Lecture Notes in Artificial Intelligence 2408

Subseries of Lecture Notes in Computer Science
Edited by J. G. Carbonell and J. Siekmann

Lecture Notes in Computer Science
Edited by G. Goos, J. Hartmanis, and J. van Leeuwen

T0145383

Springer

Berlin
Heidelberg
New York
Barcelona
Hong Kong
London
Milan
Paris
Tokyo

Antonis C. Kakas   Fariba Sadri (Eds.)

Computational Logic: Logic Programming and Beyond

Essays in Honour of Robert A. Kowalski
Part II

 Springer

Series Editors

Jaime G. Carbonell,Carnegie Mellon University, Pittsburgh, PA, USA
Jörg Siekmann, University of Saarland, Saarbrücken, Germany

Volume Editors

Antonis C. Kakas
University of Cyprus, Department of Computer Science
75 Kallipoleos St., 1678 Nicosia, Cyprus
E-mail:antonis@ucy.ac.cy

Fariba Sadri
Imperial College of Science, Technology and Medicine
Department of Computing, 180 Queen's Gate
London SW7 2BZ, United Kingdom
E-mail: fs@doc.ic.ac.uk

Cataloging-in-Publication Data applied for

Die Deutsche Bibliothek - CIP-Einheitsaufnahme

Computational logic: logig programming and beyond : essays in honour of Robert
A. Kowalski / Antonis C. Kakas ; Fariba Sadri (ed.). - Berlin ; Heidelberg ; New
York ; Barcelona ; Hong Kong ; London ; Milan ; Paris ; Tokyo : Springer Pt. 2 . -
(2002)   (Lecture notes in computer science ; Vol. 2408 : Lecture notes in   artificial
intelligence)   ISBN 3-540-43960-9

CR Subject Classification (1998): I.2.3, D.1.6, I.2, F.4, I.1

ISSN 0302-9743

ISBN 3-540-43960-9 Springer-Verlag Berlin Heidelberg New York

Springer-Verlag Berlin Heidelberg New York
a member of BertelsmannSpringer Science+Business Media GmbH

http://www.springer.de

© Springer-Verlag Berlin Heidelberg 2002
Printed in Germany

Typesetting: Camera-ready by author, data conversion by Boller Mediendesign
Printed on acid-free paper   SPIN 10873683   06/3142   5 4 3 2 1 0

Foreword

Alan Robinson

This set of essays pays tribute to Bob Kowalski on his 60th birthday, an anniversary which gives his friends and colleagues an excuse to celebrate his career as an original thinker, a charismatic communicator, and a forceful intellectual leader. The logic programming community hereby and herein conveys its respect and thanks to him for his pivotal role in creating and fostering the conceptual paradigm which is its raison d'être.

The diversity of interests covered here reflects the variety of Bob's concerns. Read on. It is an intellectual feast. Before you begin, permit me to send him a brief personal, but public, message: *Bob, how right you were, and how wrong I was*.

I should explain. When Bob arrived in Edinburgh in 1967 resolution was as yet fairly new, having taken several years to become at all widely known. Research groups to investigate various aspects of resolution sprang up at several institutions, the one organized by Bernard Meltzer at Edinburgh University being among the first. For the half-dozen years that Bob was a leading member of Bernard's group, I was a frequent visitor to it, and I saw a lot of him. We had many discussions about logic, computation, and language. By 1970, the group had zeroed in on three ideas which were soon to help make logic programming possible: the specialized inference rule of linear resolution using a selection function, together with the plan of restricting it to Horn clauses ("LUSH resolution"); the adoption of an operational semantics for Horn clauses; and a marvellously fast implementation technique for linear resolution, based on structure-sharing of syntactic expressions. Bob believed that this work now made it possible to use the predicate calculus as a programming language. I was sceptical. My focus was still on the original motivation for resolution, to build better theorem provers.

I worried that Bob had been sidetracked by an enticing illusion. In particular because of my intellectual investment in the classical semantics of predicate logic I was quite put off by the proposed operational semantics for Horn clauses. This seemed to me nothing but an adoption of MIT's notorious "Planner" ideology of computational inference. I did try, briefly, to persuade Bob to see things my way, but there was no stopping him. Thank goodness I could not change his mind, for I soon had to change mine.

In 1971, Bob and Alain Colmerauer first got together. They pooled their thinking. The rest is history. The idea of using predicate logic as a programming language then really boomed, propelled by the rush of creative energy generated by the ensuing Marseilles-Edinburgh synergy. The merger of Bob's and Alain's independent insights launched a new era. Bob's dream came true, confirmed by the spectacular practical success of Alain's Prolog. My own doubts were swept away. In the thirty years since then, logic programming has developed into a jewel of computer science, known all over the world.

Happy 60th birthday, Bob, from all of us.

Preface

Bob Kowalski together with Alain Colmerauer opened up the new field of Logic Programming back in the early 1970s. Since then the field has expanded in various directions and has contributed to the development of many other areas in Computer Science. Logic Programming has helped to place logic firmly as an integral part of the foundations of Computing and Artificial Intelligence. In particular, over the last two decades a new discipline has emerged under the name of Computational Logic which aims to promote logic as a unifying basis for problem solving. This broad role of logic was at the heart of Bob Kowalski's work from the very beginning as expounded in his seminal book "Logic for Problem Solving." He has been instrumental both in shaping this broader scientific field and in setting up the Computational Logic community.

This volume commemorates the 60th birthday of Bob Kowalski as one of the founders of and contributors to Computational Logic. It aspires to provide a landmark of the main developments in the field and to chart out its possible future directions. The authors were encouraged to provide a critical view of the main developments of the field together with an outlook on the important emerging problems and the possible contribution of Computational Logic to the future development of its related areas.

The articles in this volume span the whole field of Computational Logic seen from the point of view of Logic Programming. They range from papers addressing problems concerning the development of programming languages in logic and the application of Computational Logic to real-life problems, to philosophical studies of the field at the other end of the spectrum. Articles cover the contribution of CL to Databases and Artificial Intelligence with particular interest in Automated Reasoning, Reasoning about Actions and Change, Natural Language, and Learning.

It has been a great pleasure to help to put this volume together. We were delighted (but not surprised) to find that everyone we asked to contribute responded positively and with great enthusiasm, expressing their desire to honour Bob Kowalski. This enthusiasm remained throughout the long process of reviewing (in some cases a third reviewing process was necessary) that the invited papers had to go through in order for the decision to be made, whether they could be accepted for the volume. We thank all the authors very much for their patience and we hope that we have done justice to their efforts. We also thank all the reviewers, many of whom were authors themselves, who exhibited the same kind of zeal towards the making of this book. A special thanks goes out to Bob himself for his tolerance with our continuous stream of questions and for his own contribution to the book – his personal statement on the future of Logic Programming.

Bob has had a major impact on our lives, as he has had on many others. I, Fariba, first met Bob when I visited Imperial College for an interview as a PhD applicant. I had not even applied for logic programming, but, somehow, I ended up being interviewed by Bob. In that very first meeting his enormous enthusiasm and energy for his subject was fully evident, and soon afterwards I found myself registered to do a PhD in logic programming under his supervision. Since then, throughout all the years, Bob has been a constant source of inspiration, guidance, friendship, and humour. For me, Antonis, Bob did not supervise my PhD as this was not in Computer Science. I met Bob well after my PhD and I became a student again. I was extremely fortunate to have Bob as a new teacher at this stage. I already had some background in research and thus I was better equipped to learn from his wonderful and quite unique way of thought and scientific endeavour. I was also very fortunate to find in Bob a new good friend.

Finally, on a more personal note the first editor wishes to thank Kim for her patient understanding and support with all the rest of life's necessities thus allowing him the selfish pleasure of concentrating on research and other academic matters such as putting this book together.

Antonis Kakas and Fariba Sadri

Table of Contents, Part II

VI Logic in Databases and Information Integration

VII Automated Reasoning

VIII Non-deductive Reasoning

Table of Contents, Part I

III Software Development

IV Extensions of Logic Programming

V Applications in Logic

Author Index

MuTACLP: A Language for Temporal Reasoning with Multiple Theories

Paolo Baldan, Paolo Mancarella, Alessandra Raffaetà, and Franco Turini

Dipartimento di Informatica, Università di Pisa
Corso Italia, 40, I-56125 Pisa, Italy
{baldan,p.mancarella,raffaeta,turini}@di.unipi.it

Abstract. In this paper we introduce MuTACLP, a knowledge representation language which provides facilities for modeling and handling temporal information, together with some basic operators for combining different temporal knowledge bases. The proposed approach stems from two separate lines of research: the general studies on meta-level operators on logic programs introduced by Brogi et al. [7,9] and Temporal Annotated Constraint Logic Programming (TACLP) defined by Frühwirth [15]. In MuTACLP atoms are annotated with temporal information which are managed via a constraint theory, as in TACLP. Mechanisms for structuring programs and combining separate knowledge bases are provided through meta-level operators. The language is given two different and equivalent semantics, a top-down semantics which exploits meta-logic, and a bottom-up semantics based on an immediate consequence operator.

1 Introduction

Interest in research concerning the handling of temporal information has been growing steadily over the past two decades. On the one hand, much effort has been spent in developing extensions of logic languages capable to deal with time (see, e.g., [14,36]). On the other hand, in the field of databases, many approaches have been proposed to extend existing data models, such as the *relational*, the *object-oriented* and the *deductive* models, to cope with temporal data (see, e.g., the books [46,13] and references therein). Clearly these two strands of research are closely related, since temporal logic languages can provide solid theoretical foundations for temporal databases, and powerful knowledge representation and query languages for them [11,17,35]. Another basic motivation for our work is the need of mechanisms for combining pieces of knowledge which may be separated into various knowledge bases (e.g., distributed over the web), and thus which have to be merged together to reason with.

This paper aims at building a framework where temporal information can be naturally represented and handled, and, at the same time, knowledge can be separated and combined by means of meta-level composition operators. Concretely, we introduce a new language, called *MuTACLP*, which is based on Temporal Annotated Constraint Logic Programming (TACLP), a powerful framework defined

A.C. Kakas, F. Sadri (Eds.): Computat. Logic (Kowalski Festschrift), LNAI 2408, pp. 1–40, 2002.

by Frühwirth in [15], where temporal information and reasoning can be naturally formalized. Temporal information is represented by *temporal annotations* which say at what time(s) the formula to which they are attached is valid. Such annotations make time explicit but avoid the proliferation of temporal variables and quantifiers of the first-order approach. In this way, MuTACLP supports quantitative temporal reasoning and allows one to represent definite, indefinite and periodic temporal information, and to work both with time points and time periods (time intervals). Furthermore, as a mechanism for structuring programs and combining different knowledge sources, MuTACLP offers a set of program composition operators in the style of Brogi et al. [7,9].

Concerning the semantical aspects, the use of meta-logic allows us to provide MuTACLP with a formal and, at the same time, executable top-down semantics based on a meta-interpreter. Furthermore the language is given a bottom-up semantics by introducing an immediate consequence operator which generalizes the operator for ordinary constraint logic programs. The two semantics are equivalent in the sense that the meta-interpreter can be proved sound and complete with respect to the semantics based on the immediate consequence operator.

An interesting aspect of MuTACLP is the fact that it integrates modularity and temporal reasoning, a feature which is not common to logical temporal languages (e.g., it is lacking in [1,2,10,12,15,16,21,28]). Two exceptions are the language Temporal Datalog by Orgun [35] and the work on amalgamating knowledge bases by Subrahmanian [45]. Temporal Datalog introduces a concept of module, which, however, seems to be used as a means for defining new nonstandard algebraic operators, rather than as a knowledge representation tool. On the other hand, the work on amalgamating knowledge bases offers a multi-theory framework, based on *annotated logics*, where temporal information can be handled, but only a limited interaction among the different knowledge sources is allowed: essentially a kind of message passing mechanism allows one to delegate the resolution of an atom to other databases.

In the database field, our approach is close to the paradigm of constraint databases [25,27]. In fact, in MuTACLP the use of constraints allows one to model temporal information and to enable efficient implementations of the language. Moreover, from a deductive database perspective, each constraint logic program of our framework can be viewed as an enriched relational database where relations are represented partly intensionally and partly extensionally. The meta-level operators can then be considered as a means of constructing views by combining different databases in various ways.

The paper is organized as follows. Section 2 briefly introduces the program composition operators for combining logic theories of [7,9] and their semantics. Section 3, after reviewing the basics of constraint logic programming, introduces the language TACLP. Section 4 defines the new language MuTACLP, which integrates the basic ideas of TACLP with the composition operators on theories. In Section 5 the language MuTACLP is given a top-down semantics by means of a meta-interpreter and a bottom-up semantics based on an immediate consequence operator, and the two semantics are shown to be equivalent. Section 6 presents some examples to clarify the use of operators on theories and to show the expressive power and the knowledge representation capabilities of the language. Section 7 compares MuTACLP with some related approaches in the literature and, finally, Section 8 outlines our future research plans. Proofs of propositions and theorems are collected in the Appendix. Due to space limitations, the proofs of some technical lemmata are omitted and can be found in [4,38]. An extended abstract of this paper has been presented at the International Workshop on Spatio-Temporal Data Models and Languages [33].

2 Operators for Combining Theories

Composition operators for logic programs have been thoroughly investigated in [7,9], where both their meta-level and their bottom-up semantics are studied and compared. In order to illustrate the basic notions and ideas of such an approach this section describes the meta-level definition of the operators, which is simply obtained by adding new clauses to the well-known vanilla meta-interpreter for logic programs. The resulting meta-interpreter combines separate programs without actually building a new program. Its meaning is straightforward and, most importantly, the meta-logical definition shows that the multi-theory framework can be expressed from inside logic programming itself. We consider two operators to combine programs: union $\cup$ and intersection $\cap$. Then the so-called *program expressions* are built by starting from a set of *plain* programs, consisting of collections of clauses, and by repeatedly applying the composition operators. Formally, the language of *program expressions Exp* is defined by the following abstract syntax:

$$Exp ::= Pname \mid Exp \cup Exp \mid Exp \cap Exp$$

where *Pname* is the syntactic category of constant names for plain programs.

Following [6], the two-argument predicate *demo* is used to represent provability. Namely, $demo(\mathcal{E}, G)$ means that the formula G is provable with respect to the program expression $\mathcal{E}$.

$$demo(\mathcal{E}, empty).$$
$$demo(\mathcal{E}, (B_1, B_2)) \leftarrow demo(\mathcal{E}, B_1),\ demo(\mathcal{E}, B_2)$$
$$demo(\mathcal{E}, A) \leftarrow clause(\mathcal{E}, A, B),\ demo(\mathcal{E}, B)$$

The unit clause states that the empty goal, represented by the constant symbol *empty*, is solved in any program expression $\mathcal{E}$. The second clause deals with conjunctive goals. It states that a conjunction (B_1, B_2) is solved in the program expression $\mathcal{E}$ if B_1 is solved in $\mathcal{E}$ and B_2 is solved in $\mathcal{E}$. Finally, the third clause deals with the case of atomic goal reduction. To solve an atomic goal A, a clause with head A is chosen from the program expression $\mathcal{E}$ and the body of the clause is recursively solved in $\mathcal{E}$.

We adopt the simple naming convention used in [29]. Object programs are named by constant symbols, denoted by capital letters like P and Q. Object level expressions are represented at the meta-level by themselves. In particular, object level variables are denoted by meta-level variables, according to the so-called *non-ground representation*. An object level program P is represented, at the meta-level, by a set of axioms of the kind $clause(P, A, B)$, one for each object level clause $A \leftarrow B$ in the program P.

Each program composition operator is represented at the meta-level by a functor, whose meaning is defined by adding new clauses to the above meta-interpreter.

$$clause(\mathcal{E}_1 \cup \mathcal{E}_2, A, B) \leftarrow clause(\mathcal{E}_1, A, B)$$
$$clause(\mathcal{E}_1 \cup \mathcal{E}_2, A, B) \leftarrow clause(\mathcal{E}_2, A, B)$$
$$clause(\mathcal{E}_1 \cap \mathcal{E}_2, A, (B_1, B_2)) \leftarrow clause(\mathcal{E}_1, A, B_1),$$
$$clause(\mathcal{E}_2, A, B_2)$$

The added clauses have a straightforward interpretation. Informally, union and intersection mirror two forms of cooperation among program expressions. In the case of union $\mathcal{E}_1 \cup \mathcal{E}_2$, whose meta-level implementation is defined by the first two clauses, either expression $\mathcal{E}_1$ or $\mathcal{E}_2$ may be used to perform a computation step. For instance, a clause $A \leftarrow B$ belongs to the meta-level representation of $\mathcal{P} \cup \mathcal{Q}$ if it belongs either to the meta-level representation of $\mathcal{P}$ or to the meta-level representation of $\mathcal{Q}$. In the case of intersection $\mathcal{E}_1 \cap \mathcal{E}_2$, both expressions must agree to perform a computation step. This is expressed by the third clause, which exploits the basic unification mechanism of logic programming and the non-ground representation of object level programs.

A program expression $\mathcal{E}$ can be queried by $demo(\mathcal{E}, G)$, where G is an object level goal.

3   Temporal Annotated CLP

In this section we first briefly recall the basic concepts of Constraint Logic Programming (CLP). Then we give an overview of Temporal Annotated CLP (TACLP), an extension of CLP suited to deal with time, which will be used as a basic language for plain programs in our multi-theory framework. The reader is referred to the survey of Jaffar and Maher [22] for a comprehensive introduction to the motivations, foundations, and applications of CLP languages, and to the recent work of Jaffar et al. [23] for the formal presentation of the semantics. A good reference for TACLP is Frühwirth's paper [15].

3.1   Constraint Logic Programming

A CLP language is completely determined by its constraint domain. A *constraint domain* $\mathcal{C}$ is a tuple $\langle S_{\mathcal{C}}, \mathcal{L}_{\mathcal{C}}, \mathcal{D}_{\mathcal{C}}, \mathcal{T}_{\mathcal{C}}, solv_{\mathcal{C}} \rangle$, where

- $S_{\mathcal{C}} = \langle \Sigma_{\mathcal{C}}, \Pi_{\mathcal{C}} \rangle$ is the constraint domain *signature*, comprising the function symbols $\Sigma_{\mathcal{C}}$ and the predicate symbols $\Pi_{\mathcal{C}}$.

- $\mathcal{L}_C$ is the class of *constraints*, a set of first-order S_C-formulae, denoted by C, possibly subscripted.
- $\mathcal{D}_C$ is the *domain of computation*, a S_C-structure which provides the intended interpretation of the constraints. The domain (or support) of $\mathcal{D}_C$ is denoted by D_C.
- $\mathcal{T}_C$ is the *constraint theory*, a S_C-theory describing the logical semantics of the constraints.
- $solv_C$ is the *constraint solver*, a (computable) function which maps each formula in $\mathcal{L}_C$ to either *true*, or *false*, or *unknown*, indicating that the formula is satisfiable, unsatisfiable or it cannot be told, respectively.

We assume that Π_C contains the predicate symbol "=", interpreted as identity in $\mathcal{D}_C$. Furthermore we assume that $\mathcal{L}_C$ contains all atoms constructed from "=", the always satisfiable constraint true and the unsatisfiable constraint false, and that $\mathcal{L}_C$ is closed under variable renaming, existential quantification and conjunction. A *primitive* constraint is an atom of the form $p(t_1, \ldots, t_n)$ where p is a predicate in Π_C and $t_1, \ldots, t_n$ are terms on Σ_C.

We assume that the solver does not take variable names into account. Also, the domain, the theory and the solver *agree* in the sense that $\mathcal{D}_C$ is a model of $\mathcal{T}_C$ and for every $C \in \mathcal{L}_C$:

- $solv_C(C) = true$ implies $\mathcal{T}_C \models \exists C$, and
- $solv_C(C) = false$ implies $\mathcal{T}_C \models \neg \exists C$.

Example 1. (REAL) The constraint domain *Real* has <, <=, =, >=, > as predicate symbols, +, -, *, / as function symbols and sequences of digits (possibly with a decimal point) as constant symbols. Examples of primitive constraints are X + 3 <= Y * 1.1 and X/2 > 10. The domain of computation is the structure with reals as domain, and where the predicate symbols <, <=, =, >=, > and the function symbols +, -, *, / are interpreted as the usual relations and functions over reals. Finally, the theory $\mathcal{T}_{Real}$ is the theory of real closed fields.

A possible constraint solver is provided by the CLP($\mathcal{R}$) system [24], which relies on Gauss-Jordan elimination to handle linear constraints. Non-linear constraints are not taken into account by the solver (i.e., their evaluation is delayed) until they become linear.

Example 2. (LOGIC PROGRAMMING) The constraint domain *Term* has = as predicate symbol and strings of alphanumeric characters as function or constant symbols. The domain of computation of *Term* is the set *Tree* of *finite trees* (or, equivalently, of finite terms), while the theory $\mathcal{T}_{Term}$ is Clark's equality theory.

The interpretation of a constant is a tree with a single node labeled by the constant. The interpretation of an n-ary function symbol f is the function $f_{Tree} : Tree^n \to Tree$ mapping the trees $t_1, \ldots, t_n$ to a new tree with root labeled by f and with $t_1, \ldots, t_n$ as children.

A constraint solver is given by the *unification* algorithm. Then CLP(*Term*) coincides with logic programming.

For a given constraint domain $\mathcal{C}$, we denote by $\mathrm{CLP}(\mathcal{C})$ the CLP language based on $\mathcal{C}$. Our results are parametric to a language L in which all programs and queries under consideration are included. The set of function symbols in L, denoted by Σ_L, coincides with $\Sigma_{\mathcal{C}}$, while the set of predicate symbols Π_L includes $\Pi_{\mathcal{C}}$.

A *constraint logic program*, or simply a *program*, is a finite set of rules of the form:

$$A \leftarrow C_1, \ldots, C_n, B_1, \ldots, B_m$$

where A and $B_1, \ldots, B_m$ ($m \geq 0$) are atoms (whose predicate symbols are in Π_L but not in $\Pi_{\mathcal{C}}$), and $C_1, \ldots, C_n$ ($n \geq 0$) are primitive constraints[1] (A is called the *head* of the clause and $C_1, \ldots, C_n, B_1, \ldots, B_m$ the *body* of the clause). If $m = 0$ then the clause is called a *fact*. A query is a sequence of atoms and/or constraints.

Interpretations and Fixpoints. A $\mathcal{C}$-interpretation for a $\mathrm{CLP}(\mathcal{C})$ program is an interpretation which agrees with $\mathcal{D}_{\mathcal{C}}$ on the interpretations of the symbols in $\mathcal{L}_{\mathcal{C}}$. Formally, a $\mathcal{C}$-*interpretation* I is a subset of $\mathcal{C}$-$base_L$, i.e. of the set

$$\{p(d_1, \ldots, d_n) \mid p \text{ predicate in } \Pi_L \setminus \Pi_{\mathcal{C}}, \ d_1, \ldots, d_n \in D_{\mathcal{C}}\}.$$

Note that the meaning of primitive constraints is not specified, being fixed by $\mathcal{C}$.

The notions of $\mathcal{C}$-model and least $\mathcal{C}$-model are a natural extension of the corresponding logic programming concepts. A valuation σ is a function that maps variables into $D_{\mathcal{C}}$. A $\mathcal{C}$-ground instance A' of an atom A is obtained by applying a valuation σ to the atom, thus producing a construct of the form $p(a_1, \ldots, a_n)$ with $a_1, \ldots, a_n$ elements in $D_{\mathcal{C}}$. $\mathcal{C}$-ground instances of queries and clauses are defined in a similar way. We denote by $ground_{\mathcal{C}}(P)$ the set of $\mathcal{C}$-ground instances of clauses from P.

Finally the immediate consequence operator for a $\mathrm{CLP}(\mathcal{C})$ program P is a function $T_P^{\mathcal{C}} : \wp(\mathcal{C}\text{-}base_L) \to \wp(\mathcal{C}\text{-}base_L)$ defined as follows:

$$T_P^{\mathcal{C}}(I) = \left\{ A \mid \begin{array}{l} A \leftarrow C_1, \ldots, C_k, B_1, \ldots, B_n, \in ground_{\mathcal{C}}(P), \\ \{B_1, \ldots, B_n\} \subseteq I, \ \mathcal{D}_{\mathcal{C}} \models C_1, \ldots, C_k \end{array} \right\}$$

The operator $T_P^{\mathcal{C}}$ is continuous, and therefore it has a least fixpoint which can be computed as the least upper bound of the ω-chain $\{(T_P^{\mathcal{C}})^i\}_{i \geq 0}$ of the iterated applications of $T_P^{\mathcal{C}}$ starting from the empty set, i.e., $(T_P^{\mathcal{C}})^{\omega} = \bigcup_{i \in \mathbb{N}} (T_P^{\mathcal{C}})^i$.

3.2   Temporal Annotated Constraint Logic Programming

Temporal Annotated Constraint Logic Programming (TACLP), proposed by Frühwirth in [15,39], has been shown to be a natural and powerful framework for formalizing temporal information and reasoning. In [15] TACLP is presented

---

[1] Constraints and atoms can be in any position inside the body of a clause, although, for the sake of simplicity, we will always assume that the sequence of constraints precedes the sequence of atoms.

as an instance of annotated constraint logic (ACL) suited for reasoning about time. ACL, which can be seen as an extension of generalized annotated programs [26,30], generalizes basic first-order languages with a distinguished class of predicates, called *constraints*, and a distinguished class of terms, called *annotations*, used to label formulae. Moreover ACL provides inference rules for annotated formulae and a constraint theory for handling annotations. An advantage of the languages in the ACL framework is that their clausal fragment can be efficiently implemented: given a logic in this framework, there is a systematic way to make a clausal fragment executable as a constraint logic program. Both an interpreter and a compiler can be generated and implemented in standard constraint logic programming languages.

We next summarize the syntax and semantics of TACLP. As mentioned above, TACLP is a constraint logic programming language where formulae can be annotated with temporal labels and where relations between these labels can be expressed by using constraints. In TACLP the choice of the temporal ontology is free. In this paper, we will consider the instance of TACLP where time points are totally ordered and labels involve convex, non-empty sets of time points. Moreover we will assume that only atomic formulae can be annotated and that clauses are negation free. With an abuse of notation, in the rest of the paper such a subset of the language will be referred to simply as TACLP.

Time can be discrete or dense. Time points are totally ordered by the relation $\leq$. We denote by D the set of time points and we suppose to have a set of operations (such as the binary operations $+$, $-$) to manage such points. We assume that the time-line is left-bounded by the number 0 and open to the future, with the symbol ∞ used to denote a time point that is later than any other. A *time period* is an interval $[r, s]$ with $r, s \in D$ and $0 \leq r \leq s \leq \infty$, which represents the convex, non-empty set of time points $\{t \mid r \leq t \leq s\}^2$. Thus the interval $[0, \infty]$ denotes the whole time line.

An *annotated formula* is of the form $A \alpha$ where A is an atomic formula and α an annotation. In TACLP, there are three kinds of annotations based on time points and on time periods. Let t be a time point and $J = [r, s]$ be a time period.

(at) The annotated formula A at t means that A holds at time point t.

(th) The annotated formula A th J means that A holds *throughout*, i.e., at *every* time point in, the time period J. The definition of a th-annotated formula in terms of at is:

$$A \, \text{th} \, J \iff \forall t \, (t \in J \rightarrow A \, \text{at} \, t).$$

(in) The annotated formula A in J means that A holds at *some* time point(s) - but we do not know exactly which - in the time period J. The definition of an in-annotated formula in terms of at is:

$$A \, \text{in} \, J \iff \exists t \, (t \in J \wedge A \, \text{at} \, t).$$

The in temporal annotation accounts for indefinite temporal information.

---

[2] The results we present naturally extend to time lines that are bounded or unbounded in other ways and to time periods that are open on one or both sides.

The set of annotations is endowed with a partial order relation $\sqsubseteq$ which turns it into a lattice. Given two annotations α and β, the intuition is that $\alpha \sqsubseteq \beta$ if α is "less informative" than β in the sense that for all formulae A, $A\beta \Rightarrow A\alpha$. More precisely, being an instance of ACL, in addition to Modus Ponens, TACLP has two further inference rules: the rule ($\sqsubseteq$) and the rule ($\sqcup$).

$$\frac{A\alpha \quad \gamma \sqsubseteq \alpha}{A\gamma} \quad rule \ (\sqsubseteq) \qquad \frac{A\alpha \quad A\beta \quad \gamma = \alpha \sqcup \beta}{A\gamma} \quad rule \ (\sqcup)$$

The rule ($\sqsubseteq$) states that if a formula holds with some annotation, then it also holds with all annotations that are smaller according to the lattice ordering. The rule ($\sqcup$) says that if a formula holds with some annotation α and the same formula holds with another annotation β then it holds with the least upper bound $\alpha \sqcup \beta$ of the two annotations.

Next, we introduce the *constraint theory for temporal annotations*. Recall that a constraint theory is a non-empty, consistent first order theory that axiomatizes the meaning of the constraints. Besides an axiomatization of the total order relation $\leq$ on the set of time points D, the constraint theory includes the following axioms defining the partial order on temporal annotations.

(at th)	$\mathsf{at}\, t = \mathsf{th}\,[t, t]$
(at in)	$\mathsf{at}\, t = \mathsf{in}\,[t, t]$
(th $\sqsubseteq$)	$\mathsf{th}\,[s_1, s_2] \sqsubseteq \mathsf{th}\,[r_1, r_2] \Leftrightarrow r_1 \leq s_1,\ s_1 \leq s_2,\ s_2 \leq r_2$
(in $\sqsubseteq$)	$\mathsf{in}\,[r_1, r_2] \sqsubseteq \mathsf{in}\,[s_1, s_2] \Leftrightarrow r_1 \leq s_1,\ s_1 \leq s_2,\ s_2 \leq r_2$

The first two axioms state that $\mathsf{th}\,I$ and $\mathsf{in}\,I$ are equivalent to $\mathsf{at}\,t$ when the time period I consists of a single time point t.[3] Next, if a formula holds at every element of a time period, then it holds at every element in all sub-periods of that period ((th $\sqsubseteq$) axiom). On the other hand, if a formula holds at some points of a time period then it holds at some points in all periods that include this period ((in $\sqsubseteq$) axiom). A consequence of the above axioms is

$$(\mathsf{in}\,\mathsf{th}\,\sqsubseteq) \qquad \mathsf{in}\,[s_1, s_2] \sqsubseteq \mathsf{th}\,[r_1, r_2] \ \Leftrightarrow\ s_1 \leq r_2, r_1 \leq s_2, s_1 \leq s_2, r_1 \leq r_2$$

i.e., an atom annotated by in holds in any time period that overlaps with a time period where the atom holds throughout.

To summarize the above explanation, the axioms defining the partial order relation on annotations can be arranged in the following chain, where it is assumed that $r_1 \leq s_1,\ s_1 \leq s_2,\ s_2 \leq r_2$:

$$\mathsf{in}\,[r_1, r_2] \sqsubseteq \mathsf{in}\,[s_1, s_2] \sqsubseteq \mathsf{in}\,[s_1, s_1] = \mathsf{at}\, s_1 = \mathsf{th}\,[s_1, s_1] \sqsubseteq \mathsf{th}\,[s_1, s_2] \sqsubseteq \mathsf{th}\,[r_1, r_2]$$

Before giving an axiomatization of the least upper bound $\sqcup$ on temporal annotations, let us recall that, as explained in [15], the least upper bound of two annotations always exists but sometimes it may be "too large". In fact, rule ($\sqcup$) is correct only if the lattice order ensures $A\alpha \wedge A\beta \wedge (\gamma = \alpha \sqcup \beta) \implies A\gamma$ whereas,

---

[3] Especially in dense time, one may disallow singleton periods and drop the two axioms. This restriction has no effects on the results we are presenting.

in general, this is not true in our case. For instance, according to the lattice, $\mathtt{th}\,[1,2] \sqcup \mathtt{th}\,[4,5] = \mathtt{th}\,[1,5]$, but according to the definition of th-annotated formulae in terms of $\mathtt{at}$, the conjunction $A\,\mathtt{th}\,[1,2] \wedge A\,\mathtt{th}\,[4,5]$ does not imply $A\,\mathtt{th}\,[1,5]$, since it does not express that $A\,\mathtt{at}\,3$ holds. From a theoretical point of view, this problem can be overcome by enriching the lattice of annotations with expressions involving $\sqcup$. In practice, it suffices to consider the least upper bound for time periods that produce another *different meaningful* time period. Concretely, one restricts to $\mathtt{th}$ annotations with overlapping time periods that do not include one another:

$(\mathtt{th}\,\sqcup)$ $\mathtt{th}\,[s_1, s_2] \sqcup \mathtt{th}\,[r_1, r_2] = \mathtt{th}\,[s_1, r_2] \iff s_1 < r_1, r_1 \leq s_2, s_2 < r_2$

Summarizing, a constraint domain for *time points* is fixed where the signature includes suitable constants for time points, function symbols for operations on time points (e.g., $+, -, \ldots$) and the predicate symbol $\leq$, modeling the total order relation on time points. Such constraint domain is extended to a constraint domain $\mathcal{A}$ for handling *annotations*, by enriching the signature with function symbols $[\cdot, \cdot], \mathtt{at}, \mathtt{th}, \mathtt{in}, \sqcup$ and the predicate symbol $\sqsubseteq$, axiomatized as described above. Then, as for ordinary constraint logic programming, a TACLP language is determined by fixing a constraint domain $\mathcal{C}$, which is required to contain the constraint domain $\mathcal{A}$ for annotations. We denote by $\mathrm{TACLP}(\mathcal{C})$ the TACLP language based on $\mathcal{C}$. To lighten the notation, in the following, the "$\mathcal{C}$" will be often omitted.

The next definition introduces the clausal fragment of TACLP that can be used as an efficient temporal programming language.

Definition 1. *A TACLP clause is of the form:*

$$A\,\alpha \leftarrow C_1, \ldots, C_n, B_1\,\alpha_1, \ldots, B_m\,\alpha_m \quad (n, m \geq 0)$$

where A is an atom (not a constraint), α and α_i are (optional) temporal annotations, the C_j's are constraints and the B_i's are atomic formulae. Constraints C_j cannot be annotated.

A TACLP program is a finite set of TACLP clauses.

4 Multi-theory Temporal Annotated Constraint Logic Programming

A first attempt to extend the multi-theory framework introduced in Section 2 to handle temporal information is presented in [32]. In that paper an object level program is a collection of annotated logic programming clauses, named by a constant symbol. An annotated clause is of the kind $A \leftarrow B_1, \ldots, B_n \,\square\, [a, b]$ where the annotation $[a, b]$ represents the period of time in which the clause holds. The handling of time is hidden at the object level and it is managed at the meta-level by intersecting or joining the intervals associated with clauses. However, this approach is not completely satisfactory, in that it does not offer mechanisms for modeling indefinite temporal information and for handling periodic data. Moreover, some problems arise when we want to extract temporal information from the intervals at the object level.

To obtain a more expressive language, where in particular the mentioned deficiencies are overcome, in this paper we consider a multi-theory framework where object level programs are taken from Temporal Annotated Constraint Logic Programming (TACLP) and the composition operators are generalized to deal with temporal annotated constraint logic programs. The resulting language, called Multi-theory Temporal Annotated Constraint Logic Programming (MuTACLP for short), thus arises as a synthesis of the work on composition operators for logic programs and of TACLP. It can be thought of both as a language which enriches TACLP with high-level mechanisms for structuring programs and for combining separate knowledge bases, and as an extension of the language of program expressions with constraints and with time-representation mechanisms based on temporal annotations for atoms.

The language of program expressions remains formally the same as the one in Section 2. However now plain programs, named by the constant symbols in *Pname*, are TACLP programs as defined in Section 3.2.

Also the structure of the time domain remains unchanged, whereas, to deal with program composition, the constraint theory presented in Section 3.2 is enriched with the axiomatization of the greatest lower bound $\sqcap$ of two annotations:

$(\mathtt{th}\sqcap)$ $\mathtt{th}\,[s_1, s_2] \sqcap \mathtt{th}\,[r_1, r_2] = \mathtt{th}\,[t_1, t_2] \Leftrightarrow s_1 \leq s_2, r_1 \leq r_2, t_1 = max\{s_1, r_1\},$
$$t_2 = min\{s_2, r_2\}, t_1 \leq t_2$$

$(\mathtt{th}\sqcap')$ $\mathtt{th}\,[s_1, s_2] \sqcap \mathtt{th}\,[r_1, r_2] = \mathtt{in}\,[t_2, t_1] \Leftrightarrow s_1 \leq s_2, r_1 \leq r_2, t_1 = max\{s_1, r_1\},$
$$t_2 = min\{s_2, r_2\}, t_2 < t_1$$

$(\mathtt{th\,in}\sqcap)$ $\mathtt{th}\,[s_1, s_2] \sqcap \mathtt{in}\,[r_1, r_2] = \mathtt{in}\,[r_1, r_2] \Leftrightarrow s_1 \leq r_2, r_1 \leq s_2, s_1 \leq s_2, r_1 \leq r_2$

$(\mathtt{th\,in}\sqcap')$ $\mathtt{th}\,[s_1, s_2] \sqcap \mathtt{in}\,[r_1, r_2] = \mathtt{in}\,[s_2, r_2] \Leftrightarrow s_1 \leq s_2, s_2 < r_1, r_1 \leq r_2$

$(\mathtt{th\,in}\sqcap'')$ $\mathtt{th}\,[s_1, s_2] \sqcap \mathtt{in}\,[r_1, r_2] = \mathtt{in}\,[r_1, s_1] \Leftrightarrow r_1 \leq r_2, r_2 < s_1, s_1 \leq s_2$

$(\mathtt{in}\sqcap)$ $\mathtt{in}\,[s_1, s_2] \sqcap \mathtt{in}\,[r_1, r_2] = \mathtt{in}\,[t_1, t_2] \Leftrightarrow s_1 \leq s_2, r_1 \leq r_2, t_1 = min\{s_1, r_1\},$
$$t_2 = max\{s_2, r_2\}$$

Keeping in mind that annotations deal with time periods, i.e., *convex*, nonempty sets of time points, it is not difficult to verify that the axioms above indeed define the greatest lower bound with respect to the partial order relation $\sqsubseteq$. For instance the greatest lower bound of two $\mathtt{th}$ annotations, $\mathtt{th}\,[s_1, s_2]$ and $\mathtt{th}\,[r_1, r_2]$, can be:

- a $\mathtt{th}\,[t_1, t_2]$ annotation if $[r_1, r_2]$ and $[s_1, s_2]$ are overlapping intervals and $[t_1, t_2]$ is their (not empty) intersection (axiom $(\mathtt{th}\sqcap)$);
- an $\mathtt{in}\,[t_1, t_2]$ annotation, otherwise, where interval $[t_1, t_2]$ is the least convex set which intersects both $[s_1, s_2]$ and $[r_1, r_2]$ (axiom $(\mathtt{th}\sqcap')$, see Fig. 1.(a)).

In all other cases the greatest lower bound is always an $\mathtt{in}$ annotation. For instance, as expressed by axiom $(\mathtt{th\,in}\sqcap')$, the greatest lower bound of two annotations th $[s_1, s_2]$ and in $[r_1, r_2]$ with disjoint intervals is given by in $[s_2, r_2]$, where interval $[s_2, r_2]$ is the least convex set containing $[r_1, r_2]$ and intersecting $[s_1, s_2]$ (see Fig. 1.(b)). The greatest lower bound will play a basic role in the definition of the intersection operation over program expressions. Notice that in TACLP it is not needed since the problem of combining programs is not dealt with.

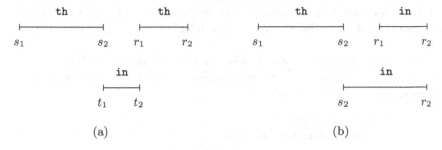

(a)   (b)

Fig. 1. Greatest lower bound of annotations.

Finally, as in TACLP we still have, in addition to Modus Ponens, the inference rules ($\sqsubseteq$) and ($\sqcup$).

Example 3. In a company there are some managers and a secretary who has to manage their meetings and appointments. During the day a manager can be busy if she/he is on a meeting or if she/he is not present in the office. This situation is modeled by the theory MANAGERS.

MANAGERS:
   $busy(M)$ th $[T_1, T_2] \leftarrow in\text{-}meeting(M)$ th $[T_1, T_2]$
   $busy(M)$ th $[T_1, T_2] \leftarrow out\text{-}of\text{-}office(M)$ th $[T_1, T_2]$

This theory is parametric with respect to the predicates *in-meeting* and *out-of-office* since the schedule of managers varies daily. The schedules are collected in a separate theory TODAY-SCHEDULE and, to know whether a manager is busy or not, such a theory is combined with MANAGERS by using the union operator.

For instance, suppose that the schedule for a given day is the following: Mr. Smith and Mr. Jones have a meeting at 9am lasting one hour. In the afternoon Mr. Smith goes out for lunch at 2pm and comes back at 3pm. The theory TODAY-SCHEDULE below represents such information.

TODAY-SCHEDULE:
   $in\text{-}meeting(mrSmith)$ th $[9am, 10am]$.
   $in\text{-}meeting(mrJones)$ th $[9am, 10am]$.
   $out\text{-}of\text{-}office(mrSmith)$ th $[2pm, 3pm]$.

To know whether Mr. Smith is busy between 9:30am and 10:30am the secretary can ask for $busy(mrSmith)$ in $[9:30am, 10:30am]$. Since Mr. Smith is in a meeting from 9am till 10am, she will indeed obtain that Mr. Smith is busy. The considered query exploits indefinite information, because knowing that Mr. Smith is busy in one instant in $[9:30am, 10:30am]$ the secretary cannot schedule an appointment for him for that period.

Example 4. At 10pm Tom was found dead in his house. The only hint is that the answering machine recorded some messages from 7pm up to 8pm. At a first glance, the doctor said Tom died one to two hours before. The detective made a further assumption: Tom did not answer the telephone so he could be already dead.

We collect all these hints and assumptions into three programs, HINTS, DOCTOR and DETECTIVE, in order not to mix firm facts with simple hypotheses that might change during the investigations.

HINTS: *found* at 10*pm*.
               ans-machine th $[7pm, 8pm]$.

DOCTOR: *dead* in $[T - 2{:}00, T - 1{:}00]$ $\leftarrow$ *found* at T

DETECTIVE: *dead* in $[T_1, T_2]$ $\leftarrow$ *ans-machine* th $[T_1, T_2]$

If we combine the hypotheses of the doctor and those of the detective we can extend the period of time in which Tom possibly died. The program expression DOCTOR $\cap$ DETECTIVE behaves as

$$dead \text{ in } [S_1, S_2] \leftarrow \text{ in } [T - 2{:}00, T - 1{:}00] \sqcap \text{ in } [T_1, T_2] = \text{ in } [S_1, S_2],$$
$$found \text{ at } T,$$
$$ans\text{-}machine \text{ th } [T_1, T_2]$$

The constraint in $[T - 2{:}00, T - 1{:}00] \sqcap$ in $[T_1, T_2] =$ in $[S_1, S_2]$ determines the annotation in $[S_1, S_2]$ in which Tom possibly died: according to axiom (in$\sqcap$) the resulting interval is $S_1 = min\{T - 2{:}00, T_1\}$ and $S_2 = max\{T - 1{:}00, T_2\}$. In fact, according to the semantics defined in the next section, a consequence of the program expression

$$\text{HINTS} \cup (\text{DOCTOR} \cap \text{DETECTIVE})$$

is just *dead* in $[7pm, 9pm]$ since the annotation in $[7pm, 9pm]$ is the greatest lower bound of in $[8pm, 9pm]$ and in $[7pm, 8pm]$.

5   Semantics of MuTACLP

In this section we introduce an operational (top-down) semantics for the language MuTACLP by means of a meta-interpreter. Then we provide MuTACLP with a least fixpoint (bottom-up) semantics, based on the definition of an immediate consequence operator. Finally, the meta-interpreter for MuTACLP is proved sound and complete with respect to the least fixpoint semantics.

In the definition of the semantics, without loss of generality, we assume all atoms to be annotated with th or in labels. In fact at t annotations can be replaced with th $[t, t]$ by exploiting the (at th) axiom. Moreover, each atom which is not annotated in the object level program is intended to be true throughout the whole temporal domain and thus it can be labelled with th $[0, \infty]$. Constraints remain unchanged.

5.1   Meta-interpreter

The extended meta-interpreter is defined by the following clauses.

$$demo(\mathcal{E}, empty). \tag{1}$$

$$demo(\mathcal{E}, (B_1, B_2)) \leftarrow demo(\mathcal{E}, B_1), demo(\mathcal{E}, B_2) \tag{2}$$

$$demo(\mathcal{E}, A \operatorname{th}[T_1, T_2]) \leftarrow S_1 \leq T_1, T_1 \leq T_2, T_2 \leq S_2, \\ clause(\mathcal{E}, A \operatorname{th}[S_1, S_2], B), demo(\mathcal{E}, B) \tag{3}$$

$$demo(\mathcal{E}, A \operatorname{th}[T_1, T_2]) \leftarrow S_1 \leq T_1, T_1 < S_2, S_2 < T_2, \\ clause(\mathcal{E}, A \operatorname{th}[S_1, S_2], B), demo(\mathcal{E}, B), \\ demo(\mathcal{E}, A \operatorname{th}[S_2, T_2]) \tag{4}$$

$$demo(\mathcal{E}, A \operatorname{in}[T_1, T_2]) \leftarrow T_1 \leq S_2, S_1 \leq T_2, T_1 \leq T_2, \\ clause(\mathcal{E}, A \operatorname{th}[S_1, S_2], B), demo(\mathcal{E}, B) \tag{5}$$

$$demo(\mathcal{E}, A \operatorname{in}[T_1, T_2]) \leftarrow T_1 \leq S_1, S_2 \leq T_2, \\ clause(\mathcal{E}, A \operatorname{in}[S_1, S_2], B), demo(\mathcal{E}, B) \tag{6}$$

$$demo(\mathcal{E}, C) \leftarrow constraint(C), C \tag{7}$$

$$clause(\mathcal{E}_1 \cup \mathcal{E}_2, A \alpha, B) \leftarrow clause(\mathcal{E}_1, A \alpha, B) \tag{8}$$

$$clause(\mathcal{E}_1 \cup \mathcal{E}_2, A \alpha, B) \leftarrow clause(\mathcal{E}_2, A \alpha, B) \tag{9}$$

$$clause(\mathcal{E}_1 \cap \mathcal{E}_2, A \gamma, (B_1, B_2)) \leftarrow clause(\mathcal{E}_1, A \alpha, B_1), \\ clause(\mathcal{E}_2, A \beta, B_2), \\ \alpha \sqcap \beta = \gamma \tag{10}$$

A clause $A \alpha \leftarrow B$ of a plain program P is represented at the meta-level by

$$clause(P, A \alpha, B) \leftarrow S_1 \leq S_2 \tag{11}$$

where $\alpha = \operatorname{th}[S_1, S_2]$ or $\alpha = \operatorname{in}[S_1, S_2]$.

This meta-interpreter can be written in any CLP language that provides a suitable constraint solver for temporal annotations (see Section 3.2 for the corresponding constraint theory). A first difference with respect to the meta-interpreter in Section 2 is that our meta-interpreter handles *constraints* that can either occur explicitly in its clauses, e.g., the constraint $s_1 \leq t_1, t_1 \leq t_2, t_2 \leq s_2$ in clause (3), or can be produced by resolution steps. Constraints of the latter kind are managed by clause (7) which passes each constraint C to be solved directly to the constraint solver.

The second difference is that our meta-interpreter implements not only Modus Ponens but also rule ($\sqsubseteq$) and rule ($\sqcup$). This is the reason why the third clause for the predicate *demo* of the meta-interpreter in Section 2 is now split into four clauses. Clauses (3), (5) and (6) implement the inference rule ($\sqsubseteq$): the atomic goal to be solved is required to be labelled with an annotation which is smaller than the one labelling the head of the clause used in the resolution step. For instance, clause (3) states that given a clause $A \, \mathtt{th} \, [s_1, s_2] \leftarrow B$ whose body B is solvable, we can derive the atom A annotated with any $\mathtt{th} \, [t_1, t_2]$ such that $\mathtt{th} \, [t_1, t_2] \sqsubseteq \mathtt{th} \, [s_1, s_2]$, i.e., according to axiom ($\mathtt{th} \sqsubseteq$), $[t_1, t_2] \subseteq [s_1, s_2]$, as expressed by the constraint $s_1 \leq t_1, t_1 \leq t_2, t_2 \leq s_2$. Clauses (5) and (6) are built in an analogous way by exploiting axioms ($\mathtt{in\,th} \sqsubseteq$) and ($\mathtt{in} \sqsubseteq$), respectively. Rule ($\sqcup$) is implemented by clause (4). According to the discussion in Section 3.2, it is applicable only to $\mathtt{th}$ annotations involving overlapping time periods which do not include one another. More precisely, clause (4) states that if we can find a clause $A \, \mathtt{th} \, [s_1, s_2] \leftarrow B$ such that the body B is solvable, and if moreover the atom A can be proved *throughout* the time period $[s_2, t_2]$ (i.e., $demo(\mathcal{E}, A \, \mathtt{th} \, [s_2, t_2])$ is solvable) then we can derive the atom A labelled with any annotation $\mathtt{th} \, [t_1, t_2] \sqsubseteq \mathtt{th} \, [s_1, t_2]$. The constraints on temporal variables ensure that the time period $[t_1, t_2]$ is a *new* time period different from $[s_1, s_2]$, $[s_2, t_2]$ and their subintervals.

Finally, in the meta-level representation of object clauses, as expressed by clause (11), the constraint $s_1 \leq s_2$ is added to ensure that the head of the object clause has a well-formed, namely non-empty, annotation.

As far as the meta-level definition of the union and intersection operators is concerned, clauses implementing the union operation remain unchanged with respect to the original definition in Section 2, whereas in the clause implementing the intersection operation a constraint is added, which expresses the annotation for the derived atom. Informally, a clause $A \, \alpha \leftarrow B$, belonging to the intersection of two program expressions $\mathcal{E}_1$ and $\mathcal{E}_2$, is built by taking one clause instance from each program expression $\mathcal{E}_1$ and $\mathcal{E}_2$, such that the head atoms of the two clauses are unifiable. Let such instances of clauses be cl_1 and cl_2. Then B is the conjunction of the bodies of cl_1 and cl_2 and A is the unified atom labelled with the greatest lower bound of the annotations of the heads of cl_1 and cl_2.

The following example shows the usefulness of clause (4) to derive *new* temporal information according to the inference rule ($\sqcup$).

Example 5. Consider the databases DB1 and DB2 containing information about people working in two companies. Jim is a consultant and he works for the first company from January 1, 1995 to April 30, 1995 and for the second company from April 1, 1995 to September 15, 1995.

DB1:
   $consultant(jim)$ th $[Jan\ 1\ 1995, Apr\ 30\ 1995]$.
DB2:
   $consultant(jim)$ th $[Apr\ 1\ 1995, Sep\ 15\ 1995]$.

The period of time in which Jim works as a consultant can be obtained by querying the union of the above theories as follows:

$$demo(\text{DB1} \cup \text{DB2}, consultant(jim)\ \text{th}\ [T_1, T_2]).$$

By using clause (4), we can derive the interval $[Jan\ 1\ 1995, Sep\ 15\ 1995]$ (more precisely, the constraints $Jan\ 1\ 1995 \leq T_1, T_1 < Apr\ 30\ 1995, Apr\ 30\ 1995 < T_2$, $T_2 \leq Sep\ 15\ 1995$ are derived) that otherwise would never be generated. In fact, by applying clause (3) alone, we can prove only that Jim is a consultant in the intervals $[Jan\ 1\ 1995, Apr\ 30\ 1995]$ and $[Apr\ 1\ 1995, Sep\ 15\ 1995]$ (or in subintervals of them) separately.

5.2 Bottom-Up Semantics

To give a declarative meaning to program expressions, we define a "higher-order" semantics for MuTACLP. In fact, the results in [7] show that the least Herbrand model semantics of logic programs does not scale smoothly to program expressions. Fundamental properties of semantics, like compositionality and full abstraction, are definitely lost. Intuitively, the least Herbrand model semantics is not compositional since it identifies programs which have different meanings when combined with others. Actually, all the programs whose least Herbrand model is empty are identified with the empty program. For example, the programs

$$\{r \leftarrow s\} \qquad\qquad \{r \leftarrow q\}$$

are both denoted by the empty model, though they behave quite differently when composed with other programs (e.g., consider the union with $\{q.\}$).

Brogi et al. showed in [9] that defining as meaning of a program P the immediate consequence operator T_P itself (rather than the least fixpoint of T_P), one obtains a semantics which is compositional with respect to several interesting operations on programs, in particular $\cup$ and $\cap$.

Along the same line, the semantics of a MuTACLP program expression is taken to be the immediate consequence operator associated with it, i.e., a function from interpretations to interpretations. The immediate consequence operator of constraint logic programming is generalized to deal with temporal annotations by considering a kind of extended interpretations, which are basically sets of annotated elements of $C\text{-}base_L$. More precisely, we first define a set of (semantical) annotations

$$Ann = \{\text{th}\,[t_1, t_2], \text{in}\,[t_1, t_2] \mid t_1, t_2\ time\ points \ \wedge\ \mathcal{D}_C \models t_1 \leq t_2\}$$

where $\mathcal{D}_C$ is the S_C-structure providing the intended interpretation of the constraints. Then the lattice of interpretations is defined as $(\wp(C\text{-}base_L \times Ann), \subseteq)$ where $\subseteq$ is the usual set-theoretic inclusion. Finally the immediate consequence operator $\mathbb{T}_{\mathcal{E}}^C$ for a program expression $\mathcal{E}$ is compositionally defined in terms of the immediate consequence operator for its sub-expressions.

Definition 2 (Bottom-up semantics). *Let $\mathcal{E}$ be a program expression, the function $\mathbb{T}_{\mathcal{E}}^C : \wp(C\text{-}base_L \times Ann) \to \wp(C\text{-}base_L \times Ann)$ is defined as follows.*

- *($\mathcal{E}$ is a plain program P)*
  $\mathbb{T}_P^C(I) =$

$$
\left\{
(A, \alpha) \;\middle|\;
\begin{array}{l}
(\alpha = \mathtt{th}\,[s_1, s_2] \;\vee\; \alpha = \mathtt{in}\,[s_1, s_2]), \\
A\,\alpha \leftarrow \bar{C}, B_1\alpha_1, \ldots, B_n\alpha_n \in ground_C(P), \\
\{(B_1, \beta_1), \ldots, (B_n, \beta_n)\} \subseteq I, \\
\mathcal{D}_C \models \bar{C}, \alpha_1 \sqsubseteq \beta_1, \ldots, \alpha_n \sqsubseteq \beta_n, s_1 \leq s_2
\end{array}
\right\}
$$

$\cup$

$$
\left\{
(A, \mathtt{th}\,[s_1, r_2]) \;\middle|\;
\begin{array}{l}
A\,\mathtt{th}\,[s_1, s_2] \leftarrow \bar{C}, B_1\alpha_1, \ldots, B_n\alpha_n \in ground_C(P), \\
\{(B_1, \beta_1), \ldots, (B_n, \beta_n)\} \subseteq I,\; (A, \mathtt{th}\,[r_1, r_2]) \in I, \\
\mathcal{D}_C \models \bar{C}, \alpha_1 \sqsubseteq \beta_1, \ldots, \alpha_n \sqsubseteq \beta_n, s_1 < r_1, r_1 \leq s_2, s_2 < r_2
\end{array}
\right\}
$$

  where $\bar{C}$ is a shortcut for $C_1, \ldots, C_k$.

- *($\mathcal{E} = \mathcal{E}_1 \cup \mathcal{E}_2$)*
  $\mathbb{T}_{\mathcal{E}_1 \cup \mathcal{E}_2}^C(I) = \mathbb{T}_{\mathcal{E}_1}^C(I) \cup \mathbb{T}_{\mathcal{E}_2}^C(I)$

- *($\mathcal{E} = \mathcal{E}_1 \cap \mathcal{E}_2$)*
  $\mathbb{T}_{\mathcal{E}_1 \cap \mathcal{E}_2}^C(I) = \mathbb{T}_{\mathcal{E}_1}^C(I) \sqcap \mathbb{T}_{\mathcal{E}_2}^C(I)$

where $I_1 \sqcap I_2 = \{(A, \gamma) \mid (A, \alpha) \in I_1,\; (A, \beta) \in I_2,\; \mathcal{D}_C \models \alpha \sqcap \beta = \gamma\}$.

Observe that the definition above properly extends the standard definition of the immediate consequence operator for constraint logic programs (see Section 3.1). In fact, besides the usual Modus Ponens rule, it captures rule ($\sqcup$) (as expressed by the second set in the definition of $\mathbb{T}_P^C$). Furthermore, also rule ($\sqsubseteq$) is taken into account to prove that an annotated atom holds in an interpretation: to derive the head $A\,\alpha$ of a clause it is not necessary to find in the interpretation exactly the atoms $B_1\,\alpha_1, \ldots, B_n\,\alpha_n$ occurring in the body of the clause, but it suffices to find atoms $B_i\,\beta_i$ which imply $B_i\,\alpha_i$, i.e., such that each β_i is an annotation stronger than α_i ($\mathcal{D}_C \models \alpha_i \sqsubseteq \beta_i$). Notice that $\mathbb{T}_{\mathcal{E}}^C(I)$ is not downward closed, namely, it is not true that if $(A, \alpha) \in \mathbb{T}_{\mathcal{E}}^C(I)$ then for all (A, γ) such that $\mathcal{D}_C \models \gamma \sqsubseteq \alpha$, we have $(A, \gamma) \in \mathbb{T}_{\mathcal{E}}^C(I)$. The downward closure will be taken only after the fixpoint of $\mathbb{T}_{\mathcal{E}}^C$ is computed. We will see that, nevertheless, no deductive capability is lost and rule ($\sqsubseteq$) is completely modeled.

The set of immediate consequences of a union of program expressions is the set-theoretic union of the immediate consequences of each program expression. Instead, an atom A labelled by γ is an immediate consequence of the intersection of two program expressions if A is a consequence of both program expressions, possibly with different annotations α and β, and the label γ is the greatest lower bound of the annotations α and β.

Let us formally define the downward closure of an interpretation.

Definition 3 (Downward closure). *The* downward closure *of an interpretation $I \subseteq \mathcal{C}\text{-}base_L \times Ann$ is defined as:*

$$\downarrow I = \{(A, \alpha) \mid (A, \beta) \in I, \ \mathcal{D}_\mathcal{C} \models \alpha \sqsubseteq \beta\}.$$

The next proposition sheds some more light on the semantics of the intersection operator, by showing that, when we apply the downward closure, the image of an interpretation through the operator $\mathbb{T}^\mathcal{C}_{\mathcal{E}_1 \cap \mathcal{E}_2}$ is the *set-theoretic intersection* of the images of the interpretation through the operators associated with $\mathcal{E}_1$ and $\mathcal{E}_2$, respectively. This property supports the intuition that the program expressions have to agree at each computation step (see [9]).

Proposition 1. *Let I_1 and I_2 be two interpretations. Then*

$$\downarrow (I_1 \Cap I_2) \ = \ (\downarrow I_1) \cap (\downarrow I_2).$$

The next theorem shows the continuity of the $\mathbb{T}^\mathcal{C}_\mathcal{E}$ operator over the lattice of interpretations.

Theorem 1 (Continuity). *For any program expression $\mathcal{E}$, the function $\mathbb{T}^\mathcal{C}_\mathcal{E}$ is continuous (over $(\wp(\mathcal{C}\text{-}base_L \times Ann), \subseteq)$).*

The fixpoint semantics for a program expression is now defined as the downward closure of the least fixpoint of $\mathbb{T}^\mathcal{C}_\mathcal{E}$ which, by continuity of $\mathbb{T}^\mathcal{C}_\mathcal{E}$, is determined as $\bigcup_{i \in \mathbb{N}} (\mathbb{T}^\mathcal{C}_\mathcal{E})^i$.

Definition 4 (Fixpoint semantics). *Let $\mathcal{E}$ be a program expression. The fixpoint semantics of $\mathcal{E}$ is defined as*

$$\mathbb{F}^\mathcal{C}(\mathcal{E}) = \downarrow (\mathbb{T}^\mathcal{C}_\mathcal{E})^\omega.$$

We remark that the downward closure is applied only once, after having computed the fixpoint of $\mathbb{T}^\mathcal{C}_\mathcal{E}$. However, it is easy to see that the closure is a continuous operator on the lattice of interpretations $\wp(\mathcal{C}\text{-}base_L \times Ann)$. Thus

$$\downarrow \left(\bigcup_{i \in \mathbb{N}} (\mathbb{T}^\mathcal{C}_\mathcal{E})^i \right) = \bigcup_{i \in \mathbb{N}} \downarrow (\mathbb{T}^\mathcal{C}_\mathcal{E})^i$$

showing that by taking the closure at each step we would have obtained the same set of consequences. Hence, as mentioned before, rule ($\sqsubseteq$) is completely captured.

5.3 Soundness and Completeness

In the spirit of [7,34] we define the semantics of the meta-interpreter by relating the semantics of an object program to the semantics of the corresponding vanilla meta-program (i.e., including the meta-level representation of the object program). When stating the correspondence between the object program and the meta-program we consider only formulae of interest, i.e., elements of C-$base_L$ annotated with labels from Ann, which are the semantic counterpart of object level annotated atoms. We show that given a MuTACLP program expression $\mathcal{E}$ (object program) for any $A \in C$-$base_L$ and any $\alpha \in Ann$, $demo(\mathcal{E}, A\,\alpha)$ is provable at the meta-level if and only if (A, α) is provable in the object program.

Theorem 2 (Soundness and completeness). *Let $\mathcal{E}$ be a program expression and let V be the meta-program containing the meta-level representation of the object level programs occurring in $\mathcal{E}$ and clauses (1)-(10). For any $A \in C$-$base_L$ and $\alpha \in Ann$, the following statement holds:*

$$demo(\mathcal{E}, A\,\alpha) \in (T_V^{\mathcal{M}})^\omega \quad \Longleftrightarrow \quad (A, \alpha) \in \mathbb{F}^{\mathcal{C}}(\mathcal{E}),$$

where $T_V^{\mathcal{M}}$ is the standard immediate consequence operator for CLP programs.

Note that V is a CLP($\mathcal{M}$) program where $\mathcal{M}$ is a multi-sorted constraint domain, including the constraint domain $Term$, presented in Example 2, and the constraint domain $\mathcal{C}$. It is worth observing that if C is a $\mathcal{C}$-ground instance of a constraint then $\mathcal{D}_{\mathcal{M}} \models C \Leftrightarrow \mathcal{D}_{\mathcal{C}} \models C$.

6 Some Examples

This section is devoted to present examples which illustrate the use of annotations in the representation of temporal information and the structuring possibilities offered by the operators. First we describe applications of our framework in the field of legal reasoning. Then we show how the intersection operator can be employed to define a kind of valid-timeslice operator.

6.1 Applications to Legal Reasoning

Laws and regulations are naturally represented in separate theories and they are usually combined in ways that are necessarily more complex than a plain merging. Time is another crucial ingredient in the definition of laws and regulations, since, quite often, they refer to instants of time and, furthermore, their validity is restricted to a fixed period of time. This is especially true for laws and regulations which concern taxation and government budget related regulations in general.

British Nationality Act. We start with a classical example in the field of legal reasoning [41], i.e. a small piece of the British Nationality Act. Simply partitioning the knowledge into separate programs and using the basic union operation, one can exploit the temporal information in an orderly way. Assume that *Jan 1 1955* is the commencement date of the law. Then statement

> x obtains the British Nationality at time t
> if x is born in U.K. at time t and
> t is after commencement and
> y is parent of x and
> y is a British citizen at time t
> or y is a British resident at time t

is modeled by the following program.

BNA:
> *get-citizenship(X)* at $T \leftarrow T \geq Jan\ 1\ 1955$, *born(X,uk)* at T,
> *parent(Y,X)* at T, *british-citizen(Y)* at T

> *get-citizenship(X)* at $T \leftarrow T \geq Jan\ 1\ 1955$, *born(X,uk)* at T,
> *parent(Y,X)* at T, *british-resident(Y)* at T

Now, the data for a single person, say John, can be encoded in a separate program.

JOHN:
> *born(john,uk)* at *Aug 10 1969*.
> *parent(bob,john)* th $[T, \infty] \leftarrow$ *born(john,_)* at T
> *british-citizen(bob)* th $[Sept\ 6\ 1940, \infty]$.

Then, by means of the union operator, one can inquire about the citizenship of John, as follows

> *demo*(BNA ∪ JOHN, *get-citizenship(john)* at T)

obtaining as result $T = Aug\ 10\ 1969$.

Movie Tickets. Since 1997, an Italian regulation for encouraging people to go to the cinema, states that on Wednesdays the ticket price is 8000 liras, whereas in the rest of the week it is 12000 liras. The situation can be modeled by the following theory BOXOFF.

BOXOFF:
> $ticket(8000, X)$ at $T \leftarrow T \geq Jan\ 1\ 1997$, *wed* at T
> $ticket(12000, X)$ at $T \leftarrow T \geq Jan\ 1\ 1997$, *non_wed* at T

The constraint $T \geq Jan\ 1\ 1997$ represents the validity of the clause, which holds from January 1, 1997 onwards.

The predicates *wed* and *non_wed* are defined in a separate theory DAYS, where w is assumed to be the last Wednesday of 1996.

DAYS: *wed* at *w*.
          wed at $T + 7 \leftarrow$ *wed* at T

          non_wed th $[w + 1, w + 6]$.
          non_wed at $T + 7 \leftarrow$ *non_wed* at T

Notice that, by means of recursive predicates one can easily express *periodic* temporal information. In the example, the definition of the predicate *wed* expresses the fact that a day is Wednesday if it is a date which is known to be Wednesday or it is a day coming seven days after a day proved to be Wednesday. The predicate *non_wed* is defined in an analogous way; in this case the unit clause states that all six consecutive days following a Wednesday are not Wednesdays.

Now, let us suppose that the owner of a cinema wants to increase the discount for young people on Wednesdays, establishing that the ticket price for people who are eighteen years old or younger is 6000 liras. By resorting to the intersection operation we can build a program expression that represents exactly the desired policy. We define three new programs CONS, DISC and AGE.

CONS: *ticket*(*8000*, X) at $T \leftarrow Y > 18$, *age*(X, Y) at T
          ticket(*12000*, X) at T.

The above theory specifies how the predicate definitions in BOXOFF must change according to the new policy. In fact to get a 8000 liras ticket now a further constraint must be satisfied, namely the customer has to be older than eighteen years old. On the other hand, no further requirement is imposed to buy a 12000 liras ticket.

DISC: *ticket*(*6000*, X) at $T \leftarrow a \leq 18$, *wed* at T, *age*(p, a) at T

The only clause in DISC states that a 6000 liras ticket can be bought on Wednesdays by a person who is eighteen years old or younger.

The programs CONS and DISC are parametric with respect to the predicate *age*, which is defined in a separate theory AGE.

AGE: *age*(X, Y) at $T \leftarrow$ *born*(X) at T_1, *year-diff*(T_1, T, Y)

At this point we can compose the above programs to obtain the program expression representing the new policy, namely

$$(\text{BOXOFF} \cap \text{CONS}) \cup \text{DISC} \cup \text{DAYS} \cup \text{AGE}.$$

Finally, in order to know how much is a ticket for a given person, the above program expression must be joined with a separate program containing the date of birth of the person. For instance, such program could be

TOM: *born*(*tom*) at *May 7 1982*.

Then the answer to the query

$$demo(((\text{BOXOFF} \cap \text{CONS}) \cup \text{DISC} \cup \text{DAYS} \cup \text{TOM}),$$
$$ticket(X, tom) \text{ at } May\ 20\ 1998)$$

is $X = 6000$ since *May 20 1998* is a Wednesday and Tom is sixteen years old.

Invim. Invim was an Italian law dealing with paying taxes on real estate transactions. The original regulation, in force since January 1, 1950, requires time calculations, since the amount of taxes depends on the period of ownership of the real estate property. Furthermore, although the law has been abolished in 1992, it still applies but only for the period antecedent to 1992.

To see how our framework allows us to model the described situation let us first consider the program INVIM below, which contains a sketch of the original body of regulations.

INVIM:

$due(Amount,X,Prop)$ th $[T_2,\infty] \leftarrow T_2 \geq Jan\ 1\ 1950,\ buys(X,Prop)$ at T_1,
$\qquad\qquad\qquad\qquad\qquad sells(X,Prop)$ at T_2,
$\qquad\qquad\qquad\qquad\qquad compute(Amount,X,Prop,T_1,T_2)$

$compute(Amount,X,Prop,T_1,T_2) \leftarrow \ldots$

To update the regulations in order to consider the decisions taken in 1992, as in the previous example we introduce two new theories. The first one includes a set of constraints on the applicability of the original regulations, while the second one is designed to embody regulations capable of handling the new situation.

CONSTRAINTS:

$due(Amount,X,Prop)$ th $[Jan\ 1\ 1993,\infty] \leftarrow$
$\qquad\qquad\qquad sells(X,Prop)$ in $[Jan\ 1\ 1950,\ Dec\ 31\ 1992]$

$compute(Amount,X,Prop,T_1,T_2)$.

The first rule specifies that the relation *due* is computed, provided that the selling date is antecedent to December, 31 1992. The second rule specifies that the rules for *compute*, whatever number they are, and whatever complexity they have, carry on unconstrained to the new version of the regulation. It is important to notice that the design of the constraining theory can be done without taking care of the details (which may be quite complicated) embodied in the original law.

The theory which handles the case of a property bought before December 31, 1992 and sold after the first of January, 1993, is given below.

ADDITIONS:

$due(Amount,X,Prop)$ th $[T_2,\infty] \leftarrow T_2 \geq Jan\ 1\ 1993,\ buys(X,Prop)$ at T_1,
$\qquad\qquad\qquad\qquad\qquad sells(X,Prop)$ at T_2,
$\qquad\qquad\qquad\qquad\qquad compute(Amount,X,Prop,T_1,Dec\ 31\ 1992)$

Now consider a separate theory representing the transactions regarding Mary, who bought an apartment on March 8, 1965 and sold it on July 2, 1997.

TRANS1:

$\qquad buys(mary,apt8)$ at $Mar\ 8\ 1965$.
$\qquad sells(mary,apt8)$ at $Jul\ 2\ 1997$.

The query

$demo(\text{INVIM} \cup \text{TRANS1}, \; due(Amount, mary, apt8) \, \text{th} \, [\_, \_])$

yields the amount, say 32.1, that Mary has to pay when selling the apartment according to the old regulations. On the other hand, the query

$demo(((\text{INVIM} \cap \text{CONSTRAINTS}) \cup \text{ADDITIONS}) \cup \text{TRANS1},$
$due(Amount, mary, apt8) \, \text{th} \, [\_, \_])$

yields the amount, say 27.8, computed according to the new regulations. It is smaller than the previous one because taxes are computed only for the period from March 8, 1965 to December 31, 1992, by using the clause in the program ADDITIONS. The clause in INVIM ∩ CONSTRAINTS cannot be used since the condition regarding the selling date ($sells(X,Prop)$ in [Jan 1 1950, Dec 31 1992]) does not hold.

In the transaction, represented by the program below, Paul buys the flat on January 1, 1995.

TRANS2:
        $buys(paul, apt9)$ at Jan 1 1995.
        $sells(paul, apt9)$ at Sep 12 1998.

$demo(\text{INVIM} \cup \text{TRANS2}, \; due(Amount, paul, apt9) \, \text{th} \, [\_, \_])$

        $Amount = 1.7$

$demo(((\text{INVIM} \cap \text{CONSTRAINTS}) \cup \text{ADDITIONS}) \cup \text{TRANS2},$
$due(Amount, paul, apt9) \, \text{th} \, [\_, \_])$

        no

If we query the theory INVIM ∪ TRANS2 we will get that Paul must pay a certain amount of tax, say 1.7, while, according to the updated regulation, he must not pay the Invim tax because he bought and sold the flat after December 31, 1992. Indeed, the answer to the query computed with respect to the theory ((INVIM ∩ CONSTRAINTS) ∪ ADDITIONS) ∪ TRANS2 is no, i.e., no tax is due.

Summing up, the union operation can be used to obtain a larger set of clauses. We can join a program with another one to provide it with definitions of its undefined predicates (e.g., AGE provides a definition for the predicate age not defined in DISC and CONS) or alternatively to add new clauses for an existing predicate (e.g., DISC contains a new definition for the predicate $ticket$ already defined in BOXOFF). On the other hand, the intersection operator provides a natural way of imposing constraints on existing programs (e.g., the program CONS constrains the definition of $ticket$ given in BOXOFF). Such constraints affect not only the computation of a particular property, like the intersection operation defined by Brogi et al. [9], but also the temporal information in which the property holds.

The use of TACLP programs allows us to represent and reason on temporal information in a natural way. Since time is explicit, at the object level we can directly access the temporal information associated with atoms. Periodic information can be easily expressed by recursive predicates (see the predicates *wed* and *non-wed* in the theory DAYS). Indefinite temporal information can be represented by using in annotations. E.g., in the program ADDITIONS the in annotation is used to specify that a certain date is within a time period (*sell(X,Prop)* in [*Jan 1 1950, Dec 31 1992*]). This is a case in which it is not important to know the precise date but it is sufficient to have an information which delimits the time period in which it can occur.

6.2   Valid-Timeslice Operator

By exploiting the features of the intersection operator we can define an operator which eases the selection of information holding in a certain interval.

Definition 5. *Let P be a plain program. For a ground interval* $[t_1, t_2]$ *we define*

$$P \Downarrow [t_1, t_2] = P \cap 1_P^{[t_1, t_2]}$$

where $1_P^{[t_1, t_2]}$ *is a program which contains a fact* "$p(X_1, \dots, X_n)$th$[t_1, t_2]$." *for all p defined in P with arity n.*

Intuitively the operator $\Downarrow$ selects only the clauses belonging to P that hold in $[t_1, t_2]$ or in a subinterval of $[t_1, t_2]$, and it restricts their validity time to such an interval. Therefore $\Downarrow$ allows us to create temporal views of programs, for instance $P \Downarrow [t, t]$ is the program P at time point t. Hence it acts as a *valid-timeslice operator* in the field of databases (see the glossary in [13]).

Consider again the Invim example of the previous section. The *whole* history of the regulation concerning Invim, can be represented by using the following program expression

(INVIM $\Downarrow$ [0, *Dec 31 1992*]) ∪ ((INVIM ∩ CONSTRAINTS) ∪ ADDITIONS)

By applying the operation $\Downarrow$, the validity of the clauses belonging to INVIM is restricted to the period from January 1, 1950 up to December 31, 1992, thus modeling the law before January 1, 1993. On the other hand, the program expression (INVIM ∩ CONSTRAINTS) ∪ ADDITIONS expresses the regulation in force since January 1, 1993, as we previously explained.

This example suggests how the operation $\Downarrow$ can be useful to model updates. Suppose that we want to represent that Frank is a research assistant in mathematics, and that, later, he is promoted becoming an assistant professor. In our formalism we can define a program FRANK that records the information associated with Frank as a research assistant.

FRANK:
   research_assistant(maths) th [*Mar 8 1993,* ∞].

On March 1996 Frank becomes an assistant professor. In order to modify the information contained in the program FRANK, we build the following program expression:

$$(\text{FRANK} \Downarrow [0, \textit{Feb 29 1996}]) \cup \{ \textit{assistant\_prof(maths)} \, \text{th} \, [\textit{Mar 1 1996}, \infty].\}$$

where the second expression is an unnamed theory. Unnamed theories, which have not been discussed so far, can be represented by the following meta-level clause:

$$clause(\{X \, \alpha \leftarrow Y\}, X \, \alpha, Y) \leftarrow T_1 \leq T_2$$

where $\alpha = \text{th} \, [T_1, T_2]$ or $\alpha = \text{in} \, [T_1, T_2]$.

The described update resembles the addition and deletion of a ground atom. For instance in $\mathcal{LDL}$++ [47] an analogous change can be implemented by solving the goal $-research\_assistant(maths), +assistant\_prof(maths)$. The advantage of our approach is that we do not change directly the clauses of a program, e.g. FRANK in the example, but we compose the old theory with a new one that represents the current situation. Therefore the state of the database before March 1, 1996 is preserved, thus maintaining the whole history. For instance, the first query below inquires the updated database before Frank's promotion whereas the second one shows how information in the database has been modified.

$$demo((\text{FRANK} \Downarrow [0, \textit{Feb 29 1996}]) \cup$$
$$\{ \textit{assistant\_prof(maths)} \, \text{th} \, [\textit{Mar 1 1996}, \infty].\},$$
$$\textit{research\_assistant(X)} \, \text{at} \, \textit{Feb 23 1994})$$
$$X = maths$$

$$demo((\text{FRANK} \Downarrow [0, \textit{Feb 29 1996}]) \cup$$
$$\{ \textit{assistant\_prof(maths)} \, \text{th} \, [\textit{Mar 1 1996}, \infty].\},$$
$$\textit{research\_assistant(X)} \, \text{at} \, \textit{Mar 12 1996})$$
$$no.$$

7 Related Work

Event Calculus by Kowalski and Sergot [28] has been the first attempt to cast into logic programming the rules for reasoning about time. In more details, Event Calculus is a treatment of time, based on the notion of event, in first-order classical logic augmented with negation as failure. It is closely related to Allen's interval temporal logic [3]. For example, let E1 be an event in which *Bob gives the Book to John* and let E2 be an event in which *John gives Mary the Book*. Assume that E2 occurs after E1. Given these event descriptions, we can deduce that there is a period started by the event E1 in which John possesses the book and that there is a period terminated by E1 in which Bob possesses the book. This situation is represented pictorially as follows:

```
     Bob has the Book        John has the Book
  <─────────────────── o ───────────────────>
                      E1
                          John has the Book        Mary has the Book
                       <─────────────────── o ───────────────────>
                                            E2
```

A series of axioms for deducing the existence of time periods and the Start and End of each time period are given by using the *Holds* predicate.

$Holds(before(e\ r))$ *if* $Terminates(e\ r)$

means that the relationship r holds in the time period *before(e r)* that denotes a time period terminated by the event e. *Holds(after(e r))* is defined in an analogous way. Event Calculus provides a natural treatment of valid time in databases, and it was extended in [43,44] to include the concept of transaction time.

Therefore Event Calculus exploits the deductive power of logic and the computational power of logic programming as in our approach, but the modeling of time is different: events are the granularity of time chosen in Event Calculus, whereas we use time points and time periods. Furthermore no provision for multiple theories is given in Event Calculus.

Kifer and Subrahmanian in [26] introduce generalized annotated logic programs (GAPs), and show how Templog [1] and an interval based temporal logic can be translated into GAPs. The annotations used there correspond to the th annotations of MuTACLP. To implement the annotated logic language, the paper proposes to use "reductants", additional clauses which are derived from existing clauses to express all possible least upper bounds. The problem is that a finite program may generate infinitely many such reductants. Then a new kind of resolution for annotated logic programs, called "ca-resolution", is proposed in [30]. The idea is to compute dynamically and incrementally the least upper bounds by collecting partial answers. Operationally this is similar to the meta-interpreter presented in Section 5.1 which relies on recursion to collect the partial answers. However, in [30] the intermediate stages of the computation may not be sound with respect to the standard CLP semantics.

The paper [26] presents also two fixpoint semantics for GAPs, defined in terms of two different operators. The first operator, called T_P, is based on interpretations which associate with each element of the Herbrand Base of a program P a *set* of annotations which is an ideal, i.e., a set downward closed and closed under *finite* least upper bounds. For each atom A, the computed ideal is the least one containing the annotations α of annotated atoms $A\,\alpha$ which are heads of (instances of) clauses whose body holds in the interpretation. The other operator, R_P, is based on interpretations which associate with each atom of the Herbrand Base a *single* annotation, obtained as the least upper bound of the set of annotations computed as in the previous case. Our fixpoint operator for MuTACLP works similarly to the T_P operator: at each step we take the closure with respect to (representable) finite least upper bounds, and, although we perform the downward closure only at the end of the computation, this does not affect the set of derivable consequences. The main difference resides in the language: MuTACLP is an extension of CLP, which focuses on temporal aspects and provides mechanisms for combining programs, taking from GAP the basic ideas for handling annotations, whereas GAP is a general language with negation and arbitrary annotations but without constraints and multiple theories.

Our temporal annotations correspond to some of the predicates proposed by Galton in [19], which is a critical examination of Allen's classical work on a theory of action and time [3]. Galton accounts for both time points and time periods in dense linear time. Assuming that the intervals I are not singletons, Galton's predicate *holds-in(A,I)* can be mapped into MuTACLP's A in I, *holds-on(A,I)* into A th I, and *holds-at(A,t)* into A at t, where A is an atomic formula. From the described correspondence it becomes clear that MuTACLP can be seen as reified FOL where annotated formulae, for example $born(john)$ at t, correspond to binary meta-relations between predicates and temporal information, for example at$(born(john), t)$. But also, MuTACLP can be regarded as a modal logic, where the annotations are seen as parameterized modal operators, e.g., $born(john)$ (at t).

Our temporal annotations also correspond to some *temporal characteristics* in the ChronoBase data model [42]. Such a model allows for the representation of a wide variety of temporal phenomena in a temporal database which cannot be expressed by using only th and in annotations. However, this model is an extension of the relational data model and, differently from our model, it is not rule-based. An interesting line of research could be to investigate the possibility of enriching the set of annotations in order to capture some other temporal characteristics, like a property that holds in an interval but not in its subintervals, still maintaining a simple and clear semantics.

In [10], a powerful temporal logic named MTL (tense logic extended by parameterized temporal operators) is translated into first order constraint logic. The resulting language subsumes Templog. The parameterized temporal operators of MTL correspond to the temporal annotations of TACLP. The constraint theory of MTL is rather complex as it involves quantified variables and implication, whose treatment goes beyond standard CLP implementations. On the other hand, MuTACLP inherits an efficient standard constraint-based implementation of annotations from the TACLP framework.

As far as the multi-theory setting is concerned, i.e. the possibility offered by MuTACLP to structure and compose (temporal) knowledge, there are few logic-based approaches providing the user with these tools. One is Temporal Datalog [35], an extension of Datalog based on a simple temporal logic with two temporal operators, namely *first* and *next*. Temporal Datalog introduces a notion of module, which however does not seem to be used as a knowledge representation tool but rather to define new non-standard algebraic operators. In fact, to query a temporal Datalog program, Orgun proposes a "point-wise extension" of the relational algebra upon the set of natural numbers, called TRA-algebra. Then he provides a mechanism for specifying generic modules, called temporal modules, which are parametric Temporal Datalog programs, with a number of input predicates (parameters) and an output predicate. A module can be then regarded as an operator which, given a temporal relation, returns a temporal relation. Thus temporal modules are indeed used as operators of TRA, through which one has access to the use of recursion, arithmetic predicates and temporal operators.

A multi-theory framework in which temporal information can be handled, based on *annotated logics*, is proposed by Subrahmanian in [45]. This is a very general framework aimed at amalgamating multiple knowledge bases which can also contain temporal information. The knowledge bases are GAPs [26] and temporal information is modeled by using an appropriate lattice of annotations. In order to integrate these programs, a so called *Mediatory Database* is given, which is a GAP having clauses of the form

$$A_0 : [m, \mu] \leftarrow A_1 : [D_1, \mu_1], \dots, A_n : [D_n, \mu_n]$$

where each D_i is a set of database names. Intuitively, a ground instance of a clause in the mediator can be interpreted as follows: "If the databases in set D_i, $1 \leq i \leq n$, (jointly) imply that the truth value of A_i is at least μ_i, then the mediator will conclude that the truth value of A_0 is at least μ". Essentially the fundamental mechanism provided to combine knowledge bases is a kind of message passing. Roughly speaking, the resolution of an atom $A_i : [D_i, \mu_i]$ is delegated to different databases, specified by the set D_i of database names, and the annotation μ_i is obtained by considering the least upper bounds of the annotations of each A_i computed in the distinct databases. Our approach is quite different because the meta-level composition operators allow us to access not only to the relation defined by a predicate but also to the definition of the predicate. For instance $P \cup Q$ is equivalent to a program whose clauses are the union of the clauses of P and Q and thus the information which can be derived from $P \cup Q$ is greater than the union of what we can derive from P and Q separately.

8   Conclusion

In this paper we have introduced MuTACLP, a language which joins the advantages of TACLP in handling temporal information with the ability to structure and compose programs. The proposed framework allows one to deal with time points and time periods and to model definite, indefinite and periodic temporal information, which can be distributed among different theories. Representing knowledge in separate programs naturally leads to use knowledge from different sources; information can be stored at different sites and combined in a modular way by employing the meta-level operators. This modular approach also favors the *reuse* of the knowledge encoded in the programs for future applications.

The language MuTACLP has been given a top-down semantics by means of a meta-interepreter and a bottom-up semantics based on an immediate consequence operator. Concerning the bottom-up semantics, it would be interesting to investigate on different definitions of the immediate consequence operator, for instance by considering an operator similar to the function R_P for generalized annotated programs [26]. The domain of interpretations considered in this paper is, in a certain sense, unstructured: interpretations are general sets of annotated atoms and the order, which is simply subset inclusion, does not take into account the order on annotations. Alternative solutions, based on different notions of interpretation, may consider more abstract domains. These domains can be obtained by endowing $C\text{-}base_L \times Ann$ with the product order (induced by the identity relation on $C\text{-}base_L$ and the order on Ann) and then by taking as elements of the domain (i.e. as interpretations) only those subsets of annotated atoms that satisfy some closure properties with respect to such an order. For instance, one can require "downward-closedness", which amounts to including subsumption in the immediate consequence operator. Another possible property is "limit-closedness", namely the presence of the least upper bound of all directed sets, which, from a computational point of view, amounts to consider computations which possibly require more than ω steps.

In [15] the language TACLP is presented as an instance of annotated constraint logic (ACL) for reasoning about time. Similarly, we could have first introduced a Multi-theory Annotated Constraint Logic (MuACL in brief), viewing MuTACLP as an instance of MuACL. To define MuACL the constructions described in this paper should be generalized by using, as basic language for plain programs, the more general paradigm of ACL where atoms can be labelled by a general class of annotations. In defining MuACL we should require that the class of annotations forms a lattice, in order to have both upper bounds and lower bounds (the latter are necessary for the definition of the intersection operator). Indeed, it is not difficult to see that, under the assumption that only atoms can be annotated and clauses are free of negation, both the meta-interpreter and the immediate consequence operator smoothly generalize to deal with general annotations.

Another interesting topic for future investigation is the treatment of negation. In the line of Frühwirth, a possible solution consists of embodying the "negation by default" of logic programming into MuTACLP by exploiting the logical equalities proved in [15]:

$$((\neg A)\,\mathtt{th}\,I) \;\Leftrightarrow\; \neg(A\,\mathtt{in}\,I) \qquad ((\neg A)\,\mathtt{in}\,I) \;\Leftrightarrow\; \neg(A\,\mathtt{th}\,I)$$

Consequently, the meta-interpreter is extended with two clauses which use such equalities:

$$demo(\mathcal{E}, (\neg A)\,\mathtt{th}\,I) \leftarrow \neg demo(\mathcal{E}, A\,\mathtt{in}\,I)$$
$$demo(\mathcal{E}, (\neg A)\,\mathtt{in}\,I) \leftarrow \neg demo(\mathcal{E}, A\,\mathtt{th}\,I)$$

However the interaction between negation by default and program composition operations is still to be fully understood. Some results on the semantic interactions between operations and negation by default are presented in [8], where, nevertheless, the handling of time is not considered.

Furthermore, it is worth noticing that in this paper we have implicitly assumed that the same unit for time is used in different programs, i.e. we have not dealt with different time *granularities*. The ability to cope with different granularities (e.g. seconds, days, etc.) is particularly relevant to support interoperability among systems. A simple way to handle this feature, is by introducing in MuTACLP a notion of *time unit* and a set of conversion predicates which transform time points into the chosen time unit (see, e.g., [5]).

We finally observe that in MuTACLP also spatial data can be naturally modelled. In fact, in the style of the constraint databases approaches (see, e.g., [25,37,20]) spatial data can be represented by using constraints. The facilities to handle time offered by MuTACLP allows one to easily establish spatio-temporal correlations, for instance time-varying areas, or, more generally, moving objects, supporting either discrete or continuous changes (see [38,31,40]).

Acknowledgments: This work has been partially supported by Esprit Working Group 28115 - DeduGIS.

References

1. M. Abadi and Z. Manna. Temporal logic programming. *Journal of Symbolic Computation*, 8:277–295, 1989.
2. J.F. Allen. Maintaining knowledge about temporal intervals. *Communications of the ACM*, 26(11):832–843, 1983.
3. J.F. Allen. Towards a general theory of action and time. *Artificial Intelligence*, 23:123–154, 1984.
4. P. Baldan, P. Mancarella, A. Raffaetà, and F. Turini. Mutaclp: A language for temporal reasoning with multiple theories. Technical report, Dipartimento di Informatica, Università di Pisa, 2001.
5. C. Bettini, X. S. Wang, and S. Jajodia. An architecture for supporting interoperability among temporal databases. In *[13]*, pages 36–55.
6. K.A. Bowen and R.A. Kowalski. Amalgamating language and metalanguage in logic programming. In K. L. Clark and S.-A. Tarnlund, editors, *Logic programming*, volume 16 of *APIC studies in data processing*, pages 153–172. Academic Press, 1982.
7. A. Brogi. *Program Construction in Computational Logic.* PhD thesis, Dipartimento di Informatica, Università di Pisa, 1993.
8. A. Brogi, S. Contiero, and F. Turini. Programming by combining general logic programs. *Journal of Logic and Computation*, 9(1):7–24, 1999.
9. A. Brogi, P. Mancarella, D. Pedreschi, and F. Turini. Modular logic programming. *ACM Transactions on Programming Languages and Systems*, 16(4):1361–1398, 1994.
10. C. Brzoska. Temporal Logic Programming with Metric and Past Operators. In *[14]*, pages 21–39.
11. J. Chomicki. Temporal Query Languages: A Survey. In *Temporal Logic: Proceedings of the First International Conference, ICTL'94*, volume 827 of *Lecture Notes in Artificial Intelligence*, pages 506–534. Springer, 1994.
12. J. Chomicki and T. Imielinski. Temporal Deductive Databases and Infinite Objects. In *Proceedings of ACM SIGACT/SIGMOD Symposium on Principles of Database Systems*, pages 61–73, 1988.
13. O. Etzion, S. Jajodia, and S. Sripada, editors. *Temporal Databases: Research and Practice*, volume 1399 of *Lecture Notes in Computer Science*. Springer, 1998.

14. M. Fisher and R. Owens, editors. *Executable Modal and Temporal Logics*, volume 897 of *Lecture Notes in Artificial Intelligence*. Springer, 1995.
15. T. Frühwirth. Temporal Annotated Constraint Logic Programming. *Journal of Symbolic Computation*, 22:555–583, 1996.
16. D. M. Gabbay. Modal and temporal logic programming. In *[18]*, pages 197–237.
17. D.M. Gabbay and P. McBrien. Temporal Logic & Historical Databases. In *Proceedings of the Seventeenth International Conference on Very Large Databases*, pages 423–430, 1991.
18. A. Galton, editor. *Temporal Logics and Their Applications*. Academic Press, 1987.
19. A. Galton. A Critical Examination of Allen's Theory of Action and Time. *Artificial Intelligence*, 42:159–188, 1990.
20. S. Grumbach, P. Rigaux, and L. Segoufin. The DEDALE system for complex spatial queries. In *Proceedings of the ACM SIGMOD International Conference on Management of Data (SIGMOD-98)*, pages 213–224, 1998.
21. T. Hrycej. A temporal extension of Prolog. *Journal of Logic Programming*, 15(1& 2):113–145, 1993.
22. J. Jaffar and M.J. Maher. Constraint Logic Programming: A Survey. *Journal of Logic Programming*, 19 & 20:503–582, 1994.
23. J. Jaffar, M.J. Maher, K. Marriott, and P.J. Stuckey. The Semantics of Constraint Logic Programs. *Journal of Logic Programming*, 37(1-3):1–46, 1998.
24. J. Jaffar, S. Michaylov, P. Stuckey, and R. Yap. The CLP(R) Language and System. *ACM Transactions on Programming Languages and Systems*, 14(3):339–395, 1992.
25. P.C. Kanellakis, G.M. Kuper, and P.Z. Revesz. Constraint query languages. *Journal of Computer and System Sciences*, 51(1):26–52, 1995.
26. M. Kifer and V.S. Subrahmanian. Theory of Generalized Annotated Logic Programming and its Applications. *Journal of Logic Programming*, 12:335–367, 1992.
27. M. Koubarakis. Database models for infinite and indefinite temporal information. *Information Systems*, 19(2):141–173, 1994.
28. R. A. Kowalski and M.J. Sergot. A Logic-based Calculus of Events. *New Generation Computing*, 4(1):67–95, 1986.
29. R.A. Kowalski and J.S. Kim. A metalogic programming approach to multi-agent knowledge and belief. In *Artificial Intelligence and Mathematical Theory of Computation*. Academic Press, 1991.
30. S.M. Leach and J.J. Lu. Computing Annotated Logic Programs. In *Proceedings of the eleventh International Conference on Logic Programming*, pages 257–271, 1994.
31. P. Mancarella, G. Nerbini, A. Raffaetà, and F. Turini. MuTACLP: A language for declarative GIS analysis. In *Proceedings of the Sixth International Conference on Rules and Objects in Databases (DOOD2000)*, volume 1861 of *Lecture Notes in Artificial Intelligence*, pages 1002–1016. Springer, 2000.
32. P. Mancarella, A. Raffaetà, and F. Turini. Knowledge Representation with Multiple Logical Theories and Time. *Journal of Experimental and Theoretical Artificial Intelligence*, 11:47–76, 1999.
33. P. Mancarella, A. Raffaetà, and F. Turini. Temporal Annotated Constraint Logic Programming with Multiple Theories. In *Tenth International Workshop on Database and Expert Systems Applications*, pages 501–508. IEEE Computer Society Press, 1999.
34. B. Martens and D. De Schreye. Why Untyped Nonground Metaprogramming Is Not (Much Of) A Problem. *Journal of Logic Programming*, 22(1):47–99, 1995.
35. M. A. Orgun. On temporal deductive databases. *Computational Intelligence*, 12(2):235–259, 1996.

36. M. A. Orgun and W. Ma. An Overview of Temporal and Modal Logic Programming. In *Temporal Logic: Proceedings of the First International Conference, ICTL'94*, volume 827 of *Lecture Notes in Artificial Intelligence*, pages 445–479. Springer, 1994.
37. J. Paredaens, J. Van den Bussche, and D. Van Gucht. Towards a theory of spatial database queries. In *Proceedings of the 13th ACM Symposium on Principles of Database Systems*, pages 279–288, 1994.
38. A. Raffaetà. *Spatio-temporal knowledge bases in a constraint logic programming framework with multiple theories*. PhD thesis, Dipartimento di Informatica, Università di Pisa, 2000.
39. A. Raffaetà and T. Frühwirth. Semantics for Temporal Annotated Constraint Logic Programming. In *Labelled Deduction*, volume 17 of *Applied Logic Series*, pages 215–243. Kluwer Academic, 2000.
40. A. Raffaetà and C. Renso. Temporal Reasoning in Geographical Information Systems. In *International Workshop on Advanced Spatial Data Management (DEXA Workshop)*, pages 899–905. IEEE Computer Society Press, 2000.
41. M. J. Sergot, F. Sadri, R. A. Kowalski, F. Kriwaczek, P. Hammond, and H. T. Cory. The British Nationality Act as a logic program. *Communications of the ACM*, 29(5):370–386, 1986.
42. S. Sripada and P. Möller. The Generalized ChronoBase Temporal Data Model. In *Meta-logics and Logic Programming*, pages 310–335. MIT Press, 1995.
43. S.M. Sripada. A logical framework for temporal deductive databases. In *Proceedings of the Very Large Databases Conference*, pages 171–182, 1988.
44. S.M. Sripada. *Temporal Reasoning in Deductive Databases*. PhD thesis, Department of Computing Imperial College of Science & Technology, 1991.
45. V. S. Subrahmanian. Amalgamating Knowledge Bases. *ACM Transactions on Database Systems*, 19(2):291–331, 1994.
46. A. Tansel, J. Clifford, S. Gadia, S. Jajodia, A. Segev, and R. Snodgrass editors. *Temporal Databases: Theory, Design, and Implementation*. Benjamin/Cummings, 1993.
47. C. Zaniolo, N. Arni, and K. Ong. Negation and aggregates in recursive rules: The LDL++Approach. In *International conference on Deductive and Object-Oriented Databases (DOOD'93)*, volume 760 of *Lecture Notes in Computer Science*. Springer, 1993.

Appendix: Proofs

Proposition 1 Let I_1 and I_2 be two interpretations. Then

$$\downarrow (I_1 \cap\!\!\!\!\cap I_2) \; = \; \downarrow I_1 \bigcap \downarrow I_2.$$

Proof. Assume $(A, \alpha) \in \downarrow (I_1 \cap\!\!\!\!\cap I_2)$. By definition of downward closure there exists γ such that $(A, \gamma) \in I_1 \cap\!\!\!\!\cap I_2$ and $\mathcal{D}_\mathcal{C} \models \alpha \sqsubseteq \gamma$. By definition of $\cap\!\!\!\!\cap$ there exist β and β' such that $(A, \beta) \in I_1$ and $(A, \beta') \in I_2$ and $\mathcal{D}_\mathcal{C} \models \beta \sqcap \beta' = \gamma$. Therefore $\mathcal{D}_\mathcal{C} \models \alpha \sqsubseteq \beta, \alpha \sqsubseteq \beta'$, by definition of downward closure we conclude $(A, \alpha) \in \downarrow I_1$ and $(A, \alpha) \in \downarrow I_2$, i.e., $(A, \alpha) \in \downarrow I_1 \bigcap \downarrow I_2$.

Vice versa assume $(A, \alpha) \in \downarrow I_1 \cap \downarrow I_2$. By definition of set-theoretic intersection and downward closure there exist β and β' such that $\mathcal{D}_\mathcal{C} \models \alpha \sqsubseteq \beta, \alpha \sqsubseteq \beta'$ and $(A, \beta) \in I_1$ and $(A, \beta') \in I_2$. By definition of $\cap\!\!\!\!\cap$, $(A, \gamma) \in I_1 \cap\!\!\!\!\cap I_2$ and $\mathcal{D}_\mathcal{C} \models \beta \sqcap \beta' = \gamma$. By property of the greatest lower bound $\mathcal{D}_\mathcal{C} \models \alpha \sqsubseteq \beta \sqcap \beta'$, hence $(A, \alpha) \in \downarrow (I_1 \cap\!\!\!\!\cap I_2)$.

Theorem 1 Let $\mathcal{E}$ be a program expression. The function $\mathbb{T}_\mathcal{E}^\mathcal{C}$ is continuous (on $(\wp(\mathcal{C}\text{-}base_L \times Ann), \subseteq)$).

Proof. Let $\{I_i\}_{i \in \mathbb{N}}$ be a chain in $(\wp(\mathcal{C}\text{-}base_L \times Ann), \subseteq)$, i.e., $I_0 \subseteq I_1 \subseteq \ldots \subseteq I_i \ldots$. Then we have to prove

$$(A, \alpha) \in \mathbb{T}_\mathcal{E}^\mathcal{C} \left(\bigcup_{i \in \mathbb{N}} I_i \right) \iff (A, \alpha) \in \bigcup_{i \in \mathbb{N}} \mathbb{T}_\mathcal{E}^\mathcal{C}(I_i).$$

The proof is by structural induction of $\mathcal{E}$.

($\mathcal{E}$ is a plain program P).

$(A, \alpha) \in \mathbb{T}_P^\mathcal{C}(\bigcup_{i \in \mathbb{N}} I_i)$
$\iff \{$definition of $\mathbb{T}_P^\mathcal{C}\}$
$((\alpha = \mathbf{th}\,[s_1, s_2] \; \vee \; \alpha = \mathbf{in}\,[s_1, s_2]) \; \wedge$
$A \; \alpha \leftarrow C_1, \ldots, C_k, B_1\alpha_1, \ldots, B_n\alpha_n \in ground_\mathcal{C}(P) \; \wedge$
$\{(B_1, \beta_1), \ldots, (B_n, \beta_n)\} \subseteq \bigcup_{i \in \mathbb{N}} I_i \; \wedge$
$\mathcal{D}_\mathcal{C} \models C_1, \ldots, C_k, \alpha_1 \sqsubseteq \beta_1, \ldots, \alpha_n \sqsubseteq \beta_n, s_1 \leq s_2) \; \vee$
$(\alpha = \mathbf{th}\,[s_1, r_2] \wedge A \; \mathbf{th}\,[s_1, s_2] \leftarrow C_1, \ldots, C_k, B_1\alpha_1, \ldots, B_n\alpha_n \in ground_\mathcal{C}(P) \; \wedge$
$\{(B_1, \beta_1), \ldots, (B_n, \beta_n)\} \subseteq \bigcup_{i \in \mathbb{N}} I_i \; \wedge \; (A, \mathbf{th}\,[r_1, r_2]) \in \bigcup_{i \in \mathbb{N}} I_i \; \wedge$
$\mathcal{D}_\mathcal{C} \models C_1, \ldots, C_k, \alpha_1 \sqsubseteq \beta_1, \ldots, \alpha_n \sqsubseteq \beta_n, s_1 < r_1, r_1 \leq s_2, s_2 < r_2)$
$\iff \{$property of set-theoretic union and $\{I_i\}_{i \in \mathbb{N}}$ is a chain. Notice that for
  ($\Longrightarrow$) j can be any element of the set $\{k \mid (B_i, \beta_i) \in I_k, i = 1, \ldots, n\}$
  which is clearly not empty$\}$
$((\alpha = \mathbf{th}\,[s_1, s_2] \; \vee \; \mathbf{in}\,[s_1, s_2]) \; \wedge$
$A \; \alpha \leftarrow C_1, \ldots, C_k, B_1\alpha_1, \ldots, B_n\alpha_n \in ground_\mathcal{C}(P) \; \wedge$
$\{(B_1, \beta_1), \ldots, (B_n, \beta_n)\} \subseteq I_j \; \wedge$
$\mathcal{D}_\mathcal{C} \models C_1, \ldots, C_k, \alpha_1 \sqsubseteq \beta_1, \ldots, \alpha_n \sqsubseteq \beta_n, s_1 \leq s_2) \; \vee$

$(\alpha = \mathtt{th}\,[s_1, r_2] \wedge A\ \mathtt{th}\,[s_1, s_2] \leftarrow C_1, \ldots, C_k, B_1\alpha_1, \ldots, B_n\alpha_n \in ground_{\mathcal{C}}(P)$
$\wedge\ \{(B_1, \beta_1), \ldots, (B_n, \beta_n)\} \subseteq I_j\ \wedge\ (A, \mathtt{th}\,[r_1, r_2]) \in I_j\ \wedge$
$\mathcal{D}_{\mathcal{C}} \models C_1, \ldots, C_k, \alpha_1 \sqsubseteq \beta_1, \ldots, \alpha_n \sqsubseteq \beta_n, s_1 < r_1, r_1 \le s_2, s_2 < r_2)$
$\Longleftrightarrow \{\text{definition of } \mathbb{T}_P^{\mathcal{C}}\}$
  $(A, \alpha) \in \mathbb{T}_P^{\mathcal{C}}(I_j)$
$\Longleftrightarrow \{\text{set-theoretic union}\}$
  $(A, \alpha) \in \bigcup_{i \in \mathbb{N}} \mathbb{T}_P^{\mathcal{C}}(I_i)$

$(\mathcal{E} = \mathcal{Q} \cup \mathcal{R})$.

  $(A, \alpha) \in \mathbb{T}_{\mathcal{Q} \cup \mathcal{R}}^{\mathcal{C}}(\bigcup_{i \in \mathbb{N}} I_i)$
$\Longleftrightarrow \{\text{definition of } \mathbb{T}_{\mathcal{Q} \cup \mathcal{R}}^{\mathcal{C}}\}$
  $(A, \alpha) \in \mathbb{T}_{\mathcal{Q}}^{\mathcal{C}}(\bigcup_{i \in \mathbb{N}} I_i) \cup \mathbb{T}_{\mathcal{R}}^{\mathcal{C}}(\bigcup_{i \in \mathbb{N}} I_i)$
$\Longleftrightarrow \{\text{inductive hypothesis}\}$
  $(A, \alpha) \in (\bigcup_{i \in \mathbb{N}} \mathbb{T}_{\mathcal{Q}}^{\mathcal{C}}(I_i)) \cup (\bigcup_{i \in \mathbb{N}} \mathbb{T}_{\mathcal{R}}^{\mathcal{C}}(I_i))$
$\Longleftrightarrow \{\text{properties of union}\}$
  $(A, \alpha) \in \bigcup_{i \in \mathbb{N}} (\mathbb{T}_{\mathcal{Q}}^{\mathcal{C}}(I_i) \cup \mathbb{T}_{\mathcal{R}}^{\mathcal{C}}(I_i))$
$\Longleftrightarrow \{\text{definition of } \mathbb{T}_{\mathcal{Q} \cup \mathcal{R}}^{\mathcal{C}}\}$
  $(A, \alpha) \in \bigcup_{i \in \mathbb{N}} \mathbb{T}_{\mathcal{Q} \cup \mathcal{R}}^{\mathcal{C}}(I_i)$

$(\mathcal{E} = \mathcal{Q} \cap \mathcal{R})$.

  $(A, \alpha) \in \mathbb{T}_{\mathcal{Q} \cap \mathcal{R}}^{\mathcal{C}}(\bigcup_{i \in \mathbb{N}} I_i)$
$\Longleftrightarrow \{\text{definition of } \mathbb{T}_{\mathcal{Q} \cap \mathcal{R}}^{\mathcal{C}}\}$
  $(A, \alpha) \in \mathbb{T}_{\mathcal{Q}}^{\mathcal{C}}(\bigcup_{i \in \mathbb{N}} I_i) \sqcap \mathbb{T}_{\mathcal{R}}^{\mathcal{C}}(\bigcup_{i \in \mathbb{N}} I_i)$
$\Longleftrightarrow \{\text{inductive hypothesis}\}$
  $(A, \alpha) \in (\bigcup_{i \in \mathbb{N}} \mathbb{T}_{\mathcal{Q}}^{\mathcal{C}}(I_i)) \sqcap (\bigcup_{i \in \mathbb{N}} \mathbb{T}_{\mathcal{R}}^{\mathcal{C}}(I_i))$
$\Longleftrightarrow \{\text{definition of } \sqcap \text{ and monotonicity of } \mathbb{T}^{\mathcal{C}}\}$
  $(A, \alpha) \in \bigcup_{i \in \mathbb{N}} (\mathbb{T}_{\mathcal{Q}}^{\mathcal{C}}(I_i) \sqcap \mathbb{T}_{\mathcal{R}}^{\mathcal{C}}(I_i))$
$\Longleftrightarrow \{\text{definition of } \mathbb{T}_{\mathcal{Q} \cap \mathcal{R}}^{\mathcal{C}}\}$
  $(A, \alpha) \in \bigcup_{i \in \mathbb{N}} \mathbb{T}_{\mathcal{Q} \cap \mathcal{R}}^{\mathcal{C}}(I_i)$

Soundness and Completeness

This section presents the proofs of the soundness and completeness results for MuTACLP meta-interpreter. Due to space limitations, the proofs of the technical lemmata are omitted and can be found in [4,38]. We first fix some notational conventions. In the following we will denote by $\mathcal{E}$, $\mathcal{N}$, $\mathcal{R}$ and $\mathcal{Q}$ generic program expressions, and by $\mathcal{C}$ the fixed constraint domain where the constraints of object programs are interpreted. Let $\mathcal{M}$ be the fixed constraint domain, where the constraints of the meta-interpreter defined in Section 5.1 are interpreted. We denote by A, B elements of $\mathcal{C}$-$base_L$, with α, β, γ annotations in Ann and by C a $\mathcal{C}$-ground instance of a constraint. All symbols may have subscripts. In the following for simplicity we will drop the reference to $\mathcal{C}$ and $\mathcal{M}$ in the name of the immediate consequence operators. Moreover we refer to the program containing the meta-level representation of object level programs and clauses (1)-(10) as "the meta-program V *corresponding* to a program expression".

We will say that an interpretation $I \subseteq C\text{-}base_L \times Ann$ *satisfies* the body of a C-ground instance $A\,\alpha \leftarrow C_1, \ldots, C_k, B_1\alpha_1, \ldots, B_n\alpha_n$ of a clause, or in symbols $I \models C_1, \ldots, C_k, B_1\alpha_1, \ldots, B_n\alpha_n$, if

1. $\mathcal{D}_C \models C_1, \ldots, C_k$ and
2. there are annotations $\beta_1, \ldots, \beta_n$ such that $\{(B_1, \beta_1), \ldots, (B_n, \beta_n)\} \subseteq I$ and $\mathcal{D}_C \models \alpha_1 \sqsubseteq \beta_1, \ldots, \alpha_n \sqsubseteq \beta_n$.

Furthermore, will often denote a sequence $C_1, \ldots, C_k$ of C-ground instances of constraints by $\bar{C}$, while a sequence $B_1\alpha_1, \ldots, B_n\alpha_n$ of annotated atoms in $C\text{-}base_L \times Ann$ will be denoted by $\bar{B}$. For example, with this convention a clause of the kind $A\,\alpha \leftarrow C_1, \ldots, C_k, B_1\alpha_1, \ldots, B_n\alpha_n$ will be written as $A\,\alpha \leftarrow \bar{C}, \bar{B}$, and, similarly, in the meta-level representation, we will write $clause(\mathcal{E}, A\,\alpha, (\bar{C}, \bar{B}))$ in place of $clause(\mathcal{E}, A\,\alpha, (C_1, \ldots, C_k, B_1\alpha_1, \ldots, B_n\alpha_n))$.

Soundness. In order to show the soundness of the meta-interpreter (restricted to the atoms of interest), we present the following easy lemma, stating that if a conjunctive goal is provable at the meta-level then also its atomic conjuncts are provable at the meta-level.

Lemma 1. *Let $\mathcal{E}$ be a program expression and let V be the corresponding meta-interpreter. For any $B_1\,\alpha_1, \ldots, B_n\,\alpha_n$ with $B_i \in C\text{-}base_L$ and $\alpha_i \in Ann$ and for any $C_1, \ldots, C_k$, with C_i a C-ground instance of a constraint, we have:*

$$For\ all\ h \quad demo(\mathcal{E}, (C_1, \ldots, C_k, B_1\,\alpha_1, \ldots, B_n\,\alpha_n)) \in T_V^h$$
$$\implies \{demo(\mathcal{E}, B_1\,\alpha_1), \ldots, demo(\mathcal{E}, B_n\,\alpha_n)\} \subseteq T_V^h \wedge \mathcal{D}_C \models C_1, \ldots, C_k.$$

The next two lemmata relate the clauses computed from a program expression $\mathcal{E}$ at the meta-level, called "virtual clauses", with the set of consequences of $\mathcal{E}$. The first lemma states that whenever we can find a virtual clause computed from $\mathcal{E}$ whose body is satisfied by I, the head $A\,\alpha$ of the clause is a consequence of the program expression $\mathcal{E}$. The second one shows how the head of a virtual clause can be "joined" with an already existing annotated atom in order to obtain an atom with a larger th annotation.

Lemma 2 (Virtual Clauses Lemma 1). *Let $\mathcal{E}$ be a program expression and V be the corresponding meta-interpreter. For any sequence $\bar{C}$ of C-ground instances of constraints, for any $A\,\alpha$, $\bar{B}$ in $C\text{-}base_L \times Ann$ and any interpretation $I \subseteq C\text{-}base_L \times Ann$, we have:*

$$clause(\mathcal{E}, A\,\alpha, (\bar{C}, \bar{B})) \in T_V^\omega \wedge I \models \bar{C}, \bar{B} \implies (A, \alpha) \in \mathbb{T}_{\mathcal{E}}(I).$$

Lemma 3 (Virtual Clauses Lemma 2). *Let $\mathcal{E}$ be a program expression and V be the corresponding meta-program. For any $A\,\text{th}\,[s_1, s_2]$, $A\,\text{th}\,[r_1, r_2]$, $\bar{B}$ in $C\text{-}base_L \times Ann$, for any sequence $\bar{C}$ of C-ground instances of constraints, and any interpretation $I \subseteq C\text{-}base_L \times Ann$, the following statement holds:*

$$clause(\mathcal{E}, A\,\text{th}\,[s_1, s_2], (\bar{C}, \bar{B})) \in T_V^\omega \ \wedge \ I \models \bar{C}, \bar{B} \ \wedge$$
$$(A, \text{th}\,[r_1, r_2]) \in I \ \wedge \ \mathcal{D}_\mathcal{C} \models s_1 < r_1, r_1 \le s_2, s_2 < r_2$$
$$\implies \ (A, \text{th}\,[s_1, r_2]) \in \mathbb{T}_\mathcal{E}(I).$$

Now, the soundness of the meta-interpreter can be proved by showing that if an annotated atom $A\,\alpha$ is provable at the meta-level from the program expression $\mathcal{E}$ then $A\,\gamma$ is a consequence of $\mathcal{E}$ for some γ such that $A\,\gamma \Rightarrow A\,\alpha$, i.e., the annotation α is less or equal to γ.

Theorem 3 (soundness). *Let $\mathcal{E}$ be a program expression and let V be the corresponding meta-program. For any $A\,\alpha$ with $A \in \mathcal{C}\text{-base}_L$ and $\alpha \in Ann$, the following statement holds:*

$$demo(\mathcal{E}, A\,\alpha) \in T_V^\omega \quad \implies \quad (A, \alpha) \in \mathbb{F}^\mathcal{C}(\mathcal{E}).$$

Proof. We first show that for all h

$$demo(\mathcal{E}, A\,\alpha) \in T_V^h \quad \implies \quad \exists \gamma : (A, \gamma) \in \mathbb{T}_\mathcal{E}^\omega \wedge \mathcal{D}_\mathcal{C} \models \alpha \sqsubseteq \gamma. \quad (12)$$

The proof is by induction on h.
(Base case). Trivial since $T_V^0 = \emptyset$.

(Inductive case). Assume that

$$demo(\mathcal{E}, A\,\alpha) \in T_V^h \quad \implies \quad \exists \gamma : (A, \gamma) \in \mathbb{T}_\mathcal{E}^\omega \wedge \mathcal{D}_\mathcal{C} \models \alpha \sqsubseteq \gamma.$$

Then:

$$demo(\mathcal{E}, A\,\alpha) \in T_V^{h+1}$$
$$\Longleftrightarrow \{\text{definition of } T_V^i\}$$
$$demo(\mathcal{E}, A\,\alpha) \in T_V(T_V^h)$$

We have four cases corresponding to clauses (3), (4), (5) and (6). We only show the cases related to clause (3) and (4) since the others are proved in an analogous way.

(clause (3)) $\{\alpha = \text{th}\,[t_1, t_2], \text{ definition of } T_V \text{ and clause (3)}\}$
   $\{clause(\mathcal{E}, A\,\text{th}\,[s_1, s_2], (\bar{C}, \bar{B})), demo(\mathcal{E}, (\bar{C}, \bar{B}))\} \subseteq T_V^h \ \wedge$
   $\mathcal{D}_\mathcal{C} \models s_1 \le t_1, t_2 \le s_2, t_1 \le t_2$
$\implies \{\text{Lemma 1 and } (\bar{C}, \bar{B}) = (C_1, \ldots, C_k, B_1\,\alpha_1, \ldots, B_n\,\alpha_n)\}$
   $clause(\mathcal{E}, A\,\text{th}\,[s_1, s_2], (C_1, \ldots, C_k, B_1\,\alpha_1, \ldots, B_n\,\alpha_n)) \in T_V^h \ \wedge$
   $\{demo(\mathcal{E}, B_1\,\alpha_1), \ldots, demo(\mathcal{E}, B_n\,\alpha_n)\} \subseteq T_V^h \ \wedge$
   $\mathcal{D}_\mathcal{C} \models C_1, \ldots, C_k \ \wedge \mathcal{D}_\mathcal{C} \models s_1 \le t_1, t_2 \le s_2, t_1 \le t_2$
$\implies \{\text{inductive hypothesis}\}$
   $\exists \beta_1, \ldots, \beta_n : clause(\mathcal{E}, A\,\text{th}\,[s_1, s_2], (C_1, \ldots, C_k, B_1\,\alpha_1, \ldots, B_n\,\alpha_n)) \in T_V^h \ \wedge$
   $\{(B_1, \beta_1), \ldots, (B_n, \beta_n)\} \subseteq \mathbb{T}_\mathcal{E}^\omega \ \wedge \ \mathcal{D}_\mathcal{C} \models \alpha_1 \sqsubseteq \beta_1, \ldots, \alpha_n \sqsubseteq \beta_n \ \wedge$
   $\mathcal{D}_\mathcal{C} \models C_1, \ldots, C_k \ \wedge \ \mathcal{D}_\mathcal{C} \models s_1 \le t_1, t_2 \le s_2, t_1 \le t_2$
$\implies \{T_V^\omega = \bigcup_{i \in \mathbb{N}} T_V^i\}$
   $clause(\mathcal{E}, A\,\text{th}\,[s_1, s_2], (C_1, \ldots, C_k, B_1\,\alpha_1, \ldots, B_n\,\alpha_n)) \in T_V^\omega \ \wedge$
   $\{(B_1, \beta_1), \ldots, (B_n, \beta_n)\} \subseteq \mathbb{T}_\mathcal{E}^\omega \ \wedge \ \mathcal{D}_\mathcal{C} \models \alpha_1 \sqsubseteq \beta_1, \ldots, \alpha_n \sqsubseteq \beta_n \ \wedge$
   $\mathcal{D}_\mathcal{C} \models C_1, \ldots, C_k \ \wedge \ \mathcal{D}_\mathcal{C} \models s_1 \le t_1, t_2 \le s_2, t_1 \le t_2$

$\implies \{\text{Lemma 2}\}$

$\quad (A, \mathbf{th}\,[s_1, s_2]) \in \mathbb{T}_{\mathcal{E}}(\mathbb{T}_{\mathcal{E}}^{\omega}) \wedge \mathcal{D}_{\mathcal{C}} \models s_1 \le t_1, t_2 \le s_2, t_1 \le t_2$

$\implies \{\mathbb{T}_{\mathcal{E}}^{\omega} \text{ is a fixpoint of } \mathbb{T}_{\mathcal{E}} \text{ and } \mathcal{D}_{\mathcal{C}} \models s_1 \le t_1, t_2 \le s_2, t_1 \le t_2\}$

$\quad (A, \mathbf{th}\,[s_1, s_2]) \in \mathbb{T}_{\mathcal{E}}^{\omega} \wedge \mathcal{D}_{\mathcal{C}} \models \mathbf{th}\,[t_1, t_2] \sqsubseteq \mathbf{th}\,[s_1, s_2]$

(clause (4)) $\{\alpha = \mathbf{th}\,[t_1, t_2], \text{ definition of } T_V \text{ and clause (4)}\}$

$\quad \{clause(\mathcal{E}, A\,\mathbf{th}\,[s_1, s_2], (\bar{C}, \bar{B})), demo(\mathcal{E}, (\bar{C}, \bar{B})), demo(\mathcal{E}, A\,\mathbf{th}\,[s_2, t_2])\} \subseteq T_V^h$

$\quad \wedge \mathcal{D}_{\mathcal{C}} \models s_1 \le t_1, t_1 < s_2, s_2 < t_2$

$\implies \{\text{Lemma 1 and } (\bar{C}, \bar{B}) = (C_1, \ldots, C_k, B_1\,\alpha_1, \ldots, B_n\,\alpha_n)\}$

$\quad clause(\mathcal{E}, A\,\mathbf{th}\,[s_1, s_2], (C_1, \ldots, C_k, B_1\,\alpha_1, \ldots, B_n\,\alpha_n)) \in T_V^h \wedge$

$\quad \{demo(\mathcal{E}, B_1\,\alpha_1), \ldots, demo(\mathcal{E}, B_n\,\alpha_n), demo(\mathcal{E}, A\,\mathbf{th}\,[s_2, t_2])\} \subseteq T_V^h \wedge$

$\quad \mathcal{D}_{\mathcal{C}} \models C_1, \ldots, C_k \wedge \mathcal{D}_{\mathcal{C}} \models s_1 \le t_1, t_1 < s_2, s_2 < t_2$

$\implies \{\text{inductive hypothesis}\}$

$\quad \exists \beta, \beta_1, \ldots, \beta_n : clause(\mathcal{E}, A\,\mathbf{th}\,[s_1, s_2], (C_1, \ldots, C_k, B_1\,\alpha_1, \ldots, B_n\,\alpha_n)) \in T_V^h \wedge$

$\quad \{(B_1, \beta_1), \ldots, (B_n, \beta_n), (A, \beta)\} \subseteq \mathbb{T}_{\mathcal{E}}^{\omega} \wedge$

$\quad \mathcal{D}_{\mathcal{C}} \models \alpha_1 \sqsubseteq \beta_1, \ldots, \alpha_n \sqsubseteq \beta_n, \mathbf{th}\,[s_2, t_2] \sqsubseteq \beta \wedge$

$\quad \mathcal{D}_{\mathcal{C}} \models C_1, \ldots, C_k \wedge \mathcal{D}_{\mathcal{C}} \models s_1 \le t_1, t_1 < s_2, s_2 < t_2.$

Since $\mathcal{D}_{\mathcal{C}} \models \mathbf{th}\,[s_2, t_2] \sqsubseteq \beta$ then $\beta = \mathbf{th}\,[w_1, w_2]$ with $\mathcal{D}_{\mathcal{C}} \models w_1 \le s_2, t_2 \le w_2$. Hence we distinguish two cases according to the relation between w_1 and s_1.

- $\mathcal{D}_{\mathcal{C}} \models w_1 \le s_1$.
  In this case we immediately conclude because $\mathcal{D}_{\mathcal{C}} \models \mathbf{th}\,[t_1, t_2] \sqsubseteq \mathbf{th}\,[w_1, w_2]$, and thus $(A, \mathbf{th}\,[w_1, w_2]) \in \mathbb{T}_{\mathcal{E}}^{\omega} \wedge \mathcal{D}_{\mathcal{C}} \models \mathbf{th}\,[t_1, t_2] \sqsubseteq \mathbf{th}\,[w_1, w_2]$.
- $\mathcal{D}_{\mathcal{C}} \models s_1 < w_1$.
  In this case $clause(\mathcal{E}, A\mathbf{th}\,[s_1, s_2], (C_1, \ldots, C_k, B_1\alpha_1, \ldots, B_n\alpha_n)) \in T_V^{\omega}$, since $T_V^{\omega} = \bigcup_{i \in \mathbb{N}} T_V^i$. Moreover, from $\mathcal{D}_{\mathcal{C}} \models s_1 < w_1, w_1 \le s_2, s_2 < t_2, t_2 \le w_2$, by Lemma 3 we obtain $(A, \mathbf{th}\,[s_1, w_2]) \in \mathbb{T}_{\mathcal{E}}(\mathbb{T}_{\mathcal{E}}^{\omega})$. Since $\mathbb{T}_{\mathcal{E}}^{\omega}$ is a fixpoint of $\mathbb{T}_{\mathcal{E}}$ and $\mathcal{D}_{\mathcal{C}} \models s_1 \le t_1, t_2 \le w_2$ we can conclude $(A, \mathbf{th}\,[s_1, w_2]) \in \mathbb{T}_{\mathcal{E}}^{\omega}$ and $\mathcal{D}_{\mathcal{C}} \models \mathbf{th}\,[t_1, t_2] \sqsubseteq \mathbf{th}\,[s_1, w_2]$.

We are finally able to prove the soundness of the meta-interpreter with respect to the least fixpoint semantics.

$\quad demo(\mathcal{E}, A\,\alpha) \in T_V^{\omega}$

$\implies \{T_V^{\omega} = \bigcup_{i \in \mathbb{N}} T_V^i\}$

$\quad \exists h : demo(\mathcal{E}, A\,\alpha) \in T_V^h$

$\implies \{\text{Statement (12)}\}$

$\quad \exists \beta : (A, \beta) \in \mathbb{T}_{\mathcal{E}}^{\omega} \wedge \mathcal{D}_{\mathcal{C}} \models \alpha \sqsubseteq \beta$

$\implies \{\text{definition of } \mathbb{F}^{\mathcal{C}}\}$

$\quad (A, \alpha) \in \mathbb{F}^{\mathcal{C}}(\mathcal{E}).$

Completeness. We first need a lemma stating that if an annotated atom $A\,\alpha$ is provable at the meta-level in a program expression $\mathcal{E}$ then we can prove at the meta-level the same atom A with any other "weaker" annotation (namely $A\,\gamma$, with $\gamma \sqsubseteq \alpha$).

Lemma 4. *Let $\mathcal{E}$ be a program expression and V be the corresponding meta-program. For any $A \in \mathcal{C}\text{-base}_L$ and $\alpha \in Ann$, the following statement holds:*

$$demo(\mathcal{E}, A\,\alpha) \in T_V^\omega \quad \Longrightarrow \quad \{demo(\mathcal{E}, A\,\gamma) \mid \gamma \in Ann, \mathcal{D}_\mathcal{C} \models \gamma \sqsubseteq \alpha\} \subseteq T_V^\omega.$$

Now the completeness result for MuTACLP meta-interpreter basically relies on two technical lemmata (Lemma 7 and Lemma 8). Roughly speaking they assert that when $\mathtt{th}$ and $\mathtt{in}$ annotated atoms are derivable from an interpretation I by using the $\mathbb{T}_\mathcal{E}$ operator then we can find corresponding virtual clauses in the program expression $\mathcal{E}$ which permit to derive the same or greater information.

Let us first introduce some preliminary notions and results.

Definition 6 (covering). *A covering for a $\mathtt{th}$-annotation $\mathtt{th}\,[t_1, t_2]$ is a sequence of annotations $\{\mathtt{th}\,[t_1^i, t_2^i]\}_{i \in \{1,\dots,n\}}$, such that $\mathcal{D}_\mathcal{C} \models \mathtt{th}\,[t_1, t_2] \sqsubseteq \mathtt{th}\,[t_1^1, t_n^2]$ and for any $i \in \{1, \dots, n\}$*

$$\mathcal{D}_\mathcal{C} \models t_1^i \leq t_2^i,\ t_1^{i+1} \leq t_2^i,\ t_1^i < t_1^{i+1}.$$

In words, a covering of a $\mathtt{th}$ annotation $\mathtt{th}\,[t_1, t_2]$ is a sequence of annotations $\{\mathtt{th}\,[t_1^i, t_2^i]\}_{i \in \{1,\dots,n\}}$ such that each of the intervals overlaps with its successor, and the union of such intervals includes $[t_1, t_2]$. The next simple lemma observes that, given two annotations and a covering for each of them, we can always build a covering for their greatest lower bound.

Lemma 5. *Let $\mathtt{th}\,[t_1, t_2]$ and $\mathtt{th}\,[s_1, s_2]$ be annotations and $\mathtt{th}\,[w_1, w_2] = \mathtt{th}\,[t_1, t_2] \sqcap \mathtt{th}\,[s_1, s_2]$. Let $\{\mathtt{th}\,[t_1^i, t_2^i]\}_{i \in \{1,\dots,n\}}$ and $\{\mathtt{th}\,[s_1^j, s_2^j]\}_{j \in \{1,\dots,m\}}$ be coverings for $\mathtt{th}\,[t_1, t_2]$ and $\mathtt{th}\,[s_1, s_2]$, respectively. Then a covering for $\mathtt{th}\,[w_1, w_2]$ can be extracted from*

$$\{\mathtt{th}\,[t_1^i, t_2^i] \sqcap \mathtt{th}\,[s_1^j, s_2^j] \mid i \in \{1, \dots n\} \ \wedge \ j \in \{1, \dots, m\}\}.$$

In the hypothesis of the previous lemma $[w_1, w_2] = [t_1, t_2] \cap [s_1, s_2]$. Thus the result of the lemma is simply a consequence of the distributivity of set-theoretical intersection with respect to union.

Definition 7. *Let $\mathcal{E}$ be a program expression, let V be the corresponding meta-program and let $I \subseteq \mathcal{C}\text{-base}_L \times Ann$ be an interpretation. Given an annotated atom $(A, \mathtt{th}\,[t_1, t_2]) \in \mathcal{C}\text{-base}_L \times Ann$, an $(\mathcal{E}, I)$-set for $(A, \mathtt{th}\,[t_1, t_2])$ is a set*

$$\{clause(\mathcal{E}, A\,\mathtt{th}\,[t_1^i, t_2^i], (\bar{C}^i, \bar{B}^i))\}_{i \in \{1,\dots,n\}} \subseteq T_V^\omega$$

such that

1. *$\{\mathtt{th}\,[t_1^i, t_2^i]\}_{i \in \{1,\dots,n\}}$ is a covering of $\mathtt{th}\,[t_1, t_2]$, and*
2. *for $i \in \{1, \dots, n\}$, $I \models \bar{C}^i, \bar{B}^i$.*

An interpretation $I \subseteq \mathcal{C}\text{-base}_L \times Ann$ is called $\mathtt{th}$-closed with respect to $\mathcal{E}$ (or $\mathcal{E}$-closed, for short) if there is an $(\mathcal{E}, I)$-set for every annotated atom $(A, \mathtt{th}\,[t_1, t_2]) \in I$.

The next lemma presents some properties of the notion of $\mathcal{E}$-closedness, which essentially state that the property of being $\mathcal{E}$-closed is invariant with respect to some obvious algebraic transformations of the program expression $\mathcal{E}$.

Lemma 6. *Let $\mathcal{E}$, $\mathcal{R}$ and $\mathcal{N}$ be program expressions and let I be an interpretation. Then the following properties hold, where $op \in \{\cup, \cap\}$*

1. *I is $(\mathcal{E}\, op\, \mathcal{E})$-closed iff I is $\mathcal{E}$-closed;*
2. *I is $(\mathcal{E}\, op\, \mathcal{R})$-closed iff I is $(\mathcal{R}\, op\, \mathcal{E})$-closed;*
3. *I is $((\mathcal{E}\, op\, \mathcal{R})\, op\, \mathcal{N})$-closed iff I is $\mathcal{E}\, op\, (\mathcal{R}\, op\, \mathcal{N})$-closed;*
4. *if I is $\mathcal{E}$-closed then I is $(\mathcal{E} \cup \mathcal{R})$-closed;*
5. *if I is $(\mathcal{E} \cap \mathcal{R})$-closed then I is $\mathcal{E}$-closed;*
6. *I is $((\mathcal{E} \cap \mathcal{R}) \cup \mathcal{N})$-closed iff I is $((\mathcal{E} \cup \mathcal{N}) \cap (\mathcal{R} \cup \mathcal{N}))$-closed.*

We next show that if we apply the $\mathbb{T}_{\mathcal{E}}$ operator to an $\mathcal{E}$-closed interpretation, then for any derived $\mathtt{th}$-annotated atom there exists an $(\mathcal{E}, I)$-set (see Definition 7). This result represents a basic step towards the completeness proof. In fact, it tells us that starting from the empty interpretation, which is obviously $\mathcal{E}$-closed, and iterating the $\mathbb{T}_{\mathcal{E}}$ then we get, step after step, $\mathtt{th}$-annotated atoms which can be also derived from the virtual clauses of the program expression at hand. For technical reasons, to make the induction work, we need a slightly stronger property.

Lemma 7. *Let $\mathcal{E}$ and $\mathcal{Q}$ be program expressions, let V be the corresponding meta-program[4] and let $I \subseteq C\text{-base}_L \times Ann$ be an $(\mathcal{E} \cup \mathcal{Q})$-closed interpretation. Then for any atom $(A, \mathtt{th}\,[t_1, t_2]) \in \mathbb{T}_{\mathcal{E}}(I)$ there exists an $(\mathcal{E} \cup \mathcal{Q}, I)$-set.*

Corollary 1. *Let $\mathcal{E}$ be any program expression and let V be the corresponding meta-program. Then for any $h \in \mathbb{N}$ the interpretation $\mathbb{T}_{\mathcal{E}}^h$ is $\mathcal{E}$-closed. Therefore $\mathbb{T}_{\mathcal{E}}^\omega$ is $\mathcal{E}$-closed.*

Another technical lemma is needed for dealing with the $\mathtt{in}$ annotations, which comes in pair with Lemma 7.

Lemma 8. *Let $\mathcal{E}$ be a program expression, let V be the corresponding meta-program and let I be any $\mathcal{E}$-closed interpretation. For any atom $(A, \mathtt{in}\,[t_1, t_2]) \in \mathbb{T}_{\mathcal{E}}(I)$ we have*

$$clause(\mathcal{E}, A\,\alpha, (\bar{C}, \bar{B})) \in T_V^\omega \wedge I \models \bar{C}, \bar{B} \wedge \mathcal{D}_C \models \mathtt{in}\,[t_1, t_2] \sqsubseteq \alpha.$$

Now we can prove the completeness of the meta-interpreter with respect to the least fixpoint semantics.

Theorem 4 (Completeness). *Let $\mathcal{E}$ be a program expression and V be the corresponding meta-program. For any $A \in C\text{-base}_L$ and $\alpha \in Ann$ the following statement holds:*

$$(A, \alpha) \in \mathbb{F}^C(\mathcal{E}) \implies demo(\mathcal{E}, A\,\alpha) \in T_V^\omega.$$

---

[4] The meta-program contains the meta-level representation of the plain programs in $\mathcal{E}$ and $\mathcal{Q}$.

Proof. We first show that for all h

$$(A, \alpha) \in \mathbb{T}_{\mathcal{E}}^{h} \qquad \Longrightarrow \qquad demo(\mathcal{E}, A\,\alpha) \in T_V^{\omega}. \qquad (13)$$

The proof is by induction on h.

(Base case). Trivial since $\mathbb{T}_{\mathcal{E}}^{0} = \emptyset$.

(Inductive case). Assume that

$$(A, \alpha) \in \mathbb{T}_{\mathcal{E}}^{h} \qquad \Longrightarrow \qquad demo(\mathcal{E}, A\,\alpha) \in T_V^{\omega}.$$

Observe that, under the above assumption,

$$\mathbb{T}_{\mathcal{E}}^{h} \models \bar{C}, \bar{B} \quad \Rightarrow \quad demo(\mathcal{E}, (\bar{C}, \bar{B})) \in T_V^{\omega}. \qquad (14)$$

In fact let $\bar{C} = C_1, \ldots, C_k$ and $\bar{B} = B_1\alpha_1, \ldots, B_n\alpha_n$. Then the notation $\mathbb{T}_{\mathcal{E}}^{h} \models \bar{C}$ amounts to say that for each i, $\mathcal{D}_{\mathcal{C}} \models C_i$ and thus $demo(\mathcal{E}, C_i) \in T_V^{\omega}$, by definition of T_V and clause (7). Furthermore $\mathbb{T}_{\mathcal{E}}^{h} \models \bar{B}$ means that for each i, $(B_i, \beta_i) \in \mathbb{T}_{\mathcal{E}}^{h}$ and $\mathcal{D}_{\mathcal{C}} \models \alpha_i \sqsubseteq \beta_i$. Hence by inductive hypothesis $demo(\mathcal{E}, B_i\,\beta_i) \in T_V^{\omega}$ and thus, by Lemma 4, $demo(\mathcal{E}, B_i\,\alpha_i) \in T_V^{\omega}$. By several applications of clause (2) in the meta-interpreter we finally deduce $demo(\mathcal{E}, (\bar{B}, \bar{C})) \in T_V^{\omega}$.

It is convenient to treat separately the cases of $\mathtt{th}$ and $\mathtt{in}$ annotations. If we assume that $\alpha = \mathtt{th}\,[t_1, t_2]$, then

$(A, \mathtt{th}\,[t_1, t_2]) \in \mathbb{T}_{\mathcal{E}}^{h+1}$
$\Longleftrightarrow \{\text{definition of } \mathbb{T}_{\mathcal{E}}^{i}\}$
$(A, \mathtt{th}\,[t_1, t_2]) \in \mathbb{T}_{\mathcal{E}}(\mathbb{T}_{\mathcal{E}}^{h})$
$\Longrightarrow \{\text{Lemma 7 and } \mathbb{T}_{\mathcal{E}}^{h} \text{ is } \mathcal{E}\text{-closed by Corollary 1}\}$
$\{clause(\mathcal{E}, A\,\mathtt{th}\,[t_1^i, t_2^i], (\bar{C}^i, \bar{B}^i))\}_{i \in \{1,\ldots,n\}} \subseteq T_V^{\omega} \wedge$
$\mathbb{T}_{\mathcal{E}}^{h} \models \bar{C}^i, \bar{B}^i \text{ for } i \in \{1, \ldots, n\} \wedge$
$\{\mathtt{th}\,[t_1^i, t_2^i]\}_{i \in \{1,\ldots,n\}} \text{ covering of } \mathtt{th}\,[t_1, t_2]$
$\Longrightarrow \{\text{previous remark (14)}\}$
$\{clause(\mathcal{E}, A\,\mathtt{th}\,[t_1^i, t_2^i], (\bar{C}^i, \bar{B}^i))\}_{i \in \{1,\ldots,n\}} \subseteq T_V^{\omega} \wedge$
$demo(\mathcal{E}, (\bar{C}^i, \bar{B}^i)) \in T_V^{\omega} \text{ for } i \in \{1, \ldots, n\} \wedge$
$\{\mathtt{th}\,[t_1^i, t_2^i]\}_{i \in \{1,\ldots,n\}} \text{ covering of } \mathtt{th}\,[t_1, t_2]$
$\Longrightarrow \{\text{definition of } T_V, \text{ clause (3) and } T_V^{\omega} \text{ is a fixpoint of } T_V\}$
$demo(\mathcal{E}, A\,\mathtt{th}\,[t_1^n, t_2^n]) \in T_V^{\omega} \wedge$
$\{clause(\mathcal{E}, A\,\mathtt{th}\,[t_1^i, t_2^i], (\bar{C}^i, \bar{B}^i))\}_{i \in \{1,\ldots,n-1\}} \subseteq T_V^{\omega} \wedge$
$demo(\mathcal{E}, (\bar{C}^i, \bar{B}^i)) \in T_V^{\omega} \text{ for } i \in \{1, \ldots, n-1\} \wedge$
$\{\mathtt{th}\,[t_1^i, t_2^i]\}_{i \in \{1,\ldots,n\}} \text{ covering of } \mathtt{th}\,[t_1, t_2]$
$\Longrightarrow \{\text{definition of } T_V, \text{ clause (4), Lemma 4 and } T_V^{\omega} \text{ is a fixpoint of } T_V\}$
$demo(\mathcal{E}, A\,\mathtt{th}\,[t_1^{n-1}, t_2^n]) \wedge \{clause(\mathcal{E}, A\,\mathtt{th}\,[t_1^i, t_2^i], (\bar{C}^i, \bar{B}^i))\}_{i \in \{1,\ldots,n-2\}} \subseteq T_V^{\omega}$
$\wedge \, demo(\mathcal{E}, (\bar{C}^i, \bar{B}^i)) \in T_V^{\omega} \text{ for } i \in \{1, \ldots, n-2\} \wedge$
$\{\mathtt{th}\,[t_1^i, t_2^i]\}_{i \in \{1,\ldots,n\}} \text{ covering of } \mathtt{th}\,[t_1, t_2]$
$\Longrightarrow \{\text{by exploiting several times clause (4) as above}\}$
$demo(\mathcal{E}, A\,\mathtt{th}\,[t_1^1, t_2^n]) \wedge \{\mathtt{th}\,[t_1^i, t_2^i]\}_{i \in \{1,\ldots,n\}} \text{ covering of } \mathtt{th}\,[t_1, t_2]$
$\Longrightarrow \{\text{by definition of covering } \mathcal{D}_{\mathcal{C}} \models \mathtt{th}\,[t_1, t_2] \sqsubseteq \mathtt{th}\,[t_1^1, t_2^n] \text{ and Lemma 4}\}$
$demo(\mathcal{E}, A\,\mathtt{th}\,[t_1, t_2]) \in T_V^{\omega}$

Instead, if $\alpha = \text{in}\,[t_1, t_2]$, then

$\quad (A, \text{in}\,[t_1, t_2]) \in \mathbb{T}_{\mathcal{E}}^{h+1}$
$\Longleftrightarrow \{\text{definition of } \mathbb{T}_{\mathcal{E}}^i\}$
$\quad (A, \text{in}\,[t_1, t_2]) \in \mathbb{T}_{\mathcal{E}}(\mathbb{T}_{\mathcal{E}}^h)$
$\Longrightarrow \{\text{Lemma 8}\}$
$\quad clause(\mathcal{E}, A\,\beta, (\bar{C}, \bar{B})) \in T_V^{\omega} \ \wedge \ \mathbb{T}_{\mathcal{E}}^h \models \bar{C}, \bar{B} \ \wedge \ \mathcal{D}_C \models \text{in}\,[t_1, t_2] \sqsubseteq \beta$
$\Longrightarrow \{\text{previous remark (14)}\}$
$\quad clause(\mathcal{E}, A\,\beta, (\bar{C}, \bar{B})) \in T_V^{\omega} \ \wedge \ demo(\mathcal{E}, (\bar{C}, \bar{B})) \in T_V^{\omega} \ \wedge \mathcal{D}_C \models \text{in}\,[t_1, t_2] \sqsubseteq \beta$
$\Longrightarrow \{\text{clause (3) or (6), and } T_V^{\omega} \text{ is a fixpoint of } T_V\}$
$\quad demo(\mathcal{E}, A\,\beta) \in T_V^{\omega} \ \wedge \ \mathcal{D}_C \models \text{in}\,[t_1, t_2] \sqsubseteq \beta$
$\Longrightarrow \{\text{Lemma 4}\}$
$\quad demo(\mathcal{E}, A\,\text{in}\,[t_1, t_2]) \in T_V^{\omega}$

We now prove the completeness of the meta-interpreter of the program expressions with respect to the least fixpoint semantics.

$\quad (A, \alpha) \in \mathbb{F}^C(\mathcal{E})$
$\Longrightarrow \{\text{definition of } \mathbb{F}^C(\mathcal{E})\}$
$\quad \exists \gamma \in Ann : \ (A, \gamma) \in \mathbb{T}_{\mathcal{E}}^{\omega} \ \wedge \ \mathcal{D}_C \models \alpha \sqsubseteq \gamma$
$\Longrightarrow \{\mathbb{T}_{\mathcal{E}}^{\omega} = \bigcup_{i \in \mathbb{N}} \mathbb{T}_{\mathcal{E}}^i\}$
$\quad \exists h : \ (A, \gamma) \in \mathbb{T}_{\mathcal{E}}^h \ \wedge \ \mathcal{D}_C \models \alpha \sqsubseteq \gamma$
$\Longrightarrow \{\text{statement (13)}\}$
$\quad demo(\mathcal{E}, A\,\gamma) \in T_V^{\omega} \ \wedge \ \mathcal{D}_C \models \alpha \sqsubseteq \gamma$
$\Longrightarrow \{\text{Lemma 4}\}$
$\quad demo(\mathcal{E}, A\,\alpha) \in T_V^{\omega}$

Description Logics for Information Integration

Diego Calvanese, Giuseppe De Giacomo, and Maurizio Lenzerini

Dipartimento di Informatica e Sistemistica
Università di Roma "La Sapienza"
Via Salaria 113, 00198 Roma, Italy
lastname @dis.uniroma1.it,
http://www.dis.uniroma1.it/∼*lastname*

Abstract. Information integration is the problem of combining the data residing at different, heterogeneous sources, and providing the user with a unified view of these data, called mediated schema. The mediated schema is therefore a reconciled view of the information, which can be queried by the user. It is the task of the system to free the user from the knowledge on where data are, and how data are structured at the sources.
In this chapter, we discuss data integration in general, and describe a logic-based approach to data integration. A logic of the Description Logics family is used to model the information managed by the integration system, to formulate queries posed to the system, and to perform several types of automated reasoning supporting both the modeling, and the query answering process. We focus, in particular, on a specific Description Logic, called $\mathcal{DLR}$, specifically designed for database applications. In the chapter, we illustrate how $\mathcal{DLR}$ is used to model a mediated schema of an integration system, to specify the semantics of the data sources, and finally to support the query answering process by means of the associated reasoning methods.

1 Introduction

Information integration is the problem of combining the data residing at different sources, and providing the user with a unified view of these data, called mediated schema. The mediated schema is therefore a reconciled view of the information, which can be queried by the user. It is the task of the data integration system to free the user from the knowledge on where data are, and how data are structured at the sources.

The interest in this kind of systems has been continuously growing in the last years. Many organizations face the problem of integrating data residing in several sources. Companies that build a Data Warehouse, a Data Mining, or an Enterprise Resource Planning system must address this problem. Also, integrating data in the World Wide Web is the subject of several investigations and projects nowadays. Finally, applications requiring accessing or re-engineering legacy systems must deal with the problem of integrating data stored in different sources.

The design of a data integration system is a very complex task, which comprises several different issues, including the following:

A.C. Kakas, F. Sadri (Eds.): Computat. Logic (Kowalski Festschrift), LNAI 2408, pp. 41–60, 2002.

1. heterogeneity of the sources,
2. relation between the mediated schema and the sources,
3. limitations on the mechanisms for accessing the sources,
4. materialized vs. virtual integration,
5. data cleaning and reconciliation,
6. how to process queries expressed on the mediated schema.

Problem (1) arises because sources are typically heterogeneous, meaning that they adopt different models and systems for storing data. This poses challenging problems in specifying the mediated schema. The goal is to design such a schema so as to provide an appropriate abstraction of all the data residing at the sources. One aspect deserving special attention is the choice of the language used to express the mediated schema. Since such a schema should mediate among different representations of overlapping worlds, the language should provide flexible and powerful representation mechanisms. We refer to [34] for a more detailed discussion on this subject. Following the work in [32,16,40], in this paper we use a formalism of the family of Description Logics to specify mediated schemas.

With regard to Problem (2), two basic approaches have been used to specify the relation between the sources and the mediated schema. The first approach, called *global-as-view* (or query-based), requires that the mediated schema is expressed in terms of the data sources. More precisely, to every concept of the mediated schema, a view over the data sources is associated, so that its meaning is specified in terms of the data residing at the sources. The second approach, called *local-as-view* (or source-based), requires the mediated schema to be specified independently from the sources. The relationships between the mediated schema and the sources are established by defining every source as a view over the mediated schema. Thus, in the local-as-view approach, we specify the meaning of the sources in terms of the concepts in the mediated schema. It is clear that the latter approach favors the extensibility of the integration system, and provides a more appropriate setting for its maintenance. For example, adding a new source to the system requires only to provide the definition of the source, and does not necessarily involve changes in the mediated schema. On the contrary, in the global-as-view approach, adding a new source typically requires changing the definition of the concepts in the mediated schema. For this reason, in the rest of the paper, we adopt the local-as-view approach. A comparison between the two approaches is reported in [51].

Problem (3) refers to the fact, that, both in the local-as-view and in the global-as-view approach, it may happen that a source presents some limitations on the types of accesses it supports. A typical example is a web source accessible through a form where one of the fields must necessarily be filled in by the user. Such a situation can be modeled by specifying the source as a relation supporting only queries with a selection on a column. Suitable notations have been proposed for such situations [44], and the consequences of these access limitations on query processing in integration systems have been investigated in several papers [44,43,27,56,55,41,42].

Problem (4) deals with a further criterion that we should take into account in the design of a data integration system. In particular, with respect to the data explicitly managed by the system, we can follow two different approaches, called *materialized* and *virtual*. In the materialized approach, the system computes the extensions of the concepts in the mediated schema by replicating the data at the sources. In the virtual approach, data residing at the sources are accessed during query processing, but they are not replicated in the integration system. Obviously, in the materialized approach, the problem of refreshing the materialized views in order to keep them up-to-date is a major issue [34]. In the following, we only deal with the virtual approach.

Whereas the construction of the mediated schema concerns the intentional level of the data integration system, problem (5) refers to a number of issues arising when considering the integration at the extensional/instance level. A first issue in this context is the interpretation and merging of the data provided by the sources. Interpreting data can be regarded as the task of casting them into a common representation. Moreover, the data returned by various sources need to be converted/reconciled/combined to provide the data integration system with the requested information. The complexity of this reconciliation step is due to several problems, such as possible mismatches between data referring to the same real world object, possible errors in the data stored in the sources, or possible inconsistencies between values representing the properties of the real world objects in different sources [28]. The above task is known in the literature as *Data Cleaning and Reconciliation*, and the interested reader is referred to [28,10,4] for more details on this subject.

Finally, problem (6) is concerned with one of the most important issues in a data integration system, i.e., the choice of the method for computing the answer to queries posed in terms of the mediated schema. While query answering in the global-as-view approach typically reduces to unfolding, an integration system based on the local-as-view approach must resort to more sophisticated query processing techniques. The main issue is that the system should be able to re-express the query in terms of a suitable set of queries posed to the sources. In this reformulation process, the crucial step is deciding how to decompose the query on the mediated schema into a set of subqueries on the sources, based on the meaning of the sources in terms of the concepts in the mediated schema. The computed subqueries are then shipped to the sources, and the results are assembled into the final answer.

In the rest of this paper, we concentrate on Problem (6), namely, query processing in a data integration system specified by means of the local-as-view approach, and we present the following contributions:

– We first provide a logical formalization of the problem. In particular, we illustrate a general architecture for a data integration system, comprising a mediated schema, a set of views, and a query. Query processing in this setting is formally defined as the problem of *answering queries using views*: compute the answer to a query only on the basis of the extension of a set of views [1,29]. We observe that, besides data integration, this problem is relevant in several fields, including data warehousing [54], query optimization [17], supporting physical data independence [50], etc.

- Then we instantiate the general framework to the case where schemas, views
  and queries are expressed by making use of a particular logical language. In
  particular:
  - The mediated schema is expressed in terms of a knowledge base consti-
    tuted by general inclusion assertions and membership assertions, formu-
    lated in an expressive Description Logic [6].
  - Queries and views are expressed as non-recursive datalog programs,
    whose predicates in the body are concepts or relations that appear in
    the knowledge base.
  - For each view, it can be specified whether the provided extension is
    sound, complete, or exact with respect to the view definition [1,11]. Such
    assumptions are used in data integration with the following meaning.
    A *sound view* corresponds to an information source which is known to
    produce only, but not necessarily all, the answers to the associated query.
    A *complete view* models a source which is known to produce all answers
    to the associated query, and maybe more. Finally, an *exact view* is known
    to produce exactly the answers to the associated query.
- We then illustrate a technique for the problem of answering queries using
  views in our setting. We first describe how to formulate the problem in
  terms of logical implication, and then we present a technique to check logical
  implication in 2EXPTIME worst case complexity.

The paper is organized as follows. Section 2 presents the general framework.
Section 3 illustrates the use of Description Logics for setting up a particular
architecture for data integration, according to the general framework. Section 4
presents the method we use for query answering using views in our architecture.
Section 5 describes other works on the problem of answering query using views.
Finally, Section 6 concludes the paper.

2   Framework

In this section we set up a logical framework for data integration. Since we
assume to work with relational databases, in the following we refer to a relational
alphabet $\mathcal{A}$, i.e., an alphabet constituted by a set of predicate and constant
symbols. Predicate symbols are used to denote the relations in the database,
whereas constant symbols denote the objects stored in relations. We adopt the
so-called *unique name assumption*, i.e., we assume that different constants denote
different objects.

A database (DB) $\mathcal{DB}$ is simply a set of relations, one for each predicate
symbol in the alphabet $\mathcal{A}$. The relation corresponding to the predicate symbol
R_i is constituted by a set of tuples of constants, which specify the objects that
satisfy the relation associated to R_i.

The main components of a data integration system are the mediated schema,
the sources, and the queries. Each component is expressed in a specific language
over the alphabet $\mathcal{A}$:

- the *mediated schema* is expressed in the schema language $\mathcal{L}_S$,
- the *sources* are modeled as views over the mediated schema, expressed in the view language $\mathcal{L}_V$,
- *queries* are issued over the mediated schema, and are expressed in the query language $\mathcal{L}_Q$.

In what follows, we provide a specification of the three components of a data integration system.

- The mediated schema S is a set of constraints, each one expressed in the language $\mathcal{L}_S$ over the alphabet $\mathcal{A}$. The language $\mathcal{L}_S$ determines the expressiveness allowed for specifying the schema of our database, i.e., the constraints that the database must satisfy. If S is constituted by the constraints $\{C_1, \ldots, C_n\}$, we say that a database $\mathcal{DB}$ satisfies S if all constraints $C_1, \ldots, C_n$ are satisfied by $\mathcal{DB}$.
- The sources are modeled in terms of a set of views $\mathcal{V} = \{V_1, \ldots, V_m\}$ over the mediated schema. Associated to each view V_i we have:
  - A definition $def(V_i)$ in terms of a query $V_i(\boldsymbol{x}) \leftarrow v_i(\boldsymbol{x}, \boldsymbol{y})$ over $\mathcal{DB}$, where $v_i(\boldsymbol{x}, \boldsymbol{y})$ is expressed in the language $\mathcal{L}_V$ over the alphabet $\mathcal{A}$. The arity of $\boldsymbol{x}$ determines the arity of the view V_i.
  - A set $ext(V_i)$ of tuples of constants, which provides the information about the extension of V_i, i.e., the content of the sources. The arity of each tuple is the same as that of V_i.
  - A specification $as(V_i)$ of which *assumption* to adopt for the view V_i, i.e., how to interpret the content of the source $ext(V_i)$ with respect to the actual set of tuples in $\mathcal{DB}$ that satisfy V_i. We describe below the various possibilities that we consider for $as(V_i)$.
- A query is expressed in the language $\mathcal{L}_Q$ over the alphabet $\mathcal{A}$, and is intended to provide the specification of which data to extract from the virtual database represented in the integration system. In general, if Q is a query and $\mathcal{DB}$ is a database satsfying S, we denote with $ans(Q, \mathcal{DB})$ the set of tuples in $\mathcal{DB}$ that satisfy Q.

The specification $as(V_i)$ determines how accurate is the knowledge on the pairs satisfying the views, i.e., how accurate is the source with respect to the specification $def(V_i)$[1]. As pointed out in several papers [1,29,37,11], the following three assumptions are relevant in a data integration system:

- *Sound Views.* When a view V_i is *sound* (denoted with $as(V_i) = sound$), its extension provides any subset of the tuples satisfying the corresponding definition. In other words, from the fact that a tuple is in $ext(V_i)$ one can conclude that it satisfies the view, while from the fact that a tuple is not in $ext(V_i)$ one cannot conclude that it does not satisfy the view. Formally, a database $\mathcal{DB}$ is *coherent with the sound view* V_i, if $ext(V_i) \subseteq ans(def(V_i), \mathcal{DB})$.

---

[1] In some papers, for example [11], different assumptions on the domain of the database are also taken into account.

- *Complete Views.* When a view V_i is *complete* (denoted with $as(V_i) = complete$), its extension provides any superset of the tuples satisfying the corresponding definition. In other words, from the fact that a tuple is in $ext(V_i)$ one cannot conclude that such a tuple satisfies the view. On the other hand, from the fact that a tuple is not in $ext(V_i)$ one can conclude that such a tuple does not satisfy the view. Formally, a database $\mathcal{DB}$ is *coherent with the complete view* V_i, if $ext(V_i) \supseteq ans(def(V_i), \mathcal{DB})$.
- *Exact Views.* When a view V_i is *exact* (denoted with $as(V_i) = exact$), its extension is exactly the set of tuples of objects satisfying the corresponding definition. Formally, a database $\mathcal{DB}$ is *coherent with the exact view* V_i, if $ext(V_i) = ans(def(V_i), \mathcal{DB})$.

The ultimate goal of a data integration system is to allow a client to extract information from the database, taking into account that the only knowledge s/he has on the database is the extension of the set of views, i.e., the content of the sources. More precisely, the problem of extracting information from the data integration system reduces to the problem of *answering queries using views.* Given

- a schema $\mathcal{S}$,
- a set of views $\mathcal{V} = \{V_1, \ldots, V_m\}$, with, for each V_i,
  • its definition $def(V_i)$,
  • its extension $ext(V_i)$, and
  • the specification $as(V_i)$ of whether it is sound, complete, or exact,
- a query Q of arity n, and
- a tuple $\boldsymbol{d} = (d_1, \ldots, d_n)$ of constants,

the problem consists in deciding whether $\boldsymbol{d} \in ans(Q, \mathcal{S}, \mathcal{V})$, i.e., deciding whether $(d_1, \ldots, d_n) \in ans(Q, \mathcal{DB})$, for each $\mathcal{DB}$ such that:

- $\mathcal{DB}$ satisfies the schema $\mathcal{S}$,
- $\mathcal{DB}$ is coherent with $V_1, \ldots, V_m$.

¿From the above definition, it is easy to see that answering queries using views is essentially an extended form of reasoning in the presence of incomplete information [53]. Indeed, when we answer the query on the basis of the views, we know only the extensions of the views, and this provides us with only partial information on the database. Moreover, since the query language may admit various forms of incomplete information (due to union, for instance), there are in general several possible databases that are coherent with the views.

The following example rephrases an example given in [1].

Example 1. Consider a relational alphabet containing (among other symbols) a binary predicate couple, and two constants Ann and Bill. Consider also two views female and male, respectively with definitions

$$\text{female}(f) \leftarrow \text{couple}(f, m)$$
$$\text{male}(m) \leftarrow \text{couple}(f, m)$$

and extensions $ext(\mathsf{female}) = \{\mathsf{Ann}\}$ and $ext(\mathsf{male}) = \{\mathsf{Bill}\}$, and assume that there are no constraints imposed by a schema.

If both views are sound, we only know that some couple has Ann as its female component and Bill as its male component. Therefore, the query $Q_c(x, y) \leftarrow \mathsf{couple}(x, y)$ asking for all couples would return an empty answer, i.e., $ans(Q_c, \mathcal{S}, \mathcal{V}) = \emptyset$. However, if both views are exact, we can conclude that all couples have Ann as their female component and Bill as their male component, and hence that $(\mathsf{Ann}, \mathsf{Bill})$ is the only couple, i.e., $ans(Q_c, \mathcal{S}, \mathcal{V}) = (\mathsf{Ann}, \mathsf{Bill})$. ∎

3   Specifying the Content of the Data Integration System

We propose here an architecture for data integration that is coherent with the framework described in Section 2, and is based on Description Logics [9,8]. In such an architecture, to specify mediated schemas, views, and queries we use the Description Logic $\mathcal{DLR}$ [6]. We first introduce $\mathcal{DLR}$, and then we illustrate how we use the logic to specify the three components of a data integration system.

3.1   The Description Logic $\mathcal{DLR}$

Description Logics[2] (DLs) have been introduced in the early 80's in the attempt to provide a formal ground to Semantic Networks and Frames. Since then they have evolved into knowledge representation languages that are able to capture virtually all class-based representation formalisms used in Artificial Intelligence, Software Engineering, and Databases. One of the distinguishing features of the work on these logics is the detailed computational complexity analysis both of the associated reasoning algorithms, and of the logical implication problem that the algorithms are supposed to solve. By virtue of this analysis, most of these logics have optimal reasoning algorithms, and practical systems implementing such algorithms are now used in several projects. In DLs, the domain of interest is modeled by means of *concepts* and *relations*, which denote classes of objects and relationships, respectively.

Here, we focus our attention on the DL $\mathcal{DLR}$ [5,6]. The basic elements of $\mathcal{DLR}$ are *concepts* (unary relations), and *n-ary relations*. We assume to deal with an alphabet $\mathcal{A}$ constituted by a finite set of atomic relations, atomic concepts, and *constants*, denoted by P, A, and a, respectively. We use R to denote arbitrary relations (of given arity between 2 and n_{max}), and C to denote arbitrary concepts, respectively built according to the following syntax:

$$R ::= \top_n \mid P \mid \$i/n : C \mid \neg R \mid R_1 \sqcap R_2$$
$$C ::= \top_1 \mid A \mid \neg C \mid C_1 \sqcap C_2 \mid \exists[\$i]R \mid (\leq k\,[\$i]R)$$

where i denotes a component of a relation, i.e., an integer between 1 and n_{max}, n denotes the *arity* of a relation, i.e., an integer between 2 and n_{max}, and k denotes a nonnegative integer. We also use the following abbreviations:

---

[2] See http://dl.kr.org for the home page of Description Logics.

$$\begin{aligned}
\mathsf{T}_n^{\mathcal{I}} &\subseteq (\Delta^{\mathcal{I}})^n \\
P^{\mathcal{I}} &\subseteq \mathsf{T}_n^{\mathcal{I}} \\
\$i/n : C^{\mathcal{I}} &= \{(d_1, \ldots, d_n) \in \mathsf{T}_n^{\mathcal{I}} \mid d_i \in C^{\mathcal{I}}\} \\
(\neg R)^{\mathcal{I}} &= \mathsf{T}_n^{\mathcal{I}} \setminus R^{\mathcal{I}} \\
(R_1 \sqcap R_2)^{\mathcal{I}} &= R_1^{\mathcal{I}} \cap R_2^{\mathcal{I}} \\
\\
\mathsf{T}_1^{\mathcal{I}} &= \Delta^{\mathcal{I}} \\
A^{\mathcal{I}} &\subseteq \Delta^{\mathcal{I}} \\
(\neg C)^{\mathcal{I}} &= \Delta^{\mathcal{I}} \setminus C^{\mathcal{I}} \\
(C_1 \sqcap C_2)^{\mathcal{I}} &= C_1^{\mathcal{I}} \cap C_2^{\mathcal{I}} \\
(\exists[\$i]R)^{\mathcal{I}} &= \{d \in \Delta^{\mathcal{I}} \mid \exists(d_1, \ldots, d_n) \in R^{\mathcal{I}}. d_i = d\} \\
(\leq k\,[\$i]R)^{\mathcal{I}} &= \{d \in \Delta^{\mathcal{I}} \mid \sharp\{(d_1, \ldots, d_n) \in R_1^{\mathcal{I}} \mid d_i = d\} \leq k\}
\end{aligned}$$

Fig. 1. Semantic rules for $\mathcal{DLR}$ (P, R, R_1, and R_2 have arity n)

- $\bot$ for $\neg\top$,
- $C_1 \sqcup C_2$ for $\neg(\neg C_1 \sqcap \neg C_2)$,
- $C_1 \Rightarrow C_2$ for $\neg C_1 \sqcup C_2$, and
- $C_1 \equiv C_2$ for $(C_1 \Rightarrow C_2) \sqcap (C_2 \Rightarrow C_1)$.

We consider only concepts and relations that are *well-typed*, which means that

- only relations of the same arity n are combined to form expressions of type $R_1 \sqcap R_2$ (which inherit the arity n), and
- $i \leq n$ whenever i denotes a component of a relation of arity n.

The semantics of $\mathcal{DLR}$ is specified as follows. An *interpretation* $\mathcal{I}$ is constituted by an *interpretation domain* $\Delta^{\mathcal{I}}$, and an *interpretation function* $\cdot^{\mathcal{I}}$ that assigns to each constant an element of $\Delta^{\mathcal{I}}$ under the unique name assumption, to each concept C a subset $C^{\mathcal{I}}$ of $\Delta^{\mathcal{I}}$, and to each relation R of arity n a subset $R^{\mathcal{I}}$ of $(\Delta^{\mathcal{I}})^n$, such that the conditions in Figure 1 are satisfied. Observe that, the "$\neg$" constructor on relations is used to express difference of relations, and not the complement [6].

3.2 Mediated Schema, Views, and Queries

We remind the reader that a mediated schema is constituted by a finite set of constraints expressed in a schema language $\mathcal{L}_\mathcal{S}$. In our setting, the schema language $\mathcal{L}_\mathcal{S}$ is based on the DL $\mathcal{DLR}$. In particular, each constraint is formulated as an *assertion* of one of the following forms:

$$R_1 \sqsubseteq R_2 \qquad\qquad C_1 \sqsubseteq C_2$$

where R_1 and R_2 are $\mathcal{DLR}$ relations of the same arity, and C_1 and C_2 are $\mathcal{DLR}$ concepts.

As we said before, a database $\mathcal{DB}$ is a set of relations, one for each predicate symbol in the alphabet $\mathcal{A}$. We denote with $R^{\mathcal{DB}}$ the relation in $\mathcal{DB}$ corresponding to the predicate symbol R (either an atomic concept, or an atomic relation). Note that a database can be seen as an interpretation for $\mathcal{DLR}$, whose domain coincides with the set of constants in the alphabet $\mathcal{A}$.

We say that a database $\mathcal{DB}$ *satisfies* an assertion $R_1 \sqsubseteq R_2$ (resp., $C_1 \sqsubseteq C_2$) if $R_1^{\mathcal{DB}} \subseteq R_2^{\mathcal{DB}}$ (resp., $C_1^{\mathcal{DB}} \subseteq C_2^{\mathcal{DB}}$). Moreover, $\mathcal{DB}$ satisfies a schema $\mathcal{S}$ if $\mathcal{DB}$ satisfies all assertions in $\mathcal{S}$.

In order to define views and queries, we now introduce the notion of query expression in our setting. We assume that the alphabet $\mathcal{A}$ is enriched with a finite set of variable symbols, simply called variables.

A *query expression* Q is a non-recursive datalog query of the form

$$Q(\boldsymbol{x}) \;\leftarrow\; conj_1(\boldsymbol{x}, \boldsymbol{y}_1) \vee \cdots \vee conj_m(\boldsymbol{x}, \boldsymbol{y}_m)$$

where each $conj_i(\boldsymbol{x}, \boldsymbol{y}_i)$ is a conjunction of *atoms*, and $\boldsymbol{x}$, $\boldsymbol{y}_i$ are all the variables appearing in the conjunct. Each atom has one of the forms $R(\boldsymbol{t})$ or $C(t)$, where $\boldsymbol{t}$ and t are variables in $\boldsymbol{x}$ and $\boldsymbol{y}_i$ or constants in $\mathcal{A}$, R is a relation, and C is a concept. The number of variables of $\boldsymbol{x}$ is called the *arity* of Q, and is the arity of the relation denoted by the query Q.

We observe that the atoms in the query expressions are arbitrary $\mathcal{DLR}$ relations and concepts, freely used in the assertions of the KB. This distinguishes our approach with respect to [22,39], where no constraints on the relations that appear in the queries can be expressed in the KB.

Given a database $\mathcal{DB}$, a query expression Q of arity n is interpreted as the set $Q^{\mathcal{DB}}$ of n-tuples of constants $(c_1, \ldots, c_n)$, such that, when substituting each c_i for x_i, the formula

$$\exists \boldsymbol{y}_1.conj_1(\boldsymbol{x}, \boldsymbol{y}_1) \vee \cdots \vee \exists \boldsymbol{y}_m.conj_m(\boldsymbol{x}, \boldsymbol{y}_m)$$

evaluates to true in $\mathcal{DB}$.

With the introduction of query expressions, we can now define views and queries. Indeed, in our setting, query expressions constitute both the view language $\mathcal{L}_\mathcal{V}$, and the query language $\mathcal{L}_\mathcal{Q}$:

- Associated to each view V_i in the set $\mathcal{V} = \{V_1, \ldots, V_m\}$ we have:
  - A definition $def(V_i)$ in terms of a query expression
  - A set $ext(V_i)$ of tuples of constants,
  - A specification $as(V_i)$ of which *assumption* to adopt for the view V_i, where each $as(V_i)$ is either *sound*, *complete*, or *exact*.
- A query is simply a query expression, as defined above.

Example 2. Consider for example the following $\mathcal{DLR}$ schema $\mathcal{S}_d$, expressing that Americans who have a doctor as relative are wealthy, and that each surgeon is also a doctor

$$\mathsf{American} \sqcap \exists [\$1](\mathsf{RELATIVE} \sqcap \$2 : \mathsf{Doctor}) \sqsubseteq \mathsf{Wealthy}$$

$$\mathsf{Surgeon} \sqsubseteq \mathsf{Doctor}$$

and two sound views V_1 and V_2, respectively with definitions

$$V_1(x) \leftarrow \mathsf{RELATIVE}(x, y) \wedge \mathsf{Surgeon}(y)$$
$$V_2(x) \leftarrow \mathsf{American}(x)$$

and extensions

$$ext(V_1) = \{\mathsf{Ann}, \mathsf{Bill}\}$$
$$ext(V_2) = \{\mathsf{Ann}, \mathsf{Dan}\}$$

Given the query $Q_w(x) \leftarrow \mathsf{Wealthy}(x)$, asking for those who are wealthy, we have that the only constant in $ans(Q_w, \mathcal{S}_d, \mathcal{V})$ is Ann. Moreover, if we add an exact view V_3 with definition $V_3(x) \leftarrow \mathsf{Wealthy}(x)$, and an extension $ext(V_3)$ not containing Bill, then, from the constraints in $\mathcal{S}_d$ and the information we have on the views, we can conclude that Bill is not American. ∎

3.3 Discussion

We observe that $\mathcal{DLR}$ is able to capture a great variety of data models with many forms of constraints [15,6]. For example, $\mathcal{DLR}$ is capable to capture formally Conceptual Data Models typically used in databases [33,24], such as the Entity-Relationship Model [18]. Hence, in our setting, query answering using views is done under the constraints imposed by a conceptual data model.

The interest in $\mathcal{DLR}$ is not confined to the expressiveness it provides for specifying data schemas. It is also equipped with effective reasoning techniques that are sound and complete with respect to the semantics. In particular, checking whether a given assertion logically follows from a set of assertions is EXPTIME-complete in $\mathcal{DLR}$ (assuming that numbers are encoded in unary), and query containment, i.e., checking whether one query is contained in another one in every model of a set of assertions, is EXPTIME-hard and solvable in 2EXP-TIME [6].

4   Query Answering

In this section we study the problem of query answering using views in the setting just defined: the schema is expressed as a $\mathcal{DLR}$ knowledge base, and queries and view definitions are espressed as $\mathcal{DLR}$ query expressions. We call the resulting problem *answering query using views in* $\mathcal{DLR}$. The technical results regarding answering query using views in $\mathcal{DLR}$ illustrated in this section are taken from [7].

The first thing to observe is that, given a schema $\mathcal{S}$ expressed in $\mathcal{DLR}$, a set of views $\mathcal{V} = \{V_1, \ldots, V_m\}$, a query Q, and a tuple $\boldsymbol{d} = (d_1, \ldots, d_n)$ of constants, verifying whether, $\boldsymbol{d}$ is in $ans(Q, \mathcal{S}, \mathcal{V})$ is essentially a form of logical implication. This observation can be made even sharper if we introduce special assertions, expressed in first-order logic with equality, that encode as logical formulas the extension of the views. In particular, for each *view* $V \in \mathcal{V}$, with $def(V) = (V(\boldsymbol{x}) \leftarrow v(\boldsymbol{x}, \boldsymbol{y}))$ and $ext(V) = \{\boldsymbol{a}_1, \ldots, \boldsymbol{a}_k\}$, we introduce the following assertions.

- If V is *sound*, then for each tuple $\boldsymbol{a}_i$, $1 \le i \le k$, we introduce the existentially quantified assertion

$$\exists \boldsymbol{y}.v(\boldsymbol{a}_i, \boldsymbol{y})$$

- If V is *complete*, then we introduce the universally quantified assertion

$$\forall \boldsymbol{x}.\forall \boldsymbol{y}.((\boldsymbol{x} \ne \boldsymbol{a}_1 \wedge \cdots \wedge \boldsymbol{x} \ne \boldsymbol{a}_k) \rightarrow \neg v(\boldsymbol{x}, \boldsymbol{y}))$$

- If V is *exact*, then, according to the definition, we treat it as a view that is both sound and complete, and introduce both types of assertions above.

Let us call $Ext(\mathcal{V})$ the set of assertions corresponding to the extension of the views $\mathcal{V}$.

Now, the problem of query answering using views in $\mathcal{DLR}$, i.e., checking whether $\boldsymbol{d} \in ans(Q, \mathcal{S}, \mathcal{V})$, can be reformulated as checking whether the following logical implication holds:

$$\mathcal{S} \cup Ext(\mathcal{V}) \models \exists \boldsymbol{y}.q(\boldsymbol{d}, \boldsymbol{y})$$

where $q(\boldsymbol{x}, \boldsymbol{y})$ is the right hand part of Q. Checking such a logical implication can in turn be rephrased as checking the unsatisfiability of

$$\mathcal{S} \cup Ext(\mathcal{V}) \cup \{\forall \boldsymbol{y}.\neg q(\boldsymbol{d}, \boldsymbol{y})\}$$

Observe that the assertion $\forall \boldsymbol{y}.\neg q(\boldsymbol{d}, \boldsymbol{y})$ has the same form as the universal assertion used for expressing extensions of complete views, except that the antecedent in the implication is empty.

The problem with the newly introduced assertions is that they are not yet expressed in a DL. The next step is to translate them in a DL. Instead of working directly with $\mathcal{DLR}$, we are going to translate the problem of query answering using views in $\mathcal{DLR}$ to reasoning in a DL, called $\mathcal{CIQ}$, that directly corresponds to a variant of Propositional Dynamic Logic [20,6].

4.1 The Description Logic $\mathcal{CIQ}$

The DL $\mathcal{CIQ}$ is obtained from $\mathcal{DLR}$ by restricting relations to be binary (such relations are called *roles* and *inverse roles*) and allowing for complex roles corresponding to regular expressions [20]. Concepts of $\mathcal{CIQ}$ are formed according to the following abstract syntax:

$$C ::= \top \mid A \mid C_1 \sqcap C_2 \mid \neg C \mid \exists R.C \mid (\le k\,Q.C)$$
$$Q ::= P \mid P^-$$
$$R ::= Q \mid R_1 \sqcup R_2 \mid R_1 \circ R_2 \mid R^* \mid R^- \mid id(C)$$

where A denotes an atomic concept, C a generic concept, P an atomic role, Q a *simple role*, i.e., either an atomic role or the inverse of an atomic role, and R a generic role. We also use the following abbreviations:

$$A^{\mathcal{I}} \subseteq \Delta^{\mathcal{I}}$$
$$\top^{\mathcal{I}} = \Delta^{\mathcal{I}}$$
$$(\neg C)^{\mathcal{I}} = \Delta^{\mathcal{I}} \setminus C^{\mathcal{I}}$$
$$(C_1 \sqcap C_2)^{\mathcal{I}} = C_1^{\mathcal{I}} \cap C_2^{\mathcal{I}}$$
$$(\exists R.C)^{\mathcal{I}} = \{d \in \Delta^{\mathcal{I}} \mid \exists (d,d') \in R^{\mathcal{I}}. d' \in C^{\mathcal{I}}\}$$
$$(\leq k\, Q.\, C)^{\mathcal{I}} = \{d \in \Delta^{\mathcal{I}} \mid \sharp\{(d,d') \in Q^{\mathcal{I}} \mid d' \in C^{\mathcal{I}}\} \leq k\}$$

$$P^{\mathcal{I}} \subseteq \Delta^{\mathcal{I}} \times \Delta^{\mathcal{I}}$$
$$(R_1 \sqcup R_2)^{\mathcal{I}} = R_1^{\mathcal{I}} \cup R_2^{\mathcal{I}}$$
$$(R_1 \circ R_2)^{\mathcal{I}} = R_1^{\mathcal{I}} \circ R_2^{\mathcal{I}}$$
$$(R^*)^{\mathcal{I}} = (R^{\mathcal{I}})^* = \bigcup_{i \geq 0} (R^{\mathcal{I}})^i$$
$$(R^-)^{\mathcal{I}} = \{(d_1, d_2) \in \Delta^{\mathcal{I}} \times \Delta^{\mathcal{I}} \mid (d_2, d_1) \in R^{\mathcal{I}}\}$$
$$id(C)^{\mathcal{I}} = \{(d, d) \in \Delta^{\mathcal{I}} \times \Delta^{\mathcal{I}} \mid d \in C^{\mathcal{I}}\}$$

Fig. 2. Semantic rules for $\mathcal{CIQ}$

- $\forall R.C$ for $\neg \exists R. \neg C$,
- $(\geq k\, Q.\, C)$ for $\neg(\leq k{-}1\, Q.\, C)$

The semantic conditions for $\mathcal{CIQ}$ are specified in Figure 2 [3].

The use of $\mathcal{CIQ}$ allows us to exploit various results established recently for reasoning in such a logic. The basis of these results lies in the correspondence between $\mathcal{CIQ}$ and a variant of Propositional Dynamic Logic [26,35] that includes converse programs and "graded modalities" [25,52] on atomic programs and their converse [47]. $\mathcal{CIQ}$ inherits from Propositional Dynamic Logics the ability of *internalizing* assertions. Indeed, one can define a role U that essentially corresponds to a *universal modality*, as the reflexive-transitive closure of all roles and inverse roles in the language. Using such a universal modality we can re-express each assertion $C_1 \sqsubseteq C_2$ as the concept $\forall U.(C_1 \Rightarrow C_2)$. This allows us to re-express logical implication as concept satisfiability [47]. Concept satisfiability (and hence logical implication) in $\mathcal{CIQ}$ is EXPTIME-complete [20].

Although $\mathcal{CIQ}$ does not have constructs for n-ary relations as $\mathcal{DLR}$, it is possible to represent n-ary relations in a *sound and complete* way wrt concept satisfiability (and hence logical implication) by means of *reification* [20]. An atomic relation P is reified by introducing a new atomic concept A_P and n functional roles $f_1, \ldots, f_n$, one for each component of P. In this way, a tuple of the relation is represented by an instance of the corresponding concept, which is linked through each of the associated roles to an object representing the component of the tuple. Performing the reification requires however some attention, since in a relation there may not be two equal tuples (i.e., constituted by the same components in the same positions) in its extension. In the reified counterpart, on the other hand, one cannot explicitly rule out (e.g., by using specific assertions) that there are two objects o_1 and o_2 "representing" the same tuple, i.e., that are connected to exactly the same objects denoting the components of

---

[3] The notation $(R^{\mathcal{I}})^i$ stands for i repetitions of $R^{\mathcal{I}}$ – i.e., $(R^{\mathcal{I}})^1 = R^{\mathcal{I}}$, and $(R^{\mathcal{I}})^i = R^{\mathcal{I}} \circ (R^{\mathcal{I}})^{i-1}$.

the tuple. However, due to the fundamental inability of $\mathcal{CIQ}$ to express that two role sequences meet in the same object, no $\mathcal{CIQ}$ concept can force such a situation. Therefore one does not need to take this constraint explicitly into account when reasoning.

Finally, we are going to make use of $\mathcal{CIQ}$ extended with object-names. An *object-name* is an atomic concept that, in each model, has as extension a single object. Object-names are not required to be disjoint, i.e, we do not make the unique name assumption on them. Disjointness can be explicitly enforced when needed through explicit assertions. In general, adding object-names to $\mathcal{CIQ}$ makes reasoning NEXPTIME-hard [49]. However our use of object-names in $\mathcal{CIQ}$ is restricted so as to keep reasoning in EXPTIME.

4.2 Reduction of Answering Queries Using Views in $\mathcal{DLR}$ to $\mathcal{CIQ}$ Unsatisfiability

We tackle answering queries using views in $\mathcal{DLR}$, by reducing the problem of checking whether $\boldsymbol{d} \in ans(Q, \mathcal{S}, \mathcal{V})$ to the problem of checking the unsatisfiability of a $\mathcal{CIQ}$ concept in which object-names appear. Object-names are then eliminated, thus obtaining a $\mathcal{CIQ}$ concept.

We translate $\mathcal{S} \cup Ext(\mathcal{V})$ into a $\mathcal{CIQ}$ concept as follows. First, we eliminate n-ary relations by means of *reification*. Then, we reformulate each assertion in $\mathcal{S}$ as a concept by internalizing assertions. Instead, representing assertions in $Ext(\mathcal{V})$ requires the following ad-hoc techniques.

We translate each existentially quantified assertion

$$\exists \boldsymbol{y}.v(\boldsymbol{a}, \boldsymbol{y})$$

as follows. We represent every constant a_i by an object-name N_{a_i}, enforcing disjointness between the object-names corresponding to different constants. We represent each existentially quantified variable y, treated as a Skolem constant, by a new object-name without disjointness constraints. We also use additional concept-names representing tuples of objects. Specifically:

- An atom $C(t)$, where C is a concept and t is a term (either a constant or a variable), is translated to

$$\forall U.(N_t \Rightarrow \sigma(C))$$

  where $\sigma(C)$ is the reified counterpart of C, N_t is the object-name corresponding to t, and U is the reflexive-transitive closure of all roles and inverse roles introduced in the reification.
- An atom $R(\boldsymbol{t})$, where R is a relation of arity n and $\boldsymbol{t} = (t_1, \ldots, t_n)$ is a tuple of terms, is translated to the conjunction of the following concepts:

$$\forall U.(N_{\boldsymbol{t}} \Rightarrow \sigma(R))$$

  where $\sigma(R)$ is the reified counterpart of R and $N_{\boldsymbol{t}}$ is an object-name corresponding to $\boldsymbol{t}$,

$$\forall U.(N_{\boldsymbol{t}} \equiv (\exists f_1.N_{t_1} \sqcap \cdots \sqcap \exists f_n.N_{t_n}))$$

and for each i, $1 \leq i \leq n$, a concept

$$\forall U.(N_{t_i} \Rightarrow ((\exists f_i^-.N_t) \sqcap (\leq 1 f_i^-.N_t)))$$

Then, the translations of the atoms are combined as in $v(\boldsymbol{a}, \boldsymbol{y})$.

To translate universally quantified assertions corresponding to the complete views and also to the query, it is sufficient to deal with assertions of the form:

$$\forall \boldsymbol{x}.\forall \boldsymbol{y}.((\boldsymbol{x} \neq \boldsymbol{a}_1 \wedge \cdots \wedge \boldsymbol{x} \neq \boldsymbol{a}_k) \rightarrow \neg conj(\boldsymbol{x}, \boldsymbol{y}))$$

Following [6], we construct for $conj(\boldsymbol{x}, \boldsymbol{y})$ a special graph, called *tuple-graph*, which reflects the dependencies between variables. Specifically, the tuple-graph is used to detect cyclic dependencies. In general, the tuple-graph is composed of $\ell \geq 1$ connected components. For the i-th connected component we build a $\mathcal{CIQ}$ concept $\delta_i(\boldsymbol{x}, \boldsymbol{y})$ as in [6]. Such a concept contains newly introduced concepts A_x and A_y, one for each x in $\boldsymbol{x}$ and y in $\boldsymbol{y}$. We have to treat variables in $\boldsymbol{x}$ and $\boldsymbol{y}$ that occur in a cycle in the tuple-graph differently from those outside of cycles. Let $\boldsymbol{x}_c$ (resp., $\boldsymbol{y}_c$) denote the variables in $\boldsymbol{x}$ (resp., $\boldsymbol{y}$) that occur in a cycle, and $\boldsymbol{x}_l$ (resp., $\boldsymbol{y}_l$) those that do not occur in cycles. We first define the concept

$$C[\boldsymbol{x}_c/\boldsymbol{s}, \boldsymbol{y}_c/\boldsymbol{t}]$$

as the concept obtained from

$$(\forall U.\neg\delta_1(\boldsymbol{x}, \boldsymbol{y})) \sqcup \cdots \sqcup (\forall U.\neg\delta_\ell(\boldsymbol{x}, \boldsymbol{y}))$$

as follows:

- for each variable x_i in $\boldsymbol{x}_c$ (resp., y_i in $\boldsymbol{y}_c$), the concept A_{x_i} (resp., A_{y_i}) is replaced by N_{s_i} (resp., N_{t_i});
- for each variable y_i in $\boldsymbol{y}_l$, the concept A_{y_i} is replaced by $\top$.

Then the concept corresponding to the universally quantified assertion is constructed as the conjunction of:

- $\forall U.C_{\boldsymbol{x}_l}$, where $C_{\boldsymbol{x}_l}$ is obtained from $\boldsymbol{x} \neq \boldsymbol{a}_1 \wedge \cdots \wedge \boldsymbol{x} \neq \boldsymbol{a}_k$ by replacing each $(x \neq a)$ with $(A_x \equiv \neg N_a)$. Observe that $(x_1, \ldots, x_n) \neq (a_1, \ldots, a_n)$ is an abbreviation for $(x_1 \neq a_1 \vee \cdots \vee x_n \neq a_n)$.
- One concept $C[\boldsymbol{x}_c/\boldsymbol{s}, \boldsymbol{y}_c/\boldsymbol{t}]$ for each possible instantiation of $\boldsymbol{s}$ and $\boldsymbol{t}$ with the constants in $Ext(\mathcal{V}) \cup \{\boldsymbol{d}\}$, with the proviso that $\boldsymbol{s}$ cannot coincide with any of the $\boldsymbol{a}_i$, for $1 \leq i \leq k$ (notice that the proviso applies only in the case where all variables in $\boldsymbol{x}$ occur in a cycle in the tuple-graph).

The critical point in the above construction is how to express a universally quantified assertion

$$\forall \boldsymbol{x}.\forall \boldsymbol{y}.((\boldsymbol{x} \neq \boldsymbol{a}_1 \wedge \cdots \wedge \boldsymbol{x} \neq \boldsymbol{a}_k) \rightarrow \neg conj(\boldsymbol{x}, \boldsymbol{y}))$$

If there are no cycles in the corresponding tuple-graph, then we can directly translate the assertion into a $\mathcal{CIQ}$ concept. As shown in the construction above, dealing with a nonempty antecedent requires some special care to correctly encode the exceptions to the universal rule. Instead, if there is a cycle, due to the fundamental inability of $\mathcal{CIQ}$ to express that two role sequences meet in the same object, no $\mathcal{CIQ}$ concept can directly express the universal assertion. The same inability, however, is shared by $\mathcal{DLR}$. Hence we can assume that the only cycles present in a model are those formed by the constants in the extension of the views or those in the tuple for which we are checking whether it is a certain answer of the query. And these are taken care of by the explicit instantiation.

As the last step to obtain a $\mathcal{CIQ}$ concept, we need to encode object-names in $\mathcal{CIQ}$. To do so we can exploit the construction used in [21] to encode $\mathcal{CIQ}$-ABoxes as concepts. Such a construction applies to the current case without any need of major adaptation. It is crucial to observe that the translation above uses object-names in order to form a sort of disjunction of ABoxes (cfr. [31]).

In [7], the following basic fact is proved for the construction presented above. Let C_{qa} be the $\mathcal{CIQ}$ concept obtained by the construction above. Then $\boldsymbol{d} \in ans(Q, \mathcal{S}, \mathcal{V})$ if and only if C_{qa} is unsatisfiable.

The size of C_{qa} is polynomial in the size of the query, of the view definitions, and of the inclusion assertions in $\mathcal{S}$, and is at most exponential in the number of constants in $ext(\mathcal{V}) \cup \{\boldsymbol{d}\}$. The exponential blow-up is due to the number of instantiations of $C[\boldsymbol{x}_c/\boldsymbol{s}, \boldsymbol{y}_c/\boldsymbol{t}]$ with constants in $ext(\mathcal{V}) \cup \{\boldsymbol{d}\}$ that are needed to capture universally quantified assertions. Hence, considering EXPTIME-completeness of satisfiability in $\mathcal{DLR}$ and in $\mathcal{CIQ}$, we get that query answering using views in $\mathcal{DLR}$ is EXPTIME-hard and can be done in 2EXPTIME.

5   Related Work

We already observed that query answering using views can be seen as a form of reasoning with incomplete information. The interested reader is referred to [53] for a survey on this subject.

We also observe that, to compute the whole set $ans(Q, \mathcal{S}, \mathcal{V})$, we need to run the algorithm presented above once for each possible tuple (of the arity of Q) of objects in the view extensions. Since we are dealing with incomplete information in a rich language, we should not expect to do much better than considering each tuple of objects separately. Indeed, in such a setting reasoning on objects, such as query answering, requires sophisticated forms of logical inference. In particular, verifying whether a certain tuple belongs to a query gives rise to a line of reasoning which may depend on the tuple under consideration, and which may vary substantially from one tuple to another. For simple languages we may indeed avoid considering tuples individually, as shown in [45] for query answering in the DL $\mathcal{ALN}$ without cyclic TBox assertions. Observe, however, that for such a DL, reasoning on objects is polynomial in both data and expression complexity [36,46], and does not require sophisticated forms of inference.

Query answering using views has been investigated in the last years in the context of simplified frameworks. In [38,44], the problem has been studied for the case of conjunctive queries (with or without arithmetic comparisons), in [2] for disjunctive views, in [48,19,30] for queries with aggregates, in [23] for recursive queries and nonrecursive views, and in [11,12] for several variants of regular path queries. Comprehensive frameworks for view-based query answering, as well as several interesting results for various query languages, are presented in [29,1].

Query answering using views is tightly related to query rewriting [38,23,51]. In particular, [3] studies rewriting of conjunctive queries using conjunctive views whose atoms are DL concepts or roles (the DL used is less expressive thatn $\mathcal{DLR}$). In general, a *rewriting* of a query with respect to a set of views is a function that, given the extensions of the views, returns a set of tuples that is contained in the answer set of the query with respect to the views. Usually, one fixes a priori the language in which to express rewritings (e.g., unions of conjunctive queries), and then looks for the best possible rewriting expressible in such a language. On the other hand, we may call *perfect* a rewriting that returns exactly the answer set of the query with respect to the views, independently of the language in which it is expressed. Hence, if an algorithm for answering queries using views exists, it can be viewed as a perfect rewriting [13,14]. The results presented here show the existence of perfect, and hence maximal, rewritings in a setting where the mediated schema, the views, and the query are expressed in $\mathcal{DLR}$.

6   Conclusions

We have illustrated a logic-based framework for data integration, and in particular for the problem of query answering using views in a data integration system. We have addressed the problem for the case of non-recursive datalog queries posed to a mediated schema expressed in $\mathcal{DLR}$. We have considered different assumptions on the view extensions (sound, complete, and exact), and we have presented a technique that solves the problem in 2EXPTIME worst case computational complexity.

We have seen in the previous section that an algorithm for answering queries using views is in fact a perfect rewriting. For the setting presented here, it remains open to find perfect rewritings expressed in a more declarative query language. Moreover it is of interest to find maximal rewritings belonging to well behaved query languages, in particular, languages with polynomial data complexity, even though we already know that such rewritings cannot be perfect [13].

Acknowledgments

The work presented here was partly supported by the ESPRIT LTR Project No. 22469 DWQ – Foundations of Data Warehouse Quality, and by MURST Cofin 2000 D2I – From Data to Integration. We wish to thank all members of the projects. Also, we thank Daniele Nardi, Riccardo Rosati, and Moshe Y. Vardi, who contributed to several ideas illustrated in the chapter.

References

1. Serge Abiteboul and Oliver Duschka. Complexity of answering queries using materialized views. In *Proc. of the 17th ACM SIGACT SIGMOD SIGART Symp. on Principles of Database Systems (PODS'98)*, pages 254–265, 1998.
2. Foto N. Afrati, Manolis Gergatsoulis, and Theodoros Kavalieros. Answering queries using materialized views with disjunction. In *Proc. of the 7th Int. Conf. on Database Theory (ICDT'99)*, volume 1540 of *Lecture Notes in Computer Science*, pages 435–452. Springer-Verlag, 1999.
3. Catriel Beeri, Alon Y. Levy, and Marie-Christine Rousset. Rewriting queries using views in description logics. In *Proc. of the 16th ACM SIGACT SIGMOD SIGART Symp. on Principles of Database Systems (PODS'97)*, pages 99–108, 1997.
4. Mokrane Bouzeghoub and Maurizio Lenzerini. Special issue on data extraction, cleaning, and reconciliation. *Information Systems*, 26(8), pages 535–536, 2001.
5. Diego Calvanese, Giuseppe De Giacomo, and Maurizio Lenzerini. Conjunctive query containment in Description Logics with *n*-ary relations. In *Proc. of the 1997 Description Logic Workshop (DL'97)*, pages 5–9, 1997.
6. Diego Calvanese, Giuseppe De Giacomo, and Maurizio Lenzerini. On the decidability of query containment under constraints. In *Proc. of the 17th ACM SIGACT SIGMOD SIGART Symp. on Principles of Database Systems (PODS'98)*, pages 149–158, 1998.
7. Diego Calvanese, Giuseppe De Giacomo, and Maurizio Lenzerini. Answering queries using views over description logics knowledge bases. In *Proc. of the 17th Nat. Conf. on Artificial Intelligence (AAAI 2000)*, pages 386–391, 2000.
8. Diego Calvanese, Giuseppe De Giacomo, Maurizio Lenzerini, Daniele Nardi, and Riccardo Rosati. Description logic framework for information integration. In *Proc. of the 6th Int. Conf. on Principles of Knowledge Representation and Reasoning (KR'98)*, pages 2–13, 1998.
9. Diego Calvanese, Giuseppe De Giacomo, Maurizio Lenzerini, Daniele Nardi, and Riccardo Rosati. Information integration: Conceptual modeling and reasoning support. In *Proc. of the 6th Int. Conf. on Cooperative Information Systems (CoopIS'98)*, pages 280–291, 1998.
10. Diego Calvanese, Giuseppe De Giacomo, Maurizio Lenzerini, Daniele Nardi, and Riccardo Rosati. Data integration in data warehousing. *Int. J. of Cooperative Information Systems*, 10(3), pages 237–271, 2001.
11. Diego Calvanese, Giuseppe De Giacomo, Maurizio Lenzerini, and Moshe Y. Vardi. Answering regular path queries using views. In *Proc. of the 16th IEEE Int. Conf. on Data Engineering (ICDE 2000)*, pages 389–398, 2000.
12. Diego Calvanese, Giuseppe De Giacomo, Maurizio Lenzerini, and Moshe Y. Vardi. Query processing using views for regular path queries with inverse. In *Proc. of the 19th ACM SIGACT SIGMOD SIGART Symp. on Principles of Database Systems (PODS 2000)*, pages 58–66, 2000.
13. Diego Calvanese, Giuseppe De Giacomo, Maurizio Lenzerini, and Moshe Y. Vardi. View-based query processing and constraint satisfaction. In *Proc. of the 15th IEEE Symp. on Logic in Computer Science (LICS 2000)*, pages 361–371, 2000.
14. Diego Calvanese, Giuseppe De Giacomo, Maurizio Lenzerini, and Moshe Y. Vardi. What is query rewriting? In *Proc. of the 7th Int. Workshop on Knowledge Representation meets Databases (KRDB 2000)*, pages 17–27. CEUR Electronic Workshop Proceedings, http://sunsite.informatik.rwth-aachen.de/Publications/CEUR-WS/Vol-29/, 2000.

15. Diego Calvanese, Maurizio Lenzerini, and Daniele Nardi. Description logics for conceptual data modeling. In Jan Chomicki and Günter Saake, editors, *Logics for Databases and Information Systems*, pages 229–264. Kluwer Academic Publisher, 1998.

16. Tiziana Catarci and Maurizio Lenzerini. Representing and using interschema knowledge in cooperative information systems. *J. of Intelligent and Cooperative Information Systems*, 2(4):375–398, 1993.

17. S. Chaudhuri, S. Krishnamurthy, S. Potarnianos, and K. Shim. Optimizing queries with materialized views. In *Proc. of the 11th IEEE Int. Conf. on Data Engineering (ICDE'95)*, Taipei (Taiwan), 1995.

18. P. P. Chen. The Entity-Relationship model: Toward a unified view of data. *ACM Trans. on Database Systems*, 1(1):9–36, March 1976.

19. Sara Cohen, Werner Nutt, and Alexander Serebrenik. Rewriting aggregate queries using views. In *Proc. of the 18th ACM SIGACT SIGMOD SIGART Symp. on Principles of Database Systems (PODS'99)*, pages 155–166, 1999.

20. Giuseppe De Giacomo and Maurizio Lenzerini. What's in an aggregate: Foundations for description logics with tuples and sets. In *Proc. of the 14th Int. Joint Conf. on Artificial Intelligence (IJCAI'95)*, pages 801–807, 1995.

21. Giuseppe De Giacomo and Maurizio Lenzerini. TBox and ABox reasoning in expressive description logics. In Luigia C. Aiello, John Doyle, and Stuart C. Shapiro, editors, *Proc. of the 5th Int. Conf. on the Principles of Knowledge Representation and Reasoning (KR'96)*, pages 316–327. Morgan Kaufmann, Los Altos, 1996.

22. Francesco M. Donini, Maurizio Lenzerini, Daniele Nardi, and Andrea Schaerf. $\mathcal{AL}$-log: Integrating Datalog and description logics. *J. of Intelligent Information Systems*, 10(3):227–252, 1998.

23. Oliver M. Duschka and Michael R. Genesereth. Answering recursive queries using views. In *Proc. of the 16th ACM SIGACT SIGMOD SIGART Symp. on Principles of Database Systems (PODS'97)*, pages 109–116, 1997.

24. Ramez A. ElMasri and Shamkant B. Navathe. *Fundamentals of Database Systems*. Benjamin and Cummings Publ. Co., Menlo Park, California, 1988.

25. M. Fattorosi-Barnaba and F. De Caro. Graded modalities I. *Studia Logica*, 44:197–221, 1985.

26. Michael J. Fischer and Richard E. Ladner. Propositional dynamic logic of regular programs. *J. of Computer and System Sciences*, 18:194–211, 1979.

27. Daniela Florescu, Alon Y. Levy, Ioana Manolescu, and Dan Suciu. Query optimization in the presence of limited access patterns. In *Proc. of the ACM SIGMOD Int. Conf. on Management of Data*, pages 311–322, 1999.

28. Helena Galhardas, Daniela Florescu, Dennis Shasha, and Eric Simon. An extensible framework for data cleaning. Technical Report 3742, INRIA, Rocquencourt, 1999.

29. Gösta Grahne and Alberto O. Mendelzon. Tableau techniques for querying information sources through global schemas. In *Proc. of the 7th Int. Conf. on Database Theory (ICDT'99)*, volume 1540 of *Lecture Notes in Computer Science*, pages 332–347. Springer-Verlag, 1999.

30. Stéphane Grumbach, Maurizio Rafanelli, and Leonardo Tininini. Querying aggregate data. In *Proc. of the 18th ACM SIGACT SIGMOD SIGART Symp. on Principles of Database Systems (PODS'99)*, pages 174–184, 1999.

31. Ian Horrocks, Ulrike Sattler, Sergio Tessaris, and Stephan Tobies. Query containment using a DLR ABox. Technical Report LTCS-Report 99-15, RWTH Aachen, 1999.

32. Michael N. Huhns, Nigel Jacobs, Tomasz Ksiezyk, Wei-Min Shen an Munindar P. Singh, and Philip E. Cannata. Integrating enterprise information models in Carnot. In *Proc. of the Int. Conf. on Cooperative Information Systems (CoopIS'93)*, pages 32–42, 1993.
33. R. B. Hull and R. King. Semantic database modelling: Survey, applications and research issues. *ACM Computing Surveys*, 19(3):201–260, September 1987.
34. Matthias Jarke, Maurizio Lenzerini, Yannis Vassiliou, and Panos Vassiliadis, editors. *Fundamentals of Data Warehouses*. Springer-Verlag, 1999.
35. Dexter Kozen and Jerzy Tiuryn. Logics of programs. In Jan van Leeuwen, editor, *Handbook of Theoretical Computer Science — Formal Models and Semantics*, pages 789–840. Elsevier Science Publishers (North-Holland), Amsterdam, 1990.
36. Maurizio Lenzerini and Andrea Schaerf. Concept languages as query languages. In *Proc. of the 9th Nat. Conf. on Artificial Intelligence (AAAI'91)*, pages 471–476, 1991.
37. Alon Y. Levy. Obtaining complete answers from incomplete databases. In *Proc. of the 22nd Int. Conf. on Very Large Data Bases (VLDB'96)*, pages 402–412, 1996.
38. Alon Y. Levy, Alberto O. Mendelzon, Yehoshua Sagiv, and Divesh Srivastava. Answering queries using views. In *Proc. of the 14th ACM SIGACT SIGMOD SIGART Symp. on Principles of Database Systems (PODS'95)*, pages 95–104, 1995.
39. Alon Y. Levy and Marie-Christine Rousset. CARIN: A representation language combining Horn rules and description logics. In *Proc. of the 12th Eur. Conf. on Artificial Intelligence (ECAI'96)*, pages 323–327, 1996.
40. Alon Y. Levy, Divesh Srivastava, and Thomas Kirk. Data model and query evaluation in global information systems. *J. of Intelligent Information Systems*, 5:121–143, 1995.
41. Chen Li and Edward Chang. Query planning with limited source capabilities. In *Proc. of the 16th IEEE Int. Conf. on Data Engineering (ICDE 2000)*, pages 401–412, 2000.
42. Chen Li and Edward Chang. On answering queries in the presence of limited access patterns. In *Proc. of the 8th Int. Conf. on Database Theory (ICDT 2001)*, 2001.
43. Chen Li, Ramana Yerneni, Vasilis Vassalos, Hector Garcia-Molina, Yannis Papakonstantinou, Jeffrey D. Ullman, and Murty Valiveti. Capability based mediation in TSIMMIS. In *Proc. of the ACM SIGMOD Int. Conf. on Management of Data*, pages 564–566, 1998.
44. Anand Rajaraman, Yehoshua Sagiv, and Jeffrey D. Ullman. Answering queries using templates with binding patterns. In *Proc. of the 14th ACM SIGACT SIGMOD SIGART Symp. on Principles of Database Systems (PODS'95)*, 1995.
45. Marie-Christine Rousset. Backward reasoning in ABoxes for query answering. In *Proc. of the 1999 Description Logic Workshop (DL'99)*, pages 18–22. CEUR Electronic Workshop Proceedings, http://sunsite.informatik.rwth-aachen. de/Publications/CEUR-WS/Vol-22/, 1999.
46. Andrea Schaerf. *Query Answering in Concept-Based Knowledge Representation Systems: Algorithms, Complexity, and Semantic Issues*. PhD thesis, Dipartimento di Informatica e Sistemistica, Università di Roma "La Sapienza", 1994.
47. Klaus Schild. A correspondence theory for terminological logics: Preliminary report. In *Proc. of the 12th Int. Joint Conf. on Artificial Intelligence (IJCAI'91)*, pages 466–471, Sydney (Australia), 1991.
48. D. Srivastava, S. Dar, H. V. Jagadish, and A. Levy. Answering queries with aggregation using views. In *Proc. of the 22nd Int. Conf. on Very Large Data Bases (VLDB'96)*, pages 318–329, 1996.

49. Stephan Tobies. The complexity of reasoning with cardinality restrictions and nominals in expressive description logics. *J. of Artificial Intelligence Research*, 12:199–217, 2000.
50. O. G. Tsatalos, M. H. Solomon, and Y. E. Ioannidis. The GMAP: A versatile tool for phyisical data independence. *Very Large Database J.*, 5(2):101–118, 1996.
51. Jeffrey D. Ullman. Information integration using logical views. In *Proc. of the 6th Int. Conf. on Database Theory (ICDT'97)*, volume 1186 of *Lecture Notes in Computer Science*, pages 19–40. Springer-Verlag, 1997.
52. Wiebe Van der Hoek and Maarten de Rijke. Counting objects. *J. of Logic and Computation*, 5(3):325–345, 1995.
53. Ron van der Meyden. Logical approaches to incomplete information. In Jan Chomicki and Günter Saake, editors, *Logics for Databases and Information Systems*, pages 307–356. Kluwer Academic Publisher, 1998.
54. Jennifer Widom. Special issue on materialized views and data warehousing. *IEEE Bulletin on Data Engineering*, 18(2), 1995.
55. Ramana Yerneni, Chen Li, Hector Garcia-Molina, and Jeffrey D. Ullman. Computing capabilities of mediators. In *Proc. of the ACM SIGMOD Int. Conf. on Management of Data*, pages 443–454, 1999.
56. Ramana Yerneni, Chen Li, Jeffrey D. Ullman, and Hector Garcia-Molina. Optimizing large join queries in mediation systems. In *Proc. of the 7th Int. Conf. on Database Theory (ICDT'99)*, pages 348–364, 1999.

Search and Optimization Problems in Datalog*

Sergio Greco[1,2] and Domenico Saccà[1,2]

[1] DEIS, Univ. della Calabria, 87030 Rende, Italy
[2] ISI-CNR, 87030 Rende, Italy
{greco,sacca}@deis.unical.it

Abstract. This paper analyzes the ability of DATALOG languages to express search and optimization problems. It is first shown that $\mathcal{NP}$ search problems can be formulated as unstratified DATALOG queries under non-deterministic stable model semantics so that each stable model corresponds to a possible solution. $\mathcal{NP}$ optimization problems are then formulated by adding a *max* (or *min*) construct to select the stable model (thus, the solution) which maximizes (resp., minimizes) the result of a polynomial function applied to the answer relation. In order to enable a simpler and more intuitive formulation for search and optimization problems, it is introduced a DATALOG language in which the use of stable model semantics is disciplined to refrain from abstruse forms of unstratified negation. The core of our language is stratified negation extended with two constructs allowing nondeterministic selections and with query goals enforcing conditions to be satisfied by stable models. The language is modular as the level of expressivity can be tuned and selected by means of a suitable use of the above constructs, thus capturing significant subclasses of search and optimization queries.

1 Introduction

DATALOG is a logic-programming language that was designed for database applications, mainly because of its declarative style and its ability to express recursive queries[3,32]. Later DATALOG has been extended along many directions (e.g., various forms of negations, aggregate predicates and set constructs) to enhance its expressive power. In this paper we investigate the ability of DATALOG languages to express search and optimization problems.

We recall that, given an alphabet Σ, a search problem is a partial multivalued function f, defined on some (not necessarily proper) subset of Σ^*, say $dom(f)$, which maps every string x of $dom(f)$ into a number of strings $y_1, \cdots, y_n$ $(n > 0)$, thus $f(x) = \{y_1, \cdots, y_n\}$. The function f is therefore represented by the following relation on $\Sigma^* \times \Sigma^*$: $graph(f) = \{(x,y) \mid x \in dom(x) \text{ and } y \in f(x)\}$. We say that $graph(f)$ is polynomially balanced if for each (x,y) in $graph(f)$, the size of y is polynomially bounded in the size of x. $\mathcal{NP}$ *search problems* are those functions

---

* Work partially supported by the Italian National Research Council (CNR) and by MURST (projects DATA-X and D2I).

A.C. Kakas, F. Sadri (Eds.): Computat. Logic (Kowalski Festschrift), LNAI 2408, pp. 61–82, 2002.
© Springer-Verlag Berlin Heidelberg 2002

f for which both $graph(f)$ is polynomially balanced and $graph(f)$ is in $\mathcal{NP}$, i.e., given $x, y \in \Sigma^*$, deciding whether $(x, y) \in graph(f)$ is in $\mathcal{NP}$.

In this paper we show that $\mathcal{NP}$ search problems can be formulated as DATALOG¬ (i.e., DATALOG with unstratified negation) queries under the nondeterministic version of total stable model semantics[11], thus the meaning of a DATALOG¬ program is given by any stable model. As an example of the language, take the *Vertex Cover* problem: given a graph $G = (V, E)$, find a vertex cover — a subset V' of V is a *vertex cover* of G if for each pair edge (x, y) in E either x or y is in V'. The problem can be formulated by the query $\langle P_{vc}, v'(X) \rangle$ where P_{vc} is the following DATALOG¬ program:

$$v'(X) \leftarrow v(X), \neg v"(X).$$
$$v"(X) \leftarrow v(X), \neg v'(X).$$
$$\text{no\_cover} \leftarrow e(X, Y), \neg v'(X), \neg v'(Y).$$
$$\text{refuse\_no\_cover} \leftarrow \text{no\_cover}, \neg \text{refuse\_no\_cover}.$$

The predicates v and e define the vertices and the edges of the graph by means of a suitable number of facts. The last rule enforces that every total stable model correspond to some vertex cover (otherwise *no_cover* would be true and, then, the atom *refuse_no_cover* would result undefined).

In order to enable a simpler and more intuitive formulation of search problems, we introduce a DATALOG language where the usage of stable model semantics is disciplined to avoid both undefinedness and unnecessary computational complexity, and to refrain from abstruse forms of unstratified negation. Thus the core of our language is stratified negation extended with two constructs (choice and subset) allowing nondeterministic selections and an additional ground goal (called *constraint goal*) in the query, enforcing conditions to be satisfied by stable models. For instance, the above query can be formulated as $\langle P_{vc'}, !\neg no\_cover, v'(X) \rangle$ where $P_{vc'}$ is the following stratified DATALOG¬ program with a subset construct to nondeterministically select a subset of the vertices:

$$v'(X) \subseteq v(X).$$
$$\text{no\_cover} \leftarrow e(X, Y), \neg v'(X), \neg v'(Y).$$

The constraint goal $!\neg no\_cover$ specifies that only those stable models by which $\neg no\_cover$ is made true are to be taken into consideration.

The expressive power (and the complexity as well) of the language gradually increases by moving from the basic language (stratified DATALOG¬) up to the whole repertoire of additional constructs. Observe that, if we do not add any constraint goal in the query, the query reduces to a stratified program with additional constructs for nondeterministic selections, which cannot be eventually retracted, thus avoiding exponential explosion of the search space. For example, the query $\langle P_{st}, st(X, Y) \rangle$, where P_{st} is defined below, computes a spanning tree of the graph G in polynomial time:

$$st(\text{nil}, X) \leftarrow v(X), \text{choice}((), (X)).$$
$$st(X, Y) \leftarrow st(Z, X), e(X, Y), st(\text{nil}, Z), Y \neq Z, Y \neq X, \text{choice}((X), (Y)).$$

The first choice selects any vertex of the graph as the root of the tree; the second choice selects one vertex y at a time to be added to the current spanning tree st so that y is connected to exactly one vertex x of st, thus preserving the tree structure. Polynomial-time computation is guaranteed since nondeterministic selections made by the choice constructs cannot be eventually discarded because there is no constraint goal to satisfy as in the example of vertex cover. Observe that also a vertex cover can be computed in polynomial time; thus we may rewrite the above query using the choice construct without constraint goal so that polynomial-time computation is guaranteed. Obviously, this kind of rewriting is not feasible for all $\mathcal{NP}$ search queries as they can be $\mathcal{NP}$ hard.

In the paper we characterize various classes of search queries, including tractable classes (for which an answer can be computed in polynomial time), and we show how such classes can be captured by a suitably disciplined usage of our DATALOG$^\neg$ language.

In the paper we also deal with the issue of formulating optimization problems. We recall that an optimization (min or max) problem, associated to a search problem f, is a function g such that $dom(g) = dom(f)$ and for each $x \in dom(g)$, $g(x) = \{y \mid y \in f(x)$ and for each other $y' \in f(x)$, $|y| \leq |y'|$ (or $|y| \geq |y'|$ if is a maximization problem)$\}$. The optimization problems associated to $\mathcal{NP}$ search problems are called $\mathcal{NP}$ optimization problems.

We show that $\mathcal{NP}$ optimization problems can be formulated as DATALOG$^\neg$ queries under the non deterministic version of total stable model semantics by using a max (or min) construct to select the model which maximizes (resp., minimizes) the cardinality of the answer relation. As an example of the language, take the *Min Vertex Cover* problem: given a graph $G = (V, E)$, find the vertex cover with minimal cardinality. The problem can be formulated by the query $\langle P_{vc'}, !\neg no\_cover, min(v'(X)) \rangle$ where $P_{vc'}$ is the above program. The goal $min(v'(X))$ further restricts the set of suitable stable models to those for which the subset of nodes v' is minimum.

The advantage of expressing $\mathcal{NP}$ search and $\mathcal{NP}$ optimization problems by using rules with built-in predicates rather than standard DATALOG$^\neg$ rules, is that the use of built-in atoms preserves simplicity and intuition in expressing problems and permits to perform query optimization. The language is 'modular' in the sense that the desired level of expressivity is achieved by enabling the constructs for non-stratified negation only when needed; in particular, if no constraint goal and min/max goal are used then polynomial time computation is guaranteed.

The paper is organized as follows. In Section 2 we introduce search and optimization queries and provide a formal ground for their classification using results from complexity theory on multivalued functions. In Section 3 we prove that $\mathcal{NP}$ search queries coincide with DATALOG$^\neg$ queries under nondeterministic total stable model semantics. We also introduce the min/max goal to capture $\mathcal{NP}$ optimization queries. In order to capture meaningful subclasses of $\mathcal{NP}$ search and optimization queries, in Section 4 we then present our language, called DATALOG$^{\neg s,c}$, and we show its ability of expressing tractable $\mathcal{NP}$ search problems. We also prove that optimization problems can be hard also when associated to tractable search problems. This explains the renewed attention [26,25,19,20,6,7] towards optimization problems, mainly with the aim of characterizing classes of problems that are constant or log approximable (i.e., there is a polynomial time algorithm that approximates the optimum value of the problem within a factor that is respectively constant or logarithmic in the size of the input). In Section 5 we introduce suitable restrictions to DATALOG$^{\neg s,c}$ in order to capture $\mathcal{NP}$ optimization subclasses that are approximable and present meaningful examples. We draw conclusions and discuss further work in Section 6.

2   Search and Optimization Queries

We assume that the reader is familiar with the basic terminology and notation of relational databases and of database queries [3,18,32].

A *relational database scheme* $\mathcal{DB}$ over a fixed countable domain U is a set of relation symbols $\{r_1, ..., r_k\}$ where each r_i has a given arity, denoted by $|r_i|$. A *database* D on $\mathcal{DB}$ is a finite structure $(A, R_1, ..., R_k)$ where $A \subseteq U$ is the *active domain* and $R_i \subseteq A^{|r_i|}$ are the (finite) *relations* of the database, one for each relation scheme r_i — we denote A by $U(D)$ and R_i by $D(r_i)$. We assume that a database is suitably encoded by a string and the recognition of whether a string represents a database on $\mathcal{DB}$ is done in polynomial time.

Definition 1. Given a database scheme $\mathcal{DB}$ and an additional relation symbol f (the query *goal*), a *search query* $NQ = \langle \mathcal{DB}, f \rangle$ is a (possibly partial) multivalued recursive function which maps every database D on $\mathcal{DB}$ to a finite, non-empty set of finite (possibly empty) relations $F \subseteq U(D)^{|f|}$ and is invariant under an isomorphism on $U - W$, where W is any finite subset of U (i.e., the function is W-*generic*). Thus $NQ(D)$ yields a set of relations on the goal, that are the *answers* of the query; the query has no answer if this set is empty or the function is not defined on D.

The class of all search queries is denoted by $\mathcal{NQ}$.   □

In classifying query classes, we shall refer to the following complexity classes of languages: the class $\mathcal{P}$ (languages that are recognized by deterministic Turing machines in polynomial time), the class $\mathcal{NP}$ (languages that are recognized by nondeterministic Turing machines in polynomial time), and the class $co\mathcal{NP}$ (the complement of $\mathcal{NP}$) — the reader can refer to [10,17,24] for excellent sources of information on this subject.

As search queries correspond to functions rather than to languages as it instead happens for boolean queries, we next introduce, for their classification, some background on complexity of functions (for a more comprehensive description of this topic we address readers to [30,31,9]).

Let a finite *alphabet* Σ with at least two elements be given. A *partial multivalued* (*MV*) *function* $f : \Sigma^* \mapsto \Sigma^*$ associates zero, one or several outcomes (*outputs*) to each input string. Let $f(x)$ stands for the set of possible results of f on an input string x; thus, we write $y \in f(x)$ if y is a value of f on the input string x. Define $dom(f) = \{x \mid \exists y(y \in f(x))\}$ and $graph(f) = \{\langle x, y \rangle \mid x \in$

$dom(f)$, $y \in f(x)$}. If $x \notin dom(f)$, we will say that f is undefined at x. It is now clear that a search query is indeed a computable MV function: the input x is a suitable encoding of a database D and each output string y encodes an answer of the query.

A computable (i.e., partial recursive) MV function f is computed by some Turing transducer, i.e., a (deterministic or not) Turing machine T which, in addition to accept any string $x \in dom(f)$, writes a string $y \in f(x)$ on an output tape before entering the accepting state. So, if $x \in dom(f)$, the set of all strings that are written in all accepting computations is $f(x)$; on the other hand, if $x \notin dom(f)$, T never enters the accepting state.

Given two MV functions f and g, define g to be a *refinement* of f if $dom(g) = dom(f)$ and $graph(g) \subseteq graph(f)$. Moreover, given a class G of MV functions, we say that $f \in_c G$ if G contains a refinement of f. For a class of MV functions F, define $F \subseteq_c G$ if, for all $f \in F$, $f \in_c G$. Since we are in general interested in finding any output of a MV function, an important practical question is whether an output can be efficiently computed by means of a polynomial-time, single-valued function. In other terms, desirable MV function classes are those which are refined by $\mathcal{PF}$, where $\mathcal{PF}$ is the class of all functions that are computed by deterministic polynomial-time transducers.

Let us now recall some important classes of MV functions. A MV function f is *polynomially balanced* if, for each x, the size of each result in $f(x)$ is polynomially bounded in the size of x. The class $\mathcal{NPMV}$ is defined as the set of all MV functions f such that (i) f is polynomially balanced, and (ii) $graph(f)$ is in $\mathcal{NP}$. By analogy, the classes $\mathcal{NPMV}_g$ and $co\mathcal{NPMV}$ are defined as the classes of all polynomially-balanced multivalued functions f for which $graph(f)$ is respectively in $\mathcal{P}$ and in $co\mathcal{NP}$. Observe that $\mathcal{NPMV}$ consists of all MV functions that are computed by nondeterministic transducers in polynomial time [30].

Definition 2.

1. $\mathcal{NQPMV}$ (resp., $\mathcal{NQPMV}_g$ and $co\mathcal{NQPMV}$) is the class of all search queries which are in $\mathcal{NPMV}$ (resp., $\mathcal{NPMV}_g$ and $co\mathcal{NPMV}$) — we shall also call the queries in this class $\mathcal{NP}$ *search queries*;
2. $\mathcal{NQPTIME}$ is the class of all queries that are computed by a nondeterministic polynomial-time transducer for which every computation path ends in an accepting state;
3. $\mathcal{NQPTIME}_g$ is equal to $\mathcal{NQPTIME} \cap \mathcal{NQPMV}_g$.   □

Observe that a query $NQ = \langle \mathcal{DB}, f \rangle$ is in $\mathcal{NQPMV}$ (resp., $\mathcal{NQPMV}_g$ and $co\mathcal{NQPMV}$) if and only if for each database D on $\mathcal{DB}$ and for each relation F on f, deciding whether F is in $NQ(D)$ is in $\mathcal{NP}$ (resp., in $\mathcal{P}$ and in $co\mathcal{NP}$).

We stress that $\mathcal{NQPMV}$ is different from the class $\mathcal{NQPTIME}$ first introduced in [1,2] — in fact, the latter class consists of all queries in $\mathcal{NQPMV}$ for which acceptance is guaranteed no matter which nondeterministic moves are guessed by the transducer.

We next present some results on whether the above query classes can be refined by $\mathcal{PF}$, thus whether a query answer in these classes can be computed in deterministic polynomial time — the results have been proven in [21].

Fact 1 *[21]*

1. $\mathcal{NQPMV}_g \subseteq (\mathcal{NQPMV} \cap co\mathcal{NQPMV})$ *and the inclusion is strict unless* $\mathcal{P} = \mathcal{NP}$;
2. *neither* $\mathcal{NQPMV} \subseteq co\mathcal{NQPMV}$ *nor* $co\mathcal{NQPMV} \subseteq \mathcal{NQPMV}$ *unless* $\mathcal{NP} = co\mathcal{NP}$;
3. $\mathcal{NQPTIME} \subset \mathcal{NQPMV}$, $\mathcal{NQPTIME} \not\subseteq co\mathcal{NQPMV}$ *unless* $\mathcal{NP} = co\mathcal{NP}$, *and* $\mathcal{NQPTIME}_g \subseteq \mathcal{NQPTIME}$ *and the inclusion is strict unless* $\mathcal{P} = \mathcal{NP}$;
4. $\mathcal{NQPTIME} \subseteq_c \mathcal{PF}$ *and* $\mathcal{NQPMV}_g \not\subseteq_c \mathcal{PF}$ *unless* $\mathcal{P} = \mathcal{NP}$. $\qquad\square$

It turns out that queries in $\mathcal{NQPTIME}$ and $\mathcal{NQPTIME}_g$ can be efficiently computed whereas queries in the other classes may not. Observe that queries in $\mathcal{NQPTIME}$ have a strange anomaly: computing an answer can be done in polynomial time, but testing whether a given relation is an answer cannot (unless $\mathcal{P} = \mathcal{NP}$). This anomaly does not occur in the class $\mathcal{NQPTIME}_g$ which, therefore, turns out to be very desirable.

Example 1. Let a database scheme $\mathcal{DB}_G = \{v, e\}$ represent a directed graph $G = (V, E)$ such that v has arity 1 and defines the nodes while e has arity 2 and defines the edges. We recall that a *kernel* is a subset V' of V such that (i) no two nodes in V' are joined by an edge and (ii) for each node x not in V', there is a node y in V' for which $(y, x) \in E$.

- NQ_{Kernel} is the query which returns the kernels of the input graph G; if the graph has no kernel then the query is not defined. The query is in $\mathcal{NQPMV}_g$, but an answer cannot be computed in polynomial time unless $\mathcal{P} = \mathcal{NP}$ since deciding whether a graph has a kernel is $\mathcal{NP}$-complete [10].
- $NQ_{SubKernel}$ is the query that, given an input graph G, returns any subset of some kernel of G. This query is in $\mathcal{NQPMV}$, but neither in $\mathcal{NQPMV}_g$ (unless $\mathcal{P} = \mathcal{NP}$) nor in $co\mathcal{NQPMV}$ (unless $\mathcal{NP} = co\mathcal{NP}$).
- $NQ_{NodeNoK}$ is the query that, given an input graph G, returns a node not belonging to any kernel of G. This query is in $co\mathcal{NQPMV}$, but not in $\mathcal{NQPMV}$ (unless $\mathcal{NP} = co\mathcal{NP}$).
- NQ_{01K} is the query that, given a graph G, returns the relation $\{0\}$ if G has no kernel, $\{1\}$ if every subset of nodes of G is a kernel, both relations $\{0\}$ and $\{1\}$ otherwise. Clearly, the query is in $\mathcal{NQPTIME}$: indeed it is easy to construct a non-deterministic polynomial-time transducer which first non-deterministically generates any subset of nodes of G and then outputs $\{1\}$ or $\{0\}$ according to whether this subset is a kernel or not. The query is not in $\mathcal{NQPTIME}_g$ otherwise we could check in polynomial time if a graph has a kernel – as the graph has a kernel iff $\{1\}$ is a result of NQ_{01K} – and, therefore, $\mathcal{P}$ would coincide with $\mathcal{NP}$.
- NQ_{CUT} is the query which returns a subset E' of the edges such that the graph $G' = (V, E')$ is 2-colorable. The query is in $\mathcal{NQPTIME}_g$. $\qquad\square$

According to Fagin's well-known result [8], a class of finite structures is $\mathcal{NP}$-recognizable iff it is definable by a second order existential formula, thus queries in $\mathcal{NQPMV}$ may be expressed as follows.

Fact 2 *Let* $NQ = \langle \mathcal{DB}, f \rangle$ *be a search query in* $\mathcal{NQPMV}$, *then there is a sequence* $\mathcal{S}$ *of relation symbols* $s_1, \ldots, s_k$, *distinct from those in* $\mathcal{DB} \cup \{f\}$, *and a closed first-order formula* $\phi(\mathcal{DB}, f, \mathcal{S})$ *such that for each database* D *on* $\mathcal{DB}$, $NQ(D) = \{ F : F \subseteq U(D)^{|f|}, S_i \subseteq U(D)^{|s_i|} (1 \leq i \leq k), \text{ and } \phi(D, F, S) \text{ is true } \}$. □

From now on, we shall formulate a query in $\mathcal{NQPMV}$ as $NQ = \{ f : (\mathcal{DB}, f, \mathcal{S}) \models \phi(\mathcal{DB}, f, \mathcal{S}) \}$.

Example 2. CUT. The query NQ_{CUT} of Example 1 can be defined as follows:

$$\{ e' : (\mathcal{DB}_G, e', s) \models (\forall x, y)[e'(x, y) \rightarrow ((e(x, y) \wedge s(x) \wedge \neg s(y))$$
$$\vee (e(x, y) \wedge \neg s(x) \wedge s(y)))] \}. \quad \square$$

Example 3. KERNEL. The query NQ_{Kernel} of Example 1 can be defined as:

$$\{ v' : (\mathcal{DB}_G, v') \models (\forall x) [(v'(x) \wedge \forall y(\neg v'(y) \vee \neg e(x, y)))$$
$$\vee (\neg v'(x) \wedge \exists y(v'(y) \wedge e(y, x)))] \} \quad \square$$

Definition 3. Given a search query $NQ = \langle \mathcal{DB}, f \rangle$, an *optimization query* $OQ = opt(NQ) = \langle \mathcal{DB}, opt(f) \rangle$, where opt is either *max* or *min*, is a search query refining NQ such that for each database D on $\mathcal{DB}$ for which NQ is defined, $OQ(D) = opt_{|F|}\{ F : F \in NQ(D) \}$ — i.e., $OQ(D)$ consists of the answers in $NQ(D)$ with the maximum or minimum (resp., if $opt = max$ or min) cardinality. The query NQ is called the *search query associated to* OQ and the relations in $NQ(D)$ are the *feasible solutions* of OQ.

The class of all optimization queries is denoted by $OPT\,\mathcal{NQ}$. Given a search class QC, the class of all queries whose search queries are in QC is denoted by $OPT\,QC$. The queries in the class $OPT\,\mathcal{NQPMV}$ are called $\mathcal{NP}$ *optimization queries*. □

Proposition 1. *Let* $OQ = \langle \mathcal{DB}, opt|f| \rangle$ *be an optimization query, then the following statements are equivalent:*

1. *OQ is in $OPT\,\mathcal{NQPMV}$.*
2. *There is a closed first-order formula* $\phi(\mathcal{DB}, f, \mathcal{S})$ *over relation symbols* $\mathcal{DB} \cup \{f\} \cup \mathcal{S}$ *such that* $OQ = opt_{|f|}\{ f : (\mathcal{DB}, f, \mathcal{S}) \models \phi(\mathcal{DB}, f, \mathcal{S}) \}$.
3. *There is a first-order formula* $\phi(\mathbf{w}, \mathcal{DB}, \mathcal{S})$, *where* $\mathbf{w}$ *is a* $a(f)$-*tuple of distinct variables, such that the relation symbols are those in* $\mathcal{DB} \cup \mathcal{S}$, *the free variables are exactly those in* $\mathbf{w}$, *and* $OQ = opt_{|w|}\{ \mathbf{w} : (\mathcal{DB}, \mathcal{S}) \models \phi(\mathbf{w}, \mathcal{DB}, \mathcal{S}) \}$.

PROOF. The equivalence of statements (1) and (2) is obvious. Clearly optimization formulae defined in Item 2 (called *feasible* in [20]) are a special case of first order optimization formulae defined in Item 3 which define the class $OPT\,PB$, of all optimization problems that can be logically defined. Moreover, in [20] it has been shown that the class $OPT\,PB$, can be expressed by means of feasible optimization first order formulae. □

The above results pinpoint that the class $OPT\,NQPMV$ corresponds to the class $OPT\,PB$ of all optimization problems that can be logically defined [19,20]. For simplicity, but without substantial loss of generality, we use as objective function the cardinality rather than a generic polynomial-time computable function. Moreover, we output the relation with the optimal cardinality rather than just the cardinality.

Example 4. MAX-CUT. The problem consists in finding the cardinality of the largest cut in the graph $G = (V, E)$. The query coincides with $max(NQ_{cut})$ (see Example 2) and can also be defined as:

$$max(\{\,(x,y) : (\mathcal{DB}_G, s) \models [(e(x,y) \wedge s(x) \wedge \neg s(y)) \vee (e(x,y) \wedge \neg s(x) \wedge s(y))]\}).$$

The query is an NP maximization query. □

Example 5. MIN-KERNEL. In this case we want to find the minimum cardinality of the kernels of a graph $G = (V, E)$. The query is $min(NQ_{kernel})$ (see Example 3) and can be equivalently defined as:

$$min(\{\,w : (\mathcal{DB}_G, v') \models v'(w) \vee \neg(\forall x)\,[\,(v'(x) \wedge \forall y(\neg v'(y) \vee \neg e(x,y)))$$
$$\vee(\neg v'(x) \wedge \exists y(v'(y) \wedge e(y,x)))]\,\})$$

This query is a NP minimization query. □

Finally, note that the query $max(NQ_{Kernel})$ equals the query $max(NQ_{SubKernel})$ although their search queries are distinct. The following results show that in general optimization queries are much harder than search queries, e.g., they cannot be solved in polynomial time even when the associated query is in $NQPTIME_g$.

Proposition 2.

1. *neither $OPT\,NQPMV \subseteq coNQPMV$ nor $OPT\,NQPMV \subseteq NQPMV$ unless $NP = coNP$;*
2. *neither $OPT\,coNQPMV \subseteq coNQPMV$ nor $OPT\,coNQPMV \subseteq NQPMV$ unless $NP = coNP$;*
3. *$OPT\,NQPMV_g \subset coNQPMV$ and $OPT\,NQPMV_g \not\subseteq NQPMV$ unless $NP = coNP$;*
4. *neither $OPT\,NQPTIME \subseteq coNPMV$ nor $OPT\,NQPTIME \subseteq NQPMV$ unless $NP = coNP$;*
5. *$OPT\,NQPTIME_g \subset coNQPMV$ and $OPT\,NQPTIME_g \not\subseteq NQPMV_g$ unless $P = NP$.*

PROOF.

1. Let $max\,Q$ be a query in $MAX\,\mathcal{NQPMV}$ — the same argument would hold also for a minimization query. Then, given a database D, to decide whether a relation f is an answer of $max\,Q(D)$, first we have to test whether f is an answer of $Q(D)$ and, then, we must verify that there is no other answer of $Q(D)$ with fewer tuples than f. As the former test is in $\mathcal{NP}$ and the latter test is in $co\mathcal{NP}$, it is easy to see that deciding whether f is an answer of $max\,Q(D)$ is neither in $\mathcal{NP}$ nor in $co\mathcal{NP}$ unless $\mathcal{NP}{=}co\mathcal{NP}$ — indeed it is in the class $\mathcal{D}^P$ [24].

2. Let us now assume that the query in the proof of part (1) is in $MAX\,co\mathcal{NQPMV}$. Then testing whether f is an answer of $Q(D)$ is in $co\mathcal{NP}$ whereas verifying that there is no other answer of $Q(D)$ with fewer tuples than f is in $co\mathcal{NP}^{\mathcal{NP}}$, that is a class at the second level of the polynomial hierarchy [24].

3. Suppose now that the query in the proof of part (1) is in $MAX\,\mathcal{NQPMV}_g$. Then testing whether f is an answer of $Q(D)$ is in $\mathcal{P}$ whereas verifying that there is no other answer of $Q(D)$ with fewer tuples than f is in $co\mathcal{NP}$.

4. Take any query $max\,Q$ in $MAX\,\mathcal{NQPMV}$. We construct the query Q' by setting $Q'(D) = Q(D) \cup \{\emptyset\}$ for each D. Then Q' is in $\mathcal{NQPTIME}$ as the transducer for Q' can now accept on every branch by eventually returning the empty relation. It is now easy to see that the complexity of finding the maximum answer for Q' is in general the same of finding the maximum answer for Q. So the results follow from part (1).

5. $OPT\,\mathcal{NQPTIME}_g \subset co\mathcal{NQPMV}$ follows from part (3) as $\mathcal{NQPTIME}_g \subset \mathcal{NQPMV}_g$ by definition. Consider now the query Q returning a maximal clique (i.e., a clique which is not contained in another one) of an undirected graph. Q is obviously in $\mathcal{NQPTIME}_g$ as a maximal clique can be constructed by selecting any node and adding additional nodes as long as the clique property is preserved. We have that $max\,Q$ is the query returning the maximum clique in a graph (i.e., the maximal clique with the maximum number of nodes) which is known to be $\mathcal{NP}$-hard.   □

3   Search and Optimization Queries in DATALOG

We assume that the reader is familiar with basic notions of logic programming and DATALOG¬ [3,22,32].

A *program* P is a finite set of rules r of the form $H(r) \leftarrow B(r)$, where $H(r)$ is an atom (*head* of the rule) and $B(r)$ is a conjunction of literals (*body* of the rule). A rule with empty body is called a *fact*. The *ground instantiation* of P is denoted by $ground(P)$; the *Herbrand universe* and the *Herbrand base* of P are denoted by U_P and B_P, respectively.

An interpretation $I \subseteq B_P$ is a *T-stable* (total stable) *model* [11] if $I = \mathbf{T}^{\infty}_{pos(P,I)}(\emptyset)$, where $\mathbf{T}$ is the classical *immediate consequence transformation* and $pos(P, I)$ denotes the positive logic program that is obtained from $ground(P)$

by (i) removing all rules r such that there exists a negative literal $\neg A$ in $B(r)$ and A is in I, and (ii) by removing all negative literals from the remaining rules. It is well-known that a program may have n T-stable models with $n \geq 0$.

Given a program P and two predicate symbols p and q, we write $p \to q$ if there exists a rule where q occurs in the head and p in the body or there exists a predicate s such that $p \to s$ and $s \to q$. A program is *stratified* if there exists no rule where a predicate p occurs in a negative literal in the body, q occurs in the head and $q \to p$, i.e. there is no recursion through negation [5]. Stratified programs have a unique stable model which coincides with the *stratified model*, obtained by partitioning the program into an ordered number of suitable subprograms (called 'strata') and computing the fixpoints of every stratum in their order [5].

A DATALOG¬ program is a logic program with negation in the rule bodies, but without functions symbols. Predicate symbols can be either extensional (i.e. defined by the facts of a database — *EDB predicate symbols*) or intensional (i.e. defined by the rules of the program — *IDB predicate symbols*). The class of all DATALOG¬ programs is simply called DATALOG¬; the subclass of all positive (resp. stratified) programs is called DATALOG (resp. DATALOG¬ˢ).

A DATALOG¬ program P has associated a relational database scheme $\mathcal{DB}_P$, which consists of all EDB predicate symbols of P. We assume that possible constants in P are taken from the same domain U of $\mathcal{DB}_P$.

Given a database D on $\mathcal{DB}_P$, the tuples of D are seen as facts added to P; so P on D yields the following logic program $P_D = P \cup \{q(t). : q \in \mathcal{DB}_P \wedge t \in D(q)\}$. Given a T-stable model M of P_D and a relation symbol r in P_D, $M(r)$ denotes the relation $\{t : r(t) \in M\}$.

Definition 4. A DATALOG¬ *search query* $\langle P, f \rangle$, where P is a DATALOG¬ program and f is an IDB predicate symbol of P, defines the query $NQ = \langle \mathcal{DB}_P, f \rangle$ such that for each D on DB_P, $NQ(D) = \{M(f) : M \text{ is a T-stable model of } P_D\}$. The set of all DATALOG¬, DATALOG or DATALOG¬ˢ search queries are denoted respectively by $search(\text{DATALOG}^\neg)$, $search(\text{DATALOG})$ and $search(\text{DATALOG}^{\neg s})$.

The DATALOG¬ *optimization query* $\langle P, opt(f) \rangle$ defines the optimization query $opt(NQ)$. The set of all DATALOG¬, DATALOG or DATALOG¬ˢ optimization queries are denoted respectively by $opt(\text{DATALOG}^\neg)$, $opt(\text{DATALOG})$ and $opt(\text{DATALOG}^{\neg s})$. □

Observe that, given a database D, if the program P_D has no stable models then both search and optimization queries are not defined on D.

Proposition 3.

1. $search(\text{DATALOG}^\neg) = \mathcal{NQPMV}$ and $opt(\text{DATALOG}^\neg) = OPT\mathcal{NQPMV}$;
2. $search(\text{DATALOG}) \subset search(\text{DATALOG}^{\neg s}) \subset \mathcal{NQPTIME}_g$.

PROOF. In [28] it has been shown that a database query NQ is defined by a query in $search(\text{DATALOG}^\neg)$ if and only if, for each input database, the answers of NQ are $\mathcal{NP}$-recognizable. Hence $search(\text{DATALOG}^\neg) = \mathcal{NQPMV}$ and $opt(\text{DATALOG}^\neg) = OPT\mathcal{NQPMV}$. Concerning part (2), observe that queries in $search(\text{DATALOG}^{\neg s})$ are a proper subset of deterministic polynomial-time queries

[5] and then $search(\text{DATALOG}^{\neg s}) \subset \mathcal{NQPTIME}_g$. Finally, the relationship $search$ $(\text{DATALOG}) \subset search(\text{DATALOG}^{\neg s})$ is well known in the literature [3].   □

Note that $search(\text{DATALOG}) = opt(\text{DATALOG})$ and $search(\text{DATALOG}^{\neg s}) = opt$ $(\text{DATALOG}^{\neg s})$ as the queries are deterministic.

Example 6. Take the queries NQ_{cut} and $max(NQ_{cut})$ of Examples 2 and 4, respectively. Consider the following DATALOG$^{\neg}$ program P_{cut}

$$\begin{aligned}
&\texttt{v}'(\texttt{X}) &&\leftarrow \texttt{v}(\texttt{X}),\ \neg\hat{\texttt{v}}'(\texttt{X}).\\
&\hat{\texttt{v}}'(\texttt{X}) &&\leftarrow \texttt{v}(\texttt{X}),\ \neg\texttt{v}'(\texttt{X}).\\
&\texttt{e}'(\texttt{X},\texttt{Y}) &&\leftarrow \texttt{e}(\texttt{X},\texttt{Y}),\ \texttt{v}'(\texttt{X}),\ \neg\texttt{v}'(\texttt{Y}).\\
&\texttt{e}'(\texttt{X},\texttt{Y}) &&\leftarrow \texttt{e}(\texttt{X},\texttt{Y}),\ \neg\texttt{v}'(\texttt{X}),\ \texttt{v}'(\texttt{Y}).
\end{aligned}$$

We have that $NQ_{cut} = \langle P_{cut}, e' \rangle$ and $max(NQ_{cut}) = \langle P_{cut}, max(e') \rangle$.   □

Example 7. Take the queries NQ_{kernel} and $min(NQ_{kernel})$ of Examples 3 and 5. Consider the following DATALOG$^{\neg}$ program P_{kernel}

$$\begin{aligned}
&\texttt{v}'(\texttt{X}) &&\leftarrow \texttt{v}(\texttt{X}),\ \neg\hat{\texttt{v}}'(\texttt{X}).\\
&\hat{\texttt{v}}'(\texttt{X}) &&\leftarrow \texttt{v}(\texttt{X}),\ \neg\texttt{v}'(\texttt{X}).\\
&\texttt{joined\_to\_v}'(\texttt{X}) &&\leftarrow \texttt{v}'(\texttt{Y}),\ \texttt{e}(\texttt{Y},\texttt{X}).\\
&\texttt{no\_kernel} &&\leftarrow \texttt{v}'(\texttt{X}),\ \texttt{joined\_to\_v}'(\texttt{X}).\\
&\texttt{no\_kernel} &&\leftarrow \hat{\texttt{v}}'(\texttt{X}),\ \neg\texttt{joined\_to\_v}'(\texttt{X}).\\
&\texttt{constraint} &&\leftarrow \neg\texttt{no\_kernel},\ \neg\texttt{constraint}.
\end{aligned}$$

We have that $NQ_{kernel} = \langle P_{kernel}, v' \rangle$ and $min(NQ_{kernel}) = \langle P_{kernel}, min(v') \rangle$. Observe that P_{kernel} has no T-stable model iff NQ_{kernel} is not defined on D (i.e., there is no kernel).   □

The problem in using DATALOG$^{\neg}$ to express search and optimization problems is that the usage of unrestricted negation in programs is often neither simple nor intuitive and, besides, it does not allow to discipline the expressive power (e.g., the classes $\mathcal{NQPTIME}$ and $\mathcal{NQPTIME}_g$ are not captured). This situation might lead to write queries that have no total stable models or whose computation is hard even though the problem is not. On the other hand, as pointed out in Proposition 3, if we just use DATALOG$^{\neg s}$ the expressive power is too low so that we cannot express simple polynomial-time problems. For instance, the query asking for a spanning tree of an undirected graph needs the use of a program with unstratified negation such as:

(1) $\texttt{reached(a)}.$
(2) $\texttt{reached(Y)} \qquad \leftarrow \texttt{spanTree(X,Y)}.$
(3) $\texttt{spanTree(X,Y)} \quad \leftarrow \texttt{reached(X)},\ \texttt{e(X,Y)},\ \texttt{Y} \neq \texttt{a},\ \neg\texttt{diffChoice(X,Y)}.$
(4) $\texttt{diffChoice(X,Y)} \leftarrow \texttt{spanTree(Z,Y)},\ \texttt{Z} \neq \texttt{X}.$

But the freedom in the usage of negation may result in meaningless programs. For instance, in the above program, in an attempt to simplify it, one could decide to modify the third rule into

(3′) spanTree(X, Y) ← reached(X), arc(X, Y), Y ≠ a, ¬reached(Y).

and remove the fourth rule. Then the resulting program will have no total stable models, thus loosing its practical meaning. Of course the risk of writing meaningless programs is present in any language, but this risk is much higher in a language with non-intuitive semantics as for unstratified negation.

In the next section we propose a language where the usage of stable model semantics is disciplined to avoid both undefinedness and unnecessary computational complexity, and to refrain from abstruse forms of unstratified negation. The core of the language is stratified DATALOG extended with only one type of non-stratified negation, hardwired into two ad-hoc constructs. The disciplined structure of negation in our language will enable us to capture interesting subclasses of $\mathcal{NQPMV}$.

4   Datalog Languages for Search and Optimization Problems

In this section we analyze the expressive power of several languages derived from DATALOG¬ by restricting the use of negation. In particular, we consider the combination of stratified negation, a nondeterministic construct, called *choice* and *subset* rules computing subsets of tuples of a given relation.

The *choice* construct is supported by several deductive database systems such as $\mathcal{LDL}++$ [33] and Coral [27], and it is used to enforce functional constraints on rules of a logic program. Thus, a goal of the form, $choice((X), (Y))$, in a rule r denotes that the set of all consequences derived from r must respect the FD $X \to Y$. In general, X can be a vector of variables — possibly an empty one denoted by "( )" — and Y is a vector of one or more variables. As shown in [29] the formal semantics of the construct can be given in terms of stable model semantics. For instance, a rule r of the form

$$r : \ p(X, Y, W) \leftarrow q(X, Y, Z, W), \ choice((X), (Y)), \ choice((Y), (X)).$$

expressing that for any stable model M, the ground instantiation of r w.r.t. M must satisfy the FDs $X \to Y$ and $Y \to X$, is rewritten into the following standard rules

$$
\begin{aligned}
r_1 : \ & p(X, Y, W) && \leftarrow q(X, Y, Z, W), \ chosen(X, Y, Z). \\
r_2 : \ & chosen(X, Y, Z) && \leftarrow q(X, Y, Z, W), \ \neg diffchoice(X, Y, Z). \\
r_3 : \ & diffchoice(X, Y, Z) \leftarrow chosen(X, Y', Z'), \ Y \neq Y'. \\
r_4 : \ & diffchoice(X, Y, Z) \leftarrow chosen(X', Y, Z'), \ Z \neq Z'.
\end{aligned}
$$

where the *choice* predicates have been substituted by the *chosen* predicate and for each *choice* predicate there is a *diffchoice* rule. The rule r will be called *choice* rule, the rule r_1 will be called *modified* rule, the rule r_2 will be called *chosen* rule and the rules r_3 and r_4 will be called *diffchoice* rules. Let P be a DATALOG¬ program with choice constructs, we denote with $sv(P)$ the program obtained by rewriting the choice rules as above — $sv(P)$ is called the *standard version* of P.

In general, the program $sv(P)$ generated by the transformation discussed above has the following properties [29,13]: 1) if P is in DATALOG or in DATALOG$^{\neg s}$ then $sv(P)$ has one or more total stable models, and 2) the **chosen** atoms in each stable model of $sv(P)$ obey the FDs defined by the choice goals. The stable models of $sv(P)$ are called *choice models* for P. The set of functional dependencies defined by choice atoms on the instances of a rule r (resp., program P) will be denoted FD_r (resp., FD_P).

A subset rule is of the form

$$\mathbf{s(X)} \subseteq \mathbf{A_1}, \dots, \mathbf{A_n}.$$

where $\mathbf{s}$ is an IDB predicate symbol not defined elsewhere in the program (*subset predicate symbol*) and all literals $\mathbf{A_1}, \dots, \mathbf{A_n}$ in the body are EDB. The rule enforces to select any subset of the relation that is derived from the body. The formal semantics of the rule is given by rewriting it into the following set of normal DATALOG$^\neg$ rules

$$\mathbf{s(X)} \leftarrow \mathbf{A_1}, \dots, \mathbf{A_n},\ \neg \mathbf{\hat{s}(X)}.$$
$$\mathbf{\hat{s}(X)} \leftarrow \mathbf{A_1}, \dots, \mathbf{A_n},\ \neg \mathbf{s(X)}.$$

where $\hat{s}$ is a new IDB predicate symbol with the same arity as s. Observe that the semantics of a subset rule can be also given in terms of choice as follows:

$$label(1).$$
$$label(2).$$
$$\mathbf{\hat{s}(1, X)} \leftarrow \mathbf{A_1}, \dots, \mathbf{A_n},\ label(L),\ choice((\mathbf{X}), (\mathbf{L})).$$
$$\mathbf{s(X)} \leftarrow \quad \mathbf{\hat{s}(1, X)}.$$

It turns out that subset rules are not necessary in our language, but we keep them in order to simplify the formulation of optimization queries.

In the following we shall denote with DATALOG$^{\neg s,c}$ the language DATALOG$^{\neg s}$ with *choice* and subset rules. More formally we say:

Definition 5. A DATALOG$^\neg$ program P with choice and subset rules is in DATALOG$^{\neg s,c}$ if P' is stratified, where P' is obtained from $sv(P")$ by removing diffchoice rules and diffchoice atoms and $P"$ is obtained from P by rewriting subset rules in terms of choice constructs.

Search and otpimization queries are denoted by $search(\text{DATALOG}^{\neg s,c})$ and $opt(\text{DATALOG}^{\neg s,c})$, respectively. Moreover, $search(\text{DATALOG}^{\neg s,c})_g$ denotes the class of queries $NQ = \langle P, f \rangle$ such that f is a relation defined by choice or subset rules and such rules are not defined in terms of other choice or subset rules; the corresponding optimization class is $opt(\text{DATALOG}^{\neg s,c})_g$. □

Proposition 4.

1. $search(\text{DATALOG}^{\neg s,c}) = \mathcal{NQPTIME}$ and
   $opt(\text{DATALOG}^{\neg s,c}) = \mathcal{OPT\,NQPTIME}$;
2. $search(\text{DATALOG}^{\neg s,c})_g = \mathcal{NQPTIME}_g$ and
   $opt(\text{DATALOG}^{\neg s,c})_g = \mathcal{OPT\,NQPTIME}_g$.

PROOF. The fact that $search(\text{DATALOG}^{\neg s,c}) = \mathcal{NQPTIME}$ has been proven in many places, e.g., in [13,21,12]. Observe now that, given any query Q in $search(\text{DATALOG}^{\neg s,c})_g$, $Q \in \mathcal{NQPTIME}$ as Q is also in $search(\text{DATALOG}^{\neg s,c})$. Moreover, for each D and for each answer of $Q(D)$, the non-deterministic choices, that are issued while executing the logic program, are kept into the answer; thus every answer contains a certificate of its recognition and, then, recognition is in $\mathcal{P}$. Hence also $Q \in \mathcal{NQPMV}_g$ and, then, $Q \in \mathcal{NQPTIME}_g$. To show that every query Q in $\mathcal{NQPTIME}_g$ is also in $search(\text{DATALOG}^{\neg s,c})_g$, we use the following characterization of $\mathcal{NQPTIME}_g$ [21]: every answer of Q can be constructed starting from the empty relation by adding one tuple at a time after a polynomial-time membership test. This construction can be easily implemented by defining a suitable query in $search(\text{DATALOG}^{\neg s,c})_g$. □

Next we show how to increase the expressive power of the language. We stress that the additional power is added in a controlled fashion so that a high level of expressivity is automatically enabled only if required by the complexity of the problem at hand.

Definition 6. Let $search(\text{DATALOG}^{\neg s,c})_!$ denote the class of queries $NQ = \langle P, !A, f \rangle$ such that $\langle P, f \rangle$ is in $search(\text{DATALOG}^{\neg s,c})$ and A is a ground literal (the *constraint goal*); for each D in $\mathcal{DB}_P$, $NQ(D) = \{M(f) : M$ is a T-stable model of P_D and either $A \in M$ if A is positive or $A \notin M$ otherwise$\}$. Accordingly, we define $opt(\text{DATALOG}^{\neg s,c})_!$, $search(\text{DATALOG}^{\neg s,c})_{g,!}$ and $opt(\text{DATALOG}^{\neg s,c})_{g,!}$. □

Proposition 5.

1. $search(\text{DATALOG}^{\neg s,c})_! = \mathcal{NQPMV}$ and
   $opt(\text{DATALOG}^{\neg s,c})_! = \mathcal{OPTNQPMV}$;
2. $search(\text{DATALOG}^{\neg s,c})_{g,!} = \mathcal{NQPMV}_g$ and
   $opt(\text{DATALOG}^{\neg s,c})_{g,!} = \mathcal{OPTNQPMV}_g$.

PROOF. Given any query $Q = \langle P, !A, f \rangle$ in $search(\text{DATALOG}^{\neg s,c})_!$, $Q \in \mathcal{NQPMV}$ since for each database D and for each relation F, testing whether $F \in Q(D)$ can be done in nondeterministic polynomial time as follows: we guess an interpretation M and, then, we check in deterministic polynomial time whether both M is a stable model and A is true in M. To prove that every query Q in $\mathcal{NQPMV}$ can be defined by some query in $search(\text{DATALOG}^{\neg s,c})_!$, we observe that Q can be expressed by a closed first-order formula by Fact 2 and that this formula can be easily translated into a query in $search(\text{DATALOG}^{\neg s,c})_!$. The proof of part (2) follows the lines of the proof of part (2) of Proposition 4. □

Example 8. The program P_{cut} of Example 6 can be replaced by the following program $P_{cut'}$:

$\text{v}'(\text{X}) \subseteq v(X).$
$\text{e}'(\text{X},\text{Y}) \leftarrow e(X,Y),\ v'(X),\ \neg v'(Y).$
$\text{e}'(\text{X},\text{Y}) \leftarrow e(X,Y),\ \neg v'(X),\ v'(Y).$

The query $\langle P_{cut'}, e' \rangle$ is in $search(\text{DATALOG}^{\neg s,c})_g$ and, therefore, the query $\langle P_{cut'}, max(e') \rangle$ is in $max(\text{DATALOG}^{\neg s,c})_g$. □

The program of the above example has been derived from the program of Example 6 by replacing the two rules with unstratified negation, defining v' with a subset rule.

Example 9. The program P_{kernel} of Example 7 can be replaced by the following program $P_{kernel'}$:

    $v'(X) \subseteq v(X)$.
    $joined\_to\_v'(X) \leftarrow v'(Y), e(Y,X)$.
    $no\_kernel \leftarrow v'(X), joined\_to\_v'(X)$.
    $no\_kernel \leftarrow \neg v'(X), \neg joined\_to\_v'(X)$.

The query $\langle P_{kernel'}, \neg no\_kernel, v' \rangle$ is in $search(\text{DATALOG}^{\neg s,c})_{g,!}$ and, therefore, the query $\langle P_{kernel'}, min|v'| \rangle$ is in $min(\text{DATALOG}^{\neg s,c})_{g,!}$.   □

The advantage of using restricted languages is that programs with built-in predicates are more intuitive and it is possible to control the expressive power.

5 Capturing Desirable Subclasses of NP Optimization Problems

We have shown that optimization queries are much harder than associated search queries. Indeed it often happens that the optimization of polynomial-time computable search queries cannot be done in polynomial time. In this section we show how to capture optimization queries for which "approximate" answers can be found in polynomial time.

Let us first recall that, as said in Proposition 1, an $\mathcal{NP}$ optimization query $opt|NQ| = \langle \mathcal{DB}, opt|f| \rangle$ corresponds to a problem in the class $OPT\,\mathcal{PB}$ that is defined as $opt|NQ| = opt_S|\{\mathbf{w} : (\mathcal{DB}, \mathcal{S}) \models \phi(\mathbf{w}, \mathcal{DB}, \mathcal{S})\}|$. In addition to the free variables $\mathbf{w}$, the first order formula ϕ may also contain quantified variables so that the general format of it is of two types:

$$(\exists \mathbf{x}_1)(\forall \mathbf{x}_2) \ldots (Q_k \mathbf{x}_k)\psi(\mathbf{w}, \mathcal{DB}, \mathcal{S}, \mathbf{x}_1, \ldots, \mathbf{x}_k), \text{ or}$$

$$(\forall \mathbf{x}_1)(\exists \mathbf{x}_2) \ldots (Q_k \mathbf{x}_k)\psi(\mathbf{w}, \mathcal{DB}, \mathcal{S}, \mathbf{x}_1, \ldots, \mathbf{x}_k),$$

where $k \geq 0$, Q_k is either $\exists$ or $\forall$, and ψ is a non-quantified formula. In the first case ϕ is a Σ_k formula while it is a Π_k formula in the latter case. (If ϕ has no quantifiers then it is both a Σ_0 and a Π_0 formula.) Accordingly, the class of all $\mathcal{NP}$ optimization problems for which the formula ϕ is a Σ_k (resp., Π_k) formula is called $OPT\,\Sigma_k$ (resp., $OPT\,\Pi_k$).

Kolaitis and Thakur [20] have introduced two hierarchies for the polynomially bounded $\mathcal{NP}$ minimization problems and for the polynomially bounded $\mathcal{NP}$ maximization problems:

$$MAX\,\Sigma_0 \subset MAX\,\Sigma_1 \subset MAX\,\Pi_1 = MAX\,\Sigma_2 \subset MAX\,\Pi_2 = MAX\,\mathcal{PB}$$
$$MIN\,\Sigma_0 = MIN\,\Sigma_1 \subset MIN\,\Pi_1 = MIN\,\Sigma_2 = MIN\,\mathcal{PB}$$

Observe that the classes $MAX\ \Sigma_0$ and $MAX\ \Sigma_1$ have been first introduced in [26] with the names MAX SNP and $MAX\ \mathcal{NP}$, respectively, whereas the class $MAX\ \Pi_1$ has been first introduced in [25].

A number of maximization problems have a desirable property: approximation. In particular, Papadimitriou and Yannakakis have shown that every problem in the class $MAX\ \Sigma_1$ is constant-approximable [26]. This is not the case for the complementary class $MIN\ \Sigma_1$ or other minimization subclasses: indeed the class $MIN\ \Sigma_0$ contains problems which are not log-approximable (unless $\mathcal{P} = \mathcal{NP}$) [20].

To single out desirable subclasses for minimization problems, Kolaitis and Thakur introduced a refinement of the hierarchies of $\mathcal{NP}$ optimization problems by means of the notion of *feasible $\mathcal{NP}$ optimization problem*, based on the fact that, as pointed out in Proposition 1, an $\mathcal{NP}$ optimization query, $opt|NQ| = \langle \mathcal{DB}, opt|f| \rangle$, can be also defined as $opt_{f,S}\{|f| : (D, f, S) \models \phi(\mathcal{DB}, f, S)\}$. Therefore, the class of all $\mathcal{NP}$ optimization problems for which the above formula ϕ is a Σ_k (resp., Π_k) formula is called $OPT\ F\Sigma_k$ (resp., $OPT\ F\Pi_k$). The following containment relations hold:

$$
\left.\begin{array}{c} MAX\ \Sigma_0 \\ MAX\ F\Sigma_1 \end{array}\right\} \subset MAX\ \Sigma_1 \subset MAX\ F\Pi_1 = MAX\ F\Sigma_2 = MAX\ \Pi_1 =
$$

$$
MAX\ \Sigma_2 \subset MAX\ F\Pi_2 = MAX\ \Pi_2 = MAX\ \mathcal{PB}
$$

$$
\left.\begin{array}{c} MIN\ \Sigma_0 = MIN\ \Sigma_1 = MIN\ F\Pi_1 \\ MIN\ F\Sigma_1 \end{array}\right\} \subset MIN\ F\Sigma_2 \subset MIN\ \Pi_1 = MIN\ \Sigma_2 =
$$

$$
MIN\ F\Pi_2 = MIN\ \Pi_2 = MIN\ \mathcal{PB}
$$

Observe that all problems in $MAX\ F\Sigma_1$ are constant-approximable since $MAX\ F\Sigma_1 \subset MAX\ \Sigma_1$.

A further refinement of feasible $\mathcal{NP}$ optimization classes can be obtained as follows. A first order formula $\phi(S)$ is *positive* w.r.t. the relation symbol S if all occurrences of S are within an even number of negation. The class of feasible $\mathcal{NP}$ minimization problems whose first order part is a positive Π_k formula ($1 \leq k \leq 2$) is denoted by $MIN\ F^+\Pi_k$. Particularly relevant is $MIN\ F^+\Pi_1$ as all optimization problems contained in this class are constant-approximable [20].

We next show that it is possible to further discipline $\mathtt{DATALOG}^{\neg s,c}$ in order to capture most of the above mentioned optimization subclasses.

First of all we point out that feasible $\mathcal{NP}$ optimization problems can be captured in $\mathtt{DATALOG}^{\neg s,c,!}$ by restricting to the class $opt(\mathtt{DATALOG}^{\neg s,c})_g$. For instance, the problem expressed by the query of Example 9 is feasible whereas the problem expressed by the query of Example 8 is not feasible.

Let P be a $\mathtt{DATALOG}^{\neg s,c}$ program, $p(\mathbf{y})$ be an atom and X a set of variables. We say that $p(\mathbf{y})$ is *free w.r.t. X* (in P) if

1. $var(p(\mathbf{y})) \subseteq X$, where $var(p(\mathbf{y}))$ is the set of variables occurring in $\mathbf{y}$, and
2. $\forall r \in P$ such that the head $H(r)$ and $p(\mathbf{y})$ unify, then $var(B(r)) \subseteq var(H(r))$ (i.e., the variables in the body also appear in the head) and for each atom $q(\mathbf{w})$ in $B(r)$, either q is an EDB predicate or $q(\mathbf{w})$ is free w.r.t. $var(q(\mathbf{w}))$.

We denote with $opt(\texttt{DATALOG}^{\neg s,c})_{\exists}$ the class of all queries $\langle P, opt|f|\rangle$ in $opt(\texttt{DATALOG}^{\neg s,c})$ such that $f(\mathbf{X})$ is free w.r.t. $\mathbf{X}$, where $\mathbf{X}$ is a list of distinct variables. Thus, $opt(\texttt{DATALOG}^{\neg s,c})_{\exists}$ denotes the class of all queries $\langle P, opt|f|\rangle$ in $opt(\texttt{DATALOG}^{\neg s,c})$, where all rules used to define (transitively) the predicate f, do not have additional variables w.r.t. to the head variables. For instance, the query of Example 8 is in $opt(\texttt{DATALOG}^{\neg s,c})_{\exists}$.

Theorem 1. $opt(\texttt{DATALOG}^{\neg s,c})_{\exists} = OPT\ \Sigma_0$.

PROOF. Let $\langle P, opt|f|\rangle$ be a query in $opt(\texttt{DATALOG}^{\neg s,c})_{\exists}$. Consider the rules that define directly or indirectly the goal f and let $\mathbf{X}$ be a list of $a(f)$ distinct variables. Since $f(\mathbf{X})$ is free w.r.t. $\mathbf{X}$ by hypothesis, it is possible to rewrite the variables in the above rules such that they are a subset of $\mathbf{X}$. It is now easy to show that the query can be written as a quantifier-free first-order formula with the free variables $\mathbf{X}$, i.e., the query is in $OPT\ \Sigma_0$. The proof that every query in $OPT\ \Sigma_0$ can be formulated as a query in $opt(\texttt{DATALOG}^{\neg s,c})_{\exists}$ is straightforward. $\square$

It turns out that all queries in $max(\texttt{DATALOG}^{\neg s,c})_{\exists}$ are constant-approximable.

Example 10. MAX CUT. Consider the program $P_{cut'}$ of Example 8. The query $\langle P_{cut'}, max(e')\rangle$ is in $MAX\ \Sigma_0$ since $e'(X,Y)$ is free w.r.t. $\langle X,Y\rangle$. $\square$

Let P be a $\texttt{DATALOG}^{\neg s,c}$ program and $p(\mathbf{y})$ be an atom. We say that P is *semipositive* w.r.t. $p(\mathbf{y})$ if

1. p is an EDB or a subset predicate symbol, or
2. $\forall r \in P$ defining p, P is semipositive w.r.t. every positive literal in the body $B(r)$ while each negative literal is EDB or subset.

We now denote with $opt(\texttt{DATALOG}^{\neg s,c})_+$ the class of all queries $\langle P, opt(f)\rangle$ in $opt(\texttt{DATALOG}^{\neg s,c})$ such that P is semipositive w.r.t. $f(\mathbf{X})$. Thus, $opt(\texttt{DATALOG}^{\neg s,c})_+$ denotes the class of all queries $\langle P, opt|f|\rangle$ in $opt(\texttt{DATALOG}^{\neg s,c})$ where negated predicates used to define (transitively) the predicate f are either EDB predicates or subset predicates. For instance, the query of Example 8 is in $opt(\texttt{DATALOG}^{\neg s,c})_+$. Moreover, since the predicate appearing in the goal is a subset predicate, the query of Example 8 is in $opt(\texttt{DATALOG}^{\neg s,c})_{g,+}$.

Theorem 2.

1. $opt(\texttt{DATALOG}^{\neg s,c})_+ = OPT\ \Sigma_1$,
2. $opt(\texttt{DATALOG}^{\neg s,c})_{g,+} = OPT\ F\Sigma_1$.

PROOF. Let $\langle P, opt|f|\rangle$ be a query in $opt(\texttt{DATALOG}^{\neg s,c})_+$ and $\mathbf{X}$ be a list of $a(f)$ distinct variables. Consider the rules that define directly or indirectly the goal f. Since P is semipositive w.r.t. $f(\mathbf{X})$ by hypothesis, it is possible to rewrite the variables in the above rules such that each of them is either in $\mathbf{X}$ or existentially quantified. It is now easy to show that the query can be formulated in the $OPT\ \Sigma_1$ format. The proof of part (2) is straightforward. $\square$

Then all queries in both $max(\texttt{DATALOG}^{\neg s,c})_+$ and $max(\texttt{DATALOG}^{\neg s,c})_{g,+}$ are constant-approximable.

Example 11. MAX SATISFIABILITY. We are given two unary relation c and a such that a fact $c(x)$ denotes that x is a clause and a fact $a(v)$ asserts that v is a variable occurring in some clause. We also have two binary relations p and n such that the facts $p(x,v)$ and $n(x,v)$ say that a variable v occurs in the clause x positively or negatively, respectively. A boolean formula, in conjunctive normal form, can be represented by means of the relations c, a, p, and n.

The maximum number of clauses simultaneously satisfiable under some truth assignment can be expressed by the query $\langle P_{sat}, max(f) \rangle$ where P_{sat} is the following program:

```
s(X) ⊆ a(X).
f(X) ← c(X), p(X,V), s(V).
f(X) ← c(X), n(X,V), ¬s(V).
```

Observe that $f(X)$ is not free w.r.t. X (indeed the query is not in *MAX* Σ_0) but P_{sat} is semipositive w.r.t. $f(X)$ so that the query is in *MAX* Σ_1. Observe now that the query goal f is not a subset predicate: indeed the query is not in *MAX* $F\Sigma_1$. □

Let $!A$ be a goal in a query in $opt(\texttt{DATALOG}^{\neg s,c})_!$ on a program P — recall that A is a positive or negative ground literal. Then a (not necessarily ground) atom C has

1. a *mark* 0 w.r.t. A if $C = A$;
2. a *mark* 1 w.r.t. A if $C = \neg A$;
3. a *mark* $k \geq 0$ w.r.t. A if there exists a rule r' in P and a substitution σ for the variables in C such that either (i) $H(r')$ has mark $(k-1)$ w.r.t. A and $C\sigma$ occurs negated in the body of r', or (ii) $H(r')$ has mark k w.r.t. A and $C\sigma$ is a positive literal in the body of r'.

Let us now define the class $opt(\texttt{DATALOG}^{\neg s,c})_{!,\nexists}$ of all queries $\langle P, !A, opt(f) \rangle$ in $opt(\texttt{DATALOG}^{\neg s,c})_!$ such that (i) $f(\mathbf{X})$ is free w.r.t. $\mathbf{X}$ and (ii) for each atom C that has an even mark w.r.t. A and for every rule r' in P, whose head unifies with C, the variables occurring in the body $B(r')$ also occur in the head $H(r')$.

We are finally able to define a subclass which captures $OPT\,F^+\Pi_1$ that is approximable when $OPT = MIN$. To this end, we define $opt(\texttt{DATALOG}^{\neg s,c})_{!,\nexists,g,+}$ as the subclass of $opt(\texttt{DATALOG}^{\neg s,c})_{!,\nexists,g}$ consisting of those queries $\langle P, !A, opt(f) \rangle$ such that there exists no subset atom $s(\mathbf{x})$ having an odd mark w.r.t. A.

Theorem 3.

1. $opt(\texttt{DATALOG}^{\neg s,c})_{!,\nexists} = OPT\,\Pi_1$;
2. $opt(\texttt{DATALOG}^{\neg s,c})_{!,\nexists,g} = OPT\,F\Pi_1$;
3. $opt(\texttt{DATALOG}^{\neg s,c})_{!,\nexists,g,+} = OPT\,F^+\Pi_1$.

PROOF. Let $\langle P, !A, opt(f) \rangle$ be a query in $opt(\texttt{DATALOG}^{\neg s,c})_{!,\nexists}$. Consider the rules that define directly or indirectly the goal f and let $\mathbf{X}$ be a list of $a(f)$ distinct variables. Since $f(\mathbf{X})$ is free w.r.t. $\mathbf{X}$ by hypothesis, it is possible to rewrite the variables in the above rules such that they are a subset of **X**. Consider now the rules that define directly or indirectly the goal !A. We can now rewrite the variables in the above rules such that they are universally quantified. It is now easy to show that the query can be written as an existential-free first-order formula with the free variables **X** and possibly additional variables universally quantified, i.e., the query is in *OPT Π_1*. The proofs of the other relationships are simple.   □

Example 12. MAX CLIQUE. In this example we want to find the cardinality of a maximum clique, i.e. a set of nodes V' such that for each pair of nodes (x, y) in V' there is an edge joining x to y. The maximum clique problem can be expressed by the query $\langle P_{clique}, !\neg no\_clique, max(v')\rangle$ where the program P_{clique} is as follows:

$\quad$ v'(X) ⊆ v(X).
$\quad$ no_clique ← v'(X), v'(Y), X ≠ Y, ¬e(X, Y).

The query is in the class $max(\mathtt{DATALOG}^{\neg s,c})_{!,\nexists,g}$ and, therefore, the optimization query is in *MAX FΠ_1* (= *MAXΠ_1*). On the other hand both atoms v'(X) and v'(Y) in the body of the rule defining the predicate no_clique have mark 1 (i.e. odd) w.r.t. the "!" goal. Therefore, the query $\langle P_{clique}, !\neg no\_clique, max(v')\rangle$ is not in the class $max(\mathtt{DATALOG}^{\neg s,\subseteq})_{!,\nexists,g,+}$, thus it is not in *MAX F$^+\Pi_1$*.   □

Example 13. MIN VERTEX COVER. As discussed in the introduction, the problem can be formulated by the query $\langle P_{vc}, !\neg no\_cover, min(v'(X))\rangle$ where P_{vc} is the following program:

$\quad$ v'(X) ⊆ v(X).
$\quad$ no_cover ← e(X, Y), ¬v'(X), ¬v'(Y).

Observe that both atoms v'(X) and v'(Y) in the rule defining no_cover have a mark 2 (i.e., even) w.r.t. the "!" goal. Therefore, the query is in $min(\mathtt{DATALOG}^{\neg s,c})_{!,\nexists,g,+}$ and, then, in *MINF$^+\Pi_1$*; so the problem is constant-approximable.   □

Additional interesting subclasses could be captured in our framework, but they are not investigated here. We just give an example of a query which is in the class *MIN F$^+\Pi_2(1)$* — this class is a subset of *MIN Π_2* where every subset predicate symbol occurs positively and at most once in every disjunction of the formula ψ. Problems in this class are log-approximable [20].

Example 14. MIN DOMINATING SET. Let $G = (V, E)$ be a graph. A subset V' of V is a dominating set if every node is either in V' or has a neighbour in V'. The query $\langle P_{ds}, !\neg no\_ds, min(v'(X))\rangle$ where P_{ds} is the following program, computes the cardinality of a minimum dominating set:

$\quad$ v'(X) ⊆ v(X).
$\quad$ q(X) ← v'(X).
$\quad$ q(X) ← e(X, Y), v'(Y).
$\quad$ no_ds ← v(X), ¬q(X).

This problem belongs to *MIN F$^+\Pi_2(1)$*.   □

Observe that the problem *min kernel* as defined in Example 8 is in the class $MIN\ F\Pi_2$, but not in $MIN\ F^+\Pi_2$, as it contains occurrences of the subset predicate v' which have an odd mark w.r.t. the "!" goal.

6   Conclusion

In this paper we have shown that $\mathcal{NP}$ search and optimization problems can be formulated as DATALOG¬ queries under non-deterministic total stable model semantics. In order to enable a simpler and more intuitive formulation of such problems, we have also introduced an extension of stratified DATALOG¬ that is able to express all $\mathcal{NP}$ search and optimization queries using a disciplined style of programming in which only simple forms of unstratified negations are supported. The core of this language, denoted by DATALOG¬$^{s,c,!}$, is stratified DATALOG¬ augmented with three types of non-stratified negations which are hardwired into ad-hoc constructs: choice predicate, subset rule and constraint goal. The former two constructs serve to issue non-deterministic selections while constructing one of possible total stable models, whereas the latter one defines some constraint that must be respected by the stable model in order to be accepted as an intended meaning of the program.

The language DATALOG¬$^{s,c,!}$ has been further refined in order to capture interesting subclasses of $\mathcal{NP}$ search queries, some of them computable in polynomial time. As for optimization queries, since in general they are not tractable also when the associated search problems are, we introduced restrictions to our language to single out classes of approximable optimization problems which have been recently introduced in the literature.

Our on-going research follows two directions:

1. efficient implementation schemes for the language, particularly to perform effective subset selections by pushing down constraints and possibly adopting 'intelligent' search strategies; this is particularly useful if one wants to find approximate solutions;
2. further extensions of the language such as (i) adding the possibility to use IDB predicates whenever an EDB predicate is required (provided that IDB definitions are only given by stratified rules), (ii) freezing, under request, nondeterministic selections to enable a "don't care" non-determinism (thus, some selections cannot be eventually retracted because of the constraint goal), and (iii) introducing additional constructs, besides to *choice* and *subset rule*, to enable nondeterministic selections satisfying predefined constraints that are tested on the fly.

References

1. Abiteboul, S., Simon, E., and Vianu, V., Non-deterministic languages to express deterministic transformations. In *Proc. ACM Symp. on Principles of Database Systems*, 1990, pp. 218-229.

2. Abiteboul, S., and Vianu, V., Non-determinism in logic-based languages. *Annals of Mathematics and Artificial Intelligence 3*, 1991, pp. 151-186.

3. Abiteboul, S., Hull, R., and Vianu, V., *Foundations of Databases*. Addison-Wesley, 1994.

4. Afrati, F., Cosmadakis, S. S., and Yannakakis, M., On Datalog vs. Polynomial Time. *Proc. ACM Symp. on Principles of Database Systems*, 1991, pp. 13-25.

5. Apt, K., Blair, H., and Walker, A., Towards a theory of declarative knowledge. In *Foundations of Deductive Databases and Logic Programming*, J. Minker (ed.), Morgan Kauffman, Los Altos, USA, 1988, 89-142.

6. Ausiello, G., Crescenzi, P., and Protasi M., Approximate solution of NP optimization problems. *Theoretical Computer Science*, No. 150, 1995, pp. 1-55.

7. Ausiello, G., Crescenzi, P., Gambosi, G., Kann, V., Marchetti-Spaccamela, A., and Protasi, M., *Complexity and Approximation - Combinatorial optimization problems and their approximability properties* Springer-Verlag, 1999.

8. Fagin, R., Generalized First-Order Spectra and Polynomial-Time Recognizable Sets. In *Complexity of Computation (R. Karp, Ed.)*, SIAM-AMS Proc., Vol. 7, 1974, pp. 43-73.

9. Fenner, S., Green, F., Homer, S., Selman, A. L., Thierauf, T. and Vollmer H., Complements of Multivalued Functions. *Chicago Journal of Theoretical Computer Science*, 1999.

10. Garey, M., and Johnson, D. S., *Computers and Intractability — A Guide to the Theory of NP-Completeness*. W.H. Freeman, New York, USA, 1979.

11. Gelfond, M., and Lifschitz, V., The Stable Model Semantics for Logic Programming. *Proc. 5th Int. Conf. on Logic Programming*, 1988, pp. 1070-1080.

12. Giannotti, F., Pedreschi, D., and Zaniolo, C., Semantics and Expressive Power of Non-Deterministic Constructs in Deductive Databases. *Journal of Computer and System Sciences, 62*, 1, 2001, pp. 15-42.

13. Giannotti, F., Pedreschi, D., Saccà, D., and Zaniolo, C., Nondeterminism in Deductive Databases. *Proc. 2nd Int. Conf. on Deductive and Object-Oriented Databases*, 1991, pp. 129-146.

14. Greco, S., Saccà, D., and Zaniolo C., Datalog with Stratified Negation and Choice: from P to D^P. *Proc. Int. Conf. on Database Theory*, 1995, pp. 574–589.

15. Greco, S., and Saccà, D., NP-Optimization Problems in Datalog. *Proc. Int. Logic Programming Symp.*, 1997, pp. 181-195.

16. Greco, S., and Zaniolo, C., Greedy Algorithms in Datalog. *Proc. Int. Joint Conf. and Symp. on Logic Programming*, 1998, pp. 294-309.

17. Johnson, D. S., A Catalog of Complexity Classes. In *Handbook of Theoretical Computer Science*, Vol. 1, J. van Leewen (ed.), North-Holland, 1990.

18. Kanellakis, P. C., Elements of Relational Database Theory. In *Handbook of Theoretical Computer Science*, Vol. 2, J. van Leewen (ed.), North-Holland, 1991.

19. Kolaitis, P. G., and Thakur, M. N., Logical Definability of NP Optimization Problems. *Information and Computation*, No. 115, 1994, pp. 321-353.

20. Kolaitis, P. G., and Thakur, M. N., Approximation Properties of NP Minimization Classes. *Journal of Computer and System Science*, No. 51, 1995, pp. 391-411.

21. Leone, N., Palopoli, L., and Saccà, D. On the Complexity of Search Queries. In *Fundamentals Of Information Systems* (T. Plle, T. Ripke, K.D. Schewe, eds), 1999, pp. 113-127.

22. Lloyd, J., *Foundations of Logic Programming.* Springer-Verlag, 1987.

23. Marek, W., and Truszczynski, M., Autoepistemic Logic. *Journal of the ACM*, Vol. 38, No. 3, 1991, pp. 588-619.

24. Papadimitriou, C. H., *Computational Complexity.* Addison-Wesley, Reading, MA, USA, 1994.

25. Panconesi, A., and Ranjan, D., Quantifiers and Approximation. *Theoretical Computer Science*, No. 1107, 1992, pp. 145-163.

26. Papadimitriou, C. H., and Yannakakis, M., Optimization, Approximation, and Complexity Classes. *Journal Computer and System Sciences*, No. 43, 1991, pp. 425-440.

27. Ramakrisnhan, R., Srivastava, D., and Sudanshan, S., CORAL — Control, Relations and Logic. In *Proc. of 18th Conf. on Very Large Data Bases*, 1992, pp. 238-250.

28. Saccà, D., The Expressive Powers of Stable Models for Bound and Unbound Queries. *Journal of Computer and System Sciences*, Vol. 54, No. 3, 1997, pp. 441-464.

29. Saccà, D., and Zaniolo, C., Stable Models and Non-Determinism in Logic Programs with Negation. In *Proc. ACM Symp. on Principles of Database Systems*, 1990, pp. 205-218.

30. Selman, A., A taxonomy of complexity classes of functions. *Journal of Computer and System Science*, No. 48, 1994, pp. 357-381.

31. A. Selman, Much ado about functions. *Proc. of the 11th Conf. on Computational Complexity*, IEEE Computer Society Press, 1996, pp. 198-212.

32. Ullman, J. K., *Principles of Data and Knowledge-Base Systems*, volume 1 and 2. Computer Science Press, New York, 1988.

33. Zaniolo, C., Arni, N., and Ong, K., Negation and Aggregates in Recursive Rules: the $\mathcal{LDL}$++ Approach. *Proc. 3rd Int. Conf. on Deductive and Object-Oriented Databases*, 1993, pp. 204-221.

The Declarative Side of Magic

Paolo Mascellani[1] and Dino Pedreschi[2]

[1] Dipartimento di Matematica, Università di Siena
via del Capitano 15, Siena - Italy
p.mascellani@dm.unipi.it
[2] Dipartimento di Informatica, Università di Pisa
Corso Italia 40, Pisa - Italy
pedre@di.unipi.it

Abstract In this paper, we combine a novel method for proving partial correctness of logic programs with a known method for proving termination, and apply them to the study of the *magic-sets* transformation. As a result, a declarative reconstruction of efficient bottom-up execution of goal-driven deduction is accomplished, in the sense that the obtained results of partial and total correctness of the transformation abstract away from procedural semantics.

1 Introduction

In the recent years, various principles and methods for the verification of logic programs have been put forward, as witnessed for instance in [11,3,16,17,13]. The main aim of this line of research is to verify the crucial properties of logic programs, notably partial and total correctness, on the basis of the declarative semantics only, or, equivalently, by abstracting away from procedural semantics.

The aim of this paper is to apply some new methods for partial correctness combined with some known methods for total correctness to a case study of clear relevance, namely bottom-up computing. More precisely, we:

- introduce a method for proving partial correctness by extending the ideas in [14],
- combine it with the approach in [6,7] for proving termination, and
- apply both to the study of the transformation techniques known as *magic-sets*, introduced for the efficient bottom-up execution of goal-driven deduction — see [9,20] for a survey.

We found the exercise stimulating, as all proofs of correctness of the magic-sets transformation(s) available in the literature are based on operational arguments, and often quite laborious. The results of partial and total correctness presented in this paper, instead, are based on purely declarative reasoning, which clarifies the natural idea underlying the magic-sets transformation. Moreover, these results are applicable under rather general assumptions, which broadly encompass the programming paradigm of deductive databases.

A.C. Kakas, F. Sadri (Eds.): Computat. Logic (Kowalski Festschrift), LNAI 2408, pp. 83–108, 2002.
© Springer-Verlag Berlin Heidelberg 2002

Preliminaries

Throughout this paper we use the standard notation of Lloyd [12] and Apt [1]. In particular, for a logic program P we denote the Herbrand Base of P by B_P, the least Herbrand model of P by M_P and the immediate consequence operator by T_P. Also, we use Prolog's convention identifying in the context of a program each string starting with a capital letter with a variable, reserving other strings for the names of constants, terms or relations. In the programs we use Prolog's list notation. Identifiers ending with "s", like xs, range over lists. Bold capital letters, like **A**, identify a possibly empty sequence (conjunction) of atoms or set of variables: the context should always be clear.

Plan of the Paper

In Section 2 we introduce a declarative method for proving the partial correctness of a logic program w.r.t. a specification. In Section 3 we use this method to obtain a declarative proof of the correctness of a particular implementation of the magic-sets transformation technique. In Section 4 is recalled the concept of acceptability, which allows to conduct declarative termination proofs for logic programs. In Section 5, we apply this concept to prove the termination of the magic programs under and some related properties. Finally, in Section 6, we provide a set of examples in order to clarify the use of the proof methods proposed and how the magic-sets transformation works.

2   Partial Correctness

Partial correctness aims at characterizing the input-output behavior of programs. The input to a logic program is a query, and the associated output is the set of *computed instances* of such a query. Therefore, partial correctness in logic programming deals with the problem of characterizing the computed instances of a query. In Apt [2,3], a notation recalling that of Hoare's triples (correctness formulas) is used. The triple:

$$\{Q\} \quad P \quad \mathcal{Q}$$

denotes the fact that $\mathcal{Q}$ is the set of computed instances of query Q. A natural question is: can we establish a correctness formula by reasoning on declarative semantics, i.e. by abstracting away from procedural semantics? The following simple result, which generalizes one from [3] tells us that this is possible in the case of ground output.

Theorem 1. *Consider the set $\mathcal{Q}$ of the correct instances of a query Q and a program P, and suppose that every query in $\mathcal{Q}$ is ground. Then:*

$$\{Q\} \quad P \quad \mathcal{Q}.$$

Proof. Clearly, every computed instance of Q is also a correct instance of Q by the Soundness of SLD-resolution. Conversely, consider a correct instance Q_1 of Q. By the Strong Completeness of SLD-resolution, there exists a computed instances Q_2 of Q such that Q_1 is an instances of Q_2. By the Soundness of SLD-resolution, Q_2 is a correct instance of Q, so it is ground. Consequently $Q_2 = Q_1$, hence Q_1 is a computed instance of Q. □

So, for programs with ground output, correct and computed instances of queries coincide, and therefore we can use directly the declarative semantics to check partial correctness. When considering one-atom queries only, the above result can be rephrased as follows: if the one-atom query A to program P admits only ground correct instances, then:

$$\{A\} \quad P \quad M_P \cap [A]. \tag{1}$$

A simple sufficient condition (taken from [3]) to check that all correct instances of a one-atom query A are ground is to show that the set $[A] \cap M_P$ is finite, i.e. that A admits a finite number of correct ground instances. So, in principle, it is possible to reason about partial correctness on the basis of the least Herbrand model only. As an example, consider the Append program:

```
append([], Ys, Ys).
append([X|Xs],Ys,[X|Zs]) ← append(Xs,Ys,Zs).
```

the interpretation:

$$I_{Append} = \{\text{append(xs,ys,zs)} \mid \text{xs,ys,zs are lists and xs * ys = zs}\} \tag{2}$$

where zs is some given list, and "*" denotes list concatenation, and the correctness formula:

$$\{\text{append(Xs, Ys, zs)}\} \quad \text{Append} \quad I_{Append}$$

We can establish such a triple provided we can show that the interpretation I_{Append} is indeed the least Herbrand model of the Append program, since the number of pairs of lists whose concatenation yields zs is clearly finite.

Unfortunately, despite the fact that the set I_{Append} is the natural intended interpretation of the Append program, it is *not* a model of Append, because the first clause does not hold in it. In fact, for many programs it is quite cumbersome to construct their least Herbrand model. Note for example that M_{Append} contains elements of the form append(s,t,u) where neither t nor u is a list. A correct definition of M_{Append} is rather intricate, and clearly, it is quite clumsy to reason about programs when even in so simple cases their semantics is defined in such a laborious way.

Why is the least Herbrand model different from the specification, or intended interpretation, of a program? The reason is that we usually design programs with reference to a class of *intended queries* which describes the admissible input for the program. As a consequence, the specification of the program is relative to the set of intended queries, whereas M_P is not. In the example, the intended queries for Append are described by the set:

$$\{\text{append}(\text{s},\text{t},\text{u}) \mid \text{s},\text{t} \text{ are lists or u is a list}\} \tag{3}$$

and it is possible to show that the specification (2) is indeed the fragment of the least Herbrand model M_{Append} restricted to the set (3) of the intended queries.

A method for identifying the intended fragment of the least Herbrand model is proposed in [4]; such a fragment is then used to establish the desired correctness formulas. This method adopts a notion of well-typedness [8,3,18], which makes it applicable to Prolog programs only, in that it exploits the left-to-right ordering of atoms within clauses.

In the next section we introduce a more general characterization of the intended fragment of the least Herbrand model, which remedies the asymmetry of the method in [4], and allows us to prove partial correctness of logic programs with no reference to control issues.

Bases

The key concept of this paper is introduced in the following:

Definition 1. *An interpretation I is called a* base *for a program P w.r.t. some model M of P iff, for every ground instance $A \leftarrow \mathbf{A}$ of every clause of P:*

$$if \ I \models A \ and \ M \models \mathbf{A}, \ then \ I \models \mathbf{A}.$$

□

The notion of a base has been designed to formalize the idea of an admissible set of "intended (one-atom) queries". Definition 1 requires that all possible clauses which allow to deduce an atom A in a base I have their bodies true in I itself. The condition that the body is true in some model of the program (obviously a necessary condition to conclude A) is used to get a weakening of the requirement. Roughly speaking, a base is required to include all possible atoms needed to deduce any atom in the base itself. The concept of a base was first introduced in [14], where it is referred to as a closed interpretation. As an example, it is readily checked that the set (3) is a base for Append.

Since a base I is assumed to describe the intended queries, the intended fragment of the least Herbrand model is $M_P \cap I$. The main motivation for introducing the notion of a base is that of obtaining a method to identify $M_P \cap I$ directly, without having to construct M_P first. To this purpose, given a base I for a program P, we define the *reduced* program, denoted P_I, as the set of ground instances of clauses from P whose heads are in I. In other words,

$$P_I = \{A \leftarrow \mathbf{A} \in ground(P) \mid A \in I\}. \tag{4}$$

The following observation is immediate:

$$T_{P_I}(X) = T_P(X) \cap I. \tag{5}$$

The following crucial result, first shown in [14], shows that the least Herbrand model of the reduced program coincides with the intended fragment (w.r.t. I) of the least Herbrand model of the program:

Theorem 2. *Let P be a program and I a base for P. Then:*

$$M_P \cap I = M_{P_I}.$$

Proof. First, we show that, for any interpretation $J \subseteq M_P$:

$$T_{P_I}(J) = T_{P_I}(J \cap I). \tag{6}$$

The ($\supseteq$) part is a direct consequence of the fact that T_{P_I} is monotonic. To establish the ($\subseteq$) part, consider $A \in T_{P_I}(J)$. Then, for some clause $A \leftarrow \mathbf{A}$ from P_I, we have $J \models \mathbf{A}$, hence $M_P \models \mathbf{A}$, which together with the fact that I is a base and $I \models A$, implies that $I \models \mathbf{A}$. Therefore $J \cap I \models \mathbf{A}$, which implies $A \in T_{P_I}(J \cap I)$.

We now show that, for all $n > 0$,

$$T_P^n(\emptyset) \cap I = T_{P_I}^n(\emptyset)$$

which implies the thesis. The proof is by induction on n. In the base case ($n = 0$), the claim is trivially true. In the induction case ($n > 0$), we calculate:

$$
\begin{aligned}
T_P^n(\emptyset) \cap I \\
= T_P(T_P^{n-1}(\emptyset)) \cap I \\
\{(5)\} \\
= T_{P_I}(T_P^{n-1}(\emptyset)) \\
\{T_P^{n-1}(\emptyset) \subseteq M_P \text{ and } (6)\} \\
= T_{P_I}(T_P^{n-1}(\emptyset) \cap I) \\
\{\text{induction hypothesis}\} \\
= T_{P_I}(T_{P_I}^{n-1}(\emptyset)) \\
= T_{P_I}^n(\emptyset).
\end{aligned}
$$

□

So, given a base I for program P, M_{P_I} is exactly the desired fragment of M_P. The reduced program P_I is a tool to construct such a desired fragment of M_P *without* constructing M_P first. Therefore, M_{P_I} directly can be used to prove correctness formulas for intended queries, i.e. queries whose ground instances are in I, as stated in the following:

Theorem 3. *Let P be program, I a base for P, and Q a one-atom query which admits ground correct instances only. Then:*

$$\{Q\} \quad P \quad M_{P_I} \cap [Q].$$

Proof. By Theorem 2, $M_{P_I} = M_P \cap I$, and $[Q] \subseteq I$ implies $M_P \cap [Q] = M_P \cap I \cap [Q]$. The result then follows immediately from (1) or, equivalently, Theorem 1. □

In the `Append` example, the intended specification (2) is indeed the least Herbrand model of the `Append` program reduced with respect to the base (3), so, using Theorem 3, we can establish the desired triple:

$$\{append(Xs, Ys, zs)\} \quad Append \quad \{append(xs, ys, zs) \mid zs = xs * ys\}$$

Later, a simple, induction-less, method for proving that a given interpretation is the least Herbrand model of certain programs is discussed.

Example 1. Consider the following program `ListSum`, computing the sum of a list of natural numbers:

```
listsum([],0) ←
listsum([X|Xs],Sum) ← listsum(Xs,PSum),sum(PSum,X,Sum)
sum(X,0,X) ←
sum(X,s(Y),s(Z)) ← sum(X,Y,Z)
```

and the Herbrand interpretations $I_{ListSum}$ and M, defined as follows:

$$I_{ListSum} = \begin{array}{l} \{listsum(xs, sum) \mid listnat(xs)\} \cup \\ \{sum(x, y, z) \mid nat(x) \wedge nat(y)\} \end{array}$$

$$M = \begin{array}{l} \{listsum(xs, sum) \mid listnat(xs) \Rightarrow nat(sum)\} \cup \\ \{sum(x, y, z) \mid nat(x) \wedge nat(y) \Rightarrow nat(z)\} \end{array}$$

where $listnat(x)$ and $nat(x)$ hold when x is, respectively, a list of natural numbers and a natural number. First, we check that M is a model of `ListSum`:

$M \models listsum([], 0)$
$M \models listsum([x|xs], sum) \Leftarrow M \models listsum(xs, psum), sum(psum, x, sum)$
$M \models sum(x, 0, x)$
$M \models sum(x, s(y), s(z)) \Leftarrow M \models sum(x, y, z)$

Next, we check that $I_{ListSum}$ is a base for `ListSum` w.r.t. M:

$I_{ListSum} \models listsum([x|xs], sum) \wedge M \models listsum(xs, psum), sum(psum, x, sum)$
$\Rightarrow I_{ListSum} \models listsum(xs, psum), sum(psum, x, sum)$

$I_{ListSum} \models sum(x, s(y), s(z)) \wedge M \models sum(x, y, z)$
$\Rightarrow I_{ListSum} \models sum(x, y, z)$

The following set is the intended interpretation of the ListSum program:

$$\begin{cases} \texttt{listsum}(\texttt{xs}, \texttt{sum}) \mid \mathit{listnat}(\texttt{xs}) \wedge \texttt{sum} = \sum_{\texttt{x} \in \texttt{xs}} \texttt{x} \\ \texttt{sum}(\texttt{x}, \texttt{y}, \texttt{z}) \mid \mathit{nat}(\texttt{x}) \wedge \mathit{nat}(\texttt{y}) \wedge \texttt{x} + \texttt{y} = \texttt{z} \end{cases} \cup \qquad (7)$$

and, although it is *not* a model of the program (the unit clause of sum does not hold in it), it is possible to prove that it is the fragment of the $M_{\texttt{ListSum}}$ restricted to the base $M_{\texttt{ListSum}}$. Therefore, by Theorem 3, provided xs is a list of natural numbers, we establish the following triple:

$$\{\texttt{listsum}(\texttt{xs}, \texttt{Sum})\} \quad \texttt{ListSum} \quad \left\{ \texttt{listsum}(\texttt{xs}, \texttt{sum}) \mid \texttt{sum} = \sum_{\texttt{x} \in \texttt{xs}} \texttt{x} \right\}.$$

□

In many examples, like the one above, bases are constructed using type information. Typically, a base is constructed by specifying the types of input positions of relations, and the model associated with a base is constructed by specifying how types propagate from input to output positions. If a decidable type definition language is adopted, such as the one proposed in [10], then checking that a given interpretation is base is fully automatazible. However, a full treatment of this aspects is outside the scope of this paper.

3 Partial Correctness and Bottom-Up Computing

Consider a naive bottom-up evaluation of the ListSum program. The sequence of approximating sets is hard to compute for several reasons.

1. The unit clause $\texttt{sum}(\texttt{X}, 0, \texttt{X}) \leftarrow$ introduces infinitely many facts at the very first step. In fact, such a clause is not *safe* in the sense of [19], i.e. variables occur in the head, which do not occur in the body.
2. Even if a safe version of the ListSum program is used, using a relation which generates natural numbers, the approximating sets grow exponentially large.
3. In any case, the bottom-up computation diverges.

In a goal-driven execution starting from the query listsum(xs,X), where xs is the input list and X is a variable, however, only a linearly increasing subset of each approximation is relevant. A more efficient bottom-up computation can be achieved using the program ListSum reduced w.r.t. an appropriate base I which includes all instances of the desired query. Indeed, Theorem 2 tells us that, in the bottom-up computation, it is equivalent to take the intersection with the base I at the limit of the computation, or at each approximation. The second option is clearly more efficient, as it allows to discard promptly all facts which are unrelevant to the purpose of answering the desired query. Therefore, the base should be chosen as small as possible, in order to minimize the size of the approximations. However, computing with the reduced program is unrealistic for two reasons. First, constructing a suitable base before the actual computation takes place is often impossible. In the ListSum example, an appropriate base should be chosen as follows:

$$I_{xs} = \begin{array}{l} \{\texttt{listsum(ys, sum)} \mid listnat(\texttt{ys}) \wedge \texttt{ys} \text{ is a suffix of } \texttt{xs}\} \cup \\ \{\texttt{sum(x, y, z)} \mid nat(\texttt{x}) \wedge nat(\texttt{y}) \wedge \texttt{z} \geq n\} \end{array}$$

where xs is the input list and n is the sum of the numbers in xs, so the expected result of the computation! Second, a reduced program is generally infinite or, at best, hopelessly large.

Nevertheless, bases and reduced programs are useful abstractions to explain the idea behind the optimization techniques like magic-sets, widely used in deductive database systems to support efficient bottom-up execution of goal-driven deduction. In fact, we shall see how the optimized *magic* program is designed to combine the construction of a base and its exploitation in an intertwined computation.

The Magic-Sets Transformation

In the literature, the problem of the efficient bottom-up execution of goal-driven computations has been tackled in a compilative way, i.e. by means of a repertoire of transformation techniques which are known under the name of *magic-sets*—see [9] or [20, Ch. 13] for a survey on this broad argument. Magic-sets is a non trivial program transformation which, given a program P and a query Q, yields a transformed program which, when executed bottom-up, mimics the top-down, Prolog-like execution of the original program P, activated on the query Q. Many variations of the basic magic-sets technique have been proposed, which however share the original idea.

All available justifications of its correctness are given by means of procedural arguments, by relating the bottom-up computation of the transformed (magic) program with the top-down computation of the original program and query. As a consequence, all known proofs of correctness of the magic-sets transformation(s) are rather complicated, although informative about the relative efficiency of the top-down and bottom-up procedures—see for instance [20, pp.836-841].

We show here how the core of the magic-sets transformation can be explained in rather natural declarative terms, by adopting the notion of a base, and the related results discussed in the previous section. Actually, we show that the "magic" of the transformation lies in computing and exploiting a base of the original program.

We provide an incremental version of the core magic-sets transformation, which allows us to compile separately each clause of the program. We need to introduce the concept of call pattern, or *mode*, which relates to that of *binding pattern* in [20]. Informally, modes indicate whether the arguments of a relation should be used either as an input or as an output, thus specifying the way a given program is intended to be queried.

Definition 2. *Consider an n-ary relation symbol p. A* mode *for p is a function:*

$$m_p : [1, n] \rightarrow \{+, -\}.$$

If $m_p(i) = '+'$, we call i an input position *of p, otherwise we call i an* output position *of p. By a* moding *we mean a collection of modes, one for each relation symbol in a program.* □

We represent modes in a compact way, writing m_p in the more suggestive form $p(m_p(1), \ldots, m_p(n))$. For instance the mode sum(+,+,-) specifies the input/output behavior of the relation sum, which is therefore expected to be queried with the two first positions filled in with ground terms.

¿From now on we assume that some fixed moding is given for any considered program. To simplify our notation, we assume, without loss of generality, that, in each relation, input positions precede output positions, so that any atom A can be viewed as $p(\mathbf{u}, \mathbf{v})$, where $\mathbf{u}$ are the terms in the input positions of p and $\mathbf{v}$ are the terms in the output positions of p. With reference to this notation, the *magic* version of an atom $A = p(\mathbf{u}, \mathbf{v})$, denoted A', is the atom $p'(\mathbf{u})$, where p' is a fresh predicate symbol (not occurring elsewhere in the program), whose arity is the number of input position of p. Intuitively, the magic atom $p'(\mathbf{u})$ represent the fact that the relation p is called with input arguments $\mathbf{u}$.

We are now ready to introduce our version of the magic-sets transformation.

Definition 3. *Consider a program P and a one-atom query Q. The* magic program O *is obtained from P and Q by the following transformation steps:*

1. *for every decomposition $A \leftarrow \mathbf{A}, B, \mathbf{B}$ of every clause from P, add a new clause $B' \leftarrow A', \mathbf{A}$;*
2. *add a new unit clause $Q' \leftarrow$;*
3. *replace each original clause $A \leftarrow \mathbf{A}$ from P with the new clause $A \leftarrow A', \mathbf{A}$.*

□

The magic program O is the optimized version of the program P w.r.t. the query Q. Observe that the transformation step (1) is performed in correspondence with every body atom of every clause in the program. Also, the only unit clause, or fact, is that introduced at step (2), also called a "seed". The collection of clauses generated at steps (1) and (2) allows to deduce all the magic atoms corresponding to the calls generated in the top-down/left-to-right execution of the original program P starting with the query Q. The declarative reading of the clause $B' \leftarrow A', \mathbf{A}$ introduced at step (1) is: "if the relation in the head of the original clause is called with input arguments as in A', and the atoms $\mathbf{A}$ preceding B in the original clause have been deduced, then the relation B is called with input arguments as in B'". Finally, the information about the calls represented by the magic atoms is exploited at step (3), where the premises of the original clauses are strengthened by an extra constraint, namely that the conclusion A is taken only if it is pertinent to some needed call, represented by the fact that A' has been deduced.

Example 2. Consider the program ListSum of Example 1 with the moding:

```
listsum(+,-)
sum(+,+,-)
```

and the query:

```
listsum([2,1,5],Sum)
```

that is consistent with the moding. The corresponding magic program is:

```
listsum([],0)  ← listsum'([])
listsum([X|Xs],Sum)  ←
     listSum'([X|Xs]),listsum(Xs,PSum),sum(PSum,X,Sum)
sum(X,0,X)  ← sum'(X,0)
sum(X,s(Y),s(Z))  ← sum'(X,s(Y)),sum(X,Y,Z)

listsum'(Xs)  ← listsum'([X|Xs])
sum'(Psum,X)  ← listsum'([X|Xs]),listsum(Xs,PSum)
sum'(X,Y)  ← sum'(X,s(Y))

listsum'([2,1,5])  ←
```

□

Partial Correctness of the Magic-Sets Transformation

We now want to show that the magic-sets transformation is correct. The correctness of the transformation is stated in natural terms in the main result of this section, which essentially says that the original and the magic program share the same logical consequences, when both are restricted to the intended query.

Theorem 4. *Let P be a program, Q be a one-atom query, and consider the magic program O. Then:*

$$M_P \cap [Q] = M_O \cap [Q].$$

Proof. The proof is organized in the following three steps:

1. the interpretation $M = \{A \in B_P \mid M_O \models A' \Rightarrow M_O \models A\}$ is a model of P;
2. the interpretation $I = \{A \in B_P \mid M_O \models A'\}$ is a base for P w.r.t. M;
3. $M_P \cap I = M_O \cap I$.

The thesis follows directly from (3), observing that $[Q] \subseteq I$ as a consequence of the fact that the magic program O contains the seed fact $Q' \leftarrow$. We now prove the facts (1), (2) and (3).

Proof of 1 Consider a ground instance $A \leftarrow \mathbf{A}$ of a clause from P: to show that M is a model of the clause, we assume:

$$M \models \mathbf{A} \tag{8}$$

$$M_O \models A' \tag{9}$$

and prove that $M_O \models A$. In turn, such conclusion is implied by $M_O \models \mathbf{A}$ as a consequence of (9) and the fact that the magic program O contains the clause $A \leftarrow A', \mathbf{A}$. To prove $M_O \models \mathbf{A}$ we proceed by induction on $\mathbf{A}$: in the base case ($\mathbf{A}$ is empty) the conclusion trivially holds. In the induction case ($\mathbf{A} = \mathbf{B}, B, \mathbf{C}$) the magic program contains the clause $B' \leftarrow A', \mathbf{B}$, and therefore $M_O \models B'$ as a consequence of (9) and the induction hypothesis. As $M \models B$ by (8), we have that $M_O \models B'$ implies $M_O \models B$, by the definition of M.

Proof of 2 Consider a ground instance $A \leftarrow \mathbf{A}$ of a clause from P, and assume:

$$I \models A \tag{10}$$

$$M \models \mathbf{A} \tag{11}$$

To obtain the desired conclusion, we prove that $I \models \mathbf{A}$ by induction on $\mathbf{A}$. In the base case ($\mathbf{A}$ is empty) the conclusion trivially holds. In the induction case ($\mathbf{A} = \mathbf{B}, B, \mathbf{C}$) the magic program O contains the clause $c : B' \leftarrow A', \mathbf{B}$. By the induction hypothesis, $I \models \mathbf{B}$, which implies $M_O \models \mathbf{B}'$ by the definition of I. This, together with (11), implies:

$$M_O \models B \tag{12}$$

by the definition of M. Next, by (10) and the definition of I, we obtain $M_O \models A'$, which, together with (12) and clause c, implies $M_O \models B'$. This directly implies $I \models B$.

Proof of 3 ($\subseteq$). First we show that M_O is a model of P_I. In fact, consider clause $A \leftarrow \mathbf{A}$ of P_I, and assume that $M_O \models \mathbf{A}$. By the definition of P_I, $I \models A$, which by the definition of I implies $M_O \models A'$. Hence, considering that $A \leftarrow A', \mathbf{A}$ is a ground instance of a cause of O, $M_O \models A$. This implies that M_O includes M_{P_I}, which, by Lemma 2, is equal to $M_P \cap I$, since I is a base for P from (i) and (ii).

($\supseteq$). Clearly $M_O \cap B_P \subseteq M_P$, as the clauses from P are strengthened in O with extra premises in the body. Hence, observing that $I \subseteq B_P$ we obtain $M_O \cap I \subseteq M_P \cap I$. □

The crucial point in this proof is the fact that the set I of atoms corresponding to the magic atoms derived in O is a base, i.e. an admissible set of intended queries, which describes all possible calls to the program originating from the top level query Q.

An Immediate consequence of Theorem 4 is the following:

Corollary 1. *Let P be a program, Q be a one-atom query, and consider the magic program O. Then, A is a ground instance of a computed answer of Q in P iff it is a ground instance of a computed answer of Q in O.* $\square$

Observe that the above equivalence result is obtained with no requirement about the fact the original program respects the specific moding, nor with any need of performing the so-called bound/free analysis. In this sense, this result is more general to the equivalence results in the literature, based on procedural reasoning. However, these results, such as that in [20] tell us more from the point of view of the relative efficiency of bottom-up and top-down computing.

As a consequence of Theorems 1 and 4, we can conclude that, for any one-atom query A which admits only ground correct instances w.r.t. a program P, the following triple holds:

$$\{A\} \quad P \quad M_O \cap [A] \tag{13}$$

i.e. the computed instances of A in P coincide with the correct instances of A in the magic program O. However, we need a syntactic condition able to guarantee that every correct instance is ground.

Well-Moded Programs

In the practice of deductive databases, the magic-sets transformation is applied to so-called *well-moded* programs, as for this programs the computational benefits of the transformation are fully exploited, in a sense which shall be clarified in the sequel.

Definition 4. *With reference to some specific, fixed moding:*

– *a one-atom query $p(\mathbf{i}, \mathbf{o})$ is called* well-moded *iff:*

$$vars(\mathbf{i}) = \emptyset;$$

– *a clause $p_0(\mathbf{o_0}, \mathbf{i_{n+1}}) \leftarrow p_1(\mathbf{i_1}, \mathbf{o_1}), \ldots, p_n(\mathbf{i_n}, \mathbf{o_n})$ is called* well-moded *if, for $i \in [1, n+1]$:*

$$vars(\mathbf{i_i}) \subseteq vars(\mathbf{o_0}) \cup \cdots \cup vars(\mathbf{o_{i-1}});$$

– *a program is called* well-moded *if every clause of it is.* $\square$

Thus, in well-moded clauses, all variables in the input positions of a body atom occur earlier in the clause, either in an output position of a preceding body atom, or in an input position of the head. Also, one-atom well-moded queries are ground at input positions. Well-modedness is a simple syntactic condition which guarantees that a given program satisfies a given moding. A well-known property of well-moded programs and queries is that they deliver ground output.

Theorem 5. *Let P be a well-moded program, and A a one-atom well-moded query. Then every computed instance of A in P is ground.*

Proof. See, for instance, [5]. The general idea of this proof is to show the following points:

1. at each step of the resolution process, the selected atom is well-moded;
2. all the output terms of a selected atom in a refutation appears in the input term of some selected atom of the refutation.

This, together with the fact that the first selected atom (the query) is well-moded, implies the claim.   □

So, well-modedness provides a (syntactic) sufficient condition to fulfill the proof obligation of triple (13).

Example 3. The program ListSum of Example 1 is well-moded w.r.t.:

    listsum(+,-)
    sum(+,+,-)

hence the following triple can be established:

$$\{\text{listsum}(xs, Sum)\} \quad \text{ListSum} \quad M_{\text{ListSum}} \cap [\text{listsum}(xs, Sum)]$$

Consider the magic program O for ListSum and listsum(xs,Sum). As a consequence of (13), we can also establish that:

$$\{\text{listsum}(xs, Sum)\} \quad \text{ListSum} \quad M_O \cap [\text{listsum}(xs, Sum)]$$

So the computed instances of the desired query can be deduced using the magic program O. This is relevant because, as we shall see later, bottom-up computing with the magic program is much easier than with the original program.   □

Moreover, well-modedness of the original program implies *safety* of the magic program, in the sense of [19]: every variable that occurs in the head of a clause of the magic program, also occurs in its body.

Theorem 6. *Let P be a well-moded program and Q a well-moded query. Then, the magic program O is safe.*

Proof. By Definition 3, there are three types of clauses in O.

Case $A \leftarrow A', \mathbf{A}$

The variables in the input positions of A occur in A', by Definition 3. By Definition 4, the variables in the output positions of A appear either in the input positions of A, and hence in A', or in the output positions of $\mathbf{A}$.

Case $Q' \leftarrow$

By the fact the Q is well-moded, Q' is ground.

Case $B' \leftarrow A', \mathbf{A}$

By Definition 3, the original clause from P is $A \leftarrow \mathbf{A}, B, \mathbf{B}$. The variables of B' are those in the input positions of B, that, by Definition 4, occur either in the input terms of A, and hence in A', or in the output terms of $\mathbf{A}$. □

Thus, despite the fact that a well-moded program, such as `ListSum` of Example 1, may not be suited for bottom-up computing, its magic version is, in the sense that the minimum requirement that finitely many new facts are inferred at each bottom-up iteration is fulfilled.

We conclude this section with some remarks about the transformation. First, observe that the optimization algorithm is modular, in the sense that each clause can be optimized separately. In particular we can obtain the optimized program transforming the program at compile time and the query, which provides the seed for the computation, at run time. Second, non-atomic queries can be dealt with easily: given a query $\mathbf{A}$, it is sufficient to add to the program a new clause $ans(\mathbf{X}) \leftarrow \mathbf{A}$, where ans is a fresh predicate and $\mathbf{X}$ are the variables in $\mathbf{A}$, and optimize the extended program w.r.t. the one-atom query $ans(\mathbf{X})$. Finally, the traditional distinction between an extensional database (EDB) and an intensional one (IDB) is immaterial to the discussion presented in this paper.

4 Total Correctness

What is the meaning of a triple $\{Q\}\ P\ \mathcal{Q}$ in the sense of total correctness? Several interpretations are possible, but the most common is to require partial correctness plus the fact that all derivations for Q in P are finite—a property which is referred to as *universal termination*. However, such a requirement would be unnecessarily restrictive if an arbitrary selection strategy is allowed in the top-down computation. For this reason, the termination analysis is usually tailored for some particular top-down strategy, such as Prolog's depth-first strategy combined with a leftmost selection rule, referred to as LD-resolution.

A proof method for termination of Prolog programs is introduced in [6,7], based on the following notion of an acceptable program.

Definition 5. *Let A be an atom and c be a clause, then:*

- *A level mapping is a function $|\ |$ from ground atoms to natural numbers.*
- *A is bounded w.r.t. $|\ |$, if $|\ |$ is bounded on the set of all ground instances of A.*

– c *is* acceptable w.r.t. $|\ |$ and *an interpretation I, if*
  • I *is a model of c,*
  • *for all ground instances $A \leftarrow \mathbf{A}, B, \mathbf{B}$ of c such that $I \models \mathbf{A}$*

$$|A| > |B|.$$

– *A program is* acceptable w.r.t. $|\ |$ and *I, if every clause of it is.*   □

The intuition behind this definition is the following. The level mapping plays the role of a termination function, and it is required to decrease from head to the body of any (ground instance of a) clause. The model I used in the notion of acceptability gives a declarative account of the leftmost selection rule of Prolog. The decreasing of the level mapping from the head A to a body atom B is required only if the body atoms to the left of B have been already refuted: in this case, by the Soundness of SLD-resolution, these atoms are true in any model of the program. In the proof method, the model I is employed to propagate inter-argument relations from left to right. The following result about acceptable programs holds.

Theorem 7. *Suppose that*

– *the program P is acceptable w.r.t. $|\ |$ and I,*
– *the one-atom query Q is bounded w.r.t. $|\ |$.*

Then all Prolog computations of Q in P are finite.

Proof. See [6,7], for a detailed proof. The general idea is to associate a multiset of integers to each query of the resolution and to show the multiset associated with a query is strictly greater than the one associated with its resolvent.   □

Moreover, it is possible to show that each terminating Prolog program P is acceptable w.r.t. the following level mapping:

$$|A| = nodes_P(A)$$

where $nodes_P$ denotes the number of nodes in the S-tree for $P \cup \{\leftarrow A\}$.

Example 4. The program `ListSum` of Example 1 is acceptable w.r.t. any model and the level mapping $|\ |$ defined as follows:

$$|\texttt{listsum}(\mathbf{xs}, \mathbf{sum})| = size(\mathbf{xs})$$
$$|\texttt{sum}(\mathbf{x}, \mathbf{y}, \mathbf{z})| = size(\mathbf{y})$$

where $size(t)$ counts the number of symbols in the (ground) term t. This can be easily checked simply observing that the number of functional symbols of every atom in the body of the clauses is strictly less than the number of functional symbols in the corresponding head.

Also, for every ground term xs and variable Sum, the query listsum(xs,Sum) is bounded, so every Prolog computation for it terminates, as a consequence of Theorem 7. In many cases, a non-trivial model is needed in the proof of termination. In the ListSum example, if the two input arguments of the relation sum in the recursive clause of listsum are swapped, then a model I is needed, such that $I \models$ listsum(xs, sum) iff $size(\text{xs}) \geq size(\text{sum})$.

Moreover, it is in general possible to use simpler level mappings, but this requires more complicate definitions: see [7,15] for details.   □

Besides its use in proving termination, the notion of acceptability makes the task of constructing the least Herbrand model of a program much easier. Call an interpretation I for a program P *supported* if for any $A \in I$ there exists a ground instance $A \leftarrow \mathbf{B}$ of a clause from P such that $I \models \mathbf{B}$. The following result from [6] holds.

Theorem 8. *Any acceptable program P has a unique supported model, which coincides with its least Herbrand model M_P.*

Proof. See [6] for details. Consider a fix-point X of T_P, strictly greater that M_P, and an element $A \in X \backslash M_P$; then, there must be a ground atom $B \in X \backslash M_P$ such that $A \leftarrow \mathbf{A}, B, \mathbf{B} \in ground(P)$. But this leads to an infinite chain of resovents, starting from A.   □

Usually, checking that an interpretation is a supported model of the program is straightforward, and does not require inductive reasoning. Also, this technique can be used with the reduced program, as reduced programs of acceptable programs are in turn acceptable.

Summarizing, the problem of establishing a triple $\{A\}\ P\ \mathcal{A}$ in the sense of total correctness, for a well-moded program P and query A, can be solved by the following steps:

1. find a base I for P such that $[A] \subseteq I$;
2. show that P is acceptable and A is bounded w.r.t. the same model and level mapping;
3. find a supported model M of P_I;
4. check that $\mathcal{A} = M \cap [A]$.

In the **Append** example of Section 2, it is easy to show that the set (2), namely {append(xs,ys,zs) | xs,ys,zs are lists and xs * ys = zs} is indeed a supported model of the program reduced by its base (3), so the desired triple can be established. In the ListSum example, it is readily checked that the set 7 from Example 2 is a supported model of the program reduced by its base I_{ListSum}.

5   Total Correctness and Bottom-Up Computing

Although a thorough study of the relative efficiency of bottom-up and top-down execution is outside the reach of our declarative methods, we are able to show the total correctness of the magic-sets transformation on the basis of the results of the previous section. In fact, we can show that if the original program is terminating in a top-down sense, then the magic program is terminating in a bottom-up sense, in a way which is made precise by the next result. Two assumptions on the original programs are necessary, namely acceptability, which implies termination, and well-modedness, which implies ground output.

Theorem 9. *Let P be a well-moded, acceptable program, and Q a one-atom well-moded, bounded query. Then the least Herbrand model of the magic program O is finite.*

Proof. Let I and $|\ |$ be the model and level mapping used in the proof of acceptability. We define a mapping of magic atoms into $\omega \cup \infty$ as follows:

$$|A'| = \max\{|B| \mid A' = B'\}.$$

Next, we show that M_O contains a finite number of magic atoms. First, we observe that, for the seed fact $Q' \in T_P(\emptyset)$, $|Q'| < \omega$, as the query Q is bounded. Consider now a magic atom B' deduced at stage $n > 1$ in the bottom-up computation, i.e. $B' \in T_O^n(\emptyset) \setminus T_O^{n-1}(\emptyset)$. By the magic transformation, there is a clause $B' \leftarrow A', \mathbf{A}$ in O such that $T_O^{n-1}(\emptyset) \models A', \mathbf{A}$. Since $T_O^{n-1}(\emptyset) \models \mathbf{A}$ implies that $\mathbf{A}$ holds in any model of P by the partial correctness Theorem 4, we have by the acceptability of P that , for each clause $A \leftarrow \mathbf{A}, B, \mathbf{B}$ in P, $|A| > |B|$, which implies $|A'| > |B'|$. Therefore, the level of newly deduced magic atoms is smaller than that of some preexisting magic atoms, which implies that finitely many magic atoms are in M_O.

To conclude the proof, we have to show that there are finitely many non-magic atoms in M_O. Observe that every non-magic atom A of M_O is a computed answer of a query B such that $M_O \models B'$. Given $A' \in M_O$, consider a query B with its output positions filled with distinct variables, and $B' = A'$. By Theorems 7 and 5, B has a finite set of ground computed answers. The thesis then follows by the fact that finitely many magic atoms are in M_O.   □

As an immediate consequence of this theorem we have that, for some $n \geq 0$:

$$T_{P_I}^n(\emptyset) = M_O$$

and therefore the bottom-up computation with O terminates. Notice that this result does not imply that the bottom-up computation with O and the top-down one with P are equally efficient, although both terminates. In [20], an extra condition on the original program is required, namely that it is *subgoal rectified*, in order to obtain that the cost of computing with the magic program is proportional to top-down evaluation.

As a final example, consider again the ListSum program of Example 1 and the query listsum(xs,Sum). By the partial correctness results, we know that:

$$\{\text{listsum}(\text{xs}, \text{Sum})\}\quad \text{ListSum}\quad M_O \cap [\text{listsum}(\text{xs}, \text{Sum})]$$

By Theorem 9 M_O is finite, so we can actually perform a bottom-up computation with O, thus obtaining M_O first, and then extract the desired computed instances from it.

6 Examples

Length of a List

Consider the program ListLen, the call pattern listlen(+,-) and the query listlen([a,b,b,a]). The optimized program is:

```
listlen([],0) ← listlen'([])
listlen([X|Xs],s(L)) ← listlen'([X|Xs]),base(X),listlen(Xs,L)

listlen'(Xs) ← base(X),listlen'([X|Xs])

listlen'([a,b,b,a]) ←
```

As we can see there is only one clause which depends from the query, namely the optimized query w.r.t. $\mathcal{C}$, and it can be easily produced at run time. The bottom-up evaluation of the optimized program is:

$$
\begin{aligned}
T_P^1(\emptyset) &= \{\text{listlen}'([a,b,b,a])\} \\
T_P^2(\emptyset) &= T_P^1(\emptyset) \cup \{\text{listlen}'([b,b,a])\} \\
T_P^3(\emptyset) &= T_P^2(\emptyset) \cup \{\text{listlen}'([b,a])\} \\
T_P^4(\emptyset) &= T_P^3(\emptyset) \cup \{\text{listlen}'([a])\} \\
T_P^5(\emptyset) &= T_P^4(\emptyset) \cup \{\text{listlen}'([])\} \\
T_P^6(\emptyset) &= T_P^5(\emptyset) \cup \{\text{listlen}([],0)\} \\
T_P^7(\emptyset) &= T_P^6(\emptyset) \cup \{\text{listlen}([b,a],s(s(0)))\} \\
T_P^8(\emptyset) &= T_P^7(\emptyset) \cup \{\text{listlen}([b,b,a],s(s(s(0))))\} \\
T_P^9(\emptyset) &= T_P^8(\emptyset) \cup \{\text{listlen}([a,b,b,a],s(s(s(s(0)))))\} \\
T_P^{10}(\emptyset) &= T_P^9(\emptyset)
\end{aligned}
$$

It can be noted that in the first part of the computation the optimized program computes the closed interpretation $I_{Listlen,[a,b,b,a]}$, and in the last one uses it in order to optimize the computation.

Sum of a List of Numbers

Consider the program ListSum, the call patterns:

```
listsum(+,-)
sum(+,+,-)
```

and the query listsum([s(0),s(s(0))], Sum). The optimized program is:

```
listsum([],0) ← listsum'([])
listsum([X|Xs],Sum) ←
    listSum'([X|Xs]),listsum(Xs,PSum),sum(PSum,X,Sum)
sum(X,0,X) ← sum'(X,0),nat(X)
sum(X,s(Y),s(Z)) ← sum'(X,s(Y)),sum(X,Y,Z)

listsum'(Xs) ← listsum'([X|Xs])
sum'(Psum,X) ← listsum'([X|Xs]),listsum(Xs,PSum)
sum'(X,Y) ← sum'(X,s(Y))

listsum'([s(0),s(s(0))]) ←
```

The bottom-up evaluation of the optimized program is:

$$
\begin{aligned}
T_P^1(\emptyset) &= & \{\texttt{listsum}'([\texttt{s}(0),\texttt{s}(\texttt{s}(0))])\} \\
T_P^2(\emptyset) &= T_P^1(\emptyset) \cup & \{\texttt{listsum}'([\texttt{s}(\texttt{s}(0))])\} \\
T_P^3(\emptyset) &= T_P^2(\emptyset) \cup & \{\texttt{listsum}'([])\} \\
T_P^4(\emptyset) &= T_P^3(\emptyset) \cup & \{\texttt{listsum}([],0)\} \\
T_P^5(\emptyset) &= T_P^4(\emptyset) \cup & \{\texttt{sum}'(0,\texttt{s}(\texttt{s}(0)))\} \\
T_P^6(\emptyset) &= T_P^5(\emptyset) \cup & \{\texttt{sum}'(0,\texttt{s}(0))\} \\
T_P^7(\emptyset) &= T_P^6(\emptyset) \cup & \{\texttt{sum}'(0,0)\} \\
T_P^8(\emptyset) &= T_P^7(\emptyset) \cup & \{\texttt{sum}(0,0,0)\} \\
T_P^9(\emptyset) &= T_P^8(\emptyset) \cup & \{\texttt{sum}(0,\texttt{s}(0),\texttt{s}(0))\} \\
T_P^{10}(\emptyset) &= T_P^9(\emptyset) \cup & \{\texttt{sum}(0,\texttt{s}(\texttt{s}(0)),\texttt{s}(\texttt{s}(0)))\} \\
T_P^{11}(\emptyset) &= T_P^{10}(\emptyset) \cup & \{\texttt{listsum}([\texttt{s}(\texttt{s}(0))],\texttt{s}(\texttt{s}(0)))\} \\
T_P^{12}(\emptyset) &= T_P^{11}(\emptyset) \cup & \{\texttt{sum}'(\texttt{s}(\texttt{s}(0)),\texttt{s}(0))\} \\
T_P^{13}(\emptyset) &= T_P^{12}(\emptyset) \cup & \{\texttt{sum}'(\texttt{s}(\texttt{s}(0)),0)\} \\
T_P^{14}(\emptyset) &= T_P^{13}(\emptyset) \cup & \{\texttt{sum}(\texttt{s}(\texttt{s}(0)),0,\texttt{s}(\texttt{s}(0)))\} \\
T_P^{15}(\emptyset) &= T_P^{14}(\emptyset) \cup & \{\texttt{sum}(\texttt{s}(\texttt{s}(0)),\texttt{s}(0),\texttt{s}(\texttt{s}(\texttt{s}(0))))\} \\
T_P^{16}(\emptyset) &= T_P^{15}(\emptyset) \cup & \{\texttt{listsum}([\texttt{s}(\texttt{s}(0)),\texttt{s}(0)],\texttt{s}(\texttt{s}(\texttt{s}(0))))\}
\end{aligned}
$$

In this case the computation of the closed interpretation is interlaced with the computation of the interesting part of the least Herbrand model.

Ancestors

Consider the following program Ancestor:

```
ancestor(X,Y) ← parent(X,Y)
ancestor(X,Y) ← parent(X,Z),ancestor(Z,Y)
```

where *Parent* is a base relation. Consider the moding ancestor(+,-) and the query ancestor(f,Y). The optimized program is:

```
ancestor(X,Y) ← ancestor'(X),parent(X,Y)
ancestor(X,Y) ← ancestor'(X),parent(X,Z),ancestor(Z,Y)

ancestor'(Y) ← parent(X,Y),ancestor'(X)
ancestor'(a) ←
```

If we suppose the following definition for the base relation `parent`:

```
parent(a,b) ←
parent(a,c) ←
parent(a,d) ←
parent(e,b) ←
parent(e,c) ←
parent(e,d) ←
parent(f,a) ←
parent(f,g) ←
parent(h,e) ←
parent(h,i) ←
```

The computation is:

$$T_P^1(\emptyset) = \{\text{ancestor}'(f)\}$$

$$T_P^2(\emptyset) = T_P^1(\emptyset) \cup \left\{ \begin{array}{l} \text{ancestor}'(a) \\ \text{ancestor}'(g) \\ \text{ancestor}(f,a) \\ \text{ancestor}(f,g) \end{array} \right\}$$

$$T_P^3(\emptyset) = T_P^2(\emptyset) \cup \left\{ \begin{array}{l} \text{ancestor}'(b) \\ \text{ancestor}'(c) \\ \text{ancestor}'(d) \\ \text{ancestor}(f,b) \\ \text{ancestor}(f,c) \\ \text{ancestor}(f,d) \end{array} \right\}$$

$$T_P^4(\emptyset) = T_P^3(\emptyset)$$

However, we obtain a different optimized program if we consider the moding `ancestor(-,+)` and the query `ancestor(X,b)`:

```
ancestor(X,Y) ← ancestor'(Y),parent(X,Y)
ancestor(X,Y) ← ancestor'(Y),parent(X,Z), ancestor(Z,Y)

ancestor'(X)  ← parent(X,Y),ancestor'(Y)

ancestor'(Y)  ←
```

The computation is:

$$T_P^1(\emptyset) = \{\texttt{ancestor'(b)}\}$$

$$T_P^2(\emptyset) = T_P^1(\emptyset) \cup \left\{ \begin{array}{l} \texttt{ancestor'(a)} \\ \texttt{ancestor'(e)} \\ \texttt{ancestor(a,b)} \\ \texttt{acenstor(e,b)} \end{array} \right\}$$

$$T_P^3(\emptyset) = T_P^2(\emptyset) \cup \left\{ \begin{array}{l} \texttt{ancestor'(f)} \\ \texttt{ancestor'(h)} \\ \texttt{ancestor(f,b)} \\ \texttt{ancestor(h,b)} \end{array} \right\}$$

$$T_P^4(\emptyset) = T_P^3(\emptyset)$$

As we can see, different call patterns generate different optimized program. In general these programs are not equivalent.

Powers

Consider now the following program Power, which computes x^y, where x and y are natural numbers:

```
power(X,0,s(0)) ←
power(X,s(Y),Z)  ← power(X,Y,W),times(X,W,Z)
times(X,0,0) ←
times(X,s(Y),Z)  ← times(X,Y,W),sum(X,W,Z)
sum(X,0,X) ←
sum(X,s(Y),s(Z))  ← sum(X,Y,Z)
```

If we consider the call patterns:

```
power(+,+,-)
times(+,+,-)
sum(+,+,-)
```

and the query:

```
power(s(s(0)),s(s(0)),Z)
```

the optimized program is:

```
power(X,0,s(0))  ← power'(X,0)
power(X,s(Y),Z)  ← power'(X,s(Y)),power(X,Y,W),times(X,W,Z)
times(X,0,0)  ← times'(X,0)
times(X,s(Y),Z)  ← times'(X,s(Y)),times(X,Y,W),sum(X,W,Z)
sum(X,0,X)  ← sum'(X,0)
sum(X,s(Y),s(Z))  ← sum'(X,s(Y)),sum(X,Y,Z)

power'(X,Y)  ← power'(X,s(Y))
```

```
times'(X,W)  ← power(X,Y,W),power'(X,s(Y))
times'(X,Y)  ← times'(X,s(Y))
sum'(X,W)  ← times(X,Y,W),times'(X,s(Y))
sum'(X,Y)  ← sum'(X,s(Y))


power'(s(s(0)),s(s(0)))  ←
```

The computation is:

$$
\begin{aligned}
T_P^1(\emptyset) &= \qquad\qquad \{\text{power}'(s(s(0)), s(s(0)))\} \\
T_P^2(\emptyset) &= T_P^1(\emptyset) \cup \{\text{power}'(s(s(0)), s(0))\} \\
T_P^3(\emptyset) &= T_P^2(\emptyset) \cup \{\text{power}'(s(s(0)), 0)\} \\
T_P^4(\emptyset) &= T_P^3(\emptyset) \cup \{\text{power}(s(s(0)), 0, s(0))\} \\
T_P^5(\emptyset) &= T_P^4(\emptyset) \cup \{\text{times}'(s(s(0)), s(0))\} \\
T_P^6(\emptyset) &= T_P^5(\emptyset) \cup \{\text{times}'(s(s(0)), 0)\} \\
T_P^7(\emptyset) &= T_P^6(\emptyset) \cup \{\text{times}(s(s(0)), 0, 0)\} \\
T_P^8(\emptyset) &= T_P^7(\emptyset) \cup \{\text{sum}'(s(s(0)), 0)\} \\
T_P^9(\emptyset) &= T_P^8(\emptyset) \cup \{\text{sum}(s(s(0)), 0, s(s(0)))\} \\
T_P^{10}(\emptyset) &= T_P^9(\emptyset) \cup \{\text{times}(s(s(0)), s(0), s(s(0)))\} \\
T_P^{11}(\emptyset) &= T_P^{10}(\emptyset) \cup \{\text{power}(s(s(0)), s(0), s(s(0)))\} \\
T_P^{12}(\emptyset) &= T_P^{11}(\emptyset) \cup \{\text{times}'(s(s(0)), s(s(0)))\} \\
T_P^{13}(\emptyset) &= T_P^{12}(\emptyset) \cup \{\text{sum}'(s(s(0)), s(s(0)))\} \\
T_P^{14}(\emptyset) &= T_P^{13}(\emptyset) \cup \{\text{sum}'(s(s(0)), s(0))\} \\
T_P^{15}(\emptyset) &= T_P^{14}(\emptyset) \cup \{\text{sum}(s(s(0)), s(0), s(s(s(0))))\} \\
T_P^{16}(\emptyset) &= T_P^{15}(\emptyset) \cup \{\text{sum}(s(s(0)), s(s(0)), s(s(s(s(0)))))\} \\
T_P^{17}(\emptyset) &= T_P^{16}(\emptyset) \cup \{\text{times}(s(s(0)), s(s(0)), s(s(s(s(0)))))\} \\
T_P^{18}(\emptyset) &= T_P^{17}(\emptyset) \cup \{\text{power}(s(s(0)), s(s(0)), s(s(s(s(0)))))\} \\
T_P^{19}(\emptyset) &= T_P^{18}(\emptyset)
\end{aligned}
$$

It is interesting to note that the computation is, in this case, really closed to that generate by a functional program with lazy evaluation.

Binary Search

Consider the following program Search, implementing the dichotomic (or binary) search on a list of pairs $(Key, Value)$ ordered with respect to Key:

```
search(N,Xs,M)  ←
    divide(Xs,Xs1,X,Y,Xs2),switch(N,X,Y,Xs1,Xs2,M)
switch(N,N,M,Xs1,Xs2,M)  ← key(N),value(M)
switch(N,X,Y,Xs1,Xs2,M)  ← greater(N,X),search(N,Xs2,M)
switch(N,X,Y,Xs1,Xs2,M)  ← greater(X,N),search(N,Xs1,M)
```

where Key and Value are base relations. Observe that the program is not completely specified, as the relations Divide, and Greater have no definition. If we consider the following call patterns:

```
search(+,+,-)
switch(+,+,+,+,+,-)
```

and the query `search(5,[(1,a),(3,b),(5,a),(10,c)],M)`, the optimized program is:

```
search(N,Xs,M)  ← search'(N,Xs),divide(Xs,Xs1,X,Y,Xs2),
                  switch(N,X,Y,Xs1,Xs2,M)
switch(N,N,M,Xs1,Xs2,M)  ← switch'(N,N,M,Xs1,Xs2),
                  key(N),value(M)
switch(N,X,Y,Xs1,Xs2,M)  ← switch'(N,X,Y,Xs1,Xs2),greater(N,X),
                  search(N,Xs2,M)
switch(N,X,Y,Xs1,Xs2,M)  ← switch'(N,X,Y,Xs1,Xs2),greater(X,N),
                  search(N,Xs1,M)

switch'(N,X,Y,Xs1,Xs2)  ← divide(Xs,Xs1,X,Y,Xs2),search'(N,Xs)
search'(N,Xs2)  ← N>X,switch'(N,X,Y,Xs1,Xs2)
search'(N,Xs1)  ← N<X,switch'(N,X,Y,Xs1,Xs2)

search'(5,[(1,a),(3,b),(5,a),(10,c)])  ←
```

The computation is the following:

$$
\begin{aligned}
T_P^1(\emptyset) &= && \{\text{search}'(5,[(1,a),(3,b),(5,a),(10,c)])\} \\
T_P^2(\emptyset) &= T_P^1(\emptyset) \cup \{\text{switch}'(5,3,b,[(1,a)],[(5,a),(10,c)])\} \\
T_P^3(\emptyset) &= T_P^2(\emptyset) \cup \{\text{search}'(5,[(5,a),(10,c)])\} \\
T_P^4(\emptyset) &= T_P^3(\emptyset) \cup \{\text{switch}'(5,5,a,[],(10,c)])\} \\
T_P^5(\emptyset) &= T_P^4(\emptyset) \cup \{\text{switch}(5,5,a,[],(10,c)],a)\} \\
T_P^6(\emptyset) &= T_P^5(\emptyset) \cup \{\text{search}(5,[(5,a),(10,c)],a)\} \\
T_P^7(\emptyset) &= T_P^6(\emptyset) \cup \{\text{switch}(5,3,b,[(1,a)],[(5,a),(10,c)],a)\} \\
T_P^8(\emptyset) &= T_P^7(\emptyset) \cup \{\text{search}(5,,[(1,a),(3,b),(5,a),(10,c)],a)\} \\
T_P^9(\emptyset) &= T_P^8(\emptyset)
\end{aligned}
$$

Fibonacci Numbers

Consider the following program, that computes the Fibonacci numbers:

```
fib(0,0)  ←
fib(s(0),s(0))  ←
fib(s(s(X)),Y)  ← fib(s(X),Y1),fib(X,Y2),sum(Y1,Y2,Y)
sum(X,0,X)  ←
sum(X,s(Y),s(Z))  ← sum(X,Y,Z)
```

with the moding:

```
fib(+,-)
sum(+,+,-)
```

and the query `fib(s(s(s(0)))),Y)`. The optimized program is:

```
fib'(s(s(s(0)))) ←
fib'(s(X))  ← fib'(s(s(X)))
fib'(X)  ← fib'(s(s(X)),fib(s(X),Y1)

sum'(Y1,Y2)  ← fib'(s(s(X))),fib(s(X),Y1),fib(X,Y2)
sum'(X,Y)  ← sum'(X,s(Y))

fib(0,0)  ← fib'(0)
fib(s(0),s(0))  ← fib'(s(0))

fib(s(s(X)),Y) ←
    fib'(s(s(X))),fib(s(X),Y1),fib(X,Y2),sum(Y1,Y2,Y)
sum(X,0,X)  ← sum'(X,0)
sum(X,s(Y),s(Z))  ← sum'(X,s(Y)),sum(X,Y,Z)
```

The computation is the following:

$$
\begin{aligned}
T_P^1(\emptyset) &= \{\texttt{fib'(s(s(s(0))))}\} \\
T_P^2(\emptyset) &= T_P^1(\emptyset) \cup \{\texttt{fib'(s(s(0)))}\} \\
T_P^3(\emptyset) &= T_P^2(\emptyset) \cup \{\texttt{fib'(s(0))}\} \\
T_P^4(\emptyset) &= T_P^3(\emptyset) \cup \{\texttt{fib'(0)},\texttt{fib(s(0),s(0))}\} \\
T_P^5(\emptyset) &= T_P^4(\emptyset) \cup \{\texttt{fib(0,0)}\} \\
T_P^6(\emptyset) &= T_P^5(\emptyset) \cup \{\texttt{sum'(s(0),0)}\} \\
T_P^7(\emptyset) &= T_P^6(\emptyset) \cup \{\texttt{sum(s(0),0,s(0))}\} \\
T_P^8(\emptyset) &= T_P^7(\emptyset) \cup \{\texttt{fib(s(s(0)),s(0))}\} \\
T_P^9(\emptyset) &= T_P^8(\emptyset) \cup \{\texttt{sum'(s(0),s(0))}\} \\
T_P^{10}(\emptyset) &= T_P^9(\emptyset) \cup \{\texttt{sum(s(0),s(0),s(s(0)))}\} \\
T_P^{11}(\emptyset) &= T_P^{10}(\emptyset) \cup \{\texttt{fib(s(s(s(0)))),s(s(0)))}\} \\
T_P^{12}(\emptyset) &= T_P^{11}(\emptyset)
\end{aligned}
$$

Here we can observe that the magic-sets transformation is suitable also for non-linear recursive programs, i.e. program with more than one mutually recursive body atoms. Once again we can see that the computation is "lazy".

7   Conclusions

In this paper, we introduced a method for proving partial correctness, revised another method for total correctness, and applied both to the case study of the magic-sets transformation for goal-driven bottom-up computing. The obtained results rely on purely declarative reasoning, abstracting away from procedural semantics, and are new under various points of view. First, partial correctness is obtained without any assumptions that the program respects the given moding. Second, termination is obtained under the only assumptions of well-moddedness, which is natural in practical bottom-up computing, and acceptability, which is a necessary and sufficient condition for top-down termination.

Moreover, both partial correctness and termination are established for logic programs in full generality, and not only for function-free Datalog programs.

Further research may be pursued on the topics of this paper. For instance, we are confident that the same kind of result can be established for other variants of the magic-sets transformation technique and also for extensions of it to general logic programs (i.e. logic program with negation in the body of the clauses). Moreover, it is interesting to investigate whether other optimization techniques may be defined using the concept of base.

Acknowledgements

Thanks are owing to Yeoshua Sagiv for useful discussions.

References

1.   K.R. Apt. Logic programming. In J. van Leeuwen, editor, *Handbook of Theoretical Computer Science*, volume B, pages 493–574. Elsevier, 1990.
2.   K. R. Apt. Declarative programming in Prolog. In D. Miller, editor, *Proc. International Symposium on Logic Programming*, pages 11–35. MIT Press, 1993.
3.   K.R. Apt. Program Verification and Prolog. In E. Börger, editor, *Specification and Validation methods for Programming languages and systems*. Oxford University Press, 1994.
4.   K.R. Apt, M. Gabbrielli, and D. Pedreschi. A Closer Look at Declarative Interpretations. *Technical Report CS-R9470*, Centre for Mathematics and Computer Science, Amsterdam, Journal of Logic Programming. 28(2): 147-180, 1996.
5.   K.R. Apt and E. Marchiori. Reasoning about Prolog programs: from modes through types to assertions. *Formal Aspects of Computing*, 6A:743–764, 1994.
6.   K.R. Apt and D. Pedreschi. Reasoning about termination of pure prolog programs. *Information and computation*, 106(1):109–157, 1993.
7.   K. R. Apt and D. Pedreschi. Modular termination proofs for logic and pure Prolog programs. In G. Levi, editor, *Advances in Logic Programming Theory*, pages 183–229. Oxford University Press, 1994.
8.   A. Bossi and N. Cocco. Verifying Correctness of Logic Programs. In J. Diaz and F. Orejas, editors, *TAPSOFT '89*, volume 352 of *Lecture Notes in Computer Science*, pages 96–110. Springer-Verlag, Berlin, 1989.
9.   C. Beeri and R. Ramakrishnan. The power of magic. In *Proc. 6th ACM-SIGMOD-SIGACT Symposium on Principles of Database systems*, pages 269–283. The Association for Computing Machinery, New York, 1987.
10.   F. Bronsard, T.K. Lakshman, and U.S. Reddy. A framework of directionality for proving termination of logic programs. In K. R. Apt, editor, *Proceedings of the Joint International Conference and Symposium on Logic Programming*, pages 321–335. MIT Press, 1992.
11.   P. Deransart. Proof methods of declarative properties of definite programs. *Theoretical Computer Science*, 118:99–166, 1993.
12.   J.W. Lloyd. *Foundations of logic programming*. Springer-Verlag, Berlin, second edition, 1987.

13.   P. Mascellani. Declarative Verification of General Logic Programs. In *Proceedings of the Student Session, ESSLLI-2000*. Birmingham UK, 2000.

14.   P. Mascellani and D. Pedreschi. Proving termination of prolog programs. In *Proceedings 1994 Joint Conf. on Declarative Programming GULP-PRODE '94*, pages 46–61, 1994.

15.   P. Mascellani and D. Pedreschi. Total correctness of prolog programs. In F.S. de Boer and M. Gabbrielli, editors, *Proceedings of the W2 Post-Conference Workshop ICLP'94*. Vrije Universiteit Amsterdam, 1994.

16.   D. Pedreschi. Verification of Logic Programs. In M. I. Sessa, editor, *Ten Years of Logic Programming in Italy*, pages 211–239. Palladio, 1995.

17.   D. Pedreschi and S. Ruggieri. Verification of Logic Programs. *Journal of Logic Programming*, 39 (1-3):125-176, April 1999

18.   S. Ruggieri. Proving (total) correctness of prolog programs. In F.S. de Boer and M. Gabbrielli, editors, *Proceedings of the W2 Post-Conference Workshop ICLP'94*. Vrije Universiteit Amsterdam, 1994.

19.   J.D. Ullman. *Principles of Database and Knowledge-base Systems, Volume I*. Principles of Computer Science Series. Computer Science Press, 1988.

20.   J.D. Ullman. *Principles of Database and Knowledge-base Systems, Volume II; The New Technologies*. Principles of Computer Science Series. Computer Science Press, 1989.

Key Constraints and Monotonic Aggregates in Deductive Databases

Carlo Zaniolo

Computer Science Department
University of California at Los Angeles
Los Angeles, CA 90095
zaniolo@cs.ucla.edu
http://www.cs.ucla.edu/~zaniolo

Abstract. We extend the fixpoint and model-theoretic semantics of logic programs to include unique key constraints in derived relations. This extension increases the expressive power of Datalog programs, while preserving their declarative semantics and efficient implementation. The greater expressive power yields a simple characterization for the notion of set aggregates, including the identification of aggregates that are monotonic with respect to set containment and can thus be used in recursive logic programs. These new constructs are critical in many applications, and produce simple logic-based formulations for complex algorithms that were previously believed to be beyond the realm of declarative logic.

1 Introduction

The basic relational data model consists of a set of tables (or base relations) and of a query language, such as SQL or Datalog, from which new relations can be derived. Unique keys can be declared to enforce functional dependency constraints on base relations, and their important role in database schema design has been recognized for a long time [1,28]. However, little attention has been paid so far to the use of unique keys, or functional dependencies, in derived relations. This paper shows that keys in derived relations increase significantly the expressive power of the query languages used to define such relations and this additional power yields considerable benefits. In particular, it produces a formal treatment of database aggregates, including user-defined aggregates, and monotonic aggregates, which can be used without restrictions in recursive queries to express complex algorithms that were previously considered problematic for Datalog and SQL.

2 Keys on Derived Relations

For example, consider a database containing relations student(Name, Major), and professor(Name, Major). In fact, let us consider the following microcollege example that only has three facts:

A.C. Kakas, F. Sadri (Eds.): Computat. Logic (Kowalski Festschrift), LNAI 2408, pp. 109–134, 2002.
© Springer-Verlag Berlin Heidelberg 2002

```
student('JimBlack', ee).        professor(ohm, ee).
                                professor(bell, ee).
```

Now, the rule is that the major of a student must match his/her advisor's main area of specialization. Then, eligible advisors can be computed as follows:

$$\text{elig\_adv}(S, P) \leftarrow \text{student}(S, \text{Majr}), \text{professor}(P, \text{Majr}).$$

Now the answer to a query ?elig_adv(S, P) is

$$\{\text{elig\_adv}('JimBlack', \text{ohm}), \text{elig\_adv}('JimBlack', \text{bell})\}$$

But, a student can only have one advisor. We can express this constraint by requiring that the first argument be a unique key for the advisor relation. We denote this constraint by the notation

$$\text{unique\_key}(\text{advisor}, [1])!$$

Thus, the first argument of unique_key specifies the predicate restricted by the key, and the second argument gives the list of the argument positions that compose the key. An empty list denotes that the derived relation can only contain a single tuple. The exclamation mark is used as the punctuation mark for key constraints. We can now write the following program for our microcollege:

Example 1. For each student select one advisor from professors in the same area

```
unique_key(advisor, [1])!

advisor(S, P) ←student(S, Majr), professor(P, Majr).

student('JimBlack', ee).
professor(ohm, ee).
professor(bell, ee).
```

Since the key condition ensures that there is only one professor in the resulting advisor table, our query has two possible answers. One is the set

$$\{\text{advisor}('JimBlack', \text{ohm})\}$$

and the other is the set:

$$\{\text{advisor}('JimBlack', \text{bell})\}$$

In the next section, we show that positive programs with keys can be characterized naturally by fixpoint semantics containing multiple canonical answers; in Section 4, we show that their meaning can also be modelled by programs with negated goals under stable models semantics.

Let us consider now some examples that provide a first illustration of the expressive power brought to logic programming by keys in derived relations. The following program constructs a spanning tree rooted in node a, for a graph stored in a binary relation g as follows:

Example 2. Computing spanning trees

```
unique_key(tree, [2])!
tree(root, a).
tree(Y, Z) ←  tree(X, Y), g(Y, Z).

g(a, b).
g(b, c).
g(a, c).
```

Two different spanning trees can be derived, as follows:

$$\{\texttt{tree(root, a)}, \texttt{tree(a, b)}, \texttt{tree(b, c)}\}$$

$$\{\texttt{tree(root, a)}, \texttt{tree(a, b)}, \texttt{tree(a, c)}\}$$

More than one key can be declared for each derived relation. For instance, let us add a second key, unique_key(tree, [1]), to the previous graph example. Then, the result may no longer be a spanning tree; instead, it is a simple path, where for each source node, there is only one sink node and vice versa:

Example 3. Computing simple paths

```
unique_key(spath, [1])!
unique_key(spath, [2])!
spath(root, X) ←g(X, Y).
spath(Y, Z) ←   spath(X, Y), g(Y, Z).
freenode ←     g(_, Y), ¬spath(_, Y).
```

The last rule in Example 3, above, detects whether any node remains free, i.e., whether there is a node not touched by the simple path. Now, a query on whether, for some simple path, there is no free node (i.e., is ¬freenode true?) can be used to decide the Hamiltonian path problem for our graph; this is an $\mathcal{NP}$-complete problem. An equivalent way to pose the same question is asking whether **freenode** is true for *all solutions*. A system that generates all possible paths and returns a positive answer when *freenode* holds for all paths implements an *all-answer* semantics. This example illustrates how exponential problems can be expressed in Datalog with keys under this semantics [14].

Polynomial time problems, however, are best treated using *single-answer* semantics, since this can be supported in polynomial time for Datalog programs with key constraints and stratified negation, as discussed later in this paper; moreover, these programs can express all the queries that are polynomial in the size of the database—i.e., the queries in the class *DB-PTIME* [1]. Under single-answer semantics, a deductive system is only expected to compute one out of the many existing canonical models for a program, and return an answer based on this particular model. For certain programs, this approach results in different query answers being returned for different canonical models computed by the system—nondeterministic queries. For other programs, however, the query answer remains the same for all canonical models—deterministic queries. This is, for instance, the case of the parity query below, which determines whether a non-empty database relation b(X) has an even number of tuples:

Example 4. Counting mod 2

$$\begin{aligned}
&\text{unique\_key}(\text{chain}, [1])! \\
&\text{unique\_key}(\text{chain}, [2])! \\
&\text{chain}(\text{nil}, X) \leftarrow \text{b}(X). \\
&\text{chain}(X, Y) \leftarrow \quad \text{chain}(\_, X), \text{b}(Y). \\
&\text{ca}(Y, \text{odd}) \qquad \text{chain}(\text{nil}, Y) \\
&\text{ca}(Y, \text{even}) \leftarrow \quad \text{ca}(X, \text{odd}), \text{chain}(X, Y). \\
&\text{ca}(Y, \text{odd}) \leftarrow \quad \text{ca}(X, \text{even}), \text{chain}(X, Y). \\
&\text{mod2}(\text{Parity}) \leftarrow \text{ca}(Y, \text{Parity}), \neg\text{chain}(Y, \_).
\end{aligned}$$

Observe that this program consists of three parts. The first part is the chain rules that enumerate the elements of b(X) one-by-one. The second part is the ca rules that perform a specific aggregate-like computation on the elements of chain—i.e., the odd/even computation for the parity query. The third part is the mod2 rule that uses negation to detect the element of the chain without a successor, and to return the aggregate value 'odd' or 'even' from that of its final element. We will later generalize this pattern to express the computation of generic aggregates.

Observe that the query in Example 4 is deterministic, inasmuch as the answer to the parity question ?mod2(even) is independent of the particular chain being constructed, and only depends on the length of this chain, which is determined by the cardinality of b(x). The parity query is a well-known polynomial query that cannot be answered by Datalog with stratified negation under the genericity assumption [1]. Furthermore, the chain predicate illustrates how the elements of a domain can be arranged in a total order; we thus conclude that negation-stratified Datalog with key constraints can express all *DB-PTIME* queries [1].

In a nutshell, key constraints under single answer semantics extend the expressive power of logic programs, and find important new applications. Of particular importance is the definition of set-aggregates. While aggregates have been used extensively in database applications, particularly in decision support and data mining applications, a general treatment of this fundamental concept had, so far, been lacking and is presented in this paper.

2.1 Basic Definitions

We assume that the reader is familiar with the relational data model and Datalog [1,36].

A logic program P/K consists of a set of rules, P, and a set of key constraints K; each such a constraint has the form unique_key(q, γ), where q is the name of the predicate in P and γ is a subset of the arguments of q. Let I be an interpretation of P; we say that I satisfies the constraint $\texttt{unique\_key}(\texttt{q}, \gamma)$, when no two atoms in I are identical in all their γ arguments. The notation $I \models K$ will be used to denote that I satisfies every key constraint in K.

The basic semantics of a positive Datalog program P consists of evaluating "in parallel" all applicable instantiations of P's rules. This semantics is formalized by the *Immediate Consequences Operator*, T_P, that defines a mapping over the (Herbrand) interpretations of P, as follows:

$$T_P(I) = \{\, A \mid A \leftarrow B_1, \ldots, B_n \in ground(P) \,\wedge\, B_1 \in I \wedge \ldots \wedge B_n \in I \,\}.$$

A rule $r \in ground(P)$ is said to be *enabled* by the interpretation I when all its goals are contained in I. Thus the operator $T_P(I)$ returns the set of the heads of rules enabled by I.

The upward powers of T_P starting from an interpretation I are defined as follows:

$$T_P^{\uparrow 0}(I) = I$$
$$T_P^{\uparrow(i+1)}(I) = T_P(T_P^{\uparrow i}(I)), \quad \text{for } i \geq 0$$
$$T_P^{\uparrow \omega}(I) = \bigsqcup_{i \geq 0} T_P^{\uparrow i}(I).$$

The semantics of a positive program is defined by the least fixpoint of T_P, denoted $lfp(T_P)$, which is also equal to the least model of P, denoted M_P [29]. The least fixpoint of T_p can be computed as the ω-power of T_P applied to the empty set: i.e., $lfp(T_p) = T_P^{\uparrow \omega}(\emptyset)$.

The *inflationary* version of the T_P operator is denoted $\mathbf{T}_P$ and defined as follows:

$$\mathbf{T}_P(I) = T_P(I) \cup I$$

For positive programs, we have:

$$T_P^{\uparrow \omega} = \mathbf{T}_P^{\uparrow \omega} = M_P = lfp(T_P) = lfp(\mathbf{T}_P)$$

The equivalence of model-theoretic and fixpoint semantics no longer holds in Datalog¬ programs, which allow the use of negated goals in rules. Various semantics have therefore been proposed for Datalog¬ programs. For instance, the *inflationary semantics*, which adopts $\mathbf{T}_P^{\uparrow \omega}$ as the meaning of a program P, can be implemented efficiently but lacks desirable logical properties [1]. On the other hand, stratified negation is widely used and combines desirable computational and logical properties [22]; however, stratified negation severely restricts the class of programs that one can write. Formal semantics for more general classes of programs are also available [10,30,2]. Because of its generality and support for nondeterminism, we will use here the *stable model* semantics, that is defined via a *stability transformation* [10], as discussed next. Given an interpretation I and a Datalog¬ program P, the stability transformation derives the positive program $ground_I(P)$ by modifying the rules of $ground(P)$ as follows:

– drop all clauses with a negative literal $\neg A$ in the body with $A \in I$, and
– drop all negative literals in the body of the remaining clauses.

Next, an interpretation M is a stable model for a Datalog$\neg$ program P iff M is the least model of the program $ground_M(P)$. In general, Datalog$\neg$ programs may have zero, one, or many stable models. We shall see how the multiplicity of stable models can be exploited to give a declarative account of non-determinism.

3   Fixpoint Semantics

We use the notation P/K to denote a logic program P constrained by the set of unique keys K.

We make no distinction between interpretations of P and interpretations of P/K; thus every $I \subseteq B_P$ is an interpretation for P/K.

Since a program with key constraints can have multiple interpretations, we will now introduce the concept of family of interpretations. A *family of interpretations* for P is defined as a non-empty set of maximal interpretations for P. More formally:

Definition 1. *Let $\Im$ be a nonempty set of interpretations for P where no element in $\Im$ is a subset of another. Then $\Im$ is called a* family of interpretations *for P.*

The set of families of interpretations for P will be denoted by $fins(P)$.

For instance, let P be the program:

a.
b ← a.

Then $fins(P)$ consists of the following families of interpretations:

1. $\{\{\}\}$
2. $\{\{a\}\}$
3. $\{\{b\}\}$
3. $\{\{a\}, \{b\}\}$
4. $\{\{a, b\}\}$

3.1   Lattice

The $fins(P)$ can be partially ordered as follows:

Definition 2. *Let $\Im_1$ and $\Im_2$ be two elements of $fins(P)$. If $\forall I_1 \in \Im_1, \exists I_2 \in \Im_2$ s.t. $I_1 \subseteq I_2$, then we say that $\Im_1$ is a subfamily of $\Im_2$ and write $\Im_1 \sqsubseteq \Im_2$.*

Now, $(\sqsubseteq, fins(P))$ is a partial order, and also a *complete lattice*, with least upper bound (lub):

$$\Im_1 \sqcup \Im_2 = \{I \in \Im_1 | \neg \exists I_2 \in \Im_2 \text{ s.t. } I_2 \supset I\} \cup \{I \in \Im_2 | \neg \exists I_1 \in \Im_1 \text{ s.t. } I_1 \supseteq I\}$$

The greatest lower bound (glb) is:

$$\Im_1 \sqcap \Im_2 = \{I_1 \cap I_2 | I_1 \in \Im_1, I_2 \in \Im_2 \ and \ \neg(\exists I' \in \Im_1, \exists I'' \in \Im_2 \ s.t. \ I' \cap I'' \supset I_1 \cap I_2)\}$$

These two operations are easily extended to families with infinitely many elements; thus we have a complete lattice, with $\{B_P\}$ as *top* and $\{\emptyset\}$ as *bottom*.

3.2 Fixpoint Semantics of Positive Programs with Keys

Let us consider first the case of positive programs P *without* key constraints, by revisiting the computation of the successive power of T_P, where T_P denotes the immediate consequence operator for P. We will also use the *inflationary* version of this operator, which was previously defined as $\mathbf{T}_P(I) = T_P(I) \cup I$.

The computation $T_P^{\uparrow \omega}(\emptyset) = \mathbf{T}_P^{\uparrow \omega}(\emptyset)$ generates an ascending chain; if I is the result obtained at the last step, the application of $\mathbf{T}_P(I)$ adds to the old I the set of new tuples $T_P(I) - I$, *all at once*. We next define an operator where the new consequences are added one by one; this will be called the *Atomic Consequence Operator (ACO)*, $\mathcal{T}_P$, which is a mapping on families of interpretations. For a singleton set $\{I\}$, $\mathcal{T}_P$ is defined as follows:

$$\mathcal{T}_P(\{I\}) = \{I' \mid \exists x \in [\mathbf{T}_P(I) - I] \ s.t. \ I' = I \cup \{x\}\} \sqcup \{I\}$$

Then, for a family of sets, $\Im$, we have

$$\mathcal{T}_P(\Im) = \bigsqcup_{I \in \Im} \mathcal{T}_P(\{I\})$$

Therefore, our new operator adds to I a single new consequence atom from $\mathbf{T}_P(I) - I$, when this is not empty; thus, it produces a family of interpretations from a singleton interpretation $\{I\}$. When $\mathbf{T}_P(I) = I$, then, by the above definition, $\mathcal{T}_P(\{I\}) = \{I\}$. The following result follows immediately from the definitions:

Proposition 1. *Let P be a positive logic program without keys. Then, $\mathcal{T}_P$ defines a mapping that is monotonic and also continuous.*

Since we have a continuous mapping in a complete lattice, the well-known Knaster-Tarski theorem, and related fixpoint results, can be used to conclude that there always exists solutions of the fixpoint equation $\Im = \mathcal{T}_P(\Im)$, and there also exists the least of such solutions, called the *least fixpoint* of $\mathcal{T}_P$. The least fixpoint of $\mathcal{T}_P$, denoted $lfp(\mathcal{T}_P)$, can be computed as the ω-power of $\mathcal{T}_P$ starting from the bottom element $\{\emptyset\}$.

Proposition 2. *Let P be a positive logic program without key constrains. Then, $\Im = \mathcal{T}_P(\Im)$ has a least fixpoint solution denoted $lfp(\mathcal{T}_P)$, where:*

$$lfp(\mathcal{T}_P) = \mathcal{T}_P^{\uparrow \omega}(\{\emptyset\}) = \bigsqcup_{0 < j} \mathcal{T}_P^{\uparrow j}(\{\emptyset\}) = \{lfp(T_P)\}$$

Thus for a positive program without keys, the least fixpoint of the $\mathcal{T}_P$ provides an equivalent characterization of the semantics of positive logic programs since the least fixpoint of $\mathcal{T}_P$ is the singleton set containing the least fixpoint of T_P.

We now consider the situation of a positive program with keys P/K. The Immediate Consequence Operator (ICO) for this program is obtained by simply ignoring the keys: $\mathbf{T}_{P/K}(I) = \mathbf{T}_P(I)$. The ACO is defined as follows:

Definition 3. *Let $T_{P/K}$ be a logic program with key constraints, and let $\{I\} \in fins(P)$ and $\Im \in fins(P)$. Then, $\mathcal{T}_{P/K}(\{I\})$ and $\mathcal{T}_{P/K}(\Im)$ are defined as follows:*

$$\mathcal{T}_{P/K}(\{I\}) = \{I' \mid \exists x \in [\mathbf{T}_P(I) - I] \ s.t. \ I' = I \cup \{x\} \ and \ I' \models K\} \sqcup \{I\}$$

$$\mathcal{T}_{P/K}(\Im) = \bigsqcup_{I \in \Im} \mathcal{T}_P(\{I\})$$

For instance, if $\mathcal{T}$ denotes the ACO for our tiny college example, then $\mathcal{T}^{\uparrow 1}(\{\emptyset\})$ is simply a family with three singleton sets, one for each fact in the program:

$$\mathcal{T}^{\uparrow 1}(\{\emptyset\}) = \{ \{professor(ohm, ee)\}, \{professor(bell, ee)\}, \{student('JimBlack', ee)\} \}$$

Thus, $\mathcal{T}^{\uparrow 2}(\{\emptyset\})$ consists of pairs taken from the three program facts:

$$\mathcal{T}^{\uparrow 2}(\{\emptyset\}) = \{ \ \{professor(bell, ee), professor(ohm, ee)\}$$
$$\{student('JimBlack', ee), professor(bell, ee)\},$$
$$\{student('JimBlack', ee), professor(ohm, ee)\}\}$$

From the first pair, above, we can only obtain a family containing the three original facts; but from the second pair and third pair we obtain two different advisors. In fact, we obtain:

$$\mathcal{T}^{\uparrow 3}(\{\emptyset\}) = \{ \ \{student('JimBlack', ee), professor(bell, ee), professor(ohm, ee)\},$$
$$\{student('JimBlack', ee), professor(bell, ee),$$
$$advisor('JimBlack', bell)\},$$
$$\{student('JimBlack', ee), professor(ohm, ee),$$
$$advisor('JimBlack', ohm)\} \}$$

In the next step, these three parallel derivations converge into the following two sets:

$$\mathcal{T}^{\uparrow 4}(\{\emptyset\}) = \{ \ \{ \ student('JimBlack', ee), professor(bell, ee), professor(ohm, ee),$$
$$advisor('JimBlack', bell)\}$$
$$\{ \ student('JimBlack', ee), professor(bell, ee), professor(ohm, ee),$$
$$advisor('JimBlack', ohm)\}\}$$

No set can be further enlarged at the next step, given that the addition of a new advisor would violate the key constraints. So we have $T^{\uparrow 5}(\{\emptyset\}) = T^{\uparrow 4}(\{\emptyset\})$, and we have reached the fixpoint.

As illustrated by this example, although the operator $T_{P/K}$ is not monotonic, the ω-power of $T_{P/K}$ has desirable characteristics that makes it the natural choice for *canonical semantics* of positive programs with keys. In fact we have the following property:

Proposition 3. *Let P/K be a positive program with key constraints. Then, $T_{P/K}^{\uparrow\omega}(\{\emptyset\})$ is a fixpoint for $T_{P/K}$, and each $\{I\} \in T_{P/K}^{\uparrow\omega}(\{\emptyset\})$ is a minimal fixpoint for $T_{P/K}$.*

Proof: The application of $T_{P/K}$ to $T_{P/K}^{\uparrow\omega}(\{\emptyset\})$ can only generate elements which were generated in the ω-derivation. Thus $T_{P/K}^{\uparrow\omega}(\{\emptyset\})$ is a fixpoint. Now, let $\{I\} \in T_{P/K}^{\uparrow\omega}(\{\emptyset\})$. Clearly, $T_{P/K}(\{I\}) = \{I\}$, otherwise the previous property does not hold. Thus $\{I\}$ is a fixpoint. To prove that it is minimal, let $J \subset I$. If we trace the derivation chain for $\{I\}$, we find a predecessor of $\{I'\}$ where I' is not a subset of J, but its immediate predecessor, I'' is. Now let $\{x\} = I' - I''$, then $J \cup \{x\}$ does not violate the key constraints (since its superset I does not), and $\{x\}$ is in $\mathbf{T}_P(J)$. Thus $\{J\}$ cannot be a fixpoint. $\qquad\square$

Therefore, under the all-answer semantics, we expect the whole family $T_{P/K}^{\uparrow\omega}(\{\emptyset\})$ to be returned as the canonical answer, whereas under a single-answer semantics any of the interpretations in $T_{P/K}^{\uparrow\omega}(\{\emptyset\})$ is accepted as a valid answer.

In the next section, we introduce an equivalent semantics for our programs with keys using the notion of stable models.

4 Stable-Model Semantics

Programs with keys have an equivalent model-theoretic semantics. We will next show that $T_{P/K}^{\uparrow\omega}(\{\emptyset\})$ corresponds to the family of stable models for the program $foe(P/K)$ obtained from P/K by expressing the key constraints by negated goals. The stable model semantics also extends naturally to stratified programs with key constraints.

4.1 Positive Programs with Key Constraints

An equivalent characterization of a positive programs P/K can be obtained by introducing negated goals in the rules of P to enforce the key constraints. The program obtained by this transformation will be denoted $foe(P/K)$, and called the *first order equivalent* of P/K. The program $foe(P/K)$ so obtained always has a formal meaning under stable model semantics [10].

Take, for instance, our advisor example; the rule in Example 1 can also be expressed as follows:

Example 5. The Advisor Example 1 Expressed Using Negation

$$\text{advisor}(S,P) \leftarrow \quad \text{student}(S,\text{Majr},\text{Year}), \text{professor}(P,\text{Majr}),$$
$$\neg\text{kviol\_advisor}(S,P).$$
$$\text{kviol\_advisor}(S,P) \leftarrow \text{advisor}(S,P'),P \neq P'.$$

Therefore, we allow a professor P to become the advisor of a student S provided that no other $P' \neq P$ is already an advisor of S. In general, if q is the name of a predicate subject to a key constraint, we use a new predicate kviol_q to denote the violation of key constraints on q; then, we add a kviol_q rule for each key declared for q. Finally, a negated kviol_q goal is added to the original rules defining q. For instance, the simple path program of Example 3 can be re-expressed in the following way:

Example 6. The simple-path program of Example 3 Expressed Using Negation

$$\text{spath}(\text{root},X) \leftarrow \quad g(X,Y), \neg\text{kviol\_spath}(\text{root},X).$$
$$\text{spath}(Y,Z) \leftarrow \quad \text{spath}(X,Y), g(Y,Z), \neg\text{kviol\_spath}(Y,Z).$$
$$\text{kviol\_spath}(X1,X2) \leftarrow \text{spath}(X1,Y2),X2 \neq Y2.$$
$$\text{kviol\_spath}(X1,X2) \leftarrow \text{spath}(Y1,X2),X1 \neq Y1.$$

Derivation of $foe(P/K)$. In general, given a program P/K constrained with keys, its first order equivalent $foe(P/K)$ is computed as follows:

1. For each rule r, with head $q(Z_1,\ldots,Z_n)$, where q is constrained by some key, add the goal $\neg\text{kviol\_q}(Z_1,\ldots,Z_n)$ to r,
2. For each unique_key(q, ArgList)! in K, where n is the arity of q, add a new rule,

$$\text{kviol\_q}(X_1,\ldots,X_n) \leftarrow q(Y_1,\ldots,Y_n),Y_1\theta_1X_1, \ldots, Y_n\theta_nX_n.$$

where θ_j denotes the equality symbol '=' for every j in ArgList, and the inequality symbol '$\neq$' for every j not in ArgList.

For instance, the *foe* of our advisor example is:

$$\text{advisor}(S,P) \leftarrow \quad \text{student}(S,\text{Majr},\text{Year}), \text{professor}(P,\text{Majr}),$$
$$\neg\text{kviol\_advisor}(S,P).$$
$$\text{kviol\_advisor}(X_1,X_2) \leftarrow \text{advisor}(Y_1,Y_2),X_1 = Y_1,X_2 \neq Y_2.$$

This transformation does in fact produce the rules of Example 6, after we replace equals with equals and eliminate all equality goals. The newly introduced predicates with the prefix kviol will be called *key-violation predicates*.

Stable models provide the formal semantics for our *foe* programs:

Proposition 4. *Let P/K be a positive logic program with keys. Then $foe(P/K)$ has one or more stable models.*

A proof for this proposition can be easily derived from [25,13], where the same transformation is used to define the formal semantics of programs with the `choice` construct.

With I an interpretation of $foe(P)$, let $pos(I)$ denote the interpretation obtained by removing all the key-violation atoms from I and leaving the others unchanged. Likewise, if $\Im$ is a family of interpretation of $foe(P)$, then we define:

$$pos(\Im) = \bigsqcup_{I \in \Im} pos(I)$$

Then, the following theorem elucidates the equivalence between the two semantics:

Proposition 5. *Let P/K be a positive program, and Σ be the set of stable models for $foe(P/K)$. Then $pos(\Sigma) = T_{P/K}^{\uparrow \omega}(\{\emptyset\})$.*

Proof: Let $I \in T_P^{\uparrow \omega}(\{\emptyset\})$, and $P_I = ground_I(foe(P/K))$ be the program produced by the stability transformation on $foe(P/K)$. It suffices to show that $T_{P_I}^{\uparrow \omega}(\{\emptyset\}) = I$, i.e., that $\{I\} = T_{P_I}^{\uparrow \omega}(\{\emptyset\})$. Now, take a derivation in $T_{P/K}^{\uparrow \omega}(\{\emptyset\})$ producing I; we can find an identical derivation in $T_{P_I}^{\uparrow \omega}(\{\emptyset\})$. This concludes our proof. □

4.2   Stratification

The notion of stratification significantly increases the expressive power of Datalog, while retaining the declarative fixpoint semantics of programs. Consider first the notion of stratification with respect to negation for programs without key constraints:

Definition 4. *Let P be a program with negated goals, and $\sigma_1, \ldots, \sigma_n$ be a partition of the predicate names in P. Then, P is said to be stratified, when for each rule $r \in P$ (with head h_r) and each goal g_r in r, the following property holds:*
 1. $stratum(h_r) > stratum(g_r)$ if g_r is a negated goal
 2. $stratum(h_r) \geq stratum(g_r)$ if g_r is a positive goal.

Therefore, a stratified program P can be viewed as a stack of rule layers, where the higher layers do not influence the lower ones. Thus the correct semantics can be assigned to a program by starting from the bottom layer and proceeding upward, with the understanding that computation for the higher layers cannot affect lower ones.

The computation can be implemented using the ICO T_P, which, in the presence of negated goals, is generalized as follows. A rule $r \in ground(P)$ is said to be *enabled* by an interpretation I when all of its positive goals are in I and none of its negated goals are in I. Then, $T_P(I)$ is defined as containing the heads of all rules in $ground(P)$ that are enabled by I. (This change automatically adjusts the definitions of **T** and $\mathcal{T}$ that are based on T_P.)

Therefore, let $I[\leq j]$ and $P[\leq j]$, respectively, denote the atoms in I and the rules in P whose head belongs to strata $\leq j$. Also let $P[j]$ denote the set of rules in P whose head belongs to stratum j. Then, we observe that for a stratified program P, the mapping defined by $P[j]$ (i.e., $T_{P[j]}$) is *monotonic with respect to* $I[j]$. Thus, if I_{j-1} is the meaning of $P[\leq j-1]$, then $\mathbf{T}_{P[j]}^{\uparrow\omega}(I_{j-1})$ is the meaning of $P[\leq j]$.

Thus, let P be a program stratified with respect to negation and without key constraints; then the following algorithm inductively constructs the iterated fixpoint for T_P (and $\mathbf{T}_P$):

Iterated Fixpoint computation for $\mathbf{T}_P$, where P is stratified with strata $\sigma_1, \ldots, \sigma_n$.

1. Let $I_0 = \emptyset$;
2. For $j = 1, \ldots, n$, let $I_j = \mathbf{T}_{P[j]}^{\uparrow\omega}(I_{j-1})$

For every $1 \leq j \leq n$, $I_j = I_n[\leq j]$ is a minimal fixpoint of $P[\leq j]$. The interpretation I_n obtained at the end of this computation is called the iterated fixpoint for T_P and defines the meaning of the program P. It is well-known that the iterated fixpoint for a stratified program P is equal to P's unique stable model [36].

These notions can now be naturally extended to programs with key constraints. A program P/K is stratified whenever its keyless counterpart P is stratified. Let $P/K[j]$ denote the rules with head in the jth stratum, along with the key constraints on their head predicates; also, let $P/K[\leq j]$ denote the rules with head in strata lower than the jth stratum, along with their applicable key constraints. Finally, let:

$$\Im[\leq j] = \bigsqcup_{I \in \Im} I[\leq j]$$

The notion of $\mathcal{T}$ can be extended in natural fashion to stratified programs. If $\Im_{j-1}$ is the meaning of $P/K[\leq j-1]$, then $\mathcal{T}_{P[j]}^{\uparrow\omega}(\Im_{j-1})$ is the natural meaning of $P/K[\leq j]$.

Thus we have the following extension of the iterated fixpoint algorithm:

Iterated Fixpoint Computation for $\mathcal{T}_{P/K}$ where P/K is stratified with strata $\sigma_1, \ldots, \sigma_n$.

1. Let $\Im_0 = \{\emptyset\}$;
2. For $j = 1, \ldots, n$, let $\Im_j = \mathcal{T}_{P/K[j]}^{\uparrow\omega}(\Im_{j-1})$

The family of interpretations $\Im_n$ obtained from this computation will be called the *iterated fixpoint* for $\mathcal{T}_{P/K}$. The iterated fixpoint for $\mathcal{T}_{P/K}$ defines the meaning of P/K; it has the property that, for each $1 \leq j \leq n$, each member in $\Im_j = \Im_n[\leq j]$ is a minimal fixpoint for $\mathcal{T}_{P/K[\leq j]}$.

Stable Model Semantics for Stratified Programs. Every program P that is stratified with respect to negation has a unique stable model that can be computed by the iterated fixpoint computation for $\mathbf{T}_P$ previously discussed. Likewise, every stratified program P/K can be expanded into its first order equivalent $foe(P/K)$. Then, it can be shown that (i) $foe(P/K)$ always has one or more stable models, and (ii) if Σ denotes the family of its stable models, then $pos(\Sigma)$ coincides with the iterated fixpoint of $\mathcal{T}_{P/K}$.

5 Single-Answer Semantics and Nondeterminism

The derivation $\mathcal{T}_{P/K}^{\uparrow\omega}(\{\emptyset\})$ can be used to compute in parallel all the stable models for a positive program $foe(P/K)$. In this computation, each application of $\mathcal{T}_{P/K}$ expands in parallel all interpretations in the current family, by the addition of a single new element to each interpretation. In [38], we discuss *condensed derivations* based on $\mathcal{T}_{P/K}$, which accelerate the derivation process by adding several new elements at each step of the computation. This ensures a faster convergence toward the final result, while still computing all stable models at once. Even with condensed derivations, the computation of all stable models requires exponential time, since the number of such models can be exponential in the size of the database. This, computational complexity might be acceptable when dealing with $\mathcal{N}\mathcal{P}$-complete problems, such as deciding the existence of an Hamiltonian path. However, in many situations involving programs with multiple stable models, *only one* such model, not all of them, is required in practice. For instance, this is the case of Example 4, where we use choice to enumerate into a chain the elements of a set one by one, with the knowledge that the even/odd parity of the whole set only depends on its cardinality, and not on the particular chain used. Therefore for Example 4, the computation of any stable model will suffice to answer correctly the parity query. Since this situation is common for many queries, we need efficient operators for computing a single stable model.

Even with $\mathcal{N}\mathcal{P}$-complete problems, it is normally desirable to generate the stable models in a serial rather than parallel fashion. For instance, for the Hamiltonian circuit problem of Example 3, we can test if the last generated model satisfies the desired property (i.e., if there is any **freenode**), and only if this test fails, proceed with the generation of another model— normally, calling on some heuristics to aid in the search for a good model. On the average, this search succeeds without having to produce an exponential number of stable models, since exponential complexity only represents the worst-case behavior for many $\mathcal{N}\mathcal{P}$-complete algorithms.

Now, the computation of a single stable model is in general $\mathcal{N}\mathcal{P}$-hard [26]; however, this computation for a program $foe(P/K)$ derived from one with key constraints can be performed in polynomial time, and, as we describe next, with minimal overhead with respect to the standard fixpoint computation. Therefore, we next concentrate on the problem of generating a single element in $\mathcal{T}_{P/K}^{\uparrow\omega}(\{\emptyset\})$, and on expressing polynomial-time queries using this single-answer semantics.

We define next the notions of *soundness* and *completeness* for nondeterministic operators to be used to compute an element in $T_P^{\uparrow\omega}(\{\emptyset\})$.

Definition 5. *Let P/K be a logic program with keys, and C be a class of functions on interpretations of P. Then we define the following two properties:*

1. Soundness. *A function $\tau \in C$ will be said to be sound for a program P/K when $\tau^{\uparrow\omega}(\emptyset) \in T_{P/K}^{\uparrow\omega}(\{\emptyset\})$. The function class C will be said to be sound when all its members are sound.*
2. Completeness. *The function class C will be said to be complete for a program $T_{P/K}$ when for each $M \in T_{P/K}^{\uparrow\omega}(\{\emptyset\})$ there exists some $\tau \in C$ such that: $\tau^{\uparrow\omega}(\emptyset) = M$.*

In situations where any answer will solve the problem at hand, there is no point in seeking completeness and we can limit ourselves to classes of functions that are sound, and efficient to compute, even if completeness is lost; *eager derivations* discussed next represent an interesting class of such functions.

Definition 6. *Let P/K be a program with key constraints, and let $\Gamma(I)$ be a function on interpretations of P. Then, $\Gamma(I)$ will be called an* eager derivation *operator for P/K if it satisfies the following three conditions:*

1. $I \subseteq \Gamma(I) \subseteq \mathbf{T}_P(I)$
2. $\Gamma(I) \models K$
3. *Every subset of $\mathbf{T}_P(I)$ that is a proper superset of $\Gamma(I)$ violates some key constraint in K.*

Let C_Γ be the class of eager derivation operators for a given program P/K. Then it is immediate to see that C_Γ is sound for all programs.

Eager derivation operators can be implemented easily. Their implementation only requires tables to memorize atoms previously derived and compare the new values against previous ones to avoid key violations. Inasmuch as table-based memorization is already part of the basic mechanism for the computation of fixpoints in deductive databases, key constraints are easy to implement.

A limitation of eager derivation operators is that they do not form a complete class for all positive programs with key constraints. This topic is discussed in [38], where classes of operators which are both sound and complete are also discussed. However, in the rest of this paper, we only use key constraints to define chain rules, such as those in Example 4; for these rules, the eager derivations are *complete*—in addition to being sound and efficiently computable.

6   Set Aggregates in Logic

The additional expressive power brought to Datalog by key constraints finds many uses; here we employ it to achieve a formal characterization of database aggregates, thus solving an important open problem in database theory and logic programming. In fact, the state-of-the-art characterization of aggregates relies on the assumption that the universe is totally ordered [36]. Using this assumption, the atoms satisfying a given predicate are chained together in ascending order, starting from the least value and ending with the largest value. Unfortunately, this solution has four serious drawbacks, since (i) it compromises data independence by violating the genericity property [1], (ii) it relies on negation, thus infecting aggregates with the nonmonotonic curse, (iii) it is often inefficient since it requires the data to be sorted before aggregation, and (iv) it cannot be applied to more advanced forms of aggregation, such as on-line aggregates and rollups, that are used in decision support and other advanced applications [33].

Online aggregation [8], in particular, cannot be expressed under the current approach that relies on a totally ordered universe to sort the elements of the set being processed, starting from its least element. In fact, at the core of on-line aggregation, there is the idea of returning partial results after visiting a proper subset of the given dataset, while the rest is still unknown. Now, it is impossible to compute the least element of a set when only part of it is known.

We next show that all these problems find a simple solution once key constraints are added to Datalog. For concreteness, we use the aggregate constructs of $\mathcal{LDL}++$ [4], but very similar syntactic constructs are used by other systems (e.g., $CORAL$ [23]), and the semantics here proposed is general and applicable to every logic-based language and database query language.

6.1 User Defined Aggregates

Consider the parity query of Example 4. To define an equivalent parity aggregate in $\mathcal{LDL}++$ the user will write the following rules:

Example 7. Definition rules for the parity aggregate mod2

```
single(mod2, _, odd).
multi(mod2, X, odd, even).
multi(mod2, X, even, odd).
freturn(mod2, _, Parity, Parity).
```

These rules have the same function as the last four rules in Example 4. The single rule specifies how to initialize the computation of the mod2 aggregate by specifying its value on a singleton set (same as the first ca rule in the example). The two multi rules instead specify how the new aggregate value (the fourth argument) should be updated for each new input value (second argument), given its previous value (third argument). (Thus these rules perform the same function as the second and the third of the ca rules in Example 4.) The freturn rule specifies (as fourth argument) the value to be returned once the last element in the set is detected (same as the last rule in Example 4). For mod2, the value returned is simply taken from the third argument, where it was left by the multi rule executed on the last element of the set. Two important observations can therefore be made:

1. We have described a very general method for defining aggregates by specifying the computation to be performed upon (i) the initial value, (ii) each
   successive value, and (iii) the final value in the set. This paradigm is very
   general, and also describes the mechanism for introducing user defined aggregates (UDAs) used by SQL3 and in the AXL system [33].
2. The correspondence between the above rules and those of Example 4 outlines
   the possibility of providing a logic semantics to UDAs by simply expanding
   the single, multi, and freturn rules into an equivalent logic program (using the chain rules) such as that of Example 4.

The rules in Example 7 are generic, and can be applied to any set of facts. To
reproduce the behavior of Example 4, they must be applied to b(X). In $\mathcal{LDL}{++}$
this is specified by the aggregate-invocation rule:

$$p(\text{mod2}\langle X\rangle) \leftarrow b(X).$$

that specifies that the result of the computation of mod2 on b(X) is returned as
the argument of a predicate, that our user has named p.

There has been much recent interest in online aggregates [8], which also find
important applications in logic programming, as discussed later in this paper.
For instance, when computing averages on non-skewed data, the aggregate often
converges toward the final value long before all the elements in the set are visited. Thus, the system should support *early returns* to allow the user to check
convergence and stop the computation as soon as the series of successive values
has converged within the prescribed accuracy [8]. UDAs with early returns can
be defined in $\mathcal{LDL}{++}$ through the use of ereturn rules.

Say, for instance, that we want to define a new aggregate myavg, and apply it
to the elements of d(Y), and view the results of this computation as a predicate
q. Then, the $\mathcal{LDL}{++}$ programmer must specify one *aggregate-application rule*,
and several *aggregate-definition rules*. For instance, the following is an aggregate
application rule:

$$r : \quad q(\text{myavg}\langle Y\rangle) \leftarrow d(Y).$$

The $\langle\ldots\rangle$ notation in the head of r denotes an aggregate; this rule specifies that
the definition rules for myavg must be applied to the stream of Y-values that
satisfy the body of the rule.

The aggregate definition rules include: (i) *single* rule(s) (ii) *multi* rule(s), (iii)
freturn rule(s) for final returns and/or (iv) *ereturn* rule(s) for early returns. All
four kinds of rules are used in the following definition of myavgr:

```
single(myavg, Y, cs(1, Y)).
multi(myavg, Y, cs(Cnt, Sum), cs(Cnt1, Sum1)) ←
                    Cnt1 = Cnt + 1, Sum1 = Sum + Y.

freturn(myavg, Y, cs(Cnt, Sum), Val) ← Val = Sum/Cnt.
```

```
ereturn(myavg, X, (Sum, Count), Avg) ←
                    Count mod 100 = 0, Avg = Sum/Count.
```

Observe that the first argument in the head of the single, multi, ereturn, and freturn rules contains the name of the aggregate: therefore, these aggregate definition rules can only be used by aggregate application rules that contain myavg⟨...⟩ in the head.

The second argument in the head of a single or multi rule holds the 'new' value from the input stream, while the last argument holds the partial value returned by the previous computation. Thus, for averages, the last argument should hold the pair cs(Count, Sum). The single rule specifies the value of the aggregate for a singleton set (containing the first value in the stream); for myavg, the singleton rule must return cs(1, Y). The multi rules prescribe an inductive computation on a set with $n + 1$ elements, by specifying how the $n + 1^{th}$ element in the stream is to be combined with the value returned (as third argument in multi) by the computation on the first n elements. For myavg, the count is increased by one and the sum is increased by the new value in the stream.

The freturn rules specify how the final value(s) of the aggregate are to be returned. For myavg, we return the ratio of sum and count. The ereturn rules specify when early returns are to be produced and what are their values. In particular for myavg, we produce early returns every 100 elements in the stream, and the value produced is the current ratio sum/count—online aggregation.

6.2 Semantics of Aggregates

In general, the semantics of an aggregate application rule r

$$\mathtt{r}: \ \mathtt{q(myavg\langle Y\rangle)} \leftarrow \mathtt{d(Y)}.$$

can be defined by expanding it into its *key-constrained equivalent* logic program, denoted $kce(r)$, which contains the following rules:

1. A *main rule*

$$\mathtt{p(Y)} \leftarrow \mathtt{results(avg, Y)}.$$

where results(avg, Y) is derived from d(Y) by a program consisting of:

2. The *chain rules* that link the elements of d(Y) into an order-inducing chain (nil is a special value not in d(Y)),

```
unique_key(chain_r, [1])!
unique_key(chain_r, [2])!
chain_r(nil, Y) ← d(Y).
chain_r(Y, Z) ←    chain_r(X, Y), d(Z).
```

3. The cagr rules that perform the inductive computation:

```
cagr(AgName, Y, New) ←    chain_r(nil, Y), Y ≠ nil, single(myagr, Y, New).
cagr(AgName, Y2, New) ← chain_r(Y1, Y2), cagr(AgName, Y1, Old),
                        multi(AgName, Y2, Old, New).
```

Thus, the `cagr` rules are used to memorize the previous results, and to apply (i) `single` to the first element of d(Y) (i.e., for the pattern chain$_r$(nil, Y)) and (ii) `multi` to the successive elements.

4. The two `results` rules, where the first rule produces *early returns* and second rule produces *final returns* as follows:

results(AgName, Y2, New) ← chain$_r$(Y1, Y2), cagr(AgName, Y1, Old),
                           ereturn(AgName, Y2, Old, Yield).
results(AgName, AgValue) ← chain$_r$(X, Y), ¬chain$_r$(Y, _),
                           cagr(AgName, Y, Old),
                           freturn(AgName, Y, Old, AgValue).

Therefore, the first `results` rule produces the early returns by applying `ereturn` to every element in the chain, and the second rule produces the final returns by applying `freturn` on the last element in the chain (i.e., the element without a successor).

In $\mathcal{LDL}$++, an implicit group-by operation is performed on the head arguments not used to apply aggregates. Thus, to compute the average salary of employees grouped by `Dno`, the user can write:

avgsal(Dno, myavg⟨Sal⟩) ← emp(Eno, Sal, Dno).

As discussed in [34], the semantics of aggregates with group-by can simply be defined by including an additional argument in the predicates chain$_r$ and `results` to hold the group-by attributes.

6.3   Applications of User Defined Aggregates

We will now discuss the use of UDAs to express polynomial algorithms in a natural and efficient way. These algorithms use aggregates in programs that yield the correct final results unaffected by the nondeterministic behavior of the aggregates. Therefore, aggregate computation here uses single-answer semantics, which assures polynomial complexity.

Let us consider first uses of nonmonotonic aggregates. For instance, say that from a set of pairs such as (Name, YearOfBirth) as input, we want to return the `Name` of the youngest person (i.e., the person born in the latest year). This computation cannot be expressed directly as an aggregate in SQL, but can be expressed by the UDA `youngest` given below (in $\mathcal{LDL}$++, a vector of n arguments $(X_1, \ldots, X_n)$ is basically treated as a n-argument function with a default name).

single(youngest, (N, Y), (N, Y)).
multi (youngest, (N, Y), (N1, Y1), (N, Y)) ←   Y ≥ Y1.
multi (youngest, (N, Y), (N1, Y1), (N1, Y1)) ← Y ≤ Y1.
freturn(youngest, (N, Y), (N1, Y1), N1).

User-defined aggregates provide a simple solution to a number of complex problems in deductive databases; due to space limitations we will here consider only simple examples—a more complete set of examples can be found in [37].

We already discussed the definition and uses of online aggregates, such as myavg that returns values every 100 samples. In a more general framework, the user would want to control how often new results are to be returned to the user, on the basis of the estimated progress toward convergence in the computation [8]. UDAs provide a natural setting for this level of control.

Applications of UDAs are too many to mention. But for an example, take the interval coalescing problem of temporal databases [35]. For instance, say that from a base relation emp(Eno, Sal, Dept, (From, To)), we project out the attribute Sal and Dept; then the same Eno appears in tuples with overlapping valid-time intervals and must be coalesced. Here we use closed intervals represented by the pair (From, To) where From is the start-time, and To is the end-time. Under the assumption that tuples are sorted by increasing start-time, we can use a special coales aggregate to perform the task in one pass through the data.

Example 8. Coalescing overlapping intervals sorted by start time.

```
empProj(Eno, coales⟨(From, To)⟩) ←      emp(Eno, _, _, (From, To)).
single(coales, (Frm, To), (Frm, To)).
multi(coales, (Nfr, Nto), (Cfr, Cto), (Cfr, Lgr)) ←    Nfr ≤ Cto,
                                                       larger(Cto, Nto, Lgr).
multi(coales, (Nfr, Nto), (Cfr, Cto), (Cfr, Nto)) ←    Nfr > Cto.
ereturn(coales, (Nfr, Nto), (Cfr, Cto), (Cfr, Cto)) ← Nfr > Cto.
freturn(coales, _, LastInt, LastInt).
larger(X, Y, X) ←    X ≥ Y.
larger(X, Y, X) ←    X < Y.
```

Thus, the single rule starts the coalescing process by setting the current interval equal to the first interval. The multi rule operates as follows: when the new interval (Nfr, Nto) overlaps the current interval (Cfr, Cto) (i.e., when Nfr ≤ Cto), the two are coalesced into an interval that begins at Cfr, and ends with the larger of Nto and Cto; otherwise, the current interval is returned and the new interval becomes the current one.

7   Monotonicity

Commercial database systems and most deductive database systems disallow the use of aggregates in recursion and require programs to be stratified with respect to aggregates. This restriction is also part of the SQL99 standards [7]. However, many important algorithms, particularly greedy algorithms, use aggregates such as count, sum, min and max in a monotonic fashion, inasmuch as previous results are never discarded. This observation has inspired a significant amount of previous work seeking efficient expression of these algorithms in logic [27,6,24,31,9,15]. At the core of this issue there is the characterization of programs where aggregates behave monotonically and can therefore be freely used in recursion. For many interesting programs, special lattices can be found in which aggregates are monotonic [24]. But the identification of such lattices cannot be automated [31], nor is the computation of fixpoints for such programs. Our newly introduced theory of aggregates provides a definitive solution to the monotonic aggregation problem, including a simple syntactic characterization to determine if an aggregate is monotonic and can thus be used freely in recursion.

7.1 Partial Monotonicity

For a program P/K, we will use the words *constrained predicates* and *free predicates* to denote predicates that are constrained by keys and those that are not. With I an interpretation, let I_c, and I_f, respectively, denote the atoms in I that are instances of constrained and free predicates; I_c will be called the *constrained component* of I, and I_f is called the *free component* of I. Then, let I and J be two interpretations such that $I \subseteq J$ and $I_c = J_c$ (thus $I_f \subseteq J_f$). Likewise, each family $\Im$ can be partitioned into the family of its constrained components, $\Im_c$, and the family of its free components, $\Im_f$.

Then, the following proposition shows that a program P/K defines a monotonic transformation with respect to the free components of families of interpretations:

Proposition 6. Partial Monotonicity: *Let $\Im$ and $\Im'$ be two families of interpretations for a program P/K. If $\Im \sqsubseteq \Im'$, while $\Im_c = \Im'_c$ then, $\mathcal{T}_{P/K}(\Im) \sqsubseteq \mathcal{T}_{P/K}(\Im')$.*

Proof. It suffices to prove the property for two singleton sets $\{I\}$ and $\{J\}$ where $I_f \subseteq J_f$, while $I_c = J_c$. Take an arbitrary $I' \in \mathcal{T}_{P/K}(\{I\})$: we need to show that there exists a $J' \in \mathcal{T}_{P/K}(\{J\})$ where $I' \subseteq J'$. If $I' \subseteq J$ the conclusion is trivial; else, let $I' = I \cup \{x\}$, $x \in T_P(I) - I$, and $I' \models K$. Since I is a subset of J but I' is not, x is not in J, and $x \in T_P(J) - J$. Also, if $J' = J \cup \{x\}$, $J' \models K$ (since $J'_c = I'_c$). Thus, $J' \in \mathcal{T}_{P/K}(\{J\})$.   □

This partial monotonicity property (i.e., monotonicity w.r.t. free predicates only) extends to the successive powers of $\mathcal{T}_{P/K}$, including its ω-power. Thus If $\Im \sqsubseteq \Im'$, while $\Im_c = \Im'_c$ then, $\mathcal{T}_{P/K}^{\uparrow \omega}(\Im) \sqsubseteq \mathcal{T}_{P/K}^{\uparrow \omega}(\Im')$. This result shows that the program P/K defines a monotonic mapping from unconstrained predicates to every other predicate in the program. It is customary in deductive databases to draw a distinction between extensional information (base relations) and intensional information (derived relations). Therefore, a program can be viewed as defining a mapping from base relations to derived relations. Therefore, the partial monotonicity property states that the mapping from database relations free of key constraints to derived relations is monotonic—i.e., the larger the base relations, the larger the derived relations.

For a base relation R that is constrained by keys, we can introduce an auxiliary input relation R_I free of key constraints, along with a copy rule that derives R from R_I. Then, we can view R_I as the input relation and R as a result of filtering R_I with the key constraints. Then, we have a monotonic mapping from the input relation R_I to the derived relations in the program.

7.2 Monotonic Aggregates

Users normally think of an aggregate application rule, such as r, as a direct mapping from r's body to r's head—a mapping which behaves according to the rules defining the aggregate. This view is also close to the actual implementation, since in a system such as $\mathcal{LDL}{+}{+}$ the execution of the rules in $kce(r)$ is already built into the system.

The *encapsulate program* for an aggregate application rule r, will be denoted $\epsilon(r)$ and contains all the rules in $kce(r)$ and the single, multi, ereturn and freturn rules defining the aggregates used in r. Then, the *transitive* mapping defined by $\epsilon(r)$ transforms families of interpretations of the body of r to families of interpretations of the heads of rules in $\epsilon(r)$. With I an interpretation of the body of r (i.e., a set of atoms from predicates in the body of r), then the mapping for $\epsilon(r)$ is equal to $T_{\epsilon(r)}^{\uparrow\omega}(\{I\})$, when there are no freturn rules, and is equal to the result of the iterated fixpoint of the stratified $\epsilon(r)$ program, otherwise.

For instance, consider the definition and application rules for an online count aggregate msum:

$$r' : \mathrm{q}(\mathrm{msum}\langle \mathrm{X}\rangle) \leftarrow \mathrm{p}(\mathrm{X}).$$

$$\mathrm{single}(\mathrm{msum}, \mathrm{Y}, \mathrm{Y}).$$
$$\mathrm{multi}(\mathrm{msum}, \mathrm{Y}, \mathrm{Old}, \mathrm{New}) \leftarrow \quad \mathrm{New} = \mathrm{Old} + \mathrm{Y}.$$
$$\mathrm{ereturn}(\mathrm{msum}, \mathrm{Y}, \mathrm{Old}, \mathrm{New}) \leftarrow \mathrm{New} = \mathrm{Old} + \mathrm{Y}.$$

The transitive mapping established by $\epsilon(r')$ can be summarized by the chain_r atoms, which describe a particular sequencing of the elements in I and the aggregate values for the sequence so generated:

$\Im$	$T_{\epsilon(r')}^{\uparrow\omega}(\Im)$
$\{\{p(3)\}\}$	$\{\{chain_r(nil, 3), q(3)\}\}$
$\{\{p(1), p(3)\}\}$	$\{\{chain_r(nil, 1), chain_r(1, 3), q(1), q(4)\},$ $\{chain_r(nil, 3), chain_r(3, 1), q(3), q(4)\}\}$
$\dots$	$\dots$

Therefore, the mapping defined by the aggregate rules is multivalued —i.e., from families of interpretations to families of interpretations. The ICO for the set of non aggregate rules P' can also be seen as a mapping between families of interpretations by simply letting $T_{P'}(\{I\}) = \{T_{P'}(I)\}$. Then, the *encapsulated consequence operator* for a program with aggregates combines the immediate consequence operator for regular rules with the transitive consequences for the aggregate rules. Because of the partial monotonicity properties of programs with key constraints, we now derive the following property:

Proposition 7. *Let P be a positive program with aggregates defined without final return rules. Then, the encapsulated consequence operator for P is monotonic in the lattice of families of interpretations.*

Therefore, aggregates defined without `freturn` rules will be called *monotonic*; thus, monotonic aggregates can be used freely in recursive programs. Aggregate computation in actual programs is very similar to the seminaive computation used to implement deductive databases [5,35], which is based on combining old values with new values according to rules obtained by the symbolic differentiation of the original rules. For aggregates, we can use the same framework with the difference that the rules for storing the old values and those for producing the results are now given explicitly by the programmer through the single/multi and ereturn/freturn rules in the definition.

7.3   Aggregates in Recursion

Our newly introduced theory of aggregates provides a definitive solution to the monotonic aggregation problem, with a simple syntactic criterion to decide if an aggregate is monotonic and can thus be used freely in recursion. The rule is as follows: *All aggregates which are defined without any* `freturn` *rule are monotonic and can be used freely in recursive rules.*

The ability of freely using aggregates with early returns in programs allows us to express concisely complex algorithms. For instance, we next define a continuous count that returns the current count after each new element but the first one (thus, it does not have a freturn since that would be redundant).

$$
\begin{aligned}
&\texttt{single(mcount, Y, 1).}\\
&\texttt{multi(mcount, Y, Old, New)} \leftarrow \quad \texttt{New = Old + 1.}\\
&\texttt{ereturn(mcount, Y, Old, New)} \leftarrow \texttt{New = Old + 1.}
\end{aligned}
$$

Using `mcount` we can now code the following applications, taken from [24].

Join the Party Some people will come to the party no matter what, and their names are stored in a `sure(Person)` relation. But others will join only after they know that at least $K = 3$ of their friends will be there. Here, `friend(P,F)` denotes that F is P's friend.

$$
\begin{aligned}
&\texttt{willcome(P)} \leftarrow \qquad\qquad \texttt{sure(P).}\\
&\texttt{willcome(P)} \leftarrow \qquad\qquad \texttt{c\_friends(P, K), K} \geq \texttt{3.}\\
&\texttt{c\_friends(P, mcount}\langle\texttt{F}\rangle\texttt{)} \leftarrow \texttt{willcome(F), friend(P, F).}
\end{aligned}
$$

Consider now a computation of these rules on the following database.

```
friend(jerry, mark).        sure(mark).
friend(penny, mark).        sure(tom).
friend(jerry, jane).        sure(jane).
friend(penny, jane).
friend(jerry, penny).
friend(penny, tom).
```

Then, the basic semi-naive computation yields:

$$\texttt{willcome(mark)}, \texttt{willcome(tom)}, \texttt{willcome(jane)},$$

$$\texttt{c\_friends(jerry, 1)}, \texttt{c\_friends(penny, 1)}, \texttt{c\_friends(jerry, 2)},$$

$$\texttt{c\_friends(penny, 2)}, \texttt{c\_friends(penny, 3)}, \texttt{willcome(penny)},$$

$$\texttt{c\_friends(jerry, 3)}, \texttt{willcome(jerry)}.$$

This example illustrates how the standard semi-naive computation can be applied to queries containing monotonic user-defined aggregates. Another interesting example is transitive ownership and control of corporations [24].

Company Control Say that owns(C1, C2, Per) denotes the percentage of shares that corporation C1 owns of corporation C2. Then, C1 controls C2 if it owns more than, say, 50% of its shares. In general, to decide whether C1 controls C3 we must also add the shares owned by corporations such as C2 that are controlled by C1. This yields the transitive control rules defined with the help of a continuous sum aggregate that returns the partial sum for each new element, but the first one.

```
control(C, C) ←          owns(C, _, _).
control(Onr, C) ←        twons(Onr, C, Per), Per > 50.
towns(Onr, C2, msum⟨Per⟩) ← control(Onr, C1), owns(C1, C2, Per).
```

Thus, every company controls itself, and a company C1 that has transitive ownership of more than 50% of C2's shares controls C2 . In the last rule, twons computes transitive ownership with the help of msum that adds up the shares of controlling companies. Observe that any pair (Onr, C2) is added at most once to control, thus the contribution of C1 to Onr's transitive ownership of C2 is only accounted once.

Bill-of-Materials (BoM) Applications BoM applications represent an important application area that requires aggregates in recursive rules. Say, for instance that assembly(P1, P2, QT) denotes that P1 contains part P2 in quantity QT. We also have elementary parts described by the relation basic_part(Part, Price). Then, the following program computes the cost of a part as the sum of the cost of the basic parts it contains.

```
part_cost(Part, 0, Cst) ←    basic_part(Part, Cst).
part_cost(Part, mcount⟨Sb⟩, msum⟨MCst⟩) ←
              part_cost(Sb, ChC, Cst), prolfc(Sb, ChC),
              assembly(Part, Sb, Mult), MCst = Cst * Mult.
```

Thus, the key condition in the body of the second rule is that a subpart Sb is counted in part_cost only when all of Sb's children have been counted. This occurs when the number of Sb's children counted so far by mcount is equal to the out-degree of this node in the graph representing assembly. This number is kept in the prolificacy table, prolfc(Part, ChC), which can be computed as follows:

```
prolfc(P1, count⟨P2⟩) ← assembly(P1, P2, _).
prolfc(P1, 0) ←          basic_part(P1, _).
```

8   Conclusions

Keys in derived relations extend the expressive power of deductive databases while retaining their declarative semantics and efficient implementations. In this paper, we have presented equivalent fixpoint and model-theoretic semantics for programs with key constraints in derived relations. Database aggregates can be easily modelled under this extension, yielding a simple characterization of monotonic aggregates. Monotonic aggregates can be freely used in recursive programs, thus providing simple and efficient expressions for optimization and greedy algorithms that had been previously considered impervious to the logic programming paradigm.

There has been a significant amount of previous work that is relevant to the results presented in this paper. In particular the $\mathcal{LDL}++$ provides the choice construct to declare functional dependency constraints in derived relations. The stable model characterization and several other results presented in this paper find a similar counterpart in properties of $\mathcal{LDL}++$ choice construct [13,37]; however, no fixpoint characterization and related results were known for $\mathcal{LDL}++$ choice. An extension of this concept to temporal logic programming was proposed by Orgun and Wadge [21], who introduced the notion of choice predicates that ensure that a given predicate is single-valued. This notion finds applications in intensional logic programming [21].

The cardinality and weight constraints proposed by Niemelä and Simons provide a powerful generalization to key constraints discussed here [20]. In fact, while the key constraint restrict the cardinality of the results to be one, the constraint that such cardinality must be restricted within a user-specified interval is supported in the mentioned work (where different weights can also be attached to atoms). Thus Niemelä and Simons (i) provide a stable model characterization for logic programs containing such constraints, (ii) propose an implementation using Smodels [19], and (ii) show how to express NP-complete problems using these constraints. The implementation approach used for Smodels is quite different from that of $\mathcal{LDL}++$; thus investigating the performance of different approaches in supporting cardinality constraints represents an interesting topic for future research.

Also left for future research, there is the topic of SLD-resolution, which (along with the fixpoint and model-theoretic semantics treated here) would provide a third semantic characterization for logic programs with key constraints [29]. Memoing techniques could be used for this purpose, and for an efficient implementation of keys and aggregates [3].

Acknowledgements

The author would like to thank the reviewers for the many improvements they have suggested, and Frank Myers for his careful proofreading of the manuscript. The author would also like to express his gratitude to Dino Pedreschi, Domenico Saccá, Fosca Giannotti and Sergio Greco who laid the seeds of these ideas during our past collaborations.

This work was supported by NSF Grant IIS-007135.

References

1. S. Abiteboul, R. Hull, and V. Vianu: *Foundations of Databases*. Addison-Wesley, 1995.
2. N. Bidoit and C. Froidevaux: General logical Databases and Programs: Default Logic Semantics and Stratification. *Information and Computation*, 91, pp. 15–54, 1991.
3. W. Chen, D. S. Warren: Tabled Evaluation With Delaying for General Logic Programs. JACM, 43(1): 20-74 (1996).
4. D. Chimenti, R. Gamboa, R. Krishnamurthy, S. Naqvi, S.Tsur and C. Zaniolo: The LDL System Prototype. *IEEE Transactions on Knowledge and Data Engineering*, 2(1), pp. 76-90, 1990.
5. S. Ceri, G. Gottlob and L. Tanca: *Logic Programming and Databases*. Springer, 1990.
6. S. W. Dietrich: Shortest Path by Approximation in Logic Programs. *ACM Letters on Programming Languages and Systems*, 1(2), pp. 119–137, 1992.
7. S. J. Finkelstein, N.Mattos, I.S. Mumick, and H. Pirahesh: Expressing Recursive Queries in SQL, ISO WG3 report X3H2-96-075, March 1996.
8. J. M. Hellerstein, P. J. Haas, H. J. Wang.: Online Aggregation. *SIGMOD 1997: Proc. ACM SIGMOD Int. Conference on Management of Data*, pp. 171-182, ACM, 1997.
9. S. Ganguly, S. Greco, and C. Zaniolo: Extrema Predicates in Deductive Databases. *JCSS* 51(2), pp. 244-259, 1995.
10. M. Gelfond and V. Lifschitz: The Stable Model Semantics for Logic Programming. *Proc. Joint International Conference and Symposium on Logic Programming*, R. A. Kowalski and K. A. Bowen (eds.), pp. 1070-1080, MIT Press, 1988.
11. F. Giannotti, D. Pedreschi, D. Saccà, C. Zaniolo: Non-Determinism in Deductive Databases. In *DOOD'91*, C. Delobel, M. Kifer, Y. Masunaga (eds.), pp. 129-146, Springer, 1991.
12. F. Giannotti, G. Manco, M. Nanni, D. Pedreschi: On the Effective Semantics of Nondeterministic, Nonmonotonic, Temporal Logic Databases. *Proceedings of 12th Int. Workshop, Computer Science Logic*, pp. 58-72, LNCS Vol. 1584, Springer, 1999.
13. F. Giannotti, D. Pedreschi, and C. Zaniolo: Semantics and Expressive Power of Non-Deterministic Constructs in Deductive Databases. *JCSS* 62, pp. 15-42, 2001.
14. Sergio Greco, Domenico Saccà: NP Optimization Problems in Datalog. *ILPS 1997: Proc. Int. Logic Programming Symposium*, pp. 181-195, MIT Press, 1997.
15. S. Greco and C. Zaniolo: Greedy Algorithms in Datalog with Choice and Negation, *Proc. 1998 Joint Int. Conference & Symposium on Logic Programming*, JCSLP'98, pp. 294-309, MIT Press, 1998.
16. R. Krishnamurthy, S. Naqvi: Non-Deterministic Choice in Datalog. In *Proc. 3rd Int. Conf. on Data and Knowledge Bases*, pp. 416-424, Morgan Kaufmann, 1988.
17. V. W. Marek and M. Truszczynski: *Nonmonotonic Logic*. Springer-Verlag, New York, 1995.
18. J. Minker: Logic and Databases: A 20 Year Retrospective. In D. Pedreschi and C. Zaniolo (eds.), *Proceedings International Workshop on Logic in Databases (LID'96)*, Springer-Verlag, pp. 5–52, 1996.
19. I. Niemelä, P. Simons and T. Syrjanen: Smodels: A System for Answer Set Programming   Proceedings of the 8th International Workshop on Non-Monotonic Reasoning, April 9-11, 2000, Breckenridge, Colorado, 4 pages. (Also see: http://www.tcs.hut.fi/Software/smodels/)

20. I. Niemelä and P. Simons: Extending the Smodels System with Cardinality and Weight Constraints. In Jack Minker (ed.): *Logic-Based Artificial Intelligence*, pp. 491-521. Kluwer Academic Publishers, 2001.
21. M.A. Orgun and W.W. Wadge,
    Towards an Unified Theory of Intensional Logic Programming.
    The Journal of Logic and Computation, 4(6), pp. 877-903, 1994.
22. T. C. Przymusinski: On the Declarative and Procedural Semantics of Stratified Deductive Databases: In J. Minker (ed.), *Foundations of Deductive Databases and Logic Programming*, pp. 193–216, Morgan Kaufmann, 1988.
23. R. Ramakrishnan, D. Srivastava, S. Sudanshan, and P. Seshadri: Implementation of the CORAL Deductive Database System. *SIGMOD'93: Proc. Int. ACM SIGMOD Conference on Management of Data*, pp. 167–176, ACM, 1993.
24. K. A. Ross and Yehoshua Sagiv: Monotonic Aggregation in Deductive Database, *JCSS* 54(1), pp. 79-97, 1997.
25. D. Saccà and C. Zaniolo: Deterministic and Non-deterministic Stable Models, *Journal of Logic and Computation*, 7(5), pp. 555-579, 1997.
26. J. S. Schlipf: Complexity and Undecidability Results in Logic Programming, *Annals of Mathematics and Artificial Intelligence*, 15, pp. 257-288, 1995.
27. S. Sudarshan and R. Ramakrishnan: Aggregation and relevance in deductive databases. *VLDB'91: Proceedings of 17th Conference on Very Large Data Bases*, pp. 501-511, Morgan Kaufmann, 1991.
28. J. D. Ullman: *Principles of Data and Knowledge-Based Systems*, Computer Science Press, New York, 1988.
29. M.H. Van Emden and R. Kowalski: The Semantics of Predicate Logic as a Programming Language. *JACM* 23(4), pp. 733-742, 1976.
30. A. Van Gelder, K. A. Ross, and J. S. Schlipf: The Well-Founded Semantics for General Logic Programs. *JACM* 38, pp. 620–650, 1991.
31. A. Van Gelder: Foundations of Aggregations in Deductive Databases. In *DOOD'93*, S. Ceri, K. Tanaka, S. Tsur (Eds.), pp. 13-34, Springer, 1993.
32. H. Wang and C. Zaniolo: User-Defined Aggregates in Object-Relational Database Systems. *ICDE 2000: International Conference on Database Engineering.* pp. 111-121, IEEE Press, 2000.
33. H. Wang and C. Zaniolo: Using SQL to Build New Aggregates and Extenders for Object-Relational Systems. *VLDB 2000: Proceedings of 26th Conference on Very Large Data Bases*, pp. 166-175, Morgan Kaufmann, 2000.
34. C. Zaniolo and H. Wang: Logic-Based User-Defined Aggregates for the Next Generation of Database Systems. In K.R. Apt, V. Marek, M. Truszczynski, D.S. Warren (eds.): *The Logic Programming Paradigm: Current Trends and Future Directions.* Springer Verlag, pp. 121-140, 1999.
35. C. Zaniolo, S. Ceri, C. Faloutzos, R. Snodgrass, V.S. Subrahmanian, and R. Zicari: *Advanced Database Systems*, Morgan Kaufmann, 1997.
36. C. Zaniolo: The Nonmonotonic Semantics of Active Rules in Deductive Databases. In *DOOD 1997*, F. Bry, R. Ramakrishnan, K. Ramamohanarao (eds.), pp. 265-282, Springer, 1997.
37. C. Zaniolo et al.: $\mathcal{LDL}$++ Documentation and Web Demo, 1988: http://www.cs.ucla.edu/ldl
38. C. Zaniolo: Key Constraints and Monotonic Aggregates in Deductive Databases. UCLA technical report, June 2001.

A Decidable CLDS
for Some Propositional Resource Logics

Krysia Broda

Department of Computing, Imperial College
180 Queens' Gate, London SW7 2BZ
kb@doc.ic.ac.uk

Abstract. The *compilation approach* for Labelled Deductive Systems
(CLDS) is a general logical framework. Previously, it has been applied
to various resource logics within natural deduction, tableaux and clausal
systems, and in the latter case to yield a decidable (first order) CLDS
for propositional Intuitionistic Logic (IL). In this paper the same clausal
approach is used to obtain a decidable theorem prover for the implication fragments of propositional substructural Linear Logic (LL) and
Relevance Logic (RL). The CLDS refutation method is based around a
semantic approach using a translation technique utilising first-order logic
together with a simple theorem prover for the translated theory using
techniques drawn from Model Generation procedures. The resulting system is shown to correspond to a standard LL(RL) presentation as given
by appropriate Hilbert axiom systems and to be decidable.

1 Introduction

Among the computational logic community no doubt there are very many people,
like me, whose enthusiasm for logic and logic programming was fired by Bob
Kowalski. In my case it led to an enduring interest in automated reasoning, and
especially the connection graph procedure. In appreciation of what Bob taught
me, this paper deals with some non-classical resource logics and uses a classical
first order theory to give a clausal theorem prover for them.

The general methodology based on Gabbay's Labelled Deductive Systems
(LDS) [9], called the Compiled Labelled Deductive Systems approach (CLDS),
is described in [5], [6]. The method allows various logics to be formalised within
a single framework and was first applied to modal logics in [14] and generally
to the multiplicative part of substructural logics in [5], [6]. The CLDS refutation method is based around a semantic approach using a translation into
first-order logic together with a simple theorem prover for the translated theory
that employs techniques drawn from Model Generation procedures. However,
one critical problem with the approach is that the resulting first order theory
is often too expressive and therefore not decidable, even when the logic being
modelled is known to be so. It was described in [4] how to construct a decidable refutation prover for the case of Intuitionistic Logic (IL); in this paper that

A.C. Kakas, F. Sadri (Eds.): Computat. Logic (Kowalski Festschrift), LNAI 2408, pp. 135–159, 2002.
© Springer-Verlag Berlin Heidelberg 2002

prover is extended, in different ways, to deal with the implication fragments of the propositional resource logics Linear Logic (LL) and Relevance Logic (RL).

The motivation for using LDS derives from the observation that many logics only differ from each other in small ways. In the family of modal logics, for example, the differences can be captured semantically through the properties of the accessibility relation, or syntactically within various side-conditions on the proof steps. In substructural logics, the differences can be captured in the syntax by means of the structural proof rules. In a CLDS, capturing differences between logics is achieved through the use of a combined language, incorporating a language for wffs and a language for terms (known as labels), called a labelling language. Elements of the two languages are combined to produce *declarative units* of the form $\alpha : \lambda$, where α is a wff and λ is a label. The interpretation of a declarative unit depends on the particular family of logics being formalised. In the case of modal logics the label λ names a possible world, whereas in substructural, or resource, logics it names a combination of resources. A theory built from declarative units is called a *configuration* and consists both of declarative units and literals stating the relationships between labels of the configuration (called R-literals). In this LDS approach applied to resource logics the declarative unit $\alpha : \lambda$ represents the statement that the "resource λ verifies the wff α". This was first exploited in [9]. Resources can be combined using the operator $\circ$ and their power of verification related by $\preceq$, where $\lambda \preceq \lambda'$ is interpreted to mean that λ' can verify everything that λ can and is thus the more powerful of the two. Depending on the properties given to $\circ$ the power of combined resources can be controlled. In RL, for example, resources can be copied; that is, $\lambda \circ \lambda \preceq \lambda$, or λ is just as powerful as multiple copies of itself. In both RL and LL the order in which resources are combined does not matter, so $\lambda \circ \lambda' \preceq \lambda' \circ \lambda$. These properties, contraction and commutativity, respectively, correspond to the structural rules of contraction and permutation of standard sequent calculi for RL and LL. In fact, in LDS, all substructural logics can be treated in a uniform way, simply by including different axioms in the labelling algebra [1].

The semantics of a CLDS is given by translating a configuration into first order logic in a particular way, the notion of semantic entailment being defined with respect to such translated configurations. An example of a configuration is the set of declarative units

$$\{p \rightarrow (p \rightarrow (q \rightarrow p)) : b, \ p : a, \ q : c, \ q \rightarrow p : b \circ a \circ a, \ p : b \circ a \circ a \circ c\}$$

and R-literals $\{a \circ a \preceq a, \ a \preceq b \circ a \circ a \circ c\}$, called *constraints* in this paper. The translation of a configuration uses a language of special monadic predicates of the form $[\alpha]^*$, one predicate for each wff α. For the above example of a configuration the translation is

$$\{[p \rightarrow (p \rightarrow (q \rightarrow p))]^*(b), [p]^*(a), [q \rightarrow p]^*(b \circ a \circ a), [p]^*(b \circ a \circ a \circ c),$$
$$a \circ a \preceq a, a \preceq b \circ a \circ a \circ c\}$$

A set of axioms to capture the meanings of the logical operators and a theory, called the *labelling algebra*, are used for manipulating labels and the relations

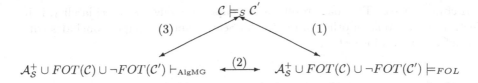

Fig. 1. Refutation CLDS

between them. The language, axiom theory and labelling algebra considered in this paper are together referred to as L_{CLDS} and R_{CLDS}, respectively, for LL and RL. An example of a semantic axiom, using the monadic predicates of the form $[\alpha]^*$, in this case that captures the meaning of the $\rightarrow$ operator, is

$$\forall x([\alpha \rightarrow \beta]^*(x) \leftrightarrow \forall y([\alpha]^*(y) \rightarrow [\beta]^*(x \circ y)))$$

For a given problem, the set of semantic axioms is implicitly instantiated for every wff that occurs in the problem; this set of instances together with a translation of the initial configuration, in which $\alpha : \lambda$ is translated as $[\alpha]^*(\lambda)$, can also be taken as a compiled form of the problem. Any standard first order theorem prover, for example Otter [12], could be used to find refutations, although not always very efficiently. In [4], a decidable refutation theorem prover based on the methods of Davis Putnam [8], Hyper-resolution [13] and model generation [11] was taken as the proof system and shown to be sound and complete with respect to the semantics. A similar approach can be taken for LL and RL, here called AlgMG, but appropriate new restrictions to retain decidability for LL and RL are required and definitions of these are the main contribution of this paper.

The CLDS approach is part of a systematic general framework that can be applied to any logic, either old or new. In case a CLDS corresponds to a known logic, the correspondence with a standard presentation of that logic must also be provided. That is, it must be shown that (i) every derivation in the chosen standard presentation of that logic can be simulated by the rules of the CLDS, in this case by the refutation theorem prover, and (ii) how to build an interpretation such that, if a formula α is not a theorem of the logic in question, then there is an appropriate model in which a suitable declarative unit constructed using α is false. It is this second part that needs care in order to obtain a decidable system for the two logics LL and RL.

The approach taken in a refutation CLDS is illustrated in Fig. 1, where C and C' are configurations and $\neg FOT(C)$ denotes the disjunction of the negations of the translated declarative units in $FOT(C)$. Arrow (2) represents the soundness and completeness of the refutation prover and arrow (1) is the definition of the semantics of a CLDS. The derived arrow (3) represents a soundness and completeness property of the refutation procedure with respect to configurations.

A fuller description of the language, labelling algebra and axioms modelling the derivation rules for the languages under consideration is given in Sect. 2, whilst Sect. 3 outlines the theorem prover and the results concerning soundness

and completeness. The main result of the paper, dealing with decidability, is in
Sect. 4, with proofs of other properties in Sect. 5 and the paper concludes with
a brief discussion in Sect. 6.

2   Refutation CLDS for Substructural Logics

The CLDS approach for the implication fragment[1] of LL and RL is now described. Definitions of the language, syntax and semantics are given, and *configurations* are introduced.

2.1   Languages and Syntax

A CLDS propositional language is defined as an ordered pair $\langle \mathcal{L}_P, \mathcal{L}_L \rangle$, where
$\mathcal{L}_L$ is a *labelling language* and $\mathcal{L}_P$ is a *propositional language*. For the implication fragment of LL and RL the language $\mathcal{L}_P$ is composed of a countable set of
proposition symbols, $\{p, q, r, \ldots\}$ and the binary connective $\rightarrow$. A special proposition symbol is $\perp$, where $\neg A$ is defined also as $A \rightarrow \perp$, so allowing negation to
be represented. (The wff $\top$ is sometimes used in place of $\perp \rightarrow \perp$.) The labelling
language $\mathcal{L}_L$ is a fragment of a first-order language composed of a binary operator $\circ$, a countable set of variables $\{x, y, z, \ldots\}$, a binary predicate $\preceq$, the set of
logical connectives $\{\neg, \wedge, \vee, \rightarrow, \leftrightarrow\}$, and the quantifiers $\forall$ and $\exists$. The first-order
language $Func(\mathcal{L}_P, \mathcal{L}_L)$ is an extension of $\mathcal{L}_L$ as follows.

Definition 1. *Let the set of all wffs in $\mathcal{L}_P$ be $\{\alpha_1, \alpha_2, \ldots\}$, then the semi-
extended labelling language $Func(\mathcal{L}_P, \mathcal{L}_L)$ comprises $\mathcal{L}_L$ extended with a set of
skolem constant symbols $\{c_{\alpha_1}, c_{\alpha_2}, \ldots\}$, also referred to as parameters.*

Terms of the semi-extended labelling language $Func(\mathcal{L}_P, \mathcal{L}_L)$ are defined inductively, as consisting of parameters and variables, together with expressions of the
form $\lambda \circ \lambda'$ for terms λ and λ', and are also called labels. Note that the parameter
c_α represents the smallest label verifying α and that all parameters will have a
special role in the semantics. There is the parameter 1 (shorthand for $c_\top$) that
represents the empty resource, since $\top$ is always provable.

To capture different classes of logics within the CLDS framework an appropri-
ate first-order theory written in the language $Func(\mathcal{L}_P, \mathcal{L}_L)$, called the *labelling
algebra*, needs to be defined. The labelling algebra is a binary first-order theory
which axiomatises (i) the binary predicate $\preceq$ as a pre-ordering relation and (ii)
the properties *identity* and *order preserving* of the commutative and associative
function symbol $\circ$. For RL, the structural property *contraction* is also included.

Definition 2. *The labelling algebra $\mathcal{A}_L$, written in $Func(\mathcal{L}_P, \mathcal{L}_L)$, is the first
order theory given by the axioms (1) - (5), where x, y and z all belong to
$Func(\mathcal{L}_P, \mathcal{L}_L)$. The algebra $\mathcal{A}_R$ is the algebra $\mathcal{A}_L$ enhanced by axiom (6).*

---

[1]   Restricted in order to keep the paper short.

1. **(identity)** $\forall x[1 \circ x \preceq x \wedge x \preceq 1 \circ x]$
2. **(order-preserving)** $\forall x, y, z[x \preceq y \rightarrow x \circ z \preceq y \circ z \wedge z \circ x \preceq z \circ y]$
3. **(pre-ordering)** $\forall x[x \preceq x]$ and $\forall x, y, z[x \preceq y \wedge y \preceq z \rightarrow x \preceq z]$
4. **(commutativity)** $\forall x, y[x \circ y \preceq y \circ x]$
5. **(associativity)** $\forall x, y, z[(x \circ y) \circ z \preceq x \circ (y \circ z)]$ and $\forall x, y, z[x \circ (y \circ z) \preceq (x \circ y) \circ z]$
6. **(contraction)** $\forall x[x \circ x \preceq x]$

The CLDS language facilitates the formalisation of two types of information, (i) what holds at particular points, given by the declarative units, and (ii) which points are in relation with each other and which are not, given by constraints ($\preceq$-literals). A declarative unit is defined as a pair *"formula:label"* expressing that a formula "holds" at a point. The label component is a ground term of the language $Func(\mathcal{L}_P, \mathcal{L}_L)$ and the formula is a wff of the language $\mathcal{L}_P$. A constraint is any ground literal in $Func(\mathcal{L}_P, \mathcal{L}_L)$ of the form $\lambda_1 \preceq \lambda_2$ or $\lambda_1 \npreceq \lambda_2$), where λ_1 and λ_2 are labels, expressing that λ_2 is, or is not, related to λ_1. In the applications considered here, little use will be made of negated constraints. In Intuitionistic Logic "related to" was interpreted syntactically as "subset of", but for L_{CLDS} it is interpreted as "has exactly the same elements as" and for R_{CLDS} as "has the same elements as, but possibly with more occurences". This combined aspect of the CLDS syntax yields a definition of a CLDS theory, called a *configuration*, which is composed of a set of constraints and a set of declarative units. An example of a configuration was given in the introduction. The formal definition of a configuration is as follows.

Definition 3. *Given a CLDS language, a configuration $\mathcal{C}$ is a tuple $\langle \mathcal{D}, \mathcal{F} \rangle$, where $\mathcal{D}$ is a finite set of constraints (referred to as a* diagram*) and $\mathcal{F}$ is a function from the set of ground terms of* $Func(\mathcal{L}_P, \mathcal{L}_L)$ *to the set of sets of wffs of $\mathcal{L}_P$. Statements of the form $\alpha \in \mathcal{F}(\lambda)$ will be written as $\alpha : \lambda \in \mathcal{C}$.*

2.2 Semantics

The model-theoretic semantics of CLDS is defined in terms of a first-order semantics using a translation method. This enables the development of a model-theoretic approach which is equally applicable to any logic also belonging to different families whose operators have a semantics which can be expressed in a first-order theory. As mentioned before, a declarative unit $\alpha : \lambda$ represents that the formula is verified (or holds) at the point λ, whose interpretation is strictly related to the type of underlying logic. These notions are expressed in terms of first-order statements of the form $[\alpha]^*(\lambda)$, where $[\alpha]^*$ is a predicate symbol. The relationships between these predicate symbols are constrained by a set of first-order axiom schemas which capture the satisfiability conditions of each type of formula α. The *extended labelling algebra* $Mon(\mathcal{L}_P, \mathcal{L}_L)$ is an extension of the language $Func(\mathcal{L}_P, \mathcal{L}_L)$ given by adding a monadic predicate symbol $[\alpha]^*$ for each wff α of $\mathcal{L}_P$. It is formally defined below.

Table 1. Basic and clausal semantic axioms for L_{CLDS} and R_{CLDS}

$$
\begin{aligned}
&\text{Ax1:} \quad \forall x \forall y (x \preceq y \wedge [\alpha]^*(x) \to [\alpha]^*(y)) \\
&\text{Ax2:} \quad \forall x ([\alpha]^*(x) \to \exists y ([\alpha]^*(y) \wedge \forall z ([\alpha]^*(z) \to y \preceq z))) \\
&\text{Ax3:} \quad \forall x ([\alpha \to \beta]^*(x) \leftrightarrow \forall y ([\alpha]^*(y) \to [\beta]^*(x \circ y))) \\
&\text{Ax2a:} \quad \forall x ([\alpha]^*(x) \to [\alpha]^*(c_\alpha)) \\
&\text{Ax2b:} \quad \forall x ([\alpha]^*(x) \to c_\alpha \preceq x) \\
&\text{Ax3a:} \quad \forall x \forall y ([\alpha \to \beta]^*(x) \wedge [\alpha]^*(y) \to [\beta]^*(x \circ y)) \\
&\text{Ax3b:} \quad \forall x ([\alpha \to \beta]^*(x) \leftarrow [\beta]^*(x \circ c_\alpha)) \\
&\text{Ax3c:} \quad \forall x ([\alpha \to \beta]^*(x) \vee [\alpha]^*(c_\alpha))
\end{aligned}
$$

Definition 4. *Let $Func(\mathcal{L}_P, \mathcal{L}_L)$ be a semi-extended labelling language. Let the ordered set of wffs of $\mathcal{L}_P$ be $\alpha_1, \ldots, \alpha_n, \ldots$, then the* extended labelling language, *called $Mon(\mathcal{L}_P, \mathcal{L}_L)$, is defined as the language $Func(\mathcal{L}_P, \mathcal{L}_L)$ extended with the set $\{[\alpha_1]^*, \ldots, [\alpha_n]^*, \ldots\}$ of unary predicate symbols.*

The *extended algebra* $\mathcal{A}_L^+$ for L_{CLDS} is a first-order theory written in $Mon(\mathcal{L}_P, \mathcal{L}_L)$, which extends the labelling algebra $\mathcal{A}_L$ with a particular set of axiom schemas. A L_{CLDS} system $\mathcal{S}$ can now be defined as $\mathcal{S} = \langle \langle \mathcal{L}_P, \mathcal{L}_L \rangle, \mathcal{A}_L^+, \text{AlgMG} \rangle$, where AlgMG is the program for processing the first order theory $\mathcal{A}_L^+$. Similarly for R_{CLDS}, but using $\mathcal{A}_R^+$ that includes the (contraction) property.

The axiom schemas are given in Table 1. There are the *basic axioms*, (Ax1) - (Ax3), and the clausal axioms, (Ax3a), (Ax3b), etc., derived from them by taking each half of the $\leftrightarrow$ in turn. The first axiom (Ax1) characterises the property that increasing labels λ and λ', such that $\lambda \preceq \lambda'$, imply that the sets of wffs verified by those labels are also increasing. The second axiom (Ax2) characterises a special property that states that, if a wff α is verified by some label, then it is verified by a "smallest" label. Both these axioms relate declarative units to constraints. The axiom (Ax3) characterises the operator $\to$. Several of the axioms have been simplified by the use of parameters, (Ax1) and (Ax2) (effectively applying Skolemisation). In (Ax2) the variable y is Skolemised to the parameter c_α. The Skolem term c_α is a constant, not depending on x, and this is the feature that eventually yields decidability. A standard Skolemisation technique would result in a function symbol depending on x, but the simpler version suffices for the following reason. Using (Ax2), any two "normal" Skolem terms, $c_\alpha(x_1)$ and $c_\alpha(x_2)$, would satisfy

$$c_\alpha(x_1) \preceq c_\alpha(x_2) \text{ and } c_\alpha(x_2) \preceq c_\alpha(x_1)$$

By (Ax1) this would allow the equivalence of $[\alpha]^*(c_\alpha(x))$ and $[\alpha]^*(c_\alpha(y))$ for any x and y. The single representative c_α is introduced in place of the "normal" Skolem terms $c_\alpha(x)$. It is not very difficult to show that, for any set S of instances of the axiom schema Skolemised in the "normal" way using Skolem symbols $c_\alpha(x)$, S is inconsistent iff the same set of instances of the axioms, together with a set of clause schema of the form $\forall x ([\alpha]^*(c_\alpha) \leftrightarrow [\alpha]^*(c_\alpha(x)))$, is inconsistent.

The Skolemised (Ax2) can also be simplified to the following equivalent version (also called (Ax2))

$$\forall x([\alpha]^*(x) \rightarrow ([\alpha]^*(c_\alpha) \wedge c_\alpha \preceq x))$$

from which (Ax2a) and (Ax2b) are derived. In the system of [4] for IL a further simplification was possible, in that (Ax3c) could be replaced by $[\alpha \rightarrow \beta]^*(1) \vee [\alpha]^*(a)$. This is not the case for L_{CLDS} or R_{CLDS}, which consequently require a slightly more complicated algorithm AlgMG. The clausal axioms in Table 1, together with the appropriate properties of the Labelling Algebra, are also called the Extended Labelling Algebra, $\mathcal{A}_L^+$ or $\mathcal{A}_R^+$. It is for finite sets of instances of these axioms that a refutation theorem prover is given in Sect. 3.

The notions of satisfiability and semantic entailment are common to any CLDS and are based on a translation method which associates syntactic expressions of the CLDS system with sentences of the first-order language $Mon(\mathcal{L}_P, \mathcal{L}_L)$, and hence associates configurations with first-order theories in the language $Mon(\mathcal{L}_P, \mathcal{L}_L)$. Each declarative unit $\alpha : \lambda$ is translated into the sentence $[\alpha]^*(\lambda)$, and constraints are translated as themselves. A formal definition is given below.

Definition 5. *Let $\mathcal{C} = \langle \mathcal{D}, \mathcal{F} \rangle$ be a configuration. The* first-order translation *of $\mathcal{C}$, $FOT(\mathcal{C})$, is a theory in $Mon(\mathcal{L}_P, \mathcal{L}_L)$ and is defined by the expression: $FOT(\mathcal{C}) = \mathcal{D} \cup \mathcal{DU}$, where $\mathcal{DU} = \{[\alpha]^*(\lambda) \mid \alpha \in \mathcal{F}(\lambda), \lambda$ is a ground term of $Func(\mathcal{L}_P, \mathcal{L}_L)\}$.*

The notion of semantic entailment for L_{CLDS} as a relation between configurations is given in terms of classical semantics using the above definition. In what follows, wherever $\mathcal{A}_L^+$ and $\models_L$ are used, $\mathcal{A}_R^+$ and $\models_R$ could also be used, assuming the additional property of (contraction) in the Labelling Algebra.[2]

Definition 6. *Let $S = \langle \langle \mathcal{L}_P, \mathcal{L}_L, \rangle, \mathcal{A}_L^+, AlgMG \rangle$ be a L_{CLDS}, $\mathcal{C} = \langle \mathcal{D}, \mathcal{F} \rangle$ and $\mathcal{C}' = \langle \mathcal{D}', \mathcal{F}' \rangle$ be two configurations of S, and $FOT(\mathcal{C}) = \mathcal{D} \cup \mathcal{DU}$ and $FOT(\mathcal{C}') = \mathcal{D}' \cup \mathcal{DU}'$ be their respective first-order translations. The configuration $\mathcal{C}$ semantically entails $\mathcal{C}'$, written $\mathcal{C} \models_L \mathcal{C}'$, iff $\mathcal{A}_L^+ \cup FOT(\mathcal{C}) \cup \neg FOT(\mathcal{C}') \models_{FOL}$.*

If δ is a declarative unit or constraint belonging to $\mathcal{C}'$ and $FOT(\delta)$ its first order translation, then $\mathcal{C} \models_L \mathcal{C}'$ implies that $\mathcal{A}_L^+ \cup FOT(\mathcal{C}) \cup \neg FOT(\delta) \models_{FOL}$, which will also be written as $\mathcal{C} \models_L \delta$.

Declarative units of the form $\alpha : 1$, such that $\mathcal{T}_\emptyset \models_L \alpha : 1$, where $\mathcal{T}_\emptyset$ is an empty configuration (i.e. $\mathcal{D}$ and $\mathcal{F}$ are both empty), are called *theorems*. In order to show that a theorem $\alpha : 1$ holds in L_{CLDS} (R_{CLDS}), appropriate instances of the axioms in $\mathcal{A}_L^+$ ($\mathcal{A}_R^+$) are first formed for each subformula of α, and then $\neg [\alpha]^*(1)$ is added. This set of clauses is refuted by AlgMG. More generally, to show that α follows from the wffs $\beta_1, \ldots, \beta_n$, the appropriate instances include those for each subformula of $\alpha, \beta_1, \ldots, \beta_n$, together with $\neg [\alpha]^*(i)$, where $i = c_{\beta_1} \circ \ldots \circ c_{\beta_n}$, together with the set $\{[\beta_j]^*(c_{\beta_j})\}$. This derives from consideration

---

[2] Recall $\neg FOT(\mathcal{C})$ means the disjunction of the negation of the literals in $FOT(\mathcal{C})$.

of the deduction theorem, namely, that $\{\beta_j\}$ implies α iff $\beta_1 \rightarrow \dots \beta_n \rightarrow \alpha$ is a theorem. Notice that, if a formula β occurs more than once, then $[\beta]^*(c_\beta)$ need only be included once in the translated data, but its label c_β is included in i as many times as it occurs.

3   A Theorem Prover for L_{CLDS} and R_{CLDS} Systems

The Extended Labelling Algebra $\mathcal{A}_L^+$ enjoys a very simple clausal form. The theorem prover AlgMG, described below as a logic program, uses an adaptation of the Model Generation techniques [11]. The axioms of the Labelling Algebra $\mathcal{A}_L$, or, including (contraction), $\mathcal{A}_R$, together with Axioms (Ax1) and (Ax2a) are incorporated into the unification algorithm, called AlgU. Axioms (Ax1), (Ax2a) and (Ax2b) were otherwise accounted for in the derivation of the remaining axioms and are not explicitly needed any further. First, some definitions are given for this particular kind of first order theory.

Note 1. In this section, a clause will either be denoted by C, or by $L \vee D$, where L is a literal and D is a disjunction of none or more literals. All variables are implicitly universally quantified. Literals are generally denoted by L or $\neg L$, but may also be denoted by: $L(x)$ or $L(y)$, when the argument is exactly the variable x or y, $L(u)$, when the argument contains no variables, $L(xu)$, when it contains a variable x and other ground terms u, in which case u is called the *ground part*, or $L(w)$ when the argument may, or may not, contain a variable. The suffices $_1$, $_2$, etc. are also used if necessary. For ease of reading and writing, label combinations such as $a \circ b \circ c$ will be written as abc. It is convenient to introduce the multi-set difference operator $-$ on labels in which every occurrence counts. For example, $aab - ab = a$ and $ab - 1 = ab$.

In the sequel, by *non-unit* parameter will be meant any parameter c_α other than $c_\top$ $(=1)$.

Definition 7. *For a given set of clauses S, the set $\mathcal{D}_S$, the* Herbrand Domain *of S, is the set $\{c_\alpha | c_\alpha$ is a non-unit parameter occurring in $S\} \cup \{1\}$. The* Herbrand Universe *of S is the set of terms formed using the operator $\circ$ applied to elements from the Herbrand Domain. A* ground instance *of a clause C or literal L (written $C\theta$ or $L\theta$) is the result of replacing each variable x_i in C or L by a ground term t_i from the Herbrand Universe, where the substitution $\theta = \{x_i := t_i\}$.*

Definition 8. *u_1 unifies with u_2 (with respect to AlgU) iff $u_1 \preceq u_2$. Notice that unification is not symmetric.*

In AlgMG it is also necessary to unify non-ground terms and details of the various cases (derived from the ground case), which are different for each of RL and LL, are given next. They are labelled (a), (b) etc. for reference.

(a) **(ground, ground + var)** u_1 unifies with xu_2 , where u_2 may implicitly be the label 1, iff there is a ground substitution θ for x such that u_1 unifies with $(xu_2)\theta$. In the case of LL there is only one possible value for θ, viz. $x := u_1 - u_2$, but in the case of RL there may be several possible values, depending on the number of implicit contraction operations applied to u_1. For example, aaa unifies with ax, with substitutions $x := 1$, $x := a$ or $x := aa$.

(b) **(ground+var, ground)** xu_1 unifies with u_2, where u_1 may implicitly be the label 1, iff there is a ground substitution θ such that $(xu_1)\theta$ unifies with u_2. The substitution θ is chosen so that $(xu_1)\theta$ is the largest possible term that unifies with u_2 (under $\preceq$). For example, in RL, ax unifies with ab with substitution $x := b$, even though other substitutions for x are possible, eg $x := abb$.[3] If $u_1 = 1$ this case reduces to $x := u_2$.

(c) **(var+ground, var+ground)** x_1u_1 unifies with x_2u_2 iff there are substitutions θ_1 and θ_2 for variables x_1 and x_2 of the form $x_1 := u_3x$ and $x_2 := u_4x$, such that u_1u_3 unifies with u_2u_4. Either or both of u_1, u_2 may implicitly be the label 1. The substitution for x_1 is maximal (under $\preceq$), in the sense that any other possible substitution for x_1 has the form $x_1 := u_5x$, where $u_5 \preceq u_3$. In LL there is only one possible substitution for x_2 of the right form, namely $x_2 := x \circ (u_1 - u_2)$. In RL there may be several possible substitutions, depending on the number of implicit contraction steps. For example, in RL, aax_1 unifies with bx_2 with both the substitutions $x_1 := bx, x_2 := ax$ or $x_1 := bx, x_2 := aax$. However, because of the presence of the variable x in the substitution for x_2, it is only necessary to use the maximal substitution, which is the first one. The reader can check the correct results are obtained if $u_1 = 1$ or $u_2 = 1$, respectively, that $x_1 = x_2u_2$ or $x_2 = u_1x_1$.

Subsumption can also be applied between literals.

Definition 9. *$L(w)$ subsumes $L(w')$ iff w unifies with w' with unifier θ and $L(w')$ is identical to $L(w)\theta$.*

This definition leads to the following cases.

(d) **(ground, ground)** $L(u_1)$ subsumes $L(u_2)$ iff $u_1 \preceq u_2$
(e) **(ground, ground+var)** $L(u_1)$ does not subsume $L(xu_2)$.
(f) **(ground+var, ground)** $L(xu_1)$ subsumes $L(u_2)$ iff there is a ground substitution θ for x such that $(xu_1)\theta$ unifies with u_2.
(g) **(ground+var, ground+var)** $L(x_1u_1)$ subsumes $L(x_2u_2)$ iff there is a substitution θ for x_1 of the form $x_1 := x_2u_3$ such that u_3u_1 unifies with u_2. For example, in RL, $P(xaa)$ subsumes $P(ay)$ and $P(aby)$, but it does not subsume $P(by)$.

Literal L subsumes clause C iff L subsumes a literal in C.

Definition 10. *Unit clause $L(w)$ resolves with $D \vee \neg L(w')$ to give $D\theta$ iff w unifies with w' with unifier θ. If D is empty and $L(w)$ and $\neg L(w')$ resolve, then they are called complements of each other. A Hyper-resolvent is a clause with no negative literals formed by resolving a clause with one or more positive unit clauses.*

---

[3] Recall that in the presence of contraction $bb \preceq b$.

Brief Overview of AlgMG. AlgMG for the implication fragment operates on sets of clauses, each of which may either be a Horn clause (including unit clauses), or a non-Horn clause of the form $\forall x([\alpha]^*(c_\alpha) \vee [\alpha \rightarrow \beta]^*(x))$. There is just one kind of negative unit clause, $\neg[\alpha]^*(i)$, derived from the initial goal, where α is the wff to be proved and $i = i_1 \circ \ldots \circ i_n$ is the label consisting of the parameters $i_1, \ldots, i_n$ that verify the formulas from which α is to be proved.

AlgMG incorporates the special unification algorithm AlgU, which is used to unify two labels x and z, where x and/or z may contain a variable, implicitly taking into account the properties of $\mathcal{A}_L^+$ or $\mathcal{A}_R^+$ and the different definitions of unifier (cases (a) to (c) above). Notice that the order of parameters in a label does not matter because of the properties (associativity) and (commutativity), so abc would match with bca, for example. By (identity), the parameter 1 is only explicitly needed in the label 1 itself, which is treated as the empty multiset. There are, in fact, only a restricted number of kinds of unification which can arise using AlgMG and these are listed after the available rules have been described.

The initial set of clauses for refuting a formula α are derived from instances of the semantic axioms appropriate for the predicates occurring in the first order translation of α (called the "appropriate set of clauses for showing α"). There are seven different rules in AlgMG, which can be applied to a finite list of clauses. Five are necessary for the operation of the algorithm and the other two, (Simplify) and (Purity), are useful for the sake of practicality; only (Simplify) is included here. The (Purity) rule serves to remove a clause if it can be detected that it cannot usefully contribute to the derivation. Unit clauses in a list, derived by the (Hyper) or (Split) rule, or given initially, are maintained as a partial model of the initial clauses. The following rules are available in AlgMG:

End A list containing an atom and its complement is marked as successfully finished. The only negative unit clause is derived from the initial goal.

Subsumption Any clause subsumed by a unit clause L is removed.

Simplify A unit clause $[\alpha]^*(x)$ can be used to remove any literal $\neg[\alpha]^*(w)$ in a clause since $[\alpha]^*(x)$ complements $\neg[\alpha]^*(w)$.

Fail A list in which no more steps are possible is marked as failed and can be used to give a model of the initial clauseset.

Hyper A hyper-resolvent (with respect to AlgU) is formed from a non-unit clause in the list and (positive) unit clauses in the list. Only hyper-resolvents that cannot immediately be subsumed are generated.

Split If L is a list of clauses containing clause $L' \vee L''$, two new lists $[L'|L^-]$ and $[L''|L^-]$ are formed, where L^- results from removing $L' \vee L''$ from L. The algorithm is then applied to each list.

The possible opportunities for unification that arise in AlgMG are as follows:

1. Unification of a label of the form xu in a positive literal, where x may be missing, with y in a negative literal in a (Hyper) step – the unifier is given as in case (a) or case (c) as appropriate.
2. Unification of a label x in a positive literal with some label w in a (Simplify) step. This always succeeds and w is unchanged. (This is a special case of (b) or (c).)

3. Unification of a label of the form xu_1 in a positive literal, where either of x or u_1 may be missing, with u_2 in the negative literal in an (End) step. This is either the ground case of unification, that is $u_1 \preceq u_2$, or case (b).
4. Unification in a (Hyper) step between a label of the form xu, where either x or u may be missing, with $c_\alpha y$. This is again either case (a) or (c).

If use of either the (Hyper) or (Simplify) rule yields a label in which there are two variables, they can be replaced by a new variable x.

The (Hyper) rule is the problem rule in AlgMG for the systems L_{CLDS} and R_{CLDS}. Its unrestricted use in a branch can lead to the generation of atoms with labels of increasing length. For example, the clause schema arising from $\alpha \to \alpha$ is $[\alpha \to \alpha]^*(x) \wedge A(y) \to A(xy)$, which, if there are atoms of the form $[\alpha \to \alpha]^*(u_1)$ and $A(u_2)$, will lead to $A(u_1u_2)$, $A(u_1u_2u_2)$ and so on, possibly none of them subsumed by earlier atoms. Therefore, without some restriction on its use, development of a branch could continue forever. The LDS tableau system in [1] and the natural deduction system in [2] both exhibited a similar problem, but its solution was not addressed in those papers. In the application to IL, due to the additional property of monotonicity in the labelling algebra, that $x \preceq x \circ y$, labels could be regarded as sets of parameters. Together with the fact that the Herbrand Domain for any particular problem was finite, there was an upper bound on the size of labels generated (i.e. on the number of occurrences of parameters in a label) and hence the number of applications of (Hyper) was finite and termination of the algorithm was assured. In the two systems L_{CLDS} and R_{CLDS} this is not so any more and a more complex bound must be used to guarantee termination. Before introducing these restrictions, an outline logic program for AlgMG is given together with some examples of its operation.

Outline Program for Algorithm AlgMG. The program is given below. A rudimentary version has been built in Prolog to check very simple examples similar to those in this paper.

```
0(start)     dp(S,F,R) :- dp1 ([ ],S,F,R).
1(fail)      dp1(M,S,M,false) :- noRulesApply(M,S).
2(end)       dp1(M,S,[],true) :- endApplies(S,M).
3(subsume)   dp1(M,S,F,R) :- subsumed(C,M,S), remove(C,S,NewS),
                 dp1( M,NewS,F,R).
4(simplify)  dp1(M,S,F,R) :- simplify(M,S,NewS), dp1(M,NewS,F,R).
5(hyper)     dp1(M,S,F,R) :- hyper(M,S,New), add(New,S,M,NewS,NewM),
                 dp1(NewM,NewS,F,R).
6(split)     dp1(M,S,F,R) :- split(M,S,NewS,S1,S2),
                 dp1([S1|M],NewS,F1,R1),dp1([S2|M],NewS,F2,R2),
                 join(F1,F2,F), and(R1,R2,R).
```

The initial call is the query dp(S, F, R), in which F and R will be variables, and S is a list of clauses appropriate for showing α and derived from a L_{CLDS} or R_{CLDS}. At termination, R will be bound either to *true* or to *false* and in the latter case F will be bound to a list of unit clauses. The list F can be used to find a finite model of S Assume that any subsumed clauses in the initial set of

Initial clauses:
(1) $P_0(a)$
(2) $\neg P_1(a)$
(3) $P_2(b) \vee P_1(x)$
(4) $P_3(bx) \to P_1(x)$
(5) $P_0(x) \wedge A(y) \to B(xy)$
(6) $P_2(x) \wedge B(y) \to C(xy)$
(7) $A(c) \vee P_3(x)$
(8) $C(cx) \to P_3(x)$

Initial translation:
$P_0(x) \qquad [\alpha \to \beta]^*(x)$

$P_1(x) \quad \begin{bmatrix} (\beta \to \gamma) \\ \to (\alpha \to \gamma) \end{bmatrix}^*(x)$

$P_2(x) \qquad [\beta \to \gamma]^*(x)$
$P_3(x) \qquad [\alpha \to \gamma]^*(x)$

Derivation:
(9)   (Split (3)) $P_2(b)$
(10)   (Split (7)) $A(c)$
(11)   (Hyper (5)) $B(ac)$

(12) (Hyper (6)) $C(abc)$
(13) (Hyper (8)) $P_3(ab)$
(14) (Hyper (4)) $P_1(a)$
(15) (End)
(16) (Split (7)) $P_3(x)$
(17) (Hyper (4)) $P_1(x)$
(18) (End)
(19) (Split (3)) $P_1(x)$
(20) (End)

Fig. 2. Refutation of $(\alpha \to \beta) \to ((\beta \to \gamma) \to (\alpha \to \gamma))$ in L_{CLDS} using AlgMG

clauses have been removed. This means that in the initial call to dp, S contains neither subsumed clauses nor tautologies - the latter because of the way the clauses are originally formed. This property will be maintained throughout. In $dp1$ the first argument is the current (recognised) set of positive unit clauses, which is assumed to be empty at the start.[4] The predicates used in the Prolog version of AlgMG can be interpreted as follows ((S, M) represents the list of all clauses in S and M):

add(New,S,M,NewS,NewM) holds iff the units in New derived from the (Hyper) rule are added to M to form $NewM$ and disjunctions in New are added to S to form $NewS$.

and(X,Y,Z) holds iff $Z = X \wedge Y$.

endApplies(S,M) holds iff (End) can be applied to (S, M).

hyper(M,S,New) holds iff New is a set of hyper-resolvents using unit clauses in M and a clause in S, that do not already occur in M. The labels of any new hyper-resolvents are subject to a size restriction (see later), in order that there are not an infinite number of hyper-resolvents.

join(F1,F2,F) holds iff F is the union of $F1$ and $F2$.

noRulesApply(M,S) holds iff there are no applicable rules to (M, S).

remove(P,S,NewS) holds iff clause P is removed from S to give $NewS$.

simplify(M,S,NewS) holds iff clauses in S can be simplified to $NewS$ by units in M.

split(M,S,NewS, S1,S2) holds iff $S1 \vee S2$ is removed from S to leave $NewS$.

subsumed(C,M,S) holds if Clause C in S is subsumed by clauses from S or M.

Examples. Two examples of refutations appear in Figs. 2 and 3, in which the LL theorem $(\alpha \to \beta) \to ((\beta \to \gamma) \to (\alpha \to \gamma))$ and the RL theorem $(\alpha \to \beta) \to$

---

[4] In case the initial goal is to be shown from some data, in the start clause this initial data would be placed in the first argument of dp1.

Initial clauses:

(1)	$P_0(a)$	(6)	$P_1(x) \wedge B(y) \rightarrow C(xy)$
(2)	$P_1(b)$	(7)	$P_2(x) \wedge P_3(y) \rightarrow P_4(xy)$
(3)	$P_2(c)$	(8)	$P_4(x) \wedge P_3(y) \rightarrow D(xy)$
(4)	$\neg D(abc)$	(9)	$A(d) \vee P_3(x)$
(5)	$P_0(x) \wedge A(y) \rightarrow B(xy)$	(10)	$C(dx) \rightarrow P_3(x)$

Initial translation:

$$P_0(x) \, [\alpha \rightarrow \beta]^*(x) \qquad\qquad P_3(x) \, [\alpha \rightarrow \gamma]^*(x)$$
$$P_1(x) \, [\beta \rightarrow \gamma]^*(x) \qquad\qquad P_4(x) \, [(\alpha \rightarrow \gamma) \rightarrow \delta]^*(x)$$
$$P_2(x) \, [(\alpha \rightarrow \gamma) \rightarrow ((\alpha \rightarrow \gamma) \rightarrow \delta)]^*(x)$$

Derivation:

(11)	(Split (9))	$A(d)$	(17)	(End)
(12)	(Hyper (5))	$B(ad)$	(18)	(Split(9))	$P_3(x)$
(13)	(Hyper (6))	$C(bad)$	(19)	(Hyper (7))	$P_4(cx)$
(14)	(Hyper (10))	$P_3(ba)$	(20)	(Hyper (8)	$D(cx)$
(15)	(Hyper (7))	$P_4(bac)$	(21)	(End)
(16)	(Hyper (8))	$D(bacba)$		

Fig. 3. Refutation in R_{CLDS} using AlgMG

$((\beta \rightarrow \gamma) \rightarrow ((\alpha \rightarrow \gamma) \rightarrow ((\alpha \rightarrow \gamma) \rightarrow \delta))) \rightarrow \delta)$ are, respectively, proved. For ease of reading, the parameters used are called $a, b, c, \ldots$ instead of having the form $c_{\alpha \rightarrow \beta}$, etc. and the predicates A, B and C are used in place of $[\alpha]^*$, $[\beta]^*$ and $[\gamma]^*$. In Fig. 2, the (translation of the) data $\alpha \rightarrow \beta$ is added as a fact and the goal is (the translation of) $(\beta \rightarrow \gamma) \rightarrow (\alpha \rightarrow \gamma)$. In Fig. 3, the initial data $\alpha \rightarrow \beta$, $\beta \rightarrow \gamma$ and $(\alpha \rightarrow \gamma) \rightarrow ((\alpha \rightarrow \gamma) \rightarrow \delta)$ are added as facts. The goal in this case is δ. These arrangements simply make the refutations a little shorter than if the initial goal had been the immediate translation of the theorem to be proved.

The calls to $dp1$ can be arranged into a tree, bifurcation occurring when the (Split) rule is used. In the derivations each line after the list of initial clauses records a derived clause. Derived unit clauses would be added to an accumulating partial model M, which is returned in case of a branch ending in failure. In Fig. 2, for example, there are three branches in the tree of calls to $dp1$, which all contain lines (1) - (8) implicitly and terminate using the (End) rule. The first branch contains lines (9) - (15), the second contains lines (9), (16) - (18), and the third contains lines (19), (20). Deletions of clauses due to purity and subsumption, and of literals due to simplify are not made, for the sake of simplicity. However, line (17) could have been achieved by a (Simplify) step instead. A possible subsumption step after line (16) is the removal of clauses (7) and (8).

Notice that, in Fig. 2 only some of the appropriate axioms have been included. It might be expected that clauses derived from both halves of the appropriate equivalence schemas would be included, resulting in the inclusion of, for instance, $P_3(x) \wedge A(y) \rightarrow C(xy)$. However, it is only necessary to include a restricted number of clauses based on the polarity of the sub-formula occurrences.

4 Main Results

4.1 Termination of AlgMG

In this section suitable termination criteria are described for the (Hyper) rule of AlgMG for the two logics in question, Linear Logic and Relevance Logic. A different condition is imposed for each of L_{CLDS} and R_{CLDS} and in such a way that termination of a branch without the use of (End) will not cause loss of soundness. That is, AlgMG will terminate a branch without (End) only if the original goal to be checked is not a theorem of LL (or RL). It is assumed that the translation of the initial goal α is initially included in the list S in AlgMG in the form $\neg[\alpha]^*(1)$. The termination conditions for the two logics are, at first sight, rather similar; however, the condition for LL uses a global restriction, whereas that for RL uses local restrictions, dependent on the particular development of the AlgMG tree.

When forming the translation of a configuration, clauses corresponding to axiom (Ax3c) for which the same wff α is involved all make use of the same parameter c_α. The number of occurences of a non-unit parameter c_α for wff α in an instance of axiom (Ax3c) is called the *relevant index* of c_α and is denoted by m_α. For example, in case an occurrence of axiom (Ax3c) is made for the two wffs $\alpha \to \beta$ and $\alpha \to \gamma$, then the two occurrences would be $[\alpha]^*(c_\alpha) \vee [\beta]^*(x)$ and $[\alpha]^*(c_\alpha) \vee [\gamma]^*(x)$ and $m_\alpha = 2$.

Definition 11. *Let L_{CLDS} be a propositional Linear LDS based on the languages $\mathcal{L}_P$ and $\mathcal{L}_L$, and S be a set of clauses appropriate for showing the wff α. The finite subset of terms in $\mathrm{Func}(\mathcal{L}_P, \mathcal{L}_L)$ that mentions only parameters in S and does not include any non-unit parameter c_α more times than its relevant index m_α is called the* restricted Linear Herbrand Universe H_L. *The restricted set of ground instances S_{H_L} is the set of ground instances of clauses in S such that every argument is in H_L. The restricted atom set B_{H_L} is the set of atoms using predicates mentioned in S and terms in H_L.*

Termination in L_{CLDS}. The criterion to ensure termination in L_{CLDS} is as follows:

Let $\mathcal{B}$ be a branch of a tree generated by AlgMG; an atom $L(w)$ may be added to $\mathcal{B}$ only if it is not subsumed by any other atom in $\mathcal{B}$ and has a ground instance $L(u)$, where $u \in H_L$.

(Notice that any atom of the form $P(ux)$, where u contains every parameter exactly m_α times, has only one ground instance, $P(w)$, such that $w \in H_L$. This instance occurs when $x = 1$ and $w = u$. This atom would therefore only be added to $\mathcal{B}$ if not already present.)

The above criterion places an upper bound on the potential size of u such that, at worst, there can be $\Pi(m_{\alpha_i} + 1)$ atoms for each predicate in any branch, where m_{α_i} are the relevant indices for non-unit parameters c_{α_i}. There is one predicate for each subformula in α, the given formula to be tested. In fact, for

LL, it is possibly simpler to use a more restrictive translation, in which a different parameter is introduced for *each* occurrence of α. Then the relevant index of any non-unit parameter is always 1, and the terms in H_L are restricted to containing any non-unit parameter at most once. The formula for the number of atoms then reduces to 2^n, where n is the number of non-unit parameters introduced by the translation. In practice there are fewer than this maximum due to subsumption.

If AlgMG is started with an initial set of sentences S appropriate for showing α and termination occurs with (End) in all branches, then, as is shown in Sect. 5, α is a theorem of LL. On the other hand, suppose termination of a branch $\mathcal{B}$ occurs without using (End), possibly because of the size restriction. Then a model of S_{H_L} can be constructed as follows: Assign true to each atom in B_{H_L} that occurs also in $\mathcal{B}$ or that is subsumed by an atom in $\mathcal{B}$, and false to all other atoms in B_{H_L}.

For illustration, if the example in Fig. 3 were redone using L_{CLDS}, then the step at line (16) would not have been generated, nor could the branch be extended further; the atoms in it can be used to obtain a finite model of the initial clauses. The following atoms would be assigned true:

$$P_0(a), P_1(b), P_2(c), A(d), B(ad), C(bad), P_3(ba), P_4(bac)$$

and all other atoms in B_L would be assigned false. It is easy to check that this is a model of the ground instances of clauses (1) - (10) whose terms all lie in H_L.

Suppose that each clause C in S is modified by the inclusion of a new condition of the form $restricted(x)$, one condition for each variable x in C. The atom $restricted(x)$ is to be interpreted as true exactly if x lies within H_L. It is easy to show that the set of modified clauses is unsatisfiable over $\mathcal{D}_S$ iff the set S_{H_L} is unsatisfiable. This property will be exploited when proving the correspondence of L_{CLDS} with LL.

Termination in R_{CLDS}. In the case of RL, the termination is complicated by the presence of contraction, illustrated in the example in Fig. 3, where the atom $D(bacba)$, derived at line (16), includes the parameter b more than m_b times $(m_b = 1)$.[5] The restriction dictating which atoms to generate by (Hyper) in R_{CLDS} uses the notion of *relevant set*, which in turn uses the notion of *full* labels. Unlike the case for LL, there is no easily stated global restriction on labels (such as that indicated by $restricted(x)$). The criterion described below was inspired by the description given in [16] for the relevant logic LR.

Definition 12. *Let R_{CLDS} be a propositional relevant LDS based on $\mathcal{L}_P$ and $\mathcal{L}_L$ and S be a set of clauses appropriate for showing α. A ground label in $\mathcal{L}_L$, that mentions only parameters in S and in which every non-unit parameter a occurs at least m_a times, is called* full. *A ground label in $\mathcal{L}_L$, that mentions only parameters in S and is not full, is called* small. *A parameter a that occurs in a*

---

[5] The inclusion of $P(b)$ in the data is due to an implied occurrence of axiom (Ax3c) and there is just one such implicit occurrence.

small label, but less than m_a times, belongs to its small part. *A parameter a that occurs in a label (either full or small) at least m_a times belongs to its* full part. *A ground atom having a predicate occurring in S that has a full/small label is also called a full/small atom.*

Definition 13. *Let R_{CLDS} be a propositional relevant LDS based on $\mathcal{L}_P$ and $\mathcal{L}_L$ and S be a set of clauses appropriate for showing α. Suppose that $\mathcal{B}$ is a branch derived from the application of AlgMG such that no subsumption steps can be made to $\mathcal{B}$ and let $P(u_1)$ be a ground atom occurring in $\mathcal{B}$. The relevant set of $P(u_1)$ (relative to $\mathcal{B}$), is the set of ground atoms $P(u_2)$ such that: only parameters occurring in S occur in u_2 and either, (i) there is at least one non-unit parameter a in $P(u_1)$ occuring k times, $0 \le k < m_a$, that also occurs in $P(u_2)$ more than k times, or, (ii) there is at least one non-unit parameter a in $P(u_1)$ occuring k times, $1 \le k$, that occurs in $P(u_2)$ zero times.*

As an example, suppose there are two parameters a and b and that $m_a = 2$ and $m_b = 3$, then the relevant set of $P(aab)$ $(=P(a^2b))$ is the set of atoms of one of the forms: $P(a^rb^2)$, $P(a^rb^3)$, $P(a^rb^p)$, where $r \ge 1$, $p \ge 4$, or $P(b^s)$, $P(a^s)$, where $s \ge 0$. The relevant set of the full atom $P(a^2b^3)$ is the set of atoms of the form $P(a^s)$ or $P(b^s)$, where $s \ge 0$.

If $P(w)$ is not ground, then the relevant set is the intersection of the relevant set of each ground instance of $P(w)$. The criterion to ensure termination in R_{CLDS} can now be stated.

> In R_{CLDS} the (Hyper) rule is restricted so that a ground atom $P(w)$ is only added to a branch $\mathcal{B}$ if (i) it is not subsumed by any literal in $\mathcal{B}$ and (ii) it belongs to the relevant set of every other P-atom in $\mathcal{B}$.

In other words, if $P(w)$ is added to a branch, then for every atom $P(z)$ in the branch, either the number of occurrences of at least one non-unit parameter a in z that occurs fewer than m_a times is increased in w, or some non-unit parameter in z is reduced to zero in w. Notice that, if there are no P-atoms in the branch, then $P(w)$ can be added vacuously according to the criterion. In case the (Hyper) rule generates a non-ground atom, then as long as it is not subsumed and some ground instance of it satisfies property (ii) above it can be added to the branch.

Although relevant sets are (countably) infinite, the impact of all relevant sets having to include any new literal in a branch is quite strict and very quickly reduces the number of possibilities to a finite number. For instance, a literal $P(u)$ in a branch with a small label $u = u_1u_2$, where u_1 is the small part of u, will prevent any other literal $P(u')$, where the small part of u' is subsumed by u_1, from being added to the branch. For instance, if $P(a^4b^2)$ belongs to a branch, and $m_a = 2$, $m_b = 3$, then no literal of the form $P(a^sb^2)$ or $P(a^sb)$, $s \ge 1$, can be added to the branch. If $m_a = m_b = 2$, then no literal of the form $P(a^sb^r)$ can be added, $s \ge 1, r \ge 1$. For any particular set of initial clauses there are only a finite number of labels that can occur as small parts of labels. This observation means that the maximum number of literals in a branch will be finite. It also allows for the following definition of measure for a branch that decreases with each new atom added to the branch.

Definition 14. *Let* $\langle \langle \mathcal{L}_P, \mathcal{L}_L \rangle, \mathcal{A}_R^+, AlgMG \rangle$ *be a* R_{CLDS} *and* S *be a set of clauses appropriate for showing* α. *The* relevant measure *of the positive atoms in a branch* $\mathcal{B}$ *derived using AlgMG, with no pairwise subsumption, is defined as the sum, over each predicate* P *in* S, *of the number of possible small parts of labels that do not occur in any* P-*literal in* $\mathcal{B}$ *or in any* P-*literal subsumed by a literal in* $\mathcal{B}$.

It is easy to see that, when a new atom $P(w)$ is added to a branch $\mathcal{B}$ by AlgMG, then the relevant measure will decrease. Eventually, either (i) (End) will be applied to $\mathcal{B}$, or (ii) the measure of $\mathcal{B}$ will have been reduced to zero, or (iii) no further steps are possible using AlgMG. For example, suppose that branch $\mathcal{B}$ includes just the atom $P(a^2b)$, that there is one predicate P and two parameters a and b each with a relevant index of 2. The relevant measure is 7, since the small parts a^2b and ab are, respectively, covered by $P(a^2b)$ and $P(ab)$, subsumed by $P(a^2b)$. If $P(a^2b^2)$ is now added then the branch measure is reduced to 5. Also, the literal $P(a^2b)$ would be subsumed.

In summary, in applying AlgMG, an atom can be added to a branch as long as it respects the following (informal) criterion:

$\mathbf{L_{CLDS}}$ An atom is added to a branch $\mathcal{B}$ only if the ground part of its label belongs to H_L and if it is not subsumed by any atom in $\mathcal{B}$.

$\mathbf{R_{CLDS}}$ An atom $P(w_1)$ is added to a branch $\mathcal{B}$ only if it has a ground instance which belongs to some relevant set of every atom in $\mathcal{B}$ and if it is not subsumed by any atom in $\mathcal{B}$. In practice, this means that $P(w_1)$ is not subsumed, and, for each atom $P(w_2)$, it must either increase the number of occurrences of at least one non-full parameter in w_2, or it must reduce the number of occurences of at least one non-unit parameter in w_2 to zero.

4.2   Properties of AlgMG.

There are several properties that hold about the relationship between the Semantics given by the Axioms in the Extended Labelling Algebra $\mathcal{A}_L^+$ and the procedure AlgMG, which are stated in Theorem 1. A proof of these properties can be made in a similar way to that given in [4] for IL. An outline is given here, including in detail the new cases for the two logics LL and RL.

Theorem 1 (Properties of AlgMG). *Let* S *be a* L_{CLDS}, α *be a propositional LL formula,* $\mathcal{A}_L^+(\alpha)$ *be the particular clauses and instances of the Semantic Axioms for showing* α *and* $\mathcal{G}_\alpha = \mathcal{A}_L^+(\alpha) \cup \{\neg[\alpha]^*(1)\}$. *Let AlgMG be initiated by the call* $dp(\mathcal{G}_\alpha, F, R)$ *for variables* F *and* R, *then the following properties hold:*

1. *If AlgMG returns* $R = true$ *then* $\mathcal{G}_\alpha \models_{FOL}$.
2. *If AlgMG returns* $R = false$ *then* F *is a partial model of* $\mathcal{G}_\alpha$, *in a way to be explained.*
3. *AlgMG terminates.*

4. If α is also a Hilbert theorem of propositional LL (i.e. α can be derived from the Hilbert Axioms for LL and Modus Ponens), then $\mathcal{G}_\alpha \models_{FOL}$.

5. If $\mathcal{G}_\alpha \models_{FOL}$ then α is a theorem of LL.

Similar properties hold for RL.

In AlgMG every step (except (Hyper)) reduces the total number of literals in $M \cup S$. However, the number of (Hyper) steps is restricted to a finite number in RL by the use of relevant sets and in LL by the restriction of terms to belong to H_L. Exactly the same proof for termination of AlgMG as in [4] can then be used.

Properties (1) and (2) are soundness and completeness results for AlgMG, in the sense that they show that the algorithm is correct with respect to finding refutations. These properties can be proved as in [4], except for the case of clause 1, the case that covers extending the resulting value of F to become a model of the clauses S, which is detailed in the proof of Lemma 1. Properties (4) and (5) show that AlgMG corresponds with LL, (4) showing it gives a refutation for any theorem of LL, and (5) showing that it only succeeds for theorems. Similarly for RL. Proofs of these properties can be made following the same proof structure as in [4], but with some changes to cope with the different logics. Lemmas 2 and 3 give the details for the two logics considered in this paper.

5   Proving the Properties of AlgMG

Proving Properties 1 and 2. Properties (1) and (2) of AlgMG are proved by showing that the following proposition, called (PROP1and2) holds for each clause of (dp1):

   if the $dp1$ conditions of the clause satisfy invariant (INV) and the other conditions are also true, then the $dp1$ conclusion of the clause satisfies (INV) also, where (INV) is

   Either, $R = $ false, $M \subseteq F$ and F can be extended to a model of S or, $R = $ true, $F = [\ ]$ and $M \cup S$ have no Herbrand models.

For the case of L_{CLDS}, when $R = $ false F is extended to be a model of the ground instances of S, taken over the domain of the initial clauses set of clauses S, S_{H_L}, which are called restricted ground instances in the Lemma below.

Note that, for the (End) clause in L_{CLDS}, when $R = $ true, it is the set of restricted ground instances of $M \cup S$ that has no models. This implies that $M \cup S$ also has no Herbrand models, for any such model would also be a model of the restricted instances. (It suffices to deal with Herbrand models since non-existence of a Herbrand model of S implies the non-existence of any model of S (see, for example, [7]).)

Lemma 1. *The* `fail` *clause of dp1 satisfies (PROP1and2).*

Proof. The details of the proof are different for each of the two logics. For LL a model of restricted ground instances is found, whereas for RL a Herbrand model is given.

R is false; all rules have been applied and $F = M$. Certainly, $M \subseteq F$. There are then two cases: for LL and for RL.

Case for Linear Logic. The set F is extended to be a model M_0 of the restricted ground instances of the clauses remaining in S as follows: Any ground atom with label in H_L that is subsumed by a literal in M is true in M_0. All other ground atoms with label in H_L are false in M_0.

The clauses left in S can only generate subsumed clauses, disallowed atoms or they are a negative unit. Assume that there is a restricted ground instance of a non-negative clause C in S that is false in M_0. That is, for some instance C', of C, its condition literals are true in M_0 and its conclusion is false in M_0. If the conclusion is a single literal then, as (Hyper) has been applied to C already, the conclusion is either true in M, and hence in M_0, or it is subsumed by a clause in M, and again is true in M_0. Both contradict the assumptions. If the conclusion is a disjunction, then (Split) must have eventually been applied and the conclusion will again be true in M, or the disjunction is subsumed by a literal in M, contradicting the assumption. In case $C = \neg L$ is a false negative unit clause in S, then some instance $C' = \neg L'$ is false, or L' is true in M_0. But in that case (End) would have been applied, a contradiction. The model obtained is a model of the clauses remaining when no more steps are possible in some chosen branch.

Case for Relevant logic. Let the set of atoms formed using predicates in the initial set of clauses S and labels drawn from the Herbrand Domain of S, $\mathcal{D}_S$, be called B_S. A model M_0 of the atoms in B_S is assigned, using the atoms in M, by the following assignment conditions:

(i) Any ground atom in B_S that is subsumed by an atom in M is true in M_0.

(ii) Any ground atom in B_S that subsumes an atom L in M by contraction of parameters in the full part of L only, is true in M_0.

(iii) All other ground atoms in B_S are false in M_0.

Assume that there is a ground instance of a non-negative clause C in S that is false in M_0. That is, for some instance C', of C, its condition literals are true in M_0 and its conclusion is false in M_0. If the conclusion is a single literal then, as (Hyper) has been applied to C already, the conclusion L is either true in M, and hence in M_0, or it is subsumed by a clause in M, and again is true in M_0, or it is disallowed. Both the first two circumstances contradict the assumption. For the third circumstance, since L is disallowed, there is some literal L', in M or subsumed by a literal in M, which is subsumed by L by contracting only parameters that occur in the full part of L'. But then by assignment condition (ii) both L' and L are assigned true, again contradicting the assumption. The remainder of the proof is as given for L_{CLDS}.

An example of a failed refutation in RL is given in Fig. 4, in which there is an attempt to show $(\alpha \rightarrow \alpha) \rightarrow (\alpha \rightarrow (\alpha \rightarrow \alpha))$. For this problem there are two parameters a and b with respective relevant indices $m_a = 1$ and $m_b = 2$. In the

Initial clauses: (7) $A(b) \vee P_2(x)$

Initial translation: (1) $\neg P_0(1)$ (8) $P_1(bx) \rightarrow P_2(x)$

$P_0(x) \quad \left[\dfrac{(\alpha \rightarrow \alpha) \rightarrow}{(\alpha \rightarrow (\alpha \rightarrow \alpha))} \right]^* (x)$ (2) $P_2(ax) \rightarrow P_0(x)$ Derivation:

(3) $P_1(a) \vee P_0(x)$ (9) (Split (3)) $P_1(a)$

$P_1(x) \qquad [\alpha \rightarrow \alpha]^*(x)$ (4) $P_1(x) \wedge A(y) \rightarrow A(xy)$ (10) (Split (5)) $A(b)$

$P_2(x) \qquad [\alpha \rightarrow (\alpha \rightarrow \alpha)]^*(x)$ (5) $A(b) \vee P_1(x)$ (11) (Hyper (4)) $A(ab)$

(6) $A(bx) \rightarrow P_1(x)$ (12) (Hyper (6)) $P_1(1)$

Fig. 4. Failed refutation in $\mathcal{A}_R$ using AlgMG

branch (9) - (12) any further literals generated using (4), such as $A(a^2b)$, are not allowed as they are not a member of the relevant set of $A(ab)$. The atoms $P_1(a)$, $P_1(1)$, $A(b)$ and $A(ab)$ are assigned true, as are $P_1(a^k)$ and $A(a^kb)$, $k \geq 2$. All others are assigned false. Note that atoms of the form $A(b^k)$, $k \geq 2$, are not assigned true by assignment condition (ii), based on atom $A(b)$, because neither b nor a occur in the full part, which is just the parameter 1. The reader can check that this is a model for the clauses (1)-(8).

The number of atoms in a branch for each predicate depends on how soon atoms with full parts are derived for that predicate. If, for example, there are two parameters a and b, $m_a = 2$ and $m_b = 3$, then if $P(aabbb)$ happened to be derived immediately, no other P atoms with both a and b in the label would be derived. Those with fewer occurrences (at least one of each parameter) would be prevented by subsumption, whereas those with more occurrences would be prevented by the termination restriction (ii). On the other hand, the worst case number of P atoms generated, with at least one occurrence of each parameter in the label, would be 6; for example, the following generation order would require all 6: $ab, a^2b, ab^2, a^2b^2, ab^3, a^2, b^3$.

5.1 Proving Correspondence of L$_{\text{CLDS}}$/R$_{\text{CLDS}}$ with LL/RL

In order to show that the refutation system L$_{\text{CLDS}}$ presented here does indeed correspond to a standard Hilbert axiom presentation for Linear Logic it is necessary to show that theorems derived within the two systems are the same (Properties 4 and 5 of Theorem 1). Similarly for R$_{\text{CLDS}}$ and Relevant Logic. The complete set of axioms used in the implication fragments of LL and RL is shown in Table 2. Axioms (I2), (I3) and (I4) correspond, respectively, to contraction, distributivity and permutation. A useful axiom, (I5), is derivable also from (I3) and (I4) and is included for convenience. All axioms are appropriate for RL, whereas (I2) is omitted for LL. Respectively, Theorems 2 and 3 state that theorems in LL and RL derived from these axioms together with the rule of Modus Ponens (MP) are also theorems of AlgMG, and that theorems of L$_{\text{CLDS}}$ and R$_{\text{CLDS}}$ are also theorems in the Hilbert System(s).

Table 2. The Hilbert axioms for I_{CLDS}

$\alpha \to \alpha$	(I1)	$(\alpha \to (\beta \to \gamma)) \to (\beta \to (\alpha \to \gamma))$	(I4)
$(\alpha \to (\alpha \to \beta)) \to (\alpha \to \beta)$	(I2)	$(\alpha \to \beta) \to ((\beta \to \gamma) \to (\alpha \to \gamma))$	(I5)
$(\alpha \to \beta) \to ((\gamma \to \alpha) \to (\gamma \to \beta))$	(I3)		

Correspondence Part I. Property (4) of AlgMG is shown in Theorem 2. An outline proof is given. For RL the appropriate Hilbert axioms are (Ax1) - (Ax5); (Ax2) is omitted for LL.

Theorem 2. *Let P be a Hilbert theorem of LL then the union of $\{\neg[P]^*(1)\}$ and the appropriate set of instances of the semantic axioms (equivalences) for $\neg[P]^*(1)$, P_S, has no models in H_L. (For RL, P_S has no models.)*

Proof. (Outline only.) The proof is essentially the same for both logics. Let P_S be the set of defining equivalences for P and its subformulas, $\forall x[[P]^*(x) \leftrightarrow R(x)]$ be the defining equivalence for $[P]^*$ and $\forall x[[P]^*(x) \leftrightarrow T_P(x)]$ be the resulting equivalence after replacing every occurrence in $R(x)$ of an atom that has a defining equivalence in P_S by the right-hand side of that equivalence. It is shown next that $T_P(1)$ is always true and hence that there are no models of P_S and $\neg[P]^*(1)$.

This property of $T_P(1)$ is shown by induction on the number of (MP) steps in the Hilbert proof of P. In case P is an axiom and uses no applications of (MP) in its proof then the property can be seen to hold by construction. For instance, in the case of the contraction axiom (I2), $T_{(I2)}(1)$ is the sentence

$$\forall y(\forall zv([\alpha]^*(z) \wedge [\alpha]^*(v) \to [\beta]^*(zyv)) \to \forall u([\alpha]^*(u) \to [\beta]^*(uy)))$$

In the case of LL, the equivalences include also the *restricted* predicate (shortened to r in the illustration below). For the permutation axiom (I4), $T_{(I4)}(1)$, after some simplification[6], is the sentence

$$\forall z \left(\begin{array}{l} \forall y([\alpha]^*(y) \to \forall v(r(zyv) \to ([\beta]^*(v) \to [\gamma]^*(zyv)))) \to \\ \forall u([\beta]^*(u) \to \forall w(r(zuw) \to ([\alpha]^*(u) \to [\gamma]^*(zuw)))) \end{array} \right)$$

Let the property hold for all theorems that have Hilbert proofs using $< n$ applications of (MP), and consider a theorem P such that its proof uses n (MP) steps, with the last step being a derivation from P' and $P' \to P$. By hypothesis, $T_{P'}(1)$ is true, and $T_{P' \to P}(1)$ is true.

Hence, since $\forall x[T_{P' \to P}(x) \leftrightarrow \forall u[T_{P'}(u) \to T_P(ux)]]$, then $T_P(1)$ is also true.

The contrapositive of Theorem 2 allows the conclusion that P is not a theorem to be drawn from the existence of a model for $\{\neg[P]^*(1)\} \cup P_S$ as found by a terminating AlgMG.

---

[6] In particular, $restricted(xy)$ implies also $restricted(x)$ and $restricted(y)$.

Correspondence Part II. To show that every formula classified as a theorem by AlgMG in R_{CLDS} or L_{CLDS} is also derivable using the appropriate Hilbert axioms and the rule of Modus Ponens, Theorem 3 is used.

Theorem 3. *Let $\mathcal{G}_\alpha$ be the set of instances of $\mathcal{A}_L^+$ for showing α (not including $\neg[\alpha]^*(1)$), then if there exists an AlgMG refutation in L_{CLDS} of $\mathcal{G}_\alpha \cup \neg[\alpha]^*(1)$ then there is a Hilbert proof in LL of α, which is therefore a theorem of LL. That is, if $\mathcal{G}_\alpha, \neg[\alpha]^*(1) \models_{FOL}$ then $\vdash_{HI} \alpha^7$. Similarly for R_{CLDS} and RL.*

Proof. Suppose $\mathcal{G}_\alpha, \neg[\alpha]^*(1) \models_{FOL}$, hence any model of $\mathcal{G}_\alpha$ is also a model of $[\alpha]^*(1)$; it is required to show $\vdash_{HI} \alpha$. Lemma 2 below states there is a model M of $\mathcal{A}_L^+$ ($\mathcal{A}_R^+$), and hence of $\mathcal{G}_\alpha$, with the property that $[\alpha]^*(1) = true$ iff $\vdash_{HI} \alpha$. Therefore, since M is a model of $\mathcal{A}_L^+$ ($\mathcal{A}_R^+$) it is a model of $[\alpha]^*(1)$ and hence $\vdash_{HI} \alpha$ is true, as required. The desired model is based on the canonical interpretation introduced in [1].

Definition 15. *The canonical interpretation for L_{CLDS} is an interpretation from $\mathrm{Mon}(\mathcal{L}_P, \mathcal{L}_L)$ onto the power set of $\mathcal{L}_P$ defined as follows:*

- $||c_\alpha|| = \{z : \vdash_{HI} \alpha \to z\}$, *for each parameter c_α;*
- $||\lambda \circ \lambda'|| = \{z : \vdash_{HI} \alpha \wedge \beta \to z\} = \{z : \vdash_{HI} \alpha \to (\beta \to z)\}$, *where $\alpha \in ||\lambda||$ and $\beta \in ||\lambda'||$;*
- $||1|| = \{z : \vdash_{HI} z\}$ *and*
- $|| \preceq || = \{(||x||, ||y||) : ||x|| \subseteq ||y||\}$;
- $||[\alpha]^*|| = \{||x|| : \alpha \in ||x||\}$;

Similarly for R_{CLDS}. For the case of L_{CLDS} an interpretation of the *restricted* predicate is also needed. This depends on the particular theorem that is to be proven, as it makes use of the relevant indices of the parameters occurring in the translated clauses. The interpretation is given by:

$$||restricted|| = \{||x|| :$$
$$\forall z (z \in ||x|| \to z \text{ is provable using } \leq m_{\alpha_i} \text{ occurrences of } \alpha_i)\}$$

In other words, $restricted(x) = true$ iff x includes $\leq m_{\alpha_i}$ occurrences of parameter α_i. (In case a new parameter is used for each instance of Axiom (Ax3c) then the definition does not depend on the particular theorem to be proven as $m_{\alpha_i} = 1$ for every c_{α_i}.)

The canonical interpretation is used to give a Herbrand model for $\mathcal{A}_I^+$ ($\mathcal{A}_R^+$), by setting $[\alpha]^*(x) = true$ iff $\alpha \in ||x||$. This means, in particular, that if $[\alpha]^*(1) = true$ then $\alpha \in ||1||$ and hence $\vdash_{HI} \alpha$. The following Lemma states that the canonical interpretation of Definition 15 is a model of $\mathcal{A}_I^+$ ($\mathcal{A}_R^+$).

Lemma 2. *The properties of the labelling algebra $\mathcal{A}_L$ ($\mathcal{A}_R$) given in Definition 2 and the semantic axioms of $\mathcal{A}_L^+$ ($\mathcal{A}_R^+$) are satisfied by the canonical interpretation for L_{CLDS} (R_{CLDS}).*

---

[7] The notation $\vdash_{HI} \gamma$ indicates that γ is provable using the appropriate Hilbert axioms.

Proof. Each of the properties of the labelling algebra is satisfied by the canonical interpretation. For R_{CLDS} the case for contraction is given here. The other cases are as given in [4]. For L_{CLDS} the case for Axiom (Ax3a) is given. The other cases are as given in [4] but modified to include the *restricted* predicate.

contraction Suppose that $\delta \in ||\lambda|| \circ ||\lambda||$. Then there is a Hilbert proof of $\alpha \to (\alpha \to \delta)$, where $\alpha \in ||\lambda||$. By axiom (I2) $\vdash_{HI} \alpha \to \delta$ and $\delta \in ||\lambda||$.

(Ax3a) Let the maximum number of parameter occurrences allowed be fixed by the global relevant indices for the particular theorem to be proved. Suppose $restricted(x)$, $restricted(y)$ and $restricted(xy)$ and that $\alpha \in ||x||$ and $\alpha \to \beta \in ||y||$. Then there are Hilbert proofs of $\delta \to \alpha$ and $\gamma \to \alpha \to \beta$ for $\delta \in ||x||$ and $\gamma \in ||y||$ such that no more than the allowed number of subformula occurrences, as given by the relevant indices for the problem, are used in the combined proofs of δ and γ. To show $\delta \to (\gamma \to \beta)$, and hence $\beta \in ||x \circ y||$, use axioms (I4) and (I5).

6   Conclusions

In this paper the method of Compiled Labelled Deductive Systems, based on the principles in [9], is applied to the two resource logics, LL and RL. The method of CLDS provides logics with a uniform presentation of their derivability relations and semantic entailments and its semantics is given in terms of a translation approach into first-order logic. The main features of a CLDS system and model theoretic semantics are described here. The notion of a configuration in a CLDS system generalises the standard notion of a theory and the notion of semantic entailment is generalised to relations between structured theories. The method is used to give presentations of L_{CLDS} and R_{CLDS}, which are seen to be generalisations, respectively, of Linear and Relevance Logic through the correspondence results in Sect. 5, which shows that there is a one-way translation of standard theories into configurations, while preserving the theorems of LL and RL.

The translation results in a compiled theory of a configuration. A refutation system based on a Model Generation procedure is defined for this theory, which, together with a particular unification algorithm and an appropriate restriction on the size of terms, yields a decidability test for formulas of propositional Linear Logic or Relevance Logic. The main contribution of this paper is to show how the translation approach into first order logic for Labelled Deductive Systems can still yield decidable theories. This meets one of the main criticisms levelled at LDS, and at CLDS in particular, that for decidable logics the CLDS representation is not decidable.

The method used in this paper can be extended to include all operators of Linear Logic, including the additive and exponential operators. For instance, the axiom for the additive disjunction operator $\vee$ in LL is

$$\forall x([\alpha \vee \beta]^*(x) \leftrightarrow \forall y(([\alpha \to \gamma]^*(y) \wedge [\beta \to \gamma]^*(y)) \to [\gamma]^*(x \circ y)))$$

From an applicative point of view, the CLDS approach provides a logic with reasoning which is closer to the needs of computing and A.I. These are in fact application areas with an increasing demand for logical systems able to represent and to reason about *structures* of information (see [9]). For example in [3] it is shown how a CLDS can provide a flexible framework for abduction.

For the automated theorem proving point of view, the translation method described in Section 2.2 facilitates the use of first-order therem provers for deriving theorems of the underlying logic. In fact, the first order axioms of a CLDS extended algebra $\mathcal{A}_S^+$ can be translated into clausal form, and so any clausal theorem proving method might be appropriate for using the axioms to automate the process of proving theorems. The clauses resulting from the translation of a particular configuration represent a partial coding of the data. A resolution refutation that simulates the application of natural deduction rules could be developed, but because of the simple structure of the clauses resulting from a subtructural CLDS theory the extended Model Generation method used here is appropriate.

References

1. M. D'Agostino and D. Gabbay. A generalisation of analytic deduction via labelled deductive systems. Part I: Basic substructural Logics. Journal of Automated Reasoning, 13:243-281, 1994.
2. K. Broda, M. Finger and A. Russo. Labelled Natural Deduction for Substructural Logics. Logic Journal of the IGPL, Vol. 7, No. 3, May 1999.
3. K. Broda and D. Gabbay. An Abductive CLDS. In Labelled Deduction, Kluwer, Ed. D. Basin et al, 1999.
4. K.Broda and D. Gabbay. A CLDS for Propositional Intuitionistic Logic. TABLEAUX-99, USA, LNAI 1617, Ed. N. Murray, 1999.
5. K. Broda and A. Russo. A Unified Compilation Style Labelled Deductive System for Modal and Substructural Logic using Natural Deduction. Technical Report 10/97. Department of Computing, Imperial College 1997.
6. K. Broda, A. Russo and D. Gabbay. A Unified Compilation Style Natural Deduction System for Modal, Substructural and Fuzzy logics, in Dicovering World with Fuzzy logic: Perspectives and Approaches to Formalization of Human-consistent Logical Systems. Eds V. Novak and I.Perfileva, Springer-Verlag 2000
7. A. Bundy. The Computer Modelling of Mathematical Reasoning. Academic Press, 1983.
8. C. L. Chang and R. Lee. Symbolic Logic and Mechanical Theorem Proving. Academic Press 1973.
9. D. Gabbay. Labelled Deductive Systems, Volume I - Foundations. OUP, 1996.
10. J. H. Gallier. Logic for Computer Science. Harper and Row, 1986.
11. R. Hasegawa, H. Fujita and M. Koshimura. MGTP: A Model Generation Theorem Prover - Its Advanced Features and Applications. In TABLEAUX-97, France, LNAI 1229, Ed. D. Galmiche, 1997.
12. W. Mc.Cune. Otter 3.0 Reference Manual and Guide. Argonne National Laboraqtory, Argonne, Illinois, 1994.
13. J.A. Robinson. Logic, Form and Function. Edinburgh Press, 1979.
14. A. Russo. Modal Logics as Labelled Deductive Systems. PhD. Thesis, Department of Computing, Imperial College, 1996.

15. R. A. Schmidt. Resolution is a decision procedure for many propositional modal logics. Advances in Modal Logic, Vol.1, CSLI, 1998.
16. P. B. Thistlethwaite, M. A. McRobbie and R. K. Meyer. Automated Theorem-Proving in Non-Classical Logics, Wiley, 1988.

A Critique of Proof Planning*

Alan Bundy

Division of Informatics,
University of Edinburgh

Abstract. Proof planning is an approach to the automation of theorem proving in which search is conducted, not at the object-level, but among a set of proof methods. This approach dramatically reduces the amount of search but at the cost of completeness. We critically examine proof planning, identifying both its strengths and weaknesses. We use this analysis to explore ways of enhancing proof planning to overcome its current weaknesses.

Preamble

This paper consists of two parts:

1. *a brief 'bluffer's guide' to proof planning[1]; and*
2. *a critique of proof planning organised as a 4x3 array.*

Those already familiar with proof planning may want to skip straight to the critique which starts at §2, p164.

1 Background

Proof planning is a technique for guiding the search for a proof in automated theorem proving, [Bundy, 1988, Bundy, 1991, Kerber, 1998, Benzmüller *et al*, 1997]. The main idea is to identify common patterns of reasoning in families of similar proofs, to represent them in a computational fashion and to use them to guide the search for a proof of conjectures from the same family. For instance, proofs by mathematical induction share the common pattern depicted in figure 1. This common pattern has been represented in the proof planners *Clam* and *λClam* and used to guide a wide variety of inductive proofs [Bundy *et al*, 1990b, Bundy *et al*, 1991, Richardson *et al*, 1998].

---

* The research reported in this paper was supported by EPSRC grant GR/M/45030. I would like to thank Andrew Ireland, Helen Lowe, Raul Monroy and two anonymous referees for helpful comments on this paper. I would also like to thank other members of the Mathematical Reasoning Group and the audiences at CIAO and Scottish Theorem Provers for helpful feedback on talks from which this paper arose.

[1] Pointers to more detail can be found at
http://dream.dai.ed.ac.uk/projects/proof_planning.html

A.C. Kakas, F. Sadri (Eds.): Computat. Logic (Kowalski Festschrift), LNAI 2408, pp. 160–177, 2002.
© Springer-Verlag Berlin Heidelberg 2002

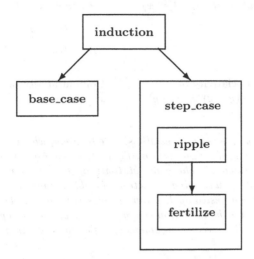

Inductive proofs start with the application of an induction *rule, which reduces the conjecture to some* base *and* step cases. *One of each is shown above. In the step case rippling reduces the difference between the* induction conclusion *and the* induction hypothesis *(see §1.2, p162 for more detail). Fertilization applies the induction hypothesis to simplify the rippled induction conclusion.*

Fig. 1. *ind_strat*: A Strategy for Inductive Proof

1.1 Proof Plans and Critics

The common patterns of reasoning are represented using *tactics*: computer programs which control proof search by applying rules of inference [Gordon *et al*, 1979]. These tactics are specified by *methods*. These methods give both the preconditions under which the tactics are applicable and the effects of their successful application. Meta-level reasoning is used to combine the tactics into a customised proof plan for the current conjecture. This meta-level reasoning matches the preconditions of later tactics to the effects of earlier ones. Examples of such customised proof plans are given in figure 2.

Proof planning has been extended to capture common causes of proof failure and ways to patch them [Ireland, 1992, Ireland & Bundy, 1996b]. With each proof method are associated some *proof critics*. Critics have a similar format to methods, but their preconditions specify situations in which the method's associated tactic will fail and instead of tactics they have instructions on patching a failed proof. Each of the critics associated with a method has a different precondition. These are used to decide on an appropriate patch. Most of the critics built to date have been associated with the ripple method, or rather with its principle sub-method, *wave*, which applies one ripple step (see §1.2, p162). Among the

$$ind\_strat(\ \boxed{x+1}^{\uparrow},x) \qquad ind\_strat(\ \boxed{x+1}^{\uparrow},x)\ then$$

$$[\ ind\_strat(\ \boxed{y+1}^{\uparrow},y)$$

$$ind\_strat(\ \boxed{y+1}^{\uparrow},y)$$

$$]$$

Associativity of $+$ Commutativity of $+$

$x+(y+z)=(x+y)+z$ $x+y=y+x$

The associativity of $+$ is an especially simple theorem, which can be proved with a single application of ind_strat from figure 1, using a one step induction rule on induction variable x. The commutativity of $+$ is a bit more complicated. ind_strat is first applied using induction variable x then in both the base and step cases there is a nested application of ind_strat using y. The first argument of ind_strat indexes the induction rule using the rippling concept of wave-fronts (see §1.2, p162). The second argument specifies the induction variable.

Fig. 2. Special-Purpose Proof Plans

---

patches these critics suggest are: a generalisation of the current conjecture, the use of an intermediate lemma, a case split and using an alternative induction rule. The use of a critic to generalise a conjecture is illustrated in figure 8.

Proof planning has been tested successfully on a wide range of inductive and other theorems. These include conjectures arising from formal methods, *i.e.* from the verification, synthesis and transformation of both software and hardware. They include, for instance: the transformation of naive into tail recursive programs [Hesketh *et al*, 1992], the verification of a microprocessor, [Cantu *et al*, 1996], the synthesis of logic programs [Kraan *et al*, 1996], decision procedures [Armando *et al*, 1996] and the rippling tactic [Gallagher, 1993], resolution completeness proofs [Kerber & Sehn, 1997], proofs of limit theorems [Melis, 1998] and diagonalization proofs [Huang *et al*, 1995, Gow, 1997]. Critics are especially useful at coming up with, so called, 'eureka' steps, *i.e.* those proof steps that usually seem to require human intervention, for instance constructing appropriate induction rules, intermediate lemmas and generalisations [Lowe *et al*, 1998] and loop invariants [Ireland & Stark, 1997].

Proof planning has also been applied outwith mathematics to the computer games of bridge [Frank *et al*, 1992] and Go [Willmott *et al*, 1999] and also to problems of configuring systems from parts, [Lowe, 1991, Lowe *et al*, 1996].

1.2 Rippling

Rippling is the key method in proof plans for inductive proof. Not only does it guide the manipulation of the induction conclusion to prepare it for the application of the induction hypothesis, but preparation for rippling suggests an appropriate induction rule and variable and different patterns of rippling failure suggest new lemmas and generalisations. Since it is also cited several times in the critique, we have included a brief introduction to rippling here.

Rippling is useful whenever there is a goal to be proved in the context of one or more 'givens'. Givens may be axioms, previously proved theorems, assumptions or hypotheses. It works by calculating the difference between the goal and the given(s) and then systematically reducing it. The similarities and differences between the goal and given(s) are marked with meta-level annotations. These annotations are shown graphically in figure 5, where the notation of rippling is explained. An example of rippling is given in figure 6.

$$rev(nil) = nil$$
$$rev(H :: T) = rev(T) <> (H :: nil)$$

$$qrev(nil, L) = L$$
$$qrev(H :: T, L) = qrev(T, H :: L)$$

rev and qrev are alternative recursive functions for reversing a list. Each is defined by a one-step list recursion using a base and step case. :: is an infix list cons and <> an infix list append. rev is a naive reverse function and qrev a more efficient, tail-recursive function. The second argument of qrev is called an accumulator. *This accumulator should be set to nil when qrev is first applied to reverse a list. Figure 4 states two theorems that relate these two functions.*

Fig. 3. Recursive Definitions of Two Reverse Functions

$$\forall k.\ rev(k) = qrev(k, nil) \tag{1}$$
$$\forall k, l.\ rev(k) <> l = qrev(k, l) \tag{2}$$

Theorem (1) shows that rev and qrev output the same result from the same input when the accumulator of qrev is initialised to nil. Theorem (2) generalises theorem (1) for all values of this accumulator. Paradoxically, the more specialised theorem (1) is harder to prove. One way to prove it is first to generalise it to theorem (2).

Fig. 4. Two Theorems about List Reversing Functions

Given: $rev(t) <> L = qrev(t, L)$

Goal: $rev(\boxed{h :: t}^{\uparrow}) <> \lfloor l \rfloor = qrev(\boxed{h :: t}^{\uparrow}, \lfloor l \rfloor)$

Wave-Rules:

$$rev(\boxed{H :: T}^{\uparrow}) \Rightarrow \boxed{rev(T) <> H :: nil}^{\uparrow} \tag{3}$$

$$qrev(\boxed{H :: T}^{\uparrow}, L) \Rightarrow qrev(T, \boxed{H :: L}^{\downarrow}) \tag{4}$$

$$(\boxed{X <> Y}^{\uparrow}) <> Z \Rightarrow X <> (\boxed{Y <> Z}^{\downarrow}) \tag{5}$$

The example is drawn from the inductive proof of theorem (2) in figure 4. The given and the goal are the induction hypothesis and induction conclusion, respectively, of this theorem. Wave-rules (3) and (4) are annotated versions of the step cases of the recursive definitions of the two list reversing functions in figure 3. Wave-rule (5) is from the associativity of <>.

The grey boxes are called wave-fronts *and the holes in them are called wave-holes. The wave-fronts in the goal indicate those places where the goal differs from the given. Those in the wave-rules indicate the differences between the left and right hand sides of the rules. The arrows on the wave-fronts indicate the direction in which rippling will move them: either outwards ($\uparrow$) or inwards ($\downarrow$). The corners, $\lfloor ... \rfloor$, around the l in the goal indicate a* sink. *A sink is one of rippling's target locations for wave-fronts; the other target is to surround an instance of the whole given with a wave-front.*

The wave-rules are used to rewrite each side of the goal. The effect is to move the wave-fronts either to surround an instance of the given or to be absorbed into a sink. An example of this process is given in figure 6

Fig. 5. The Notation of Rippling

---

2 Critique

Our critique of proof planning is organised along two dimensions. On the first dimension we consider four different aspects of proof planning: (1) its potential for advance formation, (2) its theorem proving power, (3) its support for interaction and (4) its methodology. On the second dimension, for each aspect of the first dimension we present: (a) the original dream, (b) the reality of current implementations and (c) the options available for overcoming obstacles and realising part of that original dream.

2.1 The Advance Formation of Plans

The Dream: In the original proposal for proof planning [Bundy, 1988] it was envisaged that the formation of a proof plan for a conjecture would precede its use to guide the search for a proof. Meta-level reasoning would be used to join general proof plans together by matching the preconditions of later ones to the

---

Given: $rev(t) <> L = qrev(t, L)$
Goal:

$$rev(\,\boxed{h :: t}^{\uparrow}\,) <> \lfloor l \rfloor = qrev(\,\boxed{h :: t}^{\uparrow}, \lfloor l \rfloor)$$

$$(\,\boxed{rev(t)} <> h :: nil^{\uparrow}\,) <> \lfloor l \rfloor = qrev(t, \lfloor h :: l \rfloor)$$

$$rev(t) <> \lfloor (h :: nil) <> l \rfloor = qrev(t, \lfloor h :: l \rfloor)$$

$$rev(t) <> \lfloor h :: l \rfloor = qrev(t, \lfloor h :: l \rfloor)$$

The example comes from the step case of the inductive proof of theorem (2) from figure 4. Note that the induction variable k becomes the constant t in the given and the wave-front $\boxed{h :: t}^{\uparrow}$ in the goal. However, the other universal variable, l, becomes a first-order meta-variable, L, in the given, but a sink, $\lfloor l \rfloor$, in the goal. We use uppercase to indicate meta-variables and lowercase for object-level variables and constants.
The left-hand wave-front is rippled-out using wave-rule (3) from figure 5, but then rippled-sideways using wave-rule (5), where it is absorbed into the left-hand sink. The right-hand wave-front is rippled-sideways using wave-rule (4) and absorbed into the right-hand sink. After the left-hand sink is simplified, using the recursive definition of <>, the contents of the two sinks are identical and the goal can be fertilized with the given, completing the proof. Note that fertilization unifies the meta-variable L with the sink h :: l.
Note that there is no point in rippling sideways unless this absorbs wave-fronts into sinks. Sinks mark the potential to unify wave-fronts with meta-variables during fertilization. Without sinks to absorb the wave-fronts, fertilization will fail. Such a failure is illustrated in figure 7

Fig. 6. An Example of Rippling

---

effects of earlier ones. A tactic would then be extracted from the customised proof plan thus constructed. A complete proof plan would be sent to a tactic-based theorem prover where it would be unpacked into a formal proof with little or no search.

The Reality: Unfortunately, in practice, this dream proved impossible to realise. The problem is due to the frequent impossibility of checking the preconditions of methods against purely abstract formulae. For instance, the preconditions of rippling include checking for the presence of wave-fronts in the current goal formula, that a wave-rule matches a sub-expression of this goal and that any new inwards wave-fronts have a wave-hole containing a sink. These preconditions cannot be checked unless the structure of the goal is known in some detail. To know this structure requires anticipating the effects of the previous methods in the current plan. The simplest way to implement this is to apply each of the tactics of the previous methods in order.

Similar arguments hold for most of the other proof methods used by proof planners. This is especially true in applications to game playing where the different counter actions of the opposing players must be explored before a response can be planned, [Willmott et al, 1999]. So the reality is an interleaving of proof planning and proof execution. Moreover, the proof is planned in a *consecutive* fashion, *i.e.* the proof steps are developed starting at one end of the proof then proceeding in order. At any stage of the planning process only an initial or final segment of the object-level proof is known.

The Options: One response to this reality is to admit defeat, abandon proof planning and instead recycle the preconditions of proof methods as preconditions for the application of tactics. Search can then be conducted in a space of condition/action production rules in which the conditions are the method preconditions and the actions are the corresponding tactics. Satisfaction of a precondition will cause the tactic to be applied thus realising the preconditions of subsequent tactics. Essentially, this strategy was implemented by Horn in the Oyster2 system [Horn, 1992]. The experimental results were comparable to earlier versions of *Clam*, *i.e.* if tactics are applied as soon as they are found to be applicable then proof planning conveys no advantage over Horn's production rule approach.

However, in subsequent developments some limited abstraction has been introduced into proof planning, in particular, the use of (usually second-order) meta-variables. In many cases the method preconditions can be checked on such partially abstract formulae. This allows choices in early stages of the proof to be delayed then made subsequently, *e.g.* as a side effect of unification of the meta-variables. We call this *middle-out reasoning* because it permits the non-consecutive development of a proof, *i.e.* instead of having to develop a proof from the top down or the bottom up we can start in the middle and work outwards. Middle-out reasoning can significantly reduce search by postponing a choice with a high branching factor until the correct branch can be determined. Figure 8 provides an example of middle-out reasoning.

Among the choices that can be successfully delayed in this way are: the witness of an existential variable, the induction rule, [Bundy et al, 1990a], an intermediate lemma and generalisation of a goal [Ireland & Bundy, 1996b, Ireland & Bundy, 1996a]. Each of these has a high branching factor – infinite in some cases. A single abstract branch containing meta-variables can simultaneously represent all the alternative branches. Incremental instantiation of the meta-variables as a side effect of subsequent proof steps will implicitly exclude some of these branches until only one remains. Even though the higher-order[2] unification required to whittle down these choices is computationally expensive the cost is far less than the separate exploration of each branch. Moreover, the wave annotation can be exploited to control higher-order unification by requiring wave-fronts to unify with wave-fronts and wave-holes to unify with wave-holes.

---

[2] Only second-order unification is required for the examples tackled so far, but higher-order unification is required in the general case.

Given: $rev(t) = qrev(t, nil)$

Goal:

$$rev(\boxed{h :: t}^{\uparrow}) = qrev(\boxed{h :: t}^{\uparrow}, nil)$$

$$(\boxed{rev(t) <> h :: nil}^{\uparrow}) = \underbrace{qrev(\boxed{h :: t}^{\uparrow}, nil)}$$

$$\textbf{blocked}$$

*The example comes from the failed step case of the inductive proof of theorem
(1) from figure 4. A particular kind of ripple failure is illustrated.*

*The left-hand wave-front can be rippled-out using wave-rule (3) and is then
completely rippled. However, the right-hand wave-front cannot be rippled-
sideways even though wave-rule (4) matches it. This is because there is no
sink to absorb the resulting inwards directed wave-front. If the wave-rule was
nevertheless applied then any subsequent fertilization attempt would fail.*

*Figure 8 shows how to patch the proof by a generalisation aimed to introduce
a sink into the appropriate place in the theorem and thus allow the ripple to
succeed.*

Fig. 7. A Failed Ripple

We have exploited this middle-out technique to especially good effect in our use
of critics, [Ireland & Bundy, 1996b].

Constraints have also been used as a least commitment mechanism in the
Ωmega proof planner [Benzmüller *et al*, 1997]. Suppose a proof requires an object with certain properties. The existence of such an object can be assumed
and the properties posted as constraints. Such constraints can be propagated
as the proof develops and their satisfaction interleaved with that proof in an
opportunistic way [Melis *et al*, 2000b, Melis *et al*, 2000a].

Middle-out reasoning recovers a small part of the original dream of advance
proof planning and provides some significant search control advantage over the
mere use of method preconditions in tactic-based production rules.

2.2   The Theorem Proving Power of Proof Planning

The Dream: One of the main aims of proof planning was to enable automatic theorem provers to prove much harder theorems than conventional theorem provers were capable of. The argument was that the meta-level planning
search space was considerably smaller than the object-level proof search space.
This reduction was partly due to the fact that proof methods only capture common patterns of reasoning, excluding many unsuccessful parts of the space. It was
also because the higher-level methods, *e.g. ind_strat*, each cover many object-
level proof steps. Moreover, the use of abstraction devices, like meta-variables,
enables more than one proof branch to be explored simultaneously. Such search
space reductions should bring much harder proofs into the scope of exhaustive
search techniques.

Schematic Conjecture: $\forall k, l.\ F(rev(k), l) = qrev(k, G(l))$
Given: $F(rev(t), L) = qrev(t, G(L))$
Goal:

$$F(rev(\boxed{h :: t}^{\uparrow}), \lfloor l \rfloor) = qrev(\boxed{h :: t}^{\uparrow}, G(\lfloor l \rfloor))$$

$$F(\boxed{rev(t) <> h :: nil}^{\uparrow}, \lfloor l \rfloor) = qrev(t, \boxed{h :: G(\lfloor l \rfloor)}^{\downarrow})$$

$$rev(t) <> (\ h :: nil <> F'(\boxed{rev(t) <> h :: nil}^{\uparrow}, \lfloor l \rfloor)\ ) = qrev(t, \boxed{h :: G(\lfloor l \rfloor)}^{\downarrow})$$

$$rev(t) <> (\ h :: F'(\boxed{rev(t) <> h :: nil}^{\uparrow}, \lfloor l \rfloor)\ ) = qrev(t, \boxed{h :: G(\lfloor l \rfloor)}^{\downarrow})$$

$$rev(t) <> (\lfloor h :: l \rfloor) = qrev(t, \lfloor h :: l \rfloor)$$

Meta-Variable Bindings:

$$\lambda u, v.\ u <> F'(u, v)/F$$
$$\lambda u, v.\ v./F'$$
$$\lambda u.\ u./G$$

Generalised Conjecture: $\forall k, l.\ rev(k) <> l = qrev(k, l)$

> The example shows how the failed proof attempt in figure 7 can be analysed using a critic and patched in order to get a successful proof. The patch generalises the theorem to be proved by introducing an additional universal variable and hence a sink. Middle-out reasoning is used to delay determining the exact form of the generalisation. This form is determined later as a side effect of higher-order unification during rippling.
>
> First a schematic conjecture is introduced. A new universal variable l is introduced, in the right-hand side, at the point where a sink was required in the failed proof in figure 7. Since we are not sure exactly how l relates to the rest of the right-hand side a second-order meta-variable G is wrapped around it. On the left-hand side a balancing occurrence of l is introduced using the meta-variable F. Note that l becomes a first-order meta-variable L in the given, but a sink $\lfloor l \rfloor$ in the goal.
>
> Induction on k, rippling, simplification and fertilization are now applied, but higher-order unification is used to instantiate F and G. If the schematic conjecture is now instantiated we see that the generalised conjecture is, in fact, theorem (2) from figure 4.

Fig. 8. Patching a Failed Proof using Middle-Out Reasoning

The Reality: This dream has been partially realised. The reduced search space does allow the discovery of proofs that would be beyond the reach of purely object-level, automatic provers: for instance, many of the proofs listed in §1.1, p161.

Unfortunately, these very search reduction measures can also *exclude* the proofs of hard theorems from the search space, making them impossible to find. The reduced plan space is *incomplete*. Hard theorems may require uncommon or even brand new patterns of reasoning, which have not been previously captured in proof methods. Or they may require existing tactics to be used in unusual ways that are excluded by their current heuristic preconditions. Indeed, it is often a characteristic of a breakthrough in mathematical proof that the proof incorporates some new kind of proof method, *cf* Gödel's Incompleteness Theorems. Such proofs will not be found by proof planning using only already known proof methods, but could potentially be stumbled upon by exhaustive search at the object-level.

The Options: Firstly, we consider ways of reducing the incompleteness of proof planning, then ways of removing it.

We should strive to ensure that the preconditions of methods are as general as possible, for instance, minimising the use of heuristic preconditions, as opposed to preconditions that are *required* for the legal application of the method's tactic. This will help ensure that the tactic is applied whenever it is appropriate and not excluded due to a failure to anticipate an unusual usage. A balance is required here since the absence of *all* heuristic preconditions may increase the search space to an infeasible size. Rather diligence is needed to design both tactics and their preconditions which generalise away from the particular examples that may have suggested the reasoning pattern in the first place.

The use of critics expands the search space by providing a proof patch when the preconditions of a method fail. In practice, critics have been shown to facilitate the proof of hard theorems by providing the 'eureka' steps, *e.g.* missing lemmas, goal generalisations, unusual induction rules, *etc*, that hard theorems often require [Ireland & Bundy, 1996b]. However, even with these additions, the plan space is still incomplete; so the problem is only postponed.

One way to *restore* completeness would be to allow arbitrary object-level proof steps, *e.g.* the application of an individual rule of inference such as rewriting, generalisation, induction, *etc*, with no heuristic limits on its application. Since such a facility is at odds with the philosophy of proof planning, its use would need to be carefully restricted. For instance, a proof method could be provided that made a single object-level proof step at random, but only when all other possibilities had been exhausted. Provided that the rest of the plan space was finite, *i.e.* all other proof methods were terminating, then this random method would occasionally be called and would have the same potential for stumbling upon new lines of proof that a purely object-level exhaustive prover does, *i.e.* we would not expect it to happen very often – if at all.

It is interesting to speculate about whether it would be possible to draw a more permanent benefit from such serendipity by learning a new proof method from the example proof. Note that this might require the invention of new meta-level concepts: consider, for instance, the learning of rippling from example object-level proofs, which would require the invention of the meta-level concepts of wave-front, wave-hole, *etc.*

Note that a first-order object-level proof step might be applied to a formula containing meta-variables. This would require the first-order step to be applied using higher-order unification, – potentially creating a larger search space than would otherwise occur. Also, some object-level proof steps require the specification of an expression, *e.g.* the witness of an existential quantifier, an induction variable and term, the generalisation of an expression. If these expressions are not provided via user interaction then infinite branching could be avoided by the use of meta-variables. So object-level rule application can introduce meta-variables even if they are not already present. These considerations further underline the need to use such object-level steps only as a last resort.

2.3   The Support for Interaction of Proof Planning

The Dream: Proof planning is not just useful for the automation of proof, it can also assist its interactive development. The language of proof planning describes the high-level structure of a proof and, hence, provides a high-level channel of communication between machine and user. This can be especially useful in a very large proof whose description at the object-level is unwieldy. The different proof methods chunk the proof into manageable pieces at a hierarchy of levels. The method preconditions and effects describe the relationships between and within each chunk and at each level. For instance, the language of rippling enables a proof state to be described in terms of differences between goals and givens, why it is important to reduce those differences and of ways to do so.

The preconditions and effects of methods and critics support the automatic analysis and patching of failed proof attempts. Thus the user can be directed to the reasons for a failed proof and the kind of steps required to remedy the situation. This orients the user within a large and complex search space and gives useful hints as to how to proceed.

The Reality: The work of Lowe, Jackson and others in the *XBarnacle* system [Lowe & Duncan, 1997] shows that proof planning can be of considerable assistance in interactive proof. For instance, in Jackson's PhD work, [Jackson, 1999, Ireland *et al*, 1999], the user assists in the provision of goal generalisations, missing lemmas, *etc.* by instantiating meta-variables. However, each of the advantages listed in the previous section brings corresponding disadvantages.

Firstly, proof planning provides an enriched language of human/computer communication but at the price of introducing new jargon for the user to understand. The user of *XBarnacle* must learn the meaning of wave-fronts, flawed inductions, fertilization, *etc.*

Secondly, and more importantly, the new channel of communication assists users at the cost of restricting them to the proof planning search space; *cf* the discussion of incompleteness in §2.2, p168. For instance, *XBarnacle* users can get an explanation of why a method or critic did or did not apply in terms of successful or failed preconditions. They can over-ride those preconditions to force or prevent a method or critic applying. But their actions are restricted to the search space of tactics and critics. If the proof lies *outside* that space then they are unable to direct *XBarnacle* to find it.

The Options: The first problem can be ameliorated in a number of ways. Jargon can be avoided, translated or explained according to the expertise and preferences of the user. For instance, "fertilization" can be avoided in favour of, or translated into, the "use of the induction hypothesis". "Wave-front", on the other hand, has no such ready translation into standard terminology and must be explained within the context of rippling. Thus, although this problem can be irritating, it can be mitigated with varying amounts of effort.

The second problem is more fundamental. Since it is essentially the same as the problem of the incompleteness of the plan space, discussed in §2.2, p168, then one solution is essentially that discussed at the end of §2.2, p169. New methods can be provided which apply object-level proof steps under user control. As well as providing an escape mechanism for a frustrated user this might also be a valuable device for system developers. It would enable them to concentrate on the parts of a proof they were interested in automating while using interaction to 'fake' the other parts.

The challenge is to integrate such object-level steps into the rest of the proof planning account. For instance, what story can we now tell about how such object-level steps exploit the effects of previous methods and enable the preconditions of subsequent ones?

2.4   The Methodology of Proof Planning

The Dream: Proof planning aims to capture common patterns of reasoning and repair in methods and critics. In [Bundy, 1991] we provide a number of criteria by which these methods and critics are to be assessed. These include expectancy[3], generality, prescriptiveness[4], simplicity, efficiency and parsimony. In particular, each method and critic should apply successfully in a wide range of situations (generality) and a few methods and critics should generate a large number of proofs (parsimony). Moreover, the linking of effects of earlier methods and critics to the preconditions of later ones should enable a good 'story' to be told about how and why the proof plan works. This 'story' enables the expectancy criterion to be met.

The Reality: It is hard work to ensure that these criteria are met. A new method or critic may originally be inspired by only a handful of examples. There is a constant danger of producing methods and critics that are too fine tuned to

---

[3] Some degree of assurance that the proof plan will succeed.

[4] The less search required the better.

these initial examples. This can arise both from a lack of imagination in generalising from the specific situation and from the temptation to get quick results in automation. Such over-specificity leads to a proliferation of methods and critics with limited applicability. Worse still, the declarative nature of methods may be lost as methods evolve into arbitrary code tuned to a particular problem set. The resulting proof planner will be brittle, *i.e.* will frequently fail when confronted with new problems. It will become increasing hard to tell an intelligible story about its reasoning. Critical reviewers will view the empirical results with suspicion, suspecting that the system has been hand-tuned to reproduce impressive results on only a handful of hard problems.

As the consequences of over-specificity manifest themselves in failed proof attempts so the methods and critics can be incrementally generalised to cope with the new situations. One can hope that this process of incremental generalisation will converge on a few robust methods and critics, so realising the original dream. However, a reviewer may suspect that this process is both infinite and non-deterministic, with each incremental improvement only increasing the range of the methods and critics by a small amount.

The opposite problem is caused by an over-general or missing precondition, permitting a method to apply in an inappropriate situation. This may occur, for instance, where a method is developed in a context in which a precondition is implicit, but then applied in a situation in which it is absent. This problem is analogous to feature interaction in telecomms or of predicting the behaviour of a society of agents.

The Options: The challenge is not only to adopt a development methodology that meets the criteria in [Bundy, 1991] but also *to be seen to do so.* This requires both diligence in the development of proof plans and the explicit demonstration of this diligence. Both aims can be achieved by experimental or theoretical investigations designed to test explicit hypotheses.

For instance, to test the criterion of generality, systematic and thorough application of proof planning systems should be conducted. This testing requires a large and diverse set of examples obtained from independent sources. The diversity should encompass the form, source and difficulty level of the examples. However, the generality of the whole system should not be obtained at the cost of parsimony, *i.e.* by providing lots of methods and critics 'hand crafted' to cope with each problematic example; so *each* of the methods and critics must be shown to be general-purpose. Unfortunately, it is not possible to test each one in isolation, since the methods and critics are designed to work as a family. However, it is possible to record how frequently each method and critic is used during the course of a large test run.

To meet the criterion of expectancy the specifications of the methods and critics should be declarative statements in a meta-logic. It should be demonstrated that the effects of earlier methods enable the preconditions of later ones and that the patches of critics invert the failed preconditions of the methods to which they are attached. Such demonstrations will deal both with the situation in which method preconditions/effects are too-specific (they will not be strong enough hypotheses) and in which they are too general (they will not be provable). The work of Gallagher [Gallagher, 1993] already shows that this kind of reasoning about method preconditions and effects can be automated.

To meet the criterion of prescriptiveness the search space generated by rival methods needs to be compared either theoretically or experimentally; the method with the smaller search space is to be preferred. However, reductions in search space should not be obtained at the cost of unacceptable reductions in success rate. So it might be shown experimentally and/or via expectancy arguments that acceptable success rates are maintained. Reduced search spaces will usually contribute to increased efficiency, but it is possible that precondition testing is computationally expensive and that this cost more than offsets the benefits of the increased prescriptiveness, so overall efficiency should also be addressed.

3   Conclusion

In this paper we have seen that some of the original dreams of proof planning have not been fully realised in practice. We have shown that in some cases it has not been possible to deliver the dream in the form in which it was originally envisaged, for instance, because of the impossibility of testing method preconditions on abstract formulae or the inherent incompleteness of the planning search space. In each case we have investigated whether and how a lesser version of the original dream *can* be realised. This investigation both identifies the important benefits of the proof planning approach and points to the most promising directions for future research. In particular, there seem to be three important lessons that have permeated the analysis.

Firstly, the main benefits of proof planning are in facilitating a non-consecutive exploration of the search space, *e.g.* by 'middle-out' reasoning. This allows the postponement of highly branching choice points using least commitment mechanisms, such as meta-variables or constraints. Parts of the search space with low branching rates are explored first and the results of this search determine the postponed choices by side-effect, *e.g.* using higher-order unification or constraint solving. This can result in dramatic search space reductions. In particular, 'eureka' steps can be made in which witnesses, generalisations, intermediate lemmas, customised induction rules, *etc*, are incrementally constructed. The main vehicle for such non-consecutive exploration is critics. *Our analysis points to the further development of critics as the highest priority in proof planning research.*

Secondly, in order to increase the coverage of proof planners in both automatic and interactive theorem proving it is necessary to combine it with more brute force approaches. For instance, *it may be necessary to have default methods in which arbitrary object-level proof steps are conducted either at random or under user control.* One might draw an analogy with simulated annealing in which it is sometimes necessary to make a random move in order to escape from a local minimum.

Thirdly, frequent and systematic rational reconstruction is necessary to off-set the tendency to develop over-specialised methods and critics. This tendency is a natural by-product of the experimental development of proof planning as specifications are tweaked and tuned to deal with challenging examples. It is necessary to clean-up non-declarative specifications, merge and generalise methods and critics and to test proof planners in a systematic and thorough way. *The assessment criteria of [Bundy, 1991] must be regularly restated and reapplied.*

Despite the limitations exposed by the analysis of this paper, proof planning has been shown to have a real potential for efficient and powerful, automatic and interactive theorem proving. Much of this potential still lies untapped and our analysis has identified the priorities and directions for its more effective realisation.

Afterword

I first met Bob Kowalski in June 1971, when I joined Bernard Meltzer's Meta-mathematics Unit as a research fellow. Bernard had assembled a world class centre in automatic theorem proving. In addition to Bob, the other research fellows in the Unit were: Pat Hayes, J Moore, Bob Boyer and Donald Kuehner; Donald was the co-author, with Bob, of SL-Resolution, which became the theoretical basis for Prolog.

Bob's first words to me were "Do you like computers? I don't!". This sentiment was understandable given the primitive computer facilities then available to us: one teletype with a 110 baud link to a shared ICL 4130 with 64k of memory. Bob went on to forsake the automation of mathematical reasoning as the main domain for theorem proving and instead pioneered logic programming: the application of theorem proving to programming. I stuck with mathematical reasoning and focussed on the problem of proof search control. However, I was one of the earliest adopters of Prolog and have been a major beneficiary of Bob's work, using logic programming both as a practical programming methodology and as a domain for formal verification and synthesis. I am also delighted to say that Bob has remained a close family friend for 30 years.

Happy 60th birthday Bob!

References

[Armando *et al*, 1996]   Armando, A., Gallagher, J., Smaill, A. and Bundy, A. (3-5 January 1996). Automating the synthesis of decision procedures in a constructive metatheory. In *Proceedings of the Fourth International Symposium on Artificial Intelligence and Mathematics*, pages 5–8, Florida. Also in the Annals of Mathematics and Artificial Intelligence, 22, pp 259–79, 1998.

[Benzmüller *et al*, 1997]   Benzmüller, C., Cheikhrouhou, L., Fehrer, D., Fiedler, A., Huang, X., Kerber, M., Kohlhase, K., Meier, A, Melis, E., Schaarschmidt, W., Siekmann, J. and Sorge, V. (1997).

Ωmega: Towards a mathematical assistant. In McCune, W., (ed.), *14th International Conference on Automated Deduction*, pages 252–255. Springer-Verlag.

[Bundy, 1988]   Bundy, A. (1988). The use of explicit plans to guide inductive proofs. In Lusk, R. and Overbeek, R., (eds.), *9th International Conference on Automated Deduction*, pages 111–120. Springer-Verlag. Longer version available from Edinburgh as DAI Research Paper No. 349.

[Bundy, 1991]   Bundy, Alan. (1991). A science of reasoning. In Lassez, J.-L. and Plotkin, G., (eds.), *Computational Logic: Essays in Honor of Alan Robinson*, pages 178–198. MIT Press. Also available from Edinburgh as DAI Research Paper 445.

[Bundy *et al*, 1990a]   Bundy, A., Smaill, A. and Hesketh, J. (1990a). Turning eureka steps into calculations in automatic program synthesis. In Clarke, S. L.H., (ed.), *Proceedings of UK IT 90*, pages 221–6. IEE. Also available from Edinburgh as DAI Research Paper 448.

[Bundy *et al*, 1990b]   Bundy, A., van Harmelen, F., Horn, C. and Smaill, A. (1990b). The Oyster-Clam system. In Stickel, M. E., (ed.), *10th International Conference on Automated Deduction*, pages 647–648. Springer-Verlag. Lecture Notes in Artificial Intelligence No. 449. Also available from Edinburgh as DAI Research Paper 507.

[Bundy *et al*, 1991]   Bundy, A., van Harmelen, F., Hesketh, J. and Smaill, A. (1991). Experiments with proof plans for induction. *Journal of Automated Reasoning*, 7:303–324. Earlier version available from Edinburgh as DAI Research Paper No 413.

[Cantu *et al*, 1996]   Cantu, Francisco, Bundy, Alan, Smaill, Alan and Basin, David. (1996). Experiments in automating hardware verification using inductive proof planning. In Srivas, M. and Camilleri, A., (eds.), *Proceedings of the Formal Methods for Computer-Aided Design Conference*, number 1166 in Lecture Notes in Computer Science, pages 94–108. Springer-Verlag.

[Frank *et al*, 1992]   Frank, I., Basin, D. and Bundy, A. (1992). An adaptation of proof-planning to declarer play in bridge. In *Proceedings of ECAI-92*, pages 72–76, Vienna, Austria. Longer Version available from Edinburgh as DAI Research Paper No. 575.

[Gallagher, 1993]   Gallagher, J. K. (1993). *The Use of Proof Plans in Tactic Synthesis*. Unpublished Ph.D. thesis, University of Edinburgh.

[Gordon *et al*, 1979]   Gordon, M. J., Milner, A. J. and Wadsworth, C. P. (1979). *Edinburgh LCF - A mechanised logic of computation*, volume 78 of *Lecture Notes in Computer Science*. Springer-Verlag.

[Gow, 1997]   Gow, J. (1997). *The Diagonalization Method in Automatic Proof*. Undergraduate project dissertation, Dept of Artificial Intelligence, University of Edinburgh.

[Hesketh *et al*, 1992]   Hesketh, J., Bundy, A. and Smaill, A. (June 1992). Using middle-out reasoning to control the synthesis of tail-recursive programs. In Kapur, Deepak, (ed.), *11th International Conference on Automated Deduction*, volume 607 of *Lecture Notes in Artificial Intelligence*, pages 310–324, Saratoga Springs, NY, USA.

[Horn, 1992]   Horn, Ch. (1992). Oyster-2: Bringing type theory into practice. *Information Processing*, 1:49–52.

[Huang et al, 1995]   Huang, X., Kerber, M. and Cheikhrouhou, L. (1995). Adapting the diagonalization method by reformulations. In Levy, A. and Nayak, P., (eds.), *Proc. of the Symposium on Abstraction, Reformulation and Approximation (SARA-95)*, pages 78–85. Ville d'Esterel, Canada.

[Ireland & Bundy, 1996a]   Ireland, A. and Bundy, A. (1996a). Extensions to a Generalization Critic for Inductive Proof. In McRobbie, M. A. and Slaney, J. K., (eds.), *13th International Conference on Automated Deduction*, pages 47–61. Springer-Verlag. Springer Lecture Notes in Artificial Intelligence No. 1104. Also available from Edinburgh as DAI Research Paper 786.

[Ireland & Bundy, 1996b]   Ireland, A. and Bundy, A. (1996b). Productive use of failure in inductive proof. *Journal of Automated Reasoning*, 16(1–2):79–111. Also available from Edinburgh as DAI Research Paper No 716.

[Ireland & Stark, 1997]   Ireland, A. and Stark, J. (1997). On the automatic discovery of loop invariants. In *Proceedings of the Fourth NASA Langley Formal Methods Workshop*. NASA Conference Publication 3356. Also available as Research Memo RM/97/1 from Dept of Computing and Electrical Engineering, Heriot-Watt University.

[Ireland, 1992]   Ireland, A. (1992). The Use of Planning Critics in Mechanizing Inductive Proofs. In Voronkov, A., (ed.), *International Conference on Logic Programming and Automated Reasoning – LPAR 92, St. Petersburg*, Lecture Notes in Artificial Intelligence No. 624, pages 178–189. Springer-Verlag. Also available from Edinburgh as DAI Research Paper 592.

[Ireland et al, 1999]   Ireland, A., Jackson, M. and Reid, G. (1999). Interactive Proof Critics. *Formal Aspects of Computing: The International Journal of Formal Methods*, 11(3):302–325. A longer version is available from Dept. of Computing and Electrical Engineering, Heriot-Watt University, Research Memo RM/98/15.

[Jackson, 1999]   Jackson, M. (1999). *Interacting with Semi-automated Theorem Provers via Interactive Proof Critics*. Unpublished Ph.D. thesis, School of Computing, Napier University.

[Kerber & Sehn, 1997]   Kerber, Manfred and Sehn, Arthur C. (1997). Proving ground completeness of resolution by proof planning. In Dankel II, Douglas D., (ed.), *FLAIRS-97, Proceedings of the 10th International Florida Artificial Intelligence Research Symposium*, pages 372–376, Daytona, Florida, USA. Florida AI Research Society, St. Petersburg, Florida, USA.

[Kerber, 1998]   Kerber, Manfred. (1998). Proof planning: A practical approach to mechanized reasoning in mathematics. In Bibel,

Wolfgang and Schmitt, Peter H., (eds.), *Automated Deduction, a Basis for Application – Handbook of the German Focus Programme on Automated Deduction*, chapter III.4, pages 77–95. Kluwer Academic Publishers, Dordrecht, The Netherlands.

[Kraan et al, 1996] Kraan, I., Basin, D. and Bundy, A. (1996). Middle-out reasoning for synthesis and induction. *Journal of Automated Reasoning*, 16(1–2):113–145. Also available from Edinburgh as DAI Research Paper 729.

[Lowe & Duncan, 1997] Lowe, H. and Duncan, D. (1997). XBarnacle: Making theorem provers more accessible. In McCune, William, (ed.), *14th International Conference on Automated Deduction*, pages 404–408. Springer-Verlag.

[Lowe, 1991] Lowe, Helen. (1991). Extending the proof plan methodology to computer configuration problems. *Artificial Intelligence Applications Journal*, 5(3). Also available from Edinburgh as DAI Research Paper 537.

[Lowe et al, 1996] Lowe, H., Pechoucek, M. and Bundy, A. (October 1996). Proof planning and configuration. In *Proceedings of the Ninth Exhibition and Symposium on Industrial Applications of Prolog*. Also available from Edinburgh as DAI Research Paper 859.

[Lowe et al, 1998] Lowe, H., Pechoucek, M. and Bundy, A. (1998). Proof planning for maintainable configuration systems. *Artificial Intelligence in Engineering Design, Analysis and Manufacturing*, 12:345–356. Special issue on configuration.

[Melis, 1998] Melis, E. (1998). The "limit" domain. In Simmons, R., Veloso, M. and Smith, S., (eds.), *Proceedings of the Fourth International Conference on Artificial Intelligence in Planning Systems*, pages 199–206.

[Melis et al, 2000a] Melis, E., Zimmer, J. and Müller, T. (2000a). Extensions of constraint solving for proof planning. In Horn, W., (ed.), *European Conference on Artificial Intelligence*, pages 229–233.

[Melis et al, 2000b] Melis, E., Zimmer, J. and Müller, T. (2000b). Integrating constraint solving into proof planning. In Ringeissen, Ch., (ed.), *Frontiers of Combining Systems, Third International Workshop, FroCoS'2000*, number 1794 in Lecture Notes on Artificial Intelligence, pages 32–46. Springer.

[Richardson et al, 1998] Richardson, J. D. C, Smaill, A. and Green, I. (July 1998). System description: proof planning in higher-order logic with Lambda-Clam. In Kirchner, Claude and Kirchner, Hélène, (eds.), *15th International Conference on Automated Deduction*, volume 1421 of *Lecture Notes in Artificial Intelligence*, pages 129–133, Lindau, Germany.

[Willmott et al, 1999] Willmott, S., Richardson, J., Bundy, A. and Levine, J. (1999). An adversarial planning approach to Go. In Jaap van den Herik, H. and Iida, H., (eds.), *Computers and Games*, pages 93–112. 1st Int. Conference, CG98, Springer. Lecture Notes in Computer Science No. 1558.

A Model Generation Based Theorem Prover MGTP for First-Order Logic

Ryuzo Hasegawa, Hiroshi Fujita, Miyuki Koshimura, and Yasuyuki Shirai

Graduate School of Information Science and Electrical Engineering
Kyushu University
6-1, Kasuga-koen, Kasuga, Fukuoka 816-8580, JAPAN
{hasegawa,fujita,koshi,shirai}@ar.is.kyushu-u.ac.jp

Abstract. This paper describes the major results on research and development of a model generation theorem prover MGTP. It exploits OR parallelism for non-Horn problems and AND parallelism for Horn problems achieving more than a 200-fold speedup on a parallel inference machine PIM with 256 processing elements. With MGTP, we succeeded in proving difficult mathematical problems that cannot be proven on sequential systems, including several open problems in finite algebra.

To enhance the pruning ability of MGTP, several new features are added to it. These include: CMGTP and IV-MGTP to deal with constraint satisfaction problems, enabling negative and interval constraint propagation, respectively, non-Horn magic set to suppress the generation of useless model candidates caused by irrelevant clauses, a proof simplification method to eliminate duplicated subproofs, and MM-MGTP for minimal model generation.

We studied several techniques necessary for the development of applications, such as negation as failure, abductive reasoning and modal logic systems, on MGTP. These techniques share a basic idea, which is to use MGTP as a meta-programming system for each application.

1 Introduction

Theorem proving is an important basic technology that gave rise to logic programming, and is acquiring a greater importance not only for reasoning about mathematical theorems but also for developing knowledge processing systems. We started research on parallel theorem provers in 1989 in the Fifth Generation Computer Systems (FGCS) project, with the aim of integrating logic programming and theorem proving technologies. The immediate goal of this research was to develop a fast theorem proving system on the parallel inference machine PIM [42], by effectively utilizing KL1 languages [55] and logic programming techniques.

MGTP [11,12] basically follows the model generation method of SATCHMO [38] which has a good property that one way unification suffices. Indeed, the method is very suited to KL1 implementation because we can use fast builtin unification without occur-check. MGTP exploits OR parallelism from non-Horn

A.C. Kakas, F. Sadri (Eds.): Computat. Logic (Kowalski Festschrift), LNAI 2408, pp. 178–213, 2002.
© Springer-Verlag Berlin Heidelberg 2002

problems by independently exploring each branch of a proof tree caused by case splitting, whereas it exploits AND parallelism from Horn problems that do not cause case splitting. Although OR parallelization of MGTP is relatively easy, it is essential to reduce the amount of inter processor communication. For this, we proposed a new method called the N-sequential method [22]. The basic idea is that we run in each processing element (PE) a sequential algorithm to traverse a proof tree depth-first and restrict the number of tasks being activated to at most the number N of available PEs.

Almost linear speedup was achieved for both Horn and non-Horn problems on a PIM/m system consisting of 256 PEs. With MGTP, we succeeded in solving some open quasigroup problems in finite algebra [13]. We also solved several hard condensed detachment problems that could not be solved by OTTER [39] with any strategy [25]. On the other hand, research on solving quasigroup problems with MGTP reveals that it lacks negative constraint propagation ability. Then, we developed CMGTP (Constraint-MGTP) [50] that can handle constraint propagations with negative atoms. As a result, CMGTP's search spaces became much smaller than the original MGTP's. Recently, we have been developing Java versions of MGTP (JavaMGTP) aiming at better efficiency as well as wider usability. JavaMGTP achieves several tens fold speedup compared to KL1 based implementations on a sequential machine.

However, in order to further improve the efficiency of model generation, several problems remain to be solved that are common to model generation based provers: redundant inference caused by clauses that are irrelevant to the given goal, duplication of the same subproof after case-splitting, and generation of nonminimal models.

To solve the first problem, we developed a method called *non-Horn magic sets* (NHM) [24,45]. NHM is a natural extension of the magic sets developed in the deductive database field, and is applicable to non-Horn problems. We showed that NHM has the same power as *relevancy testing* in SATCHMORE [36], although they take completely different approaches.

For the second problem, we came up with a method that combines the relevancy testing realized by NHM and SATCHMORE with *folding-up* proposed by Letz [34], within a single framework [32]. The method has not only an effect similar to relevancy testing that suppresses useless model extensions with irrelevant clauses, but also a folding-up function to eliminate duplicate subproofs. These can be achieved by computing relevant literals that contribute to closing a branch.

The third problem is how to avoid generating nonminimal models that are redundant and thus would cause inefficiency. To this end, we proposed an efficient method that employs *branching assumptions and lemmas* so as to prune branches that lead to nonminimal models, and to reduce minimality tests on obtained models [23]. Then, we have implemented MM-MGTP based on the method. Experimental results with MM-MGTP show a remarkable speedup compared to the original MGTP.

Regarding applications, MGTP can be viewed as a meta-programming system. We can build various reasoning systems on MGTP by writing the inference rules used for each system as MGTP input clauses. Along this idea, we developed several techniques and reasoning systems necessary for AI applications. They include a method to incorporate negation as failure into MGTP [29], abductive reasoning systems [30], and modal logic systems [31]. In particular, MGTP has actually been used as a rule-based engine for the argumentation and negotiation support system in the legal area.

2   An Abstract MGTP Procedure

MGTP is a theorem proving system for first-order logic. An input for MGTP is given as a set of clauses of the implicational form:

$$\Pi \to \Sigma$$

where, normally, the antecedent Π is a conjunction of atoms and the consequent Σ is a disjunction of atoms[1]. A clause is said to be positive if its antecedent is empty or **true**, negative if its consequent is empty or **false**, and mixed otherwise. A clause is called a Horn clause if it has at most one atom in its consequent, otherwise it is called a non-Horn clause. A clause is said to be *range-restricted* if every variable in the consequent of the clause appears in the antecedent, and *violated* under a model candidate M if it holds that $M \models \Pi\sigma$ and $M \not\models \Sigma\sigma$ with some substitution σ.

A generic algorithm of a standard MGTP procedure is sketched in Fig. 1. The task of MGTP is to try to construct a model for a given set of clauses, by extending the current model candidate M so as to satisfy violated clauses under M (*model extension*). The function MG takes as an initial input positive Horn clauses U_0, positive non-Horn clauses D_0, and an empty model candidate M, and returns **true/false** (SAT/UNSAT) as a proof result. MG also outputs a model every time it is found. It works as follows:

(1) As long as the unit buffer U is not empty, MG picks up an atom u from U, tests whether $M \not\models u$ (*subsumption test*), and extends a model candidate M with u (*Horn extension*). Then, the conjunctive matching procedure $CJM(M, u)$ is invoked to search for clauses whose antecedents Π are satisfied by $M \cup \{u\}$ under some substitution σ. If such nonnegative clauses are found, their consequents $\Sigma\sigma$ are added to U or the disjunction buffer D according to the form of a consequent. When the antecedent of a negative clause is satisfied by $M \cup \{u\}$ in $CJM(M, u)$, MG rejects M and returns **false** (*model rejection*).

(2) When U becomes empty, and if D is not empty, MG picks up a disjunction d from D. If d is not satisfied by M, MG recursively calls itself to expand M with each disjunct $L_j \in d$ (*non-Horn extension*).

(3) When both U and D become empty, MG outputs M and returns **true**.

---

[1] This is the primitive form of a clause in a standard MGTP, which will be extended in several ways in MGTP descendants.

```
procedure MGTP :
  begin
    U₀ ← positive Horn clauses; D₀ ← positive non-Horn clauses;
    output MG(U₀, D₀, ∅);
  end ;
boolean function MG(buffer U, buffer D, buffer M) :
  begin
    while (U ≠ ∅) begin                                        ···(1)
      U ← U \ {u ∈ U};
      if (M ⊭ u) then begin
        M ← M ∪ {u}; CJM(M, u);
        if (M is rejected) then return false ;
      end
    end ;
    if (D ≠ ∅) then begin                                      ···(2)
      D ← D \ {d ∈ D}; (where d = (L₁ ∨ ... ∨ Lₙ))
      if (M ⊭ d) then
                 n
        return  ⋁ MG(U ∪ {Lⱼ}, D, M);
                j=1
    end
    else begin output M; return true ; end                    ···(3)
  end .
```

$$U \leftarrow U \setminus \{u \in U\}$$
$$M \not\models u$$
$$M \leftarrow M \cup \{u\}; \ CJM(M,u);$$
$$D \leftarrow D \setminus \{d \in D\}; \ (\text{where } d = (L_1 \vee \ldots \vee L_n))$$
$$M \not\models d$$
$$\text{return } \bigvee_{j=1}^{n} MG(U \cup \{L_j\}, D, M);$$

Fig. 1. A standard MGTP procedure

The standard MGTP procedure might be modified in several ways. For instance, each disjunct of Σ is allowed to be a conjunction of literals. This is especially useful, in fact, for implementing a negation as failure mechanism [29]. We can also extend the procedure to deal with negative literals by introducing two additional operations: unit refutation and unit simplification. This extension yields CMGTP [28,50] which is meant for solving constraint satisfaction problems more efficiently, and MM-MGTP [23] for minimal model generation. Although the procedure apparently looks sequential, it can be parallelized by exploiting parallelism inherent in it. These issues will be described in detail in subsequent sections.

3 Parallel Implementation

There are several ways to parallelize the proving process in MGTP. These are to exploit parallelism in conjunctive matching, subsumption tests, and case splitting. For ground non-Horn cases, it is most promising to exploit OR parallelism induced by case splitting. Here we use OR parallelism to seek multiple models, which produce multiple solutions, in parallel. For Horn clauses, we have to exploit AND parallelism during the traversal of a single branch. The main source of AND parallelism is conjunctive matching and subsumption testing.

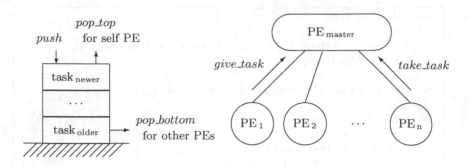

Fig. 2. Task stack

Fig. 3. Process diagram for OR parallelization

3.1 OR Parallelization

For ground non-Horn clauses, it is relatively easy for MGTP to exploit OR parallelism by exploring different branches (model candidates) in different processing elements (PEs) independently. However, inter-PE communication increases rapidly as the number of branching combinatorially explodes and a large amount of data structures, e.g. model candidates and model extending candidates, is copied to one PE after another. Conventional PE allocation methods, such as cyclic and probabilistic allocation, are based on the principle that whenever tasks are created in own PE, all of them but one are to be thrown to other PEs. Although this scheme is easy to implement, the amount of inter-PE communication is at least proportional to the number of tasks created in the whole computation.

To overcome this, we proposed a new method called the N-sequential method [22]. The basic idea is that we run in each PE a sequential algorithm to traverse a proof tree depth-first and restrict the number of activated tasks at any time to at most the number N of available PEs. In this method, a PE can move an unsolved task to other idle PE only when requested from it. When the number of created tasks exceeds the number of free PEs, the excess of tasks are executed sequentially within their current PE. Each PE maintains a task stack as shown in Fig. 2 for use in the sequential traversal of multiple unsolved branches. Created tasks are pushed onto the stack, then popped up from the top of stack (*pop_top*) when the current task has been completed. On receipt of a request from the other PE, a task at the bottom is given to it (*pop_bottom*). We provide a master process as shown in Fig. 3 which acts as a matchmaker between task-requesting (*take_task*) and task-offering (*give_task*) PEs. The task stack process and the master process are written in KL1 and incorporated to the MGTP program.

OR Parallel Performance. The performance of OR parallel MGTP was evaluated on a PIM/m with 128 PEs and a Sun Enterprise 10000 (SE10k) with 24 PEs. For the latter we used the Klic system which compiles KL1 programs into C codes and makes it possible to run them on a single machine or parallel ma-

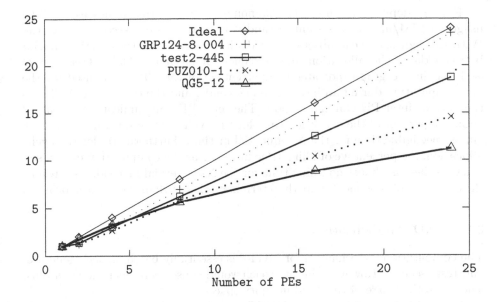

Fig. 4. Speedup ratio by OR parallelization on SE10k(1–24PE)

chines like SE10k. The experimental results show significant speedups on both systems.

Figure 4 shows a speedup ratio by OR parallel execution for non-Horn problems using the N-sequential method on SE10k. Here, GRP124-8.004 and PUZ010-1 are problems taken from the TPTP library [53], QG5-12 is a quasigroup problem to be explained in later sections, and test2-445 is an artificial benchmark spanning a trinary tree. A satisfactory speedup is attained for such problem as GRP124-8.004 and test2-445 in which the number of non-Horn extensions dominates that of Horn extensions. The reason why the PUZ010-1 and QG5-12 degrade the speedup is that they require a significant number of Horn-extensions, whereas they do only a small number of non-Horn extensions.

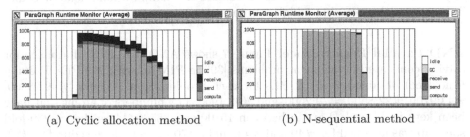

(a) Cyclic allocation method   (b) N-sequential method

Fig. 5. Snapshot of "xamonitor" on PIM/m

Figure 5 depicts snapshots of a "xamonitor" window that indicates the CPU usage on PIM/m which is sampled and displayed at every second of interval. With this figure, one can observe clear distinction of the characteristic behavior between the cyclic allocation and N-sequential methods. The lighter part of each bar in the graph indicates the percentage of the CPU time used for the net computation during an interval (one second), and the darker part indicates the rate of inter-PE communication. The inter-PE communication consumed about 20 percent of the execution time for the cyclic allocation, whereas it took almost negligible time for the N-sequential method. Furthermore, for the cyclic allocation, the percentage of idling time increases as the computation progresses, whereas there is almost no idling time for the N-sequential method. As a result, the execution of N-sequential method terminates faster than the cyclic allocation.

3.2   AND Parallelization

The computational mechanism for MGTP is essentially based on the "generate-and-test" scheme. However, this approach would cause over-generation of atoms, leading to the waste of time and memory spaces.

In the AND parallelization of MGTP, we adopted the *lazy model generation* method [26] that induces a demand-driven style of computation. In this method, a generator process to perform model extension generates a specified number of atoms only when required by the tester process to perform rejection testing. The lazy mechanism can avoid over-generation of atoms in model extension, and provides flexible control to maintain a high running rate in a parallel environment.

Figure 6 shows a process diagram for AND parallel MGTP. It consists of generator(G), tester(T), and master(M) processes. In our implementation, several G and T processes are allocated to each PE. G(T) processes perform conjunctive matching with mixed(negative) clauses. Atoms created by a G process are stored in a *New* buffer in the G, and are sent via the Master to T processes to perform rejection testing. The M process prevents G processes from generating too many atoms by monitoring the number of atoms stored in *New* buffers and by keeping that number in a moderate range. This number indicates the difference between the number of atoms generated by G processes and the number of atoms tested by T processes. By simply controlling G and T processes with the buffering mechanism mentioned above, the idea of lazy model generation can be implemented. This also enables us to balance the computational load of G and T processes, thus keeping a high running rate.

AND Parallel Performance. Figure 7 shows AND parallel performance for solving condensed detachment problems [39] on PIM/m with 256 PEs. Proving time (sec) obtained with 32 PEs for each problem is as follows: #49:18600, #44:9700, #22:8600, #79:2500, and #82:1900. The numbers of atoms that have been kept in M and D are in between 15100 and 36500. More than a 230-fold speedup was attained for #49 and #44, and a 170 to 180-fold speedup for #22, #79 and #82.

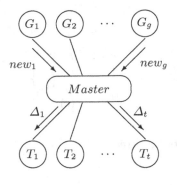 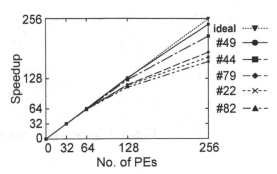

Fig. 6. Process diagram for AND parallelization

Fig. 7. Speedup ratio

To verify the effectiveness of an AND parallel MGTP, we challenged 12 hard condensed detachment problems. These problems could not be solved by OTTER with any strategy proposed in [39]. 7 of 12 problems were solved within an hour except for problem #23, in which the maximum number of atoms being stored in M and D was 85100. The problems we failed to solve were such that this size exceeds 100000 and more than 5 hours are required to solve them.

3.3   Java Implementation of MGTP

While MGTP was originally meant for parallel theorem proving based on parallel logic programming technology, Java implementations of it (JavaMGTP) [20,21] have been developed aiming at more pervasive use of MGTP through the Java technology. Here, we will briefly describe these recent implementations and results for interested readers.

The advantages of JavaMGTP's over the previous implementations with logic languages include platform independence, friendly user interfaces, and ease of extension. Moreover, JavaMGTP achieved the best performance on conventional machines among a family of model generation based provers. This achievement is brought by several implementation techniques that include a special mechanism called *A-cells* for handling multiple contexts, and an efficient term indexing. It is also a key to the achievement that we effectively utilize Java language facilities such as sophisticated class hierarchies, method overriding, and automatic memory management (garbage collection), as well as destructive assignment.

A-cells. Finding a clause $\Gamma \rightarrow B_1 \vee \ldots \vee B_m$ *violated* under a current model candidate M, i.e., $(M \models \Gamma) \wedge (\forall j_{(1 \leq j \leq m)}. \ M \not\models B_j)$ holds, MGTP extends M to $M \cup \{B_1\}, \ldots, M \cup \{B_m\}$. Repeating such extension forms a tree of model candidates, called an MG tree. Thus, each model candidate M_i comprises a sequence $< M_i^0, \ldots, M_i^j, \ldots, M_i^{k_i} >$ of sets of literals, where j is a serial number given to

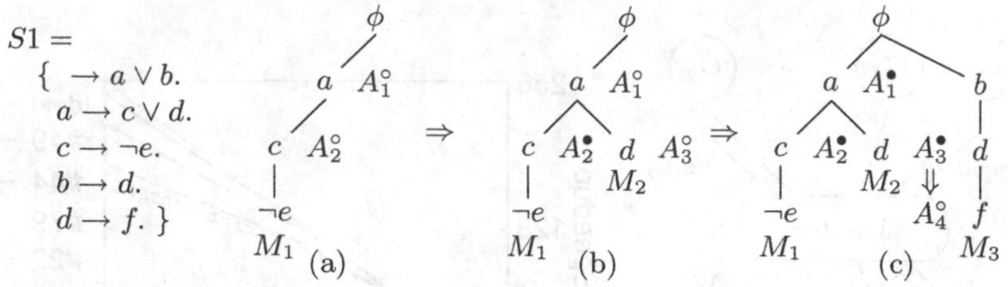

$$S1 =$$

$$\{ \ \to a \lor b.$$
$$a \to c \lor d.$$
$$c \to \neg e.$$
$$b \to d.$$
$$d \to f. \}$$

Fig. 8. Clause set S1 and its MG-tree

a context, i.e., a branch extended with a disjunct literal B^j, and M_i^j contains B^j and literals used for Horn extension that follow B^j. The most frequent operation in MGTP is to check if a ground literal L belongs to the current model candidate M. For this, we introduce an *Activation-cell* (A-cell) [21]. For each M_i^j above, we allocate an A-cell A_j containing a boolean flag **act**. When M_i^j gets included in the current model candidate M, the **act** flag of the associated A-cell A_j is set **true** (denoted by A_j°), indicating M_i^j is *active*. When M_i^j becomes no longer a member of M, the **act** of A_j is set **false** (denoted by $A_j^\bullet$), indicating M_i^j is *inactive*. On the other hand, we allocate for each atom P two variables called **pac** and **nac**, and assign a pointer to A_j to pac(nac) when $P(\neg P)$ becomes a member of M_i^j. Note that all occurrences of P and its complement $\neg P$ are represented with a unique object for P in the system. Thus, whether $P(\neg P) \in M_i^j$ can be checked simply by looking into A_j via pac(nac) of P. This A-cell mechanism reduces the complexity of the membership test to $\mathcal{O}(1)$ from $\mathcal{O}(|M|)$ which would be taken if it were naively implemented.

Figure 8 (a) shows an MG tree when a model M_1 is generated, in which pac of a refers to an A-cell A_1°, and both pac of c and nac of e refer to A_2°. In Fig. 8 (b), the current model candidate has moved from M_1 to M_2, so that the A-cell A_2° is inactivated (changed to $A_2^\bullet$), which means that neither c nor $\neg e$ belongs to the current model candidate $M_2 = \{a, d\}$. In Fig. 8 (c), the current model candidate is now $M_3 = \{b, d, f\}$, and the fact is easily recognized by looking into pac fields of b, d, and f. Note that d's pac field was updated from A_3° to A_4°. It is also easily seen that none of the other "old" literals a, c, and $\neg e$ belongs to M_3, since their pac or nac field refers to the inactivated A-cell $A_1^\bullet$ or $A_2^\bullet$.

Graphics. A JavaMGTP provides users with a graphical interface called Proof Tree Visualizer (PTV) for visualizing and controlling the proof process, which is especially useful for debugging and educational purpose. Several kinds of graphical representation for a proof tree can be chosen in PTV, e.g., a standard tree and a radial tree (see Fig. 9). The available graphical functions on a proof tree include: zooming up/down, marking/unmarking nodes, and displaying statistical information on each node. All these graphical operations are performed in concurrent with the proving process by using the multi-threading facility of Java.

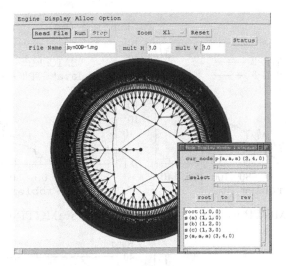

Fig. 9. A snapshot of PTV window

Moreover, one can pause/resume a proving process via the mouse control on the graphic objects for a proof tree.

Performance of a JavaMGTP. We compared JavaMGTP written in JDK1.2 (+JIT) with a Klic version of MGTP (KlicMGTP) written in KLIC v3.002 and the fastest version [49] of SATCHMO [38] written in ECL^{i}PS^e v3.5.2, on a SUN UltraSPARC10 (333MHz, 128MB). 153 range-restricted problems are taken from TPTP v2.1.1 [53], of which 42 satisfiable problems were run in all-solution mode. In Fig. 10–13, the problems are arranged and numbered in an ascending order of their execution times taken by JavaMGTP. In Fig. 12,13, a black bar shows the runtime ratio for a propositional problem, while a white bar for a first-order problem. A gap between bars (ratio zero) indicates the problems for which the two systems gave different proofs.

JavaMGTP Vs. KlicMGTP. Regarding the problems after #66 for which JavaMGTP takes more than one millisecond, JavaMGTP is 12 to 26 times (except #142) as fast as KlicMGTP for propositional cases, while 5 to 20 times for first-order cases (Fig. 12). This difference in performance is explained as follows. In JavaMGTP, CJM of ground literals like $p, q(a)$ is performed with A-cells, while CJM of a nonground literal like $r(X, Y)$ is performed with a term memory (TM) [51] rather heavier than A-cells. On the other hand, KlicMGTP always utilizes a TM for CJM, which contains some portions to be linearly scanned. Moreover, since in KlicMGTP, the TM has to be copied every time case splitting occurs, this overhead degrades the performance more significantly as the problem becomes harder.

Runtime (ms)

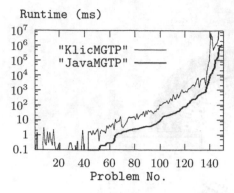

Runtime (ms)

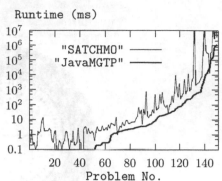

Fig. 10. JavaMGTP vs. KlicMGTP

Fig. 11. JavaMGTP vs. SATCHMO

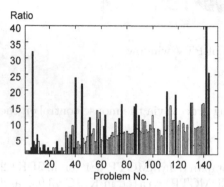

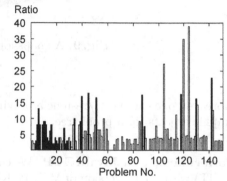

Fig. 12. JavaMGTP vs. KlicMGTP

Fig. 13. JavaMGTP vs. SATCHMO

JavaMGTP Vs. SATCHMO. SATCHMO solved three problems faster than JavaMGTP, while it failed to solve some problems due to memory overflow. This is because the proofs given by the two systems differ for such problems. For the other problems, SATCHMO gives the same proofs as JavaMGTP. Observe the problems after #66 in Fig. 13. JavaMGTP is 8 to 23 times as fast as SATCHMO for propositional cases. As for first-order cases, JavaMGTP achieves 27- to 38-fold speedup compared to SATCHMO for some problems, although its speedup gain is about 3 to 5 for most problems. In SATCHMO, since a model candidate M is maintained by using *assert/retract* of Prolog, the complexity of CJM is always $\mathcal{O}(|M|)$. On the other hand, JavaMGTP can perform CJM of ground literals in $\mathcal{O}(1)$ with A-cells. Consequently, a remarkable effect brought by this is seen for propositional problems as well as in Fig. 12. The difference in runtime for first-order problems is mainly caused by that in speed between *match*-TM and linear-search based *findall* operations, employed in JavaMGTP and SATCHMO, respectively. To get an instance of a literal, the latter takes time proportional to the number N of asserted literals, while the former a constant time w.r.t. N.

4   Extensions of MGTP Features

4.1   Extension for Constraint Solving

In this section, we present two types of extensions of the MGTP system in terms of constraint solving. Those extensions aimed at solving constraint satisfaction problems in MGTP efficiently. MGTP presents a general framework to represent and solve first order clauses, but sometimes it lacks the ability of constraint propagation using the problem (or domain) structures.

We consider, as an example, quasigroup (QG) existence problems in finite algebra [3]. This problem can be defined as finite-domain constraint satisfaction problems. In solving these problems, we found that the negative information should be propagated explicitly to prune redundant branches. This ability has been realized in the extension of MGTP, called CMGTP. Another example we consider here is channel routing problems in VLSI design. For these problems, it is needed to propagate interval constraint information as well as negative information. This additional propagation ability has been realized in the other extension of MGTP, called IV-MGTP.

CMGTP. In 1992, MGTP succeeded in solving several open quasigroup (QG) problems on a parallel inference machine PIM/m consisting of 256 processors [13]. Later, other theorem provers or constraint solvers such as DDPP, FINDER, and Eclipse solved other new open problems more efficiently than the original MGTP. Those researches have revealed that the original MGTP lacked negative constraint propagation ability. This motivated us to develop CMGTP [28,50] that allows negated atoms in the MGTP clause to enable it to propagate negative constraints explicitly.

Quasigroup Problems. A quasigroup is a pair $\langle Q, \circ \rangle$ where Q is a finite set, $\circ$ a binary operation on Q and for any $a, b, c \in Q$,

$$a \circ b = a \circ c \Rightarrow b = c$$
$$a \circ c = b \circ c \Rightarrow a = b.$$

The multiplication table of this binary operation $\circ$ forms a Latin square (shown in Fig. 14).

QG problems we tried to solve are classified to 7 categories (called QG1, QG2, ..., QG7), each of which is defined by adding some constraints to original quasigroup constraints. For example, QG5 constraint is defined as $\forall X, Y \in Q$. $((Y \circ X) \circ Y) \circ Y = X$. This constraint is represented with an MGTP clause:

$$p(Y, X, A) \wedge p(A, Y, B) \wedge p(B, Y, C), X \neq C \rightarrow . \tag{1}$$

From the view point of constraint propagation, rule (1) can be rewritten as follows[2]:

$$p(Y, X, A) \wedge p(A, Y, B) \rightarrow p(B, Y, X). \tag{2}$$

---

[2] In addition, we assume functionality in the arguments of p.

∘	1	2	3	4	5
1	1	3	2	5	4
2	5	2	4	3	1
3	4	5	3	1	2
4	2	1	5	4	3
5	3	4	1	2	5

Fig. 14. Latin square (order 5)

$$p(Y, X, A) \wedge p(B, Y, X) \rightarrow p(A, Y, B). \tag{3}$$
$$p(B, Y, X) \wedge p(A, Y, B) \rightarrow p(Y, X, A). \tag{4}$$

These rules are still in the MGTP representation. To generate negative constraints, we add extra rules containing negative atoms to the original MGTP rule, by making contrapositives of it. For example, rule (2) can be augmented by the following rules:

$$p(Y, X, A) \wedge \neg p(B, Y, X) \rightarrow \neg p(A, Y, B). \tag{5}$$
$$p(A, Y, B) \wedge \neg p(B, Y, X) \rightarrow \neg p(Y, X, A). \tag{6}$$

Each of the above rules is logically equivalent to (2), but has a different operational meaning, that is, if a negative atom is derived, it can simplify the current disjunctive clauses in the disjunction buffer D. This simplification can reduce the number of redundant branches significantly.

CMGTP Procedure. The structure of the model generation processes in CMGTP is basically the same as MGTP. The differences between CMGTP and MGTP are in the unit refutation processes and the unit simplification processes with negative atoms. We can use negative atoms explicitly in CMGTP to represent constraints. If there exist P and $\neg P$ in the current model candidate M, then *false* is derived by the unit refutation mechanism. If for a unit clause $\neg P_i \in M(P_i \in M)$, there exists a disjunction which includes $P_i(\neg P_i)$, then $P_i(\neg P_i)$ is removed from that disjunction by the unit simplification mechanism.

The refutation and simplification processes added to MGTP guarantee that for any atom $P \in M$, P and $\neg P$ are not in the current M simultaneously, and disjunctions in the current D have already been simplified by all unit clauses in M.

Experimental Results. Table 1 compares the experimental results for QG problems on CP, CMGTP and other systems. CP is an experimental program written in SICStus Prolog, that is dedicated to QG problems [50]. In CP, the domain variable and its candidates to be assigned are represented with shared variables.

The number of failed branches generated by CP and CMGTP are almost equal to DDPP and less than those from FINDER and MGTP. In fact, we

Table 1. Comparison of experimental results for QG5

Order	DDPP	FINDER	MGTP	CP	CMGTP	IV-MGTP
			Failed Branches			
9	15	40	239	15	15	15
10	50	356	7026	38	38	52
11	136	1845	51904	117	117	167
12	443	13527	2749676	372	372	320

confirmed that CP and CMGTP have the same pruning ability as DDPP by comparing the proof trees generated by these systems. The slight differences in the number of failed branches were caused by the different selection functions used.

For general performance, CP was superior to the other systems in almost every case. In particular, we obtained a new result in October 1993 that no model exists for QG5.16 by running CP on a SPARCstation-10 for 21 days. On the other hand, CMGTP is about 10 times slower than CP. The difference in speed is mainly caused by the term memory manipulation necessary for CMGTP.

IV-MGTP. In MGTP (CMGTP), interpretations (called *model candidates*) are represented as finite sets of ground atoms (literals). In many situations this turns out being too redundant. Take, for example, variables I, J ranging over the domain $\{1, \ldots, 4\}$, and interpret $\leq$, $+$ naturally. A rule like "$p(I) \wedge \{I + J \leq 4\} \to q(J)$" splits into three model extensions: $q(1)$, $q(2)$, $q(3)$, if $p(1)$ is present in the current model candidate. Now assume we have the rule "$q(I) \wedge q(J) \wedge \{I \neq J\} \to .$" saying that q is functional in its argument and, say, $q(4)$ is derived from another rule. Then all three branches must be refuted separately.

Totally ordered, finite domains occur naturally in many problems. In such problems, situations such as the one just sketched are common. Thus we developed an IV-MGTP system [19] to enhance MGTP with mechanisms to deal with them efficiently.

Introducing Constrained Atoms into MGTP. In order to enhance MGTP with totally ordered, finite domain constraints, we adopt the notation: $p(t_1, \ldots, t_r, S_1, \ldots, S_m)$ for what we call a *constrained atom*. This notation is motivated from the viewpoint of signed formula logic programming (SFLP) [37] and constraint logic programming (CLP) over finite domains [41].

Constrained atoms explicitly stipulate subsets of domains and thus are in *solved form*. The language of IV-MGTP needs to admit other forms of atoms, in order to be practically useful in solving problems with totally ordered domains. An IV-MGTP atom is an expression $p(t_1, \ldots, t_r, \kappa_1, \ldots, \kappa_m)$, where the κ_i has one of the following forms:

1. $\{i_1, \ldots, i_r\}$, where $i_j \in N$ for $1 \leq j \leq r$ (κ_i is in solved form);
2. $]\iota_1, \iota_2[$, where $\iota_j (j = 1, 2) \in N \cup CVar$; the intended meaning is $]\iota_1, \iota_2[= \{i \in N \mid i < \iota_1 \text{ or } i > \iota_2\}$;

3. $[\iota_1, \iota_2]$, where $\iota_j (j = 1, 2) \in N \cup CVar$; the intended meaning is $[\iota_1, \iota_2] = \{i \in N \mid \iota_1 \leq i \leq \iota_2\}$;
4. $U \in DVar$.

where $CVar$ is a set of *constraint variables* which hold elements from a domain N, and $DVar$ is a set of *domain variables* which hold subsets of a domain N. In this framework, since intervals play a central role, we gave the name IV-MGTP to the extension of MGTP.

For each predicate p with constrained arguments, an IV-MGTP program contains a *declaration* line of the form "declare $p(t, \ldots, t, j_1, \ldots, j_m)$". If the i-th place of p is t, then the i-th argument of p is a standard term; if the i-th place of p is a positive integer j, then the i-th argument of p is a constraint over the domain $\{1, \ldots, j\}$.

Each IV-MGTP atom $p(t_1, \ldots, t_r, \kappa_1, \ldots, \kappa_m)$ consists of two parts: the standard *term part* $p(t_1, \ldots, t_r)$ and the *constraint part* $\langle \kappa_1, \ldots, \kappa_m \rangle$. Each of r and m can be 0. The latter, $m = 0$, is in particular the case for a predicate that has no declaration. By this convention, every MGTP program is an IV-MGTP program. If $m = 1$ and the domain of κ_1 is $\{1, 2\}$, the IV-MGTP programs are equivalent to CMGTP programs where $\{1\}$ is interpreted as positive and $\{2\}$ as negative. Hence, every CMGTP program is also an IV-MGTP program.

Model Candidates in IV-MGTP. While the deduction procedure for IV-MGTP is almost the same as for CMGTP, model candidates are treated differently. In MGTP, a list of current model candidates that represent Herbrand interpretations is kept during the deduction process, and model candidates can be simply identified with sets of ground atoms. The same holds in IV-MGTP, only that some places of a predicate contain a ground constraint in solved form (that is: a subset of a domain) instead of a ground term. Note that, while in MGTP one model candidate containing ground atoms $\{L_1, \ldots, L_r\}$ trivially represents exactly one possible interpretation of the set of atoms $\{L_1, \ldots, L_r\}$, in IV-MGTP one model candidate represents many IV-MGTP interpretations which differ in the constraint parts.

Thus, model candidates can be conceived as sets of constrained atoms of the form $p\ (t_1, \ldots, t_r, S_1, \ldots, S_m)$, where the S_i are subsets of the appropriate domain. If M is a model candidate, $p(t_1, \ldots, t_r)$ the ground term part, and $\langle S_1, \ldots, S_m \rangle$ the constraint part in M, then define $M\ (\ p(t_1, \ldots, t_r)\ ) = \langle S_1, \ldots, S_m \rangle$. We say that a ground constrained atom $L = p\ (t_1, \ldots, t_r, i_1, \ldots, i_m)$ is *satisfied* by M ($M \models L$) iff there are domain elements $s_1 \in i_1, \ldots, s_m \in i_m$ such that $\langle s_1, \ldots, s_m \rangle \in M(p(t_1, \ldots, t_r))$.

Formally, a *model candidate M* is a *partial* function that maps ground instances of the term part of constrained atoms which is declared as "$p(t, \ldots, t, j_1, \ldots, j_m)$" into $(2^{\{1, \ldots, j_1\}} - \{\emptyset\}) \times \cdots \times (2^{\{1, \ldots, j_m\}} - \{\emptyset\})$. Note that $M(p(t_1, \ldots, t_r))$ can be undefined.

Besides rejection, subsumption, and extension of a model candidate, in IV-MGTP there is a fourth possibility not present in MGTP, that is, *model can-*

didate update. We see that model candidate update is really a combination of subsumption and rejection. Consider the following example.

Example 1. Let $C = p(\{1, 2\})$ be the consequent of an IV-MGTP rule and assume $M(p) = \langle\{2, 3\}\rangle$. Neither is the single atom in C inconsistent with M nor is it subsumed by M. Yet the information contained in C is not identical to that in M and it can be used to refine M to $M(p) = \langle\{2\}\rangle$.

Channel Routing Problems. Channel routing problems in VLSI design can be represented as constraint satisfaction problems, in which connection requirements (what we call *nets*) between terminals must be solved under the condition that each net has a disjoint path from all others. For these problems, many specialized solvers employing heuristics were developed. Our experiments are not primarily intended to compare IV-MGTP with such solvers, but to show the effectiveness of the interval/extraval representation and its domain calculation in the IV-MGTP procedure.

We consider a multi-layer channel which consists of multiple layers, each of which has multiple tracks. We assume in addition, to simplify the problem, that each routing path makes no detour and contains only one track. By this assumption, the problem can be formalized to determine the layer and the track numbers for each net with the help of constraints that express the two binary relations: *not equal* (*neq*) and *above*. $neq(N_1, N_2)$ means that the net N_1 and N_2 do not share the same track. $above(N_1, N_2)$ means that if N_1 and N_2 share the same layer, the track number of N_1 must be larger than that of N_2.

For example, *not equal* constraints for nets N_1 and N_2 are represented in IV-MGTP as follows:

$$p(N_1, [L, L], [T_1, T_1]) \wedge p(N_2, [L, L], [T_{21}, T_{22}]) \wedge neq(N_1, N_2)$$
$$\rightarrow p(N_2, [L, L],]T_1, T_1[)$$

where the predicate p has two constraint domains: layer number L and track number T_i.

Experimental Results. We developed an IV-MGTP prototype system in Java and made experiments on a Sun Ultra 5 under JDK 1.2. The results are compared with those on the same problems formulated and run with CMGTP [50] (also written in Java [21]). We experimented with problems consisting of 6, 8, 10, and 12 net patterns on the 2 layers channel each of which has 3 tracks. The results are shown in Table 2.

IV-MGTP reduces the number of models considerably. For example, we found the following model in a 6-net problem:

$$\{ p(1, [1, 1], [3, 3]), p(2, [1, 1], [1, 1]), p(3, [1, 1], [2, 2]),$$
$$p(4, [2, 2], [2, 3]), p(5, [2, 2], [1, 2]), p(6, [1, 1], [2, 3]) \},$$

which contains 8 ($= 1 \times 1 \times 1 \times 2 \times 2 \times 2$) CMGTP models. The advantage of using IV-MGTP is that the different feasible track numbers can be represented as

Table 2. Experimental results for the channel routing problem

Number of Nets = 6

	IV-MGTP	CMGTP
models	250	840
branches	286	882
runtime(msec)	168	95

Number of Nets = 8

	IV-MGTP	CMGTP
models	1560	10296
branches	1808	10302
runtime(msec)	706	470

Number of Nets = 10

	IV-MGTP	CMGTP
models	4998	51922
branches	6238	52000
runtime(msec)	2311	3882

Number of Nets = 12

	IV-MGTP	CMGTP
models	13482	538056
branches	20092	539982
runtime(msec)	7498	31681

interval constraints. In CMGTP, the above model is split into 8 different models. Obviously, as the number of nets increases, the reduction ratio of the number of models becomes larger. We conclude that IV-MGTP can effectively suppress unnecessary case splitting by using interval constraints, and hence, reduce the total size of proofs.

Because CMGTP program can be transferred to IV-MGTP program, QG problems can be transferred into IV-MGTP program. IV-MGTP, however, cannot solve QG problems more efficiently than CMGTP, that is, QG problems do not receive the benefit of IV-MGTP representation and deduction process. The efficiency or advantage by using IV-MGTP depends on the problem domain how beneficial the effect of interval/extraval constraints on performance is. For problems where the ordering of the domain elements has no significance, such as the elements of a QG problem (whose numeric elements are considered strictly as symbolic values, not arithmetic values), CMGTP and IV-MGTP have essentially the same pruning effect. However, where reasoning on the arithmetic ordering between the elements is important, such as in channel routing problems, IV-MGTP outperforms CMGTP.

Completeness. MGTP provides a sound and complete procedure in the sense of standard Herbrand interpretation. The extensions, CMGTP and IV-MGTP described above, however, lost completeness [19]. The reason is essentially the same as for incompleteness of resolution and hypertableaux with unrestricted selection function [18].

It can be demonstrated with the simple example $P = \{\to p, \ \neg q \to \neg p, \ q \to\}$. The program P is unsatisfiable, yet deduction procedures based on selection of only antecedent (or only consequent) literals cannot detect this. Likewise, the incomplete treatment of negation in CMGTP comes up with the incorrect model $\{p\}$ for P. The example can be transferred to IV-MGTP [3]. Assume p and q are

---

[3] We discuss only about IV-MGTP in the rest of this section, because CMGTP can be considered as a special case of IV-MGTP. It is sufficient to say about IV-MGTP.

defined "declare $p(2)$" and "declare $q(2)$". The idea is to represent a positive literal p with $p(\{2\})$ and a negative literal $\neg p$ with $p(\{1\})$. Consider

$$P' = \{\rightarrow p(\{2\}),\ q(\{1\}) \rightarrow p(\{1\}),\ q(\{2\}) \rightarrow\} \tag{7}$$

which is unsatisfiable (recall that p and q are functional), but has an IV-MGTP model, where $M(p) = \langle\{2\}\rangle$, and $M(q)$ is undefined.

In order to handle such cases, we adopt a non-standard semantics called *extended interpretations* which is suggested in SFLP [37]. The basic idea underlying extended interpretations (e-interpretations) is to introduce the disjunctive information inherent to constraints into the interpretations themselves. In e-interpretations, an e-interpretation of a predicate p is a partial function $\mathbf{I}$ mapping ground instances of the term part of p into its constraint part. This means that the concepts introduced for model candidates can be used for e-interpretations.

An extended interpretation $\mathbf{I}$ does *e-satisfy* an IV-MGTP ground atom $L = p(t_1, \ldots, t_r, S_1, \ldots, S_m)$ iff $\mathbf{I}(p(t_1, \ldots, t_r))$ is defined, has the value $\langle S_1', \ldots, S_m'\rangle$, and $S_i' \subseteq S_i$ for all $1 \leq i \leq m$.

Using the above definition, we have proved the following completeness theorem [19].

Theorem 1 (Completeness). *An IV-MGTP program P having an IV-MGTP model M is e-satisfiable by M (viewed as an e-interpretation).*

Simple conversion of this theorem and proof makes the case of CMGTP trivial.

4.2   Non-Horn Magic Set

The basic behaviors of model generation theorem provers, such as SATCHMO and MGTP, are to detect a violated clause under some interpretation, called a model candidate, and to extend the model candidate so that the clause is satisfied.

However, when there are several violated clauses, a computational cost may greatly differ according to the order in which those clauses are evaluated. Especially when a non-Horn clause irrelevant to the given goal is selected, many interpretations generated with the clause would become useless. Thus, in the model generation method, it is necessary to develop a method to suppress the generation of useless interpretations.

To this end, Loveland *et al.* proposed a method, called *relevancy testing* [56,36], to restrict the selecting of a violated clause to only those whose consequent literals are all relevant to the given goal ("totally relevant"). Then they implemented this idea in SATCHMORE (SATCHMO with RElevancy).

Let HC be a set of Horn clauses, and I be a current model candidate. A relevant literal is defined as a goal called in a failed search to prove $\perp$ from $HC \cup I$ or a goal called in a failed search to prove the antecedent of a non-Horn clause by Prolog execution.

The relevancy testing can avoid useless model extension with irrelevant violated clauses. However, there is some overhead, because it computes relevant literals dynamically by utilizing Prolog over Horn clauses whenever violated non-Horn clauses are detected.

On the other hand, compared to top-down provers, a model generation prover like SATCHMO or MGTP can avoid solving duplicate subgoals because it is based on bottom-up evaluation. However, it also has the disadvantage of generating irrelevant atoms to prove the given goal. Thus it is necessary to combine bottom-up with top-down proving to use goal information contained in negative clauses, and to avoid generating useless model candidates. For this purpose, several methods such as magic sets, Alexander templates, and bottom-up meta-interpretation have been proposed in the field of deductive databases [9].

All of these transform the given Horn intentional databases to efficient Horn intentional databases, which generate only ground atoms relevant to the given goal in extensional databases. However, these were restricted to Horn programs.

To further extend these methods, we developed a new transformation method applicable to non-Horn clauses. We call it the non-Horn magic set (NHM) [24]. NHM is a natural extension of the magic set yet works within the framework of the model generation method. Another extension for non-Horn clauses has been proposed, which simulates top-down execution based on the model elimination procedure within a forward chaining paradigm [52].

In the NHM method, each clause in a given clause set is transformed into two types of clauses. One is used to simulate backward reasoning and the other is to control inferences in forward reasoning. The set of transformed clauses is proven by bottom-up theorem provers.

There are two kinds of transformation methods: the *breadth-first NHM* and the *depth-first NHM*. The former simulates breadth-first backward reasoning, and the latter simulates depth-first backward reasoning.

Breadth-first NHM. For the breadth-first NHM method, a clause $A_1 \wedge \cdots \wedge A_n \rightarrow B_1 \vee \cdots \vee B_m$ in the given clause set S is transformed into the following (extended) clauses:

$$T_B^1 : goal(B_1) \wedge \ldots \wedge goal(B_m) \rightarrow goal(A_1) \wedge \ldots \wedge goal(A_n).$$
$$T_B^2 : goal(B_1) \wedge \ldots \wedge goal(B_m) \wedge A_1 \wedge \ldots \wedge A_n \rightarrow B_1 \vee \ldots \vee B_m.$$

In this transformation, for $n = 0$ (a positive clause), the first transformed clause T_B^1 is omitted. For $m = 0$ (a negative clause), the conjunction of $goal(B_1)$, $\ldots$, $goal(B_m)$ becomes *true*. For $n \neq 0$, two clauses T_B^1 and T_B^2 are obtained by the transformation.

Here, the meta-predicate $goal(A)$ represents that the atom A is relevant to the goal and it must be solved. The clause T_B^1 simulates top-down evaluation. Intuitively, T_B^1 means that when it is necessary to solve the consequent $B_1, \ldots, B_m$ of the original clause, it is necessary to solve the antecedent $A_1, \ldots, A_n$ before doing that. The n antecedent literals are solved in parallel. On the other hand, the clause T_B^2 simulates relevancy testing. T_B^2 means that a model extension with the consequent is performed only when $A_1, \ldots, A_n$ are satisfied by the current model candidate and all the consequent atoms $B_1, \ldots, B_m$ are relevant to the given goal. That is, the original clause is not used for model extension if there exists any consequent literal B_j such that B_j is not a goal.

Depth-first NHM. For the depth-first NHM transformation, a clause $A_1 \wedge \cdots \wedge A_n \to B_1 \vee \cdots \vee B_m$ in S is transformed into $n+1$ (extended) clauses:

$$T_D^1 : goal(B_1) \wedge \ldots \wedge goal(B_m) \to goal(A_1) \wedge cont_{k,1}(V_k).$$
$$T_D^2 : cont_{k,1}(V_k) \wedge A_1 \to goal(A_2) \wedge cont_{k,2}(V_k).$$
$$\vdots$$
$$T_D^n : cont_{k,(n-1)}(V_k) \wedge A_{n-1} \to goal(A_n) \wedge cont_{k,n}(V_k).$$
$$T_D^{n+1} : cont_{k,n}(V_k) \wedge A_n \to B_1 \vee \ldots \vee B_m.$$

where k is the clause identifier of the original clause, V_k is the tuple of all variables appearing in the original clause. The transformed clauses are interpreted as follows: If all consequent literals $B_1, \cdots, B_m$ are goals, we first attempt to solve the first atom A_1. At that time, the variable bindings obtained in the satisfiability checking of the antecedent are propagated to the next clause T_D^2 by the continuation literal $cont_{k,1}(V_k)$. If atom A_1 is solved under $cont_{k,1}(V_k)$, then we attempt to solve the second atom A_2, and so on.

Unlike the breadth-first NHM transformation, n antecedent atoms are being solved sequentially from A_1 to A_n. During this process, the variable binding information is propagated from A_1 to A_n in this order.

Several experimental results obtained so far suggest that the NHM and relevancy testing methods have a similar or the same pruning ability. To clarify this, we defined the concept of *weak relevancy testing* that mitigates the condition of relevancy testing, and then proved that the NHM method is equivalent to the weak relevancy testing in terms of the ability to prune redundant branches [45]. However, significant differences between NHM and SATCHMORE can be admitted. First, SATCHMORE performs the relevancy testing dynamically during proof, while NHM is based on the static analysis of input clauses and transforms them as a preprocessing of proof. Second, the relevancy testing by SATCHMORE repeatedly calls Prolog to compute relevant literals backward whenever a new violated clause is found. This process often results in re-computation of the same relevant literals. In contrast, for NHM, *goal* literals are computed forward and their re-computation is avoided.

4.3 Eliminating Redundant Searches by Dependency Analysis

There are two types of redundancies in model generation: One is that the same subproof tree may be generated at several descendants after a case-splitting occurs. Another is caused by unnecessary model candidate extensions.

Folding-up is a well known technique for eliminating duplicate subproofs in a tableaux framework [34]. In order to embed folding-up into model generation,

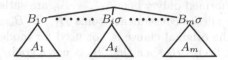

Fig. 15. Model extension

we have to analyze dependency in a proof for extracting lemmas from proven subproofs. Lemmas are used for pruning other remaining subproofs. Dependency analysis makes unnecessary parts visible because such parts are independent of essential parts in the proof. In other words, we can separate unnecessary parts from the proof according to dependency analysis.

Identifying unnecessary parts and eliminating them are considered as *proof simplification*. The computational mechanism for their elimination is essentially the same as that for *proof condensation* [46] and *level cut* [2]. Taking this into consideration, we implemented not only folding-up but also proof condensation by embedding a single mechanism, i.e. proof simplification, into model generation [32].

In the following, we consider the function MG in Fig. 1 to be a builder of proof trees in which each leaf is labeled with $\bot$ (for a failed branch, that is, UNSAT) or $\top$ (for a success branch, that is, SAT), and each non-leaf node is labeled with an atom used for model extension.

Definition 1 (Relevant atom). *Let P be a finite proof tree. A set $Rel(P)$ of relevant atoms of P is defined as follows:*

1. *If $P = \overset{|}{\bot}$ and $A_1\sigma \wedge \ldots \wedge A_n\sigma \rightarrow$ is the negative clause used for building P, then $Rel(P) = \{A_1\sigma, \ldots, A_n\sigma\}$.*
2. *If $P = \overset{|}{\top}$, then $Rel(P) = \emptyset$.*
3. *If P is in the form depicted in Fig. 15, $A_1\sigma \wedge \ldots \wedge A_n\sigma \rightarrow B_1\sigma \vee \ldots \vee B_m\sigma$ is the mixed or positive clause used for forming the root of P and*
   (a) $\forall i(1 \leq i \leq m)B_i\sigma \in Rel(P_i)$, then $Rel(P) = \cup_{i=1}^{m}(Rel(P_i) \setminus \{B_i\sigma\}) \cup \{A_1\sigma, \ldots, A_n\sigma\}$
   (b) $\exists i(1 \leq i \leq m)B_i\sigma \notin Rel(P_i)$, then $Rel(P) = Rel(P_{i_0})$ (where i_0 is the minimal index satisfying $1 \leq i_0 \leq m$ and $B_{i_0}\sigma \notin Rel(P_{i_0})$)

Informally, relevant atoms of a proof tree P are atoms which contribute to building P and appear as ancestors of P if P does not contain $\top$. If P contains $\top$, the set of relevant atoms of P is $\emptyset$.

Definition 2 (Relevant model extension). *A model extension with a clause $A_1\sigma \wedge \ldots \wedge A_n\sigma \rightarrow B_1\sigma \vee \ldots \vee B_m\sigma$ is relevant to the proof if the model extension yields the proof tree in the form depicted is Fig. 15 and either $\forall i(1 \leq i \leq m)B_i\sigma \in Rel(P_i)$ or $\exists i(1 \leq i \leq m)(P_i$ contains $\top)$ holds.*

We can eliminate irrelevant model extensions as follows. Let P be a proof tree in the form depicted in Fig. 15. If there exists a subproof tree P_i $(1 \leq i \leq m)$ such that $B_i\sigma \notin Rel(P_i)$ and P_i does not contain $\top$, we can conclude that the model extension forming the root of P is unnecessary because $B_i\sigma$ does not contribute to P_i. Therefore, we can delete other subproof trees $P_j(1 \leq j \leq m, j \neq i)$ and take P_i to be a simplified proof tree of P. When P contains $\top$, we see that the model extension forming the root of P is necessary from a model finding point of view.

Performing proof simplification *during* the proof, instead of *after* the proof has been completed, makes the model generation procedure more efficient. Let assume that we build a proof tree P (in the form depicted in Fig. 15) in a left-first manner and check whether $B_i\sigma \in Rel(P_i)$ after P_i is built. If $B_i\sigma \notin Rel(P_i)$ holds, we can ignore building the proofs $P_j(i < j \leq m)$ because the model extension does not contribute to the proof P_i. Thus $m - i$ out of m branches are eliminated after i branches have been explored. This proof elimination mechanism is essentially the same as the proof condensation [46] and the level cut [2] facilities. We can make use of a set of relevant atoms not only for proof condensation but also for generating lemmas.

Theorem 2. *Let S be a set of clauses, M a set of ground atoms and $P = MG(U_0, D_0, \emptyset)$. Note that MG in Fig. 1 is modified to return a proof tree. If all leaves in P are labeled with $\bot$, i.e. P does not contain $\top$, then $S \cup Rel(P)$ is unsatisfiable.*

This theorem says that a set of relevant atoms can be considered as a lemma. Consider the model generation procedure shown in Fig. 1. Let M be a current model candidate and P be a subproof tree which was previously obtained and does not contain $\top$. If $M \supset Rel(P)$ holds, we can reject M without further proving because $S \cup M$ is unsatisfiable where S is a clause set to be proven. This rejection mechanism can reduce search spaces by orders of magnitude. However, it is expensive to test whether $M \supset Rel(P)$. Thus, we restrict the usage of the rejection mechanism.

Definition 3 (Context unit lemma). *Let S be a set of clauses and P a proof tree of S in the form depicted in Fig. 15. When $B_i\sigma \in Rel(P_i)$, $Rel(P_i) \setminus \{B_i\sigma\} \models_S \neg B_i\sigma$ is called a context unit lemma[4] extracted from P_i. We call $Rel(P_i) \setminus \{B_i\sigma\}$ the context of the lemma.*

Note that $B_i\sigma \in Rel(P_i)$ implies $Rel(P_i)$ is not empty. Therefore, P_i does not contain $\top$. Thus, $S \cup Rel(P_i)$ is unsatisfiable according to Theorem 2.

The context of the context unit lemma extracted from $P_i(1 \leq i \leq m)$ is satisfied in model candidates of sibling proofs $P_j(j \neq i, 1 \leq j \leq m)$, that is, the lemma is available in P_j. Furthermore, the lemma can be lifted to the nearest ancestor's node which does not satisfy the context (in other words, which is

---

[4] $\Gamma \models_S L$ is an abbreviation of $S \cup \Gamma \models L$ where Γ is a set of ground literals, S is a set of clauses, and L is a literal.

labeled with an atom in the context) and is available in its descendant's proofs. Lifting context unit lemmas to appropriate nodes and using them for pruning proof tree is an implementation of folding-up [34] for model generation.

In this way, not only folding-up but also proof condensation can be achieved by calculating sets of relevant atoms of proofs. We have already implemented the model generation procedure with folding-up and proof condensation and experienced their pruning effects on some typical examples. For all non-Horn problems (1984 problems) in the TPTP library [53] version 2.2.1, the overall success rate was about 19% (cf., pure model generation 16%, Otter(v3.0.5) 27%[5]) for a time limit of 10 minutes on a Sun Ultra1 (143MHz, 256MB, Solaris2.5.1) workstation.

4.4   Minimal Model Generation

The notion of minimal models is important in a wide range of areas such as logic programming, deductive databases, software verification, and hypothetical reasoning. Some applications in such areas would actually need to generate Herbrand minimal models of a given set of first-order clauses.

A model generation algorithm can generate all minimal Herbrand models if they are finite, though it may generate non-minimal models [10]. Bry and Yahya proposed a sound (in the sense that it generates only minimal models) and complete (in the sense that it generates all minimal models) minimal model generation prover MM-SATCHMO [10]. It uses *complement splitting* (or *folding-down* in [34]) for pruning some branches leading to nonminimal models and *constrained search* for eliminating non-minimal models. Niemelä also presented a propositional tableaux calculus for minimal model reasoning [43], where he introduced the *groundedness test* which substitutes for constrained searches.

The following theorem says that a model being eliminated by factorization [34] in the model generation process is not minimal. This implies that model generation with factorization is complete for generating minimal models. It is also known that factorization is more flexible than complement splitting for pruning the redundant search spaces [34].

Theorem 3. *Let P be a proof tree of a set S of clauses. We assume that N_1 and N_2 are sibling nodes in P, N_i is labeled with a literal L_i, and P_i is a subproof tree under $N_i(i = 1, 2)$ shown in Fig. 16(a).*

If there is a node N_3, descended from N_2, labeled with L_1, then for each model M found in proof tree P_3, there exists a model M' found in P_1 such that $M' \subset M$ where P_3 is a subproof tree under N_3 (Fig. 16(b)).

To avoid a circular argument, the proof tree has to be supplied with an additional factorization dependency relation.

---

[5]   This measurement is obtained by our experiment with just Otter (not Otter+MACE).

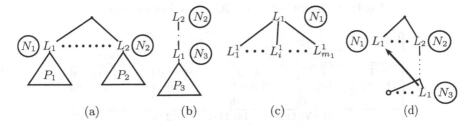

Fig. 16. Proof trees explaining Theorem 3, 4 and Definition 5

Definition 4 (Factorization dependency relation). *A factorization dependency relation on a proof tree is a strict partial ordering $\prec$ relating sibling nodes in the tree ($N_1 \prec N_2$ means that searching minimal models under N_2 is delegated to that under N_1).*

Definition 5 (Factorization). *Given a proof tree P and a factorization dependency relation $\prec$ on P. First, select a node N_3 labeled with literal L_1 and another node N_1 labeled with the same literal L_1 such that (1) N_3 is a descendant of N_2 which is the sibling node of N_1, and (2) $N_2 \not\prec N_1$.*

Then, mark N_3 with N_1 and modify $\prec$ by first adding the pair of nodes $\langle N_1, N_2 \rangle$ and then forming the transitive closure of the relation. We say that N_3 has been factorized with N_1. Marking N_3 with N_1 indicates finding models under N_3 is delegated to that under N_1. The situation is depicted in Fig. 16(d).

Corollary 1. *Let S be a set of clauses. If a minimal model M of S is built by model generation, then M is also built by model generation with factorization.*

We can replace $L_1 \vee L_2 \vee \ldots \vee L_n$ used for non-Horn extension with an augmented one $(L_1 \wedge \neg L_2 \wedge \ldots \wedge \neg L_n) \vee (L_2 \wedge \neg L_3 \wedge \ldots \wedge \neg L_n) \vee \ldots \vee L_n$, which corresponds to complement splitting. Here a negated literal is called a *branching assumption*. If none of branching assumptions $\neg L_{i+1}, \ldots, \neg L_n$ is used in a branch expanded below L_i, we can use $\neg L_i$ as a unit lemma in the proof of L_j ($i+1 \leq j \leq n$). The unit lemma is called a *branching lemma*.

We consider model generation with complement splitting as pre-determining factorization dependency relation on sibling nodes $N_1, \ldots, N_m$ as follows: $N_j \prec N_i$ if $i < j$ for all i and j ($1 \leq i, j \leq m$). According to this consideration, complement splitting is a restricted way of implementing factorization.

We have proposed a minimal model generation procedure [23] that employs branching assumptions and lemmas. We consider model generation with branching assumptions and lemmas as arranging factorization dependency relation on sibling nodes $N_1, \ldots, N_m$ as follows: For each i ($1 \leq i \leq m$), $N_j \prec N_i$ for all j ($i < j \leq m$) if $N_{j_0} \prec N_i$ for some j_0 ($i < j_0 \leq m$) and otherwise $N_i \prec N_j$ for all j ($i < j \leq m$). Performing branching assumptions and lemmas can still be taken as a restricted implementation of factorization. Nevertheless, it provides

Table 3. Results of MM-MGTP and other systems

| Problem | MM-MGTP | | MM- | |
	Rcmp	Mchk	SATCHMO	MGTP
ex1	0.271	0.520	8869.950	0.199
(N=5)	100000	100000	100000	100000
	0	0	0	0
ex1	34.150	OM (>144)	OM (>40523)	19.817
(N=7)	10000000	–	–	10000000
	0	–	–	0
ex2	0.001	0.001	1107.360	9.013
(N=14)	1	1	1	1594323
	26	26	1594323	0
ex3	19.816	5.076	OM (>2798)	589.651
(N=16)	65536	65536	–	86093442
	1	1	–	0
ex3	98.200	26.483	OM (>1629)	5596.270
(N=18)	262144	262144	–	774840978
	1	1	–	0
ex4	0.002	0.002	0.3	0.004
	341	341	341	501
	96	96	284	0
ex5	0.001	0.001	0.25	0.001
	17	17	17	129
	84	84	608	0

top: time(sec), middle: No. of models,
bottom: No. of failed branches, OM: Out of memory.
MM-MGTP and MGTP: run on Java (Solaris_JDK_1.2.1_03)
MM-SATCHMO: run on ECL^{i}PS^s Prolog Version 3.5.2
All programs were run on Sun Ultra10 (333MHz, 128MB)

an efficient way of applying factorization to minimal model generation, since it is unnecessary to compute the transitive closure of the factorization dependency relation.

In order to make the procedure sound in the sense that it generates only minimal models, it is necessary to test whether a generated model is minimal or not. The following theorem gives a necessary condition for a generated model to be nonminimal.

Theorem 4. *Let S be a set of clauses and P a proof tree of S obtained by the model generation with factorization. We assume that N_1 and N_2 are sibling nodes in P, P_i a subproof tree under N_i, and M_i a model found in $P_i(i = 1, 2)$. If $N_2 \not\prec N_1$, then $M_1 \not\subset M_2$.*

Theorem 4 says that we have to test whether $M_1 \subset M_2$ only when M_i is found under a node N_i $(i = 1, 2)$ such that $N_2 \prec N_1$.

We implemented a minimal model generation prover called MM-MGTP with branching assumptions and lemmas on Java [23]. The implementation takes Theorem 4 into account. It is applicable to first-order clauses as well as MM-SATCHMO. Table 3 shows experimental results on MM-MGTP, MM-SATCH-MO, and MGTP. There are two versions of MM-MGTP: *model checking* (Mchk) and *model re-computing* (Rcmp). The former is based on constrained search and the latter on the groundedness test.

Although the model checking MM-MGTP is similar to MM-SATCHMO, the way of treating model constraints differs somewhat. Instead of dynamically adding model constraints (negative clauses) to the given clause set, MM-MGTP retains them in the form of a model tree consisting of only models. Thus, the constrained search for minimal models in MM-SATCHMO is replaced by a *model tree traversal* for minimality testing.

In the model re-computing version, a re-computation procedure for minimality testing is invoked instead of a model tree traversal. The procedure is the same as *MG* except that some routines are modified for restarting the execution. It returns UNSAT if the current model is minimal, otherwise SAT. Experimental results show remarkable speedup compared to MM-SATCHMO. See [23] for a detailed consideration on the experiment.

5   Applications

A model generation theorem prover has a general reasoning power in various AI applications. In particular, we first implemented a propositional modal tableaux system on MGTP, by representing each rule of tableaux with MGTP input clauses. This approach has lead to research on logic programming with negation as failure [29], abductive reasoning [30], modal logic systems [31], mode analysis of FGHC programs [54], and legal reasoning [44,27], etc. In the following sections, we focus on the issue of implementing negation as failure within a framework of model generation, and describe how this feature is used to build a legal reasoning system.

5.1   Embedding Negation as Failure into MGTP

Negation as failure is one of the most important techniques developed in the logic programming field, and logic programming supporting this feature can be a powerful knowledge representation tool. Accordingly, declarative semantics such as the *answer set* semantics have been given to extensions of logic programs containing both negation as failure (*not*) and classical negation ($\neg$), where the negation as failure operator is considered to be a non-monotonic operator [16].

However, for such extended classes of logic programs, the top-down approach cannot be used for computing the answer set semantics because there is no local property in evaluating programs. Thus, we need bottom-up computation for correct evaluation of negation as failure formulas. For this purpose, we use the framework of MGTP, which can find the answer sets as the fixpoint of model candidates.

Here, we introduce a method to transform any logic program (with negation as failure) into a *positive disjunction program* (without negation as failure) [40] for which MGTP can compute the minimal models [29].

Translation into MGTP Rules. A *positive disjunctive program* is a set of rules of the form:

$$A_1 \mid \ldots \mid A_l \leftarrow A_{l+1}, \ldots, A_m \qquad (8)$$

where $m \geq l \geq 0$ and each A_i is an atom.

The meaning of a positive disjunctive program $\mathcal{P}$ can be given by the *minimal models* of $\mathcal{P}$ [40]. The minimal models of positive disjunctive programs can be computed using MGTP. We represent each rule of the form (8) in a positive disjunctive program with the following MGTP input clauses:

$$A_{l+1} \wedge \ldots \wedge A_m \rightarrow A_1 \vee \ldots \vee A_l \qquad (9)$$

General and Extended Logic Programs. MGTP can also compute the stable models of a general logic program [15] and the answer sets of an extended disjunctive program [16] by translation into positive disjunctive programs.

An *extended logic program* is a set of rules of the form:

$$L_1 \mid \ldots \mid L_l \leftarrow L_{l+1}, \ldots, L_m, \, not\, L_{m+1}, \ldots, \, not\, L_n \qquad (10)$$

where $n \geq m \geq l \geq 0$ and each L_i is a literal. This logic program is called a general logic program if $l \leq 1$, and each L_i is an atom.

While a general logic program contains negation-as-failure but does not contain classical negation, an extended disjunctive program contains both of them.

In evaluating $not\, L$ in a bottom-up manner, it is necessary to interpret $not\, L$ with respect to a fixpoint of the computation, because even if L is not currently proved, L might be proved in subsequent inferences. When we have to evaluate $not\, L$ in a current model candidate, we split the model candidate into two: (1) the model candidate where L is assumed not to hold, and (2) the model candidate where it is necessary that L holds. Each negation-as-failure formula $not\, L$ is thus translated into negative and positive literals with a modality expressing belief, i.e., "disbelieve L" (written as $\neg \mathsf{K} L$) and "believe L" (written as $\mathsf{K} L$).

Based on the above discussion, we translate each rule of the form (10) to the following MGTP rule:

$$L_{l+1} \wedge \ldots \wedge L_m \rightarrow H_1 \vee \ldots \vee H_l \vee \mathsf{K} L_{m+1} \vee \ldots \vee \mathsf{K} L_n \qquad (11)$$

where $H_i \equiv \neg \mathsf{K} L_{m+1} \wedge \ldots \wedge \neg \mathsf{K} L_n \wedge L_i \, (i = 1, \ldots, l)$

For any MGTP rule of the form (11), if a model candidate M satisfies $L_{l+1}, \ldots, L_m$, then M is split into $n - m + l \, (n \geq m \geq 0, \, 0 \leq l \leq 1)$ model candidates.

In order to reject model candidates when their guesses turn out to be wrong, the following two schemata (integrity constraints) are introduced:

$$\neg \mathsf{K}L \wedge L \;\rightarrow \qquad \text{for every literal } L \in \mathcal{L}. \tag{12}$$

$$\neg \mathsf{K}L \wedge \mathsf{K}L \;\rightarrow \qquad \text{for every literal } L \in \mathcal{L}. \tag{13}$$

Added to the schemata above, we need the following 3 schemata to deal with classical negation. Below, $\overline{L}$ is the literal complement to a literal L.

$$L \wedge \overline{L} \;\rightarrow \qquad \text{for every literal } L \in \mathcal{L}. \tag{14}$$

$$\mathsf{K}L \wedge \overline{L} \;\rightarrow \qquad \text{for every literal } L \in \mathcal{L}. \tag{15}$$

$$\mathsf{K}L \wedge \mathsf{K}\overline{L} \;\rightarrow \qquad \text{for every literal } L \in \mathcal{L}. \tag{16}$$

Next is the condition to guarantee stability at a fixpoint that all of the guesses made so far in a model candidate M are correct.

For every ground literal L, if $\mathsf{K}L \in M$, then $L \in M$.

The above computation by the MGTP is sound and complete with respect to the answer set semantics. This technique is simply based on a bottom-up model generation method together with integrity constraints over K-literals expressed by object-level schemata on the MGTP.

Compared with other approaches, the proposed method has several computational advantages: put simply, it can find all minimal models for every class of groundable logic program or disjunctive database, incrementally, without backtracking, and in parallel.

This method has been applied to a legal reasoning system [44].

5.2   Legal Reasoning

As an real application, MGTP has been applied to a legal reasoning system [44,27]. Since legal rules imply uncertainty and inconsistency, we have to introduce other language rather than the MGTP input language, for users to represent law and some judicial precedents. In this section, we show an extended logic programming language, and a method to translate it into the MGTP input clauses to solve legal problems automatically using MGTP.

Extended Logic Programming Language. In our legal reasoning system, we adopted the extended logic programming language defined below to represent legal knowledge and judicial precedents. We consider rules of the form:

$$R :: L_0 \leftarrow L_1 \wedge \ldots \wedge L_m \wedge not\ L_{m+1} \wedge \ldots \wedge not\ L_n. \tag{17}$$

$$R ::\leftarrow L_1 \wedge L_2. \tag{18}$$

$$R :: L_0 \Leftarrow L_1 \wedge \ldots \wedge L_m \wedge not\ L_{m+1} \wedge \ldots \wedge not\ L_n. \tag{19}$$

where $L_i(0 \leq i \leq n)$ represents a literal, *not* represents negation as failure (NAF), and R is a rule identifier, which has all variables occurring in $L_i(0 \leq i \leq n)$ as its arguments.

(17) is called an *exact rule*, in which if all literals in the rule body are assigned **true**, then the rule head is assigned **true** without any exception. (18) is called an *integrity constraint* which means the constraint that L_1 and L_2 must not be assigned **true** in the same context. (19) is called a *default rule*, in which if all literals in the rule body are assigned **true**, then the rule head is assigned **true** unless it causes a conflict or destroys an integrity constraints.

Example:

$$r1(X) :: fly(X) \Leftarrow bird(X) \wedge not \; baby(X).$$
$$r2(X) :: \neg fly(X) \Leftarrow penguin(X).$$
$$r3(X) :: bird(X) \leftarrow penguin(X).$$
$$f1 :: bird(a).$$
$$f2 :: penguin(a).$$
$$f3 :: baby(b).$$

In this example, $r1(X)$ can derive $fly(a)$, that is inconsistent with $\neg fly(a)$ derived from $r2(X)$. Since $r1(X)$ and $r2(X)$ are represented with default rules, we cannot conclude whether a flies or a does not fly. If $r2(X)$, however, were defined as a more specific rule than $r1(X)$, that is, $r2(X)$ is preferred to $r1(X)$, $\neg fly(a)$ could defeat $fly(a)$. In order to realize such reasoning about rule preference, we introduce another form of literal representation: $R_1 < R_2$ which means "rule R_2 is preferred to R_1" (where R_1 and R_2 are rule identifiers with arguments). For example, the following rule represents that $r2(X)$ is preferred to $r1(X)$ when X is a bird:

$$r4(X) :: r1(X) < r2(X) \leftarrow bird(X).$$

If we recognize it as a default rule, we can replace $\leftarrow$ with $\Leftarrow$. The rule preference defined as above is called *dynamic* in the sense that the preference is determined according to its arguments.

Semantics of the Rule Preference. A lot of semantics for a rule preference structure have been proposed: introducing the predicate preference relation into circumscription [35,17], introducing the rule preference relation into the default theory [4,8,1,5,6], using literal preference relation [7,48], defining its semantics as translation rules [33,47].

Among these, our system adopted the approach presented in [33], because it can be easily applied to legal reasoning and is easy to translate into MGTP input clauses.

Translation into the MGTP Input Clauses. Assume we have the default rule as:

$$R_1 :: L_0^1 \Leftarrow L_1^1 \wedge \ldots, L_m^1 \wedge not \; L_{m+1}^1 \wedge \ldots \wedge not \; L_n^1.$$

If we have the following default rule:

$$R_2 :: L_0^2 \Leftarrow L_1^2 \wedge \ldots \wedge L_k^2 \wedge not\ L_{k+1}^2 \wedge \ldots \wedge not\ L_q^2.$$

then R_1 is translated to:

$$L_0^1 \leftarrow L_1^1 \wedge \ldots \wedge L_m^1 \wedge not\ L_{m+1}^1 \wedge \ldots \wedge not\ L_n^1 \wedge not\ defeated(R_1).$$

This translation shows the interpretation of our default rules, that is, the rule head can be derived if the rule body is satisfied and there is no proof that R_1 can be defeated. The predicate *defeated* is newly introduced and defined as the following rules:

$$defeated(R_2\theta) \leftarrow$$
$$L_1^1\theta \wedge \ldots \wedge L_m^1\theta \wedge not\ L_{m+1}^1\theta \wedge \ldots \wedge not\ L_n^1\theta \wedge not\ defeated(R_1\theta) \wedge$$
$$L_1^2\theta \wedge \ldots \wedge L_k^2\theta \wedge not\ L_{k+1}^2\theta \wedge \ldots \wedge not\ L_q^2\theta \wedge not\ R_1\theta < R_2\theta.$$

where θ is a most general unifier that satisfies the following condition: *There exists the unifier θ such that $L_0^1\theta = \neg L_0^2\theta$, or there exists the unifier θ such that for some integrity constraint $\leftarrow L_1 \wedge L_2$, $L_1\theta = L_0^1\theta$ and $L_2\theta = L_0^2\theta$, or $L_2\theta = L_0^1\theta$ and $L_1\theta = L_0^2\theta$.*

In this way, default rules with rule preference relations are translated to the rule with NAF, The deduction process in MGTP for those rule set is based on [29].

Introducing Modal Operator. For each NAF literal in a rule, a modal operator K is introduced. If we have the following clause:

$$A_l \leftarrow A_{l+1} \wedge \ldots \wedge A_m \wedge not\ A_{m+1} \wedge \ldots \wedge not\ A_n$$

then we translate it with modal operators into:

$$A_{l+1} \wedge \ldots \wedge A_m \rightarrow (-KA_{m+1} \wedge \ldots \wedge -KA_n, A_l) \vee KA_{m+1} \vee \ldots \vee KA_n$$

In addition, we provide the integrity constraint for K such as $P \wedge \neg KP \rightarrow$, which enables MGTP to derive the stable models for the given input clauses. These integrity constraints are built in the MGTP deduction process with slight modification.

Extracting Stable Models. The derived models from MGTP contain not only all possible stable models but also the models which are constructed only by hypotheses. A stable model must satisfy the following condition called *T-condition*. T-condition is a criteria to extract final stable models from the derived models from MGTP.

T-Condition. If $KP \in M$, then $P \in M$.

If the proof structure included in a stable model also occurs in all the other stable models, we call it a *justified argument*, otherwise a *plausible argument*. Justified arguments are sound for any attacks against them, while plausible arguments are not sound for some attacks, that is, they might be attacked by some arguments and cause a conflict.

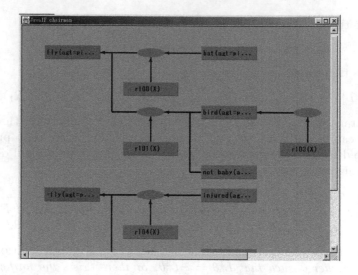

Fig. 17. The interface window in the argumentation support system

System and Experiments. We have developed an argumentation support system [27] including the legal reasoning system by MGTP. The system is written in Java and works on each client machine which is connected with other client via a TCP/IP network. Each participant (including parties concerned and a judge if needed) makes argument diagrams according to his/her own assertion by hand or sometimes automatically, and sends them to all others. Figure 17 shows an example of argument diagrams on the user interface window. The system maintains the current status of each node, that is, agreed by all, disagreed by someone, attacked by some nodes or attacking some nodes, etc. Based on these status, the judge, if necessary, intervenes their arguments and undertakes mediation.

As an experiment, we implemented a part of Japanese civil law on the system. More than 10 legal experts used the system, investigated the arguments which were automatically derived from the legal reasoning system, and had high opinions of the ability about: representation of the extended logic programming language, negotiation protocol adopted, and efficiency of reasoning.

6   Conclusion

We have reviewed research and development of the model generation theorem prover MGTP, including our recent activities around it.

MGTP is one of successful application systems developed at the FGCS project. MGTP achieved more than a 200-fold speedup on a PIM/m consisting of 256 PEs for many theorem proving benchmarks. By using parallel MGTP systems, we succeeded in solving some hard mathematical problems such as condensed detachment problems and quasigroup existence problems in finite algebra.

In the current parallel implementation, however, we have to properly use an AND parallel MGTP for Horn problems and an OR parallel MGTP for non-Horn problems separately. Thus, it is necessary to develop a parallel version of MGTP which can combine AND- and OR-parallelization for proving a set of general clauses. In addition, when running MGTP (written in Klic [14]) on other commercial parallel computers, it is difficult for them to attain such a good parallel performance as PIM for problems that require fine-grain concurrency. At present, the N-sequential method to exploit coarse-grain concurrency with low communication costs would be a practical solution for this. Recent results with Java versions of MGTP (JavaMGTP) shows several tens fold speedup compared to Klic versions. This achievement is largely due to the new *A-cell* mechanism for handling multiple contexts and several language facilities of Java including destructive assignment to variables.

To enhance the MGTP's pruning ability, we extended the MGTP features in several ways. NHM is a key technology for making MGTP practical and applicable to several applications such as disjunctive databases and abductive reasoning. The essence of the NHM method is to simulate a top-down evaluation in a framework of bottom-up computation by static clause transformation to propagate goal (negative) information, thereby pruning search spaces. This propagation is closely related to the technique developed in CMGTP to manipulate (negative) constraints. Thus, further research is needed to clarify whether the NHM method can be incorporated to CMGTP or its extended version, IV-MGTP.

It is also important in real applications that MGTP avoids duplicating the same subproofs and generating nonminimal models. The proof simplification based on dependency analysis is a technique to embed both folding-up and proof condensation in a model generation framework, and has a similar effect to NHM. Although the proof simplification is weaker than NHM in the sense that relevancy testing is performed after a model extension occurs, it is compensated by the folding-up function embedded. Incorporating this method into a minimal model generation prover MM-MGTP would enhance its pruning ability furthermore.

Lastly, we have shown that the feature of negation as failure, which is a most important invention in logic programming, can be easily implemented on MGTP, and have presented a legal reasoning system employing the feature. The basic idea behind this is to translate formulas with special properties, such as non-monotonicity and modality, into first order clauses on which MGTP works as a meta-interpreter. The manipulation of these properties is thus reduced to generate-and-test problems for model candidates. These can then be handled by the MGTP very efficiently through case-splitting of disjunctive consequences and rejection of inconsistent model candidates.

A family of MGTP systems is available at `http://ss104.is.kyushu-u.ac.jp/software/`.

Acknowledgment

We would like to thank Prof. Kazuhiro Fuchi of Keio University, the then director of ICOT, and Prof. Koichi Furukawa of Keio University, the then deputy director of ICOT, who have given us continuous support and helpful comments during the Fifth Generation Computer Systems Project. Thanks are also due to members of the MGTP research group including Associate Prof. Katsumi Inoue of Kobe University and Prof. Katsumi Nitta of Tokyo Institute of Technology for their fruitful discussions and cooperation.

References

1. Franz Baader and Bernhard Hollunder. How to prefer more specific defaults in terminological default logic. In *Proc. International Joint Conference on Artificial Intelligence*, pages 669–674, 1993.
2. Peter Baumgartner, Ulrich Furbach, and Ilkka Niemelä. Hyper Tableaux. In José Júlio Alferes, Luís Moniz Pereira, and Ewa Orłowska, editors, *Proc. European Workshop: Logics in Artificial Intelligence, JELIA*, volume 1126 of *Lecture Notes in Artificial Intelligence*, pages 1–17. Springer-Verlag, 1996.
3. Frank Bennett. Quasigroup Identities and Mendelsohn Designs. *Canadian Journal of Mathematics*, 41:341–368, 1989.
4. Gerhard Brewka. Preferred subtheories : An extended logical framework for default reasoning. In *Proc. International Joint Conference on Artificial Intelligence*, pages 1043–1048, Detroit, MI, USA, 1989.
5. Gerhard Brewka. Adding priorities and specificity to default logic . In *Proc. JELIA 94*, pages 247–260, 1994.
6. Gerhard Brewka. Reasoning about priorities in default logic. In *Proc. AAAI 94*, pages 940–945, 1994.
7. Gerhard Brewka. Well-founded semantics for extended logic programs with dynamic preference. *Journal of Artificial Intelligence Research*, 4:19–36, 1996.
8. Gerhard Brewka and Thomas F. Gordon. How to Buy a Porsche: An Approach to defeasible decision making. In *Proc. AAA94 workshop on Computational Dialectics*, 1994.
9. François Bry. Query evaluation in recursive databases: bottom-up and top-down reconciled. *Data & Knowledge Engineering*, 5:289–312, 1990.
10. François Bry and Adnan Yahya. Minimal Model Generation with Positive Unit Hyper-Resolution Tableaux. In *Proc. 5th International Workshop, TABLEAUX'96*, volume 1071 of *Lecture Notes in Artificial Intelligence*, pages 143–159, Terrasini, Palermo, Italy, May 1996. Springer-Verlag.
11. Hiroshi Fujita and Ryuzo Hasegawa. A Model-Generation Theorem Prover in KL1 Using Ramified Stack Algorithm. In *Proc. 8th International Conference on Logic Programming*, pages 535–548. The MIT Press, 1991.
12. Masayuki Fujita, Ryuzo Hasegawa, Miyuki Koshimura, and Hiroshi Fujita. Model Generation Theorem Provers on a Parallel Inference Machine. In *Proc. International Conference on Fifth Generation Computer Systems*, volume 1, pages 357–375, Tokyo, Japan, June 1992.
13. Masayuki Fujita, John Slaney, and Frank Bennett. Automatic Generation of Some Results in Finite Algebra. In *Proc. International Joint Conference on Artificial Intelligence*, 1993.

14. Tetsuro Fujita, Takashi Chikayama, Kazuaki Rokuwasa, and Akihiko Nakase. KLIC: A Portable Implementation of KL1. In *Proc. International Conference on Fifth Generation Computer Systems*, pages 66–79, Tokyo, Japan, December 1994.

15. Michael Gelfond and Vladimir Lifschitz. The Stable Model Semantics for Logic Programming. In *Proc. 5th International Conference and Symposium on Logic Programming*, pages 1070–1080. MIT Press, 1988.

16. Michael Gelfond and Vladimir Lifschitz. Classical Negation in Logic Programs and Disjunctive Databases. *New Generation Computing*, 9:365–385, 1991.

17. Benjamin Grosof. Generalization Prioritization. In *Proc. 2nd Conference on Knowledge Representation and Reasoning*, pages 289–300, 1991.

18. Reiner Hähnle. Tableaux and related methods. In Alan Robinson and Andrei Voronkov, editors, *Handbook of Automated Reasoning*, volume I. North-Holland, 2001.

19. Reiner Hähnle, Ryuzo Hasegawa, and Yasuyuki Shirai. Model Generation Theorem Proving with Finite Interval Constraints. In *Proc. First International Conference on Computational Logic (CL2000)*, 2000.

20. Ryuzo Hasegawa and Hiroshi Fujita. Implementing a Model-Generation Based Theorem Prover MGTP in Java. *Research Reports on Information Science and Electrical Engineering*, 3(1):63–68, 1998.

21. Ryuzo Hasegawa and Hiroshi Fujita. A new Implementation Technique for a Model-Generation Theorem Prover to Solve Constraint Satisfaction Problems. *Research Reports on Information Science and Electrical Engineering*, 4(1):57–62, 1999.

22. Ryuzo Hasegawa, Hiroshi Fujita, and Miyuki Koshimura. MGTP: A Parallel Theorem-Proving System Based on Model Generation. In *Proc. 11th International Conference on Applications of Prolog*, pages 34–41, Tokyo, Japan, September 1998.

23. Ryuzo Hasegawa, Hiroshi Fujita, and Miyuki Koshimura. Efficient Minimal Model Generation Using Branching Lemmas. In *Proc. 17th International Conference on Automated Deduction*, volume 1831 of *Lecture Notes in Artificial Intelligence*, pages 184–199, Pittsburgh, Pennsylvania, USA, June 2000. Springer-Verlag.

24. Ryuzo Hasegawa, Katsumi Inoue, Yoshihiko Ohta, and Miyuki Koshimura. Non-Horn Magic Sets to Incorporate Top-down Inference into Bottom-up Theorem Proving. In *Proc. 14th International Conference on Automated Deduction*, volume 1249 of *Lecture Notes in Artificial Intelligence*, pages 176–190, Townsville, North Queensland, Australia, July 1997. Springer-Verlag.

25. Ryuzo Hasegawa and Miyuki Koshimura. An AND Parallelization Method for MGTP and Its Evaluation. In *Proc. First International Symposium on Parallel Symbolic Computation*, Lecture Notes Series on Computing, pages 194–203. World Scientific, September 1994.

26. Ryuzo Hasegawa, Miyuki Koshimura, and Hiroshi Fujita. Lazy Model Generation for Improving the Efficiency of Forward Reasoning Theorem Provers. In *Proc. International Workshop on Automated Reasoning*, pages 221–238, Beijing, China, July 1992.

27. Ryuzo Hasegawa, Katsumi Nitta, and Yasuyuki Shirai. The Development of an Argumentation Support System Using Theorem Proving Technologies. In *Research Report on Advanced Software Enrichment Program 1997*, pages 59–66. Information Promotion Agency, Japan, 1999. (in Japanese).

28. Ryuzo Hasegawa and Yasuyuki Shirai. Constraint Propagation of CP and CMGTP: Experiments on Quasigroup Problems. In *Proc. Workshop 1C (Automated Reasoning in Algebra), CADE-12*, Nancy, France, 1994.

29. Katsumi Inoue, Miyuki Koshimura, and Ryuzo Hasegawa. Embedding Negation as Failure into a Model Generation Theorem Prover. In *Proc. 11th International Conference on Automated Deduction*, volume 607 of *Lecture Notes in Artificial Intelligence*, pages 400–415, Saratoga Springs, NY, USA, 1992. Springer-Verlag.

30. Katsumi Inoue, Yoshihiko Ohta, Ryuzo Hasegawa, and Makoto Nakashima. Bottom-Up Abduction by Model Generation. In *Proc. International Joint Conference on Artificial Intelligence*, pages 102–108, 1993.

31. Miyuki Koshimura and Ryuzo Hasegawa. Modal Propositional Tableaux in a Model Generation Theorem Prover. In *Proc. 3rd Workshop on Theorem Proving with Analytic Tableaux and Related Methods*, pages 145–151, May 1994.

32. Miyuki Koshimura and Ryuzo Hasegawa. Proof Simplification for Model Generation and Its Applications. In *Proc. 7th International Conference, LPAR 2000*, volume 1955 of *Lecture Notes in Artificial Intelligence*, pages 96–113. Springer-Verlag, November 2000.

33. Robert A. Kowalski and Francesca Toni. Abstract Argumentation. *Artificial Intelligence and Law Journal*, 4:275–296, 1996.

34. Reinhold Letz, Klaus Mayr, and Christoph Goller. Controlled Integration of the Cut Rule into Connection Tableau Calculi. *Journal of Automated Reasoning*, 13:297–337, 1994.

35. Vladimir Lifschitz. Computing Circumscription. In *Proc. International Joint Conference on Artificial Intelligence*, pages 121–127, Los Angeles, CA, USA, 1985.

36. Donald W. Loveland, David W. Reed, and Debra S. Wilson. Satchmore: Satchmo with RElevancy. *Journal of Automated Reasoning*, 14(2):325–351, April 1995.

37. James J. Lu. Logic Programming with Signs and Annotations. *Journal of Logic and Computation*, 6(6):755–778, 1996.

38. Rainer Manthey and Franqois Bry. SATCHMO: a theorem prover implemented in Prolog. In *Proc. 9th International Conference on Automated Deduction*, volume 310 of *Lecture Notes in Computer Science*, pages 415–434, Argonne, Illinois, USA, May 1988. Springer-Verlag.

39. William McCune and Larry Wos. Experiments in Automated Deduction with Condensed Detachment. In *Proc. 11th International Conference on Automated Deduction*, volume 607 of *Lecture Notes in Artificial Intelligence*, pages 209–223, Saratoga Springs, NY, USA, 1992. Springer-Verlag.

40. Jack Minker. On indefinite databases and the closed world assumption. In *Proc. 6th International Conference on Automated Deduction*, volume 138 of *Lecture Notes in Computer Science*, pages 292–308, Courant Institute, USA, 1982. Springer-Verlag.

41. Ugo Montanari and Francesca Rossi. Finite Domain Constraint Solving and Constraint Logic Programming. In *Constraint Logic Programming: Selected Research*, pages 201–221. The MIT press, 1993.

42. Hiroshi Nakashima, Katsuto Nakajima, Seiichi Kondo, Yasutaka Takeda, Yū Inamura, Satoshi Onishi, and Kanae Matsuda. Architecture and Implementation of PIM/m. In *Proc. International Conference on Fifth Generation Computer Systems*, volume 1, pages 425–435, Tokyo, Japan, June 1992.

43. Ilkka Niemelä. A Tableau Calculus for Minimal Model Reasoning. In *Proc. 5th International Workshop, TABLEAUX'96*, volume 1071 of *Lecture Notes in Artificial Intelligence*, pages 278–294, Terrasini, Palermo, Italy, May 1996. Springer-Verlag.

44. Katsumi Nitta, Yoshihisa Ohtake, Shigeru Maeda, Masayuki Ono, Hiroshi Ohsaki, and Kiyokazu Sakane. HELIC-II: A Legal Reasoning System on the Parallel Inference Machine. In *Proc. International Conference on Fifth Generation Computer Systems*, volume 2, pages 1115–1124, Tokyo, Japan, June 1992.

45. Yoshihiko Ohta, Katsumi Inoue, and Ryuzo Hasegawa. On the Relationship Between Non-Horn Magic Sets and Relevancy Testing. In *Proc. 15th International Conference on Automated Deduction*, volume 1421 of *Lecture Notes in Artificial Intelligence*, pages 333–349, Lindau, Germany, July 1998. Springer-Verlag.
46. Franz Oppacher and E. Suen. HARP: A Tableau-Based Theorem Prover. *Journal of Automated Reasoning*, 4:69–100, 1988.
47. Henry Prakken and Giovanni Sartor. Argument-based Extended Logic Programming with Defeasible Priorities. *Journal of Applied Non-Classical Logics*, 7:25–75, 1997.
48. Chiaki Sakama and Katsumi Inoue. Representing Priorities in Logic Programs. In *Proc. International Conference and Symposium on Logic Programming*, pages 82–96, 1996.
49. Heribert Schütz and Tim Geisler. Efficient Model Generation through Compilation. In *Proc. 13th International Conference on Automated Deduction*, volume 1104 of *Lecture Notes in Artificial Intelligence*, pages 433–447. Springer-Verlag, 1996.
50. Yasuyuki Shirai and Ryuzo Hasegawa. Two Approaches for Finite-domain Constraint Satisfaction Problem - CP and MGTP -. In *Proc. 12th International Conference on Logic Programming*, pages 249–263. MIT Press, 1995.
51. Mark Stickel. The Path-Indexing Method For Indexing Terms. Technical Note 473, AI Center, SRI, 1989.
52. Mark E. Stickel. Upside-Down Meta-Interpretation of the Model Elimination Theorem-Proving Procedure for Deduction and Abduction. *Journal of Automated Reasoning*, 13(2):189–210, October 1994.
53. Geoff Sutcliffe, Christian Suttner, and Theodor Yemenis. The TPTP Problem Library. In *Proc. 12th International Conference on Automated Deduction*, volume 814 of *Lecture Notes in Artificial Intelligence*, pages 252–266, Nancy, France, 1994. Springer-Verlag.
54. Evan Tick and Miyuki Koshimura. Static Mode Analyses of Concurrent Logic Programs. *Journal of Programming Languages*, 2:283–312, 1994.
55. Kazunori Ueda and Takashi Chikayama. Design of the Kernel Language for the Parallel Inference Machine. *Computer Journal*, 33:494–500, December 1990.
56. Debra S. Wilson and Donald W. Loveland. Incorporating Relevancy Testing in SATCHMO. Technical Reports CS-1989-24, Department of Computer Science, Duke University, Durham, North Carolina, USA, 1989.

A 'Theory' Mechanism for a Proof-Verifier Based on First-Order Set Theory*

Eugenio G. Omodeo[1] and Jacob T. Schwartz[2]

[1] University of L'Aquila, Dipartimento di Informatica
omodeo@univaq.it
[2] University of New York, Department of Computer Science,
Courant Institute of Mathematical Sciences
schwartz@cs.nyu.edu

We often need to associate some highly compound meaning with a symbol. Such a symbol serves us as a kind of container carrying this meaning, always with the understanding that it can be opened if we need its content.

(Translated from [12, pp. 101–102])

Abstract. We propose classical set theory as the core of an automated proof-verifier and outline a version of it, designed to assist in proof development, which is indefinitely expansible with function symbols generated by Skolemization and embodies a modularization mechanism named 'theory'. Through several examples, centered on the finite summation operation, we illustrate the potential utility in large-scale proof-development of the 'theory' mechanism: utility which stems in part from the power of the underlying set theory and in part from Skolemization.

Key words: Proof-verification technology, set theory, proof modularization.

1   Introduction

Set theory is highly versatile and possesses great expressive power. One can readily find terse set-theoretic equivalents of established mathematical notions and express theorems in purely set-theoretic terms.

Checking any deep fact (say the Cauchy integral theorem) using a proof-verifier requires a large number of logical statements to be fed into the system. These must formalize a line of reasoning that leads from bare set rudiments to the specialized topic of interest (say, functional analysis) and then to a target theorem. Such an enterprise can only be managed effectively if suitable modularization constructs are available.

* E.G. Omodeo enjoyed a Short-term mobility grant of the Italian National Research Council (CNR) enabling him to stay at the University of New York during the preparation of this work.

A.C. Kakas, F. Sadri (Eds.): Computat. Logic (Kowalski Festschrift), LNAI 2408, pp. 214–230, 2002.
© Springer-Verlag Berlin Heidelberg 2002

This paper outlines a version of the Zermelo-Fraenkel theory designed to assist in automated proof-verification of mathematical theorems. This system incorporates a technical notion of "theory" designed, for large-scale proof-development, to play a role similar to the notion of object class in large-scale programming. Such a mechanism can be very useful for "proof-engineering".

The theories we propose, like procedures in a programming language, have lists of formal parameters. Each "theory" requires its parameters to meet a set of assumptions. When "applied" to a list of actual parameters that have been shown to meet the assumptions, a theory will instantiate several additional "output" set, predicate, and function symbols, and then supply a list of theorems initially proved explicitly by the user inside the theory itself. These theorems will generally involve the new symbols.

Such use of "theories" and their application adds a touch of second-order logic capability to the first-order system which we describe. Since set theory has full multi-tier power, this should be all the second-order capability that is needed.

We illustrate the usefulness of the proposed theory notion via examples ranging from mere "utilities" (e.g. the specification of ordered pairs and associated projections, and the thinning of a binary predicate into a global single-valued map) to an example which characterizes a very flexible recursive definition scheme. As an application of this latter scheme, we outline a proof that a finite summation operation which is insensitive to operand rearrangement and grouping can be associated with any commutative-associative operation. This is an intuitively obvious fact (seldom, if ever, proved explicitly in algebra texts), but nevertheless it must be verified in a fully formalized context. Even this task can become unnecessarily challenging without an appropriate set-theoretic support, or without the ability to indefinitely extend the formal language with new Skolem symbols such as those resulting from "theory" invocations.

Our provisional assessment of the number of "proofware" lines necessary to reach the Cauchy integral theorem in a system like the one which we outline is 20–30 thousand statements.

2   Set Theory as the Core of a Proof-Verifier

A fully satisfactory formal logical system should be able to digest 'the whole of mathematics', as this develops by progressive extension of mathematics-like reasoning to new domains of thought. To avoid continual reworking of foundations, one wants the formal system taken as basic to remain unchanged, or at any rate to change only by extension as such efforts progress. In any fundamentally new area work and language will initially be controlled more by guiding intuitions than by entirely precise formal rules, as when Euclid and his predecessors first realized that the intuitive properties of geometric figures in 2 and 3 dimensions, and also some familiar properties of whole numbers, could be covered by modes of reasoning more precise than those used in everyday life. But mathematical developments during the last two centuries have reduced the intuitive content of geometry, arithmetic, and calculus ('analysis') in set-theoretic terms. The geometric notion of 'space' maps into 'set of all pairs (or triples) of real numbers', allowing consideration of the 'set of all n-tuples of real numbers' as 'n-dimensional space', and of more general related constructs as 'infinite dimensional' and 'functional' spaces. The 'figures' originally studied in geometry map, via the 'locus' concept, into sets of such pairs, triples, etc. Dedekind reduced 'real number x' to 'set x of rational numbers, bounded above, such that every rational not in x is larger than every rational in x'. To eliminate everything but set theory from the formal foundations of mathematics, it only remained (since 'fractions' can be seen as pairs of numbers) to reduce the notion of 'integer' to set-theoretic terms. This was done by Cantor and Frege: an integer is the class of all finite sets in 1-1 correspondence with any one such set. Subsequently Kolmogorov modeled 'random' variables as functions defined on an implicit set-theoretic measure space, and Laurent Schwartz interpreted the initially puzzling 'delta functions' in terms of a broader notion of generalized function systematically defined in set-theoretic terms. So all of these concepts can be digested without forcing any adjustment of the set-theoretic foundation constructed for arithmetic, analysis, and geometry. This foundation also supports all the more abstract mathematical constructions elaborated in such 20th century fields as topology, abstract algebra, and category theory. Indeed, these were expressed set-theoretically from their inception. So (if we ignore a few ongoing explorations whose significance remains to be determined) set theory currently stands as a comfortable and universal basis for the whole of mathematics—cf. [5].

It can even be said that set theory captures a set of reality-derived intuitions more fundamental than such basic mathematical ideas as that of number. Arithmetic would be very different if the real-world process of counting did not return the same result each time a set of objects was counted, or if a subset of a finite set S of objects proved to have a larger count than S. So, even though Peano showed how to characterize the integers and derive many of their properties using axioms free of any explicit set-theoretic content, his approach robs the integers of much of their intuitive significance, since in his reduced context they cannot be used to count anything. For this and the other reasons listed above, we prefer to work with a thoroughly set-theoretic formalism, contrived to mimic the language and procedures of standard mathematics closely.

3   Set Theory in a Nutshell

Set theory is based on the handful of very powerful ideas summarized below. All notions and notation are more or less standard (cf. [16]).[1]

- The dyadic *Boolean operations* $\cap, \setminus, \cup$ are available, and there is a *null set*, $\emptyset$, devoid of elements. The *membership* relation $\in$ is available, and set *nesting* is

---

[1] As a notational convenience, we usually omit writing universal quantifiers at the beginning of a sentence, denoting the variables which are ruled by these understood quantifiers by single uppercase Italic letters.

made possible via the singleton operation $X \mapsto \{X\}$. Derived from this, we have single-element addition and removal, and useful increment/decrement operations:

$$X \text{ with } Y := X \cup \{Y\}, \quad X \text{ less } Y := X \setminus \{Y\}, \quad \mathsf{next}(X) := X \text{ with } X.$$

Unordered lists $\{t_1, \ldots, t_n\}$ and ordered tuples $[t_1, \ldots, t_n]$ are definable too: in particular, $\{X_1, \ldots, X_n\} := \{X_1\} \cup \cdots \cup \{X_n\}$.

- *'Sets whose elements are the same are identical'*: Following a step $\ell \neq r$ in a proof, one can introduce a new constant b subject to the condition $b \in \ell \leftrightarrow b \notin r$; no subsequent conclusions where b does not appear will depend on this condition. Negated set inclusion $\not\subseteq$ can be treated similarly, since $X \subseteq Y := X \setminus Y = \emptyset$.

- *Global choice:* We use an operation **arb** which, from any non-null set X, deterministically extracts an element which does not intersect X. Assuming $\mathbf{arb}\, \emptyset = \emptyset$ for definiteness, this means that

$$\mathbf{arb}\, X \in \mathsf{next}(X) \ \& \ X \cap \mathbf{arb}\, X = \emptyset$$

for all X.

- *Set-formation:* By (possibly transfinite) element- or subset-iteration over the sets represented by the terms $t_0, t_1 \equiv t_1(x_0), \ldots, t_n \equiv t_n(x_0, \ldots, x_{n-1})$, we can form the set

$$\{\, e \,:\, x_0\, C_0\, t_0,\ x_1\, C_1\, t_1, \ldots, x_n\, C_n\, t_n \mid \varphi \,\},$$

where each C_i is either $\in$ or $\subseteq$, and where $e \equiv e(x_0, \ldots, x_n)$ and $\varphi \equiv \varphi(x_0, \ldots, x_n)$ are a set-term and a condition in which the p.w. distinct variables x_i can occur free (similarly, each t_{j+1} may involve $x_0, \ldots, x_j$). Many operations are readily definable using setformers, e.g.

$$\bigcup Y := \{\, x_2 \,:\, x_1 \in Y, x_2 \in x_1 \,\}, \quad Y \times Z := \{\, [x_1, x_2] \,:\, x_1 \in Y,\ x_2 \in Z \,\},$$
$$\mathscr{P}(Y) := \{\, x \,:\, x \subseteq Y \,\}, \quad\quad\quad \mathsf{pred}(X) := \mathbf{arb}\, \{\, y \in X \mid \mathsf{next}(y) = X \,\},$$

where if the condition φ is omitted it is understood to be **true**, and if the term e is omitted it is understood to be the same as the first variable inside the braces.

- *$\in$-recursion:* ("Transfinite") recursion over the elements of any set allows one to introduce global set operations; e.g.,

$$\mathsf{Ult\_membs}(S) := S \cup \bigcup\{\, \mathsf{Ult\_membs}(x) \,:\, x \in S \,\} \text{ and}$$
$$\mathsf{rank}(S) := \bigcup\{\, \mathsf{next}(\,\mathsf{rank}(x)\,) \,:\, x \in S \,\},$$

which respectively give the set of all "ultimate members" (i.e. elements, elements of elements, etc.) of S and the maximum "depth of nesting" of sets inside S.

- *'Infinite sets exist'*: There is at least one **s_inf** satisfying

$$\mathbf{s\_inf} \neq \emptyset \ \& \ (\forall x \in \mathbf{s\_inf})(\{x\} \in \mathbf{s\_inf}),$$

so that the p.w. distinct elements $b, \{b\}, \{\{b\}\}, \{\{\{b\}\}\}, \ldots$ belong to **s_inf** for each b in **s_inf**.

The historical controversies concerning the *choice* and *replacement* axioms of set theory are all hidden in our use of setformers and in our ability, after a statement of the form $\exists y \, \psi(X_1, \ldots, X_n, y)$ has been proved, to introduce a Skolem function $f(X_1, \ldots, X_n)$ satisfying the condition $\psi(X_1, \ldots, X_n, f(X_1, \ldots, X_n))$.

In particular, combined use of **arb** and of the setformer construct lets us write the choice set of any set X of non-null pairwise disjoint sets simply as $\{\, \mathbf{arb}\, y \,:\, y \in X \,\}$.[2]

To appreciate the power of the above formal language, consider von Neumann's elegant definition of the predicate 'X is a (possibly transfinite) ordinal', and the characterization of $\mathbb{R}$, the set of real numbers, as the set of Dedekind cuts (cf. [17]):

$$\mathsf{Ord}(X) \;:=\; X \subseteq \mathscr{P}(X) \,\&\, (\forall y, z \in X)(y \in z \lor y = z \lor z \in y)\,,$$
$$\mathbb{R} \;:=\; \bigl\{\, c \subseteq \mathbb{Q} \mid (\forall y \in c)(\exists z \in c)(y < z) \,\& $$
$$(\forall y \in c)(\forall z \in \mathbb{Q})(z < y \;\rightarrow\; z \in c) \,\bigr\} \setminus \{\emptyset, \mathbb{Q}\};$$

here the ordered field $\mathbb{Q}, <$ of rational numbers is assumed to have been defined before $\mathbb{R}$.[3]

4   Theories in Action: First Examples

Here is one of the most obvious theories one can think of:

> **THEORY** ordered_pair()
> ==>(opair, car, cdr)
> $\mathsf{car}\bigl(\,\mathsf{opair}(X, Y)\,\bigr) = X$
> $\mathsf{cdr}\bigl(\,\mathsf{opair}(X, Y)\,\bigr) = Y$
> $\mathsf{opair}(X, Y) = \mathsf{opair}(U, V) \;\rightarrow\; X = U \,\&\, Y = V$
> **END** ordered_pair.

This **THEORY** has no input parameters and no assumptions, and returns three global functions: a pairing function and its projections. To start its construction, the user simply has to

> **SUPPOSE_THEORY** ordered_pair()
> ==>
> **END** ordered_pair,

then to **ENTER_THEORY** ordered_pair, and next to define e.g.

$$\mathsf{opair}(X, Y) \;:=\; \bigl\{\, \{X\}, \{\{X\}, \{Y, \{Y\}\}\} \,\bigr\}\,,$$
$$\mathsf{car}(P) \;:=\; \mathbf{arb}\,\mathbf{arb}\, P\,,$$
$$\mathsf{cdr}(P) \;:=\; \mathsf{car}\bigl(\, \mathbf{arb}\, (P \setminus \{\mathbf{arb}\, P\}) \setminus \{\mathbf{arb}\, P\} \,\bigr)\,.$$

---

[2] Cf. [18, p. 177]. Even in the more basic framework of first-order predicate calculus, the availability of choice constructs can be highly desirable, cf. [1].

[3] For an alternative definition of real numbers which works very well too, see E.A. Bishop's adaptation of Cauchy's construction of $\mathbb{R}$ in [2, pp. 291–297].

This makes it possible to prove such intermediate lemmas as

$$\mathbf{arb}\,\{U\} \;=\; U\,,$$
$$V \in Z \;\rightarrow\; \mathbf{arb}\,\{V,Z\} = V\,,$$
$$\mathsf{car}\big(\,\{\,\{X\},\{\{X\},W\,\}\,\}\,\big) \;=\; X\,,$$
$$\mathbf{arb}\;\mathsf{opair}(X,Y) = \{X\}\,,$$
$$\mathsf{cdr}\big(\,\mathsf{opair}(X,Y)\,\big) \;=\; \mathsf{car}\big(\,\{\,\{Y,\{Y\}\}\,\}\,\big) = Y\,.$$

Once these intermediate results have been used to prove the three theorems listed earlier, the user can indicate that they are the ones he wants to be externally visible, and that the return-parameter list consists of opair, car, cdr (the detailed definitions of these symbols, as well as the intermediate lemmas, have hardly any significance outside the **THEORY** itself[4]). Then, after re-entering the main **THEORY**, which is set_theory, the user can

APPLY(opair, head, tail) ordered_pair() ==>
   $\mathsf{head}\big(\,\mathsf{opair}(X,Y)\,\big) = X$
   $\mathsf{tail}\big(\,\mathsf{opair}(X,Y)\,\big) = Y$
   $\mathsf{opair}(X,Y) = \mathsf{opair}(U,V) \;\rightarrow\; X = U\,\&\,Y = V,$

thus importing the three theorems into the main proof level. As written, this application also changes the designations 'car' and 'cdr' into 'head' and 'tail'.

Fig.1 shows how to take advantage of the functions just introduced to define notions related to maps that will be needed later on.[5]

$\mathsf{is\_map}(F) \quad := F = \{[\mathsf{head}(x), \mathsf{tail}(x)] : x \in F\}$	
$\mathsf{Svm}(F) \qquad := \mathsf{is\_map}(F)\,\&\,(\forall x,y \in F)\big(\mathsf{head}(x) = \mathsf{head}(y) \rightarrow x = y\big)$	
$\mathsf{1\_1\_map}(F) := \mathsf{Svm}(F)\,\&\,(\forall x,y \in F)\big(\mathsf{tail}(x) = \mathsf{tail}(y) \rightarrow x = y\big)$	
$F^{-1} \qquad\quad := \{[\mathsf{tail}(x), \mathsf{head}(x)] : x \in F\}$	
$\mathsf{domain}(F) := \{\mathsf{head}(x) : x \in F\}$	$\mathsf{range}(F) := \{\mathsf{tail}(x) : x \in F\}$
$F\{X\} := \{y \in \mathsf{range}(F) \mid [X,y] \in F\}$	$F_{\mid S} := F \cap \big(S \times \mathsf{range}(F)\big)$
$\mathsf{Finite}(S) \quad := \neg\exists f\big(\mathsf{1\_1\_map}(f)\,\&\,S = \mathsf{domain}(f) \neq \mathsf{range}(f) \subseteq S\big)$	

Fig. 1. Notions related to maps, single-valued maps, and 1-1 maps

For another simple example, suppose that the theory

THEORY setformer0(e, s, p)
==>
   $\mathsf{s} \neq \emptyset \rightarrow \{e(x) : x \in \mathsf{s}\} \neq \emptyset$
   $\{x \in \mathsf{s} \mid \mathsf{p}(x)\} \neq \emptyset \rightarrow \{e(x) : x \in \mathsf{s} \mid \mathsf{p}(x)\} \neq \emptyset$
END setformer0

---

[4] A similar remark on Kuratowski's encoding of an ordered pair as a set of the form $\{\{x,y\},\{x\}\}$ is made in [14, pp. 50–51].
[5] We subsequently return to the notation $[X,Y]$ for $\mathsf{opair}(X,Y)$.

has been proved, but that its user subsequently realizes that the reverse implications could be helpful too; and that the formulae

$$s \subseteq T \;\rightarrow\; \{\, \mathsf{e}(x) \,:\, x \in \mathsf{s} \mid \mathsf{p}(x)\,\} \subseteq \{\, \mathsf{e}(x) \,:\, x \in T \mid \mathsf{p}(x)\,\},$$
$$s \subseteq T \;\&\; (\forall x \in T \setminus \mathsf{s})\neg \mathsf{p}(x) \;\rightarrow\; \{\, \mathsf{e}(x) \,:\, x \in \mathsf{s} \mid \mathsf{p}(x)\,\} = \{\, \mathsf{e}(x) \,:\, x \in T \mid \mathsf{p}(x)\,\}$$

are also needed. He can then re-enter the **THEORY** setformer0, strengthen the implications already proved into bi-implications, and add the new results: of course he must then supply proofs of the new facts.

Our next sample **THEORY** receives as input a predicate $\mathsf{P} \equiv \mathsf{P}(X,V)$ and an "exception" function $\mathsf{xcp} \equiv \mathsf{xcp}(X)$; it returns a global function $\mathsf{img} \equiv \mathsf{img}(X)$ which, when possible, associates with its argument X some Y such that $\mathsf{P}(X,Y)$ holds, and otherwise associates with X the "fictitious" image $\mathsf{xcp}(X)$. The **THEORY** has an assumption, intended to guarantee non-ambiguity of the fictitious value:

> **THEORY** fcn_from_pred(P, xcp)
> $\neg\, \mathsf{P}\big(X, \mathsf{xcp}(X)\big)$ -- *convenient "guard"*
> ==>(img)
> $\mathsf{img}(X) \neq \mathsf{xcp}(X) \;\leftrightarrow\; \exists v\, \mathsf{P}(X,v)$
> $\mathsf{P}(X,V) \;\rightarrow\; \mathsf{P}\big(X, \mathsf{img}(X)\big)$
> **END** fcn_from_pred.

To construct this **THEORY** from its assumption, the user can simply define

$$\mathsf{img}(X) := \mathbf{if}\; \mathsf{P}\big(X, \mathsf{try}(X)\big)\; \mathbf{then}\; \mathsf{try}(X)\; \mathbf{else}\; \mathsf{xcp}(X)\; \mathbf{end\ if}\,,$$

where try results from Skolemization of the valid first-order formula

$$\exists y\, \forall v\big(\, \mathsf{P}(X,v) \;\rightarrow\; \mathsf{P}(X,y)\big)\,,$$

after which the proofs of the theorems of fcn_from_pred pose no problems.

As an easy example of the use of this **THEORY**, note that it can be invoked in the special form

$$\mathbf{APPLY}(\mathsf{img})\; \mathsf{fcn\_from\_pred}\big(\, \mathsf{P}(X,Y) \mapsto Y \in X\; \&\; Q(Y),$$
$$\mathsf{xcp}(X) \mapsto X \qquad\qquad\quad \big) ==> \cdots$$

for any monadic predicate Q (because $\in$ is acyclic); without the condition $Y \in X$ such an invocation would instead result in an error indication, except in the uninteresting case in which one has proved that $\forall x \,\neg\, Q(x)$.

Here is a slightly more elaborate example of a familiar **THEORY**:

> **THEORY** equivalence_classes(s, Eq)
> $(\forall x \in \mathsf{s})\big(\mathsf{Eq}(x,x)\big)$
> $(\forall x,y,z \in \mathsf{s})\big(\mathsf{Eq}(x,y) \;\rightarrow\; \big(\mathsf{Eq}(y,z) \;\leftrightarrow\; \mathsf{Eq}(x,z)\big)\big)$
> ==>(quot, cl_of) -- *"quotient"-set and globalized "canonical embedding"*
> $(\forall x,y \in \mathsf{s})\big(\mathsf{Eq}(x,y) \;\leftrightarrow\; \mathsf{Eq}(y,x)\big)$

$$(\forall x \in s)(\text{cl\_of}(x) \in \text{quot})$$
$$(\forall b \in \text{quot})(\text{arb } b \in s \ \& \ \text{cl\_of}(\text{arb } b) = b)$$
$$(\forall y \in s)(\text{Eq}(x, y) \ \leftrightarrow \ \text{cl\_of}(x) = \text{cl\_of}(y))$$
END equivalence_classes.

Suppose that this **THEORY** has been established, and that $\mathbb{N}, \mathbb{Z}$, and the multiplication operation $*$ have been defined already, where $\mathbb{N}$ is the set of natural numbers, and $\mathbb{Z}$, intended to be the set of signed integers, is defined (somewhat arbitrarily) as

$$\mathbb{Z} := \{[n, m] \ : \ n, m \in \mathbb{N} \mid n = 0 \vee m = 0\}.$$

Here the position of 0 in a pair serves as a sign indication, and the restriction of $*$ to $\mathbb{Z} \times \mathbb{Z}$ is integer multiplication (but actually, $x * y$ is always defined, whether or not $x, y \in \mathbb{Z}$). Then the set Fr of fractions and the set $\mathbb{Q}$ of rational numbers can be defined as follows:

Fr $:= \{[x, y] \ : \ x, y \in \mathbb{Z} \mid y \neq [0, 0]\}$,
Same_frac$(F, G) := (\text{head}(F) * \text{tail}(G) = \text{tail}(F) * \text{head}(G))$,
APPLY$(\mathbb{Q}, \text{Fr\_to\_}\mathbb{Q})$ equivalence_classes$(s \mapsto \text{Fr}$,
$\qquad\qquad\qquad\qquad\qquad \text{Eq}(F, G) \mapsto \text{Same\_frac}(F, G)) ==> \cdots$

Before **APPLY** can be invoked, one must prove that the restriction of Same_frac to Fr meets the **THEORY** assumptions, i.e. it is an equivalence relation. Then the system will not simply return the two new symbols $\mathbb{Q}$ and Fr_to_$\mathbb{Q}$, but will provide theorems insuring that these represent the standard equivalence-class reduction Fr/Same_frac and the canonical embedding of Fr into this quotient. Note as a curiosity —which however hints at the type of hiding implicit in the **THEORY** mechanism— that a $\mathbb{Q}$ satisfying the conclusions of the **THEORY** is not actually forced to be the standard partition of Fr but can consist of singletons or even of supersets of the equivalence classes (which is harmless).

5   A Final Case Study: Finite Summation

Consider the operation $\Sigma(F)$ or, more explicitly,

$$\sum_{x \in \text{domain}(F)} \ \sum_{[x,y] \in F} y$$

available for any *finite* map F (and in particular when $\text{domain}(F) = d \in \mathbb{N}$, so that $x \in d$ amounts to saying that $x = 0, 1, \ldots, d - 1$) such that $\text{range}(F) \subseteq$ abel, where abel is a set on which a given operation $+$ is associative and commutative and has a unit element u. Most of this is captured formally by the following **THEORY**:

THEORY sigma_add(abel, +, u)
$\quad (\forall\, x, y \,\in\,$ abel$)(x{+}y \in$ abel &$\qquad\qquad$ -- *closure w.r.t.* ...
$\qquad\qquad\qquad x{+}y = y{+}x)\qquad\qquad$ -- *...commutative operation*
$\quad$ u $\in$ abel & $(\forall\, x\, \in\,$ abel$)(x{+}$u$ = x)\qquad$ -- *designated unit element*
$\quad (\forall\, x, y, z\, \in\,$ abel$)\big((x{+}y){+}z = x{+}(y{+}z) \big)$-- *associativity*
==>$(\Sigma)\qquad\qquad\qquad\qquad\qquad\qquad\qquad$ -- *summation operation*
$\quad \Sigma(\emptyset) = $ u & $(\forall\, x \in \mathbb{N})(\forall\, y \in$ abel$)\big(\Sigma(\{[x,y]\}) = y \,\big)$
$\quad$ is_map(F) & Finite(F) & range$(F) \subseteq$ abel & domain$(F) \subseteq \mathbb{N} \;\rightarrow$
$\qquad \Sigma(F) = \Sigma(F \cap G) + \Sigma(F \setminus G)\quad$ -- *additivity*
END sigma_add.

We show below how to construct this **THEORY** from its assumptions, and how to generalize it into a **THEORY** gen_sigma_add in which the condition domain$(F) \subseteq \mathbb{N}$ is dropped, allowing the condition $(\forall\, x \in \mathbb{N})(\forall\, y \in$ abel$)\big(\Sigma(\{[x,y]\}) = y \,\big)$ to be simplified into $(\forall\, y \in$ abel$)\big(\Sigma(\{[X,y]\}) = y \,\big)$. After this, we will sketch the proof of a basic property ('rearrangement of terms') of this generalized summation operation.

5.1   Existence of a Finite Summation Operation

In order to tackle even the simple sigma_add, it is convenient to make use of recursions somewhat different (and actually simpler) than the fully general transfinite $\in$-recursion axiomatically available in our version of set theory. Specifically, we can write

$$\Sigma(F) := \text{if } F = \emptyset \text{ then } \text{u} \text{ else tail}(\text{arb } F) + \Sigma(F \text{ less arb } F) \text{ end if},$$

which is a sort of "tail recursion" based on set inclusion.

$\qquad$To see why such constructions are allowed we can use the fact that strict inclusion is a well-founded relation between finite sets, and in particular that it is well-founded over $\{ f \subseteq \mathbb{N} \times$ abel $|$ Finite$(f) \}$: this makes the above form of recursive definition acceptable.

$\qquad$In preparing to feed this definition —or something closely equivalent to it— into our proof-verifier, we can conveniently make a *détour* through the following **THEORY** (note that in the following formulae Ord(X) designates the predicate 'X is an ordinal'—see end of Sec.3):

THEORY well_founded_set(s, Lt)
$\quad (\forall t \subseteq$ s$)\big(t \neq \emptyset \rightarrow (\exists\, m \in t)(\forall u \in t)\neg\, \text{Lt}(u, m) \big)$
$\quad$ -- Lt *is thereby assumed to be irreflexive and well-founded on* s
==>(orden)
$\quad (\forall\, x, y \,\in\,$ s$)\big(\big(\text{Lt}(x,y) \rightarrow \neg\,\text{Lt}(y,x) \big) \,\&\, \neg\,\text{Lt}(x,x) \big)$
$\quad$ s $\subseteq \{\,$orden$(y) : y \in X \,\} \leftrightarrow$ orden$(X) = $ s
$\quad$ orden$(X) \neq$ s $\leftrightarrow$ orden$(X) \in$ s
$\quad$ Ord(U) & Ord(V) & orden$(U) \neq$ s $\neq$ orden$(V) \rightarrow$
$\qquad\qquad \big(\text{Lt}\big(\text{orden}(U), \text{orden}(V) \big) \rightarrow U \in V \big)$

$$\{\, u \in s \,:\, \mathsf{Lt}(\, u, \mathsf{orden}(V)\,)\,\} \subseteq \{\, \mathsf{orden}(x) \,:\, x \in V\,\}$$
$$\mathsf{Ord}(U) \,\&\, \mathsf{Ord}(V) \,\&\, \mathsf{orden}(U) \neq s \neq \mathsf{orden}(V) \,\&\, U \neq V \;\rightarrow$$
$$\mathsf{orden}(U) \neq \mathsf{orden}(V)$$
$$\exists o\Big(\, \mathsf{Ord}(o) \,\&\, s = \{\, \mathsf{orden}(x) \,:\, x \in o\,\} \,\&$$
$$\mathsf{1\_1\_map}\big(\, \{[x, \mathsf{orden}(x)] \,:\, x \in o\}\,\big)\Big)$$
END well_founded_set.

Within this **THEORY** and in justification of it, orden can be defined in two steps:

$$\mathsf{Minrel}(T) \;:=\; \textbf{if}\; \emptyset \neq T \subseteq s \;\textbf{then}\; \textbf{arb}\,\{\, m \in T \mid (\forall x \in T)\neg\,\mathsf{Lt}(x, m)\,\}$$
$$\textbf{else}\; s \;\textbf{end if},$$
$$\mathsf{orden}(X) \;:=\; \mathsf{Minrel}\big(\, s \setminus \{\, \mathsf{orden}(y) \,:\, y \in X\}\,\big),$$

after which the proof of the output theorems of the **THEORY** just described will take approximately one hundred lines.

Next we introduce a **THEORY** of recursion on well-founded sets. Even though the definition of Σ only requires much less, other kinds of recursive definition benefit if we provide a generous scheme like the following:

THEORY recursive_fcn(dom, Lt, a, b, P)
$$(\forall t \subseteq \mathsf{dom})\big(\, t \neq \emptyset \;\rightarrow\; (\exists m \in t)(\forall u \in t)\neg\,\mathsf{Lt}(u, m)\,\big)$$
-- Lt *is thereby assumed to be irreflexive and well-founded on* dom
==>(rec)
$$(\forall v \in \mathsf{dom})\big(\, \mathsf{rec}(v) =$$
$$\mathsf{a}\big(\, v, \{\, \mathsf{b}(\, v, w, \mathsf{rec}(w)\,) \,:\, w \in \mathsf{dom} \mid \mathsf{Lt}(w, v) \,\&\, \mathsf{P}(\, v, w, \mathsf{rec}(w)\,)\,\}\,\big)\,\big)$$
END recursive_fcn.

The output symbol rec of this **THEORY** is easily definable as follows:

$$\mathsf{G}(X) \;:=\; \mathsf{a}\big(\, \mathsf{orden}(X), \{\, \mathsf{b}(\, \mathsf{orden}(X), \mathsf{orden}(y), \mathsf{G}(y)\,) \,:\, y \in X \mid$$
$$\mathsf{Lt}(\, \mathsf{orden}(y), \mathsf{orden}(X)\,) \,\&\, \mathsf{P}(\, \mathsf{orden}(X), \mathsf{orden}(y), \mathsf{G}(y)\,)\,\}\,\big),$$
$$\mathsf{rec}(V) \;:=\; \mathsf{G}(\, \mathsf{index\_of}(V)\,);$$

here orden results from an invocation of our previous **THEORY** well_founded_set, namely

APPLY(orden) well_founded_set$\big(\, s \mapsto \mathsf{dom},\; \mathsf{Lt}(X, Y) \mapsto \mathsf{Lt}(X, Y)\,\big)$==>$\cdots$;

also, the restriction of index_to to dom is assumed to be the local inverse of the function orden. Note that the recursive characterization of rec in the theorem of recursive_fcn is thus ultimately justified in terms of the very general form of $\in$-recursion built into our system, as appears from the definition of G.

Since we cannot take it for granted that we have an inverse of orden, a second auxiliary **THEORY**, invokable as

APPLY(index_of) bijection$\big(\, \mathsf{f}(X) \mapsto \mathsf{orden}(X), \mathsf{d} \mapsto o1, \mathsf{r} \mapsto \mathsf{dom}\,\big)$==>$\cdots$, is useful. Here o1 results from Skolemization of the last theorem in well_founded_set. The new **THEORY** used here can be specified as follows:

> **THEORY** bijection(f, d, r)
>   1_1_map$\big(\{[x, f(x)] : x \in d\} \big) \,\&\, r = \{f(x) : x \in d\}$
> $f(X) \in r \;\to\; X \in d$ -- *convenient "guard"*
> ==>(finv)
> $Y \in r \;\to\; f\big(finv(Y)\big) = Y$
> $Y \in r \;\to\; finv(Y) \in d$
> $X \in d \;\leftrightarrow\; f(X) \in r$
> $X \in d \;\to\; finv\big(f(X)\big) = X$
> $\big(finv(Y) \in d \,\&\, \exists x\big(f(x) = Y\big)\big) \;\leftrightarrow\; Y \in r$
> $d = \{finv(y) : y \in r\} \,\&\,$ 1_1_map$\big(\{[y, finv(y)] : y \in r\}\big)$
> **END** bijection.

This little digression gives us one more opportunity to show the interplay between theories, because one way of defining finv inside bijection would be as follows:

> **APPLY**(finv) fcn_from_pred$\big($
> $P(Y, X) \mapsto f(X) = Y \,\&\, d \neq \emptyset$,
> $e(Y) \mapsto$ **if** $Y \in r$ **then** d **else arb** d **end if** $\big)$ ==> $\cdots$,

where fcn_from_pred is as shown in Sec.4.

We can now recast our first-attempt definition of Σ as

> **APPLY**(Σ) recursive_fcn$\big($
>   dom $\mapsto \{f \subseteq \mathbb{N} \times$ abel $|$ is_map$(f) \,\&\,$ Finite$(f)\}$,
>   Lt$(W, V) \mapsto W \subseteq V \,\&\, W \neq V$,
> $a(V, Z) \mapsto$ **if** $V = \emptyset$ **then** u **else** tail$($**arb** $V) +$ **arb** Z **end if**,
> $b(V, W, Z) \mapsto Z$,
> $P(V, W, Z) \mapsto W = V$ less **arb** V $\big)$ ==> $\cdots$,

whose slight intricacy is the price being paid to our earlier decision to keep the recursive definition scheme very general.

We skip the proofs that $\Sigma(\emptyset) = $ u and $(\forall x \in \mathbb{N})(\forall y \in$ abel$)\big(\Sigma(\{[x, y]\}) = y \big)$, which are straightforward. Concerning additivity, assume by absurd hypothesis that f is a finite map with domain$(f) \subseteq \mathbb{N}$ and range$(f) \subseteq$ abel such that $\Sigma(f) \neq \Sigma(f \cap g) + \Sigma(f \setminus g)$ holds for some g, and then use the following tiny but extremely useful **THEORY** (of induction over the subsets of any finite set)

> **THEORY** finite_induction(n, P)
>   Finite$(n) \,\&\, P(n)$
> ==>(m)
> $m \subseteq n \,\&\, P(m) \,\&\, (\forall k \subseteq m)\big(k \neq m \;\to\; \neg P(k)\big)$
> **END** finite_induction,

to get an inclusion-minimal such map, f0, by performing an

APPLY(f0) finite_induction$\Big(\mathsf{n} \mapsto \mathsf{f},$
$$\mathsf{P}(F) \mapsto \exists g\big(\Sigma(F) \neq \Sigma(F \cap g) + \Sigma(F \setminus g) \big) \Big) ==> \cdots .$$

Reaching a contradiction from this is very easy.

5.2   Generalized Notion of Finite Summation

Our next goal is to generalize the finite summation operation $\Sigma(F)$ to any finite map F with $\mathsf{range}(F) \subseteq \mathsf{abel}$. To do this we can use a few basic theorems on ordinals, which can be summarized as follows. Define

$$\mathsf{min\_el}(T, S) \; := \; \mathbf{if}\; S \subseteq T \;\mathbf{then}\; S \;\mathbf{else}\; \mathbf{arb}\,(S \setminus T) \;\mathbf{end\;if},$$
$$\mathsf{enum}(X, S) \; := \; \mathsf{min\_el}\big(\{\, \mathsf{enum}(y) \,:\, y \in X\}, S \big),$$

for all sets S, T (a use of $\in$-recursion quite similar to the construction used inside the **THEORY** well_founded_set![6]). Then the following *enumeration theorem* holds:

$$\exists o \Big(\mathsf{Ord}(o) \; \& \; S = \big\{ \mathsf{enum}(x, S) \,:\, x \in o \big\}$$
$$\& \; (\forall x, y \in o)\big(x \neq y \; \rightarrow \; \mathsf{enum}(x, S) \neq \mathsf{enum}(y, S) \big) \Big).$$

From this one gets the function ordin by Skolemization.

Using the predicate Finite of Fig.1, and exploiting the infinite set **s_inf** axiomatically available in our version of set theory, we can give the following definition of natural numbers:

$$\mathbb{N} := \mathbf{arb}\,\big\{\, x \in \mathsf{next}(\,\mathsf{ordin}(\mathbf{s\_inf})\,) \,\mid\, \neg\,\mathsf{Finite}(x)\,\big\}.$$

These characterizations of Finite and $\mathbb{N}$ yield

$$X \in \mathbb{N} \;\leftrightarrow\; \mathsf{ordin}(X) = X \;\&\; \mathsf{Finite}(X),$$
$$\mathsf{Finite}(X) \;\leftrightarrow\; \mathsf{ordin}(X) \in \mathbb{N},$$
$$\mathsf{Finite}(F) \;\rightarrow\; \mathsf{Finite}(\,\mathsf{domain}(F)\,) \;\&\; \mathsf{Finite}(\,\mathsf{range}(F)\,).$$

Using these results and working inside the **THEORY** gen_sigma_add, we can obtain the generalized operation Σ by first invoking

APPLY(σ) sigma_add$(\, \mathsf{abel} \mapsto \mathsf{abel}, + \mapsto +, \mathsf{u} \mapsto \mathsf{u}\,) ==> \cdots$

and then defining:

$$\Sigma(F) := \sigma\Big(\big\{\, [x, y] \,:\, x \in \mathsf{ordin}(\,\mathsf{domain}(F)\,),\; y \in \mathsf{range}(F)$$
$$\mid\, [\,\mathsf{enum}(\,x, \mathsf{domain}(F)\,), y\,] \in F \,\big\} \Big).$$

We omit the proofs that $\Sigma(\emptyset) = \mathsf{u}$, $(\forall y \in \mathsf{abel})\big(\Sigma(\{[X, y]\}) = y \big)$, and $\Sigma(F) = \Sigma(F \cap G) + \Sigma(F \setminus G)$, which are straightforward.

---

[6] This is more than just an analogy: we could exploit the well-foundedness of $\in$ to hide the details of the construction of enum into an invocation of the **THEORY** well_founded_set.

5.3   Rearrangement of Terms in Finite Summations

To be most useful, the **THEORY** of Σ needs to encompass various strong statements of the additivity property. Writing

$$\Phi(F) \equiv \text{is\_map}(F) \ \& \ \text{Finite}(\text{domain}(F)) \ \& \ \text{range}(F) \subseteq \text{abel},$$
$$\Psi(P, X) \equiv X = \bigcup P \ \& \ (\forall b, v \in P)(b \neq v \ \rightarrow \ b \cap v = \emptyset)$$

for brevity, much of what is wanted can be specified e.g. as follows:

> **THEORY** gen_sigma_add(abel, +, u)
> $\quad (\forall x, y \in \text{abel})(x{+}y \in \text{abel} \ \& \qquad$ -- *closure w.r.t.* ...
> $\qquad\qquad\qquad x{+}y = y{+}x) \qquad$ -- ... *commutative operation*
> $\quad \text{u} \in \text{abel} \ \& \ (\forall x \in \text{abel})(x{+}\text{u} = x) \quad$ -- *designated unit element*
> $\quad (\forall x, y, z \in \text{abel})((x{+}y){+}z = x{+}(y{+}z))$-- *associativity*
> ==>(Σ) $\qquad\qquad\qquad\qquad\qquad\qquad$ -- *summation operation*
> $\quad \Sigma(\emptyset) = \text{u} \ \& \ (\forall y \in \text{abel})(\Sigma(\{[X, y]\}) = y)$
> $\quad \Phi(F) \ \rightarrow \ \Sigma(F) \in \text{abel}$
> $\quad \Phi(F) \rightarrow \Sigma(F) = \Sigma(F \cap G) + \Sigma(F \setminus G)$-- *additivity*
> $\quad \Phi(F) \ \& \ \Psi(P, F) \ \rightarrow \ \Sigma(F) = \Sigma(\{ [g, \Sigma(g)] : g \in P \})$
> $\quad \Phi(F) \ \& \ \Psi(P, \text{domain}(F)) \ \rightarrow \ \Sigma(F) = \Sigma(\{ [b, \Sigma(F_{|b})] : b \in P \})$
> $\quad \Phi(F) \ \& \ \text{Svm}(G) \ \& \ \text{domain}(F) = \text{domain}(G) \ \rightarrow$
> $\qquad\qquad \Sigma(F) = \Sigma(\{ [x, \Sigma(F_{|G^{-1}\{x\}})] : x \in \text{range}(G) \})$
> **END** gen_sigma_add.

A proof of the last of these theorems, which states that Σ is insensitive to operand rearrangement and grouping, is sketched below.

   Generalized additivity is proved first: starting with the absurd hypothesis that specific f, p exist for which

$$\Phi(\text{f}) \ \& \ \Psi(\text{p}, \text{f}) \ \& \ \Sigma(\text{f}) \neq \Sigma(\{ [g, \Sigma(g)] : g \in \text{p} \})$$

holds, one can choose an inclusion-minimal such p referring to the same f and included in the p chosen at first, by an invocation

> **APPLY**(p0) finite_induction$\Big(n \mapsto p,$
> $\qquad P(Q) \mapsto \Psi(Q, \text{f}) \ \& \ \Sigma(\text{f}) \neq \Sigma(\{ [g, \Sigma(g)] : g \in Q \}))\Big)$==>$\cdots$.

From this, a contradiction is easily reached.

   The next theorem, namely

$$\Phi(F) \ \& \ \Psi(P, \text{domain}(F)) \ \rightarrow \ \Sigma(F) = \Sigma(\{ [b, \Sigma(F_{|b})] : b \in P \})$$

follows since $\Psi(P, \text{domain}(F))$ implies $\Psi(\{F_{|b} : b \in P\}, F)$.

   Proof of the summand rearrangement theorem seen above is now easy, because

$$\text{Svm}(G) \ \& \ D = \text{domain}(G) \ \rightarrow \ \Psi(\{ G^{-1}\{x\} : x \in \text{range}(G) \}, D)$$

holds for any D and hence in particular for $D = \text{domain}(F)$.

The above line of proof suggests a useful preamble is to construct the following theory of Ψ:

THEORY is_partition(p, s)
==>(flag) -- *this indicates whether or not* s *is partitioned by* p
      flag $\leftrightarrow$ s $= \bigcup$ p & $(\forall b, v)(b \neq v \ \rightarrow \ b \cap v = \emptyset)$
      flag & Finite(s) $\rightarrow$ Finite(p)
      flag & s $=$ domain(F) & $Q = \{ F_{|b} \ : \ b \in$ p $\} \ \rightarrow \ F = \bigcup Q$ &
                              $(\forall f, g \in Q)(f \neq g \ \rightarrow \ f \cap g = \emptyset)$
      Svm(G) & s $=$ domain(G) & p $= \big\{ G^{-1}\{y\} \ : \ y \in$ range(G) $\big\} \ \rightarrow$ flag
END is_partition.

6 Related Work

To support software design and specification, rapid prototyping, theorem proving, user interface design, and hardware verification, various authors have proposed systems embodying constructs for modularization which are, under one respect or another, akin to our **THEORY** construct. Among such proposals lies the OBJ family of languages [15], which integrates specification, prototyping, and verification into a system with a single underlying equational logic.

In the implementation OBJ3 of OBJ, a module can either be an *object* or a *theory*: in either case it will have a set of equations as its body, but an object is executable and has a fixed standard model whereas a theory describes non-executable properties and has loose semantics, namely a variety of admissible models. As early as in 1985, OBJ2 [13] was endowed with a generic module mechanism inspired by the mechanism for parameterized specifications of the Clear specification language [3]; the interface declarations of OBJ2 generics were not purely syntactic but contained semantic requirements that actual modules had to satisfy before they could be meaningfully substituted.

The use of OBJ for theorem-proving is aimed at providing mechanical assistance for proofs that are needed in the development of software and hardware, more than at mechanizing mathematical proofs in the broad sense. This partly explains the big emphasis which the design of OBJ places on equational reasoning and the privileged role assigned to universal algebra: equational logic is in fact sufficiently powerful to describe any standard model within which one may want to carry out computations.

We observe that an equational formulation of set theory can be designed [11], and may even offer advantages w.r.t. a more traditional formulation of Zermelo-Fraenkel in limited applications where it is reasonable to expect that proofs can be found in fully automatic mode; nevertheless, overly insisting on equational reasoning in the realm of set theory would be preposterous in light of the highly interactive proof-verification environment which we envision.

We like to mention another ambitious project, closer in spirit to this paper although based on a sophisticated variant of Church's typed lambda-calculus [6]: the Interactive Mathematical Proof System (IMPS) described in [10]. This

228 Eugenio G. Omodeo and Jacob T. Schwartz

system manages a database of mathematics, represented as a collection of interconnected axiomatic *"little theories"* which span graduate-level parts of analysis (about 25 theories: real numbers, partial orders, metric spaces, normed spaces, etc.), some algebra (monoids, groups, and fields), and also some theories more directly relevant to computer science (concerning state machines, domains for denotational semantics, and free recursive datatypes). The initial library caters for some fragments of set theory too: in particular, it contains theorems about cardinalities. Mathematical analysis is regarded as a significant arena for testing the adequacy of formalizations of mathematics, because analysis requires great expressive power for constructing proofs.

The authors of [10] claim that IMPS supports a view of the axiomatic method based on "little theories" tailored to the diverse fields of mathematics as well as the "big theory" view in which all reasoning is performed within a single powerful and highly expressive set theory. Greater emphasis is placed on the former approach, anyhow: with this approach, links — "conduits", so to speak, to pass results from one theory to another— play a crucial role. To realize such links, a syntactic device named "theory interpretation" is used in a variety of ways to translate the language of a source theory to the language of a target theory so that the image of a theorem is always a theorem: this method enables reuse of mathematical results "transported" from relatively abstract theories to more specialized ones.

One main difference of our approach w.r.t. that of IMPS is that we are willing to invest more on the "big theory" approach and, accordingly, do not feel urged to rely on a higher-order logic where functions are organized according to a type hierarchy. It may be contended that the typing discipline complies with everyday mathematical practice, and perhaps gives helpful clues to the automated reasoning mechanisms so as to ensure better performance; nevertheless, a well-thought type-free environment can be conceptually simpler.

Both OBJ and IMPS attach great importance to interconnections across theories, inheritance to mention a most basic one, and "theory ensembles" to mention a nice feature of IMPS which enables one to move, e.g., from the formal theory of a metric space to a family of interrelated replicas of it, which also caters for continuous mappings between metric spaces. As regards theory interconnections, the proposal we have made in this paper still awaits being enriched.

The literature on the OBJ family and on the IMPS system also stresses the kinship between the activity of proving theorems and computing in general; even more so does the literature on systems, such as Nuprl [8] or the Calculus of Constructions [9], which rely on a constructive foundation, more or less close to Martin-Löf's intuitionistic type theory [19]. Important achievements, and in particular the conception of declarative programming languages such as Prolog, stem in fact from the view that proof-search can be taken as a general paradigm of computation. On the other hand, we feel that too little has been done, to date, in order to exploit a "proof-by-computation" paradigm aimed at enhancing theorem-proving by means of the ability to perform symbolic computations efficiently in specialized contexts of algebra and analysis (a step in this direction was moved with [7]). Here is an issue that we intend to deepen in a forthcoming paper.

7   Conclusions

We view the activity of setting up detailed formalized proofs of important theorems in analysis and number theory as an essential part of the feasibility study that must precede the development of any ambitious proof-checker. In mathematics, set theory has emerged as the standard framework for such an enterprise, and full computer-assisted certification of a modernized version of *Principia Mathematica* should now be possible. To convince ourselves of a verifier system's ability to handle large-scale mathematical proofs —and such proofs cannot always be avoided in program-correctness verification—, it is best to follow the royal road paved by the work of Cauchy, Dedekind, Frege, Cantor, Peano, Whitehead–Russell, Zermelo–Fraenkel–von Neumann, and many others.

Only one facet of our work on large-scale proof scenarios is presented in this paper. Discussion on the nature of the basic inference steps a proof-verifier should (and reasonably can) handle has been omitted to focus our discussion on the issue of proof modularization. The obvious goal of modularization is to avoid repeating similar steps when the proofs of two theorems are closely analogous. Modularization must also conceal the details of a proof once they have been fed into the system and successfully certified.

When coupled to a powerful underlying set theory, indefinitely expansible with new function symbols generated by Skolemization, the technical notion of "theory" proposed in this paper appears to meet such proof-modularization requirements. The examples provided, showing how often the **THEORY** construct can be exploited in proof scenarios, may convince the reader of the utility of this construct.

Acknowledgements

We thank Ernst-Erich Doberkat (Universität Dortmund, D), who brought to our attention the text by Frege cited in the epigraph of this paper. We are indebted to Patrick Cegielski (Université Paris XII, F) for helpful comments.

References

1. A. Blass and Y. Gurevich. The logic of choice. *J. of Symbolic Logic*, 65(3):1264–1310, 2000.
2. D. S. Bridges. *Foundations of real and abstract analysis*. Springer-Verlag, Graduate Texts in Mathematics vol.174, 1997.
3. R. Burstall and J. Goguen. Putting theories together to make specifications. In R. Reddy, ed, Proc. 5th *International Joint Conference on Artificial Intelligence*. Cambridge, MA, pp. 1045–1058, 1977.

4. R. Caferra and G. Salzer, editors. *Automated Deduction in Classical and Non-Classical Logics*. LNCS 1761 (LNAI). Springer-Verlag, 2000.
5. P. Cegielski. Un fondement des mathématiques. In M. Barbut et al., eds, *La recherche de la vérité*. ACL – Les éditions du Kangourou, 1999.
6. A. Church. A formulation of the simple theory of types. *J. of Symbolic Logic*, 5:56–68, 1940.
7. E. Clarke and X. Zhao. Analytica—A theorem prover in Mathematica. In D. Kapur, ed, *Automated Deduction—CADE-11*. Springer-Verlag, LNCS vol. 607, pp. 761–765, 1992.
8. R. L. Constable, S. F. Allen, H. M. Bromley, W. R. Cleaveland, J. F. Cremer, R. W. Harper, D. J. Howe, T. B. Knoblock, N. P. Mendler, P. Panangaden, J. T. Sasaki, and S. F. Smith. *Implementing mathematics with the Nuprl development system*. Prentice-Hall, Englewood Cliffs, NJ, 1986.
9. Th. Coquand and G. Huet. The calculus of constructions. *Information and Computation*, 76(2/3):95–120, 1988.
10. W. M. Farmer, J. D. Guttman, F. J. Thayer. IMPS: An interactive mathematical proof system. *J. of Automated Reasoning*, 11:213–248, 1993.
11. A. Formisano and E. Omodeo. An equational re-engineering of set theories. In Caferra and Salzer [4, pp. 175–190].
12. G. Frege. Logik in der Mathematik. In *G. Frege, Schriften zur Logik und Sprachphilosophie*. Aus dem Nachlaß herausgegeben von G. Gabriel. Felix Meiner Verlag, Philosophische Bibliothek, Band 277, Hamburg, pp. 92–165, 1971.
13. K. Futatsugi, J. A. Goguen, J.-P. Jouannaud, J. Meseguer. Principles of OBJ2. Proc. *12th annual ACM Symp. on Principles of Programming Languages* (POPL'85), pp. 55-66, 1985.
14. R. Godement. *Cours d'algèbre*. Hermann, Paris, Collection Enseignement des Sciences, 3rd edition, 1966.
15. J. A. Goguen and G. Malcolm. *Algebraic semantics of imperative programs*. MIT, 1996.
16. T. J. Jech. *Set theory*. Springer-Verlag, Perspectives in Mathematical Logic, 2nd edition, 1997.
17. E. Landau. *Foundation of analysis. The arithmetic of whole, rational, irrational and complex numbers*. Chelsea Publishing Co., New York, 2nd edition, 1960.
18. A. Levy. *Basic set theory*. Springer-Verlag, Perspectives in Mathematical Logic, 1979.
19. P. Martin-Löf. *Intuitionistic type theory*. Bibliopolis, Napoli, Studies in Proof Theory Series, 1984.

An Open Research Problem: Strong Completeness of R. Kowalski's Connection Graph Proof Procedure

Jörg Siekmann[1] and Graham Wrightson[2]

[1] Universität des Saarlandes, Stuhlsatzenhausweg, D-66123 Saarbrücken, Germany.
siekmann@dfki.de
[2] Department of Computer Science and Software Engineering, The University of Newcastle, NSW 2308, Australia.
graham@cs.newcastle.edu.au

Abstract. The connection graph proof procedure (or clause graph resolution as it is more commonly called today) is a theorem proving technique due to Robert Kowalski. It is a negative test calculus (a refutation procedure) based on resolution. Due to an intricate deletion mechanism that generalises the well-known purity principle, it substantially refines the usual notions of resolution-based systems and leads to a largely reduced search space. The dynamic nature of the clause graph upon which this refutation procedure is based, poses novel meta-logical problems previously unencountered in logical deduction systems. Ever since its invention in 1975 the soundness, confluence and (strong) completeness of the procedure have been in doubt in spite of many partial results. This paper provides an introduction to the problem as well as an overview of the main results that have been obtained in the last twenty-five years.

1 Introduction to Clause Graph Resolution

We assume the reader to be familiar with the basic notions of resolution-based theorem proving (see, for example, Alan Robinson [1965], Chang, C.-L. and Lee, R.C.-T. [1973] or Don Loveland [1978]). Clause graphs introduced a new ingenious development into the field, the central idea of which is the following: In standard resolution two resolvable literals must first be found in the set of sets of literals before a resolution step can be performed, where a set of literals represents a clause (i.e. a disjunction of these literals) and a statement to be refuted is represented as a set of clauses. Various techniques were developed to carry out this search. However, Robert Kowalski [1975] proposed an enhancement to the basic data structure in order to make possible resolution steps explicit, which — as it turned out in subsequent years — not only simplified the search, but also introduced new and unexpected logical problems. This enhancement was gained by the use of so-called links between complementary literals, thus turning the *set notation* into a *graph-like structure*. The new approach allowed in particular for the removal of a link after the corresponding resolution step and

A.C. Kakas, F. Sadri (Eds.): Computat. Logic (Kowalski Festschrift), LNAI 2408, pp. 231–252, 2002.
© Springer-Verlag Berlin Heidelberg 2002

a clause that contains a literal which is no longer connected by a link may be removed also (generalised purity principle). An important side effect was that this link removal had the potential to cause the disappearance of even more clauses from the current set of clauses (avalanche effect).

Although this effect could reduce the search space drastically it also had a significant impact on the underlying logical foundations. To quote Norbert Eisinger from his monograph on Kowalski's clause graphs [1991]:

> "Let S and S' be the current set of formulae before and after a deduction step $S \vdash S'$. A step of a classical calculus and a resolution step both simply add a formula following from S. Thus, each interpreted as the conjunction of its members, S and S' are always equivalent. For clause graph resolution, however, S may contain formulae missing in S', and the removed formulae are not necessarily consequences of those still present in S'. While this does not affect the forward implication, S does in general no longer ensue from S'. In other words, it is possible for S' to possess more models than S. But, when S is unsatisfiable, so must be S', i.e. S' must not have more models than S, if soundness, unsatisfiability and hence refutability, is to be preserved."

This basic problem underlying all investigations of the properties of the clause graph procedure will be made more explicit in the following.

2   Clause Graph Resolution: The Problem

The standard resolution principle, called *set resolution* in the following, assumes the axioms and the negated theorem to be represented as a *set of clauses*. In contrast, the clause graph proof procedure represents the initial set of clauses as a *graph* by drawing a link between pairs of literal occurrences to denote that some relation holds between these two literals. If this relation is *"complementarity"* (it may denote other relations as well, see e.g. Christoph Walter [1981], but this is the standard case and the basic point of interest in this paper) of the two literals, i.e. resolvability of the respective clauses, then an initial clause graph for the set

$$S = \{\{ -P(z,c,z), -P(z,d,z)\}, \{P(a,x,a), -P(a,b,c)\},$$
$$\{P(a,w,c), P(w,y,w)\}, \{P(u,d,u), -P(b,u,d), P(u,b,b)\},$$
$$\{-P(a,b,b)\}, \{-P(c,b,c), P(v,a,d), P(a,v,b)\}\}$$

is the graph in Figure 1. Here P is a ternary predicate symbol, letters from the beginning of the alphabet $a, b, c, \ldots$ denote constants, letters from the end of the alphabet $x, y, z, v, \ldots$ denote variables and $-P(\ldots)$ denotes the negation of $P(\ldots)$.

Example 1.

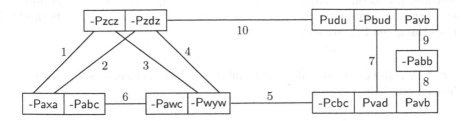

Fig. 1.

An appropriate most general unifier is associated with each link (not shown in the example of Figure 1). We use the now standard notation that adjacent boxes denote a clause, i.e. the conjunction of the literals in the boxes.

So far such a clause graph is just a data structure without commitment to a particular proof procedure and in fact there have been many proposals to base an automated deduction procedure on some graph-like notion (e.g. Andrews [1976], Andrews [1981], Bibel [1981b], Bibel [1982], Chang and Slagle [1979], Kowalski [1975], Shostak [1976], Shostak [1979], Sickel [1976], Stickel [1982], Yates and Raphael and Hart [1970], Omodeo [1982], Yarmush [1976], Murray and Rosenthal [1993], Murray and Rosenthal [1985]).

Kowalski's procedure uses a graph-like data structure as well, but its impact is more fundamental since it operates now as follows: suppose we want to perform the resolution step represented by link 6 in Figure 1 based on the unifier $\sigma = \{w \to b\}$. Renaming the variables appropriately we obtain the resolvent $\{P(a, x', a), P(b, y', b)\}$ which is inserted into the graph and if now all additional links are set this yields the graph:

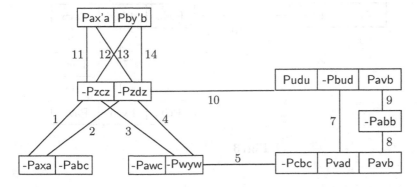

Fig. 2.

Now there are three essential operations:

1. The new links don't have to be recomputed by comparing every pair of literals again for complementarity, but this information can instead be inherited from the given link structure.

2. The link resolved upon is deleted to mark the fact that this resolution step has already been performed,

3. Clauses that contain a literal with no link connecting it to the rest of the graph may be deleted (generalised purity principle).

While the first point is the essential ingredient for the computational attractiveness of the clause graph procedure, the second and third points show the ambivalence between gross logical and computational advantages versus severe and novel theoretical problems. Let us turn to the above example again. After resolution upon link 6 we obtain the graph in Figure 2 above. Now since link 6 has been resolved upon we have it deleted it according to rule (2). But now the two literals involved become pure and hence the two clauses can be deleted as well leading to the following graph:

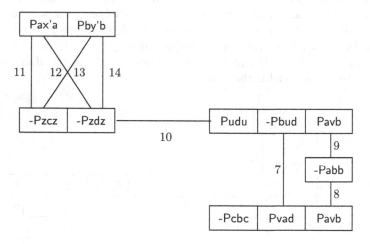

Fig. 3.

But now the literal $-P(c, b, c)$ in the bottom clause becomes pure as well and hence we have the graph:

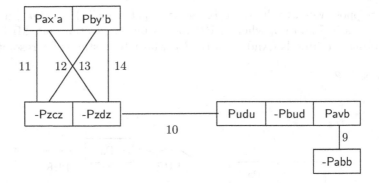

Fig. 4.

This removal causes the only literal $-P(a, b, b)$ in the bottom clause to become pure and hence, after a single resolution step followed by all these purity deletions, we arrive at the final graph:

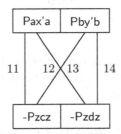

Fig. 5.

It is this strong feature that reduces redundancy in the complementary set of clauses, that marks the fascination for this proof procedure (see Ohlbach [1985], Ohlbach [1983], Bläsius [1986] and [1987], Eisinger et al. [1989], Ohlbach and Siekmann [1991], Bläsius et al. [1981], Eisinger [1981], Eisinger and Siekmann and Unvericht [1979], Ohlbach [1987], Ramesh et al. [1997], Murray and Rosenthal [1993], Siekmann and Wrightson [1980]). It can sometimes even reduce the initial redundant set to its essential contradictory subset (subgraph). But this also marks its problematical theoretical status: *how do we know that we have not deleted too many clauses?* Skipping the details of an exact definition of the various inheritance mechanisms (see e.g. Eisinger [1991] for details) the following example demonstrates the problem.

Suppose we have the refutable set $S = \{\{P(a), P(a)\}, \{-Pa\}\}$ and its initial graph as in Figure 6, where PUR means purity deletion and MER stands for merging two literals (Andrews [1968]), whilst RES stands for resolution.

Example 2.

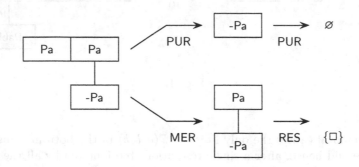

Fig. 6.

Thus in two steps we would arrive either at the empty set $\emptyset$, which stands for satisfiability, or in the lower derivation we arrive at the empty clause $\{\Box\}$, which stands for unsatisfiability.

This example would seem to show that the procedure:

(i)   is not confluent, as defined below
(ii) is not sound (correct), and
(iii) is not refutation complete (at least not in the strong sense as defined below),

and hence would be useless for all practical purposes.

But here we can spot the flaw immediately: the process did not start with the full initial graph, where all possible links are set. If, instead, all possible links are drawn in the initial graph, the example in Figure 6 fails to be a counterexample. On the other hand, after a few initial steps we always have a graph with some links deleted, for example because they have been resolved upon. So how can we be sure that the same disastrous phenomenon, as in the above example, will not occur again later on in the derivation?

These problems have been called the *confluence*, the *soundness* and the *(strong) completeness* problem of the clause graph procedure and it can be shown that for the original formulation of the procedure in Kowalski [1975] (with full subsumption and tautology removal) all these three essential properties unfortunately do not hold in general. However, for suitable remedies (of subsumption and tautology removal) the first two properties hold, whereas the third property has been open ever since.

3   Properties and Results for the Clause Graph Proof Procedure

In order to capture the strange and novel properties of logical graphs let us fix the following notions: A *clause graph of a set of clauses* S consists of a set of nodes labelled by the literal occurrences in S and a set of links that connect complementary literals. There are various possibilities to make this notion precise (e.g. Siekmann and Stephan [1976] and [1980], Brown [1976], Eisinger [1986] and [1991], Bibel [1980], Smolka [1982a,b,c] Bibel and Eder [1997], Hähnle *et al.* [2001], Murray and Rosenthal [1985]).

Let INIT(S) be the full initial clause graph for S with all possible links set. This is called a full connection graph in Bibel and Eder [1997], a total graph in Eisinger [1991] and in Siekmann, Stephan [1976] and a complete graph in Brown [1976].

Definition 1. *Clause graph resolution is called*

refutation sound   *if* $INIT(S) \xrightarrow{*} \{\Box\}$ *then* S *is unsatisfiable;*

refutation complete   *if* S *is unsatisfiable then there exists a derivation*
$$INIT(S) \xrightarrow{*} \{\Box\};$$

refutation confluent   *if* S *is unsatisfiable, and,*
if $INIT(S) \xrightarrow{*} G_1$ *and* $INIT(S) \xrightarrow{*} G_2$
then there exists $G_1 \xrightarrow{*} G'$ *and* $G_2 \xrightarrow{*} G'$ *for some* G';

affirmation sound   *if* $INIT(S) \xrightarrow{*} \varnothing$ *then* S *is satisfiable;*

affirmation complete *if* S *is satisfiable then there exists a derivation*
$$INIT(S) \xrightarrow{*} \varnothing;$$

affirmation confluent if S *is satisfiable, and,*
if $INIT(S) \xrightarrow{*} G_1$ *and* $INIT(S) \xrightarrow{*} G_2$
then there exists $G_1 \xrightarrow{*} G'$ *and* $G_2 \xrightarrow{*} G'$, *for some* G'.

The state of knowledge about the clause graph proof procedure at the end of the 1980's can be summarised by the following major theorems. There are some subtleties involved when subsumption and tautology removal are involved (see Eisinger [1991] for a thorough exposition; the discovery of the problems with subsumption and tautology removal and an appropriate remedy for these problems is due to Wolfgang Bibel).

Theorem 1 (Bibel, Brown, Eisinger, Siekmann, Stephan). *Clause graph resolution is refutation sound.*

Theorem 2 (Bibel). *Clause graph resolution is refutation complete.*

Theorem 3 (Eisinger, Smolka, Siekmann, Stephan). *Clause graph resolution is refutation confluent.*

Theorem 4 (Eisinger). *Clause graph resolution is affirmation sound.*

Theorem 5 (Eisinger). *Clause graph resolution is not affirmation confluent.*

Theorem 6 (Smolka). *For the unit refutable class, clause graph resolution with an unrestricted tautology rule is refutation complete, refutation confluent, affirmation sound, (and strongly complete).*

The important notion of strong completeness is introduced below.

Theorem 7 (Eisinger). *Clause graph resolution with an unrestricted tautology rule is refutation complete, but neither refutation confluent nor affirmation sound.*

As important and essential as the above-mentioned results may be, they are not enough for the practical usefulness of the clause graph procedure: the principal requirement for a proof procedure is not only to know that there exists a refutation, but even more importantly that the procedure can actually find it after a finite number of steps. These two notions, called *refutation completeness* and *strong refutation completeness* in the following, essentially coincide for *set resolution* but unfortunately they do not do so for the *clause graph procedure*.

This can be demonstrated by the example, in Figure 7, where we start with the graph G_0 and derive G_1 from G_0 by resolution upon the link marked ☞. The last graph G_2 contains a subgraph that is isomorphic to the first, hence the corresponding inference steps can be repeated over and over again and the procedure will not terminate with the empty clause. Note that a refutation, i.e. the derivation of the empty clause, could have been obtained by resolving upon the leftmost link between P and $-P$.

Example 3 (adapted from Eisinger [1991]).

G_0

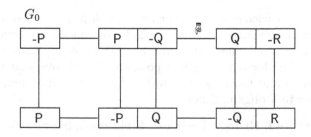

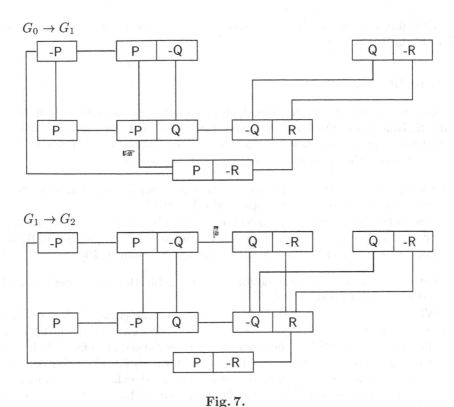

Fig. 7.

Examples of this nature gave rise to the strong completeness conjecture, which in spite of numerous attacks has remained an open problem now for over twenty years:

> *How can we ensure for an unsatisfiable graph that the derivation stops after finitely many steps with a graph that contains the empty clause?*

If this crucial property cannot be ascertained, the whole procedure would be rendered useless for all practical purposes, as we would have to backtrack to some earlier state in the derivation, and hence would have to store all intermediate graphs.

The theoretical problems and strange counter intuitive facts that arise from the (graphical) representation were first discovered by Jörg Siekmann and Werner Stephan and reported independently in Siekmann and Stephan [1976] and [1980] and by Frank Brown in Brown [1976]. They suggested a remedy to the problem: the obvious flaw in the above example can be attributed to the fact that the proof procedure never selects the essential link for the refutation (the link between $-P$ and P).

This, of course, is a property which a control strategy should have, i.e. it should be fair in the sense that every link is eventually selected. However this is a subtle property in the dynamic context of the clause graph procedure as we shall see in the following.

Control Strategies

In order to capture the strange metalogical properties of the clause graph procedure, Siekmann and Stephan [1976] and [1980] introduced two essential notions in order to capture the above-mentioned awkward phenomenon. These two notions have been the essence of all subsequent investigations:

(i) the notion of a *kernel*. This is now sometimes called the minimal refutable subgraph of a graph, e.g. in Bibel and Eder [1997];

(ii) several notions of *covering*, called fairness in Bibel and Eder [1997], exhaustiveness in Brown [1976], fairness-one and fairness-two in Eisinger [1991] and covering-one, two and three in Siekmann and Stephan [1976].

Let us have a look at these notions in turn, using the more recent and advanced notation of Eisinger [1991].

Why is it not enough to simply prove refutation completeness as in the case of clause set resolution? Ordinary refutation completeness ensures that if the initial set of clauses is unsatisfiable, then there exists a refutation, i.e. a finite derivation of the empty clause. Of course, there is a control strategy for which this would be sufficient for clause graph resolution as well, namely an exhaustive enumeration of all possible graphs, as in Figure 8, where we assume that the initial graph G_0 has n links. However such a strategy is computationally infeasible and far too expensive and would make the whole approach useless.

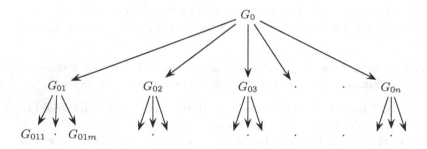

Fig. 8.

We know by Theorem 2 that the clause graph procedure is refutation complete, i.e. that there exists a subgraph from which the derivation can be obtained. Could we not use this information from a potential derivation we know to exist in order to guide the procedure in general?

Many strategies for clause graphs are in fact based on this very idea (Andrews [1981], Antoniou and Ohlbach [1983], Bibel [1981a], Bibel [1982], Chang and Slagle [1979], Sickel [1976]). However, in general, finding the appropriate subgraph essentially amounts to finding a proof in the first place and we might as well use a standard resolution-based proof procedure to find the derivation and then use this information to guide the clause graph procedure.

So let us just assume in the abstract that every full (i.e. a graph where every possible link is set) and unsatisfiable graph contains a subgraph, called a *kernel* (the shaded area in Figure 9), from which an actual refutation can be found in a finite number of steps.

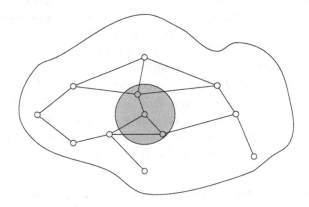

Fig. 9.

We know from Theorem 2 above and from the results in Siekmann and Stephan [1976] and [1980] that every resolution step upon a link within the kernel eventually leads to the empty clause and thus to the desired refutation. If we can ensure that:

1. resolution steps involving links outside of the kernel do not destroy the kernel, and
2. every link in the kernel is eventually selected,

then we are done. This has been the line of attack ever since. Unfortunately the second condition turned out to be more subtle and rather difficult to establish. So far no satisfactory solution to this problem has been found.

So let us look at these concepts a little closer.

Definition 2. *A filter for an inference system is a unary predicate* F *on the set of finite sequences of states. The notation* $S_0 \xrightarrow{*} S_n$ *with* F *stands for a derivation* $S_0 \xrightarrow{*} S_n$ *where* $F(S_0 \ldots S_n)$ *holds. For an infinite derivation,* $S_0 \to \ldots \to S_n \to \ldots$ *with* F *means that* $F(S_0 \ldots S_n \ldots)$ *holds for each* n.

This notion is due to Gert Smolka in [1982b] and Norbert Eisinger in [1991] and it is now used in several monographs on deduction systems (see e.g. K. Bläsius and H. J. Bürckert [1992]). Typical examples for a filter are the usual restriction and ordering strategies in automated theorem proving, such as set-of-support by Wos and Robinson and Carson [1965], linear refutation by Loveland [1970], merge resolution by Andrews [1968], unit resolution by Wos [1964], or see Kowalski [1970].

Definition 3. *A filter* F *for clause graph resolution is called*
refutation sound: $INIT(S) \xrightarrow{*} \{\square\}$ *with* F *then* S *is unsatisfiable;*
refutation complete: if S *is unsatisfiable then there exists*
$$INIT(S) \xrightarrow{*} \{\square\} \text{ with } F;$$
refutation confluent: Let S *be unsatisfiable,*
For $INIT(S) \xrightarrow{*} G_1$ with F and $INIT(S) \xrightarrow{*} G_2$
with F then there exists $G_1 \xrightarrow{*} G'$ with F and
$G_2 \xrightarrow{*} G'$ with F, for some G';
strong refutation   *for an unsatisfiable* S *there does not exist an infinite*
completeness: *derivation* $INIT(S) \to G_1 \to G_1 \to \ldots \to G_n \to \ldots$
with F.

Note that $\to$ *with* F need not be transitive, hence the special form of confluence, also note that the procedure terminates with $\{\square\}$ or with $\varnothing$.

The most important and still open question is now: can we find a general property for a filter that turns the clause graph proof procedure into a strongly complete system? Obviously the filter has to make sure that every link (in particular every link in some fixed kernel) is eventually selected for resolution and not infinitely postponed.

Definition 4. *A filter* F *for clause graph resolution is called covering, if the following holds: Let* G_0 *be an initial graph, let* $G_0 \xrightarrow{*} G_n$ *with* F *be a derivation, and let* λ *be a link in* G_n. *Then there is a finite number* $n(\lambda)$, *such that for any derivation* $G_0 \xrightarrow{*} G_n \xrightarrow{*} G$ *with* F *extending the given one by at least* $n(\lambda)$ *steps,* λ *is not in* G.

This is the weakest notion, called "coveringthree" in Siekmann and Stephan [1976], exhaustiveness in Brown [1976] and fairness in Bibel and Eder [1997]. It is well-known and was already observed in Siekmann and Stephan [1976] that the strong completeness conjecture is false for this notion of covering.

The problem is that a link can disappear without being resolved upon, namely by purity deletion, as the examples from the beginning demonstrate. Even the original links in the kernel can be deleted without being resolved upon, but may reappear after the copying process.

For this reason stronger notions of fairness are required: apparently even essential links can disappear without being resolved upon and reappear later due to the copying process. Hence we have to make absolutely sure that every link in the kernel is eventually resolved upon. To this end imagine that each initial link bears a distinct colour and that each descendant of a coloured link inherits the ancestor's colour:

Definition 5. *An ordering filter* F *for clause graph resolution is called coveringtwo, if it is a covering and at least one link of each colour must have been resolved upon after at most finitely many steps.*

At first sight this definition now seems to capture the essence, but how do we know that the "right" descendant (as there may be more than one) of the coloured ancestor has been operated upon? Hence the strongest definition of fairness for a filter:

Definition 6. *A filter* F *for clause graph resolution is called coveringone, if each colour must have disappeared after at most finitely many steps.*

While the strong completeness conjecture can be shown in the positive for the latter notion of covering (see Siekmann and Stephan [1980]), hardly any of the practical and standard filters actually fulfill this property (except for some obvious and exotic cases).

So the strong completeness conjecture boils down to finding:

1. a proof or a refutation that a covering filter is strongly complete, for the appropriate notions of coveringone, -two, and -three, and
2. strong completeness results for subclasses of the full first-order predicate calculus, or
3. an alternative notion of covering for which strong completeness can be shown.

The first two problems were settled by Norbert Eisinger and Gerd Smolka.

Theorem 8 (Smolka). *For the unit refutable class the strong completeness conjecture is true, i.e. the conjunction of a covering filter with any refutation complete and refutation confluent restriction filter is refutation complete, refutation confluent, and Noetherian, i.e. it terminates.*

This theorem, whose essential contribution is due to Gerd Smolka [1982a] accounts for the optimism at the time. After all the unit refutable class of clauses (Horn clauses) turned out to be very important for many practical purposes, includng logic programming, and the theorem shows that all the essential properties of a useful proof procedure now hold for the clause graph procedure. Based on an ingenious construction, Norbert Eisinger showed however the following devastating result which we will look at again in more detail in Section 4.

Theorem 9 (Eisinger). *In general the strong completeness conjecture is false, even for a restriction filter based on the coveringtwo definition.*

This theorem destroyed once and for all the hope of finding a solution to the problem based on the notion of fairness, as it shows that even for the strongest possible form of fairness, strong completeness cannot be obtained.

So attention turned to the third of the above options, namely of finding alternative notions of a filter for which strong completeness can be shown. Early results are in Wrightson [1989], Eisinger [1991] and more recent results are Hähnle *et al.* [2001], Meagher and Hext [1998].

Let us now look at the proof of Theorem 9 in more detail.

4   The Eisinger Example

This example is taken from Eisinger [1991], p. 158, Example 7.4.7. It shows a cyclic coveringtwo derivation, i.e. it shows that the clause graph proof procedure does not terminate even for the strong notion of a coveringtwo filter, hence in particular not for the notion of coveringthree either.

Let $S = \{PQ, -PQ, -Q - R, RS, R - S\}$ and $\text{INIT}(S) = G_0$.

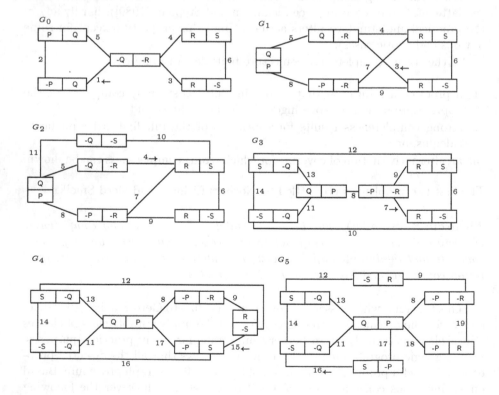

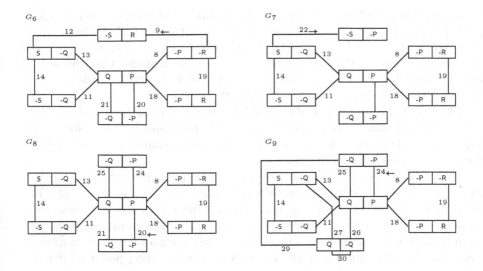

G_8 includes two copies of $-Q - P$, one of which might be removed by subsumption. To make sure that the phenomenon is not just a variation of the notorious subsumption problem described earlier in his monograph, Norbert Eisinger does not subsume, but performs the corresponding resolution steps for both clause nodes in succession.

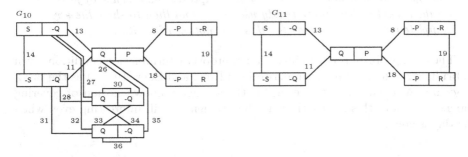

G_{10} contains two tautologies and all links which are possible among its clause nodes. In other words, it is the initial clause graph of $\{S - Q, -S - Q, QP, -P - R, -PR, Q - Q, Q - Q\}$. So far only resolution steps and purity removals were performed; now apply two tautology removals to obtain G_{11}.

G_{11} has the same structure as G_0, from which it can be obtained by applying the literal permutation $\pi : \pm Q \mapsto \mp Q, \pm P \mapsto \pm S \mapsto \mp R \mapsto \pm P$. Since $\pi^6 = \mathrm{id}$, five more "rounds" with the analogous sequence of inference steps will reproduce G_0 as G_{66}, thus after sixty-six steps we arrive at a graph isomorphic to G_0.

The only object of G_0 still present in G_{11} is the clause node labelled PQ. In particular, all initial links disappeared during the derivation. Hence G_0 and G_{66} have no object in common, which implies that the derivation is covering. The following classes of link numbers represent the "colours" introduced for the coveringtwo concept in Definition 5; the numbers of links resolved upon are asterisked: $\{1*\}$, $\{2, 8, 17, 18, 20*, 23, 24*\}$, $\{3*, 9*, 19\}$, $\{4*, 7*\}$, $\{5, 11, 13, 21, 25, 26, \ldots, 36\}$, $\{6, 10, 12, 14, 15*, 16*, 22*\}$. Only the colour $\{5, 11, \ldots, 36\}$ was never selected for resolution during the first round, and it just so happens that the second round starts with a resolution on link 11, which bears the critical colour. Hence the derivation also belongs to the coveringtwo class.

This seminal example was discovered in the autumn of 1986 and has since been published and quoted many times. It has once and for all destroyed all hope of a positive result for the strong completeness conjecture based only on the notion of covering or fairness.

The consequence of this negative result has been compared to the most unfortunate fact that the halting problem of a Turing machine is unsolvable. The (weak) analogy is in the following sense: all the work on deduction systems rests upon the basic result that the predicate calculus is semidecidable, i.e. if the theorem to be shown is in fact valid then this can be shown after a finite number of steps, provided the uniform proof procedure carries out *every* possible inference step.

Yet, here we have a uniform proof procedure — clause graph resolution — which by any intuitive notion of fairness ("carries out every possible inference step eventually") runs forever even on a valid theorem — hence is not even semidecidable.

In summary:

> *The open problem is to find a filter that captures the essence of fairness on the kernel which is practically useful[1] — and then to show the strong completeness property holds for this new notion of a filter.*

The open problem is not to invent an appropriate termination condition (even as brilliant as the decomposition criteria of Bibel and Eder [1987][2]) as the proof procedure will not terminate even for the strongest known notion of covering (fairness) — and this is exactly why the problem is still interesting even when the day is gone.

---

[1] This is important, as there are strategies which are known to be complete (for example to take a standard resolution theorem prover to find a proof and then use this information for clause-graph resolution). Hence these strategies are either based on some strange notion, or else on some too specific property.

[2] The weak notion of fairness as defined by W. Bibel and E. Eder [1987] can easily be refuted by much simpler examples (see e.g. Siekmann and Stephan [1976]) and Norbert Eisinger's construction above refutes a much stronger conjecture. The proof in the Bibel and Eder paper not only contains an excusable technical error, which we all are unfortunately prone to (the flaw is on page 336, line 29, where they assume that the fairness condition forces the procedure to resolve upon every link in the minimal complementary submatrix, here called the kernel), but unfortunately misses the very nature of the open problem (see also Siekmann and Wrightson [2001]).

5   Lifting

All of the previous results and counterexamples apply to the propositional case or ground level as it is called in the literature on deduction systems.

The question is, if and how these ground results can be lifted to the general case of the predicate calculus.

While lifting is not necessarily the wrong approach for the connection graph, the proof techniques known so far are too weak: the problem is more subtle and requires much stronger machinery for the actual lifting.

The standard argument is as follows: first the result is established for the ground case, and there is now a battery of proof techniques[3] known in order to do so. After that the result is "lifted" to the general case in the following sense: Let S be an unsatisfiable set of clauses, then by Herbrand's theorem we know that there exists a finite truth-functionally contradictory set S' of ground instances of S. Now since we have the completeness result for this propositional case we know there exists a (resolution style) derivation. Taking this derivation, we observe that all the clauses involved are just instances of the clauses at the general level and hence "lifting" this derivation amounts to exhibiting a mirror image of this derivation at the general level, as the following figures shows:

$$
\begin{array}{ccc}
S & \vdash & \{\square\} \\
\Downarrow & & \Uparrow \\
S' & \vdash_{\text{ground}} & \{\square\}
\end{array}
$$

This proof technique is due to Alan Robinson [1965].

Unfortunately this is not enough for the clause graph procedure, as we have the additional graph-like structure: not only has the ground proof to be lifted to the general level as usual, it has also to be shown that an isomorphic (or otherwise sufficient) *graph structure* can be mirrored from the ground level graph $\text{INIT}(S')$ to the graph at the general level $\text{INIT}(S)$, such that the derivation can actually be carried out within this graph structure as well:

$$
\begin{array}{ccc}
\text{INIT}(S) & \vdash & \{G(\square)\} \\
\Downarrow & & \Uparrow \\
\text{INIT}(S') & \vdash_{\text{ground}} & \{G'(\square)\}
\end{array}
$$

where $G(\square)$ is a clause graph that contains the empty clause $\square$.

This turned out to be more difficult than expected in the late 1970's, when most of this work got started. However by the end of the 1980's it was well-known that standard lifting techniques fail: the non-standard graph-oriented

---

[3] Such as induction on the excess-literal-number, which is due to W. Bledsoe (see Loveland [1978]).

lifting results in Siekmann and Stephan [1980] turned out to be false. Similarly the lifting results in Bibel [1982] and in Bibel and Eder [1997], theorem 5.4 are also false.

To quote from Norbert Eisinger's monograph ([1991], p. 125) on clause graphs

"Unfortunately the idea (of lifting a refutation) fails for an intricate difficulty which is *the* central problem in lifting graph theoretic properties. A resolution step on a link in G (the general case) requires elimination of all links in G' (the ground refutation) that are mapped to the link in G. ... Such a side effect can forestall the derivation of the successor."

This phenomenon seems to touch upon a new and fundamental problem, namely, the lifting technique has to take the topological structure of the two graphs (the ground graph and the general clause graph) into account as well, and several additional graph-theoretical arguments are asked for.

The ground case part essentially develops a strategy which from any ground initial state leads to a final state. In the clause graph resolution system any such strategy has to willy-nilly distinguish between "good" steps and "bad" steps from each ground state, because there are ground case examples where an inappropriate choice of inference steps leads to infinite derivations that do not reach a final state. Eliminating or reducing the number of links with a given atom are sample criteria for "good" steps in different strategies. The lifting part then exploits the fact that it suffices to consider the conjunction of finitely many ground instances of a given first order formula, and show how to lift the steps of a derivation existing for the ground formula to the first order level. Clause graph resolution faces the problem that a single resolution step on the general level couples different ground level steps together in a way that may be incompatible with a given ground case strategy, because "bad" steps have to be performed as a side effect of "good" steps.

That this is not always straightforward and may fail in general is shown by several (rather complex) examples (pp.123–130 in Eisinger [1991]), which we shall omit here. The interested reader may consult the monograph itself, which still represents most of what is known about the theoretical properties of clause graphs today.

To be sure, there is a very simple way to solve this problem: just add to the inference system an unrestricted copy rule and use it to insert sufficiently many variants.

However to introduce an unrestricted copy rule, as, for example, implicitly assumed in the Bibel [1982] monograph, completely destroys the practical advantages of the clause graph procedure. It is precisely the advantage of the strong redundancy removal which motivated so many practical systems to employ this rather complicated machinery (see e.g. Ohlbach and Siekmann [1991]). Otherwise we may just use ordinary resolution instead.

We feel that maybe the lifting technique should be abandoned altogether for clause graph refutation systems: the burden of mapping the appropriate graph structure (and taking its dynamically changing nature into account) seems to outweigh its advantages and a direct proof at the most general level with an appropriate technique appears far more promising. But only the future will tell.

6   Conclusion

The last twenty-five years have seen many attempts and partial results about so far unencountered theoretical problems that marred this new proof procedure, but it is probably no unfair generalisation to say, that almost every paper (including ours) on the problems has had technical flaws or major errors and the main problem — strong completeness — has been open ever since 1975 when clause graph resolution was first introduced to the scholarly community.

Why is that so?

One reason may be methodological. Clause graph resolution is formulated within three different conceptual frameworks: the usual clausal logic, the graph-theoretic properties and finally the algorithmic aspects, which account for its nonmonotonic nature. So far most of the methodological effort has been spent on the graphtheoretical notions (see e.g. Eisinger [1991]) in order to obtain a firm theoretical basis. The hope being that once these graphtheoretical properties have a sound mathematical foundation, the rest will follow suit. But this may have been a misconception: it is — after all — the *metalogical* properties of the proof procedure we are after and hence the time may have come to question the whole approach.

In (Gabbay, Siekmann [2001]) we try to turn the situation back from its (graphtheroetical) head to standing on its (logical) feet, by showing a logical encoding of the proof procedure without explicit reference to graphtheoretical properties.

Mathematics, it is said, advances through conjectures and refutations and this is a social process often carried out over more than one generation. Theoretical computer science and artificial intelligence apparently are no exceptions to this general rule.

Acknowledgements

This paper has been considerably improved by critical comments and suggestions from the anonymous referees and from Norbert Eisinger, Christoph Walther and Dov Gabbay.

The authors would like to thank Oxford University Press for their kind permissin to reprint this paper, which is appearing in the *Logic Journal of the IGPL*.

References

Andrews, P. B.: Resolution with Merging. *J. ACM* **15** (1968) 367–381.
Andrews, P. B.: Refutations by Matings. *IEEE Trans. Comp. C-25*, (1976) **8**, 801–807.

Andrews, P.B.: Theorem Proving via General Matings. *J. ACM* **28** (1981) 193–214.

Antoniuo, G., Ohlbach, H.J.: Terminator. *Proceedings 8th IJCAI*, Karlsruhe, (1983) 916–919.

Bibel, W.: A Strong Completeness Result for the Connection Graph Proof Procedure. Bericht ATP-3-IV-80, Institut für Informatik, Technische Universität, München (1980)

Bibel, W.: On the completeness of connection graph resolution. In *German Workshop on Artificial Intelligence*. J.Siekmann, ed. Informatik Fachberichte 47, Springer, Berlin, Germany (1981a) pp.246–247

Bibel, W.: On matrices with connections. *J.ACM*, **28** (1981b) 633–645

Bibel, W.: *Automated Theorem Proving*. (1982) Vieweg. Wiesbaden.

Bibel, W.: Matings in matrices. *Commun. ACM*, **26**, (1983) 844–852

Bibel, W., Eder, E.: Decomposition of tautologies into regular formula and strong completeness of connection-graph resolution *J. ACM* **44** (1997) 320–344

Bläsius, K. H.: Construction of equality graphs. SEKI report SR-86-01 (1986) Univ. Karlsruhe, Germany

Bläsius, K. H.: Equality reasoning based on graphs. SEKI report SR-87-01 (1987) Univ. Karlsruhe, Germany

Bläsius, K. H., Bürckert, H. J.: *Deduktions Systeme*, (1992) Oldenbourg Verlag. Also in English: Ellis Horwood, 1989

Bläsius, K. H., Eisinger, N., Siekmann, J., Smolka, G., Herald A., Walter, C. The Markgraf Karl refutation procedure. *Proc 7th IJCAI*, Vancouver (1981)

Brown, F. Notes on Chains and Connection Graphs. Personal Notes, Dept. of Computation and Logic, University of Edinburgh (1976)

Chang, C.-L., Lee, R.C.-T.: *Symbolic Logic and Mechanical Theorem Proving*, Academic Press (1973)

Chang, C.-L., Slagle, J.R.: Using Rewriting Rules for Connection Graphs to Prove Theorems. *Artificial Intelligence* **12** (1979) 159–178.

Eisinger, N.: What you always wanted to know about clause graph resolution. In *Proc of 8th Conf. on Automated Deduction* Oxford (1986) LNCS 230, Springer

Eisinger, N.: Subsumption for connection graphs. *Proc 7th IGCAI*, Vancouver (1981)

Eisinger, N.: Completeness, Confluence, and Related Properties of Clause Graph Resolution. Ph.D. dissertation, Universität Kaiserslautern (1988)

Eisinger, N.: *Completeness, Confluence, and Related Properties of Clause Graph Resolution*. Pitman, London, Morgan Kaufmann Publishers,Inc., San Mateo,California (1991)

Eisinger, N., Siekmann, J., Unvericht, E.: The Markgraf Karl refutation procedure. *Proc of Conf on Automated Deduction*, Austin, Texas (1979)

Eisinger, N., Ohlbach, H. J., Präcklein, A.: Elimination of redundancies in clause sets and clause graphs (1989) SEKI report, SR-89-01, University of Karlsruhe

Gabbay, D., Siekmann, J.: Logical encoding of the clause graph proof procedure, 2002, forthcoming

Hähnle, R., Murray, N. V., Rosenthal, E.: Ordered resolution versus connection graph resolution. In: R. Goré, A. Leitsch, T. Nipkow *Automated Reasoning, Proc of IJCAR 2001* (2001) LNAI 2083, Springer

Kowalski, R.: Search Strategies for Theorem Proving. *Machine Intelligence* (B.Meltzer and D.Michie, eds.), **5** Edinburgh University Press, Edinburgh, (1970) 181–201

Kowalski, R.: . A proof procedure using connection graphs. *J.ACM* **22** (1975) 572–595

Loveland, D. W.: A Linear Format for Resolution. *Proc. of Symp. on Automatic Demonstration*. Lecture Notes in Math 125, Springer Verlag, Berlin, (1970) 147–162. Also in Siekmann and Wrightson [1983b], 377–398

Loveland, D. W.: *Automated Theorem Proving: A Logical Basis* North- Holland, New York (1978)

Meagher D., Hext, J.: Link deletion in resolution theorem proving (1998) unpublished manuscript

Murray, N. V., Rosenthal, E.: Path resolution with link deletion. *Proc. of 9th IJCAII* Los Angeles (1985)

Murray, N. V., Rosenthal, E.: Dissolution: making paths vanish. *J. ACM* **40** (1993)

Ohlbach, H. J.: Ein regelbasiertes Klauselgraph Beweisverfahren. *Proc. of German Conference on AI, GWAI-83* (1983) Springer Verlag IFB vol 76

Ohlbach, H. J.: Theory unification in abstract clause graphs. *Proc. of German Conf. on AI GWAI-85* (1985) Springer Verlag IFB vol 118

Ohlbach, H. J.: Link inheritance in abstract clause graphs *J. Autom. Reasoning* **3** (1987)

Ohlbach, H. J., Siekmann, J.: The Markgraf Karl refutation procedure. In: J. L. Lassez, G. Plotkin, *Computational Logic* (1991) MIT Press, Cambridge MA

Omodeo, E. G.: The linked conjunct method for automatic deduction and related search techniques. *Computers and Mathematics with Applications* **8** (1982) 185–203

Ramesh, A., Beckert, B., Hähnle, R., Murray, N. V.: Fast subsumption checks using anti-links *J. Autom. Reasoning* **18** (1997) 47–83

Robinson, J.A.: A machine-oriented logic based on the resolution principle. *J.ACM* **12** (1965) 23–41

Shostak, R.E.: Refutation Graphs. *J. Artificial Intelligence* **7**, (1976), 51–64

Shostak, R.E.: A Graph-Theoretic View of Resolution Theorem-Proving. Report SRI International, Menlo Park (1979)

Sickel, S.: A Search Technique for Clause Interconnectivity Graphs. *IEEE Trans. Comp.* **C-25** (1976) 823–835

Siekmann, J. H., Stephan, W.: Completeness and Soundness of the Connection Graph Proof Procedure. Bericht 7/76, Fakultät Informatik, Universität Karlsruhe (1976). Also in *Proceedings of AISB/GI Conference on Artificial Intelligence*, Hamburg (1978)

Siekmann, J. H., Stephan, W.: Completeness and Consistency of the Connection Graph Proof Procedure. Interner Bericht Institut I, Fakultät Informatik, Universität Karlsruhe (1980)

Siekmann, J. H., Wrightson, G.: Paramodulated connection graphs *Acta Informatica* **13** (1980)

Siekmann, J. H., Wrightson, G.: *Automation of Reasoning.* Springer- Verlag, Berlin, Heidelberg, New York. Vol 1 and vol 2 (1983)

Siekmann, J. H., Wrightson, G.: Erratum: A counterexample to W. Bibel's and E. Eder's strong completeness result for connection graph resolution. *J. ACM* **48** (2001) 145

Smolka, G.: Completeness of the connection graph proof procedure for unit refutable clause sets. In *Proceedings of GWAI-82.* Informatik Fachberichte, vol. 58. Springer-Verlag, Berlin, Germany (1982a) 191-204.

Smolka, G.: Einige Ergebnisse zur Vollständigkeit der Beweisprozedur von Kowalski. Diplomarbeit, Fakultät Informatik, Universität Karlsruhe (1982b)

Smolka, G.: Completeness and confluence properties of Kowalksi's clause graph calculus (1982c) SEKI report SR-82-03, University of Karlsruhe, Germany

Stickel, M.: A Non-Clausal Connection-Graph Resolution Theorem-Proving Program. *Proceedings AAAI-82*, Pittsburgh (1982) 229–233

Walthe, Chr.: Elimination of redundant links in extended connection graphs. *Proc of German Workshop on AI, GWAI-81* (1981) Springer Verlag, Fachberichte vol 47

Wos, L.T., Carson, D.F., Robinson, G.A.: The Unit Preference Strategy in Theorem Proving. *AFIPS Conf. Proc. 26*, (1964) Spartan Books, Washington. Also in Siekmann and Wrightson [1983], 387–396.

Wos, L.T., Robinson, G.A., Carson, D.F.: Efficiency and Completeness of the Set of Support Strategy in Theorem Proving. *J.ACM* **12**, (1965) 536–541. Also in Siekmann and Wrightson [1983], 484–492

Wos, L. T, *et al.*: *Automated Reasoning: Introduction and Applications* (1984) Englewood Cliffs, new Jersey, Prentice-Hall

Wrightson, G.: A pragmatic strategy for clause graphs or the strong completeness of connection graphs. Report 98-3, Dept Comp. Sci., Univ of Newcastle, Australia (1989)

Yarmush, D. L.: The linked conjunct and other algorithms for mechanical theorem-proving. Technical Report IMM 412, Courant Institute of Mathematical Sciences, New York University (1976)

Yates, R. A., Raphael, B., Hart, T. P.: Resolution Graphs. *Artificial Intelligence* **1** (1970) 257–289.

Meta-reasoning: A Survey

Stefania Costantini

Dipartimento di Informatica,
Università degli Studi di L'Aquila,
via Vetoio Loc. Coppito, I-67100 L'Aquila, Italy
stefcost@univaq.it

Abstract We present the basic principles and possible applications of
systems capable of meta-reasoning and reflection. After a discussion of
the seminal approaches, we outline our own perception of the state of the
art, mainly but not only in computational logic and logic programming.
We review relevant successful applications of meta-reasoning, and the
basic underlying semantic principles.

1   Introduction

The meaning of the term "meta-reasoning" is "reasoning about reasoning". In a
computer system, this means that the system is able to reason about its own operation. This is different from performing object-level reasoning, which refers in
some way to entities external to the system. A system capable of meta-reasoning
may be able to reflect, or introspect, i.e. to shift from meta-reasoning to object-
level reasoning and vice versa.

We present the main principles and the possible applications of meta-
reasoning and reflective systems. After a review of the relevant approaches,
mainly in computational logic and logic programming, we discuss the state of
the art and recent interesting applications of meta-reasoning. Finally, we briefly
summarize the semantic foundations of meta-reasoning. We necessarily express
our own partial point of view on the field and provide the references that we
consider the most important.

There are previous good reviews on this subject, to which we are indebted
and to which we refer the reader for a wider perspective and a careful discussion
of problems, foundations, languages, approaches, and systems. We especially
mention [1], [2], [3]. Also, the reader may refer, for the computational logic
aspects, to the Proceedings of the Workshops on Meta-Programming in Logic
[4], [5], [6], [7], [8]. Much significant work on Meta-Programming was carried out
in the Esprit funded European projects Compulog I and II. Some of the results
of this work are discussed in the following sections. For a wider report we refer
the reader to [9].

More generally, about meta-reasoning in various kinds of paradigms, including object-oriented, functional and imperative languages, the reader may refer
to [10] [11], [12].

A.C. Kakas, F. Sadri (Eds.): Computat. Logic (Kowalski Festschrift), LNAI 2408, pp. 253–288, 2002.

Research about meta-reasoning and reflection in computer science has its roots in principles and techniques developed in logic, since the fundamental work of Gödel and Tarski, for which it may be useful to refer to the surveys [13], [14]. In meta-level approaches, knowledge about knowledge is represented by admitting sentences to be arguments of other sentences, without abandoning the framework of first-order logic.

An alternative important approach to formalize knowledge about knowledge is the modal approach that has initially been developed by logicians and philosophers and then has received a great deal of attention in the field of Artificial Intelligence. It aims at formalizing knowledge by a logic language augmented by a modal operator, interpreted as knowledge or belief. Thus, sentences can be expressed to represent properties of knowledge (or belief). The most common modal systems adopt a *possible world* semantics [15]. In this semantics, knowledge and belief are regarded as propositions specifying the relationship between knowledge expressed in the theory and the external world. For a review of modal and meta-languages, focused on their expressivity, on consistency problems and on the possibility of translating modal languages into a meta-level setting, the reader may refer to [16].

2   Meta-programming and Meta-reasoning

Whatever the underlying computational paradigm, every piece of software included in any system (in the following, we will say *software component*) manipulates some kind of data, organized in suitable data structures. Data can be used in various ways: for producing results, sending messages, performing actions, or just updating the component's internal state.

Data are often assumed to denote entities which are *external* to the software component. Whenever the computation should produce effects that are visible in the external environment, it is necessary to assume that there exists a *causal connection* between the software system and the environment, in the sense that the intended effect is actually achieved, by means of suitable interface devices. This means, if the software component performs an action in order, for instance, either to print some text, or to send an e-mail message, or to switch a light on, causal connection should guarantee that this is what actually happens.

There are software components however that take other programs as data. An important well-known example is a compiler, which manipulates data structures representing the source program to be translated. A compiler can be written in the language it is intended to translate (for instance, a C compiler can be written in C), or in a different language as well. It is important to notice that in any case there is no mixture between the compiler and the source program. The compiler performs a computation whose outcome is some transformed form of the source program. The source program is just text, recorded in a suitable data structure, that is step by step transformed into other representations. In essence, a compiler accepts and manipulates a *description* of the source program.

In logic, a language that takes sentences of another language as its objects of discourse is called a meta-language. The other language is called the object language. A clear separation between the object language and the meta-language is necessary: namely, it consists in the fact that sentences written in the meta-language can refer to sentences written in the object language only by means of some kind of *description,* or *encoding,* so that sentences written in the object language are treated as data. As it is well-known, Kurt Gödel developed a technique (gödelization) for coding the formulas of the theory of arithmetic by means of numbers (gödel numbers). Thus, it became possible to write formulas for manipulating other formulas, the latter represented by the corresponding gödel numbers.

In this view a compiler is a *meta-program,* and writing a compiler is more than just programming: it is meta-programming. The language in which the compiler is written acts as a meta-language. The language in which the source program is written acts as the object language. More generally, all tools for program analysis, debugging and transformation are meta-programs. They perform a kind of meta-programming that can be called *syntactic* meta-programming.

Syntactic meta-programming can be particularly useful for theorem proving. In fact, as first stressed in [17] and [18], many lemmas and theorems are actually meta-theorems, asserting the validity of a fact by simply looking at its syntactic structure. In this case a software component, namely the theorem prover, consists of two different parts: one, that we call the object level, where proofs are performed by repeatedly applying the inference rules; another one, that we call the meta-level, where meta-theorems are stated.

We may notice that a theorem prover is an "intelligent" system that performs deduction, which is a form of (mechanized) "reasoning". Then, we can say that the theorem prover at the object level performs "object-level reasoning". Meta-theorems take as arguments the description of object-level formulas and theorems, and meta-level proofs manipulate these descriptions. Then, at the meta-level the system performs reasoning about entities that are internal to the system, as opposed to object-level reasoning that concerns entities denoting elements of some external domain. This is why we say that at the meta-level the theorem prover performs "meta-level reasoning", or shortly meta-reasoning.

Meta-theorems are a particular kind of meta-knowledge, i.e. knowledge about properties of the object-level knowledge.

The object and the meta-level can usefully interact: meta-theorems can be used in order to shorten object-level proofs, thus improving the efficiency of the theorem prover, which can derive proofs more easily. In this view, meta-theorems may constitute *auxiliary inference rules* that enhance (in a pragmatic view) the "deductive power" of the system [19] [20]. Notice that, at the meta-level, new meta-theorems can also be proved, by applying suitable inference rules.

As pointed out in [21], most software components implicitly incorporate some kind of meta-knowledge: there are pieces of object-level code that "do" something in accordance to what meta-knowledge states. For instance, an object-level planner program might "know" that $near(b,a)$ holds whenever $near(a,b)$ holds, while this is not the case for $on(a,b)$. A planner with a meta-level could explicitly encode a meta-rule stating that whenever a relation $\mathcal{R}$ is symmetric, then $\mathcal{R}(a, b)$ is equivalent to $\mathcal{R}(b, a)$ and whenever instead a relation is antisymmetric this is never the case. So, at the meta-level, there could be statements that *near* is symmetric and *on* is antisymmetric.

The same results could then be obtained by means of explicit meta-reasoning, instead of implicit "knowledge" hidden in the code. The advantage is that the meta-reasoning can be performed in the same way for *any* symmetric and antisymmetric relation that one may have. Other properties of relations might be encoded at the meta-level in a similar way, and such a meta-level specification (which is independent of the specific object-level knowledge or application domain) might be reused in future applications.

There are several possible architectures for meta-knowledge and meta-reasoning, and many applications. Some of them are reviewed later. For a wider perspective however, the reader may refer to [22], [23], [24], [25], [20], [26], [27], [28], [29], [30], [31], [32], [33] where various specific architectures, applications and systems are discussed.

3   Reification

Meta-level rules manipulate a *representation* of object-level knowledge. Since knowledge is represented in some kind of language, meta-rules actually manipulate a representation of syntactic expressions of the object-level language.

In analogy with natural language, such a representation is usually called a *name* of the syntactic expression. The difference between a word of the language, such as for instance *flower,* and a name, like *"flower",* is the following: the word is used to denote an entity of the domain/situation we are talking about; the name denotes the word, so that we can say that "flower" is composed of six characters, is expressed in English and its translation into Italian is "fiore". That is, a word can be used, while a name can be inspected (for instance to count the characters) and manipulated (for instance translated).

An expression in a formal language may have different kinds of names that allow different kinds of meta-reasoning to be made on that expression. Names are expressions of the meta-language.

Taking for instance an equation such as

$$a = b - 2$$

we may have a very simple name, like in natural language, i.e.

$$\text{``}a = b - 2\text{''}$$

This kind of name, called *quotation mark name,* is usually intended as a constant of the meta-language.

A name may be instead a complex term, such as:

equation

$(left\_hand\_side(variable(``a")),$

$(right\_hand\_side$

$(binop(minus, firstop(variable(``b")), secondop(constant(``2")))))$

This term describes the equation in terms of its left-hand side and right-hand side and then describes the right-hand side as the application of a binary operator (*binop*) on two operands (*firstop* and *secondop*) where the first operand is a variable and the second one a constant. "a", "b" and "2" are constants of the meta-language, they are the names of the expressions a, b and 2 of the object language.

This more complex name, called a *structural description name,* makes it easier to inspect the expression (for instance to see whether it contains variables) and to manipulate it (for instance it is possible to transform this name into the name of another equation, by modifying some of the composing terms).

Of course, many variations are possible in how detailed names are, and what kind of detail they express. Also, many choices can be made about what names should be: for instance, the name of a variable can be a meta-constant, but can also be a meta-variable. For a discussion of different possibilities, with their advantages and disadvantages, see [34], [35], [36].

The definition of names, being a relation between object-level expressions and meta-level expressions that play the role of names, is usually called *naming relation.*

Which naming relation to choose? In general, it depends upon the kind of meta-reasoning one wants to perform. In fact, a meta-theory can only reason about the properties of object-level expressions made explicit by the naming relation. We may provide names to any language expression, from the simplest, to the more complex ones. In a logic meta-language, we may have names for variables, constants, function and predicate symbols, terms and atoms and even for entire theories: the meta-level may in principle encode and reason about the description of several object-level theories. In practice, there is a trade-off between expressivity and simplicity. In fact, names should be kept as simple as possible, to reduce the complexity (and improve the readability) of meta-level expressions. Starting from these considerations, [37] argues that the naming relation should be adapted to each particular case and therefore should be definable by the user.

In [38] it is shown that two different naming relations can coexist in the same context, for different purposes, also providing operators for transforming one representation into the other one.

The definition of a naming relation implies the definition of two operation: the first one, to compute the name of a given language expression. The second one, to compute the expression a given name stands for. The operation of obtaining the name of an object-level expression is called *reification* or *referentiation* or *quoting.* The inverse operation is called *dereification* or *dereferentiation* or *unquoting.* These are built-in operations, whose operational semantics consists in applying the naming relation in the two directions.

In [39] it is shown how the naming relation can be a sort of input parameter for a meta-language. That is, a meta-language may be, if carefully designed, to a large extent independent of the syntactic form of names, and of the class of expressions that are named. Along this line, in [36] and [33] a full theory of definable naming relations is developed, where a naming relation (with some basic properties) can be defined as a set of equations, with the associated rewrite system for applying referentiation/dereferentiation.

4   Introspection and Reflection

The idea that meta-knowledge and meta-reasoning could be useful for improving the reasoning performed at the object level (for instance by exploiting properties of relations, like symmetry), suggests that the object and the meta-level should interact. In fact, the object and the meta-level can be seen as different software components that interact by passing the control to each other.

At the object level, the operation of referentiation allows an expression to be transformed into its name and this name can be given as input argument to a meta-level component. This means that object-level computation gives place to meta-level computation. This computational step is called *upward reflection,* or *introspection,* or *shift up. Upward* because the meta-level is considered to be a "higher level" with respect to the object level. *Reflection,* or *introspection,* because the object level component suspends its activity, in order to initiate a meta-level one. This is meant to be in analogy with the process by which people become conscious (at the meta-level of mind) of mental states they are currently in (at the object level).

The inverse action, that consists in going back to the object-level activity, is called *downward reflection, or shift down.* The object-level activity can be resumed from where it had been suspended, or can be somehow restarted. Its state (if any) can be the same as before, or can be altered, according to the meta-level activity that has been performed. Downward reflection may imply that some name is dereferenced and the resulting expression ("extracted" from the name) given as input argument to the resumed or restarted object-level activity.

In logical languages, upward and downward reflection can be specified by means of special inference rules (reflection rules) or axioms (reflection axioms), that may also state what kind of knowledge is exchanged.

In functional and procedural languages, part of the run-time state of the object-level ongoing computation can be reified and passed to a meta-level function/procedure that can inspect and modify this state. When this function terminates, object-level computation resumes on a possibly modified state.

A *reflection act,* that shifts the level of the activity between the object and the meta-level, may be: *explicit,* in the sense that it is either invoked by the user (in interactive systems) or determined by some kind of specification explicitly present in the text of the theory/program; *implicit,* in the sense that it is automatically performed upon occurrence of certain predefined conditions. Explicit and implicit reflection may co-exist.

Both forms of reflection rely on the requirement of *causal connection* or, equivalently, of *introspective fidelity:* that is, the recommendations of the meta-level must be always followed at the object level. For instance, in the procedural case, the modifications to the state performed at the meta-level are effective and have a corresponding impact on the object-level computation. The usefulness of reflection consists exactly in the fact that the overall system (object + meta-levels) not only reasons about itself, but is also properly affected by the results of that reasoning.

In summary, a meta-level architecture for building software components has to provide the possibility of defining a meta-level that by means of a naming relation can manipulate the representation of object-level expressions. Notice that the levels may be several: beyond the meta-level there may be a meta-meta-level that uses a naming relation representing meta-level expressions. Similarly, we can have a meta-meta-meta-level, and so on. Also, we may have one object level and several independent meta-levels with which the object level may be from time to time associated, for performing different kinds of meta-reasoning.

The architecture may provide a reflection mechanism that allows the different levels to interact. If the reflection mechanism is not provided, then the computation is performed at the meta-level, that simulates the object-level formulas through the naming relation and simulates the object-level inference rules by means of meta-level axioms. As discussed later, this is the case in many of the main approaches to meta-reasoning.

The languages in which the object level and the meta-level(s) are expressed may be different, or they may coincide. For instance, we may have a meta-level based on a first-order logic language, were meta-reasoning is performed about an object level based on a functional or imperative language. Sometimes the languages coincide: the object language and the meta-language may be in fact the same one. In this case, this language is expressive enough as to explicitly represent (some of) its own syntactic expressions, i.e. the language is capable of *self-reference.* An interesting deep discussion about languages with self-reference can be found in [40] and [41]. The role of introspection in reasoning is discussed in [42] and [43]. An interesting contribution about reflection and its applications is [44].

5 Seminal Approaches

5.1 FOL

FOL [19], standing for *First Order Logic,* has been (to the best of our knowledge) the first reflective system appeared in the literature. It is a proof checker based on natural deduction, where knowledge and meta-knowledge are expressed in different contexts. The user can access these contexts both for expressing and for inferring new facts.

The FOL system consists of a set of theories, called contexts, based on a first-order language with sorts and conditional expressions.

A special context named META describes the proof theory and some of the model theory of FOL contexts. Given a specific context C that we take as the object theory, the naming relation is defined by attachments, which are user-defined explicit definitions relating symbols and terms in META with their interpretation in C.

The connection between C and META is provided by a special linking rule that is applicable in both directions:

$$\frac{Theorem(\text{``}W\text{''})}{W}$$

where W is any formula in the object theory C, "W" is its name, and $Theorem(\text{``}W\text{''})$ is a fact in the meta-theory. By means of a special primitive, called REFLECT, the linking rule can be explicitly applied by the user. Its effect is either that of *reflecting up* a formula W to the meta-theory, to derive meta-theorems involving "W", or vice versa that of *reflecting down* a meta-theorem "W", so that W becomes a theorem of the theory. Meta-theorems can therefore be used as subsidiary deduction rules.

Interesting applications of the FOL system to mathematical problems can be found in [17], [45].

5.2   Amalgamating Language and Meta-language in Logic Programming

A seminal approach to reflection in the context of the Horn clause language is MetaProlog, proposed by Bowen and Kowalski [46]. The proposal is based on representing Horn clause syntax and provability in the logic itself, by means of a meta-interpreter, i.e. an interpreter of the Horn clause language written in the Horn clause language itself. Therefore, also in this case the object language and the meta-language coincide.

The concept (and the first implementation) of a meta-interpreter was introduced by John McCarthy for the LISP programming language [47]. McCarthy in particular defined a universal function, written in LISP, which represents the basic features of a LISP interpreter. In particular, the universal function is able to: (i) accept as input the definition of a LISP function, together with the list of its arguments; (ii) evaluate the given function on the given arguments. Bowen and Kowalski, with MetaProlog, have developed this powerful and important idea in the field of logic programming, where the inference process is based on building proofs from a given theory, rather than on evaluating functions.

The Bowen and Kowalski meta-interpreter is specified via a predicate *demo*, that is defined by a set of meta-axioms Pr, where the relevant aspects of Horn-clause provability are made explicit. The *Demo* predicate takes as first argument the representation (name) of an object-level theory T and the representation (name) of a goal A. *Demo*("T", "A") means that the goal A is provable in the theory T.

With the above formulation, we might have an approach where inference is performed at the meta-level (via invocation of *Demo*) and the object level is simulated, by providing *Demo* with a suitable description "*T*" of an object theory *T*.

The strength and originality of MetaProlog rely instead in the *amalgamation* between the object level and the meta-level. It consists in the introduction of the following *linking rules* for upward and downward reflection:

$$\frac{T \vdash_L A}{Pr \vdash_M Demo(\text{``}T\text{''}, \text{``}A\text{''})} \qquad \frac{Pr \vdash_M Demo(\text{``}T\text{''}, \text{``}A\text{''})}{T \vdash_L A}$$

where $\vdash_M$ means provability at the meta-level M and $\vdash_L$ means provability at the object level L.

The application of the linking rules coincides, in practice, with the invocation of *Demo*, i.e., reflection is *explicit*. Amalgamation allows mixed sentences: there can be object-level sentences where the invocation of *Demo* determines a shift up to the meta-level, and meta-level sentences where the invocation of *Demo* determines a shift down to the object level. Since moreover the theory in which deduction is performed is an input argument of *Demo*, several object-level and meta-level theories can co-exist and can be used in the same inference process.

Although the extension is conservative, i.e. all theorems provable in $L+M$ are provable either in L or in M alone, the gain of expressivity, in practical terms, is great. Many traditional problems in knowledge representation find here a natural formulation.

The extension can be made non-conservative, whenever additional rules are added to *Demo*, to represent auxiliary inference rules and deduction strategies. Additional arguments can be added to *Demo* for integrating forms of *control* in the basic definition of provability. For instance it is possible to control the amount of resources consumed by the proof process, or to make the structure of the proof explicit.

The semantics of the *Demo* predicate is, however, not easy to define (see e.g. [35], [48], [49], [50]), and holds only if the meta-theory and the linking rules provide an extension to the basic Horn clause language which is conservative, i.e., only if *Demo* is a faithful representation of Horn clause provability. Although the amalgamated language is far more expressive than the object language alone, enhanced meta-interpreters are (semantically) ruled out, since in that case the extension is non-conservative.

In practice, the success of the approach has been great: enhanced meta-interpreters are used everywhere in logic programming and artificial intelligence (see for instance [51], or any other logic programming textbook). This seminal work has initiated the whole field of meta-programming in logic programming and computational logic. Problems and promises of this field are discussed by Kowalski himself in [52], [53]. The approach of meta-interpreters and other relevant applications of meta-programming are discussed in the next section.

5.3   3-LISP

3–Lisp [54] is another important example of a reflective architecture where the object language and meta-language coincide. 3–Lisp is a meta-interpreter for Lisp (and therefore it is an elaboration of McCarthy's original proposal) where (the interesting aspects of) the state of the program that is being interpreted are not stored, but are passed by as an argument of all the functions that are internal to the meta-interpreter. Then, each of these procedures takes the state as argument, makes some modification and passes the modified state to another internal procedure. These procedures call each other tail-recursively (i.e. the next procedure call is the last action they make) so as the state remains always explicit. Such a meta-interpreter is called a meta-circular interpreter. If one assumes that the meta-circular interpreter is itself executed by another meta-circular interpreter and so on, one can imagine a potentially infinite *tower* of interpreters, the lowest one executing the object level program (see the summary and formalization of this approach presented in [55]).

Here, the meta-level is accessible from the object level at run-time through a reflection act represented by a special kind of function invocation. Whenever the object-level program invokes any function f in this special way, f receives as an additional parameter a representation of the state of the program itself. Then, f can inspect and/or modify the state, before returning control to object-level execution. A reflective act implies therefore the reification of the state and the execution of f as if it were a procedure internal to the interpreter. Since f might in turn contain a reflection act, the meta-circular interpreter is able to reify its own state and start a brand-new copy of itself. In this approach one might in principle perform, via reflection, an *infinite regress* on the reflective tower of interpreters.

A program is thus able to interrupt its computation, to change something in its own state, and to continue with a modified interpretation process. This kind of mechanism is called *computational reflection*. The semantics of computational reflection is procedural, however, rather than declarative. A reflective architecture conceptually similar to 3-Lisp has been proposed for the Horn clause language and has been fully implemented [56].

Although very procedural in nature, and not easy to understand in practice, computational reflection has been having a great success in the last few years, especially in the context of imperative and object-oriented programming [11], [12]. Some authors even propose computational reflection as the basis of a new programming paradigm [57].

Since computational reflection can be perceived as the only way of performing meta-reasoning in non-logical paradigms, this success enlights once more how important meta-reasoning is, especially for complex applications.

5.4   Other Important Approaches

The amalgamated approach has been experimented by Attardi and Simi in Omega [58]. Omega is an object-oriented formalism for knowledge representation which can deal with meta-theoretical notions by providing objects that describe Omega objects themselves and derivability in Omega.

A non-amalgamated approach in logic programming is that of the Gödel language, where object theory and meta-theory are distinct. Gödel provides a (conservative) provability predicate, and an explicit form of reflection. The language has been developed and experimented in the context of the Compulog European project. It is described in the book [59]. In [60] a contribution to meta-programming in Gödel is proposed, on two aspects: on the one hand, a programming style for efficient meta-programming is outlined; on the other hand, modifications to the implementation are proposed, in order to improve the performance of meta-programs.

A project that extends and builds on both FOL and 3–Lisp is GETFOL [61],[62]. It is developed on top of a novel implementation of FOL (therefore the approach is not amalgamated: the object theory and meta-theory are distinct). GETFOL is able to introspect its own code (lifting), to reason deductively about it in a declarative meta-theory and, as a result, to produce new executable code that can be pushed back to the underlying interpretation (flattening).

The architecture is based on a sharp distinction between deduction (FOL style) and computation (3–Lisp style). Reflection in GETFOL gives access to a meta-theory where many features of the system are made explicit, even the code that implements the system itself.

The main objective of GETFOL is that of implementing theorem-provers, given its ability of implementing flexible control strategies to be adapted (via computational reflection) to the particular situation. Similarly to FOL, the kind of reasoning performed in GETFOL consists in: (i) performing some reasoning at the meta-level; (ii) using the results of this reasoning to assert facts at the object level.

An interesting extension is that of applying this concept to a system with multiple theories and multiple languages (each theory formulated in its own language) [63], where the two steps are reinterpreted as (i) doing some reasoning in one theory and (ii) jumping into another theory to do some more reasoning on the basis of what has been derived in the previous theory. These two deductions are concatenated by the application of *bridge rules,* which are inference rules where the premises belong to the language of the former theory, and the conclusion belongs to the language of the latter.

A different concept of reflection is embodied in Reflective Prolog [39] [64] [65], a self-referential Horn clause language with logical reflection. The objective of this approach is that of developing a more expressive and powerful language, while preserving the essential features of logic programming: Horn clause syntax, model-theoretic semantics, resolution via unification as procedural semantics, correctness and completeness properties.

In Reflective Prolog, Horn clauses are extended with self-reference and resolution is extended with logical reflection, in order to achieve greater expressive and inference power. The reflection mechanism is *implicit,* i.e., the interpreter of the language automatically reflects upwards and downwards by applying suitable linking rules called *reflection principles*. This allows reasoning and meta-reasoning to interleave without user's intervention, so as to exploit both knowledge and meta-knowledge in proofs: in most of the other approaches instead, there is one level which is "first–class", where deduction is actually performed, and the other level which plays a secondary role.

Reflection principles are embedded in both the procedural and the declarative semantics of the language, that is, in the extended resolution procedure which is used by the interpreter and in the construction of the models which give meanings to programs.

Procedurally, this implies that there is no need to axiomatize provability in the meta-theory. Object level reasoning is not simulated by meta-interpreters, but directly executed by the language interpreter, thus avoiding unnecessary inefficiency. Semantically, a theory composed of an object level and (one or more) meta-levels is regarded as an enhanced theory, enriched by new axioms which are entailed by the given theory and by the reflection principles interpreted as axiom schemata. Therefore, in Reflective Prolog, language and metalanguage are amalgamated in a non-conservative extension.

Reflection in Reflective Prolog gives access to a meta-theory where various kinds of meta-knowledge can be expressed, either about the application domain or about the behavior of the system. Deduction in Reflective Prolog means using at each step either meta-level or object level knowledge, in a continuous interleaving between levels. Meta-reasoning in Reflective Prolog implies a declarative definition of meta-knowledge, which is automatically integrated into the inference process. The relation between meta-reasoning in Reflective Prolog and modal logic has been discussed in [66].

An interpreter of Reflective Prolog has been fully implemented [67]. It is interesting to notice that Reflective Prolog has been implemented by means of computational reflection. This is another demonstration that computational reflection can be a good (although low-level) implementation tool.

An approach that has been successful in the context of object-oriented languages, including the most recent ones like Java, is the *meta-object protocol*. A meta-object protocol [68] [69] gives every object a corresponding meta-object that is an instance of a meta-class. Then, the behavior of an object becomes the behavior of the object/meta-object pair. At the meta-level, important aspects such as the operational semantics of inheritance, instantiation and method invocation can be defined. A meta-object protocol constitutes a flexible mean of modifying and extending an object-oriented language.

This approach has been applied to logic programming, in the ObjVProlog language [70] [71]. In addition to the above-mentioned meta-class capabilities, this language preserves the Prolog capabilities of manipulating clauses in the language itself, and provides a provability predicate.

As an example of more recent application of this approach, a review of Java reflective implementations can be found in [72].

A limitation is that only aspects directly related to objects can be described in a meta-object. Properties of sets of objects, or of the overall system, cannot be *directly* expressed. Nevertheless, some authors [72] argue that non-functional requirements such as security, fault-tolerance, atomicity, can be implemented by *implicit* reflection to the meta-object before and after the invocation of *every* object method.

6 Applications of Meta-reasoning

Meta-reasoning has been widely used for a variety of purposes, and recently the interest in new potential applications of meta-reasoning and reflection has been very significant. In this section, we provide our (necessarily partial and limited) view of some of the more relevant applications in the field.

6.1 Meta-interpreters

After the seminal work of Bowen and Kowalski [46], the most common application of meta-logic in computational logic is to define and to implement meta-interpreters. This technique has been especially used in Prolog (which is probably the most popular logic programming language) for a variety of purposes.

The basic version of a meta-interpreter for propositional Horn clause programs, reported in [53], is the following.

$demo(T, P) \leftarrow demo(T, P \leftarrow Q), demo(T, Q).$
$demo(T, P \wedge Q) \leftarrow demo(T, P), demo(T, Q).$

In the above definition, '$\wedge$' names conjunction and '$\leftarrow$' names '$\leftarrow$' itself. A theory can be named by a list containing the names of its sentences. In the propositional case, formulas and their names may coincide without the problems of ambiguity (discussed below), that arise in presence of variables. If a theory is represented by a list, then the meta-interpreter must be augmented by the additional meta-axiom:

$demo(T, P) \leftarrow member(T, P).$

For instance, query $?q$ to program

$q \leftarrow p, s.$
$p.$
$s.$

can be simulated by query $?demo([q \leftarrow p \wedge s, p, s], q)$ to the above meta-interpreter. Alternatively, it is possible to use a constant symbol to name a theory. In this case, the theory, say $t1$, can be defined by the following meta-level axioms:

$demo(t1, q \leftarrow p \wedge s).$
$demo(t1, p).$
$demo(t1, s).$

and the query becomes $?demo(t1, q).$

The meta-axioms defining *demo* can be themselves regarded as a theory that can be named, by either a list or a constant (say *d*). Thus, it is possible to write queries like *?demo(d, demo(t1, q))* which means to ask whether we can derive, by the meta-interpreter *d*, that the goal *q* can be proved in theory *t1*.

In many Prolog applications however, the theory argument is omitted, as in the so-called "Vanilla" meta-interpreter [35]. The standard declarative formulation of the Vanilla meta-interpreter in Prolog is the following (where ':–' is the Prolog counterpart of '←' and '&' indicates conjunction):

demo(empty).
demo(X) :–clause(X, Y), demo(Y).
demo(X&Y) :–demo(X), demo(Y).

For the above object-level program, we should add to the meta-interpreter the unit clauses:

clause(q, p&s).
clause(p, empty).
clause(s, empty)..

and the query would be :– *demo(q)*.

The vanilla meta-interpreter can be used for propositional programs, as well as for programs containing variables. In the latter case however, there is an important ambiguity concerning variables. In fact, variables in the object-level program are meant to range (as usual) over the domain of the program. These variables are instantiated to object-level terms. Instead, the variables occurring in the definition of the meta-interpreter, are intended to range over object-level atoms. Then, in a correct approach these are meta-variables (for an accurate discussion of this problem see [34]).

In [35], a typed version of the Vanilla meta-interpreter is advocated and its correctness proved. In [46] and [65], suitable naming mechanisms are proposed to overcome the problem.

Since however it is the untyped version that is generally used in Prolog practice, some researchers have tried to specify a formal account of the Vanilla meta-interpreter as it is. In particular, a first-order logic with ambivalent syntax has been proposed to this purpose [73], [74] and correctness results have been obtained [75].

The Vanilla meta-interpreter can be enhanced in various ways, often by making use of built-in Prolog meta-predicates that allow Prolog to act as a meta-language of itself. These predicates in fact are aimed at inspecting, building and modifying goals and at inspecting the instantiation status of variables.

First, more aspects of the proof process can be made explicit. In the above formalization, unification is implicitly demanded to the underlying Prolog interpreter and so is the order of execution of subgoals in conjunctions. Below is a formulation where these two aspects become explicit. Unification is performed by a *unify* procedure and *reorder* rearranges subgoals of the given conjunction.

$demo(empty).$
$demo(X) :-clause(H, Y), unify(H, X, Y, Y1), demo(Y1).$
$demo(X \& Y) :-reorder(X \& Y, X1 \& Y1), demo(X1), demo(Y1).$

Second, extra arguments can be added to *demo*, to represent for instance: the maximum number of steps that *demo* is allowed to perform; the actual number of steps that *demo* has performed; the proof tree; an explanation to be returned to a user and so on. Clearly, the definition of the meta-interpreter will be suitably modified according to the use of the extra arguments.

Third, extra rules can enhance the behavior of the meta-interpreter, by specifying auxiliary deduction rules. For instance, the rule

$demo(X) :-ask(X, yes).$

states that we consider X to be true, if the user answers "yes" when explicitly asked about X. In this way, the meta-interpreter exhibits an interactive behavior. The auxiliary deduction rules may be several and may interact.

In Reflective Prolog, [65] one specifies the additional rules *only,* while the definition of standard provability remains implicit. In the last example for instance, on failure of goal X, a goal $demo(X)$ would be automatically generated (this is an example of implicit upward reflection), thus employing the additional rule to query the user about X.

An interesting approach to meta-interpreters is that of [76], [77], where a binary predicate *demo* may answer queries with uninstantiated variables, which represent arbitrary fragments of the program currently being executed.

The reader may refer to [51] for an illustration of the meta-interpreter programming techniques and of their applications, including the specification of Expert Systems in Prolog.

6.2 Theory Composition and Theory Systems

Theory construction and combination is an important tool of software engineering, since it promotes modularity, software reuse and programming-in-the-large. In [53] it is observed that theory-construction can be regarded as a meta-linguistic operation. Within the Compulog European projects, two meta-logic approaches to working with theories have been proposed.

In the Algebra of Logic Programs, proposed in [78] and [79], a *program expression* defines a combination of object programs (that can be seen as theories, or modules) through a set of composition operators. The provability of a query with respect to a composition of programs can be defined by meta-axioms specifying the intended meaning of the various composition operations.

Four basic operations for composing logic programs are introduced: encapsulation (denoted by $*$), union ($\cup$), intersection ($\cap$) and import ($\lhd$).

Encapsulation copes with the requirement that a module can import from another one only its functionality, without caring of the implementation. This kind of behavior can be realized by encapsulation and union: if P is the "main program" and S is a module, the combined program is:

$$P \cup S^*$$

Intersection yields a combined theory where both the original theories are forced to agree during deduction, on every single partial conclusion.

The operation $\lhd$ builds a module $P \lhd Q$ out of two modules P and Q, where P is the *visible part* and Q the *hidden part* of the resulting module.

The usefulness of these operators for knowledge representation and reasoning is shown in [78]. The meta-logical definition of the operations is given in [79], by extending the Vanilla meta-interpreter. Two alternative implementations using the Gödel programming language are proposed and discussed in [80]. One extends the untyped Vanilla meta-interpreter. The other one exploits the meta-programming facilities offered by the language, thus using names and typed variables. The second, cleaner version seems to the authors themselves more suitable than the first one, for implementing program composition operations requiring a fine-grained manipulation of the object programs.

In the Alloy language, proposed in [81] and [82], a theory system is a collection of interdependent theories, some of which stand in a meta/object relationship, forming an arbitrary number of meta-levels. Theory systems are proposed for a meta-programming based software engineering methodology aimed at specifying, for instance, reasoning agents, programs to be manipulated, programs that manipulate them, etc. The meta/object relationship between theories provides the inspection and control facilities needed in these applications.

The basic language of theory systems is a definite clause language, augmented with ground names for every well-formed expression of the language. Each theory is named by a ground *theory term*. A theory system can be defined out of a collection of theories by using the following tools.

1. The symbol '$\vdash$' for relating theory terms and sentences. A *theoremhood statement,* like for instance $t_1 \vdash \lceil u_1 \vdash \Psi \rceil$ where t_1 and u_1 are theory terms, says that $\lceil u_1 \vdash \Psi \rceil$ is a theorem of theory t_1.
2. The distinguishes function symbol '$\diamond$', where $t_1 \diamond t_2$ means that t_1 is a meta-theory of t_2.
3. The *coincidence statement* $t_1 \equiv t_2$, expressing that t_1 and t_2 have exactly the same theorems.

The behavior of the above operators is defined by reflection principles (in the form of meta-axioms) that are suitably integrated in the declarative and proof-theoretic semantics.

6.3   The Event Calculus

Representing and reasoning about actions and temporally-scoped relations has been for years one of the key research topics in knowledge representation [83]. The Event Calculus (EC) has been proposed by Kowalski and Sergot [84] as a system for reasoning about time and actions in the framework of Logic Programming. In particular, the Event Calculus adapts the ontology of McCarthy and Hayes's *Situation Calculus* [85] i.e., actions and fluents [1], to a new task: *assimilating a narrative*, which is the description of a course of events. The essential

---

[1] It is interesting to notice that the fluent/fluxion terminology dates back to Newton

idea is to have terms, called *fluents,* which are names of time-dependent relations. Kowalski and Sergot however write $holds(r(x, y), t)$ which is understood as "fluent $r(x, y)$ is true at time t", instead of $r(x, y, t)$ like in situation calculus.

It is worthwhile to discuss the connection between Kowalski's work on meta-programming and the definition of the Event Calculus. In the logic programming framework it comes natural to recognize the higher-order nature of time-dependent propositions and to try to represent them at the meta-level. Kowalski in fact [86] considers McCarthy's Situation Calculus and comments:

> Thus we write
>
> $Holds(possess(Bob, Book1), S0)$
>
> instead of the weaker but also adequate
>
> $Possess(Bob, Book1, S0).$
>
> In the first formulation, $possess(Bob, Book1)$ is a term which names a relationship. In the second, $Possess(Bob, Book1, S0)$ is an atomic formula. Both representations are expressed within the formalism of first-order classical logic. However, the first allows variables to range over relationships whereas the second does not. If we identify relationships with atomic variable-free sentences, then we can regard a term such as $possess(Bob, Book1)$ as the name of a sentence. In this case $Holds$ is a meta-level predicate [...]

There is a clear advantage with reification from the computational point of view: by reifying, we need to write only one frame axiom, or inertia law, saying that truth of any relation does not change in time unless otherwise specified. Negation-as-failure is a natural choice for implementing the default inertia law. In a simplified, time points-oriented version, default inertia can be formulated as follows:

$$Holds(f, t) \leftarrow Happens(e),$$
$$initiates(e, f),$$
$$Date(e, t_s),$$
$$t_s < t,$$
$$not\ Clipped(t_s, f, t)$$

where $Clipped(t_s, f, t)$ is true when there is record of an event happening between t_s and t that terminates the validity of f. In other words, $Holds(f, t)$ is derivable whenever in the interval between the initiation of the fluent and the time the query is about, no terminating events has happened.

It is easy to see $Holds$ as a specialization of $Demo$. Kowalski and Sadri [87] [88], discuss in depth how an Event Calculus program can be specified and assumptions on the nature of the domain accommodated, by manipulating the usual Vanilla meta-interpreter definition.

Since the first proposal, a number of improved formalization have steamed, in order to adapt the calculus to different tasks, such as abductive planning, diagnosis, temporal database and models of legislation. All extensions and applications cannot be accounted for here, but the reader may for instance refer to [89], [90], and [91].

6.4 Logical Frameworks

A logical framework [92] is a formal system that provides tools for experimenting with deductive systems. Within a logical framework, a user can invent a new deductive system by defining its syntax, inference rules and proof-theoretic semantics. This specification is executable, so as the user can make experiments with this new system. A logical framework however cannot reasonably provide tools for defining any possible deductive system, but will stay within a certain class.

Formalisms with powerful meta-level features and strong semantic foundations have the possibility of evolving towards becoming logical frameworks.

The Maude system for instance [93] is a particular implementation of the meta-theory of rewriting logic. It provides the predefined functional module META-LEVEL, where Maude terms can be reified and where: the process of reducing a term to a normal form is represented by a function *meta-reduce*; the default interpreter is represented by a function *meta-rewrite*; the application of a rule to a term by *meta-apply*.

Recently, a reflective version of Maude has been proposed [94], based on the formalization of computational reflection proposed in [95]. The META-LEVEL module has been made more flexible, so as to allow a user to define the syntax of her own logic language L by means of meta-rules. The new language must however consist in an addition/variation to the basic syntax of the Maude language. Reflection is the tool for integrating the user-defined syntax into the proof procedure of Maude. In particular, whenever a piece of user-defined syntax is found, a reflection act to the META-LEVEL module happens, so as to apply the corresponding syntactic meta-rules. Then, the rewriting system Maude has evolved into a logical framework for logic languages based on rewriting.

The *RCL* (Reflective Computational Logic) logical framework [33] is an evolution of the Reflective Prolog metalogic language. The implicit reflection of Reflective Prolog has a semantic counterpart [39] in adding to the given theory a set of new axioms called *reflection axioms*, according to axiom schemata called *reflection principles*. Reflection principles can specify not only the shift between levels, but also many other meta-reasoning principles. For instance, reflection principles can define forms of analogical reasoning [96], and synchronous communication among logical agents [97].

RCL has originated from the idea that, more generally, reflection principles may be used to express the inference rules of user-defined deductive systems. The deductive systems that can be specified in *RCL* are however evolutions of the Horn clause language, based on a predefined enhanced syntax. A basic version of naming is provided in the enhanced Horn clause language, formalized through an equational theory.

The specification of a new deductive system DS in RCL is accomplished through the following four steps.

Step I Definition of the naming device (encoding) for DS. The user definition must extend the predefined one. RCL leaves significant freedom in the representation of names.

Step II After defining the naming convention, the user of RCL has to provide a corresponding unification algorithm (again by suitable additions to the predefined one).

Step III Representation of the axioms of DS, in the form of enhanced Horn clauses.

Step IV Definition of the inference rules of DS as reflection principles.

In particular, the user is required to express each inference rule R as a function $\mathcal{R}$, from clauses, which constitute the antecedent of the rule, to sets of clauses, which constitute the consequent.

Then, given a theory T of DS consisting of a set of axioms A and a reflection principle $\mathcal{R}$, a theory T' containing T is obtained as the deductive closure of $A \cup A'$, where A' is the set of additional axioms generated by $\mathcal{R}$. Consequently, the model-theoretic and fixed point semantics of T under $\mathcal{R}$ are obtained as the model-theoretic and fixed point semantics of T'. RCL does not actually generate T'. Rather, given a query for T, RCL dynamically generates the specific additional axioms usable to answer the query according to the reflection principle $\mathcal{R}$, i.e., according to the inference rule R of DS.

6.5   Logical Agents

In the area of intelligent software agents there are several issues that require the integration of some kind of meta-reasoning ability into the system. In fact, most existing formalisms, systems and frameworks for defining agents incorporate, in different forms, a meta-component.

An important challenge in this area is that of interconnecting several agents that are *heterogeneous* in the sense that they are not necessarily uniform in the implementation, in the knowledge they possess and in the behavior they exhibit. Any framework for developing multi-agent systems must provide a great deal of flexibility for integrating heterogeneous agents and assembling communities of independent service providers. Flexibility is required in structuring cooperative interactions among agents, and for creating more accessible and intuitive user interfaces.

Meta-reasoning is essential for obtaining such a degree of flexibility. Meta-reasoning can either be performed within the single agent, or special meta-agents can be designed, to act as meta-theories for sets of other agents. Meta-reasoning can help: (i) in the interaction among agents and with the user; (ii) in the implementation suitable strategies and plans for responding to requests. These strategies can be either domain-independent, or rely on domain- and application-specific knowledge or reasoning (auxiliary inference rules, learning algorithms, planning, and so forth)

Meta-rules and meta-programming may be particularly useful for coping with some aspects of the ontology problem: meta-rules can switch between descriptions that are syntactically different though semantically equivalent, and can help fill the gap between descriptions that are not equivalent. Also, meta-reasoning can be used for managing incomplete descriptions or requests.

The following are relevant examples of approaches to developing agent systems that make use of some form of meta-reasoning.

In the Open Agent Architecture^{TM} [98], which is meant for integrating a community of heterogeneous software agents, there are specialized server agents, called *facilitators,* that perform reasoning (and, more or less explicitly, meta-reasoning) about the agent interactions necessary for handling a complex expression. There are also *meta–agents,* that perform more complex meta-reasoning so as to assist the facilitator agent in coordinating the activities of the other agents.

In the constraint logic programming language CaseLP, there are *logical agents,* which show capabilities of complex reasoning, and *interface agents,* which provide an interface with external modules. There are no meta-agents, but an agent has *meta–goals* that trigger meta-reasoning to guide the planning process.

There are applications where agents may have objectives and may need to reason about their own as well as other agents' beliefs and about the actions that agents may take. This is the perspective of the BDI formalization of multi-agent systems proposed in [99] and [100], where BDI stands for "Belief, Desire, Intentions".

The approach of Meta-Agents [101] allow agents to reason about other agents' state, beliefs, and potential actions by introducing powerful meta-reasoning capabilities. Meta-Agents are a specification tool, since for efficient implementation they are translated into ordinary agent programs, plus some integrity constraints.

In logic programming, research on multi-agent systems starts, to the best of our knowledge, from the work by Kim and Kowalski in [102], [103]. The amalgamation of language and meta-language and the *demo* predicate with theories named by constants are used for formalizing reasoning capabilities in multi-agent domains. In this approach, the *demo* predicate is interpreted as a belief predicate and thus agents can reason, like in the BDI approach, about beliefs.

In the effort of obtaining logical agents that are rational, but also reactive (i.e. logical reasoning agents capable of timely response to external events) a more general approach has been proposed in [82], by Kowalski, and in [104] and [105] by Kowalski and Sadri. A meta-logic program defines the "observe-think-act" cycle of an agent. Integrity constraints are used to generate actions in response to updates from the environment.

In the approach of [97], agents communicate via the two meta-level primitives *tell/told.* An agent is represented by a theory, i.e. by a set of clauses prefixed with the corresponding theory name. Communication between agents is formalized by the following reflection principle $\mathcal{R}_{com}$:

$$T:told(\text{``}S\text{''},\ \text{``}A\text{''})\Leftarrow_{\mathcal{R}_{com}} S:tell(\text{``}T\text{''},\ \text{``}A\text{''}).$$

The intuitive meaning is that every time an atom of the form $tell(\text{``}T\text{''},\text{``}A\text{''})$ can be derived from a theory S (which means that agent S wants to communicate proposition A to agent T), the atom $told(\text{``}S\text{''},\text{``}A\text{''})$ is consequently derived in theory T (which means that proposition A becomes available to agent T).

The objective of this formalization is that each agent can specify, by means of clauses defining the predicate $tell$, the modalities of interaction with the other agents. These modalities can thus vary with respect to different agents or different conditions. For instance, let P be a program composed of three agents, a and b and c, defined as follows.

$a:tell(X, \text{``}ciao\text{''}):\text{-}\ friend(X).$
$a:friend(\text{``}b\text{''}).$

$b:happy:\text{-}told(\text{``}a\text{''},\ \text{``}ciao\text{''}).$

$c:happy:\text{-}told(\text{``}a\text{''},\ \text{``}ciao\text{''}).$

Agent a says "ciao" to every other agent X that considers to be its friend. In the above definition, the only friend is b. Agents b and c are happy if a says "ciao" to them. The conclusion $happy$ can be derived in agent b, while it cannot be derived in agent c. In fact, we get $a:tell(\text{``}b\text{''},\text{``}ciao\text{''})$ from $a:friend(\text{``}b\text{''})$; instead, $a:tell(\text{``}c\text{''},\text{``}ciao\text{''})$ is not a conclusion of agent a.

In [106], Dell'Acqua, Sadri and Toni propose an approach to logic-based agents as a combination of the above approaches, i.e. the approach to agents by Kowalski and Sadri [105] and the approach to meta-reasoning by Costantini et al. [65], [97]. Similarly to Kowalski and Sadri's agents, the agents in [106] are *hybrid* in that they exhibit both *rational* (or *deliberative*) and *reactive* behavior. The reasoning core of these agents is a proof procedure that combines forward and backward reasoning. Backward reasoning is used primarily for deliberative activities. Forward reasoning is used primarily for reactivity to the environment, possibly including other agents. The proof procedure is executed within an "observe-think-act" cycle that allows the agent to be alert to the environment and react to it, as well as think and devise plans. The proof procedure (IFF proof procedure proposed by Fung and Kowalski in [107]) treats both inputs from the environment and agents' actions as *abducibles* (hypotheses). Moreover, by adapting the techniques proposed in [97], the agents are capable of reasoning about their own beliefs and the beliefs of other agents.

In [108], the same authors extend the approach by providing agents with *proactive* communication capabilities. Proactive agents are able to communicate on their own initiative, not only in response to stimula. In the resulting framework reactive, rational or hybrid agents can reason about their own beliefs as well as the beliefs of other agents and can communicate proactively with each other. The agents' behavior can be regulated by condition-action rules. In this approach, there are two primitives for communication, $tell$ and ask, treated as abducibles within the "observe-think-act" cycle of the agent architecture. The predicate *told* is used to express both passive reception of messages from other agents and reception of information in response to an active request.

The following example is taken by [108] and is aimed at illustrating the basic features of the approach. Let Ag be represented by the abductive logic program $\langle P, \mathcal{A}, I \rangle$ with:

$$P = \left\{ \begin{array}{l} \texttt{told}(A, X) \leftarrow \texttt{ask}(A, X) \wedge \texttt{tell}(A, X) \\ \texttt{told}(A, X) \leftarrow \texttt{tell}(A, X) \\ \texttt{solve}(X) \leftarrow \texttt{told}(A, X) \\ \texttt{desire}(y) \leftarrow y = \texttt{car} \\ \texttt{good\_price}(p, x) \leftarrow p = 0 \end{array} \right\}$$

$$\mathcal{A} = \left\{ \texttt{tell}, \texttt{ask}, \texttt{offer} \right\}$$

$$I = \left\{ \begin{array}{l} \texttt{desire}(x) \wedge \texttt{told}(B, \text{'good\_price}(\text{'}p, \text{'}x)) \\ \Rightarrow \texttt{tell}(B, \text{'offer}(\text{'}p, \text{'}x)) \end{array} \right\}.$$

The first two clauses in P state that Ag may be told something, say X, by another agent A either because A has been explicitly asked about X (first clause) or because A tells X proactively (second clause). The third clause in P says that Ag believes anything it is told. The fourth and fifth clauses in P say, respectively, that the agent desires a car and that anything that is free is at a good price. The integrity constraint says that, if the agent desires something and it is told (by some other agent B) of a good price for it, then it makes an offer to B, by telling it.

The logic programming language DALI [109], is indebted to all previously mentioned approaches to logical agents. DALI introduces explicit reactive and proactive rules at the object level. Thus, reactivity and proactivity are modeled in the basic logic language of the agent In fact, declarative semantics is very close to that of the standard Horn clause language. Procedural semantics relies on an extended resolution. The language incorporates *tell/told* primitives, integrity constraints and *solve* rules. An "observe-think-act" cycle can of course been implemented in a DALI agent, but it is no longer necessary for modeling reactivity and proactivity.

Below is a simplified fragment of a DALI agent representing the waiter of a pub, that tries to serve a customer that enters. The customer wants some X. This request is an *external event* (indicated with 'E') that arrives to the agent. The event triggers a reactive rule (indicated with ':>' instead of usual ':-'), and determines the body of the rule to be executed. This is very much like any other goal: only, computation is not initiated by a query, but starts on reception of the event.

During the execution of the body of the reactive rule, the waiter first checks whether X is one of the available drinks. If so, the waiter serves the drink: the predicate *serve_drink* is in fact an action (indicated with 'A'). Otherwise, the waiter checks whether the request is expressed in some foreign language, for which a translation is available (this is a simple example of coping with one aspect of the ontology problem). If this is not the case, the waiter asks the customer for explanation about X: it expects to be *told* that X is actually an Y, in order to try to serve this Y.

Notice that the predicate *translate* is symmetric, where symmetry is managed by the *solve* rule. To understand the behavior, one can assume this rule to be an additional rule of a basic meta-interpreter that is not explicitly reported. A subgoal like *translate(beer, V)* is automatically transformed into a call to the meta-interpreter, of the form *solve("translate"("beer", "V"))* (formally, this is implicit upward reflection). Then, since *symmetric("translate")* succeeds, *solve("translate"("beer", "V"))* is attempted, and automatically reflected at the object level (formally, this is implicit downward reflection). Finally, the unquoted subgoal *translate(beer, V)* succeeds with V instantiated to *birra*.

$$Waiter$$
$$request(Customer, \text{``}X\text{''})_E :> serve(Customer, X).$$

$$serve(C,X) :- drink(X), serve\_drink(C,X)_A.$$
$$serve(C,X) :- translate(X,Y),$$
$$drink(Y),$$
$$serve\_drink(C,Y)_A.$$
$$serve(C,X) :- ask(C,X,Y), serve(C,Y).$$

$$ask(C,X,Y) :- ask\_for\_explanation(C, \text{``}X\text{''}), told(C, \text{``}Y\text{''}).$$

$$drink(beer).$$
$$drink(coke).$$
$$translate(birra, beer).$$
$$translate(cocacola, coke).$$
$$symmetric(\text{``}translate\text{''}).$$

$$solve(\text{``}P\text{''}(\text{``}X\text{''}, \text{``}Y\text{''})) :- symmetric(\text{``}P\text{''}), solve(\text{``}P\text{''}(\text{``}Y\text{''}, \text{``}X\text{''})).$$

Agents that interact with other agents and/or with an external environment, may expand and modify their knowledge base by incorporating new information. In a dynamic setting, the knowledge base of an agent can be seen as the set of *beliefs* of the agent, that may change over time. An agent may reach a stage where its beliefs have become inconsistent, and actions must be taken to regain consistency. The theory of belief revision aims at modeling how an agent updates its state of belief as a result of receiving new information [110], [111]. Belief revision is, in our opinion, another important issue related to intelligent agents where meta-reasoning can be usefully applied.

In [32] a model-based diagnosis system is presented, capable of revision of the description of the system to be diagnosed if inconsistencies arise from observations. Revision strategies are implemented by means of meta-programming and meta-reasoning methods.

In [112], a framework is proposed where rational, reactive agents can dynamically change their own knowledge bases as well as their own goals. In particular, an agent can make observations, learn new facts and new rules from the environment (even in contrast with its current knowledge) and then update its knowledge accordingly. To solve contradictions, techniques of contradiction removal and preferences among several sources can be adopted [113].

In [114] it is pointed out that most existing approaches to intelligent agents have difficulties to model the way agents revise their beliefs, because new information always come together certain meta-information: e.g., where the new information comes from? Is the source reliable? and so on. Then, the agent has to reason about this meta-information, in order to revise its beliefs. This leads to the proposal of a new approach, where this meta-information can be explicitly represented and reasoned about, and revision strategies can be defined in a declarative way.

7 Semantic Issues

In computational logic, meta-programming and meta-reasoning capabilities are mainly based on self-reference, i.e. on the possibility of describing language expressions in the language itself. In fact, in most of the relevant approaches the object language and the meta-language coincide.

The main tool for self-reference is a naming mechanism. An alternative form of self-reference has been proposed by McCarthy [115], who suggests that introducing function symbols denoting concepts (rather than quoted expressions) might be sufficient for most forms of meta-reasoning. But Perlis [40] observes:

> "The last word you just said" is an expression that although representable as a function still refers to a particular word, not to a concept. Thus quotation seems necessarily involved at some point if we are to have a self-describing language. It appears we must describe specific expressions as carriers of (the meaning of) concepts.

The issue of appropriate language facilities for naming is addressed by Hill and Lloyd in [35]. They point out the distinction between two possible representation schemes: the *non-ground* representation, in which an object-level variable is represented by a meta-level variable, and the *ground* representation, in which object-level expressions are represented by ground (i.e. variable free) terms at the meta-level. In the ground representation, an object level variable may be represented by a meta-level constant, or by any other ground term.

The problem with the non-ground representation is related to meta-level predicates such as the Prolog $var(X)$, which is true if the variable X is not instantiated, and is false otherwise. As remarked in [35]:

To see the difficulty, consider the goals:

$$:-var(X) \land solve(p(X))$$

and

$$:-solve(p(X)) \land var(X)$$

If the object program consists solely of the clause $p(a)$, then (using the "leftmost literal" computation rule) the first goal succeeds, while the second goal fails.

Hill and Lloyd propose a ground representation of expressions of a first-order language $\mathcal{L}$ in another first-order language $\mathcal{L}'$ with three types ω, μ and η.

Definition 1 (Hill and Lloyd ground representation). *Given a constant a in $\mathcal{L}$, there is a corresponding constant a' of type ω in $\mathcal{L}'$. Given a variable x in $\mathcal{L}$, there is a corresponding constant x' of type ω in $\mathcal{L}'$. Given an n-ary function symbol f in $\mathcal{L}$, there is a corresponding n-ary function symbol f' of type $\omega \times \ldots \omega \longrightarrow \omega$ in $\mathcal{L}'$. Given an n-ary predicate symbol p in $\mathcal{L}$, there is a corresponding n-ary function symbol f' of type $\omega \times \ldots \omega \longrightarrow \mu$ in $\mathcal{L}'$. The language $\mathcal{L}'$ has a constant empty of type μ. The mappings $a \longrightarrow a'$, $x \longrightarrow x'$, $f \longrightarrow f'$ and $p \longrightarrow p'$ are all injective.*

Moreover, $\mathcal{L}'$ contains some function and predicate symbols useful for declaratively redefining the "impure" features of Prolog and the Vanilla meta-interpreter. For instance we will have:

$constant(a'_1).$
$\ldots$
$constant(a'_n).$
$\forall_\omega x \ nonvar(x) \leftarrow constant(x).$
$\forall_\omega x \ var(x) \leftarrow \neg nonvar(x).$

The above naming mechanism is used in [35] for providing a declarative semantics to a meta-interpreter that implements SLDNF resolution [116] for normal programs and goals. This approach has then evolved into the meta-logical facilities of the Gödel language [59]. Notice that, since names of predicate symbols are function symbols, properties of predicates (e.g. symmetry) cannot be explicitly stated. Since levels in Gödel are separated rather than amalgamated, this naming mechanism does not provide operators for referentiation/dereferentiation.

An important issue raised in [40] is the following:

Now, it is essential to have also an un-naming device that would return a quoted sentence to its original (assertive) form, together with axioms stating that that is what naming and un-naming accomplish.

Along this line, the approach of [36], developed in detail in [117], proposes to name an atom of the form $\alpha_0(\alpha_1, \ldots, \alpha_n)$ as $[\beta_0, \beta_1, \ldots, \beta_n]$, where each β_i is the name of α_i. The name of the name of $\alpha_0(\alpha_1, \ldots, \alpha_n)$ is the name term $[\gamma_0, \gamma_1, \ldots, \gamma_n]$, where each γ_i is the name of β_i, etc. Requiring names of compound expressions to be compositional allows one to use unification for constructing name terms and accessing their components.

In this approach, we are able to express properties of predicates by using their names. For instance, we can say that predicate p is binary and predicate q is symmetric, by asserting $binary\_pred(p^1)$ and $symmetric(q^1)$.

Given a term t and a name term s, the expression $\uparrow t$ indicates the result of quoting t and the expression $\downarrow s$ indicates the result of unquoting s. The following axioms for the operators $\uparrow$ and $\downarrow$ formalize the relationship between terms and the corresponding name terms. They form an equality theory, called NT and first defined in [118], for the basic compositional encoding outlined above. Enhanced encodings can be obtained by adding axioms to this theory. NT states that there exist names of names (each term can be referenced n times, for any $n \geq 0$) and that the name of a compound term is obtained from the names of its components.

Definition 2 (Basic encoding NT). *Let* **NT** *be the following equality theory.*

- *For every constant or meta-constant c^n, $n \geq 0$,*
  $\uparrow c^n = c^{n+1}$.
- *For every function or predicate symbol f of arity k,*
  $\forall x_1 \ldots \forall x_k \ \uparrow (f(x_1, \ldots, x_k)) = [f^1, \uparrow x_1, \ldots, \uparrow x_k]$.
- *For every compound name term $[x_0, x_1, \ldots, x_k]$*
  $\forall x_0 \ldots \forall x_k \ \uparrow [x_0, x_1, \ldots, x_k] = [\uparrow x_0, \uparrow x_1, \ldots, \uparrow x_k]$.
- *For every term t $\downarrow\uparrow t = t$.*

The above set of axioms admits an associated convergent rewrite system UN. Then, a corresponding extended unification algorithm (E-unification algorithm) $UA(UN)$ can be defined, that deals with name terms in addition to usual terms. In [118] it is shown that:

Proposition 1 (Unification Algorithm for NT). *The E-unification algorithm $UA(UN)$ is sound for NT, terminates and converges.*

The standard semantics of the Horn clause language can be adapted, so as to include the naming device. Precisely, the technique of quotient universes by Jaffar et al. [119] can be used to this purpose.

Definition 3 (Quotient Universe). *Let R be a congruence relation. The quotient universe of U with respect to R, indicated as U/R, is the set of the equivalence classes of U under R, i.e., the partition given by R in U.*

By taking R as the finest congruence relation corresponding to UN (that always exists) we get the standard semantics of the Horn clause language [116], modulo the naming relation. The naming relation can be extended according to the application domain at hand, by adding new axioms to NT and by correspondingly extending UN and $UA(UN)$, provided that their nice formal properties are preserved. What is important is that, as advocated in [37], the approach to meta-programming and the approach to naming become independent.

It is important to observe that, as shown in [36], any (ground or non-ground) encoding providing names for variables shows in an amalgamated language the same kind of problems emphasized in [35]. In fact, let P be the following definite program, x an object-level variable and Y a meta-variable:

$$p(x) :- Y = \uparrow x, q(Y)$$
$$q(a^1).$$

Goal :-$p(a)$ succeeds by first instantiating Y to a^1 and then proving $q(a^1)$. In contrast, the goal :-$p(x)$ fails, as Y is instantiated to the name of x, say x^1, and subgoal $q(x^1)$ fails, x^1 and a^1 being distinct. Therefore, if choosing naming mechanisms providing names for variables, on the one hand terms can be inspected with respect to variable instantiation, on the other hand however important properties are lost.

A ground naming mechanism is used in [49] for providing a declarative semantics to the (conservative) amalgamation of language and meta-language in logic programming.

A naming mechanism where each well-formed expression can act as a name of itself is provided by the ambivalent logic AL of Jiang [73]. It is based on the assumption that each expression can be interpreted as a formula, as a term, as a function and as a predicate, where predicates and functions have free arity.

Unification must be extended accordingly, with the following results:

Theorem 1 (Termination of AL Unification Algorithm). *The unification algorithm for ambivalent logic terminates.*

Theorem 2 (Correctness of AL Unification Algorithm). *If the unification algorithm for ambivalent logic terminates successfully, then it provides an ambivalent unifier. If the algorithm halts with failure, then no ambivalent unifier exists.*

The limitation is that ambivalent unifiers are less general than traditional unifiers.

Theorem 3 (Properties of Resolution for AL). *Resolution is a sound and complete inference method for AL.*

Ambivalent logic has been used in [75] for proving correctness of the Vanilla meta-interpreter, also with respect to the (conservative) amalgamation of object language and meta-language. Let P be the object program, $\mathcal{L}_P$ the language of P, V_P the Vanilla meta-interpreter and $\mathcal{L}_{V_P}$ the language of V_P. Let M_P be the least Herbrand model of P, M_{V_P} be the least Herbrand model of V_P, and $M_{V_P \cup P}$ be the least Herbrand model of $V_P \cup P$. We have:

Theorem 4 (Properties of Vanilla Meta-Interpreter under AL). *For all (ground) A in $\mathcal{L}_{V_P}$, $demo(A) \in M_{V_P}$ iff $demo(A) \in M_{V_P \cup P}$; for all (ground) A in $\mathcal{L}_P$, $demo(A) \in M_P$ iff $demo(A) \in M_{V_P \cup P}$*

A similar result is obtained by Martens and De Schreye in [120] and [50] for the class of *language independent programs*. They use a non-ground representation with overloading of symbols, so as the name of an atom is a term, identical to the atom itself. Language independent programs can be characterized as follows:

Proposition 2 (Language Independence). *Let P be a definite program. Then P is language independent iff for any definite goal G, all (SLD) computed answers for $P \cup G$ are ground.*

Actually however, the real practical interest lies in enhanced meta-interpreters. Martens and De Schreye extend their results to meta-interpreters without additional clauses, but with additional arguments. An additional argument can be for instance an explicit theory argument, or an argument denoting the proof tree. The amalgamation is still conservative, but more expressivity is achieved.

The approach to proving correctness of the Vanilla meta-interpreter proposed by Levi and Ramundo in [48] uses the S-semantics introduced by Falaschi et al. in [121]. In order to fill the gap between the procedural and declarative interpretations of definite programs, the S-least Herbrand model M_P^S of a program P contains not only ground atoms, but all atoms $Q(T)$ such that $t = x\,\theta$, where θ is the computed answer substitution for $P \cup \{\leftarrow Q(x)\}$. The S-semantics is obtained as a variation of the standard semantics of the Horn clause language. Levi and Ramundo [48] and Martens and De Schreye prove (independently) that $demo(p(t)) \in M_{V_P}^S$ iff $p(t) \in M_P^S$.

In the approach of Reflective Prolog, axiom schemata are defined at the meta-level, by means of a distinguished predicate *solve* and of a naming facility. Deduction is performed at any level where there are applicable axioms. This means, conclusions drawn in the basic theory are available (by implicit reflection) at the meta-level, and vice versa. The following definition of RSLD-resolution [65] (SLD-resolution with reflection) is independent of the naming mechanism, provided that a suitable unification algorithm is supplied.

Definition 4 (RSLD-resolution). *Let G be a definite goal $\leftarrow A_1, \dots, A_k$, let A_m be the selected atom in G and let C be a definite clause.*
The goal $(\leftarrow A_1, \dots, A_{m-1}, B_1, \dots, B_q, A_{m+1}, \dots, A_k)\theta$ is derived from G and C using mgu θ iff one of the following conditions holds:

  i. *C is $A \leftarrow B_1, \dots, B_q$*
     θ is a mgu of A_m and A
 ii. *C is $solve(\alpha) \leftarrow B_1, \dots, B_q$*
     $A_m \neq solve(\delta)$
     $\uparrow A_m = \alpha'$
     θ is a mgu of α' and α

iii. A_m *is solve*(α)
   C *is* $A \leftarrow B_1, \ldots, B_q$
   $\downarrow \alpha = A'$
   θ *is a mgu of* A' *and* A

If the selected atom A_m is an object-level atom (e.g $p(a, b)$), it can be resolved in two ways. First, by using as usual the clauses defining the corresponding predicate (case (i)); for instance, if A_m is $p(a, b)$, by using the clauses defining the predicate p. Second, by using the clauses defining the predicate *solve* (case (ii), *upward reflection*) if the name $\uparrow A_m$ of A_m and α unify with mgu θ; for instance, referring to the *NT* naming relation defined above, we have $\uparrow p(a, b) = [p^1, a^1, b^1]$ and then a clause with conclusion *solve*($[p^1, v, w]$) can be used, with $\theta = \{v/a^1, w/b^1\}$.

If the selected atom A_m is *solve*(α) (e.g *solve*($[q^1, c^1, d^1]$)), again it can be resolved in two ways. First, by using the clauses defining the predicate *solve* itself, similarly to any other goal (case (i)). Second, by using the clauses defining the predicate corresponding to the atom denoted by the argument α of *solve* (case (iii), *downward reflection*); for instance, if α is $[q^1, c^1, d^1]$ and thus $\downarrow \alpha = q(c, d)$, by using the clauses defining the predicate q can be used.

In the declarative semantics of Reflective Prolog, upward and downward reflection are modeled by means of axiom schemata called *reflection principles*. The Least Reflective Herbrand Model RM_P of program P is the Least Herbrand Model of the program itself, augmented by all possible instances of the reflection principles. RM_P is the least fixed point of a suitably modified version of operator T_P.

Theorem 5 (Properties of RSLD-Resolution). *RSLD-resolution is correct and complete w.r.t. RM_P*

8 Conclusions

In this paper we have discussed the meta-level approach to knowledge representation and reasoning that has its roots in the work of logicians and has played a fundamental role in computer science. We believe in fact that meta-programming and meta-reasoning are essential ingredients for building any complex application and system.

We have tried to illustrate to a broad audience what are the main principles meta-reasoning is based upon and in which way these principles have been applied in a variety of languages and systems. We have illustrated how sentences can be arguments of other sentences, by means of naming devices. We have distinguished between amalgamated and separated approaches, depending on whether the meta-expressions are defined in (an extension of) a given language, or in a separate language. We have shown that the different levels of knowledge can interact by reflection.

In our opinion, the choice of logic programming as a basis for meta-programming and meta-reasoning has several theoretical and practical advantages. ¿From the theoretical point of view, all fundamental issues (including reflection) can be coped with on a strong semantic basis. In fact, the usual framework of first-order logic can be suitably modified and extended, as demonstrated by the various existing meta-logic languages. ¿From the practical point of view, in logic programming the meta-level mechanisms are understandable and easy-to-use and this has given rise to several successful applications. We have in fact tried (although necessarily shortly) to revise some of the important applications of meta-programming and meta-reasoning.

At the end of this survey, I wish to explicitly acknowledge the fundamental, deep and wide contribution that Robert A. Kowalski has given to this field. Robert A. Kowalski initiated meta-programming in logic programming, as well as many of its successful applications, including meta-interpreters, event calculus, logical agents. With his enthusiasm he has given constant encouragement to research in this field, and to researchers as well, including myself.

9   Acknowledgements

I wish to express my gratitude to Gaetano Aurelio Lanzarone, who has been the mentor of my research work on meta-reasoning and reflection. I gratefully acknowledge Pierangelo Dell'Acqua for his participation to this research and for the important contribution to the study of naming mechanisms and reflective resolution. I also wish to mention Jonas Barklund, for the many interesting discussions and the fruitful cooperation on these topics.

Many thanks are due to Luigia Carlucci Aiello, for her careful review of the paper, constructive criticism and useful advice. Thanks to Alessandro Provetti for his help. Thanks also to the anonymous referees, for their useful comments and suggestions. Any remaining errors or misconceptions are of course my entire responsibility.

References

1. Hill, P.M., Gallagher, J.: Meta-programming in logic programming. In Gabbay, D., Hogger, C.J., Robinson, J.A., eds.: Handbook of Logic in Artificial Intelligence and Logic Programming, Vol. 5, Oxford University Press (1995)
2. Barklund, J.: Metaprogramming in logic. In Kent, A., Williams, J.G., eds.: Encyclopedia of Computer Science and Technology. Volume 33. M. Dekker, New York (1995) 205–227
3. Lanzarone, G.A.: Metalogic programming. In Sessa, M.I., ed.: 1985–1995 Ten Years of Logic Programming in Italy. Palladio (1995) 29–70
4. Abramson, H., Rogers, M.H., eds.: Meta-Programming in Logic Programming, Cambridge, Mass., THE MIT Press (1989)
5. Bruynooghe, M., ed.: Proc. of the Second Workshop on Meta-Programming in Logic, Leuven (Belgium), Dept. of Comp. Sci., Katholieke Univ. Leuven (1990)
6. Pettorossi, A., ed.: Meta-Programming in Logic. LNCS 649, Berlin, Springer-Verlag (1992)
7. Fribourg, L., Turini, F., eds.: Logic Program Synthesis and Transformation – Meta-Programming in Logic. LNCS 883, Springer-Verlag (1994)

8. Barklund, J., Costantini, S., van Harmelen, F., eds.: Proc. Workshop on Meta Programming and Metareasonong in Logic, post-JICSLP96 workshop, Bonn (Germany), UPMAIL technical Report No. 127 (Sept. 2, 1996), Computing Science Dept., Uppsala Univ. (1996)
9. Apt, K., Turini, F., eds.: Meta-Logics and Logic Programming. The MIT Press, Cambridge, Mass. (1995)
10. Maes, P., Nardi, D., eds.: Meta-Level Architectures and Reflection, Amsterdam, North-Holland (1988)
11. Kiczales, G., ed.: Meta-Level Architectures and Reflection, Proc. Of the First Intnl. Conf. Reflection 96, Xerox PARC (1996)
12. Cointe, A., ed.: Meta-Level Architectures and Reflection, Proc. Of the Second Intnl. Conf. Reflection 99. LNCS 1616, Berlin, Springer-Verlag (1999)
13. Smorinski, C.: The incompleteness theorem. In Barwise, J., ed.: Handbook of Mathematical Logic. North-Holland (1977) 821–865
14. Smullyan, R.: Diagonalization and Self-Reference. Oxford University Press (1994)
15. Kripke, S.A.: Semantical considerations on modal logic. In: Acta Philosophica Fennica. Volume 16. (1963) 493–574
16. Carlucci Aiello, L., Cialdea, M., Nardi, D., Schaerf, M.: Modal and meta languages: Consistency and expressiveness. In Apt, K., Turini, F., eds.: Meta-Logics and Logic Programming. The MIT Press, Cambridge, Mass. (1995) 243–266
17. Aiello, M., Weyhrauch, L.W.: Checking proofs in the metamathematics of first order logic. In: Proc. Fourth Intl. Joint Conf. on Artificial Intelligence, Tbilisi, Georgia, Morgan Kaufman Publishers (1975) 1–8
18. Bundy, A., Welham, B.: Using meta-level inference for selective application of multiple rewrite rules in algebraic manipulation. Artificial Intelligence **16** (1981) 189–212
19. Weyhrauch, R.W.: Prolegomena to a theory of mechanized formal reasoning. Artificial Intelligence (1980) 133–70
20. Carlucci Aiello, L., Cecchi, C., Sartini, D.: Representation and use of metaknowledge. Proc. of the IEEE **74** (1986) 1304–1321
21. Carlucci Aiello, L., Levi, G.: The uses of metaknowledge in AI systems. In: Proc. European Conf. on Artificial Intelligence. (1984) 705–717
22. Davis, R., Buchanan, B.: Meta-level knowledge: Overview and applications. In: Procs. Fifth Int. Joint Conf. On Artificial Intelligence, Los Altos, Calif., Morgan Kaufmann (1977) 920–927
23. Maes, P.: Computational Reflection. PhD thesis, Vrije Universiteit Brussel, Faculteit Wetenschappen, Dienst Artificiele Intelligentie, Brussel (1986)
24. Genesereth, M.R.: Metalevel reasoning. In: Logic-87-8, Logic Group, Stanford University (1987)
25. Carlucci Aiello, L., Levi, G.: The uses of metaknowledge in AI systems. In Maes, P., Nardi, D., eds.: Meta-Level Architectures and Reflection. North-Holland, Amsterdam (1988) 243–254
26. Carlucci Aiello, L., Nardi, D., Schaerf, M.: Yet Another Solution to the Three Wisemen Puzzle. In Ras, Z.W., Saitta, L., eds.: Methodologies for Intelligent Systems 3: ISMIS-88, Elsevier Science Publishing (1988) 398–407
27. Carlucci Aiello, L., Nardi, D., Schaerf, M.: Reasoning about Knowledge and Ignorance. In: Proceedings of the International Conference on Fifth Generation Computer Systems 1988: FGCS-88, ICOT Press (1988) 618–627
28. Genesereth, M.R., Nilsson, J.: Logical Foundations of Artificial Intelligence. Morgan Kaufmann, Los Altos, California (1987)

29. Russell, S.J., Wefald, E.: Do the right thing: studies in limited rationality (Chapter 2: Metareasoning Architectures). The MIT Press (1991)
30. Carlucci Aiello, L., Cialdea, M., Nardi, D.: A meta level abstract description of diagnosis in Intelligent Tutoring Systems. In: Proceedings of the Sixth International PEG Conference, PEG-91. (1991) 437–442
31. Carlucci Aiello, L., Cialdea, M., Nardi, D.: Reasoning about Student Knowledge and Reasoning. Journal of Artificial Intelligence and Education 4 (1993) 397–413
32. Damásio, C., Nejdl, W., Pereira, L.M., Schroeder, M.: Model-based diagnosis preferences and strategies representation with logic meta-programming. In Apt, K., Turini, F., eds.: Meta-Logics and Logic Programming. The MIT Press, Cambridge, Mass. (1995) 267–308
33. Barklund, J., Costantini, S., Dell'Acqua, P., Lanzarone, G.A.: Reflection Principles in Computational Logic. Journal of Logic and Computation 10 (2000)
34. Barklund, J.: What is a meta-variable in Prolog? In Abramson, H., Rogers, M.H., eds.: Meta-Programming in Logic Programming. The MIT Press, Cambridge, Mass. (1989) 383–98
35. Hill, P.M., Lloyd, J.W.: Analysis of metaprograms. In Abramson, H., Rogers, M.H., eds.: Meta-Programming in Logic Programming, Cambridge, Mass., THE MIT Press (1988) 23–51
36. Barklund, J., Costantini, S., Dell'Acqua, P., Lanzarone, G.A.: Semantical properties of encodings in logic programming. In Lloyd, J.W., ed.: Logic Programming – Proc. 1995 Intl. Symp., Cambridge, Mass., MIT Press (1995) 288–302
37. van Harmelen, F.: Definable naming relations in meta-level systems. In Pettorossi, A., ed.: Meta-Programming in Logic. LNCS 649, Berlin, Springer-Verlag (1992) 89–104
38. Cervesato, I., Rossi, G.: Logic meta-programming facilities in 'Log. In Pettorossi, A., ed.: Meta-Programming in Logic. LNCS 649, Berlin, Springer-Verlag (1992) 148–161
39. Costantini, S.: Semantics of a metalogic programming language. Intl. Journal of Foundation of Computer Science 1 (1990)
40. Perlis, D.: Languages with self-reference I: foundations (or: we can have everything in first-order logic!). Artificial Intelligence 25 (1985) 301–322
41. Perlis, D.: Languages with self-reference II. Artificial Intelligence 34 (1988) 179–212
42. Konolige, K.: Reasoning by introspection. In Maes, P., Nardi, D., eds.: Meta-Level Architectures and Reflection. North-Holland, Amsterdam (1988) 61–74
43. Genesereth, M.R.: Introspective fidelity. In Maes, P., Nardi, D., eds.: Meta-Level Architectures and Reflection. North-Holland, Amsterdam (1988) 75–86
44. van Harmelen, F., Wielinga, B., Bredeweg, B., Schreiber, G., Karbach, W., Reinders, M., Voss, A., Akkermans, H., Bartsch-Spörl, B., Vinkhuyzen, E.: Knowledge-level reflection. In: Enhancing the Knowledge Engineering Process – Contributions from ESPRIT. Elsevier Science, Amsterdam, The Netherlands (1992) 175–204
45. Carlucci Aiello, L., Weyhrauch, R.W.: Using Meta-theoretic Reasoning to do Algebra. Volume 87 of Lecture Notes in Computer Science., Springer Verlag (1980) 1–13
46. Bowen, K.A., Kowalski, R.A.: Amalgamating language and metalanguage in logic programming. In Clark, K.L., Tärnlund, S.Å., eds.: Logic Programming. Academic Press, London (1982) 153–172
47. McCarthy, J.e.a.: (The LISP 1.5 Programmer's Manual)

48. Levi, G., Ramundo, D.: A formalization of metaprogramming for real. In Warren, D.S., ed.: Logic Programming - Procs. of the Tenth International Conference, Cambridge, Mass., The MIT Press (1993) 354–373

49. Subrahmanian, V.S.: Foundations of metalogic programming. In Abramson, H., Rogers, M.H., eds.: Meta-Programming in Logic Programming, Cambridge, Mass., The MIT Press (1988) 1–14

50. Martens, B., De Schreye, D.: Why untyped nonground metaprogramming is not (much of) a problem. J. Logic Programming **22** (1995)

51. Sterling, L., Shapiro, E.Y., eds.: The Art of Prolog. The MIT Press, Cambridge, Mass. (1986)

52. Kowalski, R.A.: Meta matters. invited presentation at Second Workshop on Meta-Programming in Logic META90 (1990)

53. Kowalski, R.A.: Problems and promises of computational logic. In Lloyd, J.W., ed.: Computational Logic. Springer-Verlag, Berlin (1990) 1–36

54. Smith, B.C.: Reflection and semantics in Lisp. Technical report, Xerox Parc ISL-5, Palo Alto (CA) (1984)

55. Lemmens, I., Braspenning, P.: A formal analysis of smithinsonian computational reflection. (In Cointe, P., ed.: Proc. Reflection '99) 135–137

56. Casaschi, G., Costantini, S., Lanzarone, G.A.: Realizzazione di un interprete riflessivo per clausole di Horn. In Mello, P., ed.: Gulp89, Proc. 4th Italian National Symp. on Logic Programming, Bologna (1989 (in italian)) 227–241

57. Friedman, D.P., Sobel, J.M.: An introduction to reflection-oriented programming. In Kiczales, G., ed.: Meta-Level Architectures and Reflection, Proc. Of the First Intnl. Conf. Reflection 96, Xerox PARC (1996)

58. Attardi, G., Simi, M.: Meta–level reasoning across viewpoints. In O'Shea, T., ed.: Proc. European Conf. on Artificial Intelligence, Amsterdam, North-Holland (1984) 315–325

59. Hill, P.M., Lloyd, J.W.: The Gödel Programming Language. The MIT Press, Cambridge, Mass. (1994)

60. Bowers, A.F., Gurr, C.: Towards fast and declarative meta-programming. In Apt, K.R., Turini, F., eds.: Meta-Logics and Logic Programming. The MIT Press, Cambridge, Mass. (1995) 137–166

61. Giunchiglia, F., Cimatti, A.: Introspective metatheoretic reasoning. In Fribourg, L., Turini, F., eds.: Logic Program Synthesis and Transformation – Meta-Programming in Logic. LNCS 883 (1994) 425–439

62. Giunchiglia, F., Traverso, A.: A metatheory of a mechanized object theory. Artificial Intelligence **80** (1996) 197–241

63. Giunchiglia, F., Serafini, L.: Multilanguage hierarchical logics, or: how we can do without modal logics. Artificial Intelligence **65** (1994) 29–70

64. Costantini, S., Lanzarone, G.A.: A metalogic programming language. In Levi, G., Martelli, M., eds.: Proc. 6th Intl. Conf. on Logic Programming, Cambridge, Mass., The MIT Press (1989) 218–233

65. Costantini, S., Lanzarone, G.A.: A metalogic programming approach: language, semantics and applications. Int. J. of Experimental and Theoretical Artificial Intelligence **6** (1994) 239–287

66. Konolige, K.: An autoepistemic analysis of metalevel reasoning in logic programming. In Pettorossi, A., ed.: Meta-Programming in Logic. LNCS 649, Berlin, Springer-Verlag (1992)

67. Dell'Acqua, P.: Development of the interpreter for a metalogic programming language. Degree thesis, Univ. degli Studi di Milano, Milano (1989 (in italian))

68. Maes, P.: Concepts and experiments in computational reflection. In: Proc. Of OOPSLA'87. ACM SIGPLAN NOTICES (1987) 147–155
69. Kiczales, G., des Rivieres, J., Bobrow, D.G.: The Art of Meta-Object Protocol. The MIT Press (1991)
70. Malenfant, J., Lapalme, G., Vaucher, G.: Objvprolog: Metaclasses in logic. In: Proc. Of ECOOP'89, Cambridge Univ. Press (1990) 257–269
71. Malenfant, J., Lapalme, G., Vaucher, G.: Metaclasses for metaprogramming in prolog. In Bruynooghe, M., ed.: Proc. of the Second Workshop on Meta-Programming in Logic, Dept. of Comp. Sci., Katholieke Univ. Leuven (1990) 272–83
72. Stroud, R., Welch, I.: the evolution of a reflective java extension. LNCS 1616, Berlin, Springer-Verlag (1999)
73. Jiang, Y.J.: Ambivalent logic as the semantic basis of metalogic programming: I. In Van Hentenryck, P., ed.: Proc. 11th Intl. Conf. on Logic Programming, Cambridge, Mass., THE MIT Press (1994) 387–401
74. Kalsbeek, M., Jiang, Y.: A vademecum of ambivalent logic. In Apt, K., Turini, F., eds.: Meta-Logics and Logic Programming. The MIT Press, Cambridge, Mass. (1995) 27–56
75. Kalsbeek, M.: Correctness of the vanilla meta-interpreter and ambivalent syntax. In Apt, K., Turini, F., eds.: Meta-Logics and Logic Programming. The MIT Press, Cambridge, Mass. (1995) 3–26
76. Christiansen, H.: A complete resolution principle for logical meta-programming languages. In Pettorossi, A., ed.: Meta-Programming in Logic. LNCS 649, Berlin, Springer-Verlag (1992) 205–234
77. Christiansen, H.: Efficient and complete demo predicates for definite clause languages. Datalogiske Skrifter, Technical Report 51, Dept. of Computer Science, Roskilde University (1994)
78. Brogi, A., Mancarella, P., Pedreschi, D., Turini, F.: Composition operators for logic theories. In Lloyd, J.W., ed.: Computational Logic. Springer-Verlag, Berlin (1990) 117–134
79. Brogi, A., Contiero, S.: Composing logic programs by meta-programming in Gödel. In Apt, K., Turini, F., eds.: Meta-Logics and Logic Programming. The MIT Press, Cambridge, Mass. (1995) 167–194
80. Brogi, A., Turini, F.: Meta-logic for program composition: Semantic issues. In Apt, K., Turini, F., eds.: Meta-Logics and Logic Programming. The MIT Press, Cambridge, Mass. (1995) 83–110
81. Barklund, J., Boberg, K., Dell'Acqua, P.: A basis for a multilevel metalogic programming language. In Fribourg, L., Turini, F., eds.: Logic Program Synthesis and Transformation – Meta-Programming in Logic. LNCS 883, Berlin, Springer-Verlag (1994) 262–275
82. Barklund, J., Boberg, K., Dell'Acqua, P., Veanes, M.: Meta-programming with theory systems. In Apt, K., Turini, F., eds.: Meta-Logics and Logic Programming. The MIT Press, Cambridge, Mass. (1995) 195–224
83. Shoham, Y., McDermott, D.: Temporal reasoning. In Encyclopedia of Artificial Intelligence (ed. Shapiro, S. C.) pp. 967–981, 1987.
84. Kowalski, R.A., Sergot, M.: A logic-based calculus of events. New Generation Computing 4 (1986) 67–95
85. McCarthy, J., Hayes, P.: Some philosophical problems from the standpoint of artificial intelligence. Machine Intelligence 4 (1969) 463–502
86. Kowalski, R.A.: Database updates in the event calculus. J. Logic Programming (1992) 121–146

87. Kowalski, R.A., Sadri, F.: The situation calculus and event calculus compared. In: Proc. 1994 Intl. Logic Programming Symp. (1994) 539–553

88. Kowalski, R.A., Sadri, F.: Reconciling the event calculus with the situation calculus. J. Logic Programming **31** (1997) 39–58

89. Provetti, A.: Hypothetical reasoning: From situation calculus to event calculus. Computational Intelligence Journal **12** (1996) 478–498

90. Díaz, O., Paton, N.: Stimuli and business policies as modeling constructs: their definition and validation through the event calculus. In: Proc. of CAiSE'97. (1997) 33–46

91. Sripada, S.: Efficient implementation of the event calculus for temporal database applications. In Lloyd, J.W., ed.: Proc. 12th Intl. Conf. on Logic Programming, Cambridge, Mass., The MIT Press (1995) 99–113

92. Pfenning, F.: The practice of logical frameworks. In Kirchner, H., ed.: Trees in Algebra and Programming - CAAP '96. LNCS 1059, Linkoping, Sweden, Springer–Verlag (1996) 119–134

93. Clavel, M.G., Eker, S., Lincoln, P., Meseguer, J.: Principles of Maude. In Proc. First Intl Workshop on Rewriting Logic, volume 4 of Electronic Notes in Th. Comp. Sc. (ed. Meseguer, J.), 1996.

94. Clavel, M.G., Duran, F., Eker, S., Lincoln, P., Marti-Oliet, N., Meseguer, J., Quesada, J.: Maude as a metalanguage. In Proc. Second Intl. Workshop on Rewriting Logic, volume 15 of Electronic Notes in Th. Comp. Sc., 1998.

95. Clavel, M.G., Meseguer, J.: Axiomatizing reflective logics and languages. In Kiczales, G., ed.: Proc. Reflection '96, Xerox PARC (1996) 263–288

96. Costantini, S., Lanzarone, G.A., Sbarbaro, L.: A formal definition and a sound implementation of analogical reasoning in logic programming. Annals of Mathematics and Artificial Intelligence **14** (1995) 17–36

97. Costantini, S., Dell'Acqua, P., Lanzarone, G.A.: Reflective agents in metalogic programming. In Pettorossi, A., ed.: Meta-Programming in Logic. LNCS 649, Berlin, Springer-Verlag (1992) 135–147

98. Martin, D.L., Cheyer, A.J., Moran, D.B.: The open agent architecture: a framework for building distributed software systems. Applied Artificial Intelligence **13(1–2)** (1999) 91–128

99. Rao, A.S., Georgeff, M.P.: Modeling rational agents within a BDI-architecture. In Fikes, R., Sandewall, E., eds.: Proceedings of Knowledge Representation and Reasoning (KR&R-91), Morgan Kaufmann Publishers: San Mateo, CA (1991) 473–484

100. Rao, A.S., Georgeff, M.: BDI Agents: from theory to practice. In: Proceedings of the First International Conference on Multi-Agent Systems (ICMAS-95), San Francisco, CA (1995) 312–319

101. J., D., Subrahmanian, V., Pick, G.: Meta-agent programs. J. Logic Programming **45** (2000)

102. Kim, J.S., Kowalski, R.A.: An application of amalgamated logic to multi-agent belief. In Bruynooghe, M., ed.: Proc. of the Second Workshop on Meta-Programming in Logic, Dept. of Comp. Sci., Katholieke Univ. Leuven (1990) 272–83

103. Kim, J.S., Kowalski, R.A.: A metalogic programming approach to multi-agent knowledge and belief. In Lifschitz, V., ed.: Artificial Intelligence and Mathematical Theory of Computation, Academic Press (1991)

104. Kowalski, R.A., Sadri, F.: Towards a unified agent architecture that combines rationality with reactivity. In: Proc. International Workshop on Logic in Databases. LNCS 1154, Berlin, Springer-Verlag (1996)

105. Kowalski, R.A., Sadri, F.: From logic programming towards multi-agent systems. In Annals of Mathematics and Artificial Intelligence, Vol. 25, pp. 391–410, 1999.
106. Dell'Acqua, P., Sadri, F., Toni, F.: Combining introspection and communication with rationality and reactivity in agents. In Dix, J., Cerro, F.D., Furbach, U., eds.: Logics in Artificial Intelligence. LNCS 1489, Berlin, Springer-Verlag (1998)
107. Fung, T.H., R. A. Kowalski, R.A.: The IFF proof procedure for abductive logic programming. J. Logic Programming **33** (1997) 151–165
108. Dell'Acqua, P., Sadri, F., Toni, F.: Communicating agents. In: Proc. International Workshop on Multi-Agent Systems in Logic Programming, in conjunction with ICLP'99, Las Cruces, New Mexico (1999)
109. Costantini, S.:   Towards active logic programming.   In Brogi, A., Hill, P., eds.: Proc. of 2nd International Workshop on Component-based Software Development in Computational Logic (COCL'99). PLI'99, Paris, France, http://www.di.unipi.it/ brogi/ ResearchActivity/COCL99/ proceedings/index.html (1999)
110. Gärdenfors, P.:   Belief revision: a vademecum.   In Pettorossi, A., ed.: Meta-Programming in Logic. LNCS 649, Berlin, Springer-Verlag (1992) 135–147
111. Gärdenfors, P., Roth, H.: Belief revision. In Gabbay, D., Hogger, C., Robinson, J., eds.: Handbook of Logic in Artificial Intelligence and Logic Programming. Volume 4. Clarendon Press (1995) 36–119
112. Dell'Acqua, P., Pereira, L.M.: Updating agents. (1999)
113. Lamma, E., Riguzzi, F., Pereira, L.M.: Agents learning in a three-valued logical setting. In Panayiotopoulos, A., ed.: Workshop on Machine Learning and Intelligent Agents, in conjunction with Machine Learning and Applications, Advanced Course on Artificial Intelligence (ACAI'99), Chania (Greece) (1999) (Also available at http://centria.di.fct.unl.pt/~lmp/).
114. Brewka, G.:   Declarative representation of revision strategies.   In Baral, C., Truszczynski, M., eds.: NMR'2000, Proc. Of the 8th Intl. Workshop on Non-Monotonic Reasoning. (2000)
115. McCarthy, J.: First order theories of individual concepts and propositions. Machine Intelligence **9** (1979) 129–147
116. Lloyd, J.W.: Foundations of Logic Programming, Second Edition. Springer-Verlag, Berlin (1987)
117. Dell'Acqua, P.: Reflection principles in computational logic. PhD Thesis, Uppsala University, Uppsala (1998)
118. Dell'Acqua, P.: SLD–Resolution with reflection. PhL Thesis, Uppsala University, Uppsala (1995)
119. Jaffar, J., Lassez, J.L., Maher, M.J.: A theory of complete logic programs with equality. J. Logic Programming **3** (1984) 211–223
120. Martens, B., De Schreye, D.: Two semantics for definite meta-programs, using the non-ground representation. In Apt, K., Turini, F., eds.: Meta-Logics and Logic Programming. The MIT Press, Cambridge, Mass. (1995) 57–82
121. Falaschi, M.and Levi, G., Martelli, M., Palamidessi, C.: A new declarative semantics for logic languages. In Kowalski, R. A.and Bowen, K.A., ed.: Proc. 5th Intl. Conf. Symp. on Logic Programming, Cambridge, Mass., MIT Press (1988) 993–1005

Argumentation-Based Proof Procedures for Credulous and Sceptical Non-monotonic Reasoning

Phan Minh Dung[1], Paolo Mancarella[2], and Francesca Toni[3]

[1] Division of Computer Science, Asian Institute of Technology, GPO Box 2754,
Bangkok 10501, Thailand
dung@cs.ait.ac.th
[2] Dipartimento di Informatica, Università di Pisa, Corso Italia 40,
I-56125 Pisa, Italy
p.mancarella@di.unipi.it
[3] Department of Computing, Imperial College of Science, Technology and Medicine,
180 Queen's Gate, London SW7 2BZ, U.K.
ft@doc.ic.ac.uk

Abstract. We define abstract proof procedures for performing credulous and sceptical non-monotonic reasoning, with respect to the argumentation-theoretic formulation of non-monotonic reasoning proposed in
[1]. Appropriate instances of the proposed proof procedures provide concrete proof procedures for concrete formalisms for non-monotonic reasoning, for example logic programming with negation as failure and default
logic. We propose (credulous and sceptical) proof procedures under different argumentation-theoretic semantics, namely the conventional stable
model semantics and the more liberal partial stable model or preferred
extension semantics. We study the relationships between proof procedures for different semantics, and argue that, in many meaningful cases,
the (simpler) proof procedures for reasoning under the preferred extension semantics can be used as sound and complete procedures for reasoning under the stable model semantics. In many meaningful cases still,
proof procedures for credulous reasoning under the preferred extension
semantics can be used as (much simpler) sound and complete procedures
for sceptical reasoning under the preferred extension semantics. We compare the proposed proof procedures with existing proof procedures in the
literature.

1 Introduction

In recent years argumentation [1,3,4,6,12,15,21,23,24,29,30,32] has played an important role in understanding many non-monotonic formalisms and their semantics, such as logic programming with negation as failure, default logic and
autoepistemic logic. In particular, Eshghi and Kowalski [9] have given an interpretation of negation as failure in Logic Programming as a form of assumption
based reasoning (abduction). Continuing this line of work, Dung [5] has given

A.C. Kakas, F. Sadri (Eds.): Computat. Logic (Kowalski Festschrift), LNAI 2408, pp. 289–310, 2002.
© Springer-Verlag Berlin Heidelberg 2002

a declarative understanding of this assumption based view, by formalizing the concept that an assumption can be safely accepted if "there is no evidence to the contrary". It has also been shown in [5] that the assumption based view provides a unifying framework for different semantics of logic programming. Later, this view has been further put forward [1,6,12] by the introduction the notions of attack and counterattacks between sets of assumptions, finally leading to an argumentation-theoretic understanding of the semantics of logic programming and nonmonotonic reasoning. In particular, Dung [6] has introduced an abstract framework of argumentation, that consists of a set of arguments and an attack relation between them. However, this abstract framework leaves open the question of how the arguments and their attack relationship are defined. Addressing this issue, Bondarenko et al. [1] has defined an abstract, argumentation-theoretic assumption-based framework to non-monotonic reasoning that can be instantiated to capture many of the existing approaches to non-monotonic reasoning, namely logic programming with negation as failure, default logic [25], (many cases of) circumscription [16], theorist [22], autoepistemic logic [18] and non-monotonic modal logics [17]. The semantics of argumentation can be used to characterize a number of alternative semantics for non-monotonic reasoning, each of which can be the basis for credulous and sceptical reasoning. In particular, three semantics have been proposed in [1,6] generalizing, respectively, the semantics of admissible scenaria for logic programming [5], the semantics of preferred extensions [5] or partial stable models [26] for logic programming, and the conventional semantics of stable models [10] for logic programming as well as the standard semantics of theorist [22], circumscription [16], default logic [25], autoepistemic logic [18] and non-monotonic modal logic [17].

More in detail, Bondarenko et al. understand non-monotonic reasoning as extending theories in some monotonic language by means of sets of assumptions, provided they are "appropriate" with respect to some requirements. These are expressed in argumentation-theoretic terms, as follows. According to the semantics of admissible extensions, a set of assumptions is deemed "appropriate" iff it does not attack itself and it attacks all sets of assumptions which attack it. According to the semantics of preferred extensions, a set of assumptions is deemed "appropriate" iff it is maximally admissible, with respect to set inclusion. According to the semantics of stable extensions, a set of assumptions is deemed "appropriate" iff it does not attack itself and it attacks every assumption which it does not belong.

Given any such semantics of extensions, credulous and sceptical non-monotonic reasoning can be defined, as follows. A given sentence in the underlying monotonic language is a credulous non-monotonic consequence of a theory iff it holds in some extension of the theory that is deemed "appropriate" by the chosen semantics. It is a sceptical non-monotonic consequence iff it holds in all extensions of the theory that are deemed "appropriate" by the chosen semantics.

In this paper we propose abstract proof procedures for performing credulous and sceptical reasoning under the three semantics of admissible, preferred and stable extensions, concentrating on the special class of *flat* frameworks. This class includes logic programming with negation as failure and default logic.

We define all proof procedures parametrically with respect to a proof procedure computing the semantics of admissible extensions. A number of such procedures have been proposed in the literature, e.g. [9,5,7,8,15].

We argue that the proof procedures for reasoning under the preferred extension semantics are "simpler" than those for reasoning under the stable extension semantics. This is an interesting argument in that, in many meaningful cases (e.g. when the frameworks are order-consistent [1]), the proof procedures for reasoning under the preferred extension semantics can be used as sound and complete procedures for reasoning under the stable model semantics.

The paper is organized as follows. Section 2 summarises the main features of the approach in [1]. Section 3 gives some preliminary definitions, used later on in the paper to define the proof procedures. Sections 4 and 5 describe the proof procedures for performing credulous reasoning under the preferred and stable extension semantics, respectively. Sections 6 and 7 describe the proof procedures for performing sceptical reasoning under the stable and preferred extension semantics, respectively. Section 8 compares the proposed proof procedures with existing proof procedures proposed in the literature. Section 9 concludes.

2   Argumentation-Based Semantics

In this section we briefly review the notion of assumption-based framework [1], showing how it can be used to extend any deductive system for a monotonic logic to a non-monotonic logic.

A *deductive system* is a pair $(\mathcal{L}, \mathcal{R})$ where

- $\mathcal{L}$ is a formal language consisting of countably many sentences, and
- $\mathcal{R}$ is a set of inference rules of the form

$$\frac{\alpha_1, \ldots, \alpha_n}{\alpha}$$

where $\alpha, \alpha_1, \ldots, \alpha_n \in \mathcal{L}$ and $n \geq 0$. If $n = 0$, then the inference rule is an axiom.

A set of sentences $T \subseteq \mathcal{L}$ is called a *theory*.

A *deduction* from a theory T is a sequence $\beta_1, \ldots, \beta_m$, where $m > 0$, such that, for all $i = 1, \ldots, m$,

- $\beta_i \in T$, or
- there exists $\dfrac{\alpha_1, \ldots, \alpha_n}{\beta_i}$ in $\mathcal{R}$ such that $\alpha_1, \ldots, \alpha_n \in \{\beta_1, \ldots, \beta_{i-1}\}$.

$T \vdash \alpha$ means that there is a deduction (of α) from T whose last element is α. $Th(T)$ is the set $\{\alpha \in \mathcal{L} \mid T \vdash \alpha\}$. Deductive systems are *monotonic*, in the sense that $T \subseteq T'$ implies $Th(T) \subseteq Th(T')$. They are also *compact*, in the sense that $T \vdash \alpha$ implies $T' \vdash \alpha$ for some finite subset T' of T.

Given a deductive system $(\mathcal{L}, \mathcal{R})$, an *argumentation-theoretic framework* with respect to $(\mathcal{L}, \mathcal{R})$ is a tuple $\langle T, Ab, \overline{} \rangle$ where

- T, $Ab \subseteq \mathcal{L}$, $Ab \neq \{\}$
- $\overline{}$ is a mapping from Ab into $\mathcal{L}$. $\overline{\alpha}$ is called the *contrary* of α.

The theory T can be viewed as a given set of beliefs, and Ab as a set of candidate *assumptions* that can be used to extend T. An *extension* of a theory T is a theory $Th(T \cup \Delta)$, for some $\Delta \subseteq Ab$. Sometimes, informally, we refer to the extension simply as $T \cup \Delta$ or Δ.

Given a deductive system $(\mathcal{L}, \mathcal{R})$ and an argumentation-theoretic framework $\langle T, Ab, \overline{} \rangle$ with respect to $(\mathcal{L}, \mathcal{R})$, the problem of determining whether a given sentence σ in $\mathcal{L}$ is a *non-monotonic consequence* of the framework is understood as the problem of determining whether there exist "appropriate" extensions $\Delta \subseteq Ab$ of T such that $T \cup \Delta \vdash \sigma$. In particular, σ is a *credulous* non-monotonic consequence of $\langle T, Ab, \overline{} \rangle$ if there exists *some* "appropriate" extension of T. Many logics for default reasoning are credulous in this same sense, differing however in the way they understand what it means for an extension to be "appropriate". Some logics, in contrast, are *sceptical*, in the sense they they require that σ belong to *all* "appropriate" extensions. However, the semantics of any of these logics can be made sceptical or credulous, simply by varying whether a sentence is deemed to be a non-monotonic consequence of a theory if it belongs to all "appropriate" extensions or if it belongs to some "appropriate" extension.

A number of notions of "appropriate" extensions are given in [1], for any argumentation-theoretic framework $\langle T, Ab, \overline{} \rangle$ with respect to $(\mathcal{L}, \mathcal{R})$. All these notions are formulated in argumentation-theoretic terms, with respect to a notion of "attack" defined as follows. Given a set of assumptions $\Delta \subseteq Ab$:

- Δ *attacks* an assumption $\alpha \in Ab$ iff $T \cup \Delta \vdash \overline{\alpha}$
- Δ *attacks* a set of assumptions $\Delta' \subseteq Ab$ iff Δ attacks an assumption α, for some $\alpha \in \Delta'$.

In this paper we will consider the notions of "stable", "admissible" and "preferred" extensions, defined below.

Let a set of assumptions $\Delta \subseteq Ab$ be *closed* iff $\Delta = \{\alpha \in Ab \mid T \cup \Delta \vdash \alpha\}$. Then, $\Delta \subseteq Ab$ is *stable* if and only if

1. Δ is closed,
2. Δ does not attack itself, and
3. Δ attacks α, for every assumption $\alpha \notin \Delta$.

Furthermore, $\Delta \subseteq Ab$ is *admissible* if and only if

1. Δ is closed,
2. Δ does not attack itself, and
3. for each closed set of assumptions $\Delta' \subseteq Ab$,
   if Δ' attacks Δ then Δ attacks Δ'.

Finally, $\Delta \subseteq Ab$ is *preferred* if and only if Δ is maximally admissible, with respect to set inclusion.

In general, every admissible extension is contained in some preferred extension. Moreover, every stable extension is preferred (and thus admissible) [1] but not vice versa. However, in many cases, e.g. for *stratified* and *order-consistent* argumentation-theoretic frameworks (see [1]), preferred extensions are always stable[1].

In this paper we concentrate on *flat* frameworks [1], namely frameworks in which every set of assumptions $\Delta \subseteq Ab$ is closed. For this kind of frameworks, the definitions of admissible and stable extensions can be simplified by dropping condition 1 and by dropping the requirement that Δ' be closed in condition 3 of the definition of admissible extension. In general, if the framework is flat, both admissible and preferred extensions are guaranteed to exist. Instead, even for flat frameworks, stable extensions are not guaranteed to exist. However, in many cases, e.g. for *stratified* argumentation-theoretic frameworks [1], stable extensions are always guaranteed to exist.

Different logics for default reasoning differ, not only in whether they are credulous or sceptical and how they interpret the notion of what it means to be an "appropriate" extension, but also in their underlying framework.

Bondarenko et al. [1] show how the framework can be instantiated to obtain theorist [22], (some cases of) circumscription [16], autoepistemic logic [18], non-monotonic modal logics [17], default logic [25], and logic programming, with respect to, e.g., the semantics of stable models [10] and partial stable models [26], the latter being equivalent [13] to the semantics of preferred extensions [5]. They also prove that the instances of the framework for default logic and logic programming are flat.

Default logic is the instance of the abstract framework $\langle T, Ab, \frown \rangle$ where the $\vdash$ is first-order logic augmented with domain-specific inference rules of the form

$$\frac{\alpha_1, \dots, \alpha_m, M\beta_1, \dots, M\beta_n}{\gamma}$$

where α_i, β_j, γ are sentences in classical logic. T is a classical theory and Ab consists of all expressions of the form $M\beta$ where β is a sentence of classical logic. The contrary $\overline{M\beta}$ of an assumption $M\beta$ is $\neg\beta$. The conventional semantics of extensions of default logic [25] corresponds to the semantics of stable extensions of the instance of the abstract framework for default logic [1]. Moreover, default logic inherits the semantics of admissible and preferred extensions, simply by being an instance of the framework.

Logic programming is the instance of the abstract framework $\langle T, Ab, \frown \rangle$ where T is a logic program, the assumptions in Ab are all negations $not\,p$ of atomic sentences p, and the contrary $\overline{not\,p}$ of an assumption is p. $\vdash$ is Horn logic provability, with assumptions, $not\,p$, understood as new atoms, p^* (see [9]). The logic programming semantics of stable models [10], admissible scenaria [5], and partial stable models [26]/preferred extensions [5] correspond to the semantics of stable, admissible and preferred extensions, respectively, of the instance of the abstract framework for logic programming [1].

---

[1] See the Appendix for the definition of stratified and order-consistent frameworks.

In the remainder of the paper we will concentrate on computing credulous and sceptical consequences under the semantics of preferred and stable extensions. We will rely upon a proof procedure for computing credulous consequences under the semantics of admissible extensions (see Sect. 8 for a review of such procedures). Note that we ignore the problem of computing sceptical consequences under the semantics of admissible extensions as, for flat frameworks, this problem reduces to that of computing monotonic consequences in the underlying deductive system. Indeed, in flat frameworks, the empty set of assumptions is always admissible.

We will propose abstract proof procedures, but, for simplicity, we will illustrate their behaviour within the concrete instance of the abstract framework for logic programming.

3   Preliminaries

In the sequel we assume that a framework is given and we omit mentioning it explicitly if clear by the context.

Let S be a set of sets. A subset B of S is called a **base** of S if for each element s in S there is an element b in B such that $b \subseteq s$.

We assume that the following procedures are defined, where α is a sentence in $\mathcal{L}$ and $\Delta \subseteq Ab$ is a set of assumptions:

- $support(\alpha, \Delta)$ computes a set of sets $\Delta' \subseteq Ab$ such that $\alpha \in Th(T \cup \Delta')$ and $\Delta' \supseteq \Delta$.
  $support(\alpha, \Delta)$ is said to be **complete** if it is a base of the set $\{\Delta' \subseteq Ab | \alpha \in Th(T \cup \Delta')$ and $\Delta' \supseteq \Delta\}$.
- $adm\_expand(\Delta)$ computes a set of sets $\Delta' \subseteq Ab$ such that $\Delta' \supseteq \Delta$ and Δ' is admissible.
  $adm\_expand(\Delta)$ is said to be **complete** if it is a base of the set of all admissible supersets of Δ.

We will assume that the above procedures are nondeterministic. We will write, e.g.

$$A := support(\alpha, \Delta)$$

meaning that the variable A is assigned, if any, a result of the procedure $support$. Such a statement represents a backtracking point, which may eventually fail if no further result can be produced by $support$.

The following example illustrates the above procedures.

Example 1. Consider the following logic program

$$p \leftarrow q, not\ r$$
$$q \leftarrow not\ s$$
$$t \leftarrow not\ h$$
$$f$$

and the sentence p. Possible outcomes of the procedure $support(p, \{\})$ are $\Delta_1 = \{not\,s, not\,r\}$ and $\Delta_2 = \{not\,s, not\,r, not\,f\}$. Possible outcomes of the procedure $adm\_expand(\Delta_1)$ are Δ_1 and $\Delta_1 \cup \{not\,h\}$. No possible outcomes exist for $adm\_expand(\Delta_2)$.

Note that different implementations for the above procedures are possible. In all examples in the remainder of the paper we will assume that $support$ and adm_expand return minimal sets. In the above example, Δ_1 is a minimal support whereas Δ_2 is not, and Δ_1 is a minimal admissible expansion of Δ_1 whereas $\Delta_1 \cup \{not\,h\}$ is not.

4   Computing Credulous Consequences under Preferred Extensions

To show that a sentence is a credulous consequence under the preferred extension semantics, we simply need to check the existence of an admissible set of assumptions which entails the desired sentence. This can be done by:

- finding a support set for the sentence
- showing that the support set can be extended into an admissible extension.

Proof procedure 4.1 (Credulous Preferred Extensions).

$CPE(\alpha)$:
$\qquad S := support(\alpha, \{\})$;
$\qquad \Delta := adm\_expand(S)$;
$\qquad$ **return** Δ

Notice that the two assignments in the procedure are backtracking points, due to the nondeterministic nature of both $support$ and adm_expand.

Example 2. Consider the following logic program

$$p \leftarrow not\,s$$
$$s \leftarrow q$$
$$q \leftarrow not\,r$$
$$r \leftarrow not\,q$$

and the sentence p. The procedure $CPE(p)$ will perform the following steps:

- first the set $S = \{not\,s\}$ is generated by $support(p, \{\})$
- then the set $\Delta = \{not\,s, not\,q\}$ is generated by $adm\_expand(S)$
- finally, Δ is the set returned by the procedure

Consider now the conjunction p, q. The procedure $CPE((p, q))^2$ would fail, since

- $S = \{not\,s, not\,r\}$ is generated by $support((p, q), \{\})$
- there exists no admissible set $\Delta \supseteq S$.

---

[2] Note that, in the instance of the framework of [1] for logic programming, conjunction of atoms are not part of the underlying deductive system. However, conjunctions can be accommodated by additional program clauses. E.g., in the given example, the logic program can be extended by $t \leftarrow p, q$, and CPE can be called for t.

Theorem 1 (Soundness and Completeness of CPE).

1. *If $CPE(\alpha)$ succeeds then there exists a preferred extension Δ such that $\alpha \in Th(T \cup \Delta)$.*
2. *If both support and adm_expand are complete then for each preferred extension E there exist appropriate selections such that $CPE(\alpha)$ returns $\Delta \subseteq E$.*

Proof.

1. It follows immediately from the fact that each admissible set of assumptions could be extended into a preferred extension.
2. Let E be a preferred extension such that $\alpha \in Th(T \cup E)$. Since $support(\alpha, \{\})$ is complete, there is a set $S \subset E$ such that S could be computed by $support(\alpha, \{\})$. From the completeness of adm_expand, it follows that there is $\Delta \subseteq E$ such that Δ is computed by $adm\_expand(S)$.   □

5   Computing Credulous Consequences under Stable Extensions

A stable model is nothing but a preferred extension which entails either α or its contrary, for each assumption α [1]. Hence, to show that a sentence is a credulous consequence under the stable model semantics, we simply need to find an admissible extension which entails the sentence and which can be extended into a stable model.

We assume that the following procedures are defined:

- $full\_cover(\Gamma)$ returns *true* iff the set of sentences Γ entails any assumption or its contrary, *false* otherwise;
- $uncovered(\Gamma)$ nondeterministically returns, if any, an assumption which is undefined, given Γ, i.e. neither the assumption nor its contrary is entailed by Γ.

In the following procedure CSM, both $full\_cover$ and $uncovered$ will be applied to sets of assumptions only.

Proof procedure 5.1 (Credulous Stable Models).

```
CSM(α):
    Δ := CPE(α);
    loop
        if full_cover(Δ)
            then  return Δ
            else  β := uncovered(Δ)
                  Δ := adm_expand(Δ ∪ {β});
        end if
    end loop
```

Note that CSM is a non-trivial extension of CPE: once an admissible extension is selected, as in CPE, CSM needs to further expand the selected admissible extension, if possible, to render it stable. This is achieved by the main loop in the procedure.

Clearly, the above procedure may not terminate if the underlying framework $\langle T, Ab, \frown \rangle$ contains infinitely many assumptions, since in this case the main loop may go on forever. In the following theorem we assume that the set of assumptions Ab is finite.

Theorem 2 (Soundness and Completeness of CSM).
Let $\langle T, Ab, \frown \rangle$ be a framework such that Ab is finite.

1. *If $CSM(\alpha)$ succeeds then there exists a stable extension Δ such that $\alpha \in Th(T \cup \Delta)$.*
2. *If both support and adm_expand are complete then for each stable extension E such that $\alpha \in Th(T \cup E)$ there exist appropriate selections such that $CSM(\alpha)$ returns E.*

Proof. The theorem follows directly from theorem 3. ☐

The CSM procedure is based on backward-chaining in contrast to the procedure of Niemelä et al. [19,20] that is based on forward-chaining. We explain the difference between the two procedures in the following example.

Example 3.

$$p \leftarrow not \ q$$
$$q \leftarrow not \ r$$
$$r \leftarrow not \ q$$

Assume that the given query is p. The CSM procedure would compute $\{not \ q\}$ as a support for p. The procedure $adm\_expand(\{not \ q\})$ will produce $\Delta = \{not \ q\}$ as its result. Since Δ covers all assumptions, Δ is the result produced by the procedure. Niemelä et. al procedure would start by picking an arbitrary element from $\{not \ p, not \ q, not \ r\}$ and start to apply the Fitting operator to it to get a fixpoint. For example, $not \ r$ may be selected. Then the set $B = \{q, not \ r\}$ is obtained. Since there is no conflict in B and B does not cover all the assumptions, $not \ p$ will be selected. Since $\{not \ p, q, not \ r\}$ covers all assumptions, a test to check whether p is implied from it is performed with $false$ as the result. Therefore backtracking will be made and $not \ q$ will be selected leading to the expected result.

A drawback of Niemelä et. al procedure is that it may have to make too many unnecessary choices as the above example shows. However forward chaining may help in getting closer to the solution more efficiently. The previous observations suggest a modification of the procedure which tries to combine both backward and forward chaining. This can be seen as an integration of ours and Niemelä et. al procedures. In the new procedure, $CSM2$, we make use of some additional procedures and notations:

- Given a set of sentences Γ, Γ^- denotes the set of assumptions contained in Γ.
- A set of sentences Γ is said to be **coherent** if Γ^- is admissible and $\Gamma \subseteq Th(T \cup \Gamma)$,
- Given a set of sentences Γ, $expand(\Gamma)$ defines a forward expansion of Γ satisfying the following conditions:
    1. $\Gamma \subseteq expand(\Gamma)$
    2. If Γ is coherent then
        (a) $expand(\Gamma)$ is also coherent, and
        (b) for each stable extension E, if $\Gamma^- \subseteq E$ then $expand(\Gamma)^- \subseteq E$.

Proof procedure 5.2 (Credulous Stable Models).

$CSM2(\alpha)$:
  $\Delta := CPE(\alpha)$;
  $\Gamma := expand(\Delta)$;
  loop
      if $full\_cover(\Gamma)$
          then return Γ^-
          else
              $\beta := uncovered(\Gamma)$;
              $\Delta := adm\_expand(\Gamma^- \cup \{\beta\})$;
              $\Gamma := expand(\Delta \cup \Gamma)$;
      end if
  end loop

As anticipated, the procedure $expand$ can be defined in various ways. If $expand$ is simply the identity function, i.e. $expand(\Delta) = \Delta$ the procedure $CSM2$ collapses down to CSM. In some other cases, $expand$ could also effectively perform forward reasoning, and try to produce the deductive closure of the given set of sentences. This can be achieved by defining $expand$ in such a way that

$$expand(\Delta) = Th(T \cup \Delta).$$

In still other cases, $expand(\Delta)$ could be extended to be closed under the Fitting's operator.

As in the case of Theorem 2, we need to assume that the set of assumptions in the underlying framework is finite, in order to prevent non termination of the main loop.

Theorem 3 (Soundness and Completeness of $CSM2$).
Let $\langle T, Ab, \longrightarrow \rangle$ be a framework such that Ab is finite.

1. If $CSM2(\alpha)$ succeeds then there exists a stable extension Δ such that $\alpha \in Th(T \cup \Delta)$.
2. If both CPE and adm_expand are complete then for each stable extension E such that $\alpha \in Th(T \cup E)$ there exist appropriate selections such that $CSM2(\alpha)$ returns E.

Proof.

1. We first prove by induction that at the beginning of each iteration of the loop, Γ is coherent. The basic step is clear since Δ is admissible.
   Inductive Step: Let Γ be coherent. From $\Delta := adm\_expand(\Gamma^- \cup \{\beta\})$, it follows that Δ is admissible. Because $\Gamma^- \subseteq \Delta$ and $\Gamma \subseteq Th(T \cup \Gamma)$, it follows that $\Gamma \subseteq Th(T \cup \Delta)$. From $(\Delta \cup \Gamma)^- = \Delta$, it follows that $\Delta \cup \Gamma$ is coherent. Therefore $expand(\Delta \cup \Gamma)$ is coherent.
   It is obvious that for any coherent set of sentences Γ such that $full\_cover(\Gamma)$ holds, Γ^- is stable.
2. Let E be a stable model such that $\alpha \in Th(T \cup E)$. Because CPE is complete, there is a selection such that executing the command $\Delta := CPE(\alpha)$ yields an admissible $\Delta \subseteq E$. From the properties of expand, it follows that Γ obtained from $\Gamma := expand(\Delta)$, is coherent and $\Gamma^- \subseteq E$. If $full\_cover(\Gamma)$ does not hold, then we can always select a $\beta \in E - \Gamma^-$. Therefore due to the completeness of adm_expand, we can get a Δ that is a subset of E. Hence Γ obtained from $\Gamma := expand(\Delta \cup \Gamma)$, is coherent and $\Gamma^- \subseteq E$. Continuing this process until termination, which is guaranteed by the hypothesis that Ab is finite, will return E as the result of the procedure. $\square$

However, if in the underlying framework every preferred extension is also stable, then CSM can be greatly simplified by dropping the main loop, namely CSM coincides with CPE. As shown in [1], this is the case if the underlying framework is order-consistent (see Appendix).

Theorem 4 (Soundness and completeness of CPE wrt stable models and order consistency).
Let the underlying framework be order-consistent.

1. *If $CPE(\alpha)$ succeeds then there exists a stable extension Δ such that $\alpha \in Th(T \cup \Delta)$.*
2. *If both support and adm_expand are complete then for each stable extension E there exist appropriate selections such that $CPE(\alpha)$ returns $\Delta \subseteq E$.*

The use of CPE instead of CSM, whenever possible, greatly simplifies the task of performing credulous reasoning under the stable semantics, in that it allows to keep the search for a stable extension "localised", as illustrated by the following example.

Example 4. Consider the following order-consistent logic program

$$p \leftarrow not\ s$$
$$q \leftarrow not\ r$$
$$r \leftarrow not\ q$$

which has two preferred (and stable) extensions containing p, corresponding to the sets of assumptions $\Delta_1 = \{not\ s, not\ r\}$ and $\Delta_2 = \{not\ s, not\ q\}$. The procedure $CPE(p)$ would compute the admissible extension $\{not\ s\}$ as a result, since $\{not\ s\}$ is a support for p and it is admissible (there are no attacks against $not\ s$). On the other hand, the procedure $CSM(p)$ would produce either Δ_1 or Δ_2, which are both stable sets extending $\{not\ s\}$.

6 Computing Sceptical Consequences under Stable Extensions

First, we define the notion of "contrary of sentences", by extending the notion of "contrary of assumptions". In all concrete instances of the abstract framework, e.g. logic programming, default logic, autoepistemic logic and non-monotonic modal logic, for each non-assumption sentence β there is a unique assumption α such that $\overline{\alpha} = \beta$, so the natural way of defining the "contrary of a sentence" β which is not an assumption is

$$\overline{\beta} = \alpha \text{ such that } \overline{\alpha} = \beta.$$

But in general, it is possible that for some non-assumption sentence β there may be no assumption α such that $\overline{\alpha} = \beta$, or there may be more than one assumption α such that $\overline{\alpha} = \beta$.

Thus, for general frameworks, we define the concept of **contrary of sentences** which are not assumptions as follows. Let β be a sentence such that $\beta \notin Ab$.

– if there exists α such that $\overline{\alpha} = \beta$ then $\overline{\beta} = \{\gamma | \overline{\gamma} = \beta\}$
– if there exists no α such that $\overline{\alpha} = \beta$ then we introduce a new assumption κ_β, not already in the language, and we define
  - $\overline{\kappa_\beta} = \beta$
  - $\overline{\beta} = \{\kappa_\beta\}$

Note that, in this way, the contrary of a sentence $\beta \notin Ab$ is a *set* of assumptions.

Let us denote by $Ab' \supseteq Ab$ the new set of assumptions. It is easy to see that the original framework, $\langle T, Ab, \overline{}\rangle$, and the extended framework, $\langle T, Ab', \overline{}\rangle$, are equivalent in the following sense:

– if $\Delta \subseteq Ab$ is admissible wrt the original framework then it is also admissible wrt the new framework;
– if $\Delta' \subseteq Ab'$ is admissible wrt the new framework then $\Delta' \cap Ab$ is admissible wrt the original framework.

Therefore from now on, we will assume that for each sentence β which is not an assumption there exists at least an assumption α such that $\overline{\alpha} = \beta$.

In order to show that a sentence β is entailed by each stable model, we can proceed as follows:

– check that β is a credulous consequence under the stable model semantics
– check that the contrary of the sentence is not a credulous consequence under the stable models semantics.

Notice that if $\beta \notin Ab$ the second step amounts to checking that each $\alpha \in \overline{\beta}$ is not a credulous consequence under the stable models semantics.

Moreover, notice that the first step of the computation cannot be omitted (as one could expect) since there may be cases in which neither β nor its contrary hold in any stable model (e.g. in the framework corresponding to the logic program $p \leftarrow not\ p$).

Lemma 1. *Let E be a stable extension. Then for each non-assumption β such that $\beta \notin Th(T \cup E)$, the following statements are equivalent:*

1. $\overline{\beta} \cap E \neq \emptyset$
2. $\overline{\beta} \subseteq E$

Proof. It is clear that the second condition implies the first. We need only to prove now that the first condition implies the second one. Let $\overline{\beta} \cap E \neq \emptyset$. Suppose that $\overline{\beta} - E \neq \emptyset$. Let $\alpha \in \overline{\beta} - E$. Then it is clear that $\overline{\alpha} \in Th(T \cup E)$. Contradiction to the condition that $\overline{\alpha} = \beta$ and $\beta \notin Th(T \cup E)$. $\qquad\square$

Proof procedure 6.1 (Sceptical Stable Models).

> $SSM(\alpha)$:
> > **if** $CSM(\alpha)$ fails **then** fail;
> > **select** $\beta \in \overline{\alpha}$;
> > **if** $CSM(\beta)$ succeeds **then** fail;

Notice that the SSM procedure makes use of the CSM procedure. To prevent non termination of CSM we need to assume that the set of assumptions Ab' of the underlying extended framework is finite. This guarantees the completeness of CSM (cfr. Theorem 2).

Theorem 5 (Soundness and Completeness of SSM). *Let CSM be complete.*

1. *If $SSM(\alpha)$ succeeds then $\alpha \in Th(T \cup \Delta)$, for every stable extension Δ.*
2. *If $\alpha \in Th(T \cup \Delta)$, for every stable extension Δ, and the set of stable extensions is not empty, then $SSM(\alpha)$ succeeds.*

Proof.

1. Let $SSM(\alpha)$ succeed. Assume now that α is not a skeptical consequence wrt stable semantics. There are two cases: $\alpha \in Ab$ and $\alpha \notin Ab$.
   Consider the first case where $\alpha \in Ab$. It follows that there is a stable extension E such that $\overline{\alpha} \in Th(T \cup E)$. Because of the completeness of CSM, it follows that $CSM(\overline{\alpha})$ succeeds. Hence $SSM(\overline{\alpha})$ fails, contradiction.
   Let $\alpha \notin Ab$. From lemma 1, it follows that there is a stable extension E such that $E \cap \overline{\alpha} \neq \emptyset$. That means $CSM(\beta)$ succeeds for some $\beta \in \overline{\alpha}$. Lemma 1 implies $CSM(\beta)$ succeeds for each $\beta \in \overline{\alpha}$. Hence $SMM(\alpha)$ fails. Contradiction.

2. Because CSM is complete, it is clear that $CSM(\alpha)$ succeeds. Also because of the soundness of CSM, $CSM(\beta)$ fails for each $\beta \in \overline{\alpha}$. Therefore it is obvious that SSM succeeds.   □

For a large class of argumentation frameworks, preferred extensions and stable models semantics coincide, e.g. if the frameworks are order-consistent [1]. In these frameworks, the procedure SSM can be simplified significantly as follows.

Proof procedure 6.2 (Sceptical Stable Models via CPE).

> $SSMPE(\alpha)$:
> > **if** $CPE(\alpha)$ fails **then** fail;
> > **select** $\beta \in \overline{\alpha}$;
> > **if** $CPE(\beta)$ succeeds **then** fail ;

The procedure is structurally the same as the earlier SSM, but it relies upon CPE rather than CSM, and is therefore "simpler" in the same way that CPE is "simpler" than CSM, as discussed earlier in Sect. 5.

Theorem 6 (Soundness and completeness of $SSMPE$ wrt sceptical stable semantics). *Let the underlying framework be order-consistent and CPE be complete.*

1. *If $SSMPE(\alpha)$ succeeds then $\alpha \in Th(T \cup \Delta)$, for every stable extension Δ.*
2. *If $\alpha \in Th(T \cup \Delta)$, for every stable extension Δ, then $SSMPE(\alpha)$ succeeds.*

Note that the second statement in the above theorem does not require the existence of stable extensions. This is due to the assumption that order-consistency always guarantees such condition.

7 Computing Sceptical Consequences under Preferred Extensions

The naive way of showing that a sentence is a sceptical consequence under the preferred extensions semantics is to consider each preferred extension in turn and check that the sentence is entailed by it.

The earlier procedure $SSMPE$ can be used as a simplification of the naive method only if every preferred extension is guaranteed to be stable. In general, however, the procedure $SSMPE$ is not sound under the preferred extensions semantics, since there might exist preferred extensions in which, for some assumption α, neither α nor its contrary hold, as the following example shows.

Example 5.

$$p \leftarrow not\ p$$
$$p \leftarrow q$$
$$q \leftarrow not\ r$$
$$r \leftarrow not\ q$$

Notice that there are two preferred extensions, namely $E_1 = \{not\ q, r\}$ and $E_2 = \{not\ r, q, p\}$. E_2 is also a stable extension, whereas E_1 is not since neither p nor $not\ p$ hold in E_1. Notice that $SSMPE(p)$ would succeed, hence giving an unsound result.

Nonetheless, in the general case, the following theorem shows that it is possible to restrict the number of preferred extensions to consider. This theorem is a variant of theorem 16 in [30], as we will discuss in Sect. 8.

Theorem 7. *Given an argumentation-theoretic framework $\langle T, Ab, \frown \rangle$ and a sentence α in its language, α is a sceptical non-monotonic consequence of T with respect to the preferred extension semantics, i.e. $\alpha \in Th(T \cup \Delta)$ for all preferred $\Delta \subseteq Ab$, iff*

1. $\alpha \in Th(T \cup \Delta_0)$, for some admissible set of assumptions $\Delta_0 \subseteq Ab$, and
2. for every set of assumptions $\Delta \subseteq Ab$,
   if Δ is admissible and Δ attacks Δ_0,
   then $\alpha \in Th(T \cup \Delta')$ for some set of assumptions $\Delta' \subseteq Ab$ such that
   (a) $\Delta' \supseteq \Delta$, and
   (b) Δ' is admissible.

Proof. The only if half is trivial.
The if half is proved by contradiction. Suppose there exists a set of assumptions Δ^* such that Δ^* is preferred and $\alpha \notin Th(T \cup \Delta^*)$. Suppose Δ_0 is the set of assumptions provided in part 1.
If $\Delta_0 = \emptyset$ then $\alpha \in Th(T)$ and therefore $\alpha \in Th(T \cup \Delta^*)$, thus contradicting the hypothesis.
Therefore, $\Delta_0 \neq \emptyset$. Consider the following two cases:

(i) $\Delta^* \cup \Delta_0$ attacks itself, or
(ii) $\Delta^* \cup \Delta_0$ does not attack itself.

Case (ii) implies that $\Delta^* \cup \Delta_0$ is admissible, thus contradicting the hypothesis that Δ^* is preferred (and therefore maximally admissible).
Case (i) implies that

(i.1) $\Delta^* \cup \Delta_0$ attacks Δ^*, or
(i.2) $\Delta^* \cup \Delta_0$ attacks Δ_0.

Assume that (i.1) holds. .
   $\Delta^* \cup \Delta_0$ attacks Δ^*
$\Rightarrow$ {by admissibility of Δ^*}
   Δ^* attacks $\Delta^* \cup \Delta_0$
$\Rightarrow$ {by admissibility, Δ^* does not attack itself}
   Δ^* attacks Δ_0
$\Rightarrow$ {by part 2 }
   $\alpha \in Th(T \cup \Delta^*)$
thus contradicting the hypothesis.

Assume now that (i.2) holds.

$\Delta^* \cup \Delta_0$ attacks Δ_0

$\Rightarrow$ {by admissibility of Δ_0}

Δ_0 attacks $\Delta^* \cup \Delta_0$

$\Rightarrow$ {by admissibility, Δ_0 does not attack itself}

Δ_0 attacks Δ^*

$\Rightarrow$ {by admissibility of Δ^*}

Δ^* attacks Δ_0

$\Rightarrow$ {by part 2 }

$\alpha \in Th(T \cup \Delta^*)$

thus contradicting the hypothesis.   □

This result can be used to define the following procedure to check whether or not a given sentence is a sceptical consequence with respect to the preferred extension semantics.

Let us assume the following procedure is defined

- $attacks(\Delta)$ computes a base of the set of all attacks against the set of assumptions Δ.

Proof procedure 7.1 (Sceptical Preferred Extensions).

> $SPE(\alpha)$:
> > $\Delta := CPE(\alpha)$;
> > **for each** $A := attacks(\Delta)$
> > > **for each** $\Delta' := adm\_expand(A)$
> > > > $\Delta'' := support(\alpha, \Delta')$;
> > > > **if** $adm\_expand(\Delta'')$ fails **then** fail **end if**
> > > **end for**
> > **end for**

The following soundness theorem is a trivial corollary of theorem 7.

Theorem 8 (Soundness and Completeness of SPE). *Let adm_expand be complete.*

1. *if $SPE(\alpha)$ succeeds, then $\alpha \in Th(T \cup \Delta)$, for every preferred extension Δ.*
2. *If CPE is complete and $\alpha \in Th(T \cup \Delta)$, for every preferred extension Δ, then $SPE(\alpha)$ succeeds.*

In many cases where the framework has exactly one preferred extension that is also stable (for example when the framework is stratified), it is obvious that the CPE procedure could be used as a procedure for skeptical preferred extension semantics.

8 Related Work

The proof procedures we propose in this paper rely upon proof procedures for computing credulous consequences under the semantics of admissible extensions. A number of such procedures have been proposed in the literature.

Eshghi and Kowalski [9] (see also the revised version proposed by Dung in [5]) propose a proof procedure for logic programming based upon interleaving abductive derivations, for the generation of negative literals to "derive" goals, and consistency derivations, to check "consistency" of negative literals with atoms "derivable" from the program. The proof procedure can be understood in argumentation-theoretic terms [12], as interleaving the generation of assumptions supporting goals or counter-attacking assumptions (abductive derivations) and the generation of attacks against any admissible support (consistency derivations), while checking that the generated support does not attack itself.

Dung, Kowalski and Toni [7] propose abstract proof procedures for computing credulous consequences under the semantics of admissible extensions, defined via logic programs.

Kakas and Toni [15] propose a number of proof procedures based on the construction of trees whose nodes are sets of assumptions, and such that nodes attack their parents, if any. The proof procedures are defined in abstract terms and, similarly to the procedures we propose in this paper, can be adopted for any concrete framework that is an instance of the abstract one. The procedures allow to compute credulous consequences under the semantics of admissible extensions as well as under semantics that we have not considered in this paper, namely the semantics of weakly stable extensions, acceptable extensions, well-founded extensions. The concrete procedure for computing credulous consequences under the semantics of admissible extensions, in the case of logic programming, corresponds to the proof procedure of [9].

Dung, Kowalski and Toni [8] also propose abstract proof procedures for computing credulous consequences under the semantics of admissible extensions, that can be instantiated to any instance of the framework of [1]. These procedures are defined in terms of trees whose nodes are assumptions, as well as via derivations as in [9].

Kakas and Dimopoulos [2] propose a proof procedure to compute credulous consequences under the semantics of admissible extensions for the argumentation framework of Logic Programming without Negation as Failure proposed in [14]. Here, negation as failure is replaced and extended by priorities over logic programs with no negation as failure but with explicit negation instead.

Other proof procedures for computing credulous consequences under the stable extension semantics and sceptical consequences under the semantics of preferred and stable extensions have been proposed.

Thielscher [30] proposes a proof procedure for computing sceptical consequences under the semantics of preferred extensions for the special case of logic programming [31]. This proof procedure is based upon a version of theorem 7 (theorem 16 in [30]). However, whereas [30] uses the notion of "conflict-free set of arguments" (which is an atomic, abstract notion), we use the notion of admissible set of assumptions. Moreover, theorem 16 in [30] replaces the condition in part 2 of theorem 7 "Δ' attacks Δ_0" by the (equivalent) condition corresponding to "$\Delta' \cup \Delta_0$ attacks itself". For a formal correspondence between the two approaches see [31].

Niemelä [19] and Niemelä and Simons [20] give proof procedures for computing credulous and sceptical consequences under stable extensions, for default logic and logic programming, respectively. As discussed in Sect. 5, their proof procedures for computing credulous consequences under stable extensions rely upon forward chaining, whereas the proof procedures we propose for the same task rely either on backward chaining (CSM) or on a combination of backward and forward chaining (CSM2).

Satoh and Iwayama [28] define a proof procedure for logic programming, computing credulous consequences under the stable extension semantics for range-restricted logic programs that admit at least one stable extension. Satoh [27] adapts the proof procedure in [28] to default logic. The proof procedure applies to consistent and propositional default theories.

Inoue et al. [11] apply the model generation theorem prover to logic programming to generate stable extensions, thus allowing to perform credulous reasoning under the stable extension semantics by forward chaining.

9   Conclusions

We have presented abstract proof procedures for computing credulous and sceptical consequences under the semantics of preferred and stable extensions for non-monotonic reasoning, as proposed in [1], relying upon any proof procedure for computing credulous consequences under the semantics of admissible extensions.

The proposed proof procedures are abstract in that they can be instantiated to any concrete framework for non-monotonic reasoning which is an instance of the abstract flat framework of [1]. These include logic programming and default logic. They are abstract also in that they abstract away from implementation details.

We have compared our proof procedures with existing, state of the art procedures defined for logic programming and default logic.

We have argued that the proof procedures for computing consequences under the semantics of preferred extensions are simpler than those for computing consequences under the semantics of stable extensions, and supported our arguments with examples. However, note that the (worst-case) computational complexity of the problem of computing consequences under the semantics of stable extensions is in general no worse than that of computing consequences under the semantics of preferred extensions, and in some cases it is considerably simpler [3,4]. In particular, in the case of autoepistemic logic, the problem of computing sceptical consequences under the semantics of preferred extensions is located at the fourth level of the polynomial hierarchy, whereas the same problem under the semantics of stable extensions is located at the second level.

Of course, these results do not contradict the expectation that in practice, in many cases, computing consequences under the semantics of preferred extensions is easier than under the semantics of stable extensions. Indeed, preferred extensions supporting a desired sentence can be constructed "locally", by restricting attention to the sentences in the language that are directly relevant to the sentence. Instead, stable extensions need to be constructed "globally", by considering all sentences in the language, whether they are directly relevant to the given sentence or not. This is due to the fact that stable extensions are not guaranteed to exist.

However, note that in all cases where stable extensions are guaranteed to exist and coincide with preferred extensions, e.g. for stratified and order-consistent frameworks [1], any proof procedure for reasoning under the latter is a correct (and simpler) computational mechanism for reasoning under the former.

Finally, the "locality" feature in the computation of consequences under the preferred extension semantics renders it a feasible alternative to the computation of consequences under the stable extension semantics in the non-propositional case, when the language is infinite. Indeed, both CPE and SPE do not require that the given framework be propositional.

Acknowledgements

This research has been partially supported by the EC KIT project "Computational Logic for Flexible Solutions to Applications". The third author has been supported by the UK EPSRC project "Logic-based multi-agent systems".

References

1. A. Bondarenko, P. M. Dung, R. A. Kowalski, F. Toni, An abstract, argumentation-theoretic framework for default reasoning. *Artificial Intelligence*, 93:63-101, 1997.
2. Y. Dimopoulos, A. C. Kakas, Logic Programming without Negation as Failure, Proceedings of the 1995 International Symposium on Logic Programming, pp. 369-383, 1995.
3. Y. Dimopoulos, B. Nebel, F. Toni, Preferred Arguments are Harder to Compute than Stable Extensions, Proc. of the Sixteenth International Joint Conference on Artificial Intelligence, IJCAI 99, (T. Dean ed.), pp. 36-43, 1999.
4. Y. Dimopoulos, B. Nebel, F. Toni, Finding Admissible and Preferred Arguments Can Be Very Hard, Proceedings of the Seventh International Conference on Principles of Knowledge Representation and Reasoning, KR 2000, (A. G. Cohn, F. Giunchiglia, B. Selman eds.), pp. 53-61, Morgan Kaufmann Publishers, 2000.
5. P. M. Dung, Negation as hypothesis: an abductive foundation for logic programming. *Proceedings of the 8th International Conference on Logic Programming*, Paris, France (K. Furukawa, ed.), MIT Press, pp. 3-17, 1991.
6. P. M. Dung, On the acceptability of arguments and its fundamental role in non-monotonic reasoning, logic programming and n-person games *Artificial Intelligence*,, 77:321-357, Elsevier, 1993.

7. P. M. Dung, R. A. Kowalski, F. Toni, Synthesis of proof procedures for default reasoning, *Proc. LOPSTR'96, International Workshop on Logic Program Synthesis and Transformation*, (J. Gallagher ed.), pp. 313–324, LNCS 1207, Springer Verlag, 1996.

8. P. M. Dung, R. A. Kowalski, F. Toni, Proof procedures for default reasoning. In preparation, 2002.

9. K. Eshghi, R. A. Kowalski, Abduction compared with negation as failure. *Proceedings of the 6th International Conference on Logic Programming*, Lisbon, Portugal (G. Levi and M. Martelli, eds), MIT Press, pp. 234–254, 1989

10. M. Gelfond, V. Lifschitz, The stable model semantics for logic programming. *Proceedings of the 5th International Conference on Logic Programming*, Washington, Seattle (K. Bowen and R. A. Kowalski, eds), MIT Press, pp. 1070–1080, 1988

11. K. Inoue, M. Koshimura, R. Hasegawa, Embedding negation as failure into a model generation theorem-prover. Proc. *CADE'92*, pp. 400-415, LNCS 607, Springer, 1992.

12. A. C. Kakas, R. A. Kowalski, F. Toni, The role of abduction in logic programming. *Handbook of Logic in Artificial Intelligence and Logic Programming* (D.M. Gabbay, C.J. Hogger and J.A. Robinson eds.), 5: 235-324, , Oxford University Press, 1998.

13. A. C. Kakas, P. Mancarella. Preferred extensions are partial stable models. *Journal of Logic Programming* 14(3,4), pp.341–348, 1993.

14. A. C. Kakas, P. Mancarella, P. M. Dung, The Acceptability Semantics for Logic Programs, Proceedings of the Eleventh International Conference on Logic Programming, pp. 504-519, 1994.

15. A. C. Kakas, F. Toni, Computing Argumentation in Logic Programming. *Journal of Logic and Computation* 9:515-562, Oxford University Press, 1999.

16. J. McCarthy, Circumscription – a form of non-monotonic reasoning. *Artificial Intelligence*, 1327–39, 1980.

17. D. McDermott, Nonmonotonic logic II: non-monotonic modal theories. *Journal of ACM* 29(1), pp. 33–57, 1982.

18. R. Moore, Semantical considerations on non-monotonic logic. *Artificial Intelligence* 25:75–94, 1985.

19. I. Niemelä, Towards efficient default reasoning. *Proc. IJCAI'95*, pp. 312–318, Morgan Kaufman, 1995.

20. I. Niemelä, P. Simons, Efficient implementation of the well-founded and stable model semantics. *Proc. JICSLP'96*, pp. 289–303, MIT Press, 1996.

21. J. L. Pollock. Defeasible reasoning. *Cognitive Science*, 11(4):481–518, 1987.

22. D. Poole, A logical framework for default reasoning. *Artificial Intelligence* 36:27–47, 1988.

23. H. Prakken and G. Sartor. A system for defeasible argumentation, with defeasible priorities. *Artificial Intelligence Today*, (M. Wooldridge and M. M. Veloso, eds.), LNCS 1600, pp. 365–379, Springer, 1999.

24. H. Prakken and G. Vreeswijk. Logical systems for defeasible argumentation. *Handbook of Philosophical Logic*, 2nd edition, (D. Gabbay and F. Guenthner eds.), Vol. 4, Kluwer Academic Publishers, 2001.

25. R. Reiter, A logic for default reasoning. *Artificial Intelligence* 13:81–132, Elsevier, 1980).

26. D. Saccà, C. Zaniolo, Stable model semantics and non-determinism for logic programs with negation. *Proceedings of the 9th ACM SIGACT-SIGMOD-SIGART Symposium on Principles of Database Systems*, ACM Press, pp. 205–217, 1990.

27. K. Satoh, A top-down proof procedure for default logic by using abduction. *Proceedings of the Eleventh European Conference on Artificial Intelligence*, pp. 65-69, John Wiley and Sons, 1994.

28. K. Satoh and N. Iwayama. A Query Evaluation Method for Abductive Logic Programming. Proceedings of the Joint International Conference and Symposium on Logic Programming, pp. 671 – 685, 1992.

29. G.R. Simari and R.P. Loui. A mathematical treatment of defeasible reasoning and its implementation. *Artificial Intelligence*, 52:125–257, 1992.

30. M. Thielscher, A nonmonotonic disputation-based semantics and proof procedure for logic programs. *Proceedings of the 1996 Joint International Conference and Symposium on Logic Programming* (M. Maher ed.), pp. 483–497, 1996.

31. F. Toni, Argumentation-theoretic proof procedures for logic programming. Technical Report, Department of Computing, Imperial College, 1997.

32. G. Vreeswijk. The feasibility of defeat in defeasible reasoning. *Proceedings of the 2nd Int. Conf. on Principles of Knowledge Representation and Reasoning (KR'91)*, (J.F. Allen, R. Fikes, E. Sandewall, eds.), pp. 526–534, 1991.

A   Stratified and Order Consistent Frameworks

We recall the definitions of stratified and order consistent flat argumentation-theoretic frameworks, and theire semantics properties, ad given in [1]. Both classes are characterized in terms of their attack relationship graphs.

The *attack relationship graph* of a flat assumption-based framework $\langle T, Ab, \,\overline{}\,\rangle$ is a directed graph whose nodes are the assumptions in Ab and such that there exists an edge from an assumption δ to an assumption α if and only if δ belongs to a minimal (with respect to set inclusion) attack Δ against α.

A flat assumption-based framework is *stratified* if and only if its attack relationship graph is well-founded, i.e. it contains no infinite path of the form $\alpha_1, \ldots, \alpha_n, \ldots$, where for every $i \geq 0$ there is an edge from α_{i+1} to α_i.

The notion of order-consistency requires some more auxiliary definitions. Given a flat assumption-based framework $\langle T, Ab, \,\overline{}\,\rangle$ let $\delta, \alpha \in Ab$.

- δ is *friendly* (resp. *hostile*) to α if and only if the attack relationship graph for $\langle T, Ab, \,\overline{}\,\rangle$ contains a path from δ to α with an even (resp. odd) number of edges.
- δ is *two-sided* to α, written $\delta \prec \alpha$, if δ is both friendly and hostile to α.

A flat assumption-based framework $\langle T, Ab, \,\overline{}\,\rangle$ is *order-consistent* if the relation $\prec$ is well-founded, i.e. there exists no infinite sequence of the form $\alpha_1, \ldots, \alpha_n, \ldots$, where for every $i \geq 0$, $\alpha_{i+1} \prec \alpha_i$.

The following proposition summarizes some of the semantics results of [1] as far as stratified and order-consistent frameworks are concerned.

Proposition 1 (see [1]).

- *for any stratified assumption-based framework there exists a unique stable set of assumptions, which coincides with the well-founded set of assumptions.*
- *for any order-consistent assumption-based framework stable sets of assumptions are preferred sets of assumptions and viceversa.*

It is worth recalling that the abstract notions of stratification and order-consistency generalize the notions of stratification and order-consistency for logic programming.

Automated Abduction

Katsumi Inoue

Department of Electrical and Electronics Engineering
Kobe University
Rokkodai, Nada, Kobe 657-8501, Japan
inoue@eedept.kobe-u.ac.jp

Abstract. In this article, I review Peirce's abduction in the context of Artificial Intelligence. First, I connect abduction from first-order theories with nonmonotonic reasoning. In particular, I consider relationships between abduction, default logic, and circumscription. Then, based on a first-order characterization of abduction, I show a design of abductive procedures that utilize automated deduction. With abductive procedures, proof procedures for nonmonotonic reasoning are also obtained from the relationship between abduction and nonmonotonic reasoning.

1   Introduction

Kowalski had a decisive impact on the research of abductive reasoning in AI. In 1979, Kowalski showed the role of abduction in information system in his seminal book *"Logic for Problem Solving"* [58]. In the book, Kowalski also pointed out some similarity between abductive hypotheses and defaults in nonmonotonic reasoning. This article is devoted to investigate such a relation in detail and to give a mechanism for automated abduction from first-order theories.

In this article, Peirce's logic of abduction is firstly reviewed in Section 2, and is then related to a formalization of *explanation* within first-order logic. To know what formulas hold in the theory augmented by hypotheses, the notion of *prediction* is also introduced. There are two approaches to nonmonotonic prediction: *credulous* and *skeptical* approaches, depending on how conflicting hypotheses are treated.

In Section 3, it is shown that abduction is related to the brave approach, in particular to the simplest subclass of *default logic* [87] for which efficient theorem proving techniques may exist. On the other hand, *circumscription* [70] is a notable example of the skeptical approach. Interestingly, the skeptical approach is shown to be realized using the brave approach.

In Section 4, computational properties of abduction are discussed in the context of first-order logic. To make abduction and nonmonotonic reasoning computable, the *consequence-finding* problem in first-order logic is reviewed, which is an important challenging problem in automated deduction [61,35,68]. The problem of consequence-finding is then modified so that only interesting clauses with a certain property (called *characteristic clauses*) are found. Then,

A.C. Kakas, F. Sadri (Eds.): Computat. Logic (Kowalski Festschrift), LNAI 2408, pp. 311–341, 2002.

abduction is formalized in terms of characteristic clauses. Two consequence-finding procedures are then introduced: one is *SOL resolution* [35], and the other is *ATMS* [14]. Compared with other resolution methods, SOL resolution generates fewer clauses to find characteristic clauses in general.

Finally, this article is concluded in Section 5, where Peirce's abduction is revisited with future work.

It should be noted that this article does not cover all aspects of abductive reasoning in AI. General considerations on abduction in science and AI are found in some recent books [50,26,67] and survey papers [56,78]. Applications of abduction in AI are also excluded in this article. This article mostly focuses on *first-order abduction*, i.e., automated abduction from first-order theories, and its relationship with nonmonotonic reasoning with first-order theories. Often however, abduction is used in the framework of logic programming, which is referred to as *abductive logic programming* [53,54,20]. This article omits details of abductive logic programming, but see [51] in this volume. Part of this article is excerpted from the author's thesis [36] and a summary paper by the author [37].

2   Logic of Abduction

Abduction is one of the three fundamental modes of reasoning characterized by Peirce [79], the others being *deduction* and *induction*. To see the differences between these three reasoning modes, let us look at the "beans" example used by Peirce [79, paragraph 623] in a syllogistic form. Abduction amounts to concluding the minor premise (*Case*) from the major premise (*Rule*) and the conclusion (*Result*):

> (*Rule*)   All the beans from this bag are white.
> (*Result*) These beans are white.
> _____
> (*Case*)   These beans are from this bag.

On the contrary, deduction amounts to concluding *Result* from *Rule* and *Case*, and induction amounts to concluding *Rule* from *Case* and *Result*. Later, Peirce wrote an inferential form of abduction as follows.

> The (surprising) fact, *C*, is observed;
> But if *A* were true, *C* would be a matter of course;
> Hence, there is reason to suspect that *A* is true.

This corresponds to the following rule of the form, called the *fallacy of affirming the consequent*:

$$\frac{C \quad A \supset C}{A}. \tag{1}$$

Sometimes *A* is called an *explanans* for an *explanandum C*. Both abduction and induction are non-deductive inference and generate hypotheses. However, hypothesis generation by abduction is distinguished from that by induction, in the sense that while induction infers something to be true through generalization of a number of cases of which the same thing is true, abduction can infer something quite different from what is observed.[1] Therefore, according to Peirce [79, paragraph 777], abduction is "the only kind of reasoning which supplies new ideas, the only kind which is, in this sense, synthetic". Since abduction can be regarded as a method to explain observations, Peirce considered it as the basic method for *scientific discovery*.

In the above sense, abduction is "ampliative" reasoning and may play a key role in the process of advanced inference. For example, analogical reasoning can be formalized by abduction plus deduction [79, paragraph 513]. Abduction is, however, only "probable" inference as it is non-deductive. That is, as Peirce argues, abduction is "a weak kind of inference, because we cannot say that we believe in the truth of the explanation, but only that it may be true". This phenomenon of abduction is preferable, since our *commonsense reasoning* also has a probable nature. In everyday life, we regularly form hypotheses, to explain how other people behave or to understand a situation, by filling in the gaps between what we know and what we observe. Thus, abduction is a very important form of reasoning in everyday life as well as in science and engineering.

Another important issue involved in abduction is the problem of *hypothesis selection*: what is the best explanation, and how can we select it from a number of possible explanations which satisfy the rule (1)? Peirce considered this problem philosophically, and suggested various preference criteria that are both qualitative and economical. One example of such criteria is the traditional maxim of *Occam's razor*, which adopts the simplest hypotheses.

In the following subsections, we give a logic of abduction studied in AI from two points of views, i.e., explanation and prediction.

2.1   Explanation

Firstly, we connect Peirce's logic of abduction with formalizations of abduction developed in AI within first-order logic. The most popular formalization of abduction in AI defines an explanation as a set of hypotheses which, together with the background theory, logically entails the given observations. This *deductive-nomological* view of explanation [33] has enabled us to have logical specifications of abduction and their proof procedures based on the *resolution principle* [89]. There are a number of proposals for resolution-based abductive systems [85,10,25,84,88,91,96,34,83,18,35,53,97,13,19,16].

According to the deductive-nomological view of explanation, we here connect Peirce's logic of abduction (1) with research on abduction in AI. To this end, we make the following assumptions.

1. Knowledge about a domain of discourse, or *background knowledge*, can be represented in a set of first-order formulas as the *proper axioms*. In the following, we denote such an axiom set by Σ, and call it a set of *facts*.

---

[1] The relation, difference, similarity, and interaction between abduction and induction are now extensively studied by many authors in [26].

2. An *observation* is also expressed as a first-order formula. Given an observation C, each explanation A of C satisfying the rule (1) can be constructed from a sub-vocabulary $\mathcal{H}$ of the representation language that contains Σ. We call each formula constructed from such a subset of the language a *hypothesis*. In general, a hypothesis constructed from $\mathcal{H}$ is a formula whose truth value is indefinite but may be assumed to be true. Sometimes $\mathcal{H}$ is the representation language itself.

3. The major premise $A \supset C$ in the rule (1) can be obtained *deductively* from Σ, either as an axiom contained in Σ or as a *logical consequence* of Σ:

$$\Sigma \models A \supset C. \tag{2}$$

4. Σ contains all the information required to judge the acceptability of each hypothesis A as an explanation of C. That is, each formula A satisfying (2) can be tested for its appropriateness without using information not contained in Σ. One of these domain-independent, necessary conditions is that A should not be contradictory to Σ, or that $\Sigma \cup \{A\}$ is *consistent*.

5. We adopt Occam's razor as a domain-independent criterion for hypothesis selection. Namely, the *simplest* explanation is preferred over any other.

These assumptions are useful particularly for domain-independent automated abduction. The first and second conditions above define a logical framework of abduction: the facts and the hypotheses are both first-order formulas. The third and fourth conditions give a logical specification of the link between observations and explanations: theories augmented with explanations should both entail observations and be consistent. Although these conditions are most common in abductive theories proposed in AI, the correctness of them from the philosophical viewpoint is still being argued. The fifth condition, *simplicity*, is also one of the most agreeable criterion to select explanations: a simpler explanation is preferred if every other condition is equal in multiple explanations. Note that these conditions are only for the definition of explanations. Criteria for *good*, *better*, or *best* explanations are usually given using meta information and domain-dependent heuristics. A number of factors should be considered in selecting the most reasonable explanation. Since there has been no concrete consensus among AI researchers or philosophers about the preference criteria, we will not discuss them further in this article.

An example of the above abductive theory can be seen in the *Theorist* system by Poole, Goebel and Aleliunas [84], which consists of a first-order theorem prover that distinguishes facts from hypotheses.

Definition 2.1 (Theorist) Let Σ be a set of facts, and Γ a set of hypotheses. We call a pair (Σ, Γ) an *abductive theory*. Given a closed formula G, a set E of ground instances of elements of Γ is an *explanation* of G from (Σ, Γ)[2] if

1. $\Sigma \cup E \models G$, and
2. $\Sigma \cup E$ is consistent.

---

[2] Some Theorist literature [81] gives a slightly different definition, where a set $\Sigma \cup E$ (called a *scenario*) satisfying the two conditions is called an explanation of G.

An explanation E of G is *minimal* if no proper subset E' of E is an explanation of G.

The first condition in the above definition reflects the fact that Theorist has been devised for automated scientific *theory formation*, which is useful for prototyping AI problem solving systems by providing a simple "hypothesize-test" framework, i.e., *hypothetical reasoning*. When an explanation is a finite set of hypotheses, $E = \{H_1, \ldots, H_n\}$, the first condition is equivalent to

$$\Sigma \models H_1 \wedge \ldots \wedge H_n \supset G$$

by deduction theorem, and thus can be written in the form of (2). The minimality criterion is a syntactical form of Occam's razor. Since for an explanation E of G, any $E' \subseteq E$ is consistent with Σ, the condition can be written as: an explanation E of G is minimal if no $E' \subset E$ satisfies $\Sigma \cup E' \models G$. Note that in Theorist, explanations are defined as a set of ground instances. A more general definition of (minimal) explanations is defined in [35], in which variables can be contained in explanations.

Example 2.2 Suppose that (Σ_1, Γ_1) is an abductive theory, where

$$\Sigma_1 = \{\ \forall x(\ Bird(x) \wedge \neg Ab(x) \supset Flies(x)\,),$$
$$\forall x(\ Penguin(x) \supset Ab(x)\,),$$
$$Bird(Tweety)\ \},$$
$$\Gamma_1 = \{\ \neg Ab(x)\ \}.$$

Here, the hypothesis $\neg Ab(x)$ means that for any ground term t, $\neg Ab(t)$ can be hypothesized. In other words, a hypothesis containing variables is shorthand for the set of its ground instances with respect to the elements from the universe of the language. Intuitively, $\neg Ab(x)$ means that anything can be assumed to be not abnormal (i.e., normal). In this case, a minimal explanation of $Flies(Tweety)$ is $\{\neg Ab(Tweety)\}$.

In Theorist, a set Γ of hypotheses can be any set of first-order formulas. Poole [81] shows a *naming* method which transforms each hypothesis in Γ into an atomic formula. The naming method converts an abductive theory (Σ, Γ) into a new abductive theory (Σ', Γ') in the following way. For every hypothesis $F(\mathbf{x})$ in Γ, where $\mathbf{x} = x_1, \ldots, x_n$ is the tuple of the free variables appearing in F, we associate a predicate symbol δ_F not appearing anywhere in (Σ, Γ), and define the following sets of formulas:

$$\Gamma' = \{\ \delta_F(\mathbf{x}) \mid F(\mathbf{x}) \in \Gamma\ \},$$
$$\Sigma' = \Sigma \cup \{\ \forall \mathbf{x}(\ \delta_F(\mathbf{x}) \supset F(\mathbf{x})\,) \mid F(\mathbf{x}) \in \Gamma\ \}.$$

Then, there is a 1-1 correspondence between the explanations of G from (Σ, Γ) and the explanations of G from (Σ', Γ') [81, Theorem 5.1].

Example 2.2 (continued) The hypothesis $\neg Ab(x)$ can be named $Normal(x)$:

$$\Sigma'_1 = \Sigma \cup \{ \, \forall x (\, Normal(x) \supset \neg Ab(x) \,) \, \},$$
$$\Gamma'_1 = \{ \, Normal(x) \, \}.$$

In this case, a minimal explanation of $Flies(Tweety)$ is $\{ \, Normal(Tweety) \, \}$, which corresponds to the explanation $\{ \, \neg Ab(Tweety) \, \}$ from the original (Σ_1, Γ_1).

Naming hypotheses is a technique commonly used in most abductive systems because hypotheses in the form of atomic formulas can be processed very easily in their implementation. Restriction of hypotheses to atoms is thus used in many abductive systems such as [25,96,52,9]. Note that when we use a resolution procedures for non-Horn clauses, we can allow for *negative* as well as positive *literals* as names of hypotheses, since both positive and negative literals can be resolved upon in the procedure. For Example 2.2, we do not have to rename the negative literal $\neg Ab(x)$ to the positive literal $Normal(x)$. This kind of negative abnormal literal was originally used by McCarthy [71], and is convenient for computing circumscription through abduction. Abductive systems that allow literal hypotheses can be seen in such as [85,10,35].

It should be noted that there are many other formalizations of abduction. For example, abduction is defined by the *set covering model* [6], is discussed at the *knowledge level* [63], and is formalized in various ways [100,5,12,65,80,1]. Levesque's [63] formulation suggests that abduction does not have to be formalized within first-order logic. There are some proposals for abductive theories based on other logical languages. In such cases, the background knowledge is often written in a nonmonotonic logic. For example, *abductive logic programming* (ALP) is an extension of logic programming, which is capable of abductive reasoning as well as nonmonotonic reasoning [52,53,38,44,13,28,54,19,20,51]. Abduction is also defined within a modal logic in [94], autoepistemic logic in [43], or default logic in [22]. Inoue and Sakama [43] point out that, in abduction from nonmonotonic theories, abductive explanations can be obtained not only by addition of new hypotheses, but also by removal of old hypotheses that become inappropriate.

2.2   Prediction

Theory formation frameworks like Theorist can be used for *prediction* as well as abduction. In [82], a distinction between explanation and prediction is discussed as follows. Let (Σ, Γ) be an abductive theory, G a formula, and E an explanation of G from (Σ, Γ) as defined by Definition 2.1.

1. In abduction, G is an observation which is known to be true. We may assume E is true because G is true.
2. In prediction, G is a formula or a query whose truth value is unknown but is expected to be true. We may assume E is true to make G hold under E.

Both of the above ways of theory formation perform hypothetical reasoning, but in different ways. In abduction, hypotheses used to explain observations are called *conjectures*, whereas, in prediction, hypotheses are called *defaults* [81,82]. In Example 2.2, if we have observed that *Tweety* was flying and we want to know why this observation could have occurred, then obtaining the explanation $E_1 = \neg Ab(Tweety)$ is abduction; but if all we know is only the facts Σ_1 and we want to know whether *Tweety* can fly or not, then finding E_1 is prediction where we can expect *Tweety* may fly by default reasoning. These two processes may occur successively: when an observation is made, we abduce possible hypotheses; from these hypotheses, we predict what else we can expect to be true. In such a case, hypotheses can be used as both conjectures and defaults. See also [91,50] for other discussions on the difference between explanation and prediction.

A hypothesis regarded as a default may be used unless there is evidence to the contrary. Therefore, defaults may be applied as many as possible unless augmented theories are inconsistent. This leads to the notion of *extensions* [81].

Definition 2.3 Given the facts Σ and the hypotheses (defaults) Γ, an *extension* of the abductive theory (Σ, Γ) is the set of logical consequences of $\Sigma \cup \mathbf{M}$ where $\mathbf{M}$ is a maximal (with respect to set inclusion) set of ground instances of elements of Γ such that $\Sigma \cup \mathbf{M}$ is consistent.

Using the notion of extensions, various alternative definitions of what should be predicted can be given [82]. They are related to the *multiple extension problem*: if G_1 holds in an extension X_1 and G_2 holds in another extension X_2, but there is no extension in which both G_1 and G_2 hold (i.e., $X_1 \cup X_2$ is inconsistent), then what should we predict? —Nothing? Both? Or just one of G_1 and G_2? The next two are the most well-known prediction policies:

1. Predict what holds in an extension of (Σ, Γ);
2. Predict what holds in all extensions of (Σ, Γ).

The first approach to default reasoning leads to multiple extensions and is called a *credulous* approach. On the other hand, the latter approach is called a *skeptical* approach. Credulous and skeptical reasoning are also called *brave* and *cautious* reasoning, respectively. In the next section, we see that credulous prediction can be directly characterized by explanation and that skeptical prediction can be represented by combining explanations.

3   Relating Abduction to Nonmonotonic Reasoning

In this section, we relate the abductive theories introduced in Section 2 to formalisms of nonmonotonic reasoning.

Since abduction is ampliative and plausible reasoning, conclusions of abductive reasoning may not be correct. Therefore, abduction is nonmonotonic. This can be easily verified for abductive theories. First, an explanation E is consistent with the facts Σ by definition, but E is not necessarily an explanation with respect to the new facts Σ' ($\supset \Sigma$) because $\Sigma' \cup E$ may not be consistent. Second, a *minimal* explanation E of G with respect to Σ may not be minimal with respect to Σ' ($\supset \Sigma$) because a subset E' of E may satisfy $\Sigma' \cup E' \models G$. Poole [82] investigates other possibilities of nonmonotonicity that may arise according to changes of facts, hypotheses, and observations.

The above discussion can also be verified by considering relationships between abduction and nonmonotonic logics. In fact, this link is bidirectional [36,56]: abduction can be formalized by a credulous form of nonmonotonic logic (default logic), and a skeptical nonmonotonic formalism (circumscription) can be represented using an abductive theory. The former relationship verifies the non-monotonicity of abduction, and the latter implies that abduction can be used for commonsense reasoning as well as scientific theory formation.

3.1   Nonmonotonic Reasoning

We here review two major formalisms for nonmonotonic reasoning: *default logic* [87] and *circumscription* [70]. Both default logic and circumscription extend the classical first-order predicate calculus, but in different ways. Default logic introduces *inference rules* referring to the *consistency* with a belief set, and uses them meta-theoretically to extend a first-order theory. Circumscription, on the other hand, augments a first-order theory with a *second-order axiom* expressing a kind of *minimization* principle, and restricts the objects satisfying a certain predicate to just those that the original theory says must satisfy that predicate.

Default Logic. Default logic, proposed by Reiter [87], is a logic for drawing plausible conclusions based on consistency. This is one of the most intuitive and natural logics for nonmonotonic reasoning. One of the most successful results derived from the studies on default logic can be seen in the fact that *logic programming with negation as failure* can be interpreted as a class of default logic [2,29]. In this article, we also see that abduction can be characterized by one of the simplest classes of default logic (Section 3.2).

A *default* is an inference rule of the form:

$$\frac{\alpha(\mathbf{x})\;:\;\mathsf{M}\,\beta_1(\mathbf{x}),\dots,\mathsf{M}\,\beta_m(\mathbf{x})}{\gamma(\mathbf{x})}\,, \tag{3}$$

where $\alpha(\mathbf{x})$, $\beta_1(\mathbf{x}),\dots,\beta_m(\mathbf{x})$, and $\gamma(\mathbf{x})$ are first-order formulas whose free variables are contained in a tuple of variables $\mathbf{x}$. $\alpha(\mathbf{x})$ is called the *prerequisite*, $\beta_1(\mathbf{x}),\dots,\beta_m(\mathbf{x})$ the *justifications*, and $\gamma(\mathbf{x})$ the *consequent* of the default. A default is *closed* if no formula in it contains a free variable; otherwise it is *open*. An open default is usually identified with the set of closed defaults obtained by replacing the free variables with ground terms. A default is *normal* if it contains only one justification ($m = 1$) that is equivalent to the consequent ($\beta_1 \equiv \gamma$). A *default theory* is a pair, (D, W), where D is a set of defaults and W is a set of first-order formulas which represents proper axioms. A default theory is *normal* if every default is normal.

The intended meaning of the default (3) is: for any tuple t of ground terms, "if $\alpha(\mathbf{t})$ is inferable and each of $\beta_1(\mathbf{t}), \ldots, \beta_m(\mathbf{t})$ is consistently assumed, then infer $\gamma(\mathbf{t})$". When a default is applied, it is necessary that each of its justifications is consistent with a "belief set". In order to express this condition formally, an acceptable "belief set" induced by reasoning with defaults (called an *extension*) is precisely defined in default logic as follows.

Definition 3.1 [87] Let (D, W) be a default theory, and X a set of formulas. X is an *extension* of (D, W) if it coincides with the smallest set Y of formulas satisfying the following three conditions:

1. $W \subseteq Y$.
2. Y is deductively closed, that is, it holds that $cl(Y) = Y$, where $cl(Y)$ is the logical closure of Y under classical first-order deduction.
3. For any ground instance of any default in D of the form (3), if $\alpha(\mathbf{t}) \in Y$ and $\neg\beta_1(\mathbf{t}), \ldots, \neg\beta_m(\mathbf{t}) \notin X$, then $\gamma(\mathbf{t}) \in Y$.

A default theory may have multiple or, even, no extensions. However, it is known that for any *normal default theory*, there is at least one extension [87, Theorem 3.1]. It is also noted that in default logic each extension is interpreted as an acceptable set of beliefs in accordance with default reasoning. Such an approach to default reasoning leads to multiple extensions and is a *credulous* approach. By credulous approaches one can get more conclusions depending on the choice of the extension so that conflicting beliefs can be supported by different extensions. This behavior is not necessarily intrinsic to a reasoner dealing with a default theory; we could define the theorems of a default theory to be the intersection of all its extensions so that we remain agnostic to conflicting information. This latter variant is a *skeptical* approach.

Circumscription. Circumscription, proposed by McCarthy [70], is one of the most "classical" and best-developed formalisms for nonmonotonic reasoning. An important property of circumscription that many other nonmonotonic formalisms lack, is that it is based on classical predicate logic.

Let T be a set of first-order formulas, and $\mathbf{P}$ and $\mathbf{Z}$ denote disjoint tuples of distinct predicate symbols in the language of T. The predicates in $\mathbf{P}$ are said to be *minimized* and those in $\mathbf{Z}$ to be *variables*; $\mathbf{Q}$ denotes the rest of the predicates in the language of T, called the *fixed* predicates (or *parameters*). We denote a theory T by $T(\mathbf{P}; \mathbf{Z})$ when we want to indicate explicitly that T mentions the predicates $\mathbf{P}$ and $\mathbf{Z}$. Adopting the formulation by Lifschitz [64], the *circumscription of* $\mathbf{P}$ *in* T *with* $\mathbf{Z}$, written $CIRC(T; \mathbf{P}; \mathbf{Z})$, is the augmentation of T with a second-order axiom expressing the minimality condition:

$$T(\mathbf{P}; \mathbf{Z}) \wedge \neg \exists \, \mathbf{pz}\, (T(\mathbf{p}; \mathbf{z}) \wedge \mathbf{p} < \mathbf{P}). \tag{4}$$

Here, $\mathbf{p}$ and $\mathbf{z}$ are tuples of predicate variables each of which has the same arity as the corresponding predicate symbol in $\mathbf{P}$ and $\mathbf{Z}$, and $T(\mathbf{p}; \mathbf{z})$ denotes a theory obtained from $T(\mathbf{P}; \mathbf{Z})$ by replacing each occurrence of $\mathbf{P}$ and $\mathbf{Z}$ with $\mathbf{p}$ and $\mathbf{z}$.

Also, $\mathbf{p} < \mathbf{P}$ stands for the conjunction of formulas each of which is defined, for every member P_i of $\mathbf{P}$ with a tuple $\mathbf{x}$ of object variables and the corresponding predicate variable p_i in $\mathbf{p}$, in the form:

$$\forall \mathbf{x}(p_i(\mathbf{x}) \supset P_i(\mathbf{x})) \wedge \neg\forall \mathbf{x}(P_i(\mathbf{x}) \supset p_i(\mathbf{x})).$$

Thus, the second-order formula in the definition (4) represents that the extension of the predicates from $\mathbf{P}$ is minimal in the sense that it is impossible to make it smaller without violating the constraint T. Intuitively, $CIRC(T; \mathbf{P}; \mathbf{Z})$ is intended to minimize the number of objects satisfying $\mathbf{P}$, even at the expense of allowing more or different objects to satisfy $\mathbf{Z}$.

The model-theoretic characterization of circumscription is based on the notion of *minimal models*.

Definition 3.2 [64] Let M_1 and M_2 be models of T with the same universe. We write $M_1 \leq_{\mathbf{P},\mathbf{Z}} M_2$ if M_1 and M_2 differ only in the way they interpret predicates from $\mathbf{P}$ and $\mathbf{Z}$, and the extension of every predicate P from $\mathbf{P}$ in M_1 is a subset of the extension of P in M_2. Then, a model M of T is $(\mathbf{P}, \mathbf{Z})$-*minimal* if, for no other model M' of T, $M' \leq_{\mathbf{P},\mathbf{Z}} M$ but $M \nleq_{\mathbf{P},\mathbf{Z}} M'$.

It is known that, for any formula F, $CIRC(T; \mathbf{P}; \mathbf{Z}) \models F$ if and only if F is satisfied by every $(\mathbf{P}, \mathbf{Z})$-minimal model of T [64]. Since each theorem of a circumscription is satisfied by all minimal models, this property makes the behavior of circumscription skeptical.

3.2   Abduction and Default Logic

Suppose that Σ is a set of facts and Γ is a set of hypotheses. In order to avoid confusion in terminology, we here call an extension of the abductive theory (Σ, Γ) given by Definition 2.3 a *Theorist extension*, and call an extension of a default theory (D, W) given by Definition 3.1 a *default extension*. Let $w(\mathbf{x})$ be a formula whose free variables are $\mathbf{x}$. For Σ and Γ, we define a normal default theory (D_Γ, Σ), where

$$D_\Gamma = \left\{ \frac{: \mathsf{M}\, w(\mathbf{x})}{w(\mathbf{x})} \;\middle|\; w(\mathbf{x}) \in \Gamma \right\}.$$

Notice that D_Γ is a set of *prerequisite-free* normal defaults, that is, normal defaults whose prerequisites are *true*. We obtain the next theorem by resluts from [81, Theorems 2.6 and 4.1].

Theorem 3.3 Let (Σ, Γ) be an abductive theory, and G a formula. The following three are equivalent:

(a) There is an explanation of G from (Σ, Γ).
(b) There is a Theorist extension of (Σ, Γ) in which G holds.
(c) There is a default extension of the default theory (D_Γ, Σ) in which G holds.

Theorem 3.3 is very important for the following reasons.

1. It is verified that each abductive explanation is contained in a possible set of beliefs. In particular, when the hypotheses Γ represent defaults for normal or typical properties, then in order to predict a formula G by *default reasoning*, it is sufficient to find an explanation of G from (Σ, Γ) [81].
2. All properties possessed by normal default theories are valid for abductive explanations and Theorist extensions. For instance, for any Σ and Γ, there is at least one Theorist extension of (Σ, Γ).
3. Computation of abduction can be given by *top-down default proofs* [87], which is an extension of *linear resolution* theorem proving procedures such as [59,7,66]. This fact holds for the following reasons. It is shown that, G holds in some default extension of a normal default theory (D, W) if and only if there is a top-down default proof of G with respect to (D, W) [87, Theorem 7.3]. Also, every top-down default proof returns a set S of instances of consequents of defaults from D with which G can be proven from W, i.e., $W \cup S \models G$. Therefore, such an S is an explanation from the corresponding abductive theory whenever $W \cup S$ is consistent.

The last point above is also very useful for designing and implementing hypothetical reasoning systems. In fact, many first-order abductive procedures [85,10,84,96,83] can be regarded as variants of Reiter's top-down default proof procedures: computation of explanations of G from (Σ, Γ) can be seen as an extension of proof-finding in linear resolution by introducing a set of hypotheses from Γ that, if they could be proven by preserving the consistency of the augmented theories, would complete the proofs of G. Alternatively, abduction can be characterized by a *consequence-finding* problem [35], in which some literals are allowed to be hypothesized (or *skipped*) instead of proven, so that new theorems consisting of only those skipped literals are derived at the end of deductions instead of just deriving the empty clause. In this sense, abduction can be implemented as an extension of deduction, in particular of a top-down, backward-chaining theorem-proving procedure. For example, Theorist [84,83] and SOL resolution [35] are extensions of the Model Elimination procedure [66].

Example 2.2 (continued) For the goal $G = Flies(Tweety)$, a version of Theorist implementation works as follows (written using a Prolog-like notation):

$$\leftarrow Flies(Tweety) \,,$$
$$\leftarrow Bird(Tweety) \wedge \neg Ab(Tweety) \,,$$
$$\leftarrow \neg Ab(Tweety) \,,$$
$$\square \quad \text{by defaults: } \{\neg Ab(Tweety)\} \,.$$

Then, the returned set of defaults $S = \{\neg Ab(Tweety)\}$ is checked for the consistency with Σ_1 by failing to prove the negation of S from Σ_1. In this case, it holds that

$$\Sigma_1 \not\models Ab(Tweety) \,,$$

thus showing that S is an explanation of G from (Σ_1, Γ_1).

Next, suppose that $Penguin(Tweety)$ is added to Σ_1, and let

$$\Sigma_2 = \Sigma_1 \cup \{\ Penguin(Tweety)\ \}.$$

We then get S again by the same top-down default proof as above, but the consistency check of S in this case results in a success proof:

$$\leftarrow Ab(Tweety)\,,$$
$$\leftarrow Penguin(Tweety)\,,$$
$$\Box\,.$$

Therefore, S is no longer an explanation of G from (Σ_2, Γ_1).

3.3   Abduction and Circumscription

A significant difference between circumscription and default logic lies in their ways to handle variables and equality. We then assume that the function symbols are the constants only and the number of constants is finite. Furthermore, in this subsection, a theory T means a set of formulas over the language including the equality axioms, and both the domain-closure assumption (DCA) and the unique-names assumption (UNA) are assumed to be satisfied by T. In this setting, the UNA represents that each pair of distinct constants denotes different individuals in the domain. The DCA implies that the theory has finite models and that every formula containing variables is equivalent to a propositional combination of ground atoms. Although these assumptions are strong, their importance is widely recognized in databases and logic programming. For circumscription, these assumptions make the universe fixed, so that the comparison with default logic becomes clear [24]. In particular, circumscription with these assumptions is essentially equivalent to the *Extended Closed World Assumption* (ECWA) [30].

Another big difference between circumscription and default logic is in their approaches to default prediction: skeptical versus credulous. The theorems of a circumscription are the formulas satisfied by every minimal model, while there are multiple default extensions in default logic. We, therefore, compare the theorems of a circumscription with the formulas contained in every default extension of a default theory.

On the relationship between circumscription and default logic, Etherington [24] has shown that, under some conditions, a formula is entailed by a circumscription plus the DCA and the UNA if and only if the formula is contained in every default extension of the corresponding default theory.

Proposition 3.4 [24] Assume that T is a theory satisfying the above conditions. Let $\mathbf{P}$ be a tuple of predicates, and $\mathbf{Z}$ the tuple of all predicates other than those in $\mathbf{P}$ in the language. Then, the formulas true in every default extension of the default theory:

$$\left(\ \left\{\ \frac{:\ \mathsf{M}\,\neg P_i(\mathbf{x})}{\neg P_i(\mathbf{x})}\ \middle|\ P_i \in \mathbf{P}\ \right\},\ T\ \right) \tag{5}$$

are precisely the theorems of $CIRC(T; \mathbf{P}; \mathbf{Z})$.

Since the default theory (5) is a prerequisite-free normal default theory, we can connect each of its default extensions with a Theorist extension using Theorem 3.3. Therefore, in the abductive theory, we hypothesize the negative occurrences of the minimized predicates **P**. The following corollary can be obtained by Theorem 3.3 and the model theory of circumscription.

Corollary 3.5 Let T, **P** and **Z** be the same as in Proposition 3.4. A (**P**, **Z**)-minimal model of T satisfies a formula F if and only if F has an explanation from the abductive theory $(T, \{ \neg P_i(\mathbf{x}) \,|\, P_i \in \mathbf{P} \})$.

The above corollary does not deal with a skeptical prediction but a credulous one. Moreover, Proposition 3.4 does not allow for the specification of *fixed predicates*. Gelfond *et al.* [30], on the other hand, show a more general result for the ECWA by allowing some predicates to be fixed. The idea of reducing circumscription to the ECWA is very important as it is helpful for designing resolution-based theorem provers for circumscription [86,31,41,32]. Earlier work for such a reduction of circumscription to a special case of the ECWA can be found in [73,3] where all predicates in the language are minimized.

To compute circumscription, we are particularly interested in two results of the ECWA obtained by Gelfond *et al.* [30, Theorems 5.2 and 5.3] with the notion of *free for negation*. These are also adopted as the basic characterizations for query answering in circumscriptive theories by Przymusinski [86, Theorems 2.5 and 2.6]. Inoue and Helft [41] express them using different terminology (*characteristic clauses*). Here, we relate these results of the ECWA with abduction. Let T be a theory as above, **P** the minimized predicates, **Q** the fixed predicates, and **Z** variables. For a tuple **R** of predicates in the language, we denote by $\mathbf{R}^+$ ($\mathbf{R}^-$) the positive (negative) occurrences of predicates from **R** in the language. Then, we define the abductive theory for circumscription, (T, Γ_{circ}), where the hypotheses are given as:

$$\Gamma_{circ} = \mathbf{P}^- \cup \mathbf{Q}^+ \cup \mathbf{Q}^- .$$

Intuitively, both positive and negative occurrences of **Q** are hypothesized as defaults to prevent the abductive theory from altering the definition of each predicate from **Q**. The next theorem can be obtained from [30, Theorems 5.2 and 5.3].

Theorem 3.6 [41]

(1) For any formula F not containing predicate symbols from **Z**,
  $CIRC(T; \mathbf{P}; \mathbf{Z}) \models F$ if and only if $\neg F$ has no explanation from (T, Γ_{circ}).
(2) For any formula F, $CIRC(T; \mathbf{P}; \mathbf{Z}) \models F$ if and only if there exist explanations $E_1, \ldots, E_n$ $(n \geq 1)$ of F from (T, Γ_{circ}) such that $\neg(E_1 \vee \ldots \vee E_n)$ has no explanation from (T, Γ_{circ}).

Using Theorem 3.6, we can reduce query answering in a circumscriptive theory to the finding of a combination of explanations of a query such that the negation of the disjunction cannot be explained. The basic intuition behind this theorem is as follows. In abduction, by Corollary 3.5, if a formula F is explained, then F holds in some default extension, that is, F is satisfied by some minimal model. In circumscription, on the other hand, F should be satisfied by every minimal model, or F should hold in all default extensions. This condition is checked by computing multiple explanations $E_1, \ldots, E_n$ of F corresponding to multiple default extensions such that those explanations cover all default extensions. Then, the disjunction $E_1 \vee \ldots \vee E_n$ is also an explenation of F, and is a skeptical but the weakest explanation of F [55]. Combining explanations is like an *argument system* [82,83,32], which consists of two processes where one tries to find explanations of the query and the other tries to find a counter argument to refute them.

Example 3.7 Consider the theory T consisting of the two formulas:

$$\neg Bird(x) \vee \neg Ab(x) \vee Flies(x),$$
$$Bird(Tweety),$$

where $\mathbf{P} = \{Ab\}$, $\mathbf{Q} = \{Bird\}$ and $\mathbf{Z} = \{Flies\}$, so that the abductive hypotheses are set to $\Gamma_{circ} = \{Ab\}^- \cup \{Bird\}^+ \cup \{Bird\}^-$. Let us consider the query

$$F = Flies(Tweety).$$

Now, $\{\neg Ab(Tweety)\}$ is an explanation of F. The negation of this explanation has no explanation. F is thus a theorem of $CIRC(T; Ab; Flies)$. Next, let

$$T' = T \cup \{ Ab(Tweety) \vee Ab(Sam) \}.$$

Then $\neg Ab(Sam)$ is an explanation of $\neg Ab(Tweety)$ from (T', Γ_{circ}). Hence, F is not a theorem of the circumscription of Ab in T'.

Skeptical prediction other than circumscription can also be characterized by credulous prediction. Instead of giving the hypotheses Γ_{circ}, any set Γ of hypotheses can be used in Theorem 3.6 as follows.

Corollary 3.8 Let (Σ, Γ) be an abductive theory. A formula F holds in every Theorist extension of (Σ, Γ) if and only if there exist explanations $E_1, \ldots, E_n$ $(n \geq 1)$ of F from (Σ, Γ) such that $\neg(E_1 \vee \ldots \vee E_n)$ has no explanation from (Σ, Γ).

3.4   Abduction and Other Nonmonotonic Formalization

Although we focused on default logic and circumscription as two major nonmonotonic formalization, abduction can also be used to represent other form of nonmonotonic reasoning. Here we briefly cite such work for reference. One of the most important results in this area is a formalization of nonmonotonic reasoning by means of *argumentation* framework [60,4,54]. In [4], an assumption-based framework $(\Sigma, \Gamma, \sim)$ is defined as a generalization of the Theorist framework. Here, like Theorist, Σ and Γ are defined as facts and hypotheses respectively, but are not restricted to first-order language. The mapping $\sim$ defines some notion of *contrary* of assumptions, and a *defeated argument* is defined as an augmented theory whose contrary is proved. Varying the underlying language of Σ and Γ and the notion of $\sim$, this framework is powerful enough to define the semantics of most nonmonotonic logics, including Theorist, default logic, *extended logic programs* [29], *autoepistemic logic* [74], other non-monotonic modal logics, and certain instances of circumscription. This framework is applied to defeasible rules in legal reasoning [60] and is related to other methods in abductive logic programming [54].

In [45], abduction is also related to autoepistemic logic and negation as failure in extended disjunctive logic programs. In particular, an autoepistemic translation of a hypothesis γ is given as

$$B\gamma \supset \gamma.$$

The set consisting of this autoepistemic formula produces two stable expansions, one containing γ and $B\gamma$, the other containing $\neg B\gamma$ but neither γ nor $\neg\gamma$. With this property, we can define the world in which γ is assumed to be true, while another world not assuming γ is also kept.

4   Computing Abduction via Automated Deduction

This section presents computational methods for abduction. In Section 2.1, we have seen that abduction can be characterized within first-order logic. Using this characterization, here we show a realization of automated abduction based on the *resolution principle*.

4.1   Consequence-Finding

As explained in Section 3.2, many abductive systems based on the resolution principle can be viewed as procedures that perform a kind of Reiter's *top-down default proofs*. Now, we see the underlying principle behind such abductive procedures from a different, purely *deductive*, viewpoint [35]. Firstly, the definition of abduction given in Section 2.1 can be represented as a consequence-finding problem, which is a problem of finding theorems of the given axiom set Σ.

The *consequence-finding* problem is firstly addressed by Lee in 1967 [61] in the context of Robinson's resolution principle [89]. Lee proved the completeness result that:

> Given a set of clauses Σ, if a clause C is a logical consequence of Σ, then the resolution principle can derive a clause D such that D implies C.

In this sense, the resolution principle is said *complete for consequence-finding*. In Lee's theorem, "D implies C" can be replaced with "D subsumes C". Later,

Slagle, Chang and Lee [95] and Minicozzi and Reiter [72] showed that "the resolution principle" can also be replaced with "semantic resolution" and "linear resolution", respectively. In practice, however, the set of theorems of an axiom set is generally infinite, and hence the complete deductive closure is neither obtainable nor desirable. Toward more practical automated consequence-finding, Inoue [35] reformulated the consequence-finding problem as follows.

> Given a set of clauses Σ and some criteria of "interesting" clauses, derive each "interesting" clause that is a logical consequence of Σ and is minimal with respect to subsumption.

Here, each interesting clause is called a *characteristic clause*. Criteria of interesting clauses are specified by a sub-vocabulary of the representation language called a *production field*. In the propositional case, each characteristic clause of Σ is a *prime implicate* of Σ.

The use of characteristic clauses enables us to characterize various reasoning problems of interest to AI, such as *nonmonotonic reasoning* [3,41,32,8], diagnosis [25,93], and *knowledge compilation* [69,15,90] as well as *abduction*. Moreover, for *inductive logic programming* (ILP), consequence-finding can be applied to generate hypothesis rules from examples and background knowledge [98,39], and is used as the theoretical background for discussing the completeness of ILP systems [76].[3] An extensive survey of consequence-finding in propositional logic is given by Marquis [68].

Now, characteristic clauses are formally defined as follows [35]. Let C and D be two clauses. C *subsumes* D if there is a substitution θ such that $C\theta \subseteq D$ and C has no more literals than D [66]. C *properly subsumes* D if C subsumes D but D does not subsume C. For a set of clauses Σ, $\mu\Sigma$ denotes the set of clauses in Σ not properly subsumed by any clause in Σ. A *production field* $\mathcal{P}$ is a pair, $\langle \mathbf{L}, Cond \rangle$, where $\mathbf{L}$ is a set of literals and is closed under instantiation, and $Cond$ is a certain condition to be satisfied. When $Cond$ is not specified, $\mathcal{P}$ is denoted as $\langle \mathbf{L} \rangle$. A clause C *belongs to* $\mathcal{P} = \langle \mathbf{L}, Cond \rangle$ if every literal in C belongs to $\mathbf{L}$ and C satisfies $Cond$. When Σ is a set of clauses, the set of logical consequence of Σ belonging to $\mathcal{P}$ is denoted as $Th_{\mathcal{P}}(\Sigma)$. Then, the *characteristic clauses* of Σ with respect to $\mathcal{P}$ are defined as:

$$Carc(\Sigma, \mathcal{P}) = \mu\, Th_{\mathcal{P}}(\Sigma).$$

Note that the empty clause $\square$ is the unique clause in $Carc(\Sigma, \mathcal{P})$ if and only if Σ is unsatisfiable. This means that proof-finding is a special case of consequence-finding.

When a new clause F is added to the set Σ of clauses, some consequences are newly derived with this new information. Such a new and "interesting" clause is called a "new" characteristic clauses. Formally, the *new characteristic clauses* of F with respect to Σ and $\mathcal{P}$ are defined as:

$$Newcarc(\Sigma, F, \mathcal{P}) = \mu\,[\,Th_{\mathcal{P}}(\Sigma \cup \{F\}) - Th(\Sigma)\,].$$

---

[3] In ILP, the completeness result of consequence-finding is often called the *subsumption theorem* [76], which was originally coined by Kowalski in 1970 [57].

The above definition is equivalent to the following [35]:

$$Newcarc(\Sigma, F, \mathcal{P}) = Carc(\Sigma \cup \{F\}, \mathcal{P}) - Carc(\Sigma, \mathcal{P}).$$

4.2 Abduction as Consequence-Finding

Now, we are ready to characterize abduction as consequence-finding. In the following, we denote the set of all literals in the representation language by $\mathcal{L}$, and a set Γ of hypotheses is defined as a subset of $\mathcal{L}$. Any subset E of Γ is identified with the conjunction of all elements in E. Also, for any set $\mathcal{T}$ of formulas, $\overline{\mathcal{T}}$ represents the set of formulas obtained by negating every formula in $\mathcal{T}$, i.e., $\overline{\mathcal{T}} = \{ \neg C \mid C \in \mathcal{T} \}$.

Let $G_1, \ldots, G_n$ be a finite number of observations, and suppose that they are all literals. We want to explain the observations $G = G_1 \wedge \ldots \wedge G_n$ from (Σ, Γ), where Σ is a set of clauses representing facts and Γ is a set of ground literals representing hypotheses. Let $E = E_1 \wedge \ldots \wedge E_k$ be any explanation of G from (Σ, Γ). Then, the following three hold:

1. $\Sigma \cup \{ E_1 \wedge \ldots \wedge E_k \} \models G_1 \wedge \ldots \wedge G_n$,
2. $\Sigma \cup \{ E_1 \wedge \ldots \wedge E_k \}$ is consistent,
3. Each E_i is an element of Γ.

These are equivallent to the next three conditions:

1'. $\Sigma \cup \{ \neg G_1 \vee \ldots \vee \neg G_n \} \models \neg E_1 \vee \ldots \vee \neg E_k$,
2'. $\Sigma \not\models \neg E_1 \vee \ldots \vee \neg E_k$,
3'. Each $\neg E_i$ is an element of $\overline{\Gamma}$.

By 1', a clause derived from the clause set Σ by adding the clause $\neg G$ is the negation of an explanation of G from (Σ, Γ), and this computation can be done as automated deduction over clauses.[4] By 2', such a derived clause must not be a consequence of Σ before adding $\neg G$. By 3', every literal appearing in such a clause must belong to $\overline{\Gamma}$. Moreover, E is a minimal explanation from (Σ, Γ) if and only if $\neg E$ is a minimal theorem from $\Sigma \cup \{\neg G\}$. Hence, the problem of abduction is reduced to the problem of seeking a clause such that (i) it is a minimal theorem of $\Sigma \cup \{\neg G\}$, but (ii) it is not a theorem of Σ alone, and (iii) it consists of literals only from $\overline{\Gamma}$. Therefore, we obtain the following result.

Theorem 4.1 [35]   Let (Σ, Γ) be an abductive theory, where $\Gamma \subseteq \mathcal{L}$. Put the production field as $\mathcal{P} = \langle \overline{\Gamma} \rangle$. Then, the set of minimal explanations of an observation G from (Σ, Γ) is:

$$\overline{Newcarc(\Sigma, \neg G, \mathcal{P})}.$$

---

[4] This way of computing hypotheses is often referred as "inverse entailment" in ILP [75,39]. Although there are some discussion against such a scheme of "abduction as deduction-in-reverse" [12], it is surely one of the most recognizable ways to construct possible hypotheses deductively.

In the above setting, we assumed that G is a conjunction of literals. Extending the form of each observation G_i to a clause is possible. When G is any formula, suppose that by converting $\neg G$ into the conjunctive normal form we obtain a formula $F = C_1 \wedge \cdots \wedge C_m$, where each C_i is a clause. In this case, $Newcarc(\Sigma, F, \mathcal{P})$ can be decomposed into m $Newcarc$ operations each of whose added new formula is a single clause [35]:

$$Newcarc(\Sigma, F, \mathcal{P}) = \mu \left[\bigcup_{i=1}^{m} Newcarc(\Sigma_i, C_i, \mathcal{P}) \right],$$

where $\Sigma_1 = \Sigma$, and $\Sigma_{i+1} = \Sigma_i \cup \{C_i\}$ for $i = 1, \ldots, m - 1$. This incremental computation can also be applied to get the characteristic clauses of Σ with respect to $\mathcal{P}$ as:

$$Carc(\Sigma, \mathcal{P}) = Newcarc(\emptyset, \Sigma, \mathcal{P}).$$

In Theorem 4.1, explanations obtained by a consequence-finding procedure are not necessarily ground and can contain variables. Note, however, that in implementing resolution-based abductive procedures, both the query G and its explanation E are usually considered as existentially quantified formulas. When G contains universally quantified variables, each of them is replaced with a new constant or function in $\neg G$ through Skolemization. Then, to get a universally quantified explanation in negating each new characteristic clause containing Skolem functions, we need to apply the *reverse Skolemization* algorithm [10]. For example, if $\neg P(x, \varphi(x), u, \psi(u))$ is a new characteristic clause where φ, ψ is a Skolem function, we get two explanations, $\exists x \forall y \exists u \forall v\, P(x, y, u, v)$ and $\exists u \forall v \exists x \forall y\, P(x, y, u, v)$ by reverse Skolemization.

Using Theorems 3.6 and 4.1, skeptical prediction can also be realized by consequence-finding procedures as follows.

Corollary 4.2 [41] Let $CIRC(\Sigma; \mathbf{P}; \mathbf{Z})$ be the circumscription of $\mathbf{P}$ in Σ with variables $\mathbf{Z}$. Put $\mathcal{P}_{circ} = \mathbf{P}^+ \cup \mathbf{Q}^+ \cup \mathbf{Q}^-$, where $\mathbf{Q}$ is the fixed predicates.

(1) For any formula F not containing literals from $\mathbf{Z}$,
    $CIRC(\Sigma; \mathbf{P}; \mathbf{Z}) \models F$ if and only if $Newcarc(\Sigma, F, \mathcal{P}_{circ}) = \emptyset$.
(2) For any formula F, $CIRC(\Sigma; \mathbf{P}; \mathbf{Z}) \models F$ if and only if there is a conjunction G of clauses from $Newcarc(\Sigma, \neg F, \mathcal{P}_{circ})$ such that $Newcarc(\Sigma, \neg G, \mathcal{P}_{circ}) = \emptyset$.

4.3 SOL Resolution

To compute new characteristic clauses, Inoue [35] defined an extension of the *Model Elimination* (ME) calculus [59,7,66] by adding the **Skip** rule to ME. The extension is called *SOL resolution*, and can be viewed either as OL resolution [7] (or *SL resolution* [59]) augmented with the Skip rule, or as a first-order generalization of Siegel's propositional *production algorithm* [93]. Note here that, although ME is complete for proof-finding (i.e., refutation-complete) [66], it is not complete for consequence-finding [72]. SOL resolution is useful for computing the (new) characteristic clauses for the following reasons.

(1) In computing $Newcarc(\Sigma, C, \mathcal{P})$, SOL resolution treats a newly added clause C as the *top clause* (or a *start clause*) input to ME. This is a desirable feature for consequence-finding since the procedure can directly derive the theorems relevant to the added information.

(2) It is easy to focus on producing only those theorems belonging to the production field. This is implemented by allowing an ME procedure to skip the selected literal belonging to $\mathcal{P}$. In other words, SOL resolution is restricted to searching only characteristic clauses.

Here, we show a definition of SOL resolution based on [35]. An *ordered* clause is a sequence of literals possibly containing *framed literals* which represent literals that have been resolved upon. A *structured* clause $\langle P, Q \rangle$ is a pair of a clause P and an ordered clause Q, whose clausal meaning is $P \cup Q$.

Definition 4.3 (SOL Resolution) Given a set of clauses Σ, a clause C, and a production field $\mathcal{P}$, an *SOL-deduction of a clause S from $\Sigma + C$ and $\mathcal{P}$* consists of a sequence of structured clauses, $D_0, D_1, \ldots, D_n$, such that:

1. $D_0 = \langle \Box, C \rangle$.
2. $D_n = \langle S, \Box \rangle$.
3. For each $D_i = \langle P_i, Q_i \rangle$, $P_i \cup Q_i$ is not a tautology.
4. For each $D_i = \langle P_i, Q_i \rangle$, Q_i is not subsumed by any Q_j with the empty substitution, where $D_j = \langle P_j, Q_j \rangle$ is a previous structured clause, $j < i$.
5. For each $D_i = \langle P_i, Q_i \rangle$, P_i belongs to $\mathcal{P}$.
6. $D_{i+1} = \langle P_{i+1}, Q_{i+1} \rangle$ is generated from $D_i = \langle P_i, Q_i \rangle$ according to the following steps:
   (a) Let l be the selected literal in Q_i. P_{i+1} and R_{i+1} are obtained by applying one of the rules:
      i. **(Skip)** If $P_i \cup \{l\}$ belongs to $\mathcal{P}$, then $P_{i+1} = P_i \cup \{l\}$ and R_{i+1} is the ordered clause obtained by removing l from Q_i.
      ii. **(Resolve)** If there is a clause B_i in $\Sigma \cup \{C\}$ such that $\neg k \in B_i$ and l and k are unifiable with mgu θ, then $P_{i+1} = P_i \theta$ and R_{i+1} is an ordered clause obtained by concatenating $B_i \theta$ and $Q_i \theta$, framing $l\theta$, and removing $\neg k\theta$.
      iii. **(Reduce)** If either
         A. P_i or Q_i contains an unframed literal k (*factoring/merge*) or
         B. Q_i contains a framed literal $\boxed{\neg k}$ (*ancestry*),
         and l and k are unifiable with mgu θ, then $P_{i+1} = P_i \theta$ and R_{i+1} is obtained from $Q_i \theta$ by deleting $l\theta$.
   (b) Q_{i+1} is obtained from R_{i+1} by deleting every framed literal not preceded by an unframed literal in the remainder (*truncation*).

When the Skip rule is applied to the selected literal in an SOL deduction, it is never solved by applying any resolution. To apply this rule, the selected literal has to belong to the production field. When a deduction with the top clause C is completed, that is, every literal is either solved or skipped, those skipped literals are collected and output. This output clause is a logical consequence of $\Sigma \cup \{C\}$

and every literal in it belongs to the production field $\mathcal{P}$. Note that when both Skip and resolution can be applied to the selected literal, these two rules are chosen non-deterministically. In [35], it is proved that SOL resolution is complete for both consequence-finding and finding (new) characteristic clauses. In [99], SOL resolution is implemented using the Weak Model Elimination method [66]. In [49], various pruning methods are introduced to enhance the efficiency of SOL resolution in a *connection-tableau* format [62]. In [16], del Val defines a variant of consequence-finding procedure for finding characteristic clauses, which is based on ordered resolution instead of Model Elimination.

Example 4.4 [35] Suppose that Σ consists of the two clauses:

$$(1) \quad \neg P(x) \vee Q(y,y) \vee R(z,x),$$
$$(2) \quad \neg Q(x,y) \vee R(x,y).$$

Suppose also that the set of hypotheses is given as

$$\Gamma = \{P\}^+.$$

Then the production field is $\mathcal{P} = \langle \overline{\Gamma} \rangle = \langle \{P\}^- \rangle$. Now, consider the query,

$$G = R(A,x),$$

where the variable x is interpreted as existentially quantified, and we want to compute its *answer substitution*. The first SOL-deduction from $\Sigma + \neg G$ and $\mathcal{P}$ is as follows:

$(3) \qquad \langle\, \square\, ,\, \neg R(A,x)\, \rangle,$ top clause

$(4) \qquad \langle\, \square\, ,\, \neg P(x) \vee Q(y,y) \vee \boxed{\neg R(A,x)}\, \rangle,$ resolution with (1)

$(5) \quad \langle\, \neg P(x)\, ,\, Q(y,y) \vee \boxed{\neg R(A,x)}\, \rangle,$ skip

$(6) \quad \langle\, \neg P(x)\, ,\, R(y,y) \vee \boxed{Q(y,y)} \vee \boxed{\neg R(A,x)}\, \rangle,$ resolution with (2)

$(7a) \quad \langle\, \neg P(A)\, ,\, \boxed{Q(A,A)} \vee \boxed{\neg R(A,A)}\, \rangle,$ ancestry

$(7b) \quad \langle\, \neg P(A)\, ,\, \square\, \rangle.$ truncation

In the above SOL-deduction, $P(A)$ is an explanation of the *answer* $R(A,A)$ from (Σ, Γ). Namely,

$$\Sigma \models P(A) \supset R(A,A).$$

The second SOL-deduction from $\Sigma + \neg G$ and $\mathcal{P}$ takes the same four steps as the above (3)–(6), but instead of applying ancestry at (7), $R(y,y)$ is resolved upon against the clause $\neg R(A,x')$, yielding

$(7a') \quad \langle\, \neg P(x)\, ,\, \boxed{R(A,A)} \vee \boxed{Q(A,A)} \vee \boxed{\neg R(A,x)}\, \rangle,$

$(7b') \quad \langle\, \neg P(x)\, ,\, \square\, \rangle.$

In this case, $\neg G$ is used twice in the SOL-deduction. Note that $P(x)$ is not an explanation of any definite answer. It represents that for any term t, $P(t)$ is an explanation of the *indefinite answer* $R(A,t) \vee R(A,A)$. Namely,

$$\Sigma \models \forall x(\, P(x) \supset R(A,x) \vee R(A,A)\,).$$

By Theorem 4.1 and the completeness result of SOL resolution, we can guarantee the completeness for finding explanations from first-order abductive theories. In contrast, the completeness does not hold for abductive procedures like [85,10], in which hypothesizing literals is allowed only when resolution cannot be applied for selected literals. The hypothesized, unresolved literals are "dead-ends" of deductions, and explanations obtained in this way are *most-specific* [96]. This kind of abductive computation can also be implemented in a variant of SOL resolution, called *SOL-R resolution* [35], by preferring resolution to Skip whenever both can be applied. On the other hand, there is another variant of SOL resolution, called *SOL-S resolution* [35], in which only Skip is applied by ignoring the possibility of resolution when the selected literal belongs to $\mathcal{P}$. Each explanation obtained by using SOL-S resolution is called a *least-specific* explanation [96]. While most-specific explanations are often useful for application to diagnosis [85,10], least-specific explanations are used in natural language understanding [96] and computing circumscription by Corollary 4.2 [41].

4.4   Bottom-Up Abduction

As shown by Reiter and de Kleer [88], an *assumption-based truth maintenance system* (ATMS) [14] is a propositional abductive system. In ATMS, facts are given as propositional Horn clauses and hypotheses are propositional atoms [63,34,92]. An extension of ATMS, which allows non-Horn propositional clauses for facts and propositional literals for hypotheses, is called a *clause management system* (CMS) [88]. The task of CMS is to compute the set of all minimal explanations of a literal G from (Σ, Γ), where Σ is a set of propositional clauses and $\Gamma \subseteq \mathcal{L}$ is a set of hypotheses. In ATMS, the minimal explanations of an atom G is called the *label* of G.

The *label updating algorithm* of ATMS [14] computes the label of every propositional atom in a bottom-up manner. This algorithm can be logically understood as a fixpoint computation of the following *semantic resolution*. Let Γ be a set of propositional atoms, and Σ be a set of propositional Horn clauses. Suppose that N is either *false* or any atom appearing in Σ, and that N_i $(1 \leq i \leq m; m \geq 0)$ is any atom and $A_{i,j}$ $(1 \leq i \leq m; 1 \leq j \leq n_i; n_i \geq 0)$ is an element of Γ. Then, a *clash* in semantic resolution of the form:

$$N_1 \wedge \ldots \wedge N_m \supset N$$
$$\underline{A_{i,1} \wedge \ldots \wedge A_{i,n_i} \supset N_i \,, \text{ for all } i = 1, \ldots, m}$$
$$\left(\bigwedge_{1 \leq i \leq m, \, 1 \leq j \leq n_i} A_{i,j} \right) \supset N$$

represents multiple applications of resolution. The label updating algorithm of ATMS takes each clause in Σ as input one by one, applies the above clash as many as possible, and incrementally computes every theorem of Σ that are not subsumed by any other theorem of Σ. Then, each resultant minimal theorem obtained by this computation yields a prime implicate of Σ. Now, let $PI(\Sigma, \Gamma)$ be the set of such prime implicates. The label of an atom N is obtained as

$$\{\{A_1, \ldots, A_k\} \subseteq \Gamma \mid \neg A_1 \vee \ldots \vee \neg A_k \vee N \in PI(\Sigma, \Gamma)\}.$$

In particular, each element in the label of $false$ is called a *nogood*, which is obtained as the negation of each negative clause from $PI(\Sigma, \Gamma)$. Nogoods are useful for recognizing forbidden combinations of hypotheses in many AI applications, and work as *integrity constraints* saying that those atoms cannot be assumed simultaneously. A typical implementation of the label updating algorithm performs the above clash computation for an atom N by: (i) generating the product of the labels of antecedent atoms of N, (ii) eliminating each element which is a superset of some nogood, and (iii) eliminating every non-minimal element from the rest. Although ATMS works for propositional abduction only, a similar clash rule that is complete for first-order abduction is also proposed in [18], and a method to simulate the above crash using hyperresolution is proposed for first-order abductive theories in [97].

Example 4.5 Let (Σ, Γ) be a propositional Horn abductive theory such that

$$\Sigma = \{\, A \wedge B \supset P, \quad C \supset P,$$
$$B \wedge C \supset Q, \quad D \supset Q,$$
$$P \wedge Q \supset R, \quad C \wedge D \supset false\,\},$$
$$\Gamma = \{\, A, B, C, D\,\}.$$

We here presuppose the existence of tautology $\alpha \supset \alpha$ in Σ for each assumption $\alpha \in \Gamma$, i.e.,

$$A \supset A, \quad B \supset B, \quad C \supset C, \quad D \supset D.$$

Then, the label of each non-assumption atom is computed as:

$$P: \quad \{\{A, B\}, \{C\}\},$$
$$Q: \quad \{\{B, C\}, \{D\}\},$$
$$R: \quad \{\{B, C\}, \{A, B, D\}\},$$
$$false: \quad \{\{C, D\}\}.$$

To compute the label of R in ATMS, we firstly construct the product of P and Q's labels as

$$\{\{A, B, C\}, \{B, C\}, \{A, B, D\}, \{C, D\}\},$$

then eliminate $\{C, D\}$ as a nogood and $\{A, B, C\}$ as a superset of $\{B, C\}$.

The above label updating method from [14] cannot be directly used when Σ contains non-Horn clauses. This is because semantic resolution in the above form is not deductively complete for non-Horn clauses. For a full CMS, the *level saturation* method is proposed in [88], which involves computation of all prime implicates of Σ. In [34], it is shown that a sound and complete procedure of CMS/ATMS can be provided using SOL resolution, without computing all prime implicates of Σ, for both label generating and label updating.

Example 4.6 Consider a propositional abductive theory (Σ, Γ), where

$$\Sigma = \{ P \vee Q, \quad \neg B \vee P \},$$
$$\Gamma = \{ A, B \}.$$

Let N be the set of all atoms appearing in Σ. We set the production field as

$$\mathcal{P}^* = \langle \overline{\Gamma} \cup G, \text{ the number of literals from } N - \overline{\Gamma} \text{ is at most one} \rangle.$$

Then, $Carc(\Sigma, \mathcal{P}^*)$ in this case is equivalent to Σ. While P has the label $\{\{B\}\}$, Q's label is empty. Now, suppose that a new clause,

$$\neg A \vee \neg P,$$

is added to Σ. Then, an updating algorithm based on SOL resolution finds Q's new label $\{\{A\}\}$, as well as a new nogood $\{A, B\}$:

$$\langle \square, \underline{\neg A} \vee \neg P \rangle,$$
$$\langle \neg A, \underline{\neg P} \rangle,$$

$$\langle \neg A, Q \vee \boxed{\neg P} \rangle, \qquad\qquad \langle \neg A, \underline{\neg B} \vee \boxed{\neg P} \rangle,$$
$$\langle \neg A \vee Q, \boxed{\neg P} \rangle, \qquad\qquad \langle \neg A \vee \neg B, \boxed{\neg P} \rangle,$$
$$\langle \neg A \vee Q, \square \rangle. \qquad\qquad \langle \neg A \vee \neg B, \square \rangle.$$

Abductive procedures based on *Clark completion* [9,55,28,47] also perform computation of abduction in a deductive manner. This kind of abductive procedures is often used in implementing *abductive logic programming*. Inoue *et al.* [42] develop a model generation procedure for bottom-up abduction based on a translation in [44], which applies the Skip rule of SOL resolution [35] in model generation. Abductive procedures that combine top-down and bottom-up approaches are also proposed in two ways: one is to achieve the goal-directedness in bottom-up procedures [77,42,97], and the other is to utilize derived lemmas in top-down methods [49]. Other than these resolution-based procedures, Cialdea Mayer and Pirri [11] propose tableau and sequent calculi for first-order abduction.

4.5 Computational Complexity

The computational complexity of abduction has been extensively studied. First, in the case that the background knowledge is expressed in first-order logic as in Section 2.1, the problem of finding an explanation that is consistent with Σ is not semi-decidable. That is, the problem of deciding the satisfiability of an axiom set is undecidable for first-order logic in general, hence computing an explanation is not decidable even if there exists an explanation. For the consequence-finding problem in Section 4.1, the set of characteristic clauses of Σ is not even recursively enumerable [48]. Similarly, the set of *new* characteristic clauses of F with respect to Σ, which is used to characterize explanations in abduction (Theorem 4.1), involves computation as whether a derived formula is *not* a logical consequence of Σ, which cannot be necessarily determined in a finite amount of time. Hence, to check if a set E of hypotheses obtained in a top-down default proof or SOL resolution is in fact consistent with Σ, we need some approximation like a procedure which makes a theory consistent whenever a refutation-complete theorem prover cannot succeed to prove $\neg E$ in a finite amount of time.

Next, in the propositional case, the computational complexity of abduction is studied in [6,92,21]. From the theory of enumerating prime implicates, it is known that the number of explanations grows exponentially as number of clauses or propositions grows. Selman and Levesque [92] show that finding even *one* explanation of an atom from a Horn theory and a set of atomic hypotheses is NP-hard. Therefore, even if we abandon the completeness of explanations, it is still intractable. However, if we do not restrict a set Γ of hypotheses and can hypothesize any atom to construct explanations, an explanation can be found in polynomial time. Hence, the restriction of abducible atoms is a source of complexity. On the other hand, as analyses by [6,21] show, the intrinsic difficulty also lies in checking the consistency of explanations, and the inclusion of negative clauses in a theory increases the complexity. Another source of complexity lies in the requirement of minimality for abductive explanations [21]. However, some tractable classes of abductive theories have also been discovered [23,17].

Thus, in propositional abduction, it is unlikely that there exists a polynomial-time algorithm for abductive explanations in general. We can consider approximation of abduction, by discarding either the consistency or the soundness. However, we should notice that showing that a logical framework of abduction or default reasoning is undecidable or intractable does not mean that it is useless. Since they are intrinsically difficult problems (consider, for instance, scientific discovery as the process of abduction), what we would like to know is that representing a problem in such a framework does not increase the computational complexity of the original problem.

5   Final Remark

5.1   Problems to Be Addressed

In this article, we observed that automated abduction involves automated deduction in some way. However, clarifying the relationship between abduction and deduction is just a first step towards a mechanization of Peirce's abduction. There are many future research topics in automated abduction, which include fundamental problems of abduction, applications of abduction, and computational problems of abduction. Some of these problems are also listed in [51] in this volume, and some philosophical problems are discussed in [26,67]. As a fundamental problem of abduction, we have not yet fully understood the human mechanism of explanation and prediction. The formalization in this article only reflects a small part of the whole. Most importantly, there are non-logical aspects of abduction, which are hard to be represented. The mechanization of *hypothesis selection* is one of the most challenging topics. Research on acquiring meta-knowledge like preference among explanations [47] and inventing new abducible hypotheses [40] is related to increase the quality of explanations in abduction.

For computational problems, this article showed a directly mechanized way to compute abduction. There are another approach for computation, which translates the abduction problem into other technologies developed in AI. For example, some classes of abductive theories can be transformed into propositional satisfiability and other nonmonotonic formalizations for which efficient solvers exist. Such indirect approaches are taken in recent applications involving assumption-based reasoning such as planning and diagnoses. One might think that nonmonotonic logic programming such as the stable model semantics or default logic is enough for reasoning under incomplete information when they are as expressive as the class of abductive theories. The question as to why we need abductive theories should be answered by considering the role of abduction in application domains. One may often understand abductive theories more easily and intuitively than theories represented in other nonmonotonic logics. For example, in diagnostic domains, background knowledge contains cause-effect relations and hypotheses are written as a set of causes. In the process of theory formation, incomplete knowledge is naturally represented in the form of hypothetical rules. We thus can use an abductive framework as a high-level description language while computation of abduction can be compiled into other technologies.

5.2   Towards Mechanization of Scientific Reasoning

Let us recall Peirce's theory of scientific reasoning. His theory of scientific discovery relies on the cycle of "experiment, observation, hypothesis generation, hypothesis verification, and hypothesis revision". Peirce mentions that this process involves all modes of reasoning; abduction takes place at the first stage of scientific reasoning, deduction follows to derive the consequences of the hypotheses that were given by abduction, and finally, induction is used to verify that those hypotheses are true. According to this viewpoint, let us review the logic of abduction:

$$(1)\ Facts \cup Explanation \models Observation.$$

$$(2)\ Facts \cup Explanation\ \ is\ consistent.$$

A possible interpretation of this form of hypothetical reasoning is now as follows. The formula (1) is the process of abduction, or the fallacy of affirming the consequent. The consistency check (2), on the other hand, is the place where deduction plays a role. Since our knowledge about the world may be incomplete, we should experiment with the consequences using an inductive manner in order to verify that the hypotheses are consistent with the knowledge base. At the same time, the process of inductive generalization or the synthesis from examples involves abduction too. This phenomenon of human reasoning is also discussed by Flach and Kakas [27] as the "cycle" of abductive and inductive knowledge development.

When we are given some examples, we first make hypotheses. While previous AI approaches for inductive generalization often enumerated all the possible forms of formulas, abduction would help to restrict the search space. Additional heuristics, once they are formalized, would also be helpful for constructing the hypotheses. Investigation on *knowledge assimilation* involving abduction, deduction and induction will become more and more important in AI research in the 21st century.

Acknowledgements.   Discussion with many researchers were very helpful in preparing this article. In particular, Bob Kowalski gave me valuable comments on an earlier draft of this article. I would also like to thank Toshihide Ibaraki, Koji Iwanuma, Chiaki Sakama, Ken Satoh, and Hiromasa Haneda for their suggestions on this work.

References

1. Chitta Baral.   Abductive reasoning through filtering.   *Artificial Intelligence*, 120:1–28, 2000.
2. Nicole Bidoit and Christine Froidevaux. Minimalism subsumes default logic and circumscription. In: *Proceedings of LICS-87*, pages 89–97, 1987.
3. Genevieve Bossu and Pierre Siegel. Saturation, nonmonotonic reasoning, and the closed-world assumption. *Artificial Intelligence*, 25:13–63, 1985.
4. A. Bondarenko, P. M. Dung, R. A. Kowalski, and F. Toni.   An abstract, argumentation-theoretic approach to default reasoning. *Artificial Intelligence*, 93:63–101, 1997.
5. Craig Boutilier and Verónica Becher. Abduction as belief revision. *Artificial Intelligence*, 77:43–94, 1995.
6. Tom Bylander, Dean Allemang, Michael C. Tanner, and John R. Josephson. The computational complexity of abduction. *Artificial Intelligence*, 49:25–60, 1991.
7. Chin-Liang Chang and Richard Char-Tung Lee. *Symbolic Logic and Mechanical Theorem Proving*. Academic Press, New York, 1973.
8. Viorica Ciorba. A query answering algorithm for Lukaszewicz' general open default theory. In: *Proceedings of JELIA '96*, Lecture Notes in Artificial Intelligence, 1126, pages 208–223, Springer, 1996.
9. Luca Console, Daniele Theseider Dupre, and Pietro Torasso. On the relationship between abduction and deduction. *Journal of Logic and Computation*, 1:661–690, 1991.
10. P.T. Cox and T. Pietrzykowski. Causes for events: their computation and applications. In: *Proceedings of the 8th International Conference on Automated Deduction*, Lecture Notes in Computer Science, 230, pages 608–621, Springer, 1986.
11. Marita Cialdea Mayer and Fiora Pirri. First order abduction via tableau and sequent calculi. *Journal of the IGPL*, 1(1):99–117, 1993.
12. Marita Cialdea Mayer and Fiora Pirri. Abduction is not deduction-in-reverse. *Journal of the IGPL*, 4(1):95–108, 1996.
13. Hendrik Decker. An extension of SLD by abduction and integrity maintenance for view updating in deductive databases. In: *Proceedings of the 1996 Joint International Conference and Symposium on Logic Programming*, pages 157–169, MIT Press, 1996.

14. Johan de Kleer. An assumption-based TMS. *Artificial Intelligence*, 28:127–162, 1986.
15. Alvaro del Val. Approximate knowledge compilation: the first order case. In: *Proceedings of AAAI-96*, pages 498–503, AAAI Press, 1996.
16. Alvaro del Val. A new method for consequence finding and compilation in restricted languages. In: *Proceedings of AAAI-99*, pages 259–264, AAAI Press, 1999.
17. Alvaro del Val. On some tractable classes in deduction and abduction. *Artificial Intelligence*, 116:297–313, 2000.
18. Robert Demolombe and Luis Fariñas del Cerro. An inference rule for hypothesis generation. In: *Proceedings of IJCAI-91*, pages 152–157, 1991.
19. Marc Denecker and Danny De Schreye. SLDNFA: an abductive procedure for abductive logic programs. *Journal of Logic Programming*, 34:111–167, 1998.
20. Marc Denecker and Antonis Kakas, editors. Special Issue: Abductive Logic Programming. *Journal of Logic Programming*, 44(1–3), 2000.
21. Thomas Eiter and George Gottlob. The complexity of logic-based abduction. *Journal of the ACM*, 42(1):3–42, 1995.
22. Thomas Eiter, George Gottlob, and Nicola Leone. Semantics and complexity of abduction from default theories. *Artificial Intelligence*, 90:177–223, 1997.
23. Kave Eshghi. A tractable class of abduction problems. In: *Proceedings of IJCAI-93*, pages 3–8, 1993.
24. David W. Etherington. *Reasoning with Incomplete Information*. Pitman, London, 1988.
25. Joseph J. Finger. Exploiting constraints in design synthesis. Ph.D. Dissertation, Technical Report STAN-CS-88-1204, Department of Computer Science, Stanford University, Stanford, CA, 1987.
26. Peter A. Flach and Antonis C. Kakas, editors. *Abduction and Induction—Essays on their Relation and Integration*. Kluwer Academic, 2000.
27. Peter A. Flach and Antonis C. Kakas. Abductive and inductive reasoning: background and issues. In: [26], pages 1–27, 2000.
28. T. H. Fung and R. Kowalski. The iff procedure for abductive logic programming. *Journal of Logic Programming*, 33:151–165, 1997.
29. Michael Gelfond and Vladimir Lifschitz. Classical negation in logic programs and disjunctive databases. *New Generation Computing*, 9:365–385, 1991.
30. Michael Gelfond, Halina Przymusinska, and Teodor Przymusinski. On the relationship between circumscription and negation as failure. *Artificial Intelligence*, 38:75–94, 1989.
31. Matthew L. Ginsberg. A circumscriptive theorem prover. *Artificial Intelligence*, 39:209–230, 1989.
32. Nicolas Helft, Katsumi Inoue, and David Poole. Query answering in circumscription. In: *Proceedings of IJCAI-91*, pages 426–431, 1991.
33. Carl Gustav Hempel. *Philosophy of Natural Science*. Prentice-Hall, New Jersey, 1966.
34. Katsumi Inoue. An abductive procedure for the CMS/ATMS. In: João P. Martins and Michael Reinfrank, editors, *Truth Maintenance Systems*, Lecture Notes in Artificial Intelligence, 515, pages 34–53, Springer, 1991.
35. Katsumi Inoue. Linear resolution for consequence finding. *Artificial Intelligence*, 56:301–353, 1992.
36. Katsumi Inoue. Studies on abductive and nonmonotonic reasoning. Doctoral Dissertation, Kyoto University, Kyoto, 1992.

37. Katsumi Inoue. Principles of abduction. *Journal of Japanese Society for Artificial Intelligence*, 7(1):48–59, 1992 (in Japanese).
38. Katsumi Inoue. Hypothetical reasoning in logic programs. *Journal of Logic Programming*, 18(3):191–227, 1994.
39. Katsumi Inoue. Induction, abduction, and consequence-finding. In: Céline Rouveirol and Michèle Sebag, editors, *Proceedings of the 11th International Conference on Inductive Logic Programming*, Lecture Notes in Artificial Intelligence, 2157, pages 65–79, Springer, 2001.
40. Katsumi Inoue and Hiromasa Haneda. Learning abductive and nonmonotonic logic programs. In: [26], pages 213–231, 2000.
41. Katsumi Inoue and Nicolas Helft. On theorem provers for circumscription. In: Peter F. Patel-Schneider, editor, *Proceedings of the 8th Biennial Conference of the Canadian Society for Computational Studies of Intelligence*, pages 212–219, Morgan Kaufmann, 1990.
42. Katsumi Inoue, Yoshihiko Ohta, Ryuzo Hasegawa, and Makoto Nakashima. Bottom-up abduction by model generation. In: *Proceedings of IJCAI-93*, pages 102–108, Morgan Kaufmann, 1993.
43. Katsumi Inoue and Chiaki Sakama. Abductive framework for nonmonotonic theory change. In: *Proceedings of IJCAI-95*, pages 204–210, Morgan Kaufmann, 1995.
44. Katsumi Inoue and Chiaki Sakama. A fixpoint characterization of abductive logic programs. *Journal of Logic Programming*, 27(2):107–136, 1996.
45. Katsumi Inoue and Chiaki Sakama. Negation as failure in the head. *Journal of Logic Programming*, 35(1):39–78, 1998.
46. Katsumi Inoue and Chiaki Sakama. Abducing priorities to derive intended conclusions. In: *Proceedings of IJCAI-99*, pages 44–49, Morgan Kaufmann, 1999.
47. Katsumi Inoue and Chiaki Sakama. Computing extended abduction through transaction programs. *Annals of Mathematics and Artificial Intelligence*, 25(3,4):339-367, 1999.
48. Koji Iwanuma and Katsumi Inoue. Minimal conditional answer computation and SOL. To appear, 2002.
49. Koji Iwanuma, Katsumi Inoue, and Ken Satoh. Completeness of pruning methods for consequence finding procedure SOL. In: Peter Baumgartner and Hantao Zhang, editors, *Proceedings of the 3rd International Workshop on First-Order Theorem Proving*, pages 89–100, Research Report 5-2000, Institute for Computer Science, University of Koblenz, Germany, 2000.
50. John R. Jpsephson and Susan G.Josephson. *Abductive Inference: Computation, Philosophy, Technology*. Cambridge University Press, 1994.
51. Antonis Kakas and Marc Denecker. Abductive logic programming. In this volume, 2002.
52. A.C. Kakas and P. Mancarella. Generalized stable models: a semantics for abduction. In: *Proceedings of ECAI-90*, pages 385–391, 1990.
53. A. C. Kakas, R. A. Kowalski, and F. Toni. Abductive logic programming. *Journal of Logic and Computation*, 2:719–770, 1992.
54. A. C. Kakas, R. A. Kowalski, and F. Toni. The role of abduction in logic programming. In: Dov M. Gabbay, C. J. Hogger, and J. A. Robinson, editors, *Handbook of Logic in Artificial Intelligence and Logic Programming*, Volume 5, pages 235–324, Oxford University Press, 1998.
55. Kurt Konolige. Abduction versus closure in causal theories. *Artificial Intelligence*, 53:255–272, 1992.

56. Kurt Konolige. Abductive theories in artificial intelligence. In: Gerhard Brewka, editor, *Principles of Knowledge Representation*, pages 129–152, CSLI Publications & FoLLI, 1996.

57. R. Kowalski. The case for using equality axioms in automated demonstration. In: *Proceedings of the IRIA Symposium on Automatic Demonstration*, Lecture Notes in Mathematics, 125, pages 112–127, Springer, 1970.

58. Robert A. Kowalski. *Logic for Problem Solving*. Elsevier, New York, 1979.

59. Robert Kowalski and Donald G. Kuehner. Linear resolution with selection function. *Artificial Intelligence*, 2:227–260, 1971.

60. Robert A. Kowalski and Francesca Toni. Abstract argumentation. *Artificial Intelligence and Law*, 4:275–296, 1996.

61. Char-Tung Lee. A completeness theorem and computer program for finding theorems derivable from given axioms. Ph.D. thesis, Department of Electrical Engineering and Computer Science, University of California, Berkeley, CA, 1967.

62. R. Letz, K. Mayer, and C. Goller. Controlled integration of the cut rule into connection tableau calculi. *Journal of Automated Reasoning*, 13(3):297–337, 1994.

63. Hector J. Levesque. A knowledge-level account of abduction (preliminary version). In: *Proceedings of IJCAI-89*, pages 1061–1067, 1989.

64. Vladimir Lifschitz. Computing circumscription. In: *Proceedings of IJCAI-85*, pages 121–127, 1985.

65. Jorge Lobo and Carlos Uzcátegui. Abductive consequence relations. *Artificial Intelligence*, 89:149–171, 1997.

66. Donald W. Loveland. *Automated Theorem Proving: A Logical Basis*. North-Holland, Amsterdam, 1978.

67. Lorenzo Magnani. *Abduction, Reason, and Science—Processes of Discovery and Explanation*. Kluwer Academic, 2001.

68. Pierre Marquis. Consequence finding algorithms. In: Dov M. Gabbay and Philippe Smets, editors, *Handbook for Defeasible Reasoning and Uncertain Management Systems*, Volume 5, pages 41–145, Kluwer Academic, 2000.

69. Philippe Mathieu and Jean-Paul Delahaye. A kind of logical compilation for knowledge bases. *Theoretical Computer Science*, 131:197–218, 1994.

70. John McCarthy. Circumscription—a form of non-monotonic reasoning. *Artificial Intelligence*, 13:27–39, 1980.

71. John McCarthy. Applications of circumscription to formalizing common-sense knowledge. *Artificial Intelligence*, 28:89–116, 1986.

72. Eliana Minicozzi and Raymond Reiter. A note on linear resolution strategies in consequence-finding. *Artificial Intelligence*, 3:175–180, 1972.

73. Jack Minker. On indefinite databases and the closed world assumption. In: *Proceedings of the 6th International Conference on Automated Deduction*, Lecture Notes in Computer Science, 138, pages 292–308, Springer, 1982.

74. Robert C. Moore. Semantical considerations on nonmonotonic logic. *Artificial Intelligence*, 25:75–94, 1985.

75. Stephen Muggleton. Inverse entailment and Progol. *New Generation Computing*, 13:245–286, 1995.

76. Shan-Hwei Nienhuys-Cheng and Ronald de Wolf. *Foundations of Inductive Logic Programming*. Lecture Notes in Artificial Intelligence, 1228, Springer, 1997.

77. Yoshihiko Ohta and Katsumi Inoue. Incorporating top-down information into bottom-up hypothetical reasoning. *New Generation Computing*, 11:401–421, 1993.

78. Gabriele Paul. AI approaches to abduction. In: Dov M. Gabbay and Philippe Smets, editors, *Handbook for Defeasible Reasoning and Uncertain Management Systems*, Volume 4, pages 35–98, Kluwer Academic, 2000.
79. Charles Sanders Peirce. *Elements of Logic*. In: Charles Hartshorne and Paul Weiss, editors, *Collected Papers of Charles Sanders Peirce*, Volume II, Harvard University Press, Cambridge, MA, 1932.
80. Ramón Pino-Pérez and Carlos Uzcátegui. Jumping to explanations versus jumping to conclusions. *Artificial Intelligence*, 111:131–169, 1999.
81. David Poole. A logical framework for default reasoning. *Artificial Intelligence*, 36:27–47, 1988.
82. David Poole. Explanation and prediction: an architecture for default and abductive reasoning. *Computational Intelligence*, 5:97–110, 1989.
83. David Poole. Compiling a default reasoning system into Prolog. *New Generation Computing*, 9:3–38, 1991.
84. David Poole, Randy Goebel, and Romas Aleliunas. Theorist: a logical reasoning system for defaults and diagnosis. In: Nick Cercone and Gordon McCalla, editors, *The Knowledge Frontier: Essays in the Representation of Knowledge*, pages 331–352, Springer, New York, 1987.
85. Harry E. Pople, Jr. On the mechanization of abductive logic. In: *Proceedings of IJCAI-73*, pages 147–152, 1973.
86. Teodor C. Przymusinski. An algorithm to compute circumscription. *Artificial Intelligence*, 38:49–73, 1989.
87. Raymond Reiter. A logic for default reasoning. *Artificial Intelligence*, 13:81–132, 1980.
88. Raymond Reiter and Johan de Kleer. Foundations of assumption-based truth maintenance systems: preliminary report. In: *Proceedings of AAAI-87*, pages 183–187, 1987.
89. J.A. Robinson. A machine-oriented logic based on the resolution principle. *Journal of the ACM*, 12:23–41, 1965.
90. Olivier Roussel and Philippe Mathieu. Exact knowledge compilation in predicate calculus: the partial achievement case. In: *Proceedings of the 14th International Conference on Automated Deduction*, Lecture Notes in Artificial Intelligence, 1249, pages 161–175, Springer, 1997.
91. Murray Shanahan. Prediction is deduction but explanation is abduction. In: *Proceedings of IJCAI-89*, pages 1055–1060, Morgan Kaufmann, 1989.
92. Bart Selman and Hector J. Levesque. Support set selection for abductive and default reasoning. *Artificial Intelligence*, 82:259–272, 1996.
93. Pierre Siegel, Représentation et utilization de la connaissance en calcul propositionnel. Thèse d'État, Université d'Aix-Marseille II, Luminy, France, 1987 (in French).
94. Pierre Siegel and Camilla Schwind. Hypothesis theory for nonmonotonic reasoning. In: *Proceedings of the Workshop on Nonstandard Queries and Nonstandard Answers*, pages 189–210, 1991.
95. J.R. Slagle, C.L. Chang, and R.C.T. Lee, Completeness theorems for semantic resolution in consequence-finding. In: *Proceedings of IJCAI-69*, pages 281–285, Morgan Kaufmann, 1969.
96. Mark E. Stickel. Rationale and methods for abductive reasoning in natural-language interpretation. In: R. Studer, editor, *Natural Language and Logic, Proceedings of the International Scientific Symposium*, Lecture Notes in Artificial Intelligence, 459, pages 233–252, Springer, 1990.

97. Mark E. Stickel. Upside-down meta-interpretation of the model elimination theorem-proving procedure for deduction and abduction. *Journal of Automated Reasoning*, 13(2):189–210, 1994.

98. Akihiro Yamamoto. Using abduction for induction based on bottom generalization. In: [26], pages 267–280, 2000.

99. Eiko Yamamoto and Katsumi Inoue. Implementation of SOL resolution based on model elimination. *Transactions of Information Processing Society of Japan*, 38(11):2112–2121, 1997 (in Japanese).

100. Wlodek Zadrozny. On rules of abduction. *Annals of Mathematics and Artificial Intelligence*, 9:387–419, 1993.

The Role of Logic in Computational Models of Legal Argument: A Critical Survey

Henry Prakken[1] and Giovanni Sartor[2]

[1] Institute of Information and Computing Sciences
Utrecht University, The Netherlands
http://www.cs.uu.nl/staff/henry.html
[2] Faculty of Law, University of Bologna, Italy
sartor@cirfid.unibo.it

Abstract. This article surveys the use of logic in computational models of legal reasoning, against the background of a four-layered view on legal argument. This view comprises a logical layer (constructing an argument); a dialectical layer (comparing and assessing conflicting arguments); a procedural layer (regulating the process of argumentation); and a strategic, or heuristic layer (arguing persuasively). Each further layer presupposes, and is built around the previous layers. At the first two layers the information base is fixed, while at the third and fourth layer it is constructed dynamically, during a dialogue or dispute.

1 Introduction

1.1 AI & Law Research on Legal Argument

This article surveys a field that has been heavily influenced by Bob Kowalski, the logical analysis of legal reasoning and legal knowledge representation. Not only has he made important contributions to this field (witness the many times his name will be mentioned in this survey) but also has he influenced many to undertake such a logical analysis at all. Our research has been heavily influenced by his work, building on logic programming formalisms and on the well-known argumentation-theoretic account of nonmonotonic logic, of which Bob Kowalski was one of the originators [Kakas *et al.*, 1992, Bondarenko *et al.*, 1997]. We feel therefore very honoured to contribute to this volume in honour of him.

The precise topic of this survey is the role of logic in computational models of legal argument. Argumentation is one of the central topics of current research in Artificial Intelligence and Law. It has attracted the attention of both logically inclined and design-oriented researchers. Two common themes prevail. The first is that legal reasoning is defeasible, i.e., an argument that is acceptable in itself can be overturned by counterarguments. The second is that legal reasoning is usually performed in a context of debate and disagreement. Accordingly, such notions are studied as argument moves, attack, dialogue, and burden of proof.

Historically, perhaps the first AI & Law attempt to address legal reasoning in an adversarial setting was McCarty's (partly implemented) Taxman

A.C. Kakas, F. Sadri (Eds.): Computat. Logic (Kowalski Festschrift), LNAI 2408, pp. 342–381, 2002.
© Springer-Verlag Berlin Heidelberg 2002

project, which aimed to reconstruct the lines of reasoning in the majority and dissenting opinions of a few leading American tax law cases (see e.g. [McCarty and Sridharan, 1981, McCarty, 1995]). Perhaps the first AI & Law system that explicitly defined notions like dispute and dialectical role was Rissland & Ashley's (implemented) HYPO system [Rissland and Ashley, 1987], which modelled adversarial reasoning with legal precedents. It generated 3-ply disputes between plaintiff and defendant in a legal case, where each dispute is an alternating series of attacks by the defendant on the plaintiff's claim, and of defences or counterattacks by the plaintiff against these attacks. This research was continued in Rissland & Skalak's CABARET project [Rissland and Skalak, 1991], and Aleven & Ashley's CATO project [Aleven and Ashley, 1997], both also in the 'design' strand. The main focus of all these projects is defining persuasive argument moves, moves which would be made by 'good' human lawyers.

By contrast, much logic-based research on legal argument has focused on defeasible *inference*, inspired by AI research on nonmonotonic reasoning and defeasible argumentation [Gordon, 1991, Kowalski and Toni, 1996, Prakken and Sartor, 1996,   Prakken, 1997,   Nitta and Shibasaki, 1997, Hage, 1997, Verheij, 1996]. Here the focus was first on reasoning with rules and exceptions and with conflicting rules. After a while, some turned their attention to logical accounts of case-based reasoning [Loui *et al.*, 1993, Loui and Norman, 1995, Prakken and Sartor, 1998]. Another shift in focus occurred after it was realised that legal reasoning is bound not only by the rules of logic but also by those of fair and effective procedure. Accordingly, logical models of legal argument have been augmented with a dynamic component, capturing that the information with which a case is decided is not somehow 'there' to be applied, but is constructed dynamically, in the course of a legal procedure (e.g. [Hage *et al.*, 1994, Gordon, 1994, Bench-Capon, 1998, Lodder, 1999, Prakken, 2001b]). In contrast to the above-mentioned work on dispute in the 'design' strand, here the focus is more on procedure and less on persuasive argument moves, i.e., more on the rules of the 'debating game' and less on how to play this game well.

In this survey we will discuss not only logical approaches but also some work from the 'design strand'. This is since, in our opinion, these approaches should not be regarded as alternatives but should complement and inspire each other. A purely logic-based approach runs the risk of becoming too abstract and ignored by the field for which it is intended, while a purely design-based approach is in danger of becoming too self-centred and ad-hoc.

1.2   A Four-Layered View on Legal Argument

How can all these research projects be compared and contrasted? We propose that models of legal argument can be described in terms of four layers.[1] The

---

[1] The combination of the first three layers was first discussed by [Prakken, 1995]. The first and third layer were also discussed by [Brewka and Gordon, 1994]. The fourth layer was added by [Prakken, 1997] and also discussed in [Sartor, 1997].

first, *logical* layer defines what arguments are, i.e., how pieces of information can be combined to provide basic support for a claim. The second, *dialectical* layer focuses on conflicting arguments: it introduces such notions as 'counterargument', 'attack', 'rebuttal' and 'defeat', and it defines, given a set of arguments and evaluation criteria, which arguments prevail. The third, *procedural* layer regulates how an actual dispute can be conducted, i.e., how parties can introduce or challenge new information and state new arguments. In other words, this level defines the possible speech acts, and the discourse rules governing them. Thus the procedural layer differs from the first two in one crucial respect. While those layers assume a fixed set of premises, at the procedural layer the set of premises is constructed dynamically, during a debate. This also holds for the final layer, the *strategic* or *heuristic* one, which provides rational ways of conducting a dispute within the procedural bounds of the third layer.

All four layers are to be integrated into a comprehensive view of argumentation: the logical layer defines, by providing a notion of arguments, the objects to be evaluated at the dialectical layer; the dialectical layer offers to the procedural and heuristic layers a judgement of whether a new argument might be relevant in the dispute; the procedural layer constrains the ways in which new inputs, supplied by the heuristic layer can be submitted to the dialectical one; the heuristic layer provides the matter which is to be processed in the system. Each layer can obviously be studied (and implemented) in abstraction from the other ones. However, a main premise of this article is that research at the individual levels would benefit if the connection with the other layers is always kept in mind. For instance, logical techniques (whether monotonic or not) have a better chance of being accepted by the AI & Law community when they can easily be embedded in procedural or heuristic layers of legal argument.

Let us illustrate the four layers with an example of a legal dispute.

P_1: I claim that John is guilty of murder.
O_1: I deny your claim.
P_2: John's fingerprints were on the knife.
 If someone stabs a person to death, his fingerprints must be on the knife, so, John has stabbed Bill to death.
 If a person stabs someone to death, he is guilty of murder, so, John is guilty of murder.
O_2: I concede your premises, but I disagree that they imply your claim: Witness X says that John had pulled the knife out of the dead body. This explains why his fingerprints were on the knife.
P_3 X's testimony is inadmissible evidence, since she is anonymous. Therefore, my claim still stands.

P_1 illustrates the procedural layer: the proponent of a claim starts a dispute by stating his claim. The procedure now says that the opponent can either accept or deny this claim. O does the latter with O_1. The procedure now assigns the burden of proof to P. P attempts to fulfil this burden with an argument for his claim (P_2). Note that this argument is not deductive since it includes an abductive inference step; whether it is constructible, is determined at the logical layer. The same holds for O's counterargument O_2, but whether it is a counterargument and has sufficient attacking strength is determined at the dialectical layer, while O's right to state a counterargument is defined by the procedure. The same remarks hold for P's counterargument P_3. In addition, P_3 illustrates the heuristic layer: it uses the heuristic that evidence can be attacked by arguing that it is inadmissible.

This paper is organised as follows. First, in Section 2 we discuss the four layers in more detail. Then in Section 3, we use them in discussing the most influential computational models of legal argument. In Section 4, we do the same for the main logical analyses of legal argument, after which we conclude.

2 Four Layers in Legal Argument

Let us now look in more detail at the four layers of legal argument. It is important to note that the first two layers comprise the subject matter of nonmonotonic logics. One type of such logics explicitly separates the two layers, viz. logical systems for defeasible argumentation (cf. [Prakken and Vreeswijk, 2002]). For this reason we will largely base our discussions on the structure of these systems. However, since [Dung, 1995] and [Bondarenko et al., 1997] have shown that essentially all nonmonotonic logics can be recast as such argument-based systems, most of what we will say also applies to other nonmonotonic logics.

2.1 The Logical Layer

The logical layer is concerned with the language in which information can be expressed, and with the rules for constructing arguments in this language.

The Logical Language

Deontic terms One ongoing debate in AI & Law is whether normative terms such as 'obligatory', 'permitted' and 'forbidden' should be formalised in (modal) deontic logics or whether they can be expressed in first-order logic; cf. e.g. [Jones and Sergot, 1992]. From our perspective this issue is not very relevant, since logics for defeasible argumentation can cope with any underlying logic. Moreover, as for the defeasibility of deontic reasoning, we think that special *deontic* defeasible logics (see e.g. [Nute, 1997]) are not very suited. It is better to embed one's preferred deontic monotonic logic in one's preferred general defeasible logic, since legal defeasibility is not restricted to deontic terms, but extends to all other kinds of legal knowledge, including definitions and evidential knowledge. Obviously, a unified treatment of defeasibility is to be preferred; cf. [Prakken, 1996].

Conceptual structures Others have focused on the formalisation of recurring conceptual legal structures. Important work in this area is McCarty's[1989] *Language of Legal Discourse*, which addresses the representation of such categories as space, time, mass, action, causation, intention, knowledge, and belief. This strand of work is, although very important for AI & Law, less relevant for our concerns, for the same reasons as in the deontic case: argument-based systems can deal with any underlying logic.

Conditional rules A topic that is more relevant for our concerns is the representation of conditional legal rules. The main issue here is whether legal rules satisfy contrapositive properties or not. Some AI & Law formalisms, e.g. Gordon's [1995] Pleadings Game, validate contraposition. However, [Prakken, 1997] has argued that contraposition makes counterarguments possible that would never be considered in actual reasoning practice. A possible explanation for why this is the case is Hage's [1996, 1997] view on legal rules as being constitutive. In this view (based on insights of analytical philosophy) a legal rule does not describe but constitutes states of affairs: for instance, a legal rule *makes* someone a thief or something a contract, it does not *describe* that this is the case. According to Hage, a legal rule must be *applied* to make things the case, and lawyers never apply rules contrapositively. This view is related to AI interpretations of defaults as *inference licences* or *inference policies* [Loui, 1998, Nute, 1992], while the invalidity of contraposition has also been defended in the context of causal reasoning; see e.g. [Geffner, 1992]. Finally, contraposition is also invalid in extended logic programming, where programs can have both weak and strong negations; cf. [Gelfond and Lifschitz, 1990].

Weak and strong negation The desire to formalise reasoning with rules and exceptions sometimes motivates the use of a nonprovability, consistency or weak negation operator, such as negation as failure in logic programming. Whether such a device should be used depends on one's particular convention for formalising rules and exceptions (see further Section 2 below).

Metalogic Features Much legal knowledge is metaknowledge, for instance, knowledge about the general validity of rules or their applicability to certain kinds of cases, priority principles for resolving conflicts between conflicting rules, or principles for interpreting legal rules. Clearly, for representing such knowledge metalogic tools are needed. Logic-based AI & Law research of legal argument has made ample use of such tools, as this survey will illustrate.

Non-logical languages Finally, non-logical languages can be used. On the one hand, there are the well-known knowledge representation formalisms, such as frames and semantic networks. In AI, their logical interpretation has been thoroughly studied. On the other hand, in AI & Law various special-purpose schemes have been developed, such as HYPO's factor-based representation of cases (see Section 3.3), ZENO's issue-position-based language [Gordon and Karaçapilidis, 1997], Room 5's encapsulated text frames

[Loui *et al.*, 1997], ArguMed's linked-boxes language [Verheij, 1999], or variants of Toulmin's [1958] well-known argument scheme [Bench-Capon, 1998]. Simple non-logical languages are especially convenient in systems for intelligent tutoring (such as CATO) or argument mediation (such as ROOM 5, ZENO and ArguMed), since users of such systems cannot be expected to formalise their arguments in logic. In formally reconstructing such systems, one issue is whether their representation language should be taken as primitive or translated into some known logical language. Argument-based logics leave room for both options.

Argument Construction As for argument construction, a minor issue is how to format arguments: as simple premises - conclusion pairs, as sequences of inferences (deductions) or as trees of inferences. The choice between these options seems a matter of convenience; for a discussion of the various options see e.g. [Prakken and Vreeswijk, 2002]. More crucial issues are whether incomplete arguments, i.e., arguments with hidden premises, should be allowed and whether nondeductive arguments should be allowed.

Incomplete Arguments In ordinary language people very often omit information that could make their arguments valid, such as in "John has killed Pete, so John is guilty of Murder". Here the hidden premise "Who kills another person is guilty of murder" is omitted. In some argument mediation applications, e.g. [Lodder, 1999], such incomplete arguments have been allowed, for instance, to give the listener the opportunity to agree with the argument, so that obvious things can be dealt with efficiently. In our opinion this makes sense, but only if a listener who does not agree with the argument has a way to challenge its validity.

Non-deductive argument types Non-deductive reasoning forms, such as inductive, abductive and analogical reasoning are clearly essential to any form of practical reasoning, so they must have a place in the four-layered view on argumentation. In legal reasoning inductive and abductive arguments play an important role in evidential reasoning, while analogical arguments are especially important in the interpretation of legal concepts.

The main issue is whether these reasoning forms should be regarded as argument construction principles (the logical layer) or as heuristics for finding new information (the heuristic layer). In [Prakken, 1995], one of us argued for the latter option. For instance, Prakken argued that an analogy is inherently unable to justify its conclusion since in the end it must always be decided whether the similarities outweigh the differences or not. However, others, e.g. [Loui *et al.*, 1993, Loui, 1998], have included analogical arguments at the logical layer on the grounds that if they are untenable, this will show itself in a rational dispute. Clearly, the latter view presupposes that the dialectical layer is embedded in the procedural layer. For a legal-theoretical discussion of the issue see [Peczenik, 1996, pp. 310–313]. Outside AI & Law, a prominent argument-based system that admits non-deductive arguments is [Pollock, 1995]'s OSCAR system.

Our present opinion is that both approaches make sense. One important factor here is whether the dialectical layer is embedded in the procedural layer. Another important factor is whether a reasoning form is used to justify a conclusion or not. For instance, some uses of analogy concern learning [Winston, 1980], while other uses concern justification (as in much AI & Law work on case-based reasoning). One thing is especially important: if non-deductive arguments are admitted at the logical layer, then the dialectical layer should provide for ways to attack the link between their premises and conclusion; cf. Pollock's [1995] undercutters of defeasible inference rules. For instance, if analogies are admitted, it should not only be possible to rebut them with counterexamples, i.e., with analogies for contradictory conclusions, but it should also be possible to undercut analogies by saying that the similarities are irrelevant, or that the differences are more important than the similarities.

2.2   The Dialectical Layer

The dialectical layer addresses three issues: when arguments are in conflict, how conflicting arguments can be compared, and which arguments survive the competition between all conflicting arguments.

Conflict In the literature, three types of conflicts between arguments are discussed. The first is when arguments have contradictory conclusions, as in 'A contract exists because there was an offer and an acceptance' and 'A contract does not exist because the offerer was insane when making the offer'. Clearly, this form of attack, often called *rebutting* an argument, is symmetric. The other two types of conflict are not symmetric. One is where one argument makes a non-provability assumption (e.g. with logic-programming's negation as failure) and another argument proves what was assumed unprovable by the first. For example, an argument 'A contract exists because there was an offer and an acceptance, and it is not provable that one of the parties was insane', is attacked by any argument with conclusion 'The offerer was insane'. In [Prakken and Vreeswijk, 2002] this is called *assumption attack*. The final type of conflict (identified by Pollock, e.g. 1995) is when one argument challenges a rule of inference of another argument. After Pollock, this is usually called *undercutting* an inference. Obviously, a rule of inference can only be undercut if it is not deductive. For example, an analogy can be undercut by saying that the similarity is insufficient to warrant the same conclusion. Note, finally, that all these senses of attack have a direct and an indirect version; indirect attack is directed against a subconclusion or a substep of an argument. For instance, indirect rebuttals contradict an intermediate conclusion of an argument.

Comparing Arguments The notion of conflicting, or attacking arguments does not embody any form of evaluation; comparing conflicting pairs of arguments, or in other words, determining whether an attack is successful, is another element of argumentation. The terminology varies: some terms that have been used are 'defeat' [Prakken and Sartor, 1996], 'attack' [Dung, 1995, Bondarenko *et al.*, 1997] and 'interference' [Loui, 1998]. In this article we shall use *defeat* for the weak notion and *strict defeat* for the strong, asymmetric notion.

How are conflicting arguments compared in the legal domain? Two main points must be stressed here. The first is that general, domain-independent standards are of little use. Lawyers use many domain-specific standards, ranging from general principles such as "the superior law overrides the inferior law" and "the later regulation overrides the earlier one" to case-specific and context-dependent criteria such as "preferring this rule promotes economic competition, which is good for society", or "following this argument would lead to an enormous increase in litigation, which should be avoided". The second main point is that these standards often conflict, so that the comparison of conflicting arguments is itself a subject of dispute. For instance, the standards of legal certainty and individual fairness often conflict in concrete situations. For logical models of legal argument this means that priority principles must be expressible in the logical language, and that their application must be modelled as defeasible reasoning.

Specificity Some special remarks are in order about the specificity principle. In AI this principle is often regarded as very important. However, in legal reasoning it is just one of the many standards that might be used, and it is often overridden by other standards. Moreover, there are reasons to doubt whether specificity of regulations can be syntactically defined at all. Consider the following imaginary example (due to Marek Sergot, personal communication).

1. All cows must have earmarks
2. Calfs need not have earmarks
3. All cows must have earmarks, whether calf or not
4. All calfs are cows

Lawyers would regard (2) as an exception to (1) because of (4) but they would certainly not regard (2) as an exception to (3), since the formulation of (3) already takes the possible exception into account. Yet logically (3) is equivalent to (1), since the addition "whether calf or not" is a tautology. In conclusion, specificity may be suitable as a notational convention for exceptions, but it cannot serve as a domain-independent conflict resolution principle.

Assessing the Status of Arguments The notion of defeat only tells us something about the relative strength of two individual conflicting arguments; it does not yet tell us with what arguments a dispute can be won. The ultimate status of an argument depends on the interaction between all available arguments. An important phenomenon here is *reinstatement*:[2] it may very well be that argument B defeats argument A, but that B is itself defeated by a third argument

---

[2] But see [Horty, 2001] for a critical analysis of the notion of reinstatement.

C; in that case C 'reinstates' A. Suppose, for instance, that the argument A that a contract exists because there there was an offer and acceptance, is defeated by the argument B that a contract does not exist because the offerer was insane when making the offer. And suppose that B is in turn (strictly) defeated by an argument C, attacking B's intermediate conclusion that the offerer was insane at the time of the offer. In that case C reinstates argument A.

The main distinction is that between *justified*, *defensible* and *overruled* arguments. The distinction between justified and defensible arguments corresponds to the well-known distinction between sceptical and credulous reasoning, while overruled arguments are those that are defeated by a justified argument. Several ways to define these notions have been studied, both in semantic and in proof-theoretic form, and both for justification and for defensibility. See [Prakken and Vreeswijk, 2002] for an overview and especially [Dung, 1995, Bondarenko *et al.*, 1997] for semantical studies. For present purposes the differences in semantics do not matter much; what is more important is that argument-based proof theories can be stated in the dialectical form of an argument game, as a dispute between a *proponent* and *opponent* of a claim. The proponent starts with an argument for this claim, after which each player must attack the other player's previous argument with a counterargument of sufficient strength. The initial argument *provably* has a certain status if the proponent has a *winning strategy*, i.e., if he can make the opponent run out of moves in whatever way she attacks. Clearly, this setup fits well with the adversarial nature of legal argument, which makes it easy to embed the dialectical layer in the procedural and heuristic ones.

To give an example, consider the two dialogue trees of in Figure 1. Assume that they contain all constructible arguments and that the defeat relations are as shown by the arrows (single arrows denote strict defeat while double arrows stand for mutual defeat). In the tree on the left the proponent has a winning strategy, since in all dialogues the opponent eventually runs out of moves; so argument A is provable. The tree on the right extends the first tree with three arguments. Here the proponent does not have a winning strategy, since one dialogue ends with a move by the opponent; so A is not provable in the extended theory.

Partial computation Above we said that the status of an argument depends on its interaction with all available arguments. However, we did not specify what 'available' means. Clearly, the arguments processed by the dialectical proof theory are based on input from the procedural layer, viz. on what has been said and assumed in a dispute. However, should only the actually stated arguments be taken into account, or also additional arguments that can be computed from the theory constructed during the dispute? And if the latter option is chosen, should *all* constructible arguments be considered, or only those that can be computed within given resource bounds? In the literature, all three options have been explored. The methods with partial and no computation have been defended by pointing at the fact that computer algorithms cannot be guaranteed to find arguments in reasonable time, and sometimes not even in finite time (see especially

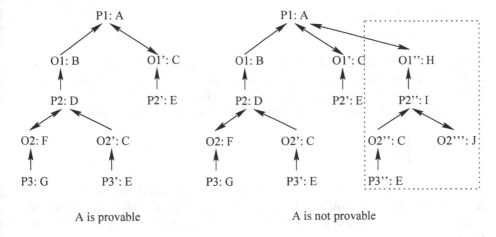

Fig. 1. Two trees of proof-theoretical dialogues.

Pollock 1995; Loui 1998). In our opinion, the choice essentially depends on the context and the intended use of the system.

Representing Exceptions Finally, we discuss the representation of exceptions to legal rules, which concerns a very common phenomenon in the law. Some exceptions are stated by statutes themselves, while others are based, for instance on the purpose of rules or on legal principles. Three different techniques have been used for dealing with exceptions. Two of them are well-known from nonmonotonic logic, while the third one is, to our knowledge, a contribution of AI & Law research.

The first general technique is the *exception clause* or *explicit-exceptions approach*, which corresponds to the use of 'unless' clauses in natural language. Logically, such clauses are captured by a nonprovability operator, which can be formalised with various well-known techniques from nonmonotonic logic or logic programming. In argument-based models the idea is that arguments concluding for the exception, thus establishing what the rule requires not to be proved, defeat arguments based upon the rule. In some formalisations, the not-to-be-proved exception is directly included in the antecedent of the rule to which it refers. So, the rule 'A if B, unless C', is (semiformally) represented as follows (where $\sim$ stands for nonprovability).

$$r_1: A \wedge \sim C \Rightarrow B$$

A more abstract and modular representation is also possible within the exception clause approach. This is achieved when the rule is formulated as requiring that no exception is proved to the rule itself. The exception now becomes the antecedent of a separate conditional.

$$r_1\colon A \wedge \sim Exc(r_1) \Rightarrow B$$
$$r_2\colon C \Rightarrow Exc(r_1)$$

While in this approach rules themselves refer to their exceptions, a variant of this technique has been developed where instead the no-exception requirement is built into the logic of rule application [Routen and Bench-Capon, 1991, Hage, 1996, Prakken and Sartor, 1996]. Semiformally this looks as follows.

$$r_1\colon A \Rightarrow B$$
$$r_2\colon C \Rightarrow Exc(r_1)$$

We shall call this the *exclusion approach*. In argument-based versions it takes the form of allowing arguments for the inapplicability of a rule defeat the arguments using that rule. Exclusion resembles Pollock's [1995] notion of undercutting defeaters.

Finally, a third technique for representing exceptions is provided by the *choice* or *implicit-exceptions approach*. As in the exclusion approach, rules do not explicitly refer to exceptions. However, unlike with exclusion, the exception is not explicitly stated as an exception. Rather it is stated as a rule with conflicting conclusion, and is turned into an exception by preference information that gives the exceptional rule priority over the general rule.

$$r_1\colon A \Rightarrow B$$
$$r_2\colon C \Rightarrow \neg B$$
$$r_1 < r_2$$

In argument-based models this approach is implemented by making arguments based on stronger rules defeat arguments based on weaker rules.

In the general study of nonmonotonic reasoning usually either only the exception-clause- or only the choice approach is followed. However, AI & Law researchers have stressed that models of legal argument should support the combined use of all three techniques, since the law itself uses all three of them.

2.3   The Procedural Layer

There is a growing awareness that there are other grounds for the acceptability of arguments besides syntactic and semantic grounds. One class of such grounds lies in the way in which a conclusion was actually reached. This is partly inspired by a philosophical tradition that emphasises the procedural side of rationality and justice; see e.g. [Toulmin, 1958, Rawls, 1972, Rescher, 1977, Habermas, 1981].

Particularly relevant for present purposes is Toulmin's [1958, pp. 7–8] advice that logicians who want to learn about reasoning in practice, should turn away from mathematics and instead study jurisprudence, since outside mathematics the validity of arguments would not depend on their syntactic form but on the disputational process in which they have been defended. According to Toulmin an argument is valid if it can stand against criticism in a properly conducted dispute, and the task of logicians is to find criteria for when a dispute has been conducted properly; moreover, he thinks that the law, with its emphasis on procedures, is an excellent place to find such criteria.

Toulmin himself has not carried out his suggestion, but others have. For instance, Rescher [1977] has sketched a dialectical model of scientific reasoning which, so he claims, explains the bindingness of inductive arguments: they must be accepted if they cannot be successfully challenged in a properly conducted scientific dispute. A formal reconstruction of Rescher's model has been given by Brewka [1994]. In legal philosophy Alexy's [1978] discourse theory of legal argumentation addresses Toulmin's concerns, based on the view that a legal decision is just if it is the outcome of a fair procedure.

Another source of the concern for procedure is AI research on resource-bounded reasoning; e.g. [Simon, 1982, Pollock, 1995, Loui, 1998]. When the available resources do not guarantee finding an optimal solution, rational reasoners have to rely on effective procedures. One kind of procedure that has been advocated as effective is dialectics [Rescher, 1977, Loui, 1998].

It is not necessary to accept the view that rationality is essentially procedural in order to see that it at least has a procedural side. Therefore, a study of procedure is of interest to anyone concerned with normative theories of reasoning.

How can formal models of legal procedure be developed? Fortunately, there already exists a formal framework that can be used. In argumentation theory, formal dialogue systems have been developed for so-called 'persuasion' or 'critical discussion'; see e.g. [Hamblin, 1971, MacKenzie, 1990, Walton and Krabbe, 1995]. According to Walton and Krabbe [1995], dialogue systems regulate four aspects of dialogues:

- *Locution rules* (what moves are possible)
- *Structural rules* (when moves are legal)
- *Commitment rules* (The effects of moves on the players' commitments);
- *Termination rules* (when dialogues terminate and with what outcome).

In persuasion, the parties in a dispute try to solve a conflict of opinion by verbal means. The dialogue systems regulate the use of speech acts for such things as making, challenging, accepting, withdrawing, and arguing for a claim. The proponent of a claim aims at making the opponent concede his claim; the opponent instead aims at making the proponent withdraw his claim. A persuasion dialogue ends when one of the players has fulfilled their aim. Logic governs the dialogue in various ways. For instance, if a participant is asked to give grounds for a claim, then in most systems these grounds have to logically imply the claim. Or if a proponent's claim is logically implied by the opponent's concessions, the opponent is forced to accept the claim, or else withdraw some of her concessions.

Most computational models of legal procedure developed so far [Hage *et al.*, 1994, Gordon, 1995, Bench-Capon, 1998, Lodder, 1999, Prakken, 2001b] have incorporated such formal dialogue systems. However, they have extended them with one interesting feature, viz. the possibility of counterargument. In argumentation-theoretic models of persuasion the only way to challenge an argument is by asking an argument for its premises. In a legal dialogue, by contrast, a party can challenge an argument even if he accepts all premises, viz. by stating a counterargument. In other words, while in the argumentation-theoretic models the underlying logic is deductive, in the AI & Law systems it is defeasible: support for a claim may be defeasible (e.g. inductive or analogical) instead of watertight, and forced or implied concession of a claim is defined in terms of defeasible instead of deductive consequence. Or in terms of our four-layered view: while the argumentation theorists only have the logical and procedural layer, the AI & Law models have added the dialectical layer in between.

In fact, clarifying the interplay between the dialectical and the procedural layer is not a trivial matter, and is the subject of ongoing logical research. See e.g. [Brewka, 2001, Prakken, 2000, Prakken, 2001c].

2.4   The Heuristic Layer

This layer (which addresses much of what is traditionally called 'rhetoric') is the most diverse one. In fact, heuristics play a role at any aspect of the other three levels: they say which premises to use, which arguments to construct, how to present them, which arguments to attack, which claims to make, concede or deny, etc. Heuristics can be divided into (at least) three kinds: *inventional heuristics*, which say how a theory can be formed (such as the classical interpretation schemes for legal rules), *selection* heuristics, which recommend a choice between various options (such as 'choose an argument with as few premises as possible, to minimise its attacking points'), and *presentation heuristics*, which tell how to present an argument (e.g. 'don't draw the conclusion yourself but invite the listener to draw it').

A keyword at the heuristic level is *persuasion*. For instance, which arguments are the most likely to make the opponent accept one's claims? Persuasiveness of arguments is not a matter of logic, however broadly conceived. Persuasiveness is not a function from a given body of information: it involves an essential nondeterministic element, viz. what the other player(s) will do in response to a player's dialectic acts. To model persuasiveness, models are needed predicting what other players (perhaps the normal, typical other player) will do. Analogous models have been studied in research on argument in negotiation [Kraus *et al.*, 1998, Parsons *et al.*, 1998].

An interesting issue is how to draw the dividing line between argument formation rules and inventional heuristics. Below we will discuss several reasoning schemes that can be reasonably regarded as of either type. We think that the criterion is whether the schemes are meant to *justify* a claim or not.

2.5   Intertwining of the Layers

The four layers can be intertwined in several ways. For instance, allocating the burden of proof is a procedural matter, usually done by the judge on the basis of procedural law. However, sometimes it becomes the subject of dispute, for instance, when the relevant procedural provisions are open-textured or ambiguous. In such a case, the judge will consider all relevant arguments for and against a certain allocation and decide which argument prevails. To this the dialectical layer applies. The result, a justified argument concerning a certain allocation, is then transferred to the procedural layer as a decision concerning the allocation.

Moreover, sometimes the question at which layer one finds himself depends on the use that is made of a reasoning scheme instead of on the reasoning scheme itself. We already mentioned analogy, which can be used in learning (heuristic layer) but also in justification (dialectical layer). Or consider, for another example, the so-called teleological interpretation scheme, i.e., the idea that law texts should usually be understood in terms of their underlying purposes. This principle may be used by a party (when it provides him with a rule which is in his interest to state) as an inventional heuristic, i.e., as a device suggesting suitable contents to be stated in his argument: interpret a law text as a rule which achieves the legislator's purposes, whenever this rule promotes your interest. If this is the use of the interpretation scheme, then a party would not input it in the dispute, but would just state the results it suggests. The principle, however, could also be viewed by a party as a justificatory premise, which the party explicitly uses to support the conclusion that a certain rule is valid, or that it prevails over alternative interpretations.

Not all inventional heuristics could equally be translatable as justificatory meta-rules. Consider for example the heuristic: interpret a text as expressing the rule that best matches the political ideology (or the sexual of racial prejudices) of the judge of your case, if this rule promotes your interest. This suggestion, even though it may be a successful heuristic, usually could not be inputted in the argumentation as a justificatory meta-rule.

3   Computational Models of Legal Argument

In the introduction we said that logic-based and design-based methods in AI & law should complement and influence each other. For this reason, we now discuss some of the most influential implemented architectures of legal argument. We do so in the light of our four-layered view.

3.1   McCarty's Work

The TAXMAN II project of McCarty (e.g. McCarty and Sridharan, 1981; McCarty, 1995) aims to model how lawyers argue for or against the application of a legal concept to a problem situation. In McCarty and Sridharan [1981] only a theoretical model is presented but in McCarty [1995] an implementation is described of most components of the model. However, their interaction in finding arguments is still controlled by the user.

Among other things, the project involves the design of a method for representing legal concepts, capturing their open-textured and dynamic nature. This method is based on the view that legal concepts have three components: firstly, a (possibly empty) set of necessary conditions for the concept's applicability; secondly, a set of instances ("exemplars") of the concept; and finally, a set of rules for transforming a case into another one, particularly for relating "prototypical" exemplars to "deformations". According to McCarty, the way lawyers typically argue about application of a concept to a new case is by finding a plausible sequence of transformations which maps a prototype, possibly via other cases, onto the new case. In our opinion, these transformations might be regarded as invention heuristics for argument construction.

3.2   Gardner

An early success of logic-based methods in AI & Law was their logical reconstruction of Gardner's [1987] program for so-called "issue spotting". Given an input case, the task of the program was to determine which legal questions involved were easy and which were hard, and to solve the easy ones. If all the questions were found easy, the program reported the case as clear, otherwise as hard. The system contained domain knowledge of three different types: legal rules, common-sense rules, and rules extracted from cases. The program considered a question as hard if either "the rules run out", or different rules or cases point at different solutions, without there being any reason to prefer one over the other. Before a case was reported as hard, conflicting alternatives were compared to check whether one is preferred over the other. For example, case law sets aside legal rules or common-sense interpretations of legal concepts.

Clearly, Gardner's program can be reconstructed as nonmonotonic reasoning with prioritised information, i.e., as addressing the dialectical layer. Reconstructions of this kind have been given by [Gordon, 1991], adapting [Poole, 1988]'s abductive model of default reasoning, and [Prakken, 1997], in terms of an argument-based logic.

3.3   HYPO

HYPO aims to model how lawyers make use of past decisions when arguing a case. The system generates 3-ply disputes between a plaintiff and a defendant of a legal claim concerning misuse of a trade secret. Each move conforms to certain rules for analogising and distinguishing precedents. These rules determine for each side which are the best cases to cite initially, or in response to the counterparty's move, and how the counterparty's cases can be distinguished. A case is represented as a set of factors pushing the case towards (pro) or against (con) a certain decision, plus a decision which resolves the conflict between the competing factors. A case is citable for a side if it has the decision wished by that side and shares with the Current Fact Situation (CFS) at least one factor which favours that decision. A citation can be countered by a counterexample, that is, a case that is at least as much on point, but has the opposite outcome. A citation may also be countered by distinguishing, that is, by indicating a factor in the CFS which is absent in the cited precedent and which supports the opposite outcome, or a factor in the precedent which is missing in the CFS, and which supports the outcome of the cited case. Finally, HYPO can create hypothetical cases by using magnitudes of factors. In evaluating the relative force of the moves, HYPO uses the set inclusion ordering on the factors that the precedents share with the CFS. However, unlike logic-based argumentation systems, HYPO does not compute an 'outcome' or 'winner' of a dispute; instead it outputs 3-ply disputes as they could take place between 'good' lawyers.

HYPO in Terms of the Four Layers Interpreting HYPO in terms of the four layers, the main choice is whether to model HYPO's analogising and distinguishing moves as argument formation rules (logical layer) or as inventional heuristics (heuristic layer). In the first interpretation, the representation language is simply as described above (a decision, and sets of factors pro and con a decision), analogising a precedent is a constructible argument, stating a counterexample is a rebutter, and distinguishing a precedent is an undercutter. Defeat is defined such that distinctions always defeat their targets, while counterarguments defeat their targets iff they are not less on point. In the second interpretation, proposed by [Prakken and Sartor, 1998], analogising and distinguishing a precedent are regarded as 'theory constructors', i.e., as ways of introducing new information into a dispute. We shall discuss this proposal below in Section 3.

Which interpretation of HYPO's argument moves is the best one is not an easy question. Essentially it asks for the nature of analogical reasoning, which is a deep philosophical question. In both interpretations HYPO has some heuristic aspects, since it defines the "best cases to cite" for each party, selecting the most-on-point cases from those allowed by the dialectical protocol. This can be regarded as a selection heuristic.

3.4   CATO

The CATO system of Aleven and Ashley [1997] applies an extended HYPO architecture for teaching case-based argumentation skills to law students, also in the trade secrets domain. CATO's main new component is a 'factor hierarchy', which expresses expert knowledge about the relations between the various factors: more concrete factors are classified according to whether they are a reason pro or con the more abstract factors they are linked to; links are given a strength (weak or strong), which can be used to solve certain conflicts. Essentially, this hierarchy fills the space between the factors and decision of a case. Thus it can be used to explain why a certain decision was taken, which in turn facilitates debates on the relevance of differences between cases.

For instance, the hierarchy positively links the factor *Security measures taken* to the more abstract concept *Efforts to maintain secrecy*. Now if a precedent contains the first factor but the CFS lacks it, then not only could a citation of the precedent be distinguished on the absence of *Security measures taken*, but also could this distinction be emphasised by saying that thus no efforts were made to maintain secrecy. However, if the CFS also contains a factor *Agreed not to disclose information*, then the factor hierarchy enables downplaying this distinction, since it also positively links this factor to *Efforts to maintain secrecy*: so the party that cited the precedent can say that in the current case, just as in the precedent, efforts were made to maintain secrecy.

The factor hierarchy is not meant to be an independent source of information from which arguments can be constructed. Rather it serves as a means to *reinterpret* precedents: initially cases are in CATO, as in HYPO, still represented as one-step decisions; the factor hierarchy can only be used to argue that the decision was in fact reached by one or more intermediate steps.

CATO in Terms of the Four Layers At the logical layer CATO adds to HYPO the generation of multi-steps arguments, exploiting the factor hierarchy. As for CATO's ability to reinterpret precedents, we do not regard this as an inventional heuristic, since the main device used in this feature, the factor hierarchy, is given in advance; instead we think that this is just the logic-layer ability to build multi-steps arguments from given information. However, CATO's way of formatting the emphasising and downplaying moves in its output can be regarded as built-in presentation heuristics.

3.5   CABARET

The CABARET system of Rissland and Skalak [1991] attempts at combining rule-based and case-based reasoning. Its case-based component is the HYPO system. The focus is on statutory interpretation, in particular on using precedents to confirm or contest the application of a rule. In [Skalak and Rissland, 1992], CABARET's model is described as a hierarchy of argument techniques including strategies, moves and primitives. A strategy is a broad characterisation of how one should argue, given one's particular viewpoint and dialectical situation. A move is a way to carry out the strategy, while a primitive is a way to implement a move. For example, when one wants to apply a rule, and not all of the rule's conditions are satisfied, then a possible strategy is to broaden the rule. This strategy can be implemented with a move that argues with an analogised precedent that the missing condition is not really necessary. This move can in turn be implemented with HYPO's ways to analogise cases. Similarly, CABARET also permits arguments that a rule which *prima facie* appears to cover the case, should not be applied to it. Here the strategy is *discrediting a rule* and the move may consist in analogising a case in which the rule's conditions were met but the rule was not applied. Again the move can be implemented with HYPO's ways to analogise cases.

CABARET in Terms of the Four Layers At the logical layer CABARET adds to HYPO the possibility to construct simple rule-based arguments, while at the dialectical layer, CABARET adds corresponding ways to attack arguments. CABARET's main feature, its model of argument strategies, clearly addresses the heuristic layer. The strategies can be seen as selection heuristics: they choose between the available attacking points, and pick up from the rule- and case-base the most relevant materials.

3.6 DART

Freeman & Farley [1996] have semi-formally described and implemented a dialectical model of argumentation. For legal applications it is especially relevant since it addresses the issue of burden of proof. Rules are divided into three epistemic categories: 'sufficient', 'evidential' and 'default', in decreasing order of priority. The rules for constructing arguments involve standard logic principles, such as modus ponens and modus tollens, but also nonstandard ones, such as for abductive reasoning ($p \Rightarrow q$ and q imply p) and a contrario reasoning ($p \Rightarrow q$ and $\neg p$ imply $\neg q$). Taken by themselves these inferences clearly are the well-known fallacies of 'affirming the consequent' and 'denying the antecedent' but this is dealt with by defining undercutters for such arguments. For instance, the above abductive argument can be undercut by providing an alternative explanation for q, in the form of a rule $r \Rightarrow q$.

The defeat relations between arguments depend both on the type of premise and on the type of inference rule. The status of arguments is defined in terms of an argument game based on a static knowledge base. DART's argument game has several variants, depending on which level of proof holds for the main claim. This is because Freeman and Farley maintain that different legal problem solving contexts require different levels of proof. For instance, for the question whether a case can be brought before court, only a 'scintilla of evidence' is required (in present terms a defensible argument), while for a decision in a case 'dialectical validity' is needed (in our terms a justified argument).

DART in Terms of the Four Layers DART essentially addresses the logical and dialectical layers, while it assumes input from the procedural layer. At the logical layer, it allows both deductive and nondeductive arguments. Freeman and Farley are well aware that this requires the definition of undercutters for the nondeductive argument types. DART's argument games are similar to dialectical proof theories for argument-based logics. However, they are not given a formal semantics. Finally, DART assumes procedural input in the form of an assignment of a level of proof to the main claim.

3.7 The Pleadings Game

Next we discuss Gordon's [1994, 1995] Pleadings Game, which is an attempt to model the procedural view on justice discussed above in Section 2.3. The legal-procedural example domain is 'civil pleading', which is the phase in Anglo-American civil procedure where the parties exchange arguments and counterarguments to identify the issues that must be decided by the court. The system is not only implemented but also formally defined. Thus this work is an excellent illustration of how logic can be used as a tool in computational models of legal argument. For this reason, and also since it clearly illustrates the relation between the first three layers, we shall discuss it in some detail.

The implemented system mediates between parties in a legal procedure: it keeps track of the stated arguments and their dialectical relations, and it checks whether the procedure is obeyed. Gordon models civil pleading as a Hamblin-MacKenzie-style dialogue game, defining speech acts for stating, conceding and denying (= challenging) a claim, and stating an argument for a claim. In addition, Gordon allows for counterarguments, thus choosing for a nonmonotonic logic as the underlying logical system. In fact, Gordon uses the argument-based proof theory of Geffner's [1992] conditional entailment.

As for the structural rules of the game, a game starts when the plaintiff states his main claim. Then the game is governed by a general rule saying that at each turn a player must respond in some permissible way to every move of the opponent that is still relevant. A move is relevant iff it concerns an issue. An issue is, very roughly, defined as a claim that dialectically matters for the main claim and has not yet been replied-to.

The other structural rules define under which conditions a move is permissible. For instance, a claim of a player may be denied by the other player iff it is an issue and is not defeasibly implied by the denier's own previous claims. And a denied claim may be defended with an argument as long as (roughly) the claim is an issue, and the argument's premises are consistent with the mover's previous claims, and (in case the other party had previously claimed them) they were conceded by the mover. If no such 'permission rule' applies, the other player is to move, except when this situation occurs at the beginning of a turn, in which case the game terminates.

The result of a terminated game is twofold: a list of issues identified during the game (i.e., the claims on which the players disagree), and a winner, if there is one. Winning is defined relative to the set of premises agreed upon during a game. If issues remain, there is no winner and the case must be decided by the court. If no issues remain, then the plaintiff wins iff its main claim is defeasibly implied by the jointly constructed theory, while the defendant wins otherwise.

An Example We now illustrate the Pleadings Game with an example. Besides illustrating this particular system, the example also illustrates the interplay between the logical, dialectical and procedural layers of legal argument. For the sake of illustration we simplify the Game on several points, and use a different (and semiformal) notation. The example, loosely based on Dutch law, concerns a dispute on offer and acceptance of contracts. The players are called plaintiff (π) and defendant (δ). Plaintiff, who had made an offer to defendant, starts the game by claiming that a contract exists. Defendant denies this claim, after which plaintiff supports it with the argument that defendant accepted his offer and that an accepted offer creates a contract.

π_1: **Claim**[(1) Contract]
δ_1: **Deny**(1)
π_2: **Argue**[(2) Offer, (3) Acceptance,
    (4) Offer $\land$ Acceptance $\Rightarrow$ Contract, so Contract]

Now defendant attacks plaintiff's supporting argument [2,3,4] by defeating its subargument that she accepted the offer. The counterargument says that defendant sent her accepting message after the offer had expired, for which reason there was no acceptance in a legal sense.

δ_2: **Concede**(2,4), **Deny**(3)
   Argue[(5) "Accept" late, (6) "Accept" late $\Rightarrow \neg$ Acceptance, so $\neg$ acceptance]

Plaintiff responds by strictly defeating δ_2 with a more specific counterargument (conditional entailment compares arguments on specificity), saying that even though defendant's accepting message was late, it still counts as an acceptance, since plaintiff had immediately sent a return message saying that he recognises defendant's message as an acceptance.

π_3: **Concede**(5), **Deny**(6),
   Argue[(5) "Accept" late, (7) "Accept" recognised,
   (8) "Accept" late $\land$ "Accept" recognised $\Rightarrow$ Acceptance,
   so Acceptance]

Defendant now attempts to leave the issues for trial by conceding π_3's argument (the only effect of this is giving up the right to state a counterargument) and its premise (8), and by denying one of the other premises, viz. (7) (she had already implicitly claimed premise (5) herself, in δ_2). Plaintiff goes along with defendant's aim by simply denying defendant's denial of (7) and not stating a supporting argument for his claim, after which the game terminates since no relevant moves are left to answer for either party.

δ_3: **Concede**(8,[5,7,8]), **Deny**(7)
π_4: **Deny**(**Deny**(7))

This game has resulted in the following dialectical graph.

π_1: [2,3,4] for Contract
δ_1: [5,6] for $\neg$ Acceptance
π_2: [5,7,8] for Acceptance

The claims in this graph that have not been conceded are

(1) Contract
(3) Acceptance
(6) "Accept" late $\Rightarrow \neg$ Acceptance
(7) "Accept" recognised

So these are the issues. Moreover, the set of premises constructed during the game, i.e. the set of conceded claims, is $\{2, 4, 5\}$. It is up to the judge whether to extend it with the issues (6) and (7). In each case conditional-entailment's proof theory must be used to verify whether the other two issues, in particular plaintiff's main claim (1), are (defeasibly) implied by the resulting premises. In fact, it is easy to see that they are entailed only if (6) and (7) are added.

The Pleadings Game in Terms of the Four Layers Clearly, the Pleadings Game explicitly models the first three layers of our model. (In fact, the game was a source of inspiration of [Prakken, 1995]'s first formulation of these layers.) Its contribution to modelling the procedural layer should be apparent from the example. Gordon has also addressed the formalisation of the dialectical layer, adapting within conditional entailment well-known AI techniques concerning naming of rules in (ab)normality predicates. Firstly, he has shown how information about properties of rules (such as validity and backing) can be expressed and, secondly, he has defined a way to express priority rules as object level rules, thus formalising disputes about rule priorities. However, a limitation of his method is that it has to accept conditional-entailment's built-in specificity principle as the highest source of priorities.

4   Logical Models of Legal Argument

Having discussed several implemented models of legal argument, we now turn to logical models. Again we will discuss them in light of our four-layers model.

4.1   Applications of Logic (Meta-)Programming

The British Nationality Act First we discuss the idea of formalising law as logic programs, viz. as a set of formulas of a logical language for which automated theorem provers exist. The underlying ideas of this approach are set out in [Sergot, 1988] and [Kowalski and Sergot, 1990], and is most closely associated with Sergot and Kowalski. The best known application is the formalisation of the British Nationality Act [Sergot et al., 1986] (but see also [Bench-Capon et al., 1987]). For present purposes the main relevance of the work of Sergot et al. is its treatment of exceptions by using negation by failure (further explored by Kowalski, 1989, 1995). To our knowledge, this was the first logical treatment of exceptions in a legal context.

In this approach, which implements the explicit-exceptions approach of Section 2, negation by failure is considered to be an appropriate translation for such locutions as 'unless the contrary is shown' or 'subject to section . . .', which usually introduce exception clauses in legislation. Consider, for example, the norm to the effect that, under certain additional conditions, an abandoned child acquires British citizenship unless it can be shown that both parents have a different citizenship. Since Kakas et al. have shown that negation as failure can be given an argument-based interpretation, where negation-as failure assumptions are defeated by proving their contrary, we can say that [Sergot et al., 1986] model reasoning with rules and exceptions at the logical and the dialectical layer.

Allen & Saxon's criticism An interesting criticism of Sergot et al.'s claim concerning exceptions was put forward by [Allen and Saxon, 1989]. They argued that the defeasible nature of legal reasoning is irreducibly procedural, so that it cannot be captured by current nonmonotonic logics, which define defeasible consequence only as a 'declarative' relation between premises and conclusion of an argument. In particular, they attacked the formalisation of 'unless shown otherwise' with negation as failure by arguing that 'shown' in this context does not mean 'logically proven from the available premises' but "shown by a process of argumentation and the presenting of evidence to an authorized decision-maker". So 'shown' would not refer to logical but to legal-procedural nonprovability.

In our opinion, Allen & Saxon are basically right, since such expressions address the allocation of the burden of proof, which in legal procedure is a matter of decision by the judge rather than of inference, and therefore primarily concerns the procedural layer rather than the dialectical one (as is Sergot *et al.*'s use of negation by failure). Note that these remarks apply not only to Sergot *et al.*'s work, but to any approach that stays within the dialectical layer. In Section 4.4 we will come back to this issue in more detail.

Applications of Logic Metaprogramming In two later projects the legal application of logic-programming was enriched with techniques from logic metaprogramming. Hamfelt [1995] uses such techniques for (among other things) representing legal collision rules and interpretation schemes. His method uses logic programming's DEMO predicate, which represents provability in the object language. Since much knowledge used in legal reasoning is metalevel knowledge, Hamfelt's approach might be a useful component of models of legal argument. However, it is not immediately clear how it can be embedded in a dialectical context, so that more research is needed.

The same holds for the work of Routen and Bench-Capon [1991], who have applied logic metaprogramming to, among other things, the representation of rules and exceptions. Their method provides a way to implement the exclusion approach of Section 2. They enrich the knowledge representation language with metalevel expressions Exception_to($rule_1, rule_2$), and ensure that their metainterpreter applies a rule only if no exceptional rule can be applied. Although this is an elegant method, it also has some restrictions. Most importantly, it is not embedded in an argument-based model, so that it cannot easily be combined with other ways to compare conflicting arguments. Thus their method seems better suited for representing coherent legal texts than for modelling legal argument.

4.2 Applications of Argument-Based Logics

Next we discuss legal applications of logics for defeasible argumentation. Several of these applications in fact use argument-based versions of logic programming.

Prakken & Sartor Prakken and Sartor [1996, 1997] have developed an argument-based logic similar to the one of [Simari and Loui, 1992], but that is expressive enough to deal with contradictory rules, rules with assumptions, inapplicability statements, and priority rules. Their system applies the well-known abstract approach to argumentation, logic programming and nonmonotonic reasoning developed by Dung [1995] and Bondarenko *et al.* [1997]. The logical language of the system is that of extended logic programming i.e., it has both negation as failure ($\sim$) and classical, or strong negation ($\neg$). Moreover, each formula is preceded by a term, its name. (In [Prakken, 1997] the system is generalised to the language of default logic.) Rules are *strict*, represented with $\rightarrow$, or else *defeasible*, represented with $\Rightarrow$. Strict rules are beyond debate; only defeasible rules can make an argument subject to defeat. Accordingly, facts are represented as strict rules with empty antecedents (e.g. $\rightarrow$ *gave-up-house*). The input information of the system, i.e., the premises, is a set of strict and defeasible rules, which is called an *ordered theory* ('ordered' since an ordering on the defeasible rules is assumed).

Arguments can be formed by chaining rules, ignoring weakly negated antecedents; each head of a rule in the argument is a conclusion of the argument. Conflicts between arguments are decided according to a binary relation of *defeat* among arguments, which is partly induced by rule priorities. An important feature of the system is that the information about these priorities is itself presented as premises in the logical language. Thus rule priorities are as any other piece of legal information established by arguments, and may be debated as any other legal issue. The results of such debates are then transported to and used by the metatheory of the system.

There are three ways in which an argument Arg_2 can defeat an argument Arg_1. The first is *assumption* defeat (in the above publications called "undercutting" defeat), which occurs if a rule in Arg_1 contains $\sim L$ in its body, while Arg_2 has a conclusion L. For instance, the argument $[r_1\colon \rightarrow p, r_2\colon p \Rightarrow q]$ (strictly) defeats the argument $[r_3\colon \sim q \Rightarrow r]$ (note that $\sim L$ reads as 'there is no evidence that L'). This way of defeat can be used to formalise the explicit-exception approach of Section 2. The other two forms of defeat are only possible if Arg_1 does not assumption-defeat Arg_2. One way is by *excluding* an argument, which happens when Arg_2 concludes for some rule r in Arg_1 that r is not applicable (formalised as $\neg appl(r)$). For instance, the argument $[r_1\colon \rightarrow p, r_2\colon p \Rightarrow \neg appl(r_3)]$ (strictly) defeats the argument $[r_3\colon \Rightarrow r]$ by excluding it. This formalises the exclusion approach of Section 2. The final way in which Arg_2 can defeat Arg_1 is by *rebutting* it: this happens when Arg_1 and Arg_2 contain rules that are in a head-to-head conflict and Arg_2's rule is not worse than the conflicting rule in Arg_1. This way of defeat supports the implicit-exception approach.

Argument status is defined with a dialectical proof theory. The proof theory is correct and complete with respect to [Dung, 1995]'s grounded semantics, as extended by Prakken and Sartor to the case with reasoning about priorities. The opponent in a game has just one type of move available, stating an argument that defeats proponent's preceding argument (here defeat is determined as if no priorities were defined). The proponent has two types of moves: the first is an argument that combines an attack on opponent's preceding argument with a priority argument that makes the attack strictly defeating opponent's argument; the second is a priority argument that neutralises the defeating force of O's last move. In both cases, if proponent uses a priority argument that is not justified by the ordered theory, this will reflect itself in the possibility of successful attack of the argument by the opponent.

We now present the central definition of the dialogue game ('*Arg*-defeat' means defeat on the basis of the priorities stated by *Arg*). The first condition says that the proponent begins and then the players take turns, while the second condition prevents the proponent from repeating a move. The last two conditions were just explained and form the heart of the definition.

A *dialogue* is a finite nonempty sequence of moves $move_i = (Player_i, Arg_i)$ $(i > 0)$, such that

1. $Player_i = P$ iff i is odd; and $Player_i = O$ iff i is even;
2. If $Player_i = Player_j = P$ and $i \neq j$, then $Arg_i \neq Arg_j$;
3. If $Player_i = P$ then Arg_i is a minimal (w.r.t. set inclusion) argument such that
   (a) Arg_i strictly Arg_i-defeats Arg_{i-1}; or
   (b) Arg_{i-1} does not Arg_i-defeat A_{i-2};
4. If $Player_i = O$ then Arg_i $\emptyset$-defeats Arg_{i-1}.

The following simple dialogue illustrates this definition. It is about a tax dispute about whether a person temporarily working in another country has changed his fiscal domicile. All arguments are citations of precedents.[3]

P_1: [f_1: *kept-house,*
      r_1: *kept-house* $\Rightarrow \neg$ *change*]

(Keeping one's old house is a reason against change of fiscal domicile.)

O_1: [f_2: $\neg$ *domestic-headquarters,*
      r_2: $\neg$ *domestic-headquarters* $\Rightarrow \neg$ *domestic-company,*
      r_3: $\neg$ *domestic-company* $\Rightarrow$ *change*]

(If the employer's headquarters are in the new country, it is a foreign company, in which case fiscal domicile has changed.)

P_2: [f_3: *domestic-property,*
      r_4: *domestic-property* $\Rightarrow$ *domestic-company,*
      f_4: r_4 *is decided by higher court than* r_2,
      r_5: r_4 *is decided by higher court than* $r_2 \Rightarrow r_2 \prec r_4$]

(If the employer has property in the old country, it is a domestic company. The court which decided this is higher than the court deciding r_2.)

The proponent starts the dialogue with an argument P_1 for $\neg$ *change*, after which the opponent attacks this argument with an argument O_1 for the opposite conclusion. O_1 defeats P_1 as required, since in our logical system two rebutting

---

[3] Facts $f_i: \rightarrow p_i$ are abbreviated as $f_i: p_i$.

arguments defeat each other if no priorities are stated. P_2 illustrates the first possible reply of the proponent to an opponent's move: it combines an object level argument for the conclusion *domestic-company* with a priority argument that gives r_4 precedence over r_2 and thus makes P_2 strictly defeat O_1. The second possibility, just stating a priority argument that neutralises the opponent's move, is illustrated by the following alternative move, which resolves the conflict between P_1 and O_1 in favour of P_1:

P_2': [f_5: r_1 *is more recent than* r_3,
$\quad\quad p'$: r_1 *is more recent than* $r_3 \Rightarrow r_3 \prec r_1$]

Kowalski & Toni Like Prakken and Sartor, Kowalski and Toni [1996] also apply the abstract approach of [Dung, 1995, Bondarenko *et al.*, 1997] to the legal domain, instantiating it with extended logic programming. Among other things, they show how priority principles can be encoded in the object language without having to refer to priorities in the metatheory of the system. We illustrate their method using the language of [Prakken and Sartor, 1996]. Kowalski and Toni split each rule $r: P \Rightarrow Q$ into two rules

$Applicable(r) \Rightarrow Q$
$P \wedge \sim Defeated(r) \Rightarrow Applicable(r)$

The predicate *Defeated* is defined as follows:

$r \prec r' \wedge Conflicting(r, r') \wedge Applicable(r') \rightarrow Defeated(r)$

Whether $r \prec r'$ holds, must be (defeasibly) derived from other information. Kowalski and Toni also define the *Conflicting* predicate in the object language.

Three Formal Reconstructions of HYPO-style Case-Based Reasoning
The dialectical nature of the HYPO system has inspired several logically inclined researchers to reconstruct HYPO-style reasoning in terms of argument-based defeasible logics. We briefly discuss three of them, and refer to [Hage, 1997] for a reconstruction in Reason-based Logic (cf. Section 4.3 below).

Loui et al. (1993) Loui *et al.* [1993] proposed a reconstruction of HYPO in the context of the argument-based logic of [Simari and Loui, 1992]. They mixed the pro and con factors of a precedent in one rule

Pro-factors $\wedge$ *Con-factors* $\Rightarrow$ *Decision*

but then implicitly extended the case description with rules containing a superset of the con factors and/or a subset of the con factors in this rule. Loui *et al.* also studied the combination of reasoning with rules and cases. This work was continued in [Loui and Norman, 1995] (discussed below in Section 4.5).

Prakken and Sartor (1998) Prakken and Sartor [1998] have modelled HYPO-style reasoning in their [1996] system, also adding additional expressiveness. As Loui *et al.* [1993] they translate HYPO's cases into a defeasible-logical theory. However, unlike Loui *et al.*, Prakken and Sartor separate the pro and con factors into two conflicting rules, and capture the case decision with a priority rule. This method is an instance of a more general idea (taken from [Loui and Norman, 1995]) to represent precedents as a set of arguments pro and con the decision, and to capture the decision by a justified priority argument that in turn makes the argument for the decision justified. In its simplest form where, as in HYPO, there are just a decision and sets of factors pro and con the decision, this amounts to having a pair of rules

r_1: *Pro-factors* $\Rightarrow$ *Decision*
r_2: *Con-factors* $\Rightarrow$ $\neg$*Decision*

and an unconditional priority rule

p: $\Rightarrow r_1 \succ r_2$

However, in general arguments can be multi-step (as suggested by [Branting, 1994]) and priorities can very well be the justified outcome of a competition between arguments.

Analogy is now captured by a 'rule broadening' heuristic, which deletes the antecedents missing in the new case. And distinguishing is captured by a heuristic which introduces a conflicting rule 'if these factors are absent, then the consequent of your broadened rule does not hold'. So if a case rule is r_1: $f_1 \wedge f_2 \Rightarrow d$, and the CFS consists of f_1 only, then r_1 is analogised by $b(r_1)$: $f_1 \Rightarrow d$, and $b(r_1)$ is distinguished by $d(b(r_1))$: $\neg f_2 \Rightarrow \neg d$. To capture the heuristic nature of these moves, Prakken and Sartor 'dynamify' their [1996] dialectical proof procedure, to let it cope with the introduction of new premises.

Finally, in [Prakken, 2002] it is, inspired by [Bench-Capon and Sartor, 2001], shown how within this setup cases can be compared not on factual similarities but on the basis of underlying values.

Horty (1999) Horty [1999] has reconstructed HYPO-style reasoning in terms of his own work on two other topics: defeasible inheritance and defeasible deontic logic. Since inheritance systems are a forerunner of logics for defeasible argumentation, Horty's reconstruction can also be regarded as argument-based. It addresses the analogical citation of cases and the construction of multi-steps arguments. To support the citation of precedents on their intermediate steps, cases are separated into 'precedent constituents', which contain a set of factors and a possibly intermediate outcome. Arguments are sequences of factor sets, starting with the current fact situation and further constructed by iteratively applying precedent constituents that share at least one factor with the set constructed thus far. Conflicting uses of precedent constituents are compared with a variant of HYPO's more-on-point similarity criterion. The dialectical status of the constructible arguments is then assessed by adapting notions from Horty's inheritance systems, such as 'preemption'.

Other Work on Argument-Based Logics Legal applications of argument-based logic programming have also been studied by Nitta and his colleagues; see e.g. [Nitta and Shibasaki, 1997]. Besides rule application, their argument construction principles also include some simple forms of analogical reasoning. However, no undercutters for analogical arguments are defined. The system also has a rudimentary dialogue game component.

Formal work on dialectical proof theory with an eye to legal reasoning has been done by Jakobovits and Vermeir [1999]. Their focus is more on technical development than on legal applications.

4.3   Reason-Based Logic

Hage [1996, 1997] and Verheij [1996] have developed a formalism for legal reasoning called 'reason-based logic' (RBL), centering around a deep philosophical account of the concept of a rule. It is meant to capture how legal (and other) principles, goals and rules give rise to reasons for and against a proposition and how these reasons can be used to draw conclusions. The underlying view on principles, rules and reasons is influenced by insights from analytical philosophy on the role of reasons in practical reasoning, especially [Raz, 1975]. Hage and Verheij stress that rule application is much more than simple modus ponens. It involves reasoning about the validity and applicability of a rule, and weighing reasons for and against the rule's consequent.

RBL's View on Legal Knowledge RBL reflects a distinction between two levels of legal knowledge. The primary level includes principles and goals, while the secondary level includes rules. Principles and goals express reasons for or against a conclusion. Without the secondary level these reasons would in each case have to be weighed to obtain a conclusion, but according to Hage and Verheij rules express the outcome of certain weighing process. Therefore, a rule does not only generate a reason for its consequent but also generates a so-called 'exclusionary' reason against applying the principles underlying the rule: the rule replaces the reasons on which it is based. This view is similar to Dworkin's [1977] well-known view that while principles are weighed against each other, rules apply in an all-or-nothing fashion. However, according to Hage [1996] and Verheij [Verheij et al., 1998] this difference is just a matter of degree: if new reasons come up, which were not taken into account in formulating the rule, then these new reasons are not excluded by the rule; the reason based on the rule still has to be compared with the reasons based on the new principles. Consequently, in RBL rules and principles are syntactically indistinguishable; their difference is only reflected in their degree of interaction with other rules and principles (but Hage [1997] somewhat deviates from this account.)

A Sketch of the Formal System To capture reasoning about rules, RBL provides the means to express properties of rules in the object language. To this end Hage and Verheij use a sophisticated naming technique, viz. *reification*, well-known from metalogic and AI [Genesereth and Nilsson, 1988, p. 13], in which every predicate constant and logical symbol is named by a function expression. For instance, the conjunction $R(a) \land S(b)$ is denoted by the infix function expression $r(a) \land' s(b)$. Unlike the naming techniques used by [Gordon, 1995] and [Prakken and Sartor, 1996], RBL's technique reflects the logical structure of the named formula.

Rules are named with a function symbol *rule*, resulting in terms like

$$\texttt{rule}(r, p(x), q(x))$$

Here r is a 'rule identifier', $p(x)$ is the rule's condition, and $q(x)$ is its consequent. RBL's object language does not contain a conditional connective corresponding to the function symbol *rule*; rules can only be stated indirectly, by assertions that they are valid, as in

$$\texttt{Valid}(\texttt{rule}(r, condition_r, conclusion_r))$$

Hage and Verheij state RBL as extra inference rules added to standard first-order logic or, in some versions, as extra semantic constraints on models of a first-order theory. We first summarise the most important rules and then give some (simplified) formalisations.

1. If a rule is valid, its conditions are satisfied and there is no evidence that it is excluded, the rule is applicable.
2. If a rule is applicable, it gives rise to a reason for its application.
3. A rule applies if and only if the set of all derivable reasons for its application outweighs the set of all derivable reasons against its application.
4. If a rule applies, it gives rise to a reason for its consequent.
5. A formula is a conclusion of the premises if and only if the reasons for the formula outweigh the reasons against the formula.

Here is how a simplified formal version of inference rule (1) looks like. Note that *condition* and *consequent* are variables, which can be instantiated with the name of any formula.

If $\texttt{Valid}(\texttt{rule}(r, condition, consequent))$ is derivable
and $\texttt{Obtains}(condition)$ is derivable
and $\texttt{Excluded}(r))$ is not derivable,
then $\texttt{Applicable}(r, \texttt{rule}(condition, consequent))$ is derivable.

Condition (4) has the following form.

If $\texttt{Applies}(r, \texttt{rule}(condition, consequent))$ is derivable,
then $\texttt{Proreason}(consequent)$ is derivable.

Finally, here is how in condition (5) the connection between object- and metalevel is made.

If $\mathtt{Outweighs}(\mathtt{Proreasons}(formula),\mathtt{Conreasons}(formula))$ is derivable, then $\mathtt{Formula}$ is derivable.

Whether the pro-reasons outweigh the con-reasons must itself be derived from the premises. The only built-in constraint is that any nonempty set outweighs the empty set. Note that while $formula$ is a variable for an object term, occurring in a well-formed formula of RBL, $\mathtt{Formula}$ is a metavariable which stands for the formula named by the term $formula$. This is how object and metalevel are in RBL connected.

In RBL the derivability of certain formulas is defined in terms of the non-derivability of other formulas. For instance, in (1) it may not be derivable that the rule is excluded. To deal with this, RBL adapts techniques of default logic, by restating the inference rules as conditions on membership of an extension.

Using RBL In RBL exceptions can be represented both explicitly and implicitly. As for explicit exceptions, since RBL has the validity and applicability requirements for rules built into the logic, the exclusion method of Section 2 can be used. RBL also supports the choice approach: if two conflicting rules both apply and do not exclude each other, then their application gives rise to conflicting reasons, which have to be weighed.

Finally, Hage and Verheij formalise legal priority principles in a similar way as [Kowalski and Toni, 1996], representing them as inapplicability rules. The following example illustrates their method with the three well known legal principles *Lex Superior* (the higher regulation overrides the lower one), *Lex Posterior* (the later rule overrides the earlier one) and *Lex Specialis* (the specificity principle). It is formalised in the language of [Prakken and Sartor, 1996]; recall that with respect to applicability, this system follows, as RBL, the exclusion approach.

The three principles can be expressed as follows.

H: x conflicts with $y \wedge y$ is inferior to $x \wedge \sim \neg\mathtt{appl}(x) \Rightarrow \neg\mathtt{appl}(y)$
T: x conflicts with $y \wedge y$ is earlier than $x \wedge \sim \neg\mathtt{appl}(x) \Rightarrow \neg\mathtt{appl}(y)$
S: x conflicts with $y \wedge x$ is more specific than $y \wedge \sim \neg\mathtt{appl}(x) \Rightarrow$
$\quad \neg\mathtt{appl}(y)$

Likewise for the ordering of these three principles:

HT: T conflicts with $H \wedge \sim \neg\mathtt{appl}(H) \Rightarrow \neg\mathtt{appl}(T)$
TS: S conflicts with $T \wedge \sim \neg\mathtt{appl}(T) \Rightarrow \neg\mathtt{appl}(S)$
HS: S conflicts with $H \wedge \sim \neg\mathtt{appl}(H) \Rightarrow \neg\mathtt{appl}(S)$

Thus the metatheory of the logic does not have to refer to priorities. However, the method contains another metareasoning feature, viz. the ability to express metalevel statements of the kind x conflicts with y.

Evaluation RBL clearly confines itself to the logical and dialectical layer of legal argument. At these layers, it is a philosophically well-motivated analysis of legal reasoning, while technically it is very expressive, supporting reasoning with rules and exceptions, with conflicting rules, and about rules and their priority relations. However, it remains to be investigated how RBL can, given its complicated technical nature and the lack of the notion of an argument, be embedded in procedural and heuristic accounts of legal argument.

4.4 Procedural Accounts of Legal Reasoning

The Pleadings Game is not the only procedural AI & Law model. We now briefly discuss some formal models of this kind.

Hage, Leenes, and Lodder At the same time when Gordon designed his system, Hage et al. [1994] developed a procedural account of Hart's distinction between clear and hard cases. They argued that whether a case is easy or hard depends on the stage of a procedure: a case that is easy at an earlier stage, can be made hard by introducing new information. This is an instance of their purely procedural view on the law, which incorporates substantive law by the judge's obligation to apply it. To formalise this account, a Hamblin-MacKenzie-style formal dialogue system with the possibility of counterargument was developed. This work was extended by [Lodder, 1999] in his DiaLaw system.

The general setup of these systems is the same as that of the Pleadings Game. For the technical differences the reader is referred to the above publications. One difference at the dialectical layer is that instead of an argument-based logic, Hage and Verheij's reason-based logic is used. Another difference in [Hage et al., 1994] is that it includes a third party, the *referee*, who is entitled to decide whether certain claims should be accepted by the parties or not. The dialogue systems also support disputes about the procedural legality of a move. Finally, arguments do not have to be logically valid; the only use of reason-based logic is to determine whether a claim of one's opponent follows from one's commitments and therefore must be accepted.

Bench-Capon Bench-Capon [1998] has also developed a dialogue game for legal argument. As the above-discussed games, it has the possibility of counterargument (although it does not incorporate a formalised account of the dialectical layer). The game also has a referee, with roughly the same role as in [Hage et al., 1994]. Bench-Capon's game is especially motivated by the desire to generate more natural dialogues than the "stilted" ones of Hamblin-MacKenzie-style systems. To this end, arguments are defined as variants of Toulmin's [1958] argument structures, containing a *claim*, *data* for this claim, a *warrant* connecting data and claim, a *backing* for the warrant, and possible *rebuttals* of the claim with an exception. The available speech acts refer to the use or attack of these items, which, according to Bench-Capon, induces natural dialogues.

Formalising Allocations of the Burden of Proof Above we supported Allen and Saxon's [1989] criticism of Sergot *et al.*'s [1986] purely logical- and dialectical-layer account of reasoning with exceptions. Additional support is provided by Prakken [2001a], who argues that allocations of burden of proof cannot be modelled by 'traditional' nonmonotonic means.

Burden of proof is one of the central notions of legal procedure, and it is clearly connected with defeasibility [Loui, 1995, Sartor, 1995]. There are two aspects of having the burden of proving a claim: the task to come with an argument for that claim, and the task to uphold this argument against challenge in a dispute. The first aspect can be formalised in Hamblin-MacKenzie-style dialogue systems (discussed above in Section 2.3). The second aspect requires a system that assesses arguments on the basis of the dialectical interactions between all available arguments. At first sight, it would seem that dialectical proof theories of nonmonotonic logics can be directly applied here. However, there is a problem, which we shall illustrate with an example from contract law.

In legal systems it is generally the case that the one who argues that a valid contract exists has the burden of proving those facts that ordinarily give rise to the contract, while the party who denies the existence of the contract has the burden of proving why, despite these facts, exceptional circumstances prevent the contract from being valid. Now suppose that plaintiff claims that a contract between him and defendant exists because plaintiff offered defendant to sell her his car, and defendant accepted. Then plaintiff has the burden of proving that there was such an offer and acceptance, while defendant has the burden of proving, for instance, that the car had a hidden defect. Suppose we formalise this in [Prakken and Sartor, 1996] as follows:

r_1: *offer* $\wedge$ *acceptance* $\wedge \sim$ *exception*(r_1) $\Rightarrow$ *contract*
r_2: *hidden defect* $\Rightarrow$ *exception*(r_1)

Suppose further that in the dispute arguments for and against *hidden defect* are exchanged, and that the judge regards them of equal strength.

What follows dialectically? If plaintiff starts with moving his argument for *contract*, then defendant can assumption-defeat this argument with her argument for *exception*(r_1). Plaintiff cannot attack this with his argument against *hidden defect* since it is of equal strength as defendant's argument, so it does not strictly defeat it. In conclusion, plaintiff's argument is not justified (but merely defensible), so the outcome of our logical reconstruction is that plaintiff has not fulfilled his burden of proof.

However, the problem with this reconstruction is that it ignores that neither has defendant fulfilled her burden of proof: she had to prove *hidden defect*, but her argument for this conclusion also is merely defensible. The problem with the (sceptical) dialectical proof theory is that plaintiff has the burden of proof with respect to all subissues of the dispute; there is no way to distribute the burden of proof over the parties, as is common in legal dispute. This problem is not confined to the particular system or knowledge representation method, but seems a fundamental problem of current 'traditional' nonmonotonic logics.

An additional problem for such logics is that in legal procedure the allocation of the burden of proof is ultimately a matter of decision by the judge, and therefore cannot be determined by logical form. Any full model of reasoning under burden of proof should leave room for such decisions.

In [Prakken, 2001a] the dialectical proof theory for grounded semantics is adapted to enable distributions of the burden of proof over the parties, which in [Prakken, 2001b] is embedded in a dialogue game model for legal procedure. The basic idea of [Prakken, 2001a] is that the required strength of a move depends on who has the burden of proof concerning the issue under attack (as decided by the judge in the dialogue game). The resulting system has no clear link with argument-based semantics in the style of [Dung, 1995, Bondarenko et al., 1997]. For logicians this is perhaps disappointing, but for others this will count as support for the view that the semantics of (legal) defeasible reasoning is essentially procedural.

ZENO's argumentation framework Another account of distributions of the burden of proof in dialectical systems is given by Gordon and Karaçapilidis [1997]. In fact, [Prakken, 2001a]'s proposal can partly be seen as a generalisation and logical formalisation of this account. Gordon and Karaçapilidis incorporate variants of Freeman and Farley's 'levels of proof' in their 'ZENO argumentation framework'. This is the dialectical-layer part of the ZENO argument mediation system: it maintains a 'dialectical graph' of the issues, the positions with respect to these issues, and the arguments pro and con these positions that have been advanced in a discussion, including positions and arguments about the strength of other arguments. Arguments are links between positions.

Part of the framework is a status assignment to positions: each position is assigned *in* or *out* depending on two factors: the required level of proof for the position, and the relative strengths of the arguments pro and con the position that themselves have antecedents that are *in*. For instance, a position with level 'scintilla of evidence' is *in* iff at least one argument pro is *in* (here they deviate from Freeman and Farley). And a position with level 'preponderance of evidence' is *in* iff the joint pro arguments that are *in* outweigh the joint con arguments that are *in*. The burden of proof can be distributed over the parties since levels of proof can be assigned to arbitrary positions instead of (as in [Freeman and Farley, 1996]) only to the initial claim of a dispute.

4.5 Formalisations of the Heuristic Layer

In logical models of legal argument the heuristic layer has so far received very little attention. Above we discussed Prakken and Sartor's [1998] logical reconstruction of HYPO-style analogising and distinguishing as premise introduction heuristics. Perhaps the most advanced formal work on the heuristic layer to date is Loui and Norman's [1995] study of the use of rationales of cases and rules in legal argument, which we will now discuss in some detail.

Loui and Norman (1995) Loui and Norman [1995] formally define a protocol for the exchange of arguments and counterarguments, and analyse within the protocol various uses of rationales of rule and cases. These uses are modelled as ways to modify a previously stated argument. Thus their various uses of rationales can be regarded as inventional heuristics.

More precisely, each move states and/or modifies one or more arguments. Newly stated arguments are added to the so-called *argument record*, which is shared by the players. Modifications modify an argument on the record moved by the other player, in order to have new ways to attack it. Each move must change the status of the main claim: if the proponent moves, the status must change to 'justified', while if the opponent moves, it must change to 'defensible' or 'overruled'. Whether a move achieves this, is tested by applying the argumentation logic of [Simari and Loui, 1992] to the argument record resulting from the move (taking only the explicitly stated arguments into account).

We now summarise the types of rationales identified by Loui and Norman and how they can be used to generate new 'attacking points'. Then we discuss the use of one type in more detail.

Compression rationales. Some rules compress a line of reasoning in a single if-then rule. For instance, the rule 'vehicles are not allowed in the park' might compress 'vehicles used for private transportation are not allowed in the park' and 'vehicles are normally used for private transportation'. Unpacking the compressed rule enables an attack on the latter rule, for instance, with 'ambulances are not used for private transport'. Semiformally: unpack your opponent's use of $A \Rightarrow B$ as $A \Rightarrow C, C \Rightarrow B$ and state an argument for $\neg C$.

Specialisation rationales. Sometimes a rule can be argued to implement a principle. For instance, the rule 'mail order buyers can cancel their order within one week' could be argued to specialise the principle 'weak contract parties should be protected', since mail order buyers (usually consumers) are weak parties and allowing them to cancel their order within a week is a way to protect them. A rationale-based attack could restate the rule as 'insofar as mail order buyers are weak parties, they can cancel their order within one week'. This enables an attack on the weakness of the party, for instance, when the buyer is a company. The logical form is: if we have a rule $B \Rightarrow C$ and a principle $W \Rightarrow P$, and we have that $B \Rightarrow W$ and $C \Rightarrow P$, then replace the rule with $W \Rightarrow C$, and attack the modified argument with an argument for $\neg W$.

Fit rationales. Sometimes a rule is defended by arguing that it explains the decisions of a given set of precedents. This rule could be modified into a rule that equally well explains the cases but that does not apply in the new case, or is susceptible to a new attack. Other forms of attack are also possible, for instance, adding a precedent to the set and arguing that a conflicting rule better explains the resulting set.

Disputation rationales. Sometimes the *ratio decidendi* of a precedent is the result of a choice between conflicting arguments. Then the case rule can be replaced by these conflicting arguments, and by showing that in the new fact situation the outcome of the dispute would have been different.

Let us illustrate this in some detail. Assume a case rule $B \wedge C \Rightarrow A$, which compresses the adjudication between the following three arguments (for notational convenience we use specificity to express the comparison of the arguments).

Arg_1: B, $B \Rightarrow A$, so A
Arg_2: C, $C \Rightarrow D$, $D \Rightarrow \neg A$, so $\neg A$
Arg_3: B, C, $B \Rightarrow F$, $F \wedge C \Rightarrow \neg D$, so $\neg D$

Loui and Norman's protocol allows the following dispute:

– P: I have an argument for A:
  • Arg_0: B, C, $B \wedge C \Rightarrow A$, so A
  Argument record $= \{Arg_0\}$
– O: Your rule compresses the adjudication between three arguments, so:
  Argument record $= \{Arg_1, Arg_2, Arg_3\}$
– O: And I attack Arg_3 with:
  • Arg_4: B, G, $B \wedge G \Rightarrow \neg F$, so $\neg F$
  Argument record $= \{Arg_1, Arg_2, Arg_3, Arg_4\}$

Applying [Simari and Loui, 1992]'s system to the argument records before and after O's move, we see that A is justified on the basis of the former record, but overruled on the basis of the latter. So O has fulfilled her task of changing the status of P's main claim.

5 Conclusion

One aim of this review has been to show that there is more to legal argument than inference (whether deductive or defeasible). Another aim has been to argue that logic is more fruitfully applied to legal reasoning if the context in which it is to be used is taken into account. Our four-layered view on legal argument is an attempt to provide the necessary context. Two main features of this context are that it is dynamic and that it is dialectical: the theory with which to reason is not given but must be constructed, in dialectical interaction with one's adversaries, and within procedural bounds.

Summarising in more detail our overview of logical research on legal argument, we can say that the dialectical layer has been largely dealt with. Adapting general techniques from nonmonotonic logic, various sophisticated methods have been developed for formalising reasoning with rules and exceptions, with rule priorities (including combining several sources of priorities), about rule priorities, and about other properties of legal rules, such as their backing, validity or applicability, and their correct interpretation. Above all, AI & Law has shown how all these elements can be integrated. Of course, the implementation of these formal models involves computational issues. However, these issues fall outside the present paper: the field of AI & Law can here rely on relevant work in other fields, such as automated theorem proving.

At the procedural layer considerable progress has been made. However, a general framework is still lacking. Such a framework is needed since, although most current procedures are carefully designed, it is often hard to see their underlying structure. This makes it hard to study their properties and also to design new procedures. A possible framework is proposed in [Prakken, 2000], leaving room for various sets of speech acts, various underlying defeasible logics, and various options on trying alternative moves.

For logicians, the study of disputational procedures opens a new range of research questions. One issue is the formalisation of 'self-modifying' procedures, i.e., the possibility to change a procedure in a dispute governed by that same procedure; cf. [Vreeswijk, 2000]. Another issue is the relation between the dialectical and procedural layer, especially when dialogue systems incorporate dialectical proof theories in their dialogue rules (as studied by [Prakken, 2001c]). Also, general principles should be studied for how to enable as many 'sensible' dialogues as possible while disallowing all 'non-sensible' dialogues.

Finally, the formalisation of the heuristic aspects of legal argument is still in its early stages. Some interesting research issues are:

- Formalisation of nondeductive types of arguments.
- Formalisation of inventional, presentation and selection heuristics.
- Formalisation of the notion of *persuasive* argumentation.
- Drawing the dividing line between argument construction rules and premise introduction heuristics.

Finally, there is the more general issue as to the limits of argument-based approaches. Perhaps more 'holistic' approaches are needed, where people exchange entire theories with each other, which are assessed on coherence; cf. e.g. [Bench-Capon and Sartor, 2001].

References

[Aleven and Ashley, 1997] V. Aleven and K.D. Ashley. Evaluating a learning environment for case-based argumentation skills. In *Proceedings of the Sixth International Conference on Artificial Intelligence and Law*, pages 170–179, New York, 1997. ACM Press.

[Alexy, 1978] R. Alexy. *Theorie der juristischen Argumentation. Die Theorie des rationalen Diskurses als eine Theorie der juristischen Begründung*. Suhrkamp Verlag, Frankfurt am Main, 1978.

[Allen and Saxon, 1989] L.E. Allen and C.S. Saxon. Relationship of expert systems to the operation of a legal system. In *Preproceedings of the III International Conference on "Logica, Informatica, Diritto" (Appendix)*, pages 1–15, Florence, 1989.

[Bench-Capon and Sartor, 2001] T.J.M. Bench-Capon and G. Sartor. Theory based explanation of case law domains. In *Proceedings of the Eighth International Conference on Artificial Intelligence and Law*, pages 12–21, New York, 2001. ACM Press.

[Bench-Capon et al., 1987] T.J.M. Bench-Capon, G.O. Robinson, T.W. Routen, and M.J. Sergot. Logic programming for large scale applications in law: a formalisation of supplementary benefit legislation. In *Proceedings of the First International Conference on Artificial Intelligence and Law*, pages 190–198, New York, 1987. ACM Press.

[Bench-Capon, 1998] T.J.M. Bench-Capon. Specification and implementation of Toulmin dialogue game. In *Legal Knowledge-Based Systems. JURIX: The Eleventh Conference*, pages 5–19, Nijmegen, 1998. Gerard Noodt Instituut.

[Bondarenko et al., 1997] A. Bondarenko, P.M. Dung, R.A. Kowalski, and F. Toni. An abstract, argumentation-theoretic approach to default reasoning. *Artificial Intelligence*, 93:63–101, 1997.

[Branting, 1994] L.K. Branting. A computational model of ratio decidendi. *Artificial Intelligence and Law*, 2:1–31, 1994.

[Brewka and Gordon, 1994] G. Brewka and T.F. Gordon. How to buy a porsche, an approach to defeasible decision making. In *Working Notes of the AAAI-94 Workshop on Computational Dialectics*, pages 28–38, Seattle, Washington, 1994.

[Brewka, 1994] G. Brewka. A logical reconstruction of Rescher's theory of formal disputation based on default logic. In *Proceedings of the Eleventh European Conference on Artificial Intelligence*, pages 366–370, 1994.

[Brewka, 2001] G. Brewka. Dynamic argument systems: a formal model of argumentation processes based on situation calculus. *Journal of Logic and Computation*, 11:257–282, 2001.

[Dung, 1995] P.M. Dung. On the acceptability of arguments and its fundamental role in nonmonotonic reasoning, logic programming, and n-person games. *Artificial Intelligence*, 77:321–357, 1995.

[Dworkin, 1977] R.M. Dworkin. Is law a system of rules? In R.M. Dworkin, editor, *The Philosophy of Law*, pages 38–65. Oxford University Press, Oxford, 1977.

[Freeman and Farley, 1996] K. Freeman and A.M. Farley. A model of argumentation and its application to legal reasoning. *Artificial Intelligence and Law*, 4:163–197, 1996.

[Gardner, 1987] A. Gardner. *Artificial Intelligence Approach to Legal Reasoning*. MIT Press, Cambridge, MA, 1987.

[Geffner, 1992] H. Geffner. *Default reasoning: causal and conditional theories*. MIT Press, Cambridge, MA, 1992.

[Gelfond and Lifschitz, 1990] M. Gelfond and V. Lifschitz. Logic programs with classical negation. In *Proceedings of the Seventh Logic Programming Conference*, pages 579–597, Cambridge, MA, 1990. MIT Press.

[Genesereth and Nilsson, 1988] M.R. Genesereth and N.J. Nilsson. *Logical Foundations of Artificial Intelligence*. Morgan Kaufmann Publishers Inc, Palo Alto, CA, 1988.

[Gordon and Karaçapilidis, 1997] T.F. Gordon and N. Karaçapilidis. The Zeno argumentation framework. In *Proceedings of the Sixth International Conference on Artificial Intelligence and Law*, pages 10–18, New York, 1997. ACM Press.

[Gordon, 1991] T.F. Gordon. An abductive theory of legal issues. *International Journal of Man-Machine Studies*, 35:95–118, 1991.

[Gordon, 1994] T.F. Gordon. The Pleadings Game: an exercise in computational dialectics. *Artificial Intelligence and Law*, 2:239–292, 1994.

[Gordon, 1995] T.F. Gordon. *The Pleadings Game. An Artificial Intelligence Model of Procedural Justice*. Kluwer Academic Publishers, Dordrecht/Boston/London, 1995.

[Habermas, 1981] J. Habermas. *Theorie des Kommunikativen Handelns*. p, Frankfurt, 1981.

[Hage et al., 1994] J.C. Hage, R.E. Leenes, and A.R. Lodder. Hard cases: a procedural approach. *Artificial Intelligence and Law*, 2:113–166, 1994.

[Hage, 1996] J.C. Hage. A theory of legal reasoning and a logic to match. *Artificial Intelligence and Law*, 4:199–273, 1996.

[Hage, 1997] J.C. Hage. *Reasoning With Rules. An Essay on Legal Reasoning and Its Underlying Logic*. Law and Philosophy Library. Kluwer Academic Publishers, Dordrecht/Boston/London, 1997.

[Hamblin, 1971] C.L. Hamblin. Mathematical models of dialogue. *Theoria*, 37:130–155, 1971.

[Hamfelt, 1995] A. Hamfelt. Formalizing multiple interpretation of legal knowledge. *Artificial Intelligence and Law*, 3:221–265, 1995.

[Horty, 1999] J. Horty. Precedent, deontic logic, and inheritance. In *Proceedings of the Seventh International Conference on Artificial Intelligence and Law*, pages 63–72, New York, 1999. ACM Press.

[Horty, 2001] J. Horty. Argument construction and reinstatement in logics for defeasible reasoning. *Artificial Intelligence and Law*, 9:1–28, 2001.

[Jakobovits and Vermeir, 1999] H. Jakobovits and D. Vermeir. Dialectic semantics for argumentation frameworks. In *Proceedings of the Seventh International Conference on Artificial Intelligence and Law*, pages 53–62, New York, 1999. ACM Press.

[Jones and Sergot, 1992] A.J.I. Jones and M.J. Sergot. Deontic logic in the representation of law: towards a methodology. *Artificial Intelligence and Law*, 1:45–64, 1992.

[Kakas et al., 1992] A.C. Kakas, R.A. Kowalski, and F. Toni. Abductive logic programming. *Journal of Logic and Computation*, 2:719–770, 1992.

[Kowalski and Sergot, 1990] R.A. Kowalski and M.J. Sergot. The use of logical models in legal problem solving. *Ratio Juris*, 3:201–218, 1990.

[Kowalski and Toni, 1996] R.A. Kowalski and F. Toni. Abstract argumentation. *Artificial Intelligence and Law*, 4:275–296, 1996.

[Kowalski, 1989] R.A. Kowalski. The treatment of negation in logic programs for representing legislation. In *Proceedings of the Second International Conference on Artificial Intelligence and Law*, pages 11–15, New York, 1989. ACM Press.

[Kowalski, 1995] R.A. Kowalski. Legislation as logic programs. In Z. Bankowski, I. White, and U. Hahn, editors, *Informatics and the Foundations of Legal Reasoning*, Law and Philosophy Library, pages 325–356. Kluwer Academic Publishers, Dordrecht/Boston/London, 1995.

[Kraus et al., 1998] S. Kraus, K. Sycara, and A. Evenchik. Reaching agreements through argumentation: a logical model and implementation. *Artificial Intelligence*, 104:1–69, 1998.

[Lodder, 1999] A.R. Lodder. *DiaLaw. On Legal Justification and Dialogical Models of Argumentation*. Law and Philosophy Library. Kluwer Academic Publishers, Dordrecht/Boston/London, 1999.

[Loui and Norman, 1995] R.P. Loui and J. Norman. Rationales and argument moves. *Artificial Intelligence and Law*, 3:159–189, 1995.

[Loui et al., 1993] R.P. Loui, J. Norman, J. Olson, and A. Merrill. A design for reasoning with policies, precedents, and rationales. In *Proceedings of the Fourth International Conference on Artificial Intelligence and Law*, pages 202–211, New York, 1993. ACM Press.

[Loui et al., 1997] R.P. Loui, J. Norman, J. Alpeter, D. Pinkard, D. Craven, J. Linsday, and M. Foltz. Progress on Room 5: A testbed for public interactive semi-formal legal argumentation. In *Proceedings of the Sixth International Conference on Artificial Intelligence and Law*, pages 207–214, New York, 1997. ACM Press.

[Loui, 1995] R.P. Loui. Hart's critics on defeasible concepts and ascriptivism. In *Proceedings of the Fifth International Conference on Artificial Intelligence and Law*, pages 21–30, New York, 1995. ACM Press.

[Loui, 1998] R.P. Loui. Process and policy: resource-bounded non-demonstrative reasoning. *Computational Intelligence*, 14:1–38, 1998.

[MacKenzie, 1990] J.D. MacKenzie. Four dialogue systems. *Studia Logica*, 51:567–583, 1990.

[McCarty and Sridharan, 1981] L.T. McCarty and N.S. Sridharan. The representation of an evolving system of legal concepts: II. Prototypes and deformations. In *Proceedings of the Seventh International Joint Conference on Artificial Intelligence*, pages 246–253, 1981.

[McCarty, 1989] L.T. McCarty. A language for legal discourse I. basic features. In *Proceedings of the Second International Conference on Artificial Intelligence and Law*, pages 180–189, New York, 1989. ACM Press.

[McCarty, 1995] L.T. McCarty. An implementation of Eisner v. Macomber. In *Proceedings of the Fifth International Conference on Artificial Intelligence and Law*, pages 276–286, New York, 1995. ACM Press.

[Nitta and Shibasaki, 1997] K. Nitta and M. Shibasaki. Defeasible reasoning in Japanese criminal jurisprudence. *Artificial Intelligence and Law*, 5:139–159, 1997.

[Nute, 1992] D. Nute. Inferences, rules, and instrumentalism. *International Journal of Expert Systems*, 5:267–274, 1992.

[Nute, 1997] D. Nute, editor. *Defeasible Deontic Logic*, volume 263 of *Synthese Library*. Kluwer Academic Publishers, Dordrecht/Boston/London, 1997.

[Parsons et al., 1998] S. Parsons, C. Sierra, and N.R. Jennings. Agents that reason and negotiate by arguing. *Journal of Logic and Computation*, 8:261–292, 1998.

[Peczenik, 1996] A. Peczenik. Jumps and logic in the law. *Artificial Intelligence and Law*, 4:297–329, 1996.

[Pollock, 1995] J.L. Pollock. *Cognitive Carpentry. A Blueprint for How to Build a Person*. MIT Press, Cambridge, MA, 1995.

[Poole, 1988] D.L. Poole. A logical framework for default reasoning. *Artificial Intelligence*, 36:27–47, 1988.

[Prakken and Sartor, 1996] H. Prakken and G. Sartor. A dialectical model of assessing conflicting arguments in legal reasoning. *Artificial Intelligence and Law*, 4:331–368, 1996.

[Prakken and Sartor, 1997] H. Prakken and G. Sartor. Argument-based extended logic programming with defeasible priorities. *Journal of Applied Non-classical Logics*, 7:25–75, 1997.

[Prakken and Sartor, 1998] H. Prakken and G. Sartor. Modelling reasoning with precedents in a formal dialogue game. *Artificial Intelligence and Law*, 6:231–287, 1998.

[Prakken and Vreeswijk, 2002] H. Prakken and G.A.W. Vreeswijk. Logics for defeasible argumentation. In D. Gabbay and F. Günthner, editors, *Handbook of Philosophical Logic*, volume 4, pages 219–318. Kluwer Academic Publishers, Dordrecht/Boston/London, second edition, 2002.

[Prakken, 1995] H. Prakken. From logic to dialectics in legal argument. In *Proceedings of the Fifth International Conference on Artificial Intelligence and Law*, pages 165–174, New York, 1995. ACM Press.

[Prakken, 1996] H. Prakken. Two approaches to the formalisation of defeasible deontic reasoning. *Studia Logica*, 57:73–90, 1996.

[Prakken, 1997] H. Prakken. *Logical Tools for Modelling Legal Argument. A Study of Defeasible Argumentation in Law*. Law and Philosophy Library. Kluwer Academic Publishers, Dordrecht/Boston/London, 1997.

[Prakken, 2000] H. Prakken. On dialogue systems with speech acts, arguments, and counterarguments. In *Proceedings of the 7th European Workshop on Logic for Artificial Intelligence (JELIA'2000)*, number 1919 in Springer Lecture Notes in AI, pages 224–238, Berlin, 2000. Springer Verlag.

[Prakken, 2001a] H. Prakken. Modelling defeasibility in law: logic or procedure? *Fundamenta Informaticae*, 48:253–271, 2001.

[Prakken, 2001b] H. Prakken. Modelling reasoning about evidence in legal procedure. In *Proceedings of the Eighth International Conference on Artificial Intelligence and Law*, pages 119–128, New York, 2001. ACM Press.

[Prakken, 2001c] H. Prakken. Relating protocols for dynamic dispute with logics for defeasible argumentation. *Synthese*, 127:187–219, 2001.

[Prakken, 2002] H. Prakken. An exercise in formalising teleological case-based reasoning. *Artificial Intelligence and Law*, 10, 2002. in press.

[Rawls, 1972] J. Rawls. *A Theory of Justice*. Oxford University Press, Oxford, 1972.

[Raz, 1975] J. Raz. *Practical Reason and Norms*. Princeton University Press, Princeton, 1975.

[Rescher, 1977] N. Rescher. *Dialectics: a Controversy-oriented Approach to the Theory of Knowledge*. State University of New York Press, Albany, N.Y., 1977.

[Rissland and Ashley, 1987] E.L. Rissland and K.D. Ashley. A case-based system for trade secrets law. In *Proceedings of the First International Conference on Artificial Intelligence and Law*, pages 60–66, New York, 1987. ACM Press.

[Rissland and Skalak, 1991] E.L. Rissland and D.B. Skalak. CABARET: statutory interpretation in a hybrid architecture. *International Journal of Man-Machine Studies*, 34:839–887, 1991.

[Routen and Bench-Capon, 1991] T. Routen and T.J.M. Bench-Capon. Hierarchical formalizations. *International Journal of Man-Machine Studies*, 35:69–93, 1991.

[Sartor, 1995] G. Sartor. Defeasibility in legal reasoning. In Z. Bankowski, I. White, and U. Hahn, editors, *Informatics and the Foundations of Legal Reasoning*, Law and Philosophy Library, pages 119–157. Kluwer Academic Publishers, Dordrecht/Boston/London, 1995.

[Sartor, 1997] G. Sartor. Logic and argumentation in legal reasoning. *Current Legal Theory*, pages 25–63, 1997.

[Sergot et al., 1986] M.J. Sergot, F. Sadri, R.A. Kowalski, F. Kriwaczek, P. Hammond, and H.T. Cory. The British Nationality Act as a logic program. *Communications of the ACM*, 29:370–386, 1986.

[Sergot, 1988] M.J. Sergot. Representing legislation as logic programs. In J.E. Hayes, D. Michie, and J. Richards, editors, *Machine Intelligence*, volume 11, pages 209–260. Oxford University Press, Oxford, 1988.

[Simari and Loui, 1992] G.R. Simari and R.P. Loui. A mathematical treatment of defeasible argumentation and its implementation. *Artificial Intelligence*, 53:125–157, 1992.

[Simon, 1982] H. Simon. *Models of Bounded Rationality*, volume 2 (collected papers). MIT Press, Cambridge, MA, 1982.

[Skalak and Rissland, 1992] D.B. Skalak and E.L. Rissland. Arguments and cases. an inevitable intertwining. *Artificial Intelligence and Law*, 1:3–44, 1992.

[Toulmin, 1958] S.E. Toulmin. *The Uses of Argument*. Cambridge University Press, Cambridge, 1958.

[Verheij et al., 1998] B. Verheij, J.C. Hage, and H.J. van der Herik. An integrated view on rules and principles. *Artificial Intelligence and Law*, 6:3–26, 1998.

[Verheij, 1996] B. Verheij. *Rules, reasons, arguments: formal studies of argumentation and defeat*. Doctoral dissertation University of Maastricht, 1996.

[Verheij, 1999] B. Verheij. Automated argument assistance for lawyers. In *Proceedings of the Seventh International Conference on Artificial Intelligence and Law*, pages 43–52, New York, 1999. ACM Press.

[Vreeswijk, 2000] G.A.W. Vreeswijk. Representation of formal dispute with a standing order. *Artificial Intelligence and Law*, 8:205–231, 2000.

[Walton and Krabbe, 1995] D.N. Walton and E.C.W. Krabbe. *Commitment in Dialogue. Basic Concepts of Interpersonal Reasoning*. State University of New York Press, Albany, NY, 1995.

[Winston, 1980] P.H. Winston. Learning and reasoning by analogy. *Communications of the ACM*, 23:689–703, 1980.

Logic Programming Updating - A Guided Approach

José Júlio Alferes and Luís Moniz Pereira

Centro de Inteligência Artificial, Fac. Ciências e Tecnologia, Univ. Nova de Lisboa,
P-2825-114 Caparica, Portugal,
Voice:+351 21 294 8533, Fax: +351 21 294 8541
jja,lmp@di.fct.unl.pt

Abstract. In this work we review and synthesize, in a selective way, a series of recent developments concerning the dynamics of the evolution of logic programs by means of updates. We do so because this comparatively new and expanding area merits the attention of more researchers and more teachers alike, though there does not exist a single integrative source to induct them to the topic.

1 Introduction

Inasmuch we have accompanied the area of logic program updating from its inception, and contributed assiduously to its growth, we assumed ourselves in a good position to promote the topic and fill-in the absence and lack of a coherent self-contained exposition. The opportunity to do so is afforded by the present chapter-length work in honour of Bob Kowalski, who has done so much to promote logic programming as a whole. Note, however, that this is not a survey. It simply brings together at this juncture, within a uniform notation, continuity of exposition, and under the same 2-valued semantics, the marrow of a series of developments on the topic of logic program updates, which have been co-authored with others. A critical survey would require a much longer work, including the introduction to each of other authors' approaches and notation. Notwithstanding, the original papers we reference contain a number of comparative and critical remarks that the reader can follow up to that effect.

We begin at the beginning, by recapitulating the seminal work of [34] on revision programs (here dubed MT-revision-programs) to specify model updates, and go on to show how they can captured a program transformation, as a result of work by [6]. Next, we present the topic and issues of program updates, a generalization of model updates, show how they can specify the result of sequences of updates known as dynamic logic programs (DLPs) [4], and illustrate their applications. Subsequently, we introduce the language LUPS [8], devised for specifying update commands which produce DLPs, and exhibit its application in a number of domains. Finally, we proffer future perspectives on logic program updating, and mention ongoing work and implementations.

A.C. Kakas, F. Sadri (Eds.): Computat. Logic (Kowalski Festschrift), LNAI 2408, pp. 382–412, 2002.

We are indebted to the co-authors of joint papers from which we have extracted or adapted much material, namely João Leite, Halina Przymusinska, Teodor Przymusinski, and Paulo Quaresma.

2 Model Updates

As the world changes so must programs that represent knowledge about it. When dealing with modifications to a knowledge base represented by a propositional theory, two kinds of abstract frameworks have been distinguished both by Keller and Winslett in [28] and by Katsuno and Mendelzon in [27]. One, theory revision, deals with incorporating new knowledge about a static world state. The other deals with changing worlds, and is known as theory update. In this work, we are concerned with theory update only, and, in particular, within the framework of logic programming.

Within the framework of logic programming, simple fact by fact updates have long been addressed [14, 23, 24]. Program updating is distinct from program revision, where a program accommodates, perhaps non-monotonically by revising assumptions, additional information about a world state. Work on logic programs revision (or contradiction removal) has received more attention (e.g. in [3, 5, 22, 40, 41]).

A key insight into the issue of updating theories is due to Winslett [39], who showed that, contrary to theory revision, one must consider the effect of an update in each of the states of the world that are consistent with our current knowledge of its state. The following realistic situation chisels the differences between program update and revision crisply.

Example 1. My secretary has just booked me on a flight from here to London on Wednesday but can't remember to which airport, Gatwick or Heathrow. Clearly, this statement can be represented by:

$$booked\_for\_gatwick \lor booked\_for\_heathrow$$

where propositions $booked\_for\_X$ mean that I have a booking for a Wednesday flight to airport X.

Now someone tells me there never are flights from here to Gatwick on Wednesday i.e., assuming that no one cannot have a booking for a non-existing flight, I'm told $\neg booked\_for\_gatwick$. I conclude that I'll be flying to Heathrow, i.e. $booked\_for\_heathrow$. This is knowledge revision. The state of the world hasn't changed with respect to the flight information, but on obtaining more information I have revised accordingly my knowledge about that state of the world.

Alternatively, I hear on the radio that all flights to Gatwick on Wednesday have been cancelled and, consequently, all possibly existing bookings for those flights have been withdrawn. In other words, the world changed such that now $\neg booked\_for\_gatwick$ holds. I'm at a loss regarding whether I still have a flight to London on Wednesday. This is knowledge update. The world of flights has changed, and refining my knowledge about its previous state is inadequate: I cannot conclude that I have a booking for a flight to Heathrow (*booked_for_heathrow*). I have obtained knowledge about the new world state but it doesn't help me to disambiguate the knowledge I had about its previous state. What I can do is pick up the phone and book a flight to Heathrow on Wednesday, on any airline. That will change my flight world and at the same time update my knowledge about that change. However, I'm now unsure whether I might not have two flights booked to Heathrow. But if my secretary suddenly remembers he had definitely booked me to Gatwick, then I will no longer believe I have two flights to Heathrow on Wednesday.

Accordingly, theory update is performed "model by model" [27, 39], where a set of formulae T_U is a *theory update* of T, following an update request U, iff the models of T_U result from updating each of the models of T by U. Thus a theory update is determined by the update of its models.

The same idea can be applied to knowledge bases represented by logic programs: a program P_U is a *program update* of P, following an update request U, iff the models of P_U (according to some logic program semantics S) are the result of updating each of the models of P (given by semantics S) by U. So, to obtain P_U, first compute all models of P according to a given semantics S; to each of these models apply the update request U to obtain a new set of models $\mathcal{M}$; P_U is then any logic program whose models are exactly $\mathcal{M}$.

In [10, 34, 35, 37], the issue of program change via updating rules has been introduced. There, a new set of models is obtained by means of the update rules, from each of the models of the given program. Any program satisfying the new set of models will count as an update of the original program. However, no procedure, for obtaining such a single program whose model are the ones resulting from the update, is set forth by the cited authors. (Except in the trivial case where the original and final programs are just sets of facts [10, 37].) Following [27, 39], it is essential to start by specifying precisely how a program's models are to change, before even attempting to specify program change.

In this section, we begin with an overview of the work on model updating, and move on to present a correct transformation on normal programs which, from an initial program, produces another program whose models enact the required change in the initial program's models, as specified by the update rules.

2.1   Marek and Truszczyński's Revision Programs

In [34], the authors introduced a language for specifying updates to knowledge bases, which they called *revision programs*. Given the set of all models of a knowledge base, a revision program specifies exactly how the models are to be changed.[1]

To avoid confusion between the concept of program (or theory) revision, whose difference to updates are reviewed in the beginning of Section 2, and the

---

[1] For more detailed motivation and background to this section of the present paper the reader is referred to [34, 37].

name "revision programs" choosen by [34] to denote the programs that specify knowledge base updates, in the sequel we will call these *MT-revision-programs*.

The language of MT-revision-programs is quite similar to that of logic programming: MT-revision-programs are collections of *update rules*, which in turn are built of atoms by means of the special operators: $\leftarrow$, *in*, *out*, and ",". The first is an implication symbol, *in* specifies that some atom is added to the models, via an update, *out* that some atom is deleted, and the comma denotes conjunction.

Definition 1 (Update rules for atoms). *Let U be a countable set of atoms. An update in-rule or, simply, an in-rule, is any expression of the form:*

$$in(p) \leftarrow in(q_1), \ldots, in(q_m), out(s_1), \ldots, out(s_n) \qquad (1)$$

where p, q_i, $1 \leq i \leq m$, and s_j, $1 \leq j \leq n$, are all in U, and $m, n \geq 0$.

An update out-rule or, simply, an out-rule, is any expression of the form:

$$out(p) \leftarrow in(q_1), \ldots, in(q_m), out(s_1), \ldots, out(s_n) \qquad (2)$$

where p_i, q_i, $1 \leq i \leq m$, and s_j, $1 \leq j \leq n$, are all in U, and $m, n \geq 0$.

Intuitively, MT-revision-programs can be regarded as operators which, given some initial interpretation I_i, produce its updated version I_u.

Example 2. Consider a knowledge base with two models $I_1 = \{gatwick\}$ and $I_2 = \{heathrow\}$. The information that flights for Gatwick have been cancelled, can be represented by the MT-revision-program UP:

$$out(gatwick)$$

stating that in the resulting knowledge base, *gatwick* is to be deleted. We will see that this MT-revision-program, applied to I_1 and I_2 produces the two models: $\{\}$ and $\{heathrow\}$. In the first, both *gatwick* and *heathrow* are false, and in the latter, *gatwick* is false and *heathrow* is true. As desired, in both of them *gatwick* was deleted, and thus is false.

It is worth noting here some similarities between these update rules and STRIPS operators [18], in that both specify what should be added and what should be deleted from the current knowledge base. However, differently from STRIPS the preconditions of update rules may depend on the models of the resulting knowledge base. With STRIPS they may only depend on the models of the previous knowledge base.

Example 3. Let UP be the MT-revision-program:

$$in(a) \leftarrow out(b)$$
$$in(b) \leftarrow out(a)$$

and an initial knowledge base whose only model is $I_i = \{\}$, where both a and b are false . Intuitively, the first rule of UP, states that if b is not true in the resulting theory (after the update) then a must be added. The second states that if a is not true, then b must be added. Thus, there are two possible update interpretations: $I_u = \{a\}$ and $I_u = \{b\}$.

In this example, differently from STRIPS, the update in a knowledge base can be conditional on the truth or falsity of some atoms in the resulting models. However, the first example shows that there are some changes that are mandatory in a MT-revision-program. In it, removing *gatwick* is not conditional on anything, and must be done in every resulting model. This notion of mandatory or necessary changes is formalized as follows:

Definition 2 (Necessary change). *Let P be a MT-revision-program with least model M. The* necessary change *determined by P is the pair (In_P, Out_P), where:*

$$In_P = \{a : in(a) \in M\}$$
$$Out_P = \{a : out(a) \in M\}$$

If $In_P \cap Out_P = \{\}$ then P is said coherent.

Intuitively, the necessary change determined by a program P specifies those atoms that must be added and those atoms that must be deleted, whatever the initial interpretation.

Example 4. Take the MT-revision-program $P = \{out(b) \leftarrow out(a); in(b); out(a)\}$. The necessary changes are irreconcilable (since b must be added, and simultaneously deleted) and P is incoherent.

To build a model of the resulting knowledge base, after the update specified by a MT-revision-program, necessary change must be considered. But, depending on the initial interpretation, other changes are in order. These changes are formalized as follows:

Definition 3 (Justified update). *Let P be a MT-revision-program and I_i and I_u two interpretations. The* reduct $P_{I_u | I_i}$ *with respect to I_i and I_u is obtained by the following operations:*

- *Removing from P all rules whose body contains some $in(a)$ and $a \notin I_u$;*
- *Removing from P all rules whose body contains some $out(a)$ and $a \in I_u$;*
- *Removing from the body of remaining rules of P all $in(a)$ such that $a \in I_i$;*
- *Removing from the body of remaining rules of P all $out(a)$ such that $a \notin I_i$.*

Whenever P is coherent, I_u is a P-justified update of I_i if the following stability condition holds: $I_u = (I_i - Out_{P_{I_u | I_i}}) \cup In_{P_{I_u | I_i}}.$

The first two operations delete rules which are useless given I_u. Due to stability, the initial interpretation is preserved as much as possible in the final one. The last two rules achieve this since any exceptions to preservation are explicitly dealt with by the union and difference operations in the two stability conditions.

Example 5. In example 2, note e.g. that $\{\}$ is a justified update of $\{gatwick\}$. In fact, the reduct operation does not change the MT-revision-program, and its necessary change is given by $In = \{\}$ and $Out = \{gatwick\}$. Stability is guaranteed because:

$$\{\} = (\{gatwick\} - \{gatwick\}) \cup \{\}$$

Example 6. In example 3, note e.g. that $\{a\}$ is a justified update of $\{\}$. The reduct operation yields the MT-revision-program with the single fact $in(a)$, and thus its necessary change is given by $In = \{a\}$ and $Out = \{\}$. Stability is guaranteed because:

$$\{a\} = (\{\} - \{\}) \cup \{a\}$$

2.2 Model Updates as Logic Programs

With MT-revision-programs, program updating is only implicitly achieved, by recourse to the updating of each of a program's models to obtain a new set of updated models, via the update rules. Following [27], any program satisfying the updated models counts as a program update. Unfortunately, such an approach suffers, in general, from several important drawbacks:

- In order to obtain the update KB' of a knowledge base KB one has to first compute all the models M of DB (typically, a daunting task) and then individually compute their (possibly multiple) updates M_U by U. An update M_U of a given interpretation M is obtained by changing the status of only those literals in M that are *"forced"* to change by the update U, while keeping all the other literals intact by *inertia*.
- The updated knowledge base KB' is not defined directly but, instead, it is indirectly characterized as a knowledge base whose models coincide with the set of all updated models M_U of KB. In general, there is therefore no natural way of computing[2] KB' because the only straightforward candidate for KB' is the typically intractably large knowledge base KB'' consisting of all clauses that are entailed by all the updated models M_U of KB.

To overcome these important drawbacks, in this section we demonstrate a program transformation path to updating, similar to that of [6], that directly obtains, from the original program, an updated program with the required models, which is similar to the first. The updated program's models will be exactly those derivable, one by one, from the original program's models through the update rules. Thus it is possible to sidetrack the model generation path.

We shall see that any MT-revision-program can be transformed into an extended logic program which, by the way, becomes part of the new updated program. This transformation is similar in character to the one in [37], which serves a different purpose though.[3] The normal program which is subjected to the MT-revision-program, has to be transformed too. The final updated program is the union of the two transformations.

Definition 4 (Translation of MT-revision-programs into extended LPs).
Given a MT-revision-program UP, its translation into the updated logic program ULP is defined as follows.

---

[2] In fact, in general such a database KB' may not exist at all.

[3] In [6] a more complex transformation is considered where the program to be updated can also be an extended logic program.

– *Each update in-rule of the form (1) translates into:*

$$p \leftarrow q_1, \ldots, q_m, \neg s_1, \ldots, \neg s_n$$

– *Each update out-rule of the form (2) translates into:*

$$\neg p \leftarrow q_1, \ldots, q_m, \neg s_1, \ldots, \neg s_n$$

The rationale for this translation can best be understood in conjunction with the next definition, for they go together. Suffice it to say that we can simply equate explicit negation $\neg$ with *out*, since the programs to be updated are normal ones, and thus devoid of explicit negation (so no confusion can arise).

Definition 5 (Update transformation of a normal program). *Given a MT-revision-program UP, consider its corresponding updated logic program ULP. For any normal logic program P, its updated program U with respect to ULP (or to UP) is obtained through the operations:*

– *The rules of ULP belong to U;*
– *The rules of P belong to U, subject to the changes below;*
– *For each atom A figuring in a head of a rule of ULP:*
  – *Replace in every rule of U originated in P all occurences of A by A_i, where A_i is a new atom;*
  – *Include in U the rules $A \leftarrow A_i, not \neg A$ and $\neg A \leftarrow not A_i, not A$.*

The purpose of the first operation is to ensure change according to the MT-revision-program. The second operation guarantees that, for inertia, rules in P remain unchanged unless they can be affected by some update rule. The third operation changes all atoms in rules originating in P which are possibly affected by some update rule, by renaming such atoms. The new name stands for the atom as defined by the P program. The fourth operation introduces inertia rules, stating that any possibly affected atom contributes to the definition of its new version, unless actually affected through being overriden by the contrary conclusion of an update rule; the $not \neg A$ and $not A$ conditions cater for this test.

Example 7. The translation of the MT-revision-program of example 2 is:

$$\neg gatwick$$

Now suppose that the initial knowledge base is given by the logic program:

$$gatwick \leftarrow not\, heathrow$$
$$heathrow \leftarrow not\, gatwick$$

whose stable models are exactly those initial models mentioned in example 2. The corresponding updated program is (with the obvious abbreviations) :

$$\neg g \qquad g_i \leftarrow not\, h \qquad g \leftarrow g_i, not\, \neg g$$
$$h \leftarrow not\, g_i \qquad \neg g \leftarrow not\, g_i, not\, g$$

Its answer-sets are $\{\neg g, g_i\}$ and $\{\neg g, h\}$.

Note that, if we restrict to the original language of P, and ignore explicit negation, these two models become exactly those that are the justified updates.

Example 8. Suppose that the initial knowledge base of example 3 is given by the empty program (whose only stable model is the empty set). The updated program of example 3 is simply:

$$a \leftarrow not\, b$$
$$b \leftarrow not\, a$$

whose answer-sets are $\{a\}$ and $\{b\}$.

Theorem 1 (Correctness of the update transformation).   *Let P be a normal logic program and UP a coherent MT-revision-program. Modulo any new atoms of the form A_i and any explicitly negated elements, the answer-sets of the updated program U of P with respect to UP, are exactly the UP-justified updates of the stable models of P.*

In general, the updated programs of normal programs (without explicit negation) are extended programs (with explicit negation). To update them in turn, and thus making it possible to iterate the update process, the issue of updating extended programs has to be addressed. This was one main motivation of [6] for defining updates for models and programs with explicit negation. For the sake of simplicity of exposition, and as in our opinion this is not a central point of the present work, we refer the interested reader to [6]. There a definition of update transformation of extended program, that allows for the iteration of updates can be found that allows for the iteration of updates.

3   Program Updates

As mention above, a key insight into the issue of updating theories is due to Winslett [39], who showed that one must consider the effect of an update in each of the states of the world that are consistent with our current knowledge of its state. MT-revision-programs, described in the previous section, follows this approach. The common intuition behind the update of a model has been based on what is referred to as the commonsense law of inertia, i.e. things do not change unless they are expressly made to, in this case the truth value of each element of the model to be updated should remain the same unless explicitly changed by the update. Suppose, for example, that we have a model in which "sunshine" is true and "rain" is false; if later we receive the information that the sun is no longer shining, we conclude that "sunshine" is false, due to the update, and that "rain" is still false by inertia.

Suppose now that our vision of the world is described instead by a logic program and we want to update it. Is updating each of its models enough? Is all the information borne by a logic program contained within its set of models?

The answer to both these questions is negative. A logic program encodes more than the set of its individual models. It encodes the relationships between the elements of a model, which are lost if we envisage updates simply on a model by model basis, as proposed by MT-revision-programs.

In fact, while the semantics of the resulting knowledge base after an update indeed represents the intended meaning when just the extensional part of the original knowledge base (the set of facts) is being updated, it leads to strongly *counter-intuitive* results when also the intensional part of the database (the set of rules) undergoes change, as the following example shows:

Example 9. Consider the logic program P :

$$P : sleep \quad \leftarrow not\, tv\_on$$
$$tv\_on \quad \leftarrow \quad\quad\quad\quad\quad (3)$$
$$watch\_tv \leftarrow tv\_on.$$

Clearly $M = \{tv\_on, watch\_tv\}$ is its only stable model. Suppose now that the update U states that there is a power failure, and if there is a power failure then the TV is no longer on, as represented by the logic program U:

$$U : \quad\quad out(tv\_on) \leftarrow in(power\_failure)$$
$$in(power\_failure) \leftarrow \quad\quad\quad\quad\quad (4)$$

With MT-revision-programs, we would obtain $M_U = \{power\_failure, watch\_tv\}$ as the only update of M by U. This is because $power\_failure$ needs to be added to the model and its addition forces us to make tv_on false. As a result, even though there is a power failure, we are still watching TV. However, by inspecting the initial program and the updating rules, we are likely to conclude that since "$watch\_tv$" was true only because "tv_on" was true, the removal of "tv_on" should implicitly make "$watch\_tv$" false by default. Moreover, one would expect "$sleep$" to become true as well.[4] Consequently, the intended model of the update of P by U is the model $M'_U = \{power\_failure, sleep\}$.

Suppose now that another update U_2 follows, described by the logic program:

$$U_2 : out(power\_failure) \leftarrow \quad\quad\quad\quad\quad (5)$$

stating that power is back up again. We should now expect the TV to be on again. Since power was restored, i.e. "$power\_failure$" is false, the rule "$out(tv\_on) \leftarrow in(power\_failure)$" of U should have no effect and the truth value of "tv_on" should be obtained by inertia from rule "$tv\_on \leftarrow$ " of the original program P.

This example illustrates that, when updating knowledge bases, it is not suffi-cient to just consider the truth values of literals figuring in the heads of its rules because the truth value of their rule bodies may also be affected by the updates of other literals. In other words, it suggests that the *principle of inertia* should

---

[4] Note the similarities between this and the "ramification problem", i.e. the problem of proliferation of implicit consequences of actions.

be applied not just to the individual literals in an interpretation but rather to the *entire rules of the knowledge base.*

Newton's first law, also known as the law of inertia, states that: *"every body remains at rest or moves with constant velocity in a straight line, unless it is compelled to change that state by an unbalanced force acting upon it"* (adapted from [36]). One often tends to interpret this law in a commonsensical way, as things keeping as they are unless some kind of force is applied to them. This is true but it doesn't exhaust the meaning of the law. It is the result of all applied forces that governs the outcome. Take a body to which several forces are applied, and which is in a state of equilibrium due to those forces canceling out. Later one of those forces is removed and the body starts to move.

The same kind of behaviour presents itself when updating programs. Before obtaining the truth value, by inertia, of those elements not directly affected by the update program, one should verify whether the truth of such elements is not indirectly affected by the updating of other elements. That is, the body of a rule may act as a force that sustains the truth of its head, but this force may be withdrawn when the body becomes false.

Another way to view program updating, and in particular the rôle of inertia, is to say that the rules of the initial program carry over to the updated program, due to inertia, instead of the truth of literals, just in case they aren't overruled by the update program. Once again this should be so because the rules encode more information than their models.

This approach was first adopted in [31], where the authors present a program transformation which, given an initial program and an update program, produces an updated program obeying the rule of inertia.

The above example also leads us to another important observation, namely, that the notion of an update KB' of one knowledge base KB by another knowledge base U should not just depend on the *semantics* of KB and U, as it is the case with interpretation updates, but that it should also depend on their *syntax*. This is best illustrated by the following, even simpler, example:

Example 10. Consider the logic program P :

$$P : innocent \leftarrow not\ found\_guilty \tag{6}$$

whose only stable model is $M = \{innocent\}$, because $found\_guilty$ is false by default. Suppose now that the update U states that the person has been found guilty:

$$U : in(found\_guilty). \tag{7}$$

Using the interpretation approach of MT-revision-programs, we would obtain $M_U = \{innocent, found\_guilty\}$ as the only update of M by U thus leading us to the counter-intuitive conclusion that the person is both innocent and guilty. This is because $found\_guilty$ must be added to the model M and yet its addition does not force us to make $innocent$ false. However, it is intuitively clear that the interpretation $M'_U = \{found\_guilty\}$, stating that the person is guilty but no longer presumed innocent, should be the only model of the updated program.

Mark that, such a desired model could not be obtained simply by imposing an integrity constraint to the effect that *innocent* and *found_guilty* cannot be simultaneously true. Such an integrity constraint would remove the undesired model but would not introduce the desired one, leaving us with a resulting update with no models at all.

Observe, however, that the program P is *semantically equivalent* to the following program P':

$$P' : innocent \leftarrow \qquad\qquad (8)$$

because the programs P and P' have exactly the same set of stable models, namely the model M. Nevertheless, while the model $\{innocent, found\_guilty\}$ is not the intended model of the update of P by U it is in fact the only reasonable model of the update of P' by U.

To overcome these drawbacks of MT-revision-programs, and consider inertia of logic program rules, [4] introduced *dynamic logic programming*. In this setting, sequences of logic programs $P_1 \oplus \ldots \oplus P_n$ are given. Intuitively a sequence may be viewed as the result of, starting with program P_1, updating it with program $P_2, \ldots$, and updating it with program P_n. Alternatively, the different P_is in the sequence can be viewed as different time points in possible future evolutions of the knowledge, or even as knowledge of ever more specific objects organized in a hierarchy (see [11] for more on this view). In DLP, newer or more specific rules (coming from new, newly acquired, or more specific knowledge) can be added at the end of the sequence, bothering not whether they conflict with the previous or less specific knowledge. The role of dynamic programming is to ensure that these added rules are in force, and that previous or less specific rules are still valid (by inertia) as far as possible, i.e. they are kept for as long as they do not conflict with newly added ones.

3.1   Dynamic Logic Programs

In this section we recapitulate Dynamic Logic Programming (DLP) [4], a framework that can be used to model the evolution of a logic program through sequences of updates.

To represent update rules, instead of using the operators *in* and *out* of MT-revision-programs, DLP directly uses the syntax of logic programs. In fact, $in(A)$ in the body of an update rule simply stands for A being true in the updated knowledge base, and $out(A)$ for A not being true in the updated knowledge base, i.e. *not* A being true. Accordingly, there is no need for these special operators, and the *in*s in bodies can be removed whilst *out*s are replaced by default negation.

Operator $in(A)$ in the head of an update rule simply stands for making A true, and thus can also be replaced by the atom itself. To represent negative information in logic programs and their updates, instead of the operator *out* of MT-revision-programs, DLP allows for the presence of default negation in rule heads. It is worth noting why, in the update setting, generalizing the language to allow default negation in rule heads (thus defining *"generalized logic programs"*) is more adequate than introducing explicit negation in programs (both in heads and bodies). Suppose we are given a rule stating that A is true whenever some condition $Cond$ is met. This is naturally represented by the rule $A \leftarrow Cond$. Now suppose we want to say, as an update, that A should no longer be the case (i.e. should be deleted or retracted), if some condition $Cond'$ is met. How do we represent this new knowledge? By using extended logic programming (with explicit negation) this could be represented by $\neg A \leftarrow Cond'$. But this rule says more than we want to. It states that A is false upon $Cond'$, and we only want to go as far as to say that the truth of A is to be deleted in that case. All is wont to be said is that, if $Cond'$ is true, then $not\ A$ should be the case, i.e. $not\ A \leftarrow Cond'$. As argued in [20], the difference between explicit and default negation is fundamental whenever the information about some atom A cannot be assumed to be complete. Under these circumstances, the former means that there is evidence for A being false, while the latter means that there is no evidence for A being true. In the deletion example, we desire the latter case.

In other words, a $not\ A$ head means A is deleted if the body holds. Deleting A means that A is no longer true, not necessarily that it is false. When the CWA is adopted as well, then this deletion causes A to be false. In the updates setting, the CWA must be explicitly encoded from the start, by making all $not\ A$ false in the initial program being updated. That is, the two concepts, deletion and CWA, are orthogonal and must be separately incorporated. In the stable models [26, 32] and well-founded semantics [12] of single generalized programs, the CWA is adopted *ab initio*, and default negation in the heads is conflated with non-provability because there is no updating and thus no deletion. Note however that, unlike with single generalized programs (cf. [26]), in updates the head not s cannot be moved freely into the body, to obtain simple denials: there is inescapable pragmatic information in specifying exactly which not literal figures in the head, namely the one being deleted when the body holds true. It is not indifferent that any other (positive) body literal in the denial would be moved to the head. Example 12 shows just that.

We now recall the semantics of *single generalized logic programs*. The class of generalized logic programs can be viewed as a special case of yet broader classes of programs, introduced earlier in [26] and in [32]. As shown in [4], their semantics coincides with the stable models semantics [19] for the special case of normal programs. Moreover, the semantics also coincides with the one in [32] (and, consequently, with the one in [26]) when the latter is restricted to the language of generalized programs.

Definition 6 (Generalized logic program). *A generalized logic program in the language $\mathcal{L}$ is a finite or infinite set of ground rules r of the form:*

$$L_0 \leftarrow L_1, \ldots, L_n. \qquad n \geq 0$$

where each L_i is a literal in $\mathcal{L}$ (i.e. an atom or a default literal $not\ A$ where A is an atom). By head(r) we mean L_0, by body(r) the set of literals $\{L_1, \ldots, L_n\}$, by $body_{pos}(r)$ the set of all atoms in body(r), and by $body_{neg}(r)$ the set of all default literals in body(r). We refer to $body_{pos}(r)$ as the prerequisites of r.

In the sequel, whenever L is of the form not A, we use not L to stand for the atom A.

The semantics of generalized logic programs is defined as a generalization of the stable models semantics [19]. Before advancing the generalized definition, let us sketch, as a first step, a definition equivalent to the stable models semantics for the case of normal logic programs.

In the fixpoint operator $\Gamma(M)$ of the stable models semantics, first one deletes every program rule whose body contains some not A where $A \in M$, and deletes too, from rule bodies every literal not A such that $A \notin M$. The least model of the so obtained program is then computed. In its stead, one may take default literals in rule bodies as new propositional variables, add a fact not A for every $A \notin M$, and then compute the least model of the resulting definite program. It is easy to check that the resulting set of atoms, not of the form not A, will be exactly the same as in $\Gamma(M)$. Moreover, for every fixpoint of $\Gamma(M)$, $A \notin M$ iff all rules of the program with head A have a false body in M. Thus, if one is only interested in fixpoints, instead of adding facts not A for every $A \notin M$, one may add not A for just every A having no rule with a true body in M. This approach views stable models as deriving not A for every atom A which is not "supported" in the program by the model.

Now, since one can have default literals in rule heads, there are more ways of deriving them. But the previous one remains, i.e. if for some A there is no rule for A whose body is true, then not A should be the case. This is the basic intuition behind the definition of stable models for generalized programs: given a model M, first add facts not A for every A with no rule with true body in M; M is a stable model if the least model obtained after such additions coincides with M, where M has been enlarged with new propositional variables not A for every $A \notin M$.

Definition 7 (Default assumptions). *Let M be a model of P. Then:*

$$Default(P, M) = \{not\, A \mid \nexists r \in P : head(r) = A \wedge M \models body(r)\}$$

Definition 8 (Stable Models of Generalized Programs). *A model M is a stable model of the generalized program P iff $M = least(P \cup Default(P, M))$*

In DLP, sequences of generalized programs $P_1 \oplus \ldots \oplus P_n$ are given. As said before, intuitively a sequence may be viewed as the result of, starting with program P_1, updating it with program P_2, ..., and updating it with program P_n. The role of dynamic programming is to ensure that the newly added rules (from latter programs) are in force, and that previous rules (from previous programs) are still valid (by inertia) as far as possible, i.e. they are kept for as long as they do not conflict with newly added ones.

The semantics of dynamic logic programs is defined according to the rationale above. Given a model M of the last program P_n, start by removing all the rules from previous programs whose head is the complement of some later rule with true body in M (i.e. by removing all rules which conflict with more recent ones). All others persist through by inertia. Then, as for the stable models of a single generalized program, add facts $not\,A$ for all atoms A which have no rule at all with true body in M, and compute the least model. If M is a fixpoint of this construction, M is a stable model of the sequence up to P_n.

Other possible views on and usage of DLP, justify slight generalizations of the above informally described language and semantics. In general, the distinguished programs represent knowledge true at some state s, where different states may stand for different stages of knowledge in the linear evolution of the knowledge base (as above), but also for different time points in possible future evolutions of the knowledge, or even for knowledge of ever more specific objects organized in a hierarchy. In the latter case, each program contains the rules that are specific to the object under consideration, and rules from programs above in the hierarchy are inherited just as long as they do not conflict with the more specific information (for more on this stance see [11]). These other views justify a tree-like structure of programs (rather than a sequence), and also that dynamic programs can be queried at any state, rather than only at the last one.

Definition 9 (Dynamic Logic Program). *Let S be an ordered set with a smallest element s_0 and with the property that every $s \in S$ other than s_0 has an immediate predecessor $s - 1$ and that $s_0 = s - n$ for some finite n. Then $\bigoplus\{P_i : i \in S\}$ is a Dynamic Logic Program, where each of the $P_i s$ is a generalized logic program.*

Definition 10 (Rejected rules). *Let $\bigoplus\{P_i : i \in S\}$ be a Dynamic Logic Program, let $s \in S$, and let M be a model of P_s. Then:*

$$Reject(s, M) = \{r \in P_i \mid \exists r' \in P_j, \; head(r) = not\,head(r') \wedge \; i < j \leq s \wedge \\ M \models body(r')\}$$

To allow for querying a dynamic program at any state s, the definition of stable model is parameterized by the state:

Definition 11 (Stable Models of a DLP at state s). *Let $\bigoplus\{P_i : i \in S\}$ be a Dynamic Logic Program, let $s \in S$, and let $\mathcal{P} = \bigcup_{i \leq s} P_i$. A model M of P_s is a stable model of $\bigoplus\{P_i : i \in S\}$ at state s iff:*

$$M = least([\mathcal{P} - Reject(s, M)] \cup Default(\mathcal{P}, M))$$

It is clear from the definitions that stable models of dynamic programs are a generalization of stable models of generalized and normal programs, i.e. if the dynamic program consists of a single generalized (resp. normal) program then its semantics is the same as that of the stable models of generalized (resp. normal) programs.

Moreover, if the union of all the programs in the sequence is consistent, then the stable models of the union carry over to the update sequence. More precisely, for a sequence of two programs:

Proposition 1. *If M is a stable model of the union $P \cup U$ of programs P and U then it is also a stable model of the update program $P \oplus U$, at state U. Thus, the semantics of the update program $P \oplus U$ is always weaker than or equal to the semantics of the union $P \cup U$ of programs P and U.*

In general, the converse of the above result does not hold. In particular, the union $P \cup U$ may be a contradictory program with no stable models. This is for example the case of the updates discussed in example 9 where $P \cup U$ is contradictory (after the adequate removal of *ins* and replacement of *outs* by negation as default).

Special cases where contradiction never appears are when one of the programs, P or U, is empty, or when both of them are normal logic programs (without negation in heads). In both these cases, the semantics of the $P \oplus U$ at U coincides with the semantics of the union of both programs:

Proposition 2. *If either P or U is empty then M is a stable model of $P \cup U$ iff M is a stable model of $P \oplus U$.*

Proposition 3. *If both P and U are normal programs (or if both have only clauses with default atoms not A in their heads) then M is a stable model of $P \cup U$ iff M is a stable model of $P \oplus U$.*

It is also shown in [4] that dynamic logic programs generalize the MT-revision-programs of [34]. In fact MT-revision-programs are updates $P \oplus U$ where P is a set of facts describing the initial interpretation and U results from an easy translation of the update program.

For this result, we identify update rules:

$$in(A) \leftarrow in(B), out(C)$$
$$out(A) \leftarrow in(B), out(C) \tag{9}$$

used in MT-revision-programs, with the following generalized logic program clauses:

$$A \leftarrow B, not\, C$$
$$not\, A \leftarrow B, not\, C. \tag{10}$$

Theorem 2 (Program updates generalize interpretation updates). *Let I be any interpretation and U any updating program in the language $\mathcal{L}$. Denote by P_I the generalized logic program in $\mathcal{L}$ defined by*

$$P_I = \{A \; \leftarrow \; : A \in I\} \; \cup \; \{not\, A \; \leftarrow \; : not\, A \in I\}.$$

Then M is a stable model of the program update $P_I \oplus U$ of the program P_I by the program U iff M is an U-justified update of I.

In [4] a transformational semantics for dynamic programs is also presented. According to this equivalent definition, a sequence of programs is translated into a single generalized program (with one new argument added to all predicates) whose stable models are in one-to-one correspondence with the stable models of the dynamic program. This transformational semantics is the basis of an existing implementation of dynamic logic programming.[5]

3.2 Examples

Example 11. Consider the DLP $P \oplus U_1 \oplus U_2$ introduced in example 9, used to specify the evolution of a knowledge base, and P, U_1 and U_2 are:

$P : sleep \leftarrow not\, tv\_on$ $U_1 : failure$ $U_2 : not\, failure$
$\quad\ tv\_on$ $\quad\ not\, tv\_on \leftarrow failure$
$\quad\ watch\_tv \leftarrow tv\_on$

Clearly, the stable models at state P coincide with the stable models of P, i.e. there is only one which is $\{tv\_on, watch\_tv\}$.

At state U_1 there is a single stable model: $M_1 = \{failure, sleep\}$. In fact, $Reject(U_1, M_1) = \{tv\_on\}$, because the rule $not\, tv\_on \leftarrow failure$ of U_1 has a complementary head, of a rule with true body at M_1, and $Default(P \cup U_1, M_1) = \{not\, watch\_tv, not\, tv\_on\}$. The least model of $P \cup U_1$ minus rejected rules plus default literals is exactly M_1, and thus it is a stable model at U_1.

One can easily check the only stable model at U_2 is $M_2 = \{tv\_on, watch\_tv\}$.

Example 12. Consider the DLP $P_1 \oplus P_2$, where P_1 and P_2 are:

$P_1 : c \leftarrow$ $P_2 : not\, a \leftarrow c$
$\quad\ a \leftarrow not\, b$

The only stable model at P_2 is $M = \{c, not\, a, not\, b\}$. In fact, $Default(P_1 \cup P_2, M) = \{not\, b\}$, $Reject_2(M) = \{a \leftarrow not\, b\}$, and:

$$M = \{c, not\, a, not\, b\} = least((P_1 \cup P_2 - \{a \leftarrow not\, b\}) \cup \{not\, b\})$$

Note here that, as mentioned in Section 3.1, in DLPs the head *not*'s cannot be moved freely into the body, to obtain denials. The rule in P_2 includes the pragmatic information that a is to be deleted if c is true, information that would be lost with the denial. Intuitively that rule makes a different statement from that of the rule $not\, c \leftarrow a$, which however yields the same denial. And this difference is reflected in the definition of stable models for DLPs. In fact, if the rule in P_2 is replaced by this other one, the only stable model at P_2 would be $\{not\, c, a, not\, b\}$ instead.

The reader can check that if the rule in P_2 is replaced by $u \leftarrow a, c, not\, u$ (which, under the stable models semantics, is equivalent to the denial) the results are also different from the ones above: with this rule instead, there is no stable model at P_2.

Example 13. To illustrate the usage of DLP to represent priority of ever more specific knowledge, consider the well-known problem of flying birds, where we have several rules with different priorities. First, the animals-do-not-fly rule, with the lowest priority; then, the birds-fly rule with a higher priority; next,

---
[5] Publicly available from: http://centria.di.fct.unl.pt/~jja/updates/

the penguins-do-not-fly rule with an even higher priority; and, finally, with the highest priority, all the is-a rules describing the actual taxonomy. This can be coded quite naturally in dynamic logic programming (where rules with variables simply stand for all their ground instances):

$$P_1 : not\, fly(X) \leftarrow animal(X) \qquad P_4 : animal(X) \leftarrow bird(X)$$
$$P_2 : fly(X) \leftarrow bird(X) \qquad\qquad bird(X) \leftarrow penguin(X)$$
$$P_3 : not\, fly(X) \leftarrow penguin(X) \qquad animal(pluto)$$
$$bird(duffy)$$
$$penguin(tweety)$$

The reader can check that, as intended, the dynamic logic program $P_1 \oplus P_2 \oplus P_3 \oplus P_4$ at state 4 has a single stable model where $fly(duffy)$ is true, and both $fly(pluto)$ and $fly(tweety)$ are false. Note how the rule $not\, fly(duffy) \leftarrow animal(duffy)$ of P_1 is rejected by the rule $fly(duffy) \leftarrow bird(duffy)$ of P_2, and the rule $fly(tweety) \leftarrow bird(tweety)$ of P_2 is rejected by the rule $not\, fly(tweety) \leftarrow penguin(tweety)$ of P_3.

Example 14. To illustrate the usage of DLP to represent knowledge about hierarchies of objects, consider the following example, adapted from [11].

There are 3 objects in a simple part-of hierarchy, o_1, o_2 and o_3, where object o_2 is part-of object o_1, and object o_3 is also part-of object o_1. Each of the objects has some security information, where users' access authorizations to objects are specified by the predicate $auth(User)$.

Now suppose we want to say, as general rules for o_1 (to be inherited by its part-of objects), that Bob is authorized just in case Ann is not authorized, and that either Ann or Tom is authorized if Alice isn't. Moreover, for object o_2 we want to say that Alice is authorized, and for object o_3 that Bob is not. This information can be represented by the DLP $\bigoplus\{P_{o_1}, P_{o_2}, P_{o_3}\}$, where $o_1 < o_2$ and $o_1 < o_3$, and where:

$$P_{o_1} : auth(bob) \leftarrow not\, auth(ann) \qquad\qquad P_{o_2} : auth(alice)$$
$$auth(ann) \leftarrow not\, auth(tom), not\, auth(alice)$$
$$auth(tom) \leftarrow not\, auth(ann), not\, auth(alice) \qquad P_{o_3} : not\, auth(bob)$$

The access authorizations for object o_1 are given by the stable models at o_1, which are: $\{auth(ann)\}$ and $\{auth(bob), auth(tom)\}$. For object o_2 there is a single stable model $\{auth(bob), auth(alice)\}$. For object o_3 there are two stable models: $\{auth(ann)\}$ and $\{auth(tom)\}$.

3.3   Other References to Program Updates

Recently, Dynamic Logic Programs have been studied by Eiter et al. in [17]. There, a syntactic redefinition of DLPs is presented, and semantical properties are investigated. In particular, a study on the DLP-verification of well known postulates of belief revision [1], iterated revision [13], of theory updates [27] is

carried out. Further structural properties of DLPs, when viewed as nonmonotonic consequence operators, are also studied in [17]. Structural properties of
logic program updates are also studied in [15].

As noted in [17], DLP makes no attempt to minimize the set of rules that are
rejected. It is argued there that this could be a natural approach for measuring
the change which some program P_1 undergoes when updated by some other
program P_2, thus functioning as a good criterion for minimality of change. In
case such a minimality criterion is desired, they refine the semantics of DLP, and
introduce minimal stable models at states. The following example illustrates the
intuition behind minimal stable models. For the formal definition, and further
motivation see [17].

Example 15. Consider that the program P_1

$$a \leftarrow not\, b \qquad not\, b$$

is updated by program $P_2 = \{b \leftarrow not\, a\}$.

There are two stable models of $P_1 \oplus P_2$ at P_2. One is $M_1 = \{a, not\, b\}$. In
fact, $Default(P_1 \cup P_2, M_1) = \{not\, b\}$, $Reject_2(M_1) = \{\}$, and:

$$M_1 = \{a, not\, b\} = least(P_1 \cup P_2 \cup \{not\, b\})$$

The other one is $M_2 = \{b, not\, a\}$. Here, $Default(P_1 \cup P_2, M_2) = \{not\, a\}$,
$Reject_2(M_2) = \{not\, b\}$, and:

$$M_2 = \{b, not\, a\} = least\,([(P_1 \cup P_2) - \{not\, b\}] \cup \{not\, a\})$$

M_2 is not a minimal stable model of $P_1 \oplus P_2$ because the set of P_1's rules
rejected in the case of that model (i.e. $\{not\, b\}$) is a proper superset of the set
of P_1's rules rejected in the case of M_1 (i.e. $\{\}$). Accordingly, the only minimal
stable model of $P_1 \oplus P_2$ is M_1. This is the only one that guarantees that only a
minimal set of rules from the initial P_1 is rejected.

Another important result of [17], is the clarification of the close relationship
between DLPs and inheritance programs [11]. Though defined with different
goals, inheritance programs share some close similarities with DLP. Inheritance
programs [11] aim at extending with inheritance disjunctive logic programming
with strong negation. In them, a hierarchy of objects (or knowledge bases) is
given, where each object has a logic program. The role of inheritance programs
is to establish the semantics at each object. Inheritance is used to inherit, as
much as possible, rules from objects higher in the hierarchy into the objects
lower down, i.e. as long as they do not conflict with the rules of the more specific
objects.

Other approaches to updates of logic programs by logic programs are presented in [25] and in [38]. Based on an abductive framework for (non-monotonic)
auto-epistemic theories, that make use of the notion of negative explanation and
anti-explanation, in [25] the authors define "autoepistemic updates". Based on

this work, in [38] they employ this new abduction framework (in this case rewritten for logic programming instead) to compute minimal programs which result from updating one logic program by another. In their framework, several updates are possible because non-deterministic contradiction removal is used to revise inconsistencies (through abduction) between an initial program and the one updating it, giving preference to the rules of the latter. In their framework, updating and revision take place simultaneously.

Yet another, independently defined, approach to logic programs updates, can be found in [42]. As in DLPs, the semantics of the update of a program by another is obtained by removing rules from the initial program which "somehow" contradict rules from the update program, and retaining all others by inertia. Additionally, at the end, prioritized logic programs are used to give preference to rules from the update program over all retained rules of the initial program. This last step leads to quite different results when compared to DLP. For example, consider a program P with the single rule $a \leftarrow not\, b$, which we want to update with a program $U = \{b \leftarrow not\, a\}$. With DLPs, the rule from P is not rejected, and the semantics of the update equal the semantics of $P \cup U$, i.e. it has two stable models: $\{a\}$ and $\{b\}$. With the approach of [42], priority is given to the rules of U, and the only resulting stable model is $\{b\}$. In our opinion, preferences and updates are different concepts, that have to be considered separately. In DLPs, when there are no conflicts among rules (which is the case in the above example) there is no difference between updating and the union of the programs. If preferences are wanted, then they should be added on top of DLP, to prefer some rules over others. This issue, of preferences and DLPs, is studied in [7]. One important drawback of the approach of [42] is that it cannot be used to consider sequences of updates, and simply captures a single update of one program by another.

4   Languages for Updates

Dynamic logic programming does not by itself provide a proper language for specifying (or programming) changes to logic programs. If knowledge is already represented by logic programs, dynamic programs simply represent the evolution of knowledge. But how is that evolving knowledge specified? What makes knowledge evolve? Since logic programs describe knowledge states, it's only fitting that logic programs describe transitions of knowledge states as well. It is natural to associate with each state a set of transition rules to obtain the next state. As a result, an interleaving sequence of states and rules of transition will be obtained. Imperative programming specifies transitions and leaves states implicit. Logic programming, up to now, could not specify state transitions. With the language of dynamic updates LUPS we make both states and their transitions declarative.

Usually updates are viewed as actions or commands that make the knowledge base evolve from one state to another. This is the classical view e.g. in relational databases: the knowledge (data) is expressed declaratively via a set of relations; updates are commands that change the data. In the previous section, updates were viewed declaratively as a given update store consisting of the sequence of programs. They were more in the spirit of state transition rules, rather than commands. Of course, one could say that the update commands were implicit. For instance, in example 9, the sequence $P \oplus U \oplus U_2$ could be viewed as the result of, starting from P, performing first, simultaneously, the update commands **assert** $not\,tv\_on \leftarrow power\_failure$ and **assert** $power\_failure$, and then the update command **assert** $not\,power\_failure$. But, if viewed as a language for (implicitly) specifying update commands, dynamic logic programming is quite poor. For instance, it does not provide any mechanism for saying that some rule (or fact) should be asserted only whenever some conditions are satisfied. This is essential in the domain of actions, to specify direct effects of actions. For example, suppose we want to state that $wake\_up$ should be added to our knowledge base whenever $alarm\_rings$ is true. As a language for specifying updates, dynamic logic programming does not provide a way of specifying such an update command. Note that the command is distinct from **assert** $wake\_up \leftarrow alarm\_rings$. With the latter, if the alarm stops ringing (i.e. if $not\,alarm\_rings$ is later asserted), $wake\_up$ becomes false. In the former, we expect $wake\_up$ to remain true (by inertia) even after the alarm stops ringing. As a matter of fact, in this case, we don't want to add the rule saying that $wake\_up$ is true whenever $alarm\_rings$ is also true. We simply want to add the fact $wake\_up$ as soon as $alarm\_rings$ is true. From there on, no connection between $wake\_up$ and $alarm\_rings$ should persist.

This simple one-rule example also highlights another limitation of dynamic logic programming as a language for specifying update commands: one must explicitly say to which program in the sequence a rule belongs. Sometimes, in particular in the domain of actions, there is no way to know a priori to which state (or program) a rule should belong to. Where should we assert the fact $wake\_up$? This is not known a priori because we don't know when $alarm\_rings$.

In this section we show a language for specifying logic program updates: LUPS – "Language of dynamic updates" [8]. The object language of LUPS is that of generalized logic programs. A sentence U in LUPS is a set of simultaneous update commands (or actions) that, given a pre-existing sequence of logic programs $P_0 \oplus \ldots \oplus P_n$ (i.e. a dynamic logic program), whose semantics corresponds to our knowledge at a given state, produces a sequence with one more program, $P_0 \oplus \ldots \oplus P_n \oplus P_{n+1}$, corresponding to the knowledge that results from the previous sequence after performing all the simultaneous commands. A program in LUPS is a sequence of such sentences.

Given a program in LUPS, its semantics is defined by means of a dynamic logic program generated by the sequence of commands. In [8], a translation of a LUPS program into a generalized logic program is presented, where stable models exactly correspond to the semantics of the original LUPS program. This translation directly provides an implementation of LUPS.

In this update framework, knowledge evolves from one knowledge state to another as a result of update commands stated in the object language. Knowledge states KS_i represent dynamically evolving states of our knowledge. They undergo change due to *update actions*. Without loss of generality (as will become clear below) we assume that the initial knowledge state, KS_0, is empty and that in it all predicates are *false* by default. This is the *default knowledge state*. Given the *current knowledge state KS*, its *successor knowledge state KS[U]* is produced as a result of the occurrence of a non-empty set U of simultaneous *updates*. Each of the updates can be viewed as a set of (parallel) *actions* and consecutive knowledge states are obtained as

$$KS_n = KS_0[U_1][U_2]...[U_n]$$

where U_i's represent consecutive sets of updates. We also denote this state by:

$$KS_n = U_1 \otimes U_2 \otimes \ldots \otimes U_n$$

So defined sequences of updates will be called *update programs*. In other words, an update program is a finite sequence $\mathcal{U} = \{U_s : s \in S\}$ of updates indexed by the set $S = \{1, 2, \ldots, n\}$. Each updates is a set of update commands. Update commands (to be defined below) specify *assertions* or *retractions* to the current knowledge state. By the current knowledge state we mean the one resulting from the last update performed.

Knowledge can be queried at any state $q \le n$, where n is the index of the current knowledge state. A query will be denoted by:

$$\textbf{holds } B_1, \ldots, B_k, not\, C_1, \ldots, not\, C_m \textbf{ at } q?$$

and is true iff the conjunction of its literals holds at the state KB_q. If $q = n$, we simply skip the state reference "**at** q".

4.1   Update Commands

Update commands cause changes to the current knowledge state leading to a new successor state. The simplest command consists of adding a rule to the current state: **assert** $L \leftarrow L_1, \ldots, L_k$. For example, when a law stating that abortion is punished by jail is approved, the knowledge state might be updated via the command: **assert** $jail \leftarrow abortion$.

In general, the addition of a rule to a knowledge state may depend upon some precondition. To allow for that, an assert command in LUPS has the form:

$$\textbf{assert } L \leftarrow L_1, \ldots, L_k \textbf{ when } L_{k+1}, \ldots, L_m \tag{11}$$

The meaning of such an assert rule is that if the precondition $L_{k+1}, \ldots, L_m$ is true in the current knowledge state, then the rule $L \leftarrow L_1, \ldots, L_k$ should belong to the successor knowledge state. Normally, the so added rule persists, or is in force, from then on by inertia, until possibly defeated by some future update or until retracted. This is the case for the assert-command above: the rule $jail \leftarrow abortion$ remains in effect by inertia from the successor state onwards unless later invalidated.

However, there are cases where this persistence by inertia should not be assumed. Take, for instance, the *alarm_ring* discussed in the introduction. This fact is a one-time event that should not persist by inertia, i.e. it is not supposed to hold by inertia after the successor state. In general, facts that denote names of events or actions should be *non-inertial*. Both are true in the state they occur, and do not persist by inertia for later states. Accordingly, the rule within the assert command may be preceded with the keyword *event*, indicating that the added rule is non-inertial. Assert commands are thus of the form (11) or of the form:[6]

$$\textbf{assert event } L \leftarrow L_1, \ldots, L_k \textbf{ when } L_{k+1}, \ldots, L_m \qquad (12)$$

While some update commands, such as **assert** *republican_congress*, represent newly incoming information, and are thus one-time non-persistent update commands (whose effect, i.e. the truth of *republican_congress*, may nevertheless persist by inertia), some other update commands are liable to be *persistent*, i.e., to remain in force until cancelled. For example, an update like:

$$\textbf{assert } jail \leftarrow abortion \textbf{ when } rep\_congress, rep\_president$$

or

$$\textbf{assert } wake\_up \textbf{ when } alarm\_sounds$$

might be always true, or at least true until cancelled. Enabling the possibility of such updates allows our system to dynamically change without any truly new (external) updates being received. For example, the persistent update command:

$$\textbf{assert } set\_hands(T) \textbf{ when } get\_hands(C) \wedge get\_time(T) \wedge (T - C) > \Delta$$

defines a perpetually operating clock whose hands move to the actual time position whenever the difference between the clock time and the actual time is sufficiently large.

In order to specify such persistent updates commands (which we call laws) we introduce the syntax:

$$\textbf{always [event] } L \leftarrow L_1, \ldots, L_k \textit{ when } L_{k+1}, \ldots, L_m \qquad (13)$$

For stopping persistent update commands, we use:

$$\textbf{cancel } L \leftarrow L_1, \ldots, L_k \textbf{ when } L_{k+1}, \ldots, L_m \qquad (14)$$

The first statement means that, in addition to any new set of arriving update commands, we are also supposed to keep executing this persistent update command. The second statement cancels this persistent update, when the conditions for cancellation are met.

The existence of persistent update commands requires a "trivial" update, which does not specify any truly new updates but simply triggers all the already defined persistent updates to fire, thus resulting in a new modified knowledge state. Such "no-operation" update ensures that the system continues to

---

[6] In both cases, if the precondition is empty we just skip the whole *when* subclause.

evolve, even when no truly new updates are specified, and may be represented by **assert** *true*. It stands for the *tick of the clock* that drives the world being modelled.

To deal with the deletion of rules, we have available the *retraction* command:

$$\textbf{retract [event] } L \leftarrow L_1, \ldots, L_k \textbf{ when } L_{k+1}, \ldots, L_m \tag{15}$$

meaning that, subject to precondition $L_{k+1}, \ldots, L_m$, the rule $L \leftarrow L_1, \ldots, L_k$ is either retracted from now on, or just retracted temporarily in the next state (non-inertial retract, i.e. an event of retraction, triggered by the *event* keyword).

The cancelling of an update command is not equivalent to the retracting of a rule. Cancelling an update just means it will no longer be added as a command to updates, it does not cancel the inertial effects of its previous application(s). However, retracting an update causes any of its inertial effects to be cancelled from now on, as well as cancelling a persistent law. Also, note that "*retract event ...*" does not mean the retracting of an event, because events persist only for one state and thus do not require retraction. It represents a temporary removal of a rule from the successor state (a temporary retraction event).

Definition 12 (LUPS). *An update program in LUPS is a finite sequence of updates, where an update is a set of commands of the form (11) to (15).*

Example 16. Consider the following scenario:

- once Republicans take over both Congress and the Presidency they establish a law stating that abortions are punishable by jail;
- once Democrats take over both Congress and the Presidency they abolish such a law;
- in the meantime, there are no changes in the law because always either the President or the Congress vetoes such changes;
- performing an abortion is an event, i.e. a non-inertial update.

Consider the following update history: (1) a Democratic Congress and a Republican President (Reagan); (2) Mary performs abortion; (3) Republican Congress is elected (Republican President remains in office: Bush); (4) Kate performs abortion; (5) Clinton is elected President; (6) Ann performs abortion; (7) G.W. Bush is elected president (8) A democrat is elected President and Democratic Congress is in place (year 2004?); (9) Susan performs abortion.

The specification in LUPS would be:

Persistent update commands:

$$\textbf{always } jail(X) \leftarrow abt(X) \textbf{ when } repC \wedge repP$$
$$\textbf{always } not\, jail(X) \leftarrow abt(X) \textbf{ when } not\, repC \wedge not\, repP$$

Alternatively, instead of the second clause, in this example, we can use a retract statement

$$\textbf{retract } jail(X) \leftarrow abt(X) \textbf{ when } not\, repC \wedge not\, repP$$

Note that, in this example, since there is no other rule implying *jail*, retracting the rule is safely equivalent to retracting its conclusion.

The above rules state that we are always supposed to update the current state with the rule $jail(X) \leftarrow abt(X)$ provided $repC$ and $repP$ hold true and that we are supposed to assert the opposite (or just retract this rule) provided $not\,repC$ and $not\,repP$ hold true. Such persistent update commands should be added to U_1.

Sequence of non-persistent update commands:

U_1 : **assert** $repP$ U_5 : **assert** $not\,repP$

     assert $not\,repC$ U_6 : **assert event** $abt(ann)$

U_2 : **assert event** $abt(mary)$ U_7 : **assert** $repP$

U_3 : **assert** $repC$ U_8 : **assert** $not\,repP$ **assert** $not\,recC$

U_4 : **assert event** $abt(kate)$ U_9 : **assert event** $abt(susan)$

Of course, in the meantime we could have a lot of trivial update events representing ticks of the clock, or any other irrelevant updates.

4.2 Semantics of LUPS

In this section we provide update programs with a meaning, by translating them into dynamic logic programs. The semantics of a LUPS program is then determined by the semantics of the so obtained dynamic program.

More precisely, the translation of a LUPS program into a dynamic program is obtainable by induction, starting from the empty program P_0, and for each update U_i, given the already built dynamic program $P_0\oplus\ldots\oplus P_{i-1}$, determining the resulting program $P_0\oplus\ldots\oplus P_{i-1}\oplus P_i$. To cope with persistent update commands we will further consider, associated with every dynamic program in the inductive construction, a set containing all currently active persistent commands, i.e. all those that were not cancelled, up to that point in the construction, from the time they were introduced. To be able to retract rules, we need to uniquely identify each such rule. This is achieved by augmenting the language of the resulting dynamic program with a new propositional variable "$rule(L \leftarrow L_1,\ldots,L_n)$" for every rule $L \leftarrow L_1,\ldots,L_n$ appearing in the original LUPS program.[7]

Definition 13 (Translation into dynamic programs). *Let* $\mathcal{U} = U_1\otimes\ldots\otimes U_n$ *be an update program. The corresponding dynamic program* $\Upsilon(\mathcal{U}) = \mathcal{P} = P_0 \oplus \ldots \oplus P_n$ *is obtained by the following inductive construction, using at each step i an auxiliary set of persistent commands* PC_i:

Base step: $P_0 = \{\}$ *with* $PC_0 = \{\}$.

Inductive step: Let $\mathcal{P}_i = P_0 \oplus \ldots \oplus P_i$ *with the set of persistent commands* PC_i *be the translation of* $\mathcal{U}_i = U_1\otimes\ldots\otimes U_i$. *The translation of* $\mathcal{U}_{i+1} = U_1\otimes\ldots\otimes U_{i+1}$ *is* $\mathcal{P}_{i+1} = P_0 \oplus \ldots \oplus P_{i+1}$ *with the set of persistent commands* PC_{i+1}, *where:*

---

[7] Note that, by definition, all such rules are ground and thus the new variable uniquely identifies the rule, where rule/1 is a reserved predicate.

$PC_{i+1} = PC_i \cup$
$\cup \{ \textbf{\textit{assert}}\ R\ \textbf{\textit{when}}\ C : \textbf{\textit{always}}\ R\ \textbf{\textit{when}}\ C \in U_{i+1} \}$
$\cup \{ \textbf{\textit{assert event}}\ R\ \textbf{\textit{when}}\ C : \textbf{\textit{always event}}\ R\ \textbf{\textit{when}}\ C \in U_{i+1} \}$
$- \{ \textbf{\textit{assert [event]}}\ R\ \textbf{\textit{when}}\ C : \textbf{\textit{cancel}}\ R\ \textbf{\textit{when}}\ D \in U_{i+1} \wedge \bigoplus_i \mathcal{P}_i \models_{sm} D \}$
$- \{ \textbf{\textit{assert [event]}}\ R\ \textbf{\textit{when}}\ C : \textbf{\textit{retract}}\ R\ \textbf{\textit{when}}\ D \in U_{i+1} \wedge \bigoplus_i \mathcal{P}_i \models_{sm} D \}$

$NU_{i+1} = U_{i+1} \cup PC_{i+1}$

$P_{i+1} = \{ R,\ rule(R) : \textbf{\textit{assert [event]}}\ R\ \textbf{\textit{when}}\ C \in NU_{i+1} \wedge \bigoplus_i \mathcal{P}_i \models_{sm} C \}$
$\cup \{ not\ rule(R) : \textbf{\textit{retract [event]}}\ R\ \textbf{\textit{when}}\ C \in NU_{i+1} \wedge \bigoplus_i \mathcal{P}_i \models_{sm} C \}$
$\cup \{ not\ rule(R) : \textbf{\textit{assert event}}\ R\ \textbf{\textit{when}}\ C \in NU_i \wedge \bigoplus_{i-1} \mathcal{P}_{i-1} \models_{sm} C \}$
$\cup \{ rule(R) : \textbf{\textit{retract event}}\ R\ \textbf{\textit{when}}\ C \in NU_i \wedge \bigoplus_{i-1} \mathcal{P}_{i-1} \models_{sm} C, rule(R) \}$

*where R denotes a generalized logic program rule, and C and D a conjunction of literals. **assert [event]** R **when** C and **retract [event]** R **when** C are used for notational convenience, and stand for either the assert or the assert-event command (resp. retract and retract-event). So, for example in the first line of the definition of P_{i+1}, R and $rule(R)$ must be added either if there exists a command **assert** R **when** C or a command **assert event** R **when** C obeying the conditions there.*

In the inductive step, if $i = 0$ the last two lines are omitted. In that case NU_i does not exist.

Definition 14 (LUPS semantics). *Let $\mathcal{U}$ be an update program. A query **holds** $L_1, \ldots, L_n$ **at** q is true in $\mathcal{U}$ iff $\bigoplus_q \Upsilon(\mathcal{U}) \models_{sm} L_1, \ldots, L_n$.*

From the results on dynamic programs, it is clear that LUPS generalizes the language of updates of MT-revision-programs presented in section 2.1:

Proposition 4 (LUPS generalizes MT-revision-programs). *Let I be an interpretation and R a MT-revision-program. Let $\mathcal{U} = U_1 \otimes U_2$ be the update program where:*

$$U_1 = \{ \textbf{\textit{assert}}\ A : A \in I \}$$

$$U_2 = \{ \textbf{\textit{assert}}\ A \leftarrow B_1, \ldots, not\ B_n : in(A) \leftarrow in(B_1), \ldots, out(B_n) \in R \}$$
$$\cup \{ \textbf{\textit{assert}}\ not\ A \leftarrow B_1, \ldots, not\ B_n : out(A) \leftarrow in(B_1), \ldots, out(B_n) \in R \}$$

Then, M is a stable model of $\Upsilon(\mathcal{U})$ iff M is an interpretation update of I by R in the sense of [34].

This definition of the LUPS semantics is based on a translation into dynamic logic programs, and is not purely syntactic. Indeed, to obtain the translated dynamic program, one needs to compute, at each step of the inductive process, the consequences of the previous one.

In [8] a translation of update programs and queries, into normal logic programs written in a meta-language is presented. That translation is purely syntactic, and is correct in the sense that a query holds in an update program iff the translation of the query holds in all stable models of the translation of the update program. It also directly provides a mechanism for implementing update programs: with a pre-processor performing the translations, query answering is reduced to that over normal logic programs.[8]

5   Application Domains

In this section we discuss and illustrate with examples the applicability of the language LUPS to several broad knowledge representation domains. The selected domains include: *theory of actions, legal reasoning, and software specification.* Additional examples and domains of application of LUPS can be found in [9].

The theory of actions (for a survey see [21]) is very closely related to knowledge updates. An action taking place at a specific moment of time may cause an effect in the form of a change of the status of some fluent. For example, an action of stepping on a sharp nail may result in severe pain. The occurrence of pain can therefore be viewed as a simple (atomic) knowledge update triggered by a given action. Similarly, a set of parallel actions can be viewed as triggering (causing) parallel atomic updates. The following *suitcase* example illustrates how LUPS can be used to handle parallel updates.

Example 17 (Suitcase). There is a suitcase with two latches which opens whenever both latches are up, and there is an action of toggling applicable to each latch [33]. This situation is represented by the three persistent rules:

$$\textbf{always } open \leftarrow up(l1), up(l2)$$
$$\textbf{always } up(L) \textbf{ when } not\ up(L), toggle(L)$$
$$\textbf{always } not\ up(L) \textbf{ when } up(L), toggle(L)$$

In the initial situation $l1$ is down, $l2$ is up, and the suitcase is closed:

$$U_1 = \{\textbf{assert } not\ up(l1), \textbf{assert } up(l2), \textbf{assert } not\ open\}$$

Suppose there are now two simultaneous toggling actions:

$$U_2 = \{\textbf{assert event } toggle(l1), \textbf{assert event } toggle(l2)\}$$

and afterwards another $l2$ toggling action: $U_3 = \{\textbf{assert event } toggle(l2)\}$. In the knowledge state 2 we will have $up(l1), not\ up(l2)$ and the suitcase is not open. Only after U_3 will latch $l2$ be up and the suitcase open.

Robert Kowalski's team did truly outstanding research work on using logic programming as a *language for legal reasoning* (see e.g. [29]). However logic programming itself lacks any mechanism for expressing dynamic changes in the

---

[8] Such a pre-processor and a meta-interpreter for query answering have been implemented and are available at: http://centria.di.fct.unl.pt/~jja/updates/

law due to revisions of the law or due to new legislation. LUPS allows us to handle such changes in a very natural way by augmenting the knowledge base only with the newly added or revised data, and automatically obtaining the updated information as a result. We illustrate this capability on the following simple example.

Example 18 (Conscientious objector). Consider a situation where someone is conscripted if he is draftable and healthy. Moreover a person is draftable when he attains a specific age. However, after some time, the law changes and a person is no longer conscripted if he is indeed a conscientious objector:

$$U_1 : \textbf{always } draftable(X) \textbf{ when } of\_age(X)$$
$$\textbf{assert } conscripted(X) \leftarrow draftable(X), healthy(X)$$
$$U_2 : \textbf{assert } healthy(a). \textbf{ assert } healthy(b). \textbf{ assert } of\_age(b).$$
$$\textbf{assert } consc\_objector(a). \textbf{ assert } consc\_objector(b)$$
$$U_3 : \textbf{assert } of\_age(a)$$
$$U_4 : \textbf{assert } not \ conscripted(X) \leftarrow consc\_objector(X)$$

In state 3, b is subject to conscription but after the last assertion his situation changes. On the other hand, a is never conscripted.

One of the most important problems in software engineering is that of choosing a suitable software specification language. The following are among the key desired properties of such a language:

1. Possibility of a concise representation of statements of natural language, commonly used in informal descriptions of various domains.
2. Availability of query answering systems which allow rapid prototyping.
3. Existence of a well developed and mathematically precise semantics of the language.
4. Ability to express conditions that change dynamically.
5. Ability to handle inconsistencies stemming from specification revisions.

It has been argued in the literature that the language of logic programming is a good potential candidate for the language of software specifications. However, it lacks simple and natural ways of expressing conditions that change dynamically and the ability to handle inconsistencies stemming from specification revisions. The following simplified banking example illustrates how LUPS can be used to represent changes in software specifications.

Example 19 (Banking transactions). Consider a software specification for performing banking transactions. Account balances are modeled by the predicate *balance(AccountNo, Balance)*. Predicates *deposit(AccountNo, Amount)*   and *withdrawal(AccountNo, Amount)* represent the actions of depositing and withdrawing money into and out of an account, respectively. A withdrawal can only be accomplished if the account has a sufficient balance. This simplified description can easily be modeled in LUPS by U_1:

always $balance(Ac, OB + Up)$ **when** $updateBal(Ac, Up), balance(Ac, OB)$
always $not \ balance(Ac, OB)$ **when** $updateBal(Ac, NB), balance(Ac, OB)$
assert $updateBal(Ac, -X) \leftarrow withdrawal(Ac, X), balance(Ac, O), O > X$
assert $updateBal(Ac, X) \leftarrow deposit(Ac, X)$

The first two rules state how to update the balance of an account, given any event of $updateBal$. By the last two rules, deposits and withdrawals are effected, causing $updateBal$.

An initial situation can be imposed via $assert$ commands. Deposits and withdrawals can be stipulated by asserting events of $deposit/2$ and $withdrawal/2$. E.g.:

U_2 : {**assert** $balance(1, 0)$, **assert** $balance(2, 50)$}
U_3 : {**assert event** $deposit(1, 40)$, **assert event** $withdrawal(2, 10)$}

causes the balance of both accounts 1 and 2 to be 40, after state 3.

Now consider the following sequence of informal specification revisions. Deposits under 50 are no longer allowed; VIP accounts may have a negative balance up to the limit specified for the account; account #1 is a VIP account with the overdraft limit of 200; deposits under 50 are allowed for accounts with negative balances. These can in turn be modeled by the sequence:

U_4 : **assert** $not \ updateBal(Ac, X) \leftarrow deposit(Ac, X), X < 50$
U_5 : **assert** $updateBal(Ac, -X) \leftarrow vip(Ac, L), withdrawal(Ac, X),$
$\qquad\qquad\qquad\qquad\qquad\quad balance(Ac, B), B + L \geq X$
U_6 : **assert** $vip(1, 200)$
U_7 : **assert** $updateBal(Ac, X) \leftarrow deposit(Ac, X), balance(Ac, B), B < 0$

6   Future Perspectives

Knowledge updating is not to be simply envisaged as taking place in the time dimension alone. Several updating dimensions may combine simultaneously, with or without the temporal one, such as specificity (as in taxonomies), strength of the updating instance (as in the legislative domain), hierarchical position of knowledge source (as in organizations), credibility of the source (as in uncertain, mined, or learnt knowledge), or opinion precedence (as in a society of agents).

What's more, updating inevitably raises issues about revising and preferring, and some work is emerging on the articulation of these distinct but highly complementary aspects. And learning is usefully seen as successive approximate change, as opposed to exact change, and combining the results of learning by multiple agents, multiple strategies, or multiple data sets, inevitably poses problems within the province of updating. Last but not least, goal directed planning can be fruitfully envisaged as abductive updating.

Thus, not only do the aforementioned topics combine naturally together – and so require precise, formal, means and tools to do so – but their combination results in turn in a nascent complex architectural basis and component for Logic Programming rational agents, which can update one another and common, structured, blackboard agents.

We surmise, consequently, that fostering this meshing of topics within the Logic Programming community is all of opportune, seeding, and fruitful. Indeed, application areas, such as software development, multi-strategy learning, abductive planning, model-based diagnosis, agent architectures, and others, are being successfully pursued employing the above outlook.

In this necessarily selective introductory guided tour, we have not delved into topics, mentioned above, which have already been the subject of research and publication, and that the captive reader may want to pursue, namely: the combination of updates with preferences [7]; the extension to multiple updating dimensions [30]; the coupling of updates and abduction in planning [2]; mutually updating agents [16]; updating postulates, structural properties, and complexity [17, 15].

A well-founded semantics for generalized programs has also been defined [9], which allows carrying over results to 3-valued updates. This semantics is actually the basis for one of our implementations under the XBS system. The other implementation, relies on a preprocessor that produces programs to be run under the DLV-system.

In the body of the references provided below the interested reader may follow up on other works and approaches, and glean in them critical appraisals and comparisons.

Acknowledgements

We thank the co-authors of joint papers from which we have extracted or adapted much material, namely João Leite, Halina Przymusinska, Teodor Przymusinski, and Paulo Quaresma. We acknowledge the support of PRAXIS projects MENTAL and ACROPOLE.

References

[1] C. Alchourrón, P. Gärdenfors, and D. Makinson. On the logic of theory change: Partial meet contraction and revision functions. *J. Symbolic Logic*, 50(2):510–530, 1985.

[2] J. Alferes, J. A. Leite, L. M. Pereira, and P. Quaresma. Planning as abductive updating. In D. Kitchin, editor, *AISB'00 Symposium on AI Planning and Intelligent Agents*, pages 1–8. AISB, 2000.

[3] J. J. Alferes, C. V. Damásio, and L. M. Pereira. A logic programming system for non-monotonic reasoning. *Journal of Automated Reasoning*, 14:93–147, 1995.

[4] J. J. Alferes, J. A. Leite, L. M. Pereira, H. Przymusinska, and T. Przymusinski. Dynamic updates of non-monotonic knowledge bases. *Journal of Logic Programming*, 45(1-3):43–70, 2000. A short version titled *Dynamic Logic Programming* appeared in A. Cohn and L. Schubert (eds.), *KR'98*, Morgan Kaufmann.

[5] J. J. Alferes and L. M. Pereira. Contradiction: when avoidance equal removal. In R. Dyckhoff, editor, *4th ELP*, volume 798 of *LNAI*. Springer–Verlag, 1994.

[6] J. J. Alferes and L. M. Pereira. Update-programs can update programs. In J. Dix, L. M. Pereira, and T. Przymusinski, editors, *NMELP'96*. Springer, 1996.

[7] J. J. Alferes and L. M. Pereira. Updates plus preferences. In M. O. Aciego et al., editor, *JELIA'00*. Springer LNAI 1919, 2000.

[8] J. J. Alferes, L. M. Pereira, H. Przymusinska, and T. Przymusinski. LUPS – a language for updating logic programs. In M. Gelfond, N. Leone, and G. Pfeifer, editors, *LPNMR'99*. Springer, 1999.

[9] J. J. Alferes, L. M. Pereira, T. Przymusinski, H. Przymusinska, and P. Quaresma. Dynamic knowledge representation and its applications. In *AIMSA'00*. Springer LNAI, 2000.

[10] C. Baral. Rule-based updates on simple knowledge bases. In *AAAI'94*, pages 136–141, 1994.

[11] F. Buccafurri, W. Faber, and N. Leone. Disjunctive logic programs with inheritance. In D. De Schreye, editor, *ICLP'99*. MIT Press, 1999.

[12] C. V. Damásio and L. M. Pereira. Default negation in the heads: why not? In R. Dyckhoff et al., editor, *ELP'96*. Springer, 1996.

[13] A. Darwiche and J. Pearl. On the logic of iterated belief revision. *Artificial Intelligence*, 89(1-2):1–29, 1997.

[14] H. Decker. Drawing updates from derivations. In *Int. Conf on Database Theory*, volume 460 of *LNCS*, 1990.

[15] M. Dekhtyar, A. Dikovsky, S. Dudakov, and N. Spyratos. Monotone expansions of updates in logical databases. In M. Gelfond, N. Leone, and G. Pfeifer, editors, *LPNMR'99*. Springer, 1999.

[16] P. Dell'Acqua and L. M. Pereira. Updating agents. In S. Rochefort, F. Sadri, and F. Toni, editors, *ICLP'99 Workshop on Multi-Agent Systems in Logic*, 1999.

[17] T. Eiter, M. Fink, G. Sabbatini, and H. Tompits. Considerations on updates of logic programs. In M. O. Aciego, I. P. de Guzmn, G. Brewka, and L. M. Pereira, editors, *JELIA'00*. Springer LNAI 1919, 2000.

[18] R. E. Fikes and N. J. Nilsson. STRIPS: a new approach to the application of theorem proving to problem solving. *Artificial Intelligence*, 2(2-3):189–208, 1971.

[19] M. Gelfond and V. Lifschitz. The stable model semantics for logic programming. In R. Kowalski and K. A. Bowen, editors, *ICLP'88*. MIT Press, 1988.

[20] M. Gelfond and V. Lifschitz. Logic programs with classical negation. In Warren and Szeredi, editors, *ICLP'90*. MIT Press, 1990.

[21] M. Gelfond and V. Lifschitz. Action languages. *Linkoping Electronic Articles in Computer and Information Science*, 3(16), 1998.

[22] L. Giordano and A. Martelli. Generalized stable models, truth maintenance and conflit resolution. In D. Warren and P. Szeredi, editors, *7th ICLP*, pages 427–441. MIT Press, 1990.

[23] A. Guessoum and J. W. Lloyd. Updating knowledge bases. *New Generation Computing*, 8(1):71–89, 1990.

[24] A. Guessoum and J. W. Lloyd. Updating knowledge bases II. *New Generation Computing*, 10(1):73–100, 1991.

[25] K. Inoue and C. Sakama. Abductive framework for nonmonotonic theory change. In *IJCAI'95*, pages 204–210. Morgan Kaufmann, 1995.

[26] K. Inoue and C. Sakama. Negation as failure in the head. *Journal of Logic Programming*, 35:39–78, 1998.

[27] H. Katsuno and A. Mendelzon. On the difference between updating a knowledge base and revising it. In J. Allen, R. Fikes, and E. Sandewall, editors, *KR'91*. Morgan Kaufmann, 1991.

[28] A. Keller and M. Winslett Wilkins. On the use of an extended relational model to handle changing incomplete information. *IEEE Trans. on Software Engineering*, 11(7):620–633, 1985.

[29] R. Kowalski. Legislation as logic programs. In *Logic Programming in Action*, pages 203–230. Springer-Verlag, 1992.

[30] J. A. Leite, J. Alferes, and L. M. Pereira. Multi-dimensional dynamic logic programming. In F. Sadri and K. Satoh, editors, *CL-2000 Workshop on Computational Logic in Multi-Agent Systems (CLIMA'00)*, 2000.

[31] J. A. Leite and L. M. Pereira. Generalizing updates: from models to programs. In *LPKR'97: ILPS'97 workshop on Logic Programming and Knowledge Representation*, 1997.

[32] V. Lifschitz and T. Woo. Answer sets in general non-monotonic reasoning (preliminary report). In B. Nebel, C. Rich, and W. Swartout, editors, *KR'92*. Morgan-Kaufmann, 1992.

[33] F. Lin. Embracing causality in specifying the indirect effects of actions. In *IJCAI'95*, pages 1985–1991. Morgan Kaufmann, 1995.

[34] V. Marek and M. Truszczyński. Revision specifications by means of programs. In C. MacNish, D. Pearce, and L. M. Pereira, editors, *JELIA'94*, volume 838 of *LNAI*, pages 122–136. Springer-Verlag, 1994.

[35] V. Marek and M. Truszczyński. Revision programming, database updates and integrity constraints. In *ICDT'95*, pages 368–382. Springer-Verlag, 1995.

[36] Isaaco Newtono. *Philosophiæ Naturalis Principia Mathematica*. Editio tertia & aucta emendata. Apud Guil & Joh. Innys, Regiæ Societatis typographos, 1726. Original quotation: *"Corpus omne perseverare in statu suo quiescendi vel movendi uniformiter in directum, nisi quatenus illud a viribus impressis cogitur statum suum mutare."*.

[37] T. Przymusinski and H. Turner. Update by means of inference rules. In V. Marek, A. Nerode, and M. Truszczyński, editors, *LPNMR'95*, volume 928 of *LNAI*, pages 156–174. Springer-Verlag, 1995.

[38] C. Sakama and K. Inoue. Updating extended logic programs through abduction. In M. Gelfond, N. Leone, and G. Pfeifer, editors, *LPNMR'99*. Springer, 1999.

[39] M. Winslett. Reasoning about action using a possible models approach. In *AAAI'88*, pages 89–93, 1988.

[40] C. Witteveen and W. Hoek. Revision by communication. In V. Marek, A. Nerode, and M. Truszczyński, editors, *LPNMR'95*, pages 189–202. Springer, 1995.

[41] C. Witteveen, W. Hoek, and H. Nivelle. Revision of non-monotonic theories: some postulates and an application to logic programming. In C. MacNish, D. Pearce, and L. M. Pereira, editors, *JELIA'94*, pages 137–151. Springer, 1994.

[42] Y. Zhang and N. Foo. Updating logic programs. In H. Prade, editor, *ECAI'98*. Morgan Kaufmann, 1998.

Representing Knowledge in A-Prolog

Michael Gelfond

Department of Computer Science
Texas Tech University
Lubbock, TX 79409, USA
mgelfond@cs.ttu.edu
http://www.cs.ttu.edu/~mgelfond

Abstract. In this paper, we review some recent work on declarative logic programming languages based on stable models/answer sets semantics of logic programs. These languages, gathered together under the name of A-Prolog, can be used to represent various types of knowledge about the world. By way of example we demonstrate how the corresponding representations together with inference mechanisms associated with A-Prolog can be used to solve various programming tasks.

1 Introduction

Understanding the basic principles which can serve as foundation for building programs capable of learning and reasoning about their environment is one of the most interesting and important challenges faced by people working in Artificial Intelligence and Computing Science. Frequently search for these principles is centered on finding efficient means of human-computer communication, i.e. on programming languages[1]. Such languages differ according to the type of information their designers want to communicate to computers. There are two basic types of languages - *algorithmic* and *declarative*. Programs in algorithmic languages describe sequences of actions for a computer to perform while declarative programs can be viewed as collections of statements describing objects of a domain and their properties. A semantic of a declarative program Π is normally given by defining its models, i.e. possible states of the world compatible with Π. Statements which are true in all such models constitute the set of valid consequences of Π. Declarative programming consists in representing knowledge, about the domain relevant to the programmer's goals, by a program Π (often called a *knowledge base*) and in reducing various programming tasks to finding models or computing consequences of Π. Normally, models are found and/or consequences are computed by general purpose reasoning algorithms often called *inference engines*. There are a number of requirements which should be satisfied by a declarative programming language. Some of these requirements are common

---

[1] In this paper by programming we mean a process of refining specifications. Consequently, the notion of a programming language is understood broadly and includes specification languages which are not necessarily executable.

A.C. Kakas, F. Sadri (Eds.): Computat. Logic (Kowalski Festschrift), LNAI 2408, pp. 413–451, 2002.

to all programming languages. For instance, there is always a need for a simple syntax and a clear definition of the meaning of a program. Among other things such a definition should provide the basis for the development of mathematical theory of the language. It is also important to have a programming methodology to guide a programmer in the process of finding a solution to his problem and in the design and implementation of this solution on a computer. These and other general requirements are well understood and frequently discussed in the programming language community. There are, however, some important requirements which seems to be pertinent mainly to declarative languages. We would like to mention those which we believe to be especially important and which will play a role in our discussion.

• A declarative language should allow construction of *elaboration tolerant* knowledge bases, i.e. the bases in which small modifications of the informal body of knowledge correspond to small modifications of the formal base representing this knowledge. It seems that this requirement is easier satisfied if we use languages with *nonmonotonic* consequence relation. (A consequence relation $\models_{\mathcal{L}}$ is called nonmonotonic if there are formulas A, B and C in $\mathcal{L}$ such that $A \models C$ but $A, B \not\models C$, i.e. addition of new information to the knowledge base of a reasoner may invalidate some of his previous conclusions [59]). This property is especially important for representing common-sense knowledge about the world. In common-sense reasoning, additions to the agent's knowledge are frequent and inferences are often based on the absence of knowledge. Modeling such reasoning in languages with a non-monotonic consequence relation seem to lead to simpler and more elaboration tolerant representations.

• Inference engines associated with the language should be sufficiently general and efficient. Notice however that, since some of the relations one needs to teach a computer about are not enumerable, such systems cannot in general be complete. Language designers should therefore look for the 'right' balance between the expressive power of the language and computational efficiency of its inference engine.

We are not sure that it is possible (and even desirable) to design a knowledge representation language suitable for all possible domains and problems. The choice of the language, its semantics and its consequence relation depends significantly on the types of statements of natural language used in the informal descriptions of programming tasks faced by the programmer.

In this paper we discuss A-Prolog – a language of logic programs under answer set (stable model) semantics [30],[31]. A-Prolog can be viewed as a purely declarative language with roots in logic programming [42,43,85], syntax and semantics of standard Prolog [18], [22], and in the work on nonmonotonic logic [73], [62]. It differs from many other knowledge representation languages by its ability to represent *defaults*, i.e. statements of the form *"Elements of a class C normally satisfy property P"*. One may learn early in life that parents normally love their children. So knowing that Mary is a mother of John he may conclude that Mary loves John and act accordingly. Later one can learn that Mary is an exception to the above default, conclude that Mary does not really like John, and use this new knowledge to change his behavior. One can argue that a substantial part of our education consists in learning various defaults, exceptions to these defaults, and the ways of using this information to draw reasonable conclusions about the world and the consequences of our actions. A-Prolog provides a powerful logical model of this process. Its syntax allows simple representation of defaults and their exceptions, its consequence relation characterizes the corresponding set of valid conclusions, and its inference mechanisms allow a program to find these conclusions in a "reasonable" amount of time.

There are other important types of statements which can be nicely expressed in A-Prolog. This includes the causal effects of actions ("statement F becomes true as a result of performing an action a"), statements expressing the lack of information ("It is not known if statement P is true or false"), various completeness assumptions, "Statements not entailed by the knowledge base are false", etc. On the negative side, A-Prolog in its current form is not adequate for reasoning with real numbers and for reasoning with complex logical formulas - the things classical logic is good at.

There is by now a comparatively large number of inference engines associated with A-Prolog. There are well known conditions which guarantee that the traditional SLDNF-resolution based goal-oriented methods of "classical" Prolog and its variants are sound with respect to various semantics of logic programming [86,27,45,1]. All of these semantics are sound with respect to the semantics of A-Prolog, i.e. if a program Π is consistent under the answer set semantics and Π entails a literal l under one of these semantics then Π entails l in A-Prolog. This property allows the use of SLDNF based methods for answering A-Prolog's queries. Similar observations hold for bottom up methods of computation used in deductive databases. The newer methods (like that of [14]) combine both, bottom-up and top-down, approaches. A more detailed discussion of these matters can be found in [48]. In the last few years we witnessed the coming of age of inference engines aimed at computing answer sets (stable models) of programs of A-Prolog [65,66,21,17]. The algorithms implemented in these engines have much in common with more traditional satisfiability algorithms. The additional power comes from the use of techniques from deductive databases, good understanding of the relationship between various semantics of logic programming and other more recent discoveries (see for instance [44]) These engines are of course applicable only to programs with finite Herbrand universes. Their efficiency and power combined with so called *answer sets programming paradigm* [64], [56] lead to the development of A-Prolog based solutions for various problems in several knowledge intensive domains [77,7,25].

This paper is an attempt to introduce the reader to some recent developments in theory and practice of A-Prolog. In section 2 we briefly review the syntax and the semantics of the basic version of A-Prolog. In section 3 A-Prolog will be used to gradually construct a knowledge base which will demonstrate some knowledge of the notion of *orphan*. The main goal of this, rather long, example is to familiarize the reader with basic methodology of representing knowledge in A-Prolog. Section 4 contains some recent results from the mathematical theory of the language. The selection, of course, strongly reflects personal taste of the author and the limitations of space and time. Many first class recent results are not even mentioned. I hope however that the amount of material is sufficient to allow the reader to form a first impression and to get some appreciation of the questions involved. Section 5 contains a brief introduction to two extensions of the basic language: A-Prolog with disjunction and A-Prolog with sets. The latter is the only part of this paper which was neither published nor discussed in a broad audience. Again the purpose is primarily to illustrate the power of the basic semantics and the ease of adding extensions to the language. Finally, section 6 deals with more advanced knowledge representation techniques and more complex reasoning problems. There are several other logical languages and reasoning methods which can be viewed as alternatives to A-Prolog (see for instance [1,12,40]). They were developed in approximately the same time frame as A-Prolog share the same roots and a number of basic ideas. The relationship and mutual fertilization between these approaches is a fascinated subject which goes beyond the natural boundaries of this paper.

2   Syntax and Semantics of the Language

In this section we give a brief introduction to the syntax and semantics of a comparatively simple variant of A-Prolog. Two more powerful dialects will be discussed in sections 5. The syntax of the language is determined by a signature σ consisting of types, $types(\sigma) = \{\tau_0, \ldots, \tau_m\}$, object constants $obj(\tau, \sigma) = \{c_0, \ldots, c_m\}$ for each type τ, and typed function and predicate constants $func(\sigma) = \{f_0, \ldots, f_k\}$ and $pred(\sigma) = \{p_0, \ldots, p_n\}$. We will assume that the signature contains symbols for integers and for the standard relations of arithmetic. Terms are built as in typed first-order languages; positive literals (or atoms) have the form $p(t_1, \ldots, t_n)$, where t's are terms of proper types and p is a predicate symbol of arity n; negative literals are of the form $\neg p(t_1, \ldots, t_n)$. In our further discussion we often write $p(t_1, \ldots, t_n)$ as $p(\bar{t})$. The symbol $\neg$ is called *classical* or *strong* negation.[2] Literals of the form $p(\bar{t})$ and $\neg p(\bar{t})$ are called contrary. By $\bar{l}$ we denote a literal contrary to l. Literals and terms not containing variables are called *ground*. The sets of all ground terms, atoms and literals over

---

[2] Logic programs with two negations appeared in [31] which was strongly influenced by the epistemic interpretation of logic programs given below. Under this view $\neg p$ can be interpreted as "believe that p is false" which explains the term "classical negation" used by the authors. Different view was advocated in [67,87] where the authors considered logic programs without negation as failure but with $\neg$. They demonstrated that in this context logic programs can be viewed as theories of a variant of intuitionistic logic with strong negation due to [63]. For more recent work on this subject see [68]. I believe that both views proved to be fruitful and continue to play an important role in our understanding of A-Prolog. A somewhat different view on the semantics of programs with two negations can be found in [1].

σ will be denoted by $terms(\sigma)$, $atoms(\sigma)$ and $lit(\sigma)$ respectively. For a set P of predicate symbols from σ, $atoms(P, \sigma)$ $(lit(P, \sigma))$ will denote the sets of ground atoms (literals) of σ formed with predicate symbols from P. Consistent sets of ground literals over signature σ, containing all arithmetic literals which are true under the standard interpretation of their symbols, are called *states* of σ and denoted by $states(\sigma)$.

A rule of A-Prolog is an expression of the form

$$l_0 \leftarrow l_1, \ldots, l_m, not\ l_{m+1}, \ldots, not\ l_n \tag{1}$$

where $n \geq 1$, l_i's are literals, l_0 is a literal or the symbol $\perp$, and not is a logical connective called *negation as failure* or *default negation*. An expression $not\ l$ says that there is no reason to believe in l. An *extended literal* is an expression of the form l or $not\ l$ where l is a literal. A rule (1) is called a *constraint* if $l_0 = \perp$.

Unless otherwise stated, we assume that the $l's$ in rules (1) are ground. Rules with variables (denoted by capital letters) will be used only as a shorthand for the sets of their ground instantiations. This approach is justified for the so called closed domains, i.e. domains satisfying the domain closure assumption [72] which asserts that *all objects in the domain of discourse have names in the language of Π*. Even though the assumption is undoubtedly useful for a broad range of applications, there are cases when it does not properly reflect the properties of the domain of discourse. Semantics of A-Prolog for open domains can be found in [8], [41].

A pair $\langle \sigma, \Pi \rangle$ where σ is a signature and Π is a collection of rules over σ is called a *logic program*. (We often denote such pair by its second element Π. The corresponding signature will be denoted by $\sigma(\Pi)$.)

The following notation will be useful for the further discussion. A set $not\ l_i, \ldots, not\ l_{i+k}$ will be denoted by $not\ \{l_i, \ldots, l_{i+k}\}$. If r is a rule of type (1) then $head(r) = \{l_0\}$, $pos(r) = \{l_1, \ldots, l_m\}$, $neg(r) = \{l_{m+1}, \ldots, l_n\}$, and $body(r) = pos(r), not\ neg(r)$. The head, $\perp$, of a constraint rule will be frequently omitted. Finally, $head(\Pi) = \bigcup_{r \in \Pi} head(r)$. Similarly, for pos and neg.

We say that a literal $l \in lit(\sigma)$ is *true* in a state X of σ if $l \in X$; l is *false* in X if $\bar{l} \in X$; Otherwise, l is unknown. $\perp$ is false in X.

The answer set semantics of a logic program Π assigns to Π a collection of *answer sets* – consistent sets of ground literals over signature $\sigma(\Pi)$ corresponding to beliefs which can be built by a rational reasoner on the basis of rules of Π. In the construction of these beliefs the reasoner is assumed to be guided by the following informal principles:

- He should satisfy the rules of Π, understood as constraints of the form: *If one believes in the body of a rule one must belief in its head.*
- He cannot believe in $\perp$ (which is understood as falsity).
- He should adhere to the *rationality principle* which says that *one shall not believe anything he is not forced to believe.*

The precise definition of answer sets will be first given for programs whose rules do not contain default negation. Let Π be such a program and let X be a state of $\sigma(\Pi)$. We say that X is *closed* under Π if, for every rule *head* $\leftarrow$ *body* of Π, head is true in X whenever *body* is true in X. (For a constraint this condition means that the body is not contained in X.)

Definition 1. *(Answer set – part one)*
A state X of $\sigma(\Pi)$ is an *answer set* for Π if X is minimal (in the sense of set-theoretic inclusion) among the sets closed under Π.

It is clear that a program without default negation can have at most one answer set. To extend this definition to arbitrary programs, take any program Π, and let X be a state of $\sigma(\Pi)$. The *reduct*, Π^X, of Π relative to X is the set of rules

$$l_0 \leftarrow l_1, \ldots, l_m$$

for all rules (1) in Π such that $l_{m+1}, \ldots, l_n \notin X$. Thus Π^X is a program without default negation.

Definition 2. *(Answer set – part two)*
A state X of $\sigma(\Pi)$ is an answer set for Π if X is an answer set for Π^X.

(The above definition differs slightly from the original definition in [31], which allowed the inconsistent answer set, $lit(\sigma)$. Answer sets defined in this paper correspond to consistent answer sets of the original version.)

Definition 3. *(Entailment)*
A program Π entails a literal l ($\Pi \models l$) if l belongs to all answer sets of Π. The Π's answer to a query l is *yes* if $\Pi \models l$, *no* if $\Pi \models \bar{l}$, and *unknown* otherwise.

Consider for instance a logic program[3]

$$\Pi_0 \begin{cases} p(a) \leftarrow not\ q(a). \\ p(b) \leftarrow not\ q(b). \\ q(a). \end{cases}$$

It has one answer set $\{q(a), p(b)\}$ and thus answers *yes* and *unknown* to queries $q(a)$ and $q(b)$ respectively. If we expand Π_0 by a rule

$$\neg q(X) \leftarrow not\ q(X). \tag{2}$$

the resulting program Π_1 would have the answer set $S = \{q(a), \neg q(b), p(b)\}$ and hence its answer to the query $q(b)$ would be *no*.

Rule (2), read as "*if there is no reason to believe that X satisfies q then it does not*" is called the *closed world assumption* for q [72]. It guarantees that the reasoner's beliefs about q are complete, i.e. for any ground term t and every answer set S of the corresponding program, $q(t) \in S$ or $\neg q(t) \in S$.

---

[3] Unless otherwise specified we assume that signature of a program consists of symbols occurring in it.

The programs may have one, many, or zero answer sets. It is easy to check for instance that programs

$$\Pi_3 = \{p \leftarrow not\ p\} \text{ and } \Pi_4 = \{p. \quad \neg p.\}$$

have no answer sets while program Π_5

$$e(0).$$
$$e(s(s(X))) \leftarrow not\ e(X).$$
$$p(s(X)) \quad \leftarrow e(X),$$
$$\qquad\qquad not\ p(X).$$
$$p(X) \qquad \leftarrow e(X),$$
$$\qquad\qquad not\ p(s(X)).$$

has an infinite collection of them.

In some cases a knowledge representation problem consists in representing a (partial) definition of new relations between objects of the domain in terms of the old, known relations. Such a definition can be mathematically described by logic programs viewed as functions from states of some input signature σ_i (given relations) into states of some output signature σ_i (defined relations).[4] More precisely [10].

Definition 4. *(lp-functions)*
A four-tuple $f = \langle \Pi(f), \sigma_i(f), \sigma_o(f), dom(f) \rangle$ where

1. $\Pi(f)$ is a logic program (with some signature σ),
2. $\sigma_i(f)$ and $\sigma_o(f)$ are sub-signatures of σ, called the input and output signatures of f respectively,
3. $dom(f)$ is a collection of states of $\sigma_i(f)$

is called *lp-function* if for any $X \in dom(f)$ program $\Pi(f) \cup X$ is *consistent*, i.e., has an answer set.

For any $X \in dom(f)$, $f(X) = \{l : l \in lit(\sigma_o(f)), \Pi(f) \cup X \models l\}$.

We finish our introduction to A-Prolog by recalling the following propositions which will be useful for our further discussion. To the best of my knowledge Proposition 1 first appeared in [54].

Proposition 1. For any answer set S of a logic program Π:

(a) For any ground instance of a rule of the type (1) from Π,

if $\{l_1, \ldots, l_m\} \subseteq S$ and $\{l_{m+1}, \ldots, l_n\} \cap S = \emptyset$ then $l_0 \in S$.

(b) If $l_0 \in S$, then there exists a ground instance of a rule of the type (1) from Π such that $\{l_1, \ldots, l_m\} \subseteq S$ and $\{l_{m+1}, \ldots, l_n\} \cap S = \emptyset$.

---

[4] This view is similar to that of databases where one of the most important knowledge representation problems consists in defining the new relations (views) in terms of the basic relations stored in the database tables. Unlike our case, however, databases normally assume the completeness of knowledge and hence only need to represent positive information. As a result, database views can be defined as functions from sets of atoms to sets of atoms.

The next proposition (a variant of a similar observation from [31]) shows how programs of A-Prolog can be reduced to *general logic programs*, i.e. programs containing neither ¬ nor ⊥. We will need the following notation:

For any predicate p occurring in Π, let p' be a new predicate of the same arity. The atom $p'(\bar{t})$ will be called the *positive form* of the negative literal $\neg p(\bar{t})$. Every positive literal is, by definition, its own positive form. The positive form of a literal l will be denoted by l^+. Π^+ stands for the general logic program obtained from Π by replacing each rule (1) by

$$l_0^+ \leftarrow l_1^+, \ldots, l_m^+, not\ l_{m+1}^+, \ldots, not\ l_n^+$$

and adding the rules

$$\leftarrow p(\bar{t}), p'(\bar{t})$$

for every atom $p(\bar{t})$ of $\sigma(\Pi)$. For any set S of literals, S^+ stands for the set of the positive forms of the elements of S.

Proposition 2. A consistent set $S \subset lit(\sigma(\Pi))$ is an answer set of Π if and only if S^+ is an answer set of Π^+.

Proposition 2 suggests the following simple way of evaluating queries in A-Prolog. To obtain an answer for query p, run queries p and p' on the program Π^+. If Π^+'s answer to p is *yes* then Π's answer to p is *yes*. If Π^+'s answer to p' is *yes* then Π's answer to p is *no*. Otherwise the answer to p is *unknown*. (The method of course works only if the corresponding inference engine terminates).

3 Defining Orphans - A Case Study

In this section we give a simple example of representing knowledge in A-Prolog. We will be dealing with a class of "personnel" systems whose background knowledge consist of collections of personal records of people. Such collections will be referred to as *databases*. There are multiple ways of designing such records. To keep a presentation concise we fix an artificially simple signature σ_i containing names of people, a special constant *nil* (read as *unknown person*), and the predicate symbols $person(P)$, $father(F, P)$, $mather(M, P)$, $child(P)$, $dead(P)$. We assume that *every person in the domain has a database record not containing false information, names of the parents of the live people are known and properly recorded, while unknown parents are represented by nil*, and that *the death records and children's records are complete*. The set of databases satisfying these assumptions will be denoted by C_0. Typical records of a database from C_0 look as follows:

$person(john)$.	$person(mike)$.	$person(kathy)$.
$father(mike, john)$.	$father(sam, mike)$.	$father(nil, kathy)$.
$mother(kathy, john)$.	$mother(mary, mike)$.	$mother(pat, kathy)$.
	$dead(mike)$.	$dead(kathy)$.
$child(john)$.		

The first record describes a child, John, whose parents are Mike and Kathy. Since the death of John is not recorded he must be alive. Similarly, we can conclude that Mike and Kathy were adults when they died, and that the name of Kathy's father is unknown.

Let us assume that we are confronted with a problem of expanding databases from C_0. In particular we need to familiarize the system with a notion of an *orphan* - a child whose parents are dead. In slightly more precise terms we need to define a function which takes a database $X \in C_0$ describing personal records of people from some domain and returns the set of the domain's orphans.

The problem can be solved by introducing an lp-function f_0 with $dom(f_0) = C_0$, $\Pi(f_0)$ consisting of rules:

$$\Pi(f_0) \begin{cases} r1. \quad orphan(P) \qquad \leftarrow child(P), \\ \qquad\qquad\qquad\qquad\quad not\ dead(P), \\ \qquad\qquad\qquad\qquad\quad parents\_dead(P). \\ \\ r2. \quad parents\_dead(P) \leftarrow father(F, P). \\ \qquad\qquad\qquad\qquad\quad mother(M, P), \\ \qquad\qquad\qquad\qquad\quad dead(F), \\ \qquad\qquad\qquad\qquad\quad dead(M). \end{cases}$$

with *father, mother, dead* and *child* being predicate symbols of $\sigma_i(f_0)$, *orphan* being the only predicate symbol of $\sigma_o(f_0)$, and both signatures sharing the same object constants. It is not difficult to convince oneself that, since X contains complete information about the live people of the domain, set $f_0(X)$ consists exactly of the domain's orphans.

Program $\Pi(f_0)$ has many attractive mathematical and computational properties. For instance it is easy to check that, for any database $X \in C_0$, the program $R_0 = \Pi(f_0) \cup X$ is *acyclic* [3], i.e. there is a function $||\ ||$ from ground atoms of $\sigma(R_0)$ to natural numbers [5] such that for any atom l occurring in the body of a rule with the head l_0, $||l_0|| > ||l||$. Acyclic general logic programs have unique answer sets which can be computed by a bottom-up evaluation [3]. Moreover, acyclicity of R_0 together with some results from [4,79] guarantee that the SLDNF resolution based interpreter of Prolog will always terminate on atomic queries and (under the 'right' interpretation) produce the intended answers. The reference to the 'right' interpretation is of course vague and deserves some comments. Suppose that, according to the database, X_0, containing records about John and Mary, John is an orphan and Mary is not. Given program R_0 the Prolog interpreter will answer queries $orphan(john)$ and $orphan(mary)$ by *yes* and *no* respectively. Since the closed world assumption is built in the semantics of "classical" logic programming, the second answer can be (correctly) interpreted as saying that Mary is not an orphan. It is important to realize however that,

---

[5] Functions from ground literals to ordinals are called *level mappings*. They often play an important role in characterizing various properties of logic programs.

from the standpoint of the semantics of A-Prolog, this interpretation is incorrect. Since neither $orphan(mary)$ nor $\neg orphan(mary)$ is entailed by the program the answer to the query $orphan(mary)$ should be *unknown*. To get the correct answer we need to complete the rules of Π_0 by explicitly defining non-orphans. This can be done by adding a simple rule encoding the corresponding closed world assumption:

$$r3. \quad \neg orphan(P) \leftarrow person(P),$$
$$not\ orphan(P).$$

It may be instructive at this point to modify our notion of a database X by explicitly defining its negative information. For relations *dead* and *child* and it easy: we just need to explicitly encode the closed world assumptions:

$$r4. \quad \neg child(P) \leftarrow person(P),$$
$$not\ child(P).$$
$$r5. \quad \neg dead(P) \leftarrow person(P),$$
$$not\ dead(P).$$

Even though we typically have complete information about the parents of people from the database this is not always the case. We can express this fact by the following default with exceptions:

$$r6. \quad \neg father(F, P) \leftarrow person(F),$$
$$person(P),$$
$$not\ father(F, P),$$
$$not\ ab(d(F, P)).$$
$$r7. \quad ab(d(F, P)) \quad \leftarrow father(nil, P).$$

Here $d(F, P)$ is used to name the default; statement $ab(d(F, P))$ says that this default is not applicable to F and P. If we assume that Bob is a person in our database we will be able to use the default to show that Bob is not the father of John. For Kathy, however, the same question will remain undecided.

Rules (r6) and (r7) can be viewed as a result of the application of the general methodology of representing defaults in A-Prolog. More detailed discussion of this methodology can be found in [8]. A more general approach which provides means for specifying priorities between defaults is discussed [36], [23], [39].

Let $X \in C_0$. Since X contains the complete records of parents of every live person p the rules (r6) and (r7) allow us to conclude that for every person r different from the father of P the answer to query $father(r, p)$ will be *no*. For dead people more negative knowledge can be extracted from the database by common-sense rules like:

$$r8. \quad \neg father(F, P) \quad \leftarrow mother(F, Q).$$
$$r9. \quad \neg father(F, P) \quad \leftarrow descendant(F, P).$$
$$r10. \quad descendent(P, P).$$
$$r11. \quad descendent(D, P) \leftarrow parent(P, C),$$
$$descendant(D, C).$$

Consider an lp-function

$$g = \langle \Pi(g), \sigma_i(f_0), \sigma_i(f_0), C_0 \rangle$$

where $\Pi(g)$ consists of rules (r4)-(r11), together with the obvious definition of relation *parent* and the rules extracting negative information for mothers. The function computes the *completion* of a database $X \in C_0$ by the corresponding negative information. By $\hat{C}_0$ we denote the collection of completions of elements of C_0. Consider

$$\Pi(f) = \Pi(f_0) \cup (r3)$$

and lp-functions

$$f = \langle \Pi(f_0) \cup (r3), \sigma_i(f_0), \sigma_o(f_0), \hat{C}_0 \rangle$$

and

$$h = \langle \Pi(f) \cup \Pi(g), \sigma_i(f_0), \sigma_o(f_0), C_0 \rangle$$

Using the Splitting Lemma (see the next section) it is not difficult to show that, for any $X \in C_0$, $h(X) = f(g(X))$, i.e. $h = f \circ g$. (Notice that, since $\Pi(g)$ is nonmonotonic, its consequences can be modified by addition of $\Pi(f)$ and so such a proof is necessary. Fortunately, it follows immediately from a fairly general theorem from [28].)

Due to the use of default negation, h is also elaboration tolerant w.r.t. *some* modifications of the background knowledge such as addition of new people and recording of deaths and changes in the adulthood status. The latter for instance can be accomplished by simply removing, say, a record *child(john)* from the background knowledge X_0 described above, at which point *John* will seize to be an orphan. A program will continue to work correctly as far as the update of the background knowledge still belongs to the class C_0.

When our knowledge of the domain cannot be captured by databases from C_0 or $\hat{C}_0$ the situation may become substantially more complex. Let us for instance consider a modification of our informal knowledge base by *removing from it the closed world assumption for property of being a child* (*cwa(child)*). Now the record of a person p can contain a statement *child(p)* or a statement $\neg child(p)$, or no information about p being a child at all. (In the latter case we say that p's age is unknown.) A new class of databases will be denoted by C_1. As before, every database X of C_1 contains atoms *alive(p)*, *father(f, p)*, *mother(m, p)* for every live person p of the domain. We still have the closed world assumption for *alive* and no false information in the X's records. Our goal is still to teach our knowledge base about the orphans, i.e. to construct an lp-function which takes a database $X \in C_1$ and returns the set of domain's orphans. It is easy to see that completions of databases from C_1 with respect to missing negative information about relations other than *child* can be defined as values of the lp-function g' obtained from g by removing *cwa(child)* from $\Pi(g)$. We denote the set of all such completions by $\hat{C}_1$. As expected, however, program f does not work correctly with databases from $\hat{C}_1$ – the closed world assumption for orphans will force the program to erroneously conclude that everyone whose age is not known is not an orphan.

The problem of finding a uniform way of modifying logic programs which would reflect the removal of some of its closed world assumptions was addressed in [10], [28]. The authors' approach is based on the notion of *interpolation* of a logic program. To be more precise we will need some additional terminology.

Let F be an lp-function, O be a set of predicate symbols from $\sigma_i(F)$ and $D = dom(F)$ be closed with respect to O, i.e., $X \in D$ contains l or $\bar{l}$ for every literal $l \in lit(O)$. For any set X of input literals we define the set $c(X, O)$ of its *covers* – $\hat{X} \in c(X, O)$ if it satisfies the following properties:

1. $\hat{X} \in D$;
2. $X \subseteq \hat{X}$;
3. for every input literal $l \notin lit(O)$, $l \in X$ iff $l \in \hat{X}$.

By $\tilde{D}$ we denote the set of states of $\sigma_i(F)$ such that

$$X = \bigcap_{\hat{X} \in c(X,O)} \hat{X}$$

Definition 5. We say that an lp-function $\tilde{F}$ is an *O-interpolation* of F if

$$dom(\tilde{F}) = \tilde{D}$$

$$\tilde{F}(X) = \bigcap_{\hat{X} \in c(X,O)} F(\hat{X})$$

$$\sigma_i(F) = \sigma_i(\tilde{F}) \text{ and } \sigma_o(F) = \sigma_o(\tilde{F})$$

Let us go back to function f from our example and consider $O = \{child\}$, program

$$\Pi(\tilde{f}) \begin{cases} 1.\ may\_be\_child(P) & \leftarrow not\ \neg child(P). \\[4pt] 2.\ parents\_dead(P) & \leftarrow father(F, P). \\ & \quad mother(M, P), \\ & \quad dead(F), \\ & \quad dead(M). \\[4pt] 3.\ orphan(P) & \leftarrow child(P), \\ & \quad not\ dead(P), \\ & \quad parents\_dead(P). \\[4pt] 4.\ may\_be\_orphan(P) & \leftarrow may\_be\_child(P), \\ & \quad not\ dead(P), \\ & \quad parents\_dead(P). \\[4pt] 5.\ \neg orphan(P) & \leftarrow not\ may\_be\_orphan(P). \end{cases}$$

and lp-function

$$\tilde{f} = \langle \Pi(\tilde{f}), \sigma_i(f), \sigma_o(f), \hat{C}_1 \rangle$$

The rules of $\Pi(\tilde{f})$ are obtained by the general algorithm from [10] which, under certain conditions, translates lp-functions described by general logic programs into their interpolations. In the next section we use the mathematical theory of A-Prolog to prove that $\tilde{f}$ is indeed a $\{child\}$-interpolation of f.

In the conclusion of this section we illustrate how A-Prolog can be used to represent

(a) simple priorities between defaults;

(b) statements about the lack of information.

To do that, let us supply our program with knowledge about some fictitious legal regulations. The first regulation says that *orphans are entitled to assistance according to special government program 1*, while the second says that all *children who are not getting any special assistance are entitled to program 0*. Legal regulations always come with exceptions and hence can be viewed as defaults. We represent both regulations by the following rules:

$$entitled(P, 1) \leftarrow orphan(P),$$
$$not\ ab(d_1(P)),$$
$$not\ \neg entitled(P, 1).$$

$$entitled(P, 0) \leftarrow child(P),$$
$$\neg dead(P),$$
$$not\ ab(d_2(P)),$$
$$not\ \neg entitled(P, 0).$$

$$ab(d_2(P)) \qquad \leftarrow orphan(P).$$

The first two rules are standard representations of defaults. The last rule says that the default d_2 is not applicable to orphans. Notice that if Joe is a child and it is not known whether he is an orphan or not then Joe will receive benefits from program 0 but not from program 1. This case of insufficient documentation can be detected by the following rule:

$$check\_status(P) \leftarrow person(P),$$
$$not\ \neg orphan(P),$$
$$not\ orphan(P).$$

Though simple, the program above illustrates many interesting features of A-Prolog: recursive rules, the use of default negation for representing defaults with exceptions, the use of both negations in formulating the closed world assumptions, the ability to discriminate between falsity and the absence of information, and to produce conclusions based on such absence. The program can be used together with various inference engines of A-Prolog, thus making it (efficiently) executable. In section 6 we will demonstrate how A-Prolog can be used to represent change and causal relations. First, however, we briefly discuss mathematical theory of A-Prolog.

4 Mathematics of A-Prolog

In this section we review several important properties of programs of A-Prolog. Our goal of course is not to give a serious introduction into the mathematics of A-Prolog. By now the theory is well developed, contains many interesting results, and probably deserves a medium size textbook. Instead we concentrate on a few important discoveries and discuss their relevance to constructing knowledge bases.

4.1 Splitting Lemma

The structure of answer sets of a program Π can sometimes be better understood by "splitting" the program into parts. We say that a set U of literals *splits* a program Π if, for every rule r of Π, $pos(r) \cup neg(r) \subseteq U$ whenever $head(r) \in U$. If U splits Π then the set of rules in Π whose heads belong to U will be called the base of Π (relative to U). We denote the base of Π by $b_U(\Pi)$. The rest of the program (called the top of Π) will be denoted by $t_U(\Pi)$.

Consider for instance a program Π_1 consisting of the rules

$$q(a) \leftarrow not\ q(b),$$
$$q(b) \leftarrow not\ q(a),$$
$$r(a) \leftarrow q(a).$$
$$r(a) \leftarrow q(b)$$

Then, $U = \{q(a), q(b)\}$ is a splitting set of Π_1, $b_U(\Pi_1)$ consists of the first two rules while $t_U(\Pi_1)$ consists of the last two.

Let U be a splitting set of a program Π and consider $X \subseteq U$. For each rule $r \in \Pi$ satisfying property

$$pos(r) \cap U \subset X \text{ and } (neg(r) \cap U) \cap X = \emptyset$$

take the rule r' such that

$$head(r') = head(r),\ pos(r') = pos(r) \setminus U,\ neg(r') = neg(r) \setminus U$$

The resulting program, $e_U(\Pi, X)$, is called *partial evaluation* of Π with respect to U and X.

A *solution* to Π with respect to U is a pair $\langle X, Y \rangle$ of sets of literals such that:

- X is an answer set for $b_U(\Pi)$;
- Y is an answer set for $e_U(t_U(\Pi), X)$;
- $X \cup Y$ is consistent.

Lemma 1. *(Splitting Lemma)*

Let U be a splitting set for a program Π. A set S of literals is a consistent answer set for Π if and only if $S = X \cup Y$ for some solution $\langle X, Y \rangle$ to Π w.r.t. U.

The Splitting Lemma has become an important tool for establishing existence and other properties of programs of A-Prolog. To demonstrate its use let us consider a class of finite *stratified* programs. A finite general logic program Π is called stratified if there is a level mapping $||\ ||$ of Π such that if $r \in \Pi$ then

1. For any $l \in pos(r)$, $||l|| \leq ||head(r)||$;
2. For any $l \in neg(r)$ $||l|| < ||head(r)||$.

This is a special case of the notion of stratified logic program introduced in [2]. The results of that paper together with those from [29] imply that a stratified program has exactly one answer set. For finite stratified programs this can be easily proven by induction on the number of levels of Π with the use of the Splitting Lemma. If Π has one level (i.e. $||l|| = 0$ for every $l \in \sigma(\Pi)$) then Π does not contain default negation and hence, by [85] has exactly one minimal Herbrand model which, by definition, coincides with the Π's answer set. If the highest level of an atom from $\sigma(\Pi)$ is $n+1$ then it suffices to notice that atoms with smaller levels form a splitting set U of Π. By inductive hypothesis, $b_U(\Pi)$ has exactly one answer set, X, $e_U(\Pi, X)$ is a program without *not* and hence has one and only answer set Y. By Splitting Lemma $X \cup Y$ is the only answer set of Π.

The Splitting Lemma can be generalized to programs with a monotone, continuous sequence of splitting sets. This more powerful version can be used to prove the uniqueness of answer set for locally stratified logic programs [69], existence of answer sets for order-consistent logic programs of [26], etc.

The above results, combined with Proposition 2 can be used to establish existence and uniqueness of answer sets of programs with $\neg$. Consider for instance lp-function h from the previous section and a set X of literals from C_0. To show that $\Pi(h) \cup X$ has the unique answer set let us notice that the corresponding general logic program $(\Pi(h) \cup X)^+$ is stratified, and therefore has the unique answer set S^+. To show that the corresponding set S is consistent we need to check that there is no atom $p(\bar{t})$ such that $p(\bar{t}), (\neg p(\bar{t}))^+ \in S^+$. By Proposition 1 we have that this could only happen if for some people f and p, $father(f, p)$ and $mother(f, p)$ or $father(f, p)$ and $descendent(f, p)$ were in S^+. It is not difficult to check that this is impossible since, according to our assumption, C_0 contains correct factual information. This, by Proposition 2, implies that $\Pi(h) \cup X$ has the unique answer set S.

The discussion in this section follows [50], in which authors gave a clear exposition of the idea of splitting in the domain of logic programs. Independently, similar results were obtained in [16]. There is a very close relationship between splitting of logic programs and splitting of autoepistemic and default theories [34,15,84].

4.2   Signed Programs

In this section we introduce the notion of *signing* of a program of A-Prolog. The notion of signing for finite general logic programs was introduced by Kunen [46], who used it as a tool in his proof that, for a certain class of program, two different semantics of logic programs coincide. Turner, in [82], extends the definition to the class of logic programs with two kinds of negation and investigates properties of signed programs. We will need some terminology.

The *absolute value* of a literal l (symbolically, $|l|$) is l if l is positive, and $\bar{l}$ otherwise.

Definition 6. A *signing* of logic program Π is a set $S \subseteq atoms(\sigma(\Pi))$ such that

1. for any rule
$$l_0 \leftarrow l_1, \ldots, l_m, not\ l_{m+1}, \ldots, not\ l_n$$
   from Π, either
$$|l_0|, \ldots, |l_m| \in S \text{ and } |l_{m+1}|, \ldots, |l_n| \notin S$$
   or
$$|l_0|, \ldots, |l_m| \notin S \text{ and } |l_{m+1}|, \ldots, |l_n| \in S$$
   ;
2. for any atom $l \in S$, $\neg l$ does not appear in Π.

A program is called *signed* if it has a signing. Obviously, programs without default negation are signed with the empty signing. Program

$$
\begin{aligned}
p(a) &\leftarrow not\ q(a). \\
p(b) &\leftarrow not\ q(b). \\
q(a). &
\end{aligned}
$$

is signed with signing $\{q(a), q(b)\}$.

Program Π_1 with signature $\sigma = \{\{a, b, c\}, \{p, q, r, ab\}\}$ and the rules

$$
\begin{aligned}
q(X) &\leftarrow p(X), \\
&\quad not\ ab(X). \\
\neg q(X) &\leftarrow r(X). \\
ab(X) &\leftarrow not\ \neg r(X).
\end{aligned}
$$

is signed with a signing $atoms(ab, \sigma)$.

Signed programs enjoy several important properties which make them attractive from the standpoint of knowledge representation. In particular,

1. Signed general logic programs are consistent, i.e. have an answer set. Simple consistency conditions can also be given for signed programs with classical negation.

2. If Π is consistent then the set of consequences of the program under answer set semantics coincides with its set of consequences under well-founded semantics [86]. Notice, that this result shows that inference engines such as SLG [14] which compute the well founded semantics of logic programs, can also be used to compute the consequences of such programs under the answer set semantics.

The following theorem gives another important property of signed programs:

An lp-function F is called *monotonic* if for any $X, Y \in dom(F)$, $F(X) \subseteq F(Y)$.

Theorem 1. *(Monotonicity Theorem, Turner)*
If an lp-function F has a signing S such that $S \cap (lit(\sigma_i(F)) \cup lit(\sigma_o(F))) = \emptyset$ then F is monotonic.

Example 1. Consider an lp-function f_1 with $\sigma_i(f_1) = \{\{a, b, c\}, \{p, r\}\}, \sigma_o(f_1) = \{\{a, b, c\}, \{q\}\}, dom(f_1)$ consisting of consistent sets of literals in σ_i, and program Π_1 above as $\Pi(f_1)$. It is easy to see that lp-function f_1 satisfies the condition of theorem 1, and hence is monotonic. It's worth noticing, that logic program $\Pi(f_1)$ is nonmonotonic. Addition of extra rules (or facts) about ab can force us to withdraw previous conclusions about q. Monotonicity is however preserved for inputs from σ_i.

Discussion of the importance of this property for knowledge representation can be found in [47].

4.3   Interpolation

In the previous section we mentioned the notion of interpolation $\tilde{F}$ of an lp-function F. The switch from F to $\tilde{F}$ reflects the removal from the informal knowledge base represented by $\Pi(F)$ some of its closed world assumptions. The notion of signing plays an important role in the following theorem (a variant of the result from [10]) which facilitates the construction of $\tilde{F}$.

Let F be an lp-function with $\Pi(F)$ not containing $\neg$, O be a set of predicate symbols from $\sigma_i(F)$ and the domain D of F be closed with respect to O, i.e., $X \in D$ contains l or $\bar{l}$ for every literal $l \in lit(O)$. By o we denote the set of predicate symbols of $\Pi(F)$ depending on O. More precisely, o is the minimal set of predicate symbols such that $O \subseteq o$ and if the body of a rule of $\Pi(F)$ with head $p(\bar{t})$ contains an atom formed by a predicate symbol from o, then $p \in o$. To define $\tilde{F}$ we expand the signature σ of $\Pi(F)$ by a new atom, m_p, (read as "may be p") for every predicate symbol p, and consider a mapping, α, from extended literals of $\Pi(F)$ into literals of the new signature $\tilde{\sigma}$:

1. if $p \in o$ then $\alpha(p(\bar{t})) = m_p(\bar{t})$ and $\alpha(not\ p(\bar{t})) = \neg p(\bar{t})$.
2. Otherwise, $\alpha(e) = e$.

If E is a set of extended literals then $\alpha(E) = \{\alpha(e) : e \in E\}$.

If r is a rule of the form

$$l_0 \leftarrow pos, not\ neg$$

then by $\alpha(r)$ we denote the rules:

$$\alpha(l_0) \leftarrow \alpha(pos), not\ neg$$
$$l_0 \leftarrow pos, \alpha(not\ neg)$$

By $\tilde{F}$ we denote the lp-function with $dom(\tilde{F}) = \tilde{dom}(F)$, $\sigma_i(\tilde{F}) = \sigma_i(F)$, $\sigma_o(\tilde{F}) = \sigma_o(F)$, and $\Pi(\tilde{F})$ consisting of the rules:

1. For any predicate symbol $p \in O$ add the rule

$$m\_p(X) \leftarrow not\ \neg p(X).$$

2. For any predicate symbol $p \in o \setminus O$ add the rule

$$\neg p(X) \leftarrow not\ m_p(X).$$

3. Replace every rule $r \in \Pi$ by $\alpha(r)$.

Theorem 2. *(Interpolation theorem)*
Let F and O be as above. If $\Pi(F)$ is signed then $\tilde{F}$ is an interpolation of F.

Let us now demonstrate how these results can be used to prove properties of the lp-functions f and $\tilde{f}$ from section 3.

Proposition 3. $\tilde{f}$ is the interpolation of f.

Proof (sketch).
(a) Let $O = \{child\}$ and $D = dom(f)$. Using definitions of $\hat{C}_1$ and $\tilde{D}$ it is not difficult to show that $\hat{C}_1 = \tilde{D}$. To check the first condition of Definition 5 we need to prove that for any $X \in \tilde{D}$, $(\Pi(\tilde{f}) \cup X)^+$ has an answer set. This follows from the fact that this program is stratified. Moreover stratifiability implies that this answer set is unique. Let us denote it by A^+. Using Proposition 1 we can check that if $orphan(p) \in A^+$ then so is $may\_be\_orphan(p)$ and therefore A^+ does not contain $(\neg orphan(p))^+$. By Proposition 2 we conclude that A is the answer set of $\Pi(\tilde{f}) \cup X$, i.e., $X \in dom(\tilde{f})$.

(b) Let U be a set of literals formed by predicate symbols of the program $\Pi(f)$ different from $orphan$ and $child$. Obviously, for any $X \in \tilde{D}$, U is a splitting set of $\Pi(f) \cup X$. The base of this program, consisting of definition of $parents\_dead$ and and $X \cap U$ contains no default negation. This, together with consistency of X, implies that the base has exactly one answer set. Let us denote it by A_U.

Now consider an lp-function r with

$$\Pi(r) = e_U(\Pi(f), A_U) \tag{3}$$

$\sigma_i(r)$ formed by predicate *child* and objects constants of the domain, $\sigma_o(r) = \sigma\_o(f)$, and the domain consisting of complete and consistent sets of literals formed by *child*. From definition of our operator $\tilde{\ }$ it is easy to see that

$$\Pi(\tilde{r}) = e_U(\Pi(\tilde{f}), A_U) \tag{4}$$

It is not difficult to check that $\Pi(r)$ is signed with a signing consisting of atoms formed by the predicate symbol "*may_be*" therefore, by Theorem 2,

$$\tilde{r}(Y) = \bigcap_{\hat{Y} \in c(Y,O)} r(\hat{Y}) \tag{5}$$

By Splitting Lemma we can conclude that

$$f(X) = \tilde{r}(Y) \text{ where } Y = X \cap lit(O) \tag{6}$$

and that, for every $\hat{X} \in c(X, O)$,

$$\tilde{f}(\hat{X}) = \tilde{r}(\hat{Y}) \text{ where } \hat{Y} = \hat{X} \cap lit(O) \tag{7}$$

Finally let us notice that

$$\hat{X} \in c(X, O) \text{ iff } \hat{Y} \in c(Y, O) \tag{8}$$

which, together with (5) – (6) implies

$$\tilde{f}(\hat{X}) = \bigcap_{\hat{X} \in c(X,O)} f(\hat{X}) \tag{9}$$

We hope that the discussion in this section will help a reader to get a feel for some of the mathematics of A-Prolog. The following sections contain several other useful mathematical results which may help to better see the variety of questions related to A-Prolog. Meanwhile we turn to the question of the extensions of the basic variant of A-Prolog.

5 Extensions of A-Prolog

There are several important extensions of A-Prolog (see for instance [38,52,66]). We will briefly discuss two of such extensions: *disjunctive A-Prolog* (DA-Prolog) [70,31], and *A-Prolog with sets* (ASET-Prolog). DA-Prolog has been studied for a substantial amount of time. It has a non-trivial theory and efficient implementation. ([61] surveys alternative ways of introducing disjunction in logic programming). ASET-Prolog is still in its developing stage.

5.1   A-Prolog with Disjunction

A program of DA-Prolog consists of rules of the form

$$l_0 \text{ or } \ldots \text{ or } l_k \leftarrow l_{k+1}, \ldots l_m, not\ l_{m+1}, \ldots, not\ l_n \qquad (10)$$

The definition of an answer set of a disjunctive program is obtained by making a small change in the definition of what it means for a set X of literals to be closed under program Π. We now say that X is *closed* under Π if, for every rule *head* $\leftarrow$ *body* of Π, at least one of literals in the head is true in X whenever *body* is true in X. The rest of definitions 1 and 2 remain unchanged. The following simple examples illustrate the definition:
A program Π_1 consisting of the rules:

p_1 **or** p_2.
$q \leftarrow p_1$.
$q \leftarrow p_2$.

has two answer sets, $\{p_1, q\}$ and $\{p_2, q\}$. The program Π_2:

p_1 **or** p_2.
$q \leftarrow not\ p_1$.
$q \leftarrow not\ p_2$.

has the same answer sets. And the program

p_1 **or** p_2.
$q \leftarrow not\ p_1$.
$q \leftarrow not\ p_2$.
$\neg p_1$.

has the answer set $\{\neg p_1, p_2, q\}$.

There are several systems capable of reasoning in DA-Prolog. Some of them use the top-down or bottom-up methods of answering queries similar to those in non-disjunctive logic programs [78,5]. A different approach is taken by the *dlv* system [20] which takes as an input a program of DA-Prolog with a finite Herbrand Universe and computes the answer sets of this program.

Knowledge Representation in DA-Prolog

The following examples demonstrate the use of disjunction for knowledge representation and reasoning.

Example 2. Let us consider the following scenario: A preliminary summer teaching schedule of a computer science department is described by a relation *teaches(prof, class)*. The preliminary character of the schedule is reflected by the following uncertainty in the database

teaches(mike, java) **or** *teaches(john, java)*.

and by the absence of the closed world assumption for *teaches*. Now assume that in summer semesters the department normally teaches at most one course on computer languages. Intuitively this implies that no course on the C language will be offered. To make such a conclusion possible we expand our database by the following information:

$lang(java).$
$lang(c).$
$offered(C) \quad \leftarrow teaches(P, C).$
$\neg offered(C1) \leftarrow lang(C1),$
$\qquad\qquad\qquad lang(C2),$
$\qquad\qquad\qquad offered(C2),$
$\qquad\qquad\qquad C1 \neq C2,$
$\qquad\qquad\qquad not\ offered(C1).$

The last statement is a standard representation of a default. The resulting program has two answer sets. In one Java is taught by Mike, in another one by John. In both cases however the C language is not offered. The example demonstrates the ability of DA-Prolog to represent reasoning by cases and to nicely combine disjunction with defaults. (For comparison of these properties with the use of disjunction in Reiter's default logic see [33]).

The next example from [13] demonstrates the expressive power of the language.

Example 3. Suppose a holding owns some companies producing a set of products. Each product is produced by at most two companies. We will use a relation $produced\_by(P, C_1, C_2)$ which holds if a product P produced by companies C_1 and C_2. The holding below consists of four companies producing four products and can be represented as follows:

$produced\_by(p1, b, s).$ $produced\_by(p2, f, b).$
$produced\_by(p3, b, b).$ $produced\_by(p4, s, p).$

This slightly artificial representation, which requires a company producing a unique product to be repeated twice (as in the case of $p3$), is used to simplify the presentation.

Suppose also that we are given a relation $controlled\_by(C_1, C_2, C_3, C_4)$ which holds if companies C_2, C_3, C_4 control company C_1. In our holding, b and s control f, which is represented by

$controlled\_by(f, b, s, s)$

Suppose now that the holding needs to sell some of the companies and that its policy in such situations is to maintain ownership of so called strategic companies, i.e. companies belonging to a minimal (with respect to the set theoretic inclusion) set S satisfying the following conditions:

1. Companies from S produce all the products.
2. S is closed under relation $controlled\_by$, i.e. if companies C_2, C_3, C_4 belong to S then so is C_1.

It is easy to see that for the holding above the set $\{b, s\}$ is not strategic while the set $\{b, s, f\}$ is.

Suppose now that we would like to write a program which, given a holding of the above form, computes sets of its strategic companies. This can be done by the rules

1. $strat(C_1)$ **or** $strat(C_2) \leftarrow produced\_by(P, C_1, C_2)$
2. $strat(C_1) \qquad\qquad\quad \leftarrow controlled\_by(C_1, C_2, C_3, C_4),$
$$strat(C_2),$$
$$strat(C_3),$$
$$strat(C_4).$$

defining the relation $strat(C)$. Let Π be a program consisting of rules (1), (2) and an input database X of the type described above. The first rule guarantees that, for every answer set A of Π and every product p, there is a company c producing p such that an atom $strat(c) \in A$. The second rule ensures that for every answer set of Π the set of atoms of the form $strat(c)$ belonging to this set is closed under the relation $controlled\_by$. Minimality of this set follows from the minimality condition in the definition of answer set. It is not difficult to check that answer sets of Π correspond one-to-one to strategic sets of the holding described by an input database. The *dlv* reasoning system can be asked to find an answer set of Π and display atoms of the form $strat$ from it.

Complexity and Expressiveness

The above problem can be viewed as an example of a classical search problem P given by a finite collection $dom(P)$ of possible input databases and a function $P(X)$ defining *solutions* of P for every input X from $dom(P)$. An algorithm solves a search problem P if for each $X \in dom(P)$ it returns *no* if $P(X)$ is empty and one of the elements of $P(X)$ otherwise. Solution of the corresponding decision problem requires an algorithm which checks if $P(X)$ is empty or not. This observation suggests the following approach to solving a search and decision problem P:

1. Encode input instances and solutions of P by collections of literals from signatures σ_i and σ_o. (Make sure that the corresponding encoding e is polynomial).
2. Construct a program Π such that for every $X \in dom(P)$ restrictions of answer sets of $\Pi \cup X$ on $lit(\sigma_o)$ correspond to $P(X)$.

If we are successful we say that Π is a *uniform logic programming solution* of P. It is natural to characterize the class of problems which can be solved by this method. First some notation: By FA-Prolog and FDA-Prolog we mean restrictions of A-Prolog and DA-Prolog to languages with finite Herbrand universes.

Theorem 3. *(Complexity results)*

1. The problem of deciding whether a program of FA-Prolog has an answer set is NP-complete [55].

2. A decision problem P can be solved by a uniform program of FA-Prolog iff it is in the class NP [75]
3. A decision problem P can be solved by a uniform program of FDA-Prolog iff it is in the complexity class Σ_2^P [13]

It is interesting to note that the problem from example 3 is Σ_2^P complete [19] and therefore the use of disjunction is essential. The above theorem shows that for decision problems we have a complete answer. The problem remains open for arbitrary search problems but it is clear that both, FA-Prolog and FDA-Prolog can capture a rather large number of such problems. For instance, according to [56], FA-Prolog can solve "all search problems , whose associated decision problems are in NP, that we considered so far".

5.2   A-Prolog with Sets

In this section we introduce a new extension, ASET-Prolog, of A-Prolog which simplifies representation and reasoning with sets of terms and with functions from such sets to natural numbers. The language does not yet have a complete implementation. Fortunately, its semantics is very close to the semantics of *choice* rules of [66], which makes it possible to run a large numbers of programs of ASET-Prolog using *smodels* reasoning system. (In fact, ASET-Prolog is an attempt to simplify and slightly generalize the original work of [66]). We start by defining the syntax and semantics of the language. To simplify the presentation we limit ourself to a language $\mathcal{L}$ without $\neg$ and assume that $\mathcal{L}$ has a finite Herbrand universe. Atoms of $\mathcal{L}$ will be called L-*atoms*. We expand $\mathcal{L}$ by two new types of atoms:

1. An *s-atom* is a statement of the form

$$\{\overline{x} \ : \ p(\overline{x})\} \subseteq \{\overline{x} \ : \ q(\overline{x})\}. \tag{11}$$

   where $\overline{x}$ is the list of all free variables occurring in the corresponding atom. The statement says that p is a subset of q.
2. An *f-atom* is a statement of the form

$$|\{\overline{x} \ : p(\overline{x})\}| \leq n \text{ or } |\{\overline{x} \ : p(\overline{x})\}| = n \tag{12}$$

   where $| \ |$ denotes the cardinality of the corresponding set. (The general description of the language allows other functions on sets except the $| \ |$.)

Let us denote the new language by S.

A *program* of ASET-Prolog (parameterized by a background language S) is a collection of rules of the form

$$l_0 \leftarrow l_1, \ldots, l_m, not\ l_{m+1}, \ldots, not\ l_n \tag{13}$$

where $l_1, \ldots, l_n$ are atoms of S and l_0 is either L-atom or s-atom of S.

To give a semantics of ASET-Prolog we generalize the notion of stable model of A-Prolog. First we need the following terminology.

Let S be a set of ground atoms of $\mathcal{S}$.

1. An L-atom atom l of $\mathcal{S}$ is true in S if $l \in S$.
2. An s-atom (11) is true in S if for any sequence $\bar{t}$ of ground terms of $\mathcal{L}$, either $p(\bar{t}) \notin S$ or $q(\bar{t}) \in S$.
3. An f-atom (12) is true in S if cardinality of the set $\{\bar{t} \; : \; p(\bar{t}) \in S\}$ satisfies the corresponding condition.

We say that S *satisfies* an atom l of S and write $S \models l$ if l is true in S; *not* l is satisfied by S if $S \not\models l$. As in the definition of stable models, we consider rules of a program Π with free variables to be schemas denoting the set of ground instances of these rules (i.e., the result of replacing free variables of Π by terms of $\mathcal{S}$). Unless stated otherwise Π is assumed to be grounded.

Definition 7. *(Stable models of ASET-Prolog)*
Let S be a collection of ground atoms. By $se(\Pi, S)$ (read as "the set elimination of Π with respect to S") we mean the program obtained from Π by:

1. removing from Π all the rules whose bodies contain s-atoms or f-atoms not satisfied by S;
2. removing all remaining s-atoms and f-atoms from the bodies of the rules;
3. replacing rules of the form $l \leftarrow \Gamma$ where l is an s-atoms not satisfied by S by rules $\leftarrow \Gamma$;
4. Replacing the remaining rules of the form: $\{\bar{x} \; : \; p(\bar{x})\} \subseteq \{\bar{x} \; : \; q(\bar{x})\} \leftarrow \Gamma$ by the rules $p(\bar{t}) \leftarrow \Gamma$ for each $p(\bar{t})$ from S.

We say that S is a stable model of Π if it is a stable model of $se(\Pi, S)$.

Let us now give several examples of the use of ASET-Prolog.

Example 4. (Computing the cardinality of sets)
We are given a complete list of statements of the form $located\_in(C, S)$ (read as "a city C is located in a state S"), e.g.,

$$located\_in(austin, tx).$$
$$located\_in(lubbock, tx).$$
$$located\_in(sacramento, ca).$$

Suppose that we need to define a relation $num(N, S)$ which holds iff N is the number of cities located in a state S. This can be done with the following rule of ASET-Prolog:

$$num(N, S) \leftarrow |\{X : located\_in(X, S)\}| = N.$$

After grounding, this rule will turn into the rules

$$num(i, tx) \leftarrow |\{X : located\_in(X, tx)\}| = i.$$
$$num(i, ca) \leftarrow |\{X : located\_in(X, ca)\}| = i.$$

where i's are integers from 0 to some maximum integer m. (Notice, that the variable X is bounded and hence it is not replaced by any term.) It is easy to check that the program has exactly one stable model A and that A contains the above facts and the atoms $num(2, tx)$ and $num(1, ca)$.

The next three examples are taken from [66]. They demonstrates the use of rules of the form:

$$\{\bar{x} \;:\; p(\bar{x})\} \subseteq \{\bar{x} \;:\; q(\bar{x}\} \leftarrow \Gamma \qquad (14)$$

with s-atoms in the heads. Rules of this form are called *selection* rules and are read as follows: " If Γ holds in a set S of beliefs of an agent then any subset of the set $\{\bar{t} \;:\; q(\bar{t}) \in S\}$ may be the extent of $p(\bar{x})$ in S"[6]. The next example demonstrates the use of selection rules:

Example 5. (Cliques)
Suppose we have a graph defined by the set of facts of the form $node(X)$ and $edge(X, Y)$

$$node(a).$$
$$node(b).$$
$$node(c).$$

$$edge(a, b)$$

We would like to define a relation $clique(X)$, i.e. to write a program Π of ASET-Prolog such that for any graph G represented as above, the set of nodes N is a clique of G iff there is a stable model S of $\Pi \cup G$ such that an atom $clique(t) \in S$ iff $t \in N$. Recall that a set of nodes of a graph G is called a clique if *every two nodes from this set are connected by an edge of G*. This can be easily expressed by the following rules:

$$\{X \;:\; clique(X)\} \subseteq \{X \;:\; node(X)\}.$$
$$\leftarrow clique(X), clique(Y), X \neq Y, not\ edge(X, Y).$$

Answer sets of the program consisting of graph G combined with the first rule correspond to arbitrary subsets of nodes of G. Adding the constraint eliminates those which do not form a clique.

The next example demonstrates how selection rules combined with cardinality constraints can allow selection of subsets of given cardinality.

Example 6. (Coloring the graphs)
Suppose we have a graph G defined by the set of facts of the form $node(X)$ and $edge(X, Y)$ as in example (5) together with a set C of colors

$$color(red).\quad color(green).\quad \ldots$$

We would like to *color the graph in a way which guarantees that no two neighboring nodes have the same color*. To this end we introduce a program Π defining a relation $colored(Node, Color)$ such that every coloring will be represented by the atoms of the form $colored(n, c)$ from some stable model of $\Pi \cup G \cup C$. Program

---

[6] By the extent of $p(\bar{x})$ in S we mean the set of ground terms such that $p(\bar{t}) \in S$.

Π will consist of the following rules:

$$\{C : colored(X, C)\} \subseteq \{C : color(C)\} \leftarrow node(X).$$
$$\leftarrow |\{C : colored(X, C)\}| = N,$$
$$N \neq 1.$$
$$\leftarrow colored(X, C),$$
$$colored(Y, C),$$
$$edge(X, Y).$$

The first rule allows the selection of arbitrary sets of colors for a given node X. The second limits the selection to one color per node. The third eliminates the selections which color neighbors by the same color. The selections left after this pruning correspond to acceptable colorings.

The next example illustrates the use of s-atoms in the body of rules.

Example 7. (Checking the course prerequisites)
Suppose that we have a record of courses passed by a student, s, given by a collection of atoms

$$passed(s, c1). \quad passed(s, c2). \quad passed(s, c3).$$

and a list of prerequisites for each class

$$prereq(c1, c4). \quad prereq(c2, c4). \quad prereq(c4, c5).$$

Our goal is to express the following rule: *A student S is allowed to take class C if he passed all the prerequisites for C and didn't pass C yet.* This rule can be written as

$$(a) \quad can\_take(S, C) \leftarrow \{X : prereq(X, C)\} \subseteq \{X : passed(S, X)\},$$
$$not \ passed(S, C).$$

It is easy to check that the stable model M of this program, Π, consists of the above facts and an atom $can\_take(s, c4)$. Indeed, after grounding the above rule will turn into rules:

$$can\_take(s, c_i) \leftarrow \{X : prereq(X, c_i)\} \subseteq \{X : passed(s, X)\},$$
$$not \ passed(s, c_i).$$

where $0 \leq i \leq 5$. (We are of course assuming that the variables are properly typed). The s-literals in the bodies of the rules are satisfied for i = 1,2,3, and 4 and are not satisfied for i=5. So $se(\Pi, M)$ consists of the facts and rules

$$can\_take(s, c_i) \leftarrow not \ passed(s, c_i).$$

where i = 1..4. It is easy to check that M is the only stable model of this program.

A careful reader probably noticed that the same example could be formalized in A-Prolog without the use of s-atoms. This can be done by introducing a new predicate symbol $not\_ready(S, C)$ read as "a student S is not yet ready to take a class C" and by replacing rule (a) above by the following two rules:

$$(b) \quad not\_ready(S, C) \leftarrow prereq(X, C)$$
$$not\ passed(S, X).$$

$$(c) \quad can\_take(S, C) \leftarrow not\ not\_ready(S, C)$$
$$not\ passed(S, C).$$

The following proposition shows that this is not an accident. First we need some notation. Let Π be a logic program over signature σ containing a rule

$$l_0 \leftarrow \Gamma_1, \{X : p(X)\} \subseteq \{X : q(X)\}, \Gamma_2 \qquad (15)$$

By Π^* we denote the program obtained from Π by replacing rule 15 by the rules

$$d \leftarrow p(X), not\ q(X). \qquad (16)$$

$$l_0 \leftarrow \Gamma_1, not\ d, \Gamma_2 \qquad (17)$$

where d is formed with a predicate symbol not belonging to σ.

Proposition 4.
1. For any stable model S of Π there is a stable model S^* of Π^* such that $S = S^* \cap lit(\sigma)$.

2. If S^* is a stable model of Π^* then $S = S^* \cap lit(\sigma)$ is a stable model of Π.

This proposition shows that allowing s-atoms in the bodies of rules does not add to the expressive power of A-Prolog. It allows however a more compact representation with fewer predicate symbols. To some extent the above proposition can help to explain why stable models of programs of A-Prolog with s-terms in the bodies of their rules do not have the anti-chain property enjoyed by stable models of "pure" A-Prolog. The following example demonstrates that this is indeed the case.

Example 8. Consider the following program Π

$$p(a).$$

$$q(a) \leftarrow \{X : p(X)\} \subseteq \{X : q(X)\}.$$

which has two models, $S_1 = \{p(a)\}$ and $S_2 = \{p(a), q(a)\}$. (Since $S_1 \subset S_2$ the set of models of Π does not form an anti-chain.) It is easy to check however that the models of Π^* are $\{p(a), d\}$ and $\{p(a), q(a)\}$ which, thanks to the presence of a new atom d, do form an anti-chain.

As mentioned before selection rules of ASET-Prolog are closely related to choice rules of [66,76] which have a form

$$m\{p(\overline{X}) : q(\overline{X})\}n \leftarrow \Gamma \qquad (18)$$

Even though the general semantics of choice rules is somewhat complicated, sometimes such rules can be viewed as a shorthand for several rules of ASET-Prolog. More precisely, let us consider a program Π containing rule (18) and assume that no other rule of Π contains p in the head. Let Π^{++} be a program obtained from Π by replacing rule (18) by rules:

$$\{X \ : \ p(X)\} \subseteq \{X \ : \ q(X)\} \leftarrow \ \Gamma$$
$$\leftarrow n < |\{X \ : \ p(X)\}|.$$
$$\leftarrow |\{X \ : \ p(X)\}| < m$$

Proposition 5. Let Π and Π^{++} be as above. Then S is a stable model of Π in the sense of [66] iff S is a stable model of ASET-Prolog program Π^{++}.

(Proofs of both propositions will appear in the forthcoming paper on ASET-Prolog.)

6   Reasoning in Dynamic Domains

Let us now consider domains containing agents capable of performing actions and reasoning about their effects. Such domains are often called *dynamic domains* or *dynamic systems*. We will base their description on the formalism of *action languages* [35], which can be thought of as formal models of the part of the natural language that are used for describing the behavior of dynamic domains. A theory in an action language normally consists of an action description and a history description [9], [51]. The former contains the knowledge about effects of actions, the latter consists of observations of an agent. Some discussion of architecture of autonomous agents build on action languages and A-Prolog can be found in [11]

6.1   Specifying Effects of Actions

An *action description language* contains propositions which describe the effects of actions on states of the system modeled by sets of *fluents* – statements whose truth depends on time. Fluent f is true in a state σ iff $f \in \sigma$. Mathematically, an *action description* – a collection of statements in an action description language – defines a *transition system* with nodes corresponding to possible states and arcs labeled by actions from the given domain. An arc (σ_1, a, σ_2) indicates that execution of an action a in state σ_1 may result in the domain moving to the state σ_2. We call an action description *deterministic* if for any state σ_1 and action a there is at most one such successor state σ_2. By a *path* of a transition system T we mean a sequence $\sigma_0, a^1, \sigma_1, \ldots, a^n, \sigma_n$ such that for any $1 \le i < n$, $(\sigma_i, a^{i+1}, \sigma_{i+1})$ is an arc of T; σ_0 and σ_n are called initial and final states of the path respectively. Due to the size of the diagram, the problem of finding its concise specification is not trivial and has been a subject of research for some time. Its solution requires the good understanding of the nature of causal effects of actions in the presence of complex interrelations between fluents. An additional level of complexity is added by the need to specify what is not changed by actions. The latter, known as the *frame problem*, is often reduced to the problem of finding a concise and accurate representation of the inertia axiom – a default which says that *things normally stay as they are* [60]. The search for such a representation substantially influenced AI research during the last twenty years. An interesting account of history of this research together with some possible solutions can be found in [74]. In this paper we limit our attention to an action description language $\mathcal{B}$ [35] which signature Σ consist of two disjoint, non-empty sets of symbols: the set $\mathbf{F}$ of fluents and the set $\mathbf{A}$ of *elementary actions*. A set $\{a_1, \ldots, a_n\}$ of elementary actions is called a *compound* action. It is interpreted as a collection of elementary actions performed simultaneously. By actions we mean both elementary and compound actions. By *fluent literals* we mean fluents and their negations. By $\bar{l}$ we denote the fluent literal complementary to l. A set S of fluent literals is called *complete* if, for any $f \in \mathbf{F}$, $f \in S$ or $\neg f \in S$. An action description of $\mathcal{B}(\Sigma)$ is a collection of propositions of the form

1. $causes(a_e, l_0, [l_1, \ldots, l_n])$,
2. $caused(l_0, [l_1, \ldots, l_n])$, and
3. $impossible\_if(a, [l_1, \ldots, l_n])$

where a_e and a are elementary and arbitrary actions respectively and $l_0, \ldots, l_n$ are fluent literals from Σ. The first proposition says that, if the elementary action a_e were to be executed in a situation in which $l_1, \ldots, l_n$ hold, the fluent literal l_0 will be caused to hold in the resulting situation. Such propositions are called *dynamic causal laws*. (The restriction on a_e being elementary is not essential and can be lifted. We require it to simplify the presentation). The second proposition, called a *static causal law*, says that, in an arbitrary situation, the truth of fluent literals, $l_1, \ldots, l_n$ is sufficient to cause the truth of l_0. The last proposition says that action a cannot be performed in any situation in which $l_1, \ldots, l_n$ hold. Notice that here a can be compound, e.g. $impossible\_if(\{a_1, a_2\}, [\,])$ means that elementary actions a_1 and a_2 cannot be performed concurrently. To define the transition diagram, T, given by an action description $\mathcal{A}$ of $\mathcal{B}$ we use the following terminology and notation. A set S of fluent literals is closed under a set Z of static causal laws if S includes the head, l_0, of every static causal law such that $\{l_1, \ldots, l_n\} \subseteq S$. The set $Cn_Z(S)$ of *consequences* of S under Z is the smallest set of fluent literals that contains S and is closed under Z. $E(a_e, \sigma)$ stands for the set of all fluent literals l_0 for which there is a dynamic causal law $causes(a_e, l_0, [l_1, \ldots, l_n])$ in $\mathcal{A}$ such that $[l_1, \ldots, l_n] \subseteq \sigma$. $E(a, \sigma) = \bigcup_{a_e \in a} E(a_e, \sigma)$. The transition system $T = \langle \mathcal{S}, \mathcal{R} \rangle$ *described* by an action description $\mathcal{A}$ is defined as follows:

1. $\mathcal{S}$ is the collection of all complete and consistent sets of fluent literals of Σ closed under the static laws of $\mathcal{A}$,
2. $\mathcal{R}$ is the set of all triples (σ, a, σ') such that $\mathcal{A}$ does not contain a proposition of the form $impossible\_if(a, [l_1, \ldots, l_n])$ such that $[l_1, \ldots, l_n] \subseteq \sigma$ and

$$\sigma' = Cn_Z(E(a, \sigma) \cup (\sigma \cap \sigma')) \tag{19}$$

where Z is the set of all static causal laws of $\mathcal{A}$. The argument of $Cn(Z)$ in (19) is the union of the set $E(a,\sigma)$ of the "direct effects" of a with the set $\sigma \cap \sigma'$ of facts that are "preserved by inertia". The application of $Cn(Z)$ adds the "indirect effects" to this union.

The above definition is from [57] and is the product of a long investigation of the nature of causality. (See for instance, [49,81].) The following theorem (a version of the result from [83]) shows the remarkable relationship between causality and beliefs of rational agents as captured by the notion of answer sets of logic programs. First we need some terminology. We start by describing an encoding τ of causal laws of $\mathcal{B}$ into a program of A-Prolog suitable for execution by *smodels*:

1. $\tau(causes(a, l_0, [l_1 \ldots l_n]))$ is the collection of atoms $d\_law(d)$, $head(d, l_0)$, $action(d, a)$ and $prec(d, i, l_i)$ for $1 \leq i \leq n$, and $prec(d, m+1, nil)$ (where d is a new term used to name the corresponding law.)
2. $\tau(caused(l_0, [l_1 \ldots l_n]))$ is the collection of atoms $s\_law(d)$, $head(d, l_0)$, $prec(d, i, l_i)$ for $1 \leq i \leq n$, and $prec(d, m+1, nil)$.
3. $\tau(impossible\_if([a_1, \ldots, a_k], [l_1 \ldots l_n]))$ is a constraint

$$\leftarrow h(l_1, T), \ldots, h(l_n, T), occurs(a_1, T), \ldots, occurs(a_k, T).$$

Here T ranges over integers, $occurs(a, t)$ says that action a occurred at moment t, and $h(l, t)$ means that fluent literal l holds at t. Finally, for any action description $\mathcal{A}$

$$\tau(\mathcal{A}) = \{\tau(law) : law \in \mathcal{A}\} \tag{20}$$

$$\phi(\mathcal{A}) = \tau(\mathcal{A}) \cup \Pi(1) \tag{21}$$

where $\Pi(1)$ is an instance of the following program

$$\Pi(N) \begin{cases} \begin{array}{ll} 1.\ h(L, T') & \leftarrow d\_law(D), \\ & head(D, L), \\ & action(D, A), \\ & occurs(A, T), \\ & prec\_h(D, T). \\ 2.\ h(L, T) & \leftarrow s\_law(D), \\ & head(D, L), \\ & prec\_h(D, T). \\ 3.\ all\_h(D, K, T) & \leftarrow prec(D, K.nil). \\ 4.\ all\_h(D, K, T) & \leftarrow prec(D, K, P), \\ & h(P, T), \\ & all\_h(D, K' \\ 5.\ prec\_h(D, T) & \leftarrow all\_h(D, 1, T). \\ 6.\ h(L, T') & \leftarrow h(L, T), \\ & not\ h(\overline{L}, T'). \end{array} \end{cases}$$

Here D, A, L are variables for the names of laws, actions, and fluent literals respectively, T, T' are consecutive time points from interval $[0, N]$ and K, K' stand for consecutive integers. The first two rules describe the meaning of dynamic and static causal laws, rules (3), (4), (5) define what it means for preconditions of law D to succeed, and rule (6) represents the inertia axiom from [60].

Theorem 4. For any action a and any state σ, a state σ' is a successor state of a on σ iff there is an answer set S of

$$\phi(\mathcal{A}) \cup \{h(l,0) : l \in \sigma\} \cup \{occurs\_at(a_i, 0) : a_i \in a\}$$

such that, $\sigma' = \{l : h(l,1) \in S\}$.

The theorem establishes a close relationship between the notion of causality and the notion of rational beliefs of an agent. The systematic study of the relationship between entailment in action theories and in A-Prolog started in [32], where the authors formulated the problem and obtained some preliminary results. For more advanced result see [83].

6.2   Planning in A-Prolog

Now we will show how the above theory can be applied to classical planning problems of AI [80], [24], [53],[58]. Let us consider for instance the *blocks world* problem which can be found in most introductory AI textbooks:

The domain consists of a set of cubic blocks sitting on a table. The blocks can be stacked, but only one block can fit directly on top of another. A robot arm can pick up a block and move it to another position, either on the table or on top of another block. The arm can only pick up one block at a time, so it cannot pick up a block that has another one on it. The goal will always be to build one or more stacks of blocks, specified in terms of what blocks are on top of what other blocks.

To build a formal representation of the domain let us introduce names $b_1, \ldots, b_n$ for blocks and t for the table. We use a fluent $on(B, L)$ to indicate that block B is on location L and an action $move(B, L)$ which moves block B to a position L. The corresponding types will be described as follows:

$$
\begin{aligned}
block(b_1). \quad &\ldots \quad block(b_n). \\
loc(t). & \\
loc(L) \leftarrow \; &block(L). \\
fluent(on(B, L)) \leftarrow \; &block(B), \quad \text{The executability conditions for action } move \\
&loc(L). \\
act(move(B, L)) \leftarrow \; &block(B), \\
&loc(L).
\end{aligned}
$$

are defined by the rules:

$impossible\_if(move(B, L), [on(A, B)]) \leftarrow block(B),$
$\qquad\qquad\qquad\qquad\qquad\qquad\qquad block(A),$
$\qquad\qquad\qquad\qquad\qquad\qquad\qquad loc(L).$
$impossible\_if(move(B_1, B_2), [on(A, B_2)]) \leftarrow block(B_1),$
$\qquad\qquad\qquad\qquad\qquad\qquad\qquad block(B_2),$
$\qquad\qquad\qquad\qquad\qquad\qquad\qquad block(A).$
$impossible\_if(move(B, B), [\,]).$

The action's direct effect is represented by a dynamic causal law,

$causes(move(B, L), on(B, L), [\,]) \leftarrow block(B),$
$\qquad\qquad\qquad\qquad\qquad\qquad loc(L).$

The static causal law,

$caused(\neg on(B, L_2), [on(B, L_1)]) \leftarrow block(B),$
$\qquad\qquad\qquad\qquad\qquad\qquad loc(L_1),$
$\qquad\qquad\qquad\qquad\qquad\qquad loc(L_2).$
$\qquad\qquad\qquad\qquad\qquad\qquad L_1 \neq L_2$

guarantees the uniqueness of a block's location. It is easy to see that the resulting program has a unique answer set, S. Atoms from S, formed by predicates $impossible\_if$, $causes$, and $caused$, form an action description, $\mathcal{A}_b$, of action language $\mathcal{B}$ which defines the transition diagram, T, of the blocks wold domain.

For simplicity we restrict ourselves to a planning problem of the following type:

Given an initial node, σ_0, of the diagram, a non-negative integer n, and a collection Σ_f of goal nodes find a path of length less than or equal to n from σ_0 to one of the elements of Σ_f. The path determines the agent's plan – a sequence of actions it needs to perform in order to achieve its goal. We will refer to such plans as solutions of the planning problem.

We assume that an initial state, σ_0, is defined by a collection I of formulas of the form $h(l, 0)$, and that the goal is given by a collection, G, of statements $g(l_0), \dots, g(l_k)$ which specifies what fluent literals must be true in a goal state. For instance, I may be

$h(on(a, t), 0). \quad h(on(b, a), 0). \quad h(on(c, t), 0).$

and the goal may be

$g(on(b, t)). \quad g(on(a, b)).$

A planning problem defined in this way will be denoted by $(\mathcal{A}, I, G, n)$.

The actual planning is done with the help of a program we call a *planning module*. The simplest planning module, PM_0, consists of the *goal constraints* and the *possible plans generator*. Constraints may be defined as follows:

$$fails(T) \quad \leftarrow 0 \leq T \leq n,$$
$$g(F),$$
$$not\ h(F,T)$$

$$succeeds(T) \leftarrow 0 \leq T \leq n$$
$$not\ fails(T).$$
$$succeeds \quad \leftarrow succeeds(T).$$

$$\leftarrow not\ succeeds.$$

$fails(T)$ holds if at least one of the fluent literals from the goal does not hold at time T; $succeeds$ holds when all the goals are satisfied at some moment of time $0 \leq T \leq n$. The last constraint requires the goal to be achievable in at most n steps.

The generator of the planning module may consists of the rules

$$0\{occurs(A,T) : act(A)\}1 \leftarrow 0 \leq T < n.$$

$$act\_occur(T) \leftarrow occurs(A,T).$$

$$\leftarrow succeeds(T),$$
$$0 \leq T_1 < T$$
$$not\ act\_occurs(T).$$
$$\leftarrow succeeds(T),$$
$$T < T_1 \leq n$$
$$act\_occurs(T).$$

We use the choice rule of [66] to guarantee that every answer set of a program containing the planning module will contain at most one occurrence of the statement $occurs(a,t)$ for every moment $0 \leq t < n$. The two constraints guarantee that at least one action occurs at any moment of time before the goal is achieved and that no actions occur afterwards. (As mentioned in section 5, the choice rule above can be viewed as a shorthand for a selection rule of ASET-Prolog or for a collection of rules of A-Prolog.) We will also need the rules

$$h(F,0) \quad \leftarrow not\ h(\neg F,0).$$
$$h(\neg F,0) \leftarrow not\ h(F,0).$$

sometimes called the *awareness axioms*, which say that for every fluent f, either f or $\neg f$ should be included in the beliefs of a reasoning agent. (If the agent's information about the initial situation is complete this axiom can be omitted).

Let $P = (\mathcal{A}, I, G, n)$ be a planning problem and consider

$$Plan(P) = \tau(\mathcal{A}) \cup I \cup G \cup PM_0 \tag{22}$$

Using theorem 4 it is not difficult to show that

Proposition 6. A sequence $a_1, \ldots, a_k$ $(0 \leq k \leq n)$ is a solution of a planning problem P with a deterministic action description iff there is an answer set S of $Plan(P)$ such that

1. for any $0 < i \le k$, $occurs(a_i, i-1) \in S$;
2. S contains no other atoms formed by $occurs$.

The proposition reduces the process of finding solutions of planning problems to that of finding answer sets of programs of A-Prolog. To apply it to our blocks world we need to show that the corresponding action description, $\mathcal{A}_b$, is deterministic. This immediately follows from the following proposition:

Proposition 7. If every static causal law of an action description $\mathcal{A}$ of $\mathcal{B}$ has at most one precondition then $\mathcal{A}$ is deterministic.

A-Prolog's inference engines like *smodels*, *dlv*, and *ccalc* are sufficiently powerful to make this method work for comparatively large applications. For instance in [7] the authors used this method for the development of a decision support system to be used by the flight controllers of the space shuttle. One of the system's goals was to find the emergency plans for performing various shuttle maneuvers in the presence of multiple failures of the system's equipment. Efficiency wise, performance of the planner was more than satisfactory (in most cases the plans were found in a matter of seconds.) This is especially encouraging since performance of A-Prolog satisfiability solvers is improving at a very high rate.

The system is implemented on top of *smodels*. It includes a knowledge base containing information about the relevant parts of the shuttle and its maneuvers, and of the actions available to the controllers. The effects of actions are given in an action description language $\mathcal{B}$. The corresponding action description contains a large number of static causal laws. It is interesting to notice that these laws, which are not available in more traditional planning languages like [71], played a very important role in the system design. We are not sure that a concise, elaboration tolerant, and clear description of the effects of controller's actions could be achieved without their use.

The system's planning module is based on the same generate and test idea as PM_0 but contains a number of constraints prohibiting certain combinations of actions. Constraints of this sort substantially improve the quality of plans as well as the efficiency of the planner [37].

To illustrate the idea let us show how this type of heuristic information can be used in blocks world planning. The rules, R, below express "do not destroy a good tower" heuristic suggested in [6]. It ensures that the moves of blocks which satisfy the planner's goal are immediately cut from its search space.

$$h(ok(t), T) \leftarrow 0 \le T \le n.$$
$$h(ok(B_1), T) \leftarrow 0 \le T \le n,$$
$$g(on(B_1, B_2)),$$
$$h(on(B_1, B_2), T),$$
$$h(ok(B_2), T).$$

$$\leftarrow occurs(move(B, L), T),$$
$$h(ok(B), T).$$

The first two rules define a fluent $ok(B)$ which holds at moment T if all the blocks of the tower with the top B have positions specified by the planner's goal. The last rule prohibits movements of the ok blocks. The planning module PM_1, consisting of PM_0 combined with the R rules above, returns better plans than PM_0 and is substantially more efficient. In a sense R-rules can be viewed as a declarative specification of control information limiting the search space of the planner. (It is interesting to note that in [6] a similar effect was achieved by expanding the original action description language with a variant of temporal logic. The use of A-Prolog makes this unnecessary).

7 Acknowledgments

There are many people who directly or indirectly contributed to this publication by creating its subject matter and influencing its author, and I would like to thank all of them. Special thanks however are due to Bob Kowalski. If it were not for his paper [42] I (almost accidentally) read in the early eighties I probably would not get interested in logic programming and would have never had a chance to become familiar with this wonderful field of research. For that, and for the opportunity to learn from Bob's work during the last twenty years, I am very grateful.

References

1. J. Alferes and L. Pereira. *Reasoning With Logic Programming*. Springer Verlag, 1996.
2. K. Apt, H. Blair, and A. Walker. Towards a theory of declarative knowledge. In J. Minker, editor, *Foundations of Deductive Databases and Logic Programming*, pages 89–148. Morgan Kaufmann, San Mateo, CA., 1988.
3. K. Apt and M. Bezem. Acyclic programs. *New Generation Computing*, 9(3,4):335–365, 1991.
4. K. Apt and A. Pellegrini. On the occur-check free logic programs. *ACM Transaction on Programming Languages and Systems*, 16(3):687–726, 1994.
5. C. Aravindan, J. Dix, and I. Niemela. Dislop: A research project on disjunctive logic programming, *AI communications*, 10 (3/4):151-165.
6. F. Bacchus and F. Kabanza. Planning for Temporally Extended Goals. *Annals of Mathematics and Artificial Intelligence*, 22:1-2, 5-27.
7. M. Balduccini, M. Barry, M. Gelfond, M. Nogueira, and R. Watson An A-Prolog decision support system for the Space Shuttle. *Lecture Notes in Computer Science - Proceedings of Practical Aspects of Declarative Languages'01*, (2001), 1990:169–183
8. C. Baral and M. Gelfond. Logic programming and knowledge representation. *Journal of Logic Programming*, 19,20:73–148, 1994.
9. C. Baral, M. Gelfond, and A. Provetti. Representing Actions: Laws, Observations and Hypothesis. *Journal of Logic Programming*, 31(1-3):201–243, May 1997.
10. C. Baral, M. Gelfond, and O. Kosheleva. Expanding queries to incomplete databases by interpolating general logic programs. *Journal of Logic Programming*, vol. 35, pp 195-230, 1998.

11. C. Baral and M. Gelfond. Reasoning agents in dynamic domains. In J Minker, editor, *Logic Based AI*. pp. 257–279, Kluwer, 2000.
12. A. Bondarenko, P.M. Dung, R. Kowalski, F. Toni, An abstract, argumentation-theoretic approach to default reasoning. *Artificial Intelligence* 93(1-2) pages 63-101, 1997.
13. M. Cadoli, T. Eiter, and G. Gottlob. Default logic as a query language. IEEE Transactions on Knowledge and Data Engineering, 9(3), pages 448–463, 1997.
14. W. Chen, T. Swift, and D. Warren. Efficient top-down computation of queries under the well-founded semantics. *Journal of Logic Programming*, 24(3):161–201, 1995.
15. P. Cholewinski. Stratified Default Logic. *In Computer Science Logic*, Springer LNCS 933, pages 456–470, 1995.
16. P. Cholewinski. Reasoning with Stratified Default Theories. *In Proc. of 3rd Int'l Conf. on Logic Programming and Nonmonotonic Reasoning*, pages 273–286, 1995.
17. P. Cholewinski, W. Marek, and M. Truszczyński. Default Reasoning System DeReS. In *Int'l Conf. on Principles of Knowledge Representation and Reasoning*, 518-528. Morgan Kauffman, 1996.
18. A. Colmerauer, H. Kanoui, R. Pasero, and P. Russel. Un systeme de communication homme-machine en francais. Technical report, Groupe de Intelligence Artificielle Universitae de Aix-Marseille, 1973.
19. T. Eiter and G. Gottlob. Complexity aspects of various semantics for disjunctive databases. In Proc. of PODS-93, pages 158-167, 1993.
20. T. Eiter, N. Leone, C. Mateis., G. Pfeifer and F. Scarcello. A deductive system for nonmonotonic reasoning, Procs of the LPNMR'97, 363–373, 1997
21. T. Eiter, W. Faber, N. Leone. Declarative problem solving in DLV. In J Minker, editor, *Logic Based AI*, 79–103 Kluwer, 2000.
22. K. Clark. Negation as failure. In Herve Gallaire and J. Minker, editors, *Logic and Data Bases*, pages 293–322. Plenum Press, New York, 1978.
23. J. Delgrande and T. Schaub. Compiling reasoning with and about preferences into default logic. In *Proc. of IJCAI 97*, 168–174, 1997.
24. Y. Dimopoulos, B. Nebel, and J. Koehler. Encoding planning problems in non-monotonic logic programs. *Lecture Notes in Artificial Intelligence - Recent Advances in AI Planning, Proc. of the 4th European Conference on Planning, ECP'97*, 1348:169–181, 1997
25. E. Erdem, V. Lifschitz, and M. Wong. Wire routing and satisfiability planning. *Proc. of CL-2000*, 822-836, 2000.
26. François Fages. Consistency of Clark's completion and existence of stable models. *Journal of Methods of Logic in Computer Science*, 1(1):51–60, 1994.
27. M. Fitting. A Kripke-Kleene semantics for logic programs. *Journal of Logic Programming*, 2(4):295–312, 1985.
28. M. Gelfond and A. Gabaldon. Building a knowledge base: an example. *Annals of mathematics and artificial Intelligence*, 25:165–199.
29. M. Gelfond. On stratified autoepistemic theories. In *Proc. AAAI-87*, pages 207–211, 1987.
30. M. Gelfond and V. Lifschitz. The stable model semantics for logic programming. In R. Kowalski and K. Bowen, editors, *Logic Programming: Proc. of the Fifth Int'l Conf. and Symp.*, pages 1070–1080, 1988.
31. M. Gelfond and V. Lifschitz. Classical negation in logic programs and disjunctive databases. *New Generation Computing*, pages 365–387, 1991.
32. M. Gelfond and V. Lifschitz. Representing Actions and Change by Logic Programs. *Journal of Logic Programming*, 17:301–323.

33. M. Gelfond, V Lifschitz, H. Przymusinska, and M. Truszczynski. Disjunctive defaults. In J. Allen, R. Fikes, and E. Sandewall, editors, *Principles of Knowledge Representation and Reasoning: Proc. of the Second Int'l Conf.*, pages 230–237, 1991.

34. M. Gelfond, H. Przymusinska. On Consistency and Completeness of Autoepistemic Theories, Fundamenta Informaticae, vol. 16, Num. 1, pp. 59-92, 1992.

35. M. Gelfond and V. Lifschitz. Action Languages. *Electronic Transactions on Artificial Intelligence*, Vol. 2, 193-210, 1998 http://www.ep.liu.se/ej/etai/1998/007

36. M. Gelfond and T. Son.   Reasoning with prioritized defaults.   In J. Dix, L. M. Pereira, T. Przymusinski, editors, *Lecture Notes in Artificial Intelligence*, 1471, pp 164-224, 1998.

37. Y. Huang, H. Kautz and B. Selman. Control Knowledge in Planning: Benefits and Tradeoffs. *16th National Conference of Artificial Intelligence (AAAI'99)*, 511–517.

38. K. Inoue and C. Sakama.   Negation as Failure in the Head.   *Journal of Logic Programming*, 35(1):39-78, 1998.

39. C. Sakama and K. Inoue Prioritized Logic Programming and its Application to Commonsense Reasoning, *Artificial Intelligence* 123(1-2):185-222, Elsevier, 2000.

40. A. C. Kakas, R. Kowalski, F. Toni, The Role of Abduction in Logic Programming, *Handbook of Logic in Artificial Intelligence and Logic Programming 5*, pages 235-324, D.M. Gabbay, C.J. Hogger and J.A. Robinson eds., Oxford University Press (1998).

41. M. Kaminski. A note on the stable model semantics of logic programs. *Artificial Intelligence*, 96(2):467–479, 1997.

42. R. Kowalski. Predicate logic as a programming language. *Information Processing 74*, pages 569–574, 1974.

43. R. Kowalski. *Logic for Problem Solving*. North-Holland, 1979.

44. C. Koch and N. Leone Stable model checking made easy. In proc. of IJCAI'99, 1999.

45. K. Kunen.   Negation in logic programming.   *Journal of Logic Programming*, 4(4):289–308, 1987.

46. K. Kunen. Signed data dependencies in logic programs. *Journal of Logic Programming*, 7(3):231–245, 1989.

47. V. Lifschitz. Restricted Monotonicity. *In proc. of AAA-93*, pages 432–437, 1993

48. V. Lifschitz. Foundations of logic programming. In Gerhard Brewka, editor, *Principles of Knowledge Representation*, pages 69–128. CSLI Publications, 1996.

49. V. Lifschitz. On the logic of causal explanation. *Artificial Intelligence*, 96:451–465, 1997.

50. V. Lifschitz and H. Turner. Splitting a logic program. In Pascal Van Hentenryck, editor, *Proc. of the Eleventh Int'l Conf. on Logic Programming*, pages 23–38, 1994.

51. V. Lifschitz, Two components of an action language. *Annals of Math and AI*, 21(2-4):305–320, 1997.

52. V. Lifschitz and L. Tang and H. Turner, Nested expressions in logic programs. *Annals of Math and AI*, Vol. 25, pages 369-389, 1999

53. V. Lifschitz. Action languages, Answer Sets, and Planning. In *The Logic Programming Paradigm: a 25-Year Perspective*, 357–353, Spring-Verlag, 1999.

54. W. Marek and V.S. Subrahmanian. The relationship between logic program semantics and non–monotonic reasoning. In G. Levi and M. Martelli, editors, *Proc. of the Sixth Int'l Conf. on Logic Programming*, pages 600–617, 1989.

55. W. Marek, and M. Truszczyński. Autoepistemic Logic. *Journal of the ACM*, 38, pages 588–619, 1991.

56. W. Marek, and M. Truszczyński. Stable models and an alternative logic programming paradigm. In *The Logic Programming Paradigm: a 25-Year Perspective*, 375–398, Spring-Verlag. 1999.

57. N. McCain and H. Turner. Causal theories of action and change. In *Proc. of AAAI*, pages 460–465, 1997.

58. N. McCain and H. Turner. Satisfiability planning with causal theories. In *Proc. of KR*, pages 212–223, 1998.

59. J. McCarthy. Programs with common sense. In *Proc. of the Teddington Conference on the Mechanization of Thought Processes*, pages 75–91, London, 1959. Her Majesty's Stationery Office.

60. J. McCarthy and P. Hayes. Some philosophical problems from the standpoint of artificial intelligence. In B. Meltzer and D. Michie, editors, *Machine Intelligence*, volume 4, pages 463–502. Edinburgh University Press, Edinburgh, 1969.

61. J. Minker Overview of disjunctive logic programming. *Annals of mathematics and artificial Intelligence*, 12:1–24, 1994.

62. R. Moore. Semantical Considerations on Nonmonotonic Logic. *Artificial Intelligence*, 25(1):75–94, 1985.

63. D. Nelson. Constructible falsity. *Journal of Symbolic Logic*, 14:16–26, 1949.

64. I. Niemela. Logic Programming with stable model semantics as a constraint programming paradigm. In proceedings of the workshop on computational aspects of nonmonotonic reasoning, pp 72–79, Trento, Italy, 1998.

65. I. Niemela and P. Simons. Smodels – an implementation of the stable model and well-founded semantics for normal logic programs. In *Proc. 4th international conference on Logic programming and non-monotonic reasoning*, pages 420–429, 1997.

66. I. Niemela and P. Simons. Extending the Smodels system with cardinality and weight constraints. In J Minker, editor, *Logic Based AI*, pp. 491–522, Kluwer, 2000.

67. D. Pearce and G. Wagner. Reasoning with negative information 1 – strong negation in logic programming. Technical report, Gruppe fur Logic, Wissentheorie and Information, Freie Universitat Berlin, 1989.

68. D. Pearce. From here to there: Stable negation in logic programming. In D. Gabbay and H. Wansing, editors. *What is negation?*, Kluwer, 1999.

69. T. Przymusinski. Perfect model semantics. *In Proc. of Fifth Int'l Conf. and Symp.*, pages 1081–1096, 1988.

70. T. Przymusinski. Stable semantics for disjunctive programs. *New generation computing*, 9(3,4):401–425, 1991.

71. E. Pednault. ADL: Exploring the middle ground between STRIPS and the situation calculus. *In Proc. of KR89*, pages 324–332, 1987.

72. R. Reiter. On closed world data bases. In H. Gallaire and J. Minker, editors, *Logic and Data Bases*, pages 119–140. Plenum Press, New York, 1978.

73. R. Reiter. A logic for default reasoning. *Artificial Intelligence*, 13(1,2):81–132, 1980.

74. M. Shanahan. *Solving the frame problem: A mathematical investigation of the commonsense law of inertia.* MIT press, 1997.

75. J. Schlipf. The expressive powers of the logic programming semantics. *Journal of the Computer Systems and Science*, 51, pages 64–86, 1995.

76. Simons, P. Extending the stable model semantics with more expressive rules. In *5th International Conference, LPNMR'99*, 305–316.

77. T. Soininen and I. Niemela. Developing a declarative rule language for applications in program configuration. *In practical aspects of declarative languages*, LNCS 1551, pages 305-319, 1999.
78. D. Seipel and H. Thone. DisLog - A system for reasoning in disjunctive deductive databases. *In proc. of DAISD'94.*
79. K. Stroetman. A Completeness Result for SLDNF-Resolution. *Journal of Logic Programming*, 15:337–355, 1993.
80. V. Subrahmanian and C. Zaniolo. Relating stable models and AI planning domains. In L. Sterling, editor, *Proc. ICLP-95*, pages 233–247. MIT Press, 1995.
81. M. Thielscher. Ramification and causality. *Artificial Intelligence*, 89(1-2):317–364, 1997.
82. H. Turner. Signed logic programs. In *Proc. of the 1994 International Symposium on Logic Programming*, pages 61–75, 1994.
83. H. Turner. Representing actions in logic programs and default theories. *Journal of Logic Programming*, 31(1-3):245–298, May 1997.
84. H. Turner. Splitting a Default Theory, *In Proc. of AAAI-96*, pages 645–651, 1996.
85. M. van Emden and R. Kowalski. The semantics of predicate logic as a programming language. *Journal of the ACM.*, 23(4):733–742, 1976.
86. A. Van Gelder, K. Ross, and J. Schlipf. The well-founded semantics for general logic programs. *Journal of ACM*, 38(3):620–650, 1991.
87. G. Wagner. Logic programming with strong negation and inexact predicates. *Journal of Logic and Computation*, 1(6):835–861, 1991.

Some Alternative Formulations of the Event Calculus

Rob Miller[1] and Murray Shanahan[2]

[1] University College London, London WC1E 6BT, U.K.
rsm@ucl.ac.uk
http://www.ucl.ac.uk/~uczcrsm/
[2] Imperial College of Science, Technology and Medicine, London SW7 2BT, U.K.
m.shanahan@ic.ac.uk
http://www-ics.ee.ic.ac.uk/~mpsha/

Abstract. The Event Calculus is a narrative based formalism for reasoning about actions and change originally proposed in logic programming form by Kowalski and Sergot. In this paper we summarise how variants of the Event Calculus may be expressed as classical logic axiomatisations, and how under certain circumstances these theories may be reformulated as "action description language" domain descriptions using the Language $\mathcal{E}$. This enables the classical logic Event Calculus to inherit various provably correct automated reasoning procedures recently developed for $\mathcal{E}$.

1 Introduction

The "Event Calculus" was originally introduced by Bob Kowalski and Marek Sergot [33] as a logic programming framework for representing and reasoning about actions (or events) and their effects, especially in database applications. Since then many alternative formulations, implementations and applications have been developed. The Event Calculus has been reformulated in various logic programming forms (e.g. [11], [12], [21], [23], [29], [53], [58], [71], [72], [73],[74]), in classical logic (e.g. [62], [42], [43]), in modal logic (e.g. [2], [3], [4], [5], [6], [7]) and as an "action description language" ([21], [22]). In one form or another it has been extended and applied, for example, in the context of planning (e.g. [15], [8], [19], [44], [45], [63], [65], [20]), cognitive robotics (e.g. [60], [61], [65], [67]), abductive reasoning (e.g. [11], [44], [45], [71] and [72]), database updates (e.g. [29], [72]), accident report processing [35], legal reasoning [30], modelling continuous change and mathematical modelling (e.g. [42], [58], [71]), modelling and reasoning about agent beliefs [35], reasoning about programming constructs [10,68], and software engineering [52].

In spite of this growing menagerie of Event Calculus formulations and applications, relatively little work has been done to show how the various versions correspond. (Indeed, much more work has been done on showing how the Event Calculus corresponds to the Situation Calculus, see e.g. [21], [31], [32], [41], [48], [49], [73], [74].) This article is an attempt to begin to address this issue. We first

A.C. Kakas, F. Sadri (Eds.): Computat. Logic (Kowalski Festschrift), LNAI 2408, pp. 452–490, 2002.

summarise recent work (e.g. [62], [42], [43]) on axiomatising the Event Calculus in classical logic, using circumscription as a method for default reasoning to solve the frame and related problems. We then describe how under certain circumstances such classical logic theories may be reformulated as "action description language" domain descriptions using the Language $\mathcal{E}$ [21,22]. This enables the classical logic Event Calculus to inherit various provably correct, logic programming and/or argumentation based automated reasoning procedures developed for $\mathcal{E}$ in [21], [22], [23] and [24].

Even if attention is restricted to classical logic formulations of the Event Calculus, there are a number of different choices or variations for the core set of axioms. The various alternatives are each geared to classes of domains with particular restrictions or features; for example to describe systems most naturally viewed as deterministic, as involving both continuous and discrete change, or which require reasoning about the future but not the past. In view of this, we first present (in Section 2) one particular (basic) form of the Event Calculus with six domain independent axioms labeled (EC1) to (EC6), and then (Section 3) list and motivate some alternatives. When describing these, possible substitutes for (for example) axiom (EC1) are labeled (EC1a), (EC1b), etc.

A central feature of the Event Calculi presented here are that they are *narrative-based*, i.e. a time structure which is independent of any action occurrences is established or assumed, and then statements about when various actions occur within this structure are incorporated in the description of the domain under consideration. The time structure is usually assumed or stated to be linear – typically the real or integer number line – although the underlying ideas can equally be applied to other (possibly branching) temporal structures. For the purposes of simplicity, unless otherwise stated we will assume in this article that time is represented either by the real numbers, the integers, the non-negative reals or the non-negative integers, and that appropriate axioms are included in the theory which establish one of these time structures.

Sections 2 and 3 of this article are mostly taken from [43].

2 A Classical Logic Event Calculus Axiomatisation

Informally, the basic idea of the Event Calculus is to state that *fluents* (time-varying properties of the world) are *true* at particular time-points if they have been *initiated* by an action occurrence at some earlier time-point, and not *terminated* by another action occurrence in the meantime. Similarly, a fluent is *false* at a particular time-point if it has been previously terminated and not initiated in the meantime. Domain dependent axioms are provided to describe which actions initiate and terminate which fluents under various circumstances, and to state which actions occur when. In the context of the Event Calculus, individual action occurrences are often referred to as "events", so that "actions" are "event types".

The Event Calculus given here is written in a sorted predicate calculus with equality, with a sort $\mathcal{A}$ for *actions* (variables $a, a_1, a_2, \ldots$), a sort $\mathcal{F}$ for fluents (variables $f, f_1, f_2, \ldots$), a sort $\mathcal{T}$ for timepoints (here either real numbers or integers, variables $t, t_1, t_2, \ldots$) and a sort $\mathcal{X}$ for domain objects (variables $x, x_1, x_2, \ldots$). To describe a very basic calculus we need five predicates (other than equality); $Happens \subseteq \mathcal{A} \times \mathcal{T}$, $HoldsAt \subseteq \mathcal{F} \times \mathcal{T}$, $Initiates \subseteq \mathcal{A} \times \mathcal{F} \times \mathcal{T}$, $Terminates \subseteq \mathcal{A} \times \mathcal{F} \times \mathcal{T}$ and $< \subseteq \mathcal{T} \times \mathcal{T}$. $Happens(A, T)$ indicates that action A occurs at time T, $HoldsAt(F, T)$ means that fluent F is true at time T, and $Initiates(A, F, T)$ (respectively $Terminates(A, F, T)$) expresses that if A occurs at T it will initiate (respectively terminate) the fluent F. "$<$" is the standard order relation for time.

It is convenient to also define auxiliary predicates $Clipped \subseteq \mathcal{T} \times \mathcal{F} \times \mathcal{T}$ and $Declipped \subseteq \mathcal{T} \times \mathcal{F} \times \mathcal{T}$ in terms of $Happens$, $Initiates$, $Terminates$, and $<$. $Clipped(T_1, F, T_2)$ (respectively $Declipped(T_1, F, T_2)$) means "the fluent F is terminated (respectively initiated) between times T_1 and T_2." The corresponding definitional axioms[1] are

$$Clipped(t_1, f, t_2) \stackrel{\text{def}}{\equiv} \exists a, t[Happens(a, t) \wedge t_1 \leq t < t_2 \qquad \qquad \text{(EC1)}$$
$$\wedge \ Terminates(a, f, t)]$$

$$Declipped(t_1, f, t_2) \stackrel{\text{def}}{\equiv} \exists a, t[Happens(a, t) \wedge t_1 \leq t < t_2 \qquad \qquad \text{(EC2)}$$
$$\wedge \ Initiates(a, f, t)]$$

We can now axiomatise the two principles stated in the introduction to this section. Fluents which have been initiated by an occurrence of an action continue to hold until an occurrence of an action which terminates them

$$HoldsAt(f, t_2) \leftarrow [Happens(a, t_1) \wedge Initiates(a, f, t_1) \qquad \qquad \text{(EC3)}$$
$$\wedge \ t_1 < t_2 \wedge \neg Clipped(t_1, f, t_2)]$$

and fluents which have been terminated by an occurrence of an action continue not to hold until an occurrence of an action which initiates them:

$$\neg HoldsAt(f, t_2) \leftarrow [Happens(a, t_1) \wedge Terminates(a, f, t_1) \qquad \qquad \text{(EC4)}$$
$$\wedge \ t_1 < t_2 \wedge \neg Declipped(t_1, f, t_2)]$$

The four axioms above capture the behaviour of fluents once initiated or terminated by an action. But we need also to describe fluents' behaviour before the occurrence of any actions which affect them. We therefore axiomatise a general principle of persistence for fluents; fluents change their truth values only via the occurrence of initiating and terminating actions:

$$HoldsAt(f, t_2) \leftarrow [HoldsAt(f, t_1) \wedge t_1 < t_2 \qquad \qquad \text{(EC5)}$$
$$\wedge \neg Clipped(t_1, f, t_2)]$$

---

[1] By $E_1 \stackrel{\text{def}}{\equiv} E_2$ we mean that expression E_1 is notational shorthand for expression E_2.

$$\neg HoldsAt(f, t_2) \leftarrow [\neg HoldsAt(f, t_1) \wedge t_1 < t_2 \qquad \text{(EC6)}$$
$$\wedge \neg Declipped(t_1, f, t_2)]$$

Definitions of the predicates *Happens*, *Initiates* and *Terminates* are given in the domain-dependent part of the theory, as illustrated in the following example.

2.1 An Example Domain Dependent Axiomatisation

As an example domain dependent theory, we axiomatise a simple scenario of a robot going outside a room by moving through a door, which can be locked and unlocked using an electronic key. For this example we will assume a real number time-line. We will use three fluents, *Inside* (the robot is inside the room), *HasKey* (the robot is holding the electronic key), and *Locked* (the door is locked), and three actions, *Insert* (insert the key in the door), *GoThrough* (move through the door), and *Pickup* (pick up the key). We assume uniqueness-of-names axioms[2] which confirm that all of these constant symbols refer to distinct fluents or actions. Inserting the key alternately locks and unlocks the door. Picking up the key causes the robot to be holding the key, and, going through the unlocked door causes the robot to swap from being inside to outside or vice-versa. We use the predicates *Initiates* and *Terminates* to express these effects:

$$Initiates(a, f, t) \equiv [[a = Pickup \wedge f = HasKey] \qquad \text{(R1)}$$
$$\vee [a = Insert \wedge f = Locked$$
$$\wedge \neg HoldsAt(Locked, t)$$
$$\wedge HoldsAt(HasKey, t)]$$
$$\vee [a = GoThrough \wedge f = Inside$$
$$\wedge \neg HoldsAt(Locked, t)$$
$$\wedge \neg HoldsAt(Inside, t)]]$$

$$Terminates(a, f, t) \equiv [[a = GoThrough \wedge f = Inside \qquad \text{(R2)}$$
$$\wedge \neg HoldsAt(Locked, t)$$
$$\wedge HoldsAt(Inside, t)]$$
$$\vee [a = Insert \wedge f = Locked$$
$$\wedge HoldsAt(Locked, t)$$
$$\wedge HoldsAt(HasKey, t)]]$$

---

[2] In this case the collection of uniqueness-of-names axioms will consist of a sentence such as $Inside \neq HasKey$ for each pair of fluent names and action names. In domains where parameterised fluents or actions are used, e.g. a $Lower(x)$ action to represent the act of lowering an object x meters, it might also typically include sentences such as $Lower(x_1) = Lower(x_2) \rightarrow x_1 = x_2$. The inclusion of such uniqueness-of-names axioms is not obligatory (we might for example wish to deliberately use two names to refer to the same action), but their omission will generally lead to unexpected results.

Let us suppose that the door is locked and the robot is inside at time 0, and that the robot picks up the key, unlocks the door and goes through the door at times 2, 4 and 6 respectively:

$$HoldsAt(Locked, 0) \; \wedge \; HoldsAt(Inside, 0) \tag{R3}$$

$$\begin{aligned} Happens(a, t) \; \equiv \; & [[a = Pickup \; \wedge \; t = 2] \; \vee \\ & [a = Insert \; \wedge \; t = 4] \; \vee \\ & [a = GoThrough \; \wedge \; t = 6]] \end{aligned} \tag{R4}$$

The reader is invited to check that from (EC1)-(EC6) and (R1)-(R4), together with uniqueness-of-names axioms for fluents and actions and an appropriate axiomatisation of the real numbers, it is for example possible to deduce that the robot is no longer inside the room at time 8, i.e. $\neg HoldsAt(Inside, 8)$.

2.2   Circumscription and the Frame Problem

In Event Calculus terms, the frame problem is the problem of expressing in a succinct and elaboration tolerant way that in most cases a given action will not initiate or terminate a given fluent. The description of which actions initiate and terminate which fluents via single biconditionals (as in axioms (R1) and (R2) above), although succinct, is rather unsatisfactory from the point of view of elaboration tolerance. For example, if new information about the initiating effects of a new action needs to be included in the robot domain (e.g. $Initiates(PressDoorBell, RingingNoise, t)$) this cannot be simply added to the theory, since it would be inconsistent with axiom (R1) (from which it is possible to infer $\neg Initiates(PressDoorBell, RingingNoise, t)$).

Hence most versions of the Event Calculus describe each fact or rule about initiation and termination in a separate axiom or clause, and provide an extra transformation or non-monotonic reasoning method to infer negative information about $Initiates$ and $Terminates$ from the collection of such rules. In the context of our classical logic Event Calculus and robot example, the individual rules would be

$$Initiates(Pickup, HasKey, t) \tag{R5}$$

$$\begin{aligned} Initiates(Insert, Locked, t) \; \leftarrow \\ [\neg HoldsAt(Locked, t) \wedge HoldsAt(HasKey, t)] \end{aligned} \tag{R6}$$

$$\begin{aligned} Initiates(GoThrough, Inside, t) \; \leftarrow \\ [\neg HoldsAt(Locked, t) \wedge \neg HoldsAt(Inside, t)] \end{aligned} \tag{R7}$$

$$\begin{aligned} Terminates(GoThrough, Inside, t) \; \leftarrow \\ [\neg HoldsAt(Locked, t) \wedge HoldsAt(Inside, t)] \end{aligned} \tag{R8}$$

$$Terminates(Insert, Locked, t) \leftarrow \qquad\qquad\qquad\qquad\qquad (R9)$$
$$[HoldsAt(Locked, t) \wedge HoldsAt(HasKey, t)]$$

Predicate completion or circumscription [39] can then be used to transform this collection of axioms into expressions such as (R1) and (R2). In this article we use the notation described in [37] to indicate circumscriptions of particular conjunctions of sentences. In particular, the circumscription

$$CIRC[(R5) \wedge (R6) \wedge (R7) \wedge (R8) \wedge (R9) \; ; \; Initiates, Terminates]$$

yields exactly (R1) and (R2). For simple domains such as the above, this type of transformation (whether described in terms of circumscription or predicate completion) is analogous to the solution to the frame problem developed by Reiter for the Situation Calculus [50].

To make useful deductions using axioms (EC1)-(EC6), it is also necessary to be able to infer both positive and negative information about *Happens* from the domain dependent part of the theory. Again the issue of elaboration tolerance arises, so that, as for *Initiates* and *Terminates*, most versions of the Event Calculus encapsulate each individual action occurrence in a separate *Happens* assertion (rather than using a biconditional such as (R4)), and then use some form of non-monotonic reasoning to infer negative information about this predicate. For example, in the case of our robot example the assertions would be

$$Happens(Pickup, 2) \qquad\qquad\qquad\qquad\qquad\qquad\qquad (R10)$$

$$Happens(Insert, 4) \qquad\qquad\qquad\qquad\qquad\qquad\qquad (R11)$$

$$Happens(GoThrough, 6) \qquad\qquad\qquad\qquad\qquad\qquad\qquad (R12)$$

The circumscription $CIRC[(R10) \wedge (R11) \wedge (R12) \; ; \; Happens]$ then gives (R4).

If we now wish to add more information about *Happens*, we can do so without altering axioms (R10)-(R12) and then reapply the circumscription operator. This information need not just be in the form of ground literals – we may have less precise information about the order or timing of action occurrences. For example, we might know that the robot pressed the door bell either just before, just after or at the same time as inserting the key, in which case we could add

$$\exists t_1.[Happens(PressDoorBell, t_1) \wedge 2 < t_1 < 6] \qquad\qquad\qquad (R13)$$

The circumscription $CIRC[(R10) \wedge (R11) \wedge (R12) \wedge (R13)$; $Happens]$ then gives

$$\exists t_1.[2 < t_1 < 6 \wedge [Happens(a,t) \equiv \tag{R14}$$
$$[[a = Pickup \wedge t=2] \vee$$
$$[a = Insert \wedge t=4] \vee$$
$$[a = GoThrough \wedge t=6] \vee$$
$$[a = PressDoorBell \wedge t=t_1]]]]$$

enabling us to deduce facts such as $\neg HoldsAt(Inside, 8)$ as before.

More generally, complete Event Calculus domain descriptions of this basic type are of the form

$$CIRC[\Sigma \; ; \; Initiates, Terminates] \wedge CIRC[\Delta \; ; \; Happens] \wedge \Omega \wedge EC$$

where Σ is a conjunction of *Initiates* and *Terminates* formulae, Δ is a conjunction of *Happens* and temporal ordering formulae, Ω is a conjunction of fluent-specific *HoldsAt* formulae such as (R3) and time-independent formulae (such as uniqueness-of-names axioms for actions and fluents), and EC is the conjunction of axioms (EC1) to (EC6) together an appropriate axiomatisation of the sort $\mathcal{T}$. The minimisation of *Initiates* and *Terminates* corresponds to the default assumption that actions have no unexpected effects, and the minimisation of *Happens* corresponds to the default assumption that there are no unexpected event occurrences. The key to this solution to the frame problem is thus the splitting of the theory into different parts, which are circumscribed separately. This technique, sometimes referred to as *forced separation*, is also employed in [9], [14] and [28], and is akin to what Sandewall calls *filtering* [56].

2.3   Narrative Information and Planning

In some circumstances it is convenient to define *Happens* in terms of other predicates representing different categories of action occurrence. For example, in the context of planning we may wish to distinguish between actions that have (definitely) happened in the past and actions that the agent will (possibly) perform in the future[3]. In this case we may include a domain independent axiom such as

$$Happens(a, t) \equiv [Occurred(a, t) \vee Perform(a, t)] \tag{EC7}$$

We can now maintain a complete definition for *Occurred* (in the same way that we previously had a complete definition for *Happens*) based on our knowledge of actions that have already taken place, whilst keeping *Perform* undefined within the theory. In this way we can formulate a deductive specification of the planning task in terms of *Perform*. For example, in the context of the robot example suppose that at time 3 we know that a *Pickup* action has already taken

---

[3] Or we may wish to distinguish between actions performed by the agent and events occurring in the environment and outside the agent's control.

place (at time 2), and wish to plan for the goal $\neg HoldsAt(Inside, 8)$. We would include the axiom

$$Occurred(a, t) \equiv [a = Pickup \wedge t = 2] \tag{R15}$$

(or the equivalent expression $CIRC[Occurred(Pickup, 2) \;\; ; \;\; Occurred]$) in the domain description, and then show that the sentence

$$Perform(a, t) \equiv [[a = Insert \wedge t = 4] \vee [a = GoThrough \wedge t = 6]] \tag{P1}$$

is a plan for $\neg HoldsAt(Inside, 8)$ in the sense that

$$[(EC1) \wedge \ldots \wedge (EC7) \wedge (R1) \wedge (R2) \wedge (R3) \wedge (R15)] \models \\ [(P1) \;\to\; \neg HoldsAt(Inside, 8)]$$

More generally, planning can be viewed as the deduction[4] of sentences of the form $[Plan \to Goal]$ from an Event Calculus domain description, where $Plan$ is a sentence such as (P1) defining the predicate $Perform$, and $Goal$ is a sentence containing just the predicates $HoldsAt$ and $<$ (we need also to establish via general theorems or a specific check that $Plan$ is consistent with the Event Calculus theory). By the Deduction Theorem (see e.g. [18]) $Theory \models [Plan \to Goal]$ is equivalent to $[Theory \wedge Plan] \models Goal$ so that planning in the context of the Event Calculus can also be understood in terms of abduction (i.e. finding plans to add to the theory so that the goal is entailed). Indeed, it is this abductive view which is taken in the majority of work on Event Calculus planning, e.g. in [15], [8], [19], [44], [45], [63], [65] and [20].

2.4 Non-determinism

In contrast to many versions of the Event Calculus, the axiomatisation described in (EC1)–(EC6) is non-deterministic, in the sense that simultaneously initiating and terminating a fluent simply gives rise to two sets of models (one in which the fluent is true immediately afterwards and one in which it is false), rather than resulting in an inconsistent theory. This is because of the requirement in axioms (EC1) and (EC2) that $t_1 \leq t$, rather than $t_1 < t$.

For example, let us suppose that the action of tossing a coin is represented as $TossCoin$, and that each occurrence of this action results in the fluent $HeadsUp$ being either true or false. We can represent this with an $Initiates$ and a $Terminates$ literal:

$$Initiates(TossCoin, HeadsUp, t) \tag{C1}$$

$$Terminates(TossCoin, HeadsUp, t) \tag{C2}$$

---

[4] That is to say, planning can be specified as a deductive task. We do not wish to claim that general purpose classical theorem provers are practical as planning systems.

Suppose the time is represented as the reals, and that a single *TossCoin* action happens at time 2:

$$Happens(TossCoin, 2) \tag{C3}$$

The theory which consists of axioms (EC1)-(EC6), $CIRC[(C1) \wedge (C2) ; Initiates, Terminates]$ and $CIRC[(C3) ; Happens]$ has four classes of models with respect to the fluent *HeadsUp* – one in which *HeadsUp* holds for all timepoints, one in which *HeadsUp* holds for no timepoints, one in which *HeadsUp* changes from true to false for all timepoints greater than 2, and one in which *HeadsUp* changes from false to true for all timepoints greater than 2. This is because using axioms (EC1) and (EC2) we can show $Clipped(2, HeadsUp, T)$ and $Declipped(2, HeadsUp, T)$ for all $T > 2$, so that (EC3) and (EC4) are trivially satisfied and the truth value of *HeadsUp* at different timepoints is constrained only by axioms (EC5) and (EC6).

The narrative-based nature of the Event Calculus (i.e. the fact that action occurrences are explicitly represented) facilitates a simple alternative to representing non-determinism. We can for example regard the action *TossCoin* as representing a choice of two deterministic actions *TossHead* and *TossTail*:

$$Happens(TossCoin, t) \rightarrow \tag{C4}$$
$$[Happens(TossHead, t) \vee Happens(TossTail, t)]$$

We can then rewrite axioms (C1) and (C2) as

$$Initiates(TossHead, HeadsUp, t) \tag{C1a}$$

$$Terminates(TossTail, HeadsUp, t) \tag{C2a}$$

The theory consisting of (EC1)-(EC6), (C5), $CIRC[(C3) \wedge (C4) ; Happens]$ and $CIRC[(C1a) \wedge (C2a) ; Initiates, Terminates]$ now also gives rise to the desired classes of models described above. (The circumscription of *Happens* eliminates models where both a *TossHead* and a *TossTail* action occur at time 2.) For tasks such as planning, it is straightforward to specify that the agent in question can attempt some actions (such as *TossCoin*) but not others (such as *TossHead* or *TossTail*).

2.5   Concurrent Actions

The syntax of the Event Calculus makes it straightforward to express that two or more actions have occurred or will occur simultaneously, since different *Happens* literals in the domain description may refer to the same timepoint.

In some domains, concurrently performed actions may cancel each others' effects, and may combine to cause effects which none of the actions performed in isolation would achieve. A standard example is that if a bowl is filled with water, lifting just the left side of the bowl or just the right side will cause the water to spill. Lifting both sides simultaneously will not cause the water to spill but will cause the bowl to be raised.

In the Event Calculus, we can describe cancellations and combinations of effects with *Happens* preconditions in the domain dependent axioms defining *Initiates* and *Terminates*. For example:

$$Initiates(LiftLeft, Spilt, t) \leftarrow \neg Happens(LiftRight, t) \tag{B1}$$

$$Initiates(LiftRight, Spilt, t) \leftarrow \neg Happens(LiftLeft, t) \tag{B2}$$

$$Initiates(LiftRight, Raised, t) \leftarrow Happens(LiftLeft, t) \tag{B3}$$

To illustrate the effect of such statements, suppose that our domain description also includes the following narrative information:

$$\neg HoldsAt(Spilt, 0) \tag{B4}$$

$$Happens(LiftLeft, 2) \tag{B5}$$

$$Happens(LiftRight, 2) \tag{B6}$$

The theory consisting of (EC1)-(EC6), (B4), $CIRC[(B1) \wedge (B2) \wedge (B3)$; *Initiates, Terminates*] and $CIRC[(B5) \wedge (B6)$; *Happens*] entails, for example, both $\neg HoldsAt(Spilt, 4)$ and $HoldsAt(Raised, 4)$.

3 Alternative and Extended Classical Logic Event Calculus Axiomatisations

The version of the Event Calculus described in Section 2 has a number of characteristics; it is geared to time-lines extending infinitely backwards as well as forwards, it is "non-deterministic" (in the sense described in Section 2.4), it regards all actions as possible under all circumstances, it regards all fluents' truth values as persisting between all relevant action occurrences, and it regards all action occurrences as instantaneous. However, the choice of which of these characteristics to include in a given Event Calculus axiomatisation is to a large extent arbitrary, and in this section we describe alternative axiomatisations which each negate one or more of these properties.

For ease of presentation, sub-sections 3.1 to 3.7 below each alter the axiomatisation (EC1)–(EC6) as little as possible to illustrate the particular point under discussion. But unless otherwise stated these alterations can be combined in a straightforward and obvious manner. For example we can combine the modifications described in sub- sections 3.2, 3.3 and 3.6 below to produce a "deterministic" Event Calculus with facilities to describe when it is impossible for particular actions to occur, and including actions of a non-zero duration.

Where in a particular sub-section no alternative to one of the axioms (EC1)–(EC6) is given, it should be assumed that the axiom in question remains unchanged.

3.1   An Alternative Axiomatisation for Non-negative Time

Where time is modeled as the non-negative reals or integers, it is often convenient to introduce two new predicates[5] $InitiallyP \subseteq \mathcal{F}$ and $InitiallyN \subseteq \mathcal{F}$ ("P" for "positive" and "N" for "negative"), and to replace axioms (EC5) and (EC6) with the following three axioms:

$$HoldsAt(f,t) \leftarrow [InitiallyP(f) \wedge \neg Clipped(0,f,t)] \qquad \text{(EC5a)}$$

$$\neg HoldsAt(f,t) \leftarrow [InitiallyN(f) \wedge \neg Declipped(0,f,t)] \qquad \text{(EC6a)}$$

$$InitiallyP(f) \vee InitiallyN(f) \qquad \text{(EC8a)}$$

Indeed, for non-negative time (EC5a), (EC6a) and (EC8a) may be deduced from (EC5) and (EC6) together with the assertion

$$[HoldsAt(f,0) \equiv InitiallyP(f)] \wedge [\neg HoldsAt(f,0) \equiv InitiallyN(f)]$$

Axioms (EC5a) and (EC6a) have an advantage over (EC5) and (EC6) in that they can readily be converted to logic program clauses without causing obvious looping problems.

However, this alternative axiomatisation is slightly weaker. For example, in the non-deterministic domain described in Sub-section 2.4 by axioms (C1), (C2) and (C3), axioms (EC5a), (EC6a) and (EC8a) would license models where $HeadsUp$ fluctuated arbitrarily between true and false at times after 2. Although this characteristic is problematic for this particular example, it can be an advantage for representing other types of domain where it is convenient to "dynamically manage the frame", i.e. to regard some fluents as having an inherent persistence during some intervals of time but not during others. Indeed, for such domains axiom (EC8a) may not be appropriate. These issues are discussed in more detail in Section 3.7.

Since this particular axiomatisation does not include the general principle of persistence encapsulated in (EC5) and (EC6) (which describes how fluents persist independently of initiating and terminating action occurrences), adding individual $HoldsAt$ literals to a given domain description (see for example axiom (R3) in Section 2.1) no longer necessarily has the same effect, particularly in axiomatisations where (EC8a) is omitted. Instead, individual observations of the form $HoldsAt(F,T)$ and $\neg HoldsAt(F,T)$ can be assimilated indirectly into the theory (perhaps automatically by a process of abduction) by appropriate

---

[5] These predicates are referred to as $InitiallyTrue$ and $InitiallyFalse$ in [42].

addition of *Happens*, *InitiallyP* and *InitiallyN* literals. Axiom (R3), for example, can be replaced by

$$InitiallyP(Locked) \wedge InitiallyP(Inside) \tag{R3a}$$

so that (R3) is now entailed by the theory consisting of (R3a), (R1), (R2), (R4), (EC1)–(EC4), (EC5a) and (EC6a).

3.2   Deterministic Event Calculus

A strictly deterministic Event Calculus (in the sense that simultaneously initiating and terminating a fluent results in inconsistency) may be formulated by replacing (EC3) and (EC4) by the following two axioms:

$$HoldsAt(f, t_2) \leftarrow [Happens(a, t_1) \wedge Initiates(a, f, t_1) \\ \wedge\ t_1 < t_2 \wedge \neg StoppedIn(t_1, f, t_2)] \tag{EC3b}$$

$$\neg HoldsAt(f, t_2) \leftarrow [Happens(a, t_1) \wedge Terminates(a, f, t_1) \\ \wedge\ t_1 < t_2 \wedge \neg StartedIn(t_1, f, t_2)] \tag{EC4b}$$

where the predicates *StoppedIn* and *StartedIn* are defined as follows:

$$StoppedIn(t_1, f, t_2) \overset{\text{def}}{\equiv} \exists a, t[Happens(a, t) \wedge t_1 < t < t_2 \\ \wedge\ Terminates(a, f, t)] \tag{EC9b}$$

$$StartedIn(t_1, f, t_2) \overset{\text{def}}{\equiv} \exists a, t[Happens(a, t) \wedge t_1 < t < t_2 \\ \wedge\ Initiates(a, f, t)] \tag{EC10b}$$

Note that *StoppedIn* and *StartedIn* are identical to *Clipped* and *Declipped* except for the inequality relations between the time-point variables. Strictly speaking (EC1) and (EC2) defining *Clipped* and *Declipped* are still required since these predicates are used in (EC5) and (EC6), or in their substitutes (EC5a) and (EC6a). But for domains using non-negative time and axioms (EC5a) and (EC6a), the definitions of *Clipped* and *Declipped* may be straightforwardly replaced by those for *StoppedIn* and *StartedIn*, provided no actions occur at time 0 which either terminate initially-positive fluents or initiate initially-negative fluents.

The effective meanings of *Initiates* and *Terminates* are slightly different in the deterministic Event Calculus (i.e. in axiomatisations including (EC3b) and (EC4b)) from their meanings in non-deterministic Event Calculus. (EC3b) and (EC4b) ensure that *Initiates*(A, F, T) can be read as "F holds immediately after an occurrence of A at time T", whereas with axioms (EC3) and (EC4) *Initiates*(A, F, T) corresponds to the slightly weaker assertion that "an occurrence of A at time T has an initiating influence on F" (which may or may not be overridden by a simultaneously occurring terminating influence).

There are several ways in which non-determinism may be reintroduced into what we have described here as deterministic Event Calculus. For example, the technique exemplified in axioms (C1a), (C2a) and (C4) (see Section 2.4) is still applicable. Other methods include the use of *determining fluents* or a *Releases* predicate (see [62] and Section 3.7).

3.3   Action Preconditions and the Qualification Problem in the Event Calculus

We have already illustrated with axioms such as (R6)-(R9) (See Section 2.2) how preconditions for particular effects of actions may be expressed within the Event Calculus. These types of precondition are often referred to as *fluent preconditions*. There are also various ways in which *action preconditions* (i.e. conditions necessary for actions to be possible at all) can be expressed. One method is to introduce a new predicate *Impossible* $\subseteq \mathcal{A} \times \mathcal{T}$ and write an appropriate definition for *Impossible* with respect to each action in the domain in question. For instance, in our example domain we may wish to express that it is impossible for the robot to pickup the key if it is not fitted with a grabber, and it is impossible for the robot to go through a locked door:

$$Impossible(Pickup, t) \leftarrow \neg HoldsAt(HasGrabber, t) \tag{R16}$$

$$Impossible(GoThrough, t) \leftarrow HoldsAt(Locked, t) \tag{R17}$$

We can regard the qualification problem (at least in part) as the problem of expressing, in a succinct and elaboration tolerant way, that under most circumstances most actions are possible. To achieve this in the Event Calculus, we can minimise the predicate *Impossible*. $CIRC[(\text{R16}) \wedge (\text{R17}) \, ; \, Impossible]$ gives

$$Impossible(a, t) \equiv [[a = Pickup \, \wedge \, \neg HoldsAt(HasGrabber, t)] \tag{R18}$$
$$\vee \, [a = GoThrough \, \wedge \, HoldsAt(Locked, t)]]$$

For narrative formalisms such as the Event Calculus, the way in which this type of knowledge is to be interpreted, and thus the way in which the domain independent axioms need to be adapted, depends to some extent on the individual domain and mode of reasoning under consideration. For tasks such as planning, which involves (hypothetical) reasoning about future events, it makes sense to regard the assertion $Impossible(A, T)$ as stating "it is impossible to predict the effects of attempting to perform action A at time T" (so that $\neg Impossible(A, T)$ can be regarded as analogous to $Poss(A, S)$ in Reiter's Situation Calculus [50]). In this case it is necessary only to block any inferences about what holds or does not hold at any time after an (attempt at an) 'impossible' action occurrence. This can be done by appropriately modifying the definitions of *Clipped* and *Declipped*:

$$Clipped\,(t_1, f, t_2) \overset{\text{def}}{\equiv} \qquad\qquad\qquad\qquad\qquad\qquad\qquad\qquad\text{(EC1c)}$$
$$[\exists a, t[Happens(a, t) \;\wedge\; t_1 \leq t < t_2 \;\wedge\; Terminates(a, f, t)]$$
$$\vee\; \exists a, t[Happens(a, t) \;\wedge\; t < t_2 \;\wedge\; Impossible(a, t)]]$$

$$Declipped\,(t_1, f, t_2) \overset{\text{def}}{\equiv} \qquad\qquad\qquad\qquad\qquad\qquad\qquad\text{(EC2c)}$$
$$[\exists a, t[Happens(a, t) \;\wedge\; t_1 \leq t < t_2 \;\wedge\; Initiates(a, f, t)]$$
$$\vee\; \exists a, t[Happens(a, t) \;\wedge\; t < t_2 \;\wedge\; Impossible(a, t)]]$$

On the other hand, if for example the domain includes certain knowledge about actions or events that have actually occurred in the past, it makes sense to regard the assertion $Impossible(A, T)$ as stating "action A could not have occurred at time T". Hence where the definition of $Happens$ is split as in axiom (EC7) (see Section 2.3), we can include additional constraints such as

$$\neg Occurred(a, t) \;\leftarrow\; Impossible(a, t) \qquad\qquad\qquad\qquad\text{(EC11c)}$$

Notice for example that from (R15), (R16) and (EC11c) we can infer the action precondition $HoldsAt(HasGrabber, 2)$ for the known occurrence of $Pickup$ at time 2. This illustrates why we would not want to include constraints analogous to (EC11c) for hypothetical future performances of actions – at time 0 we would not for example want $Perform(Pickup, 1)$ to constitute a plan for the goal $HoldsAt(HasGrabber, 1)$. We could however safely state that nothing happens when an agent attempts to perform an impossible action, by replacing (EC7) with

$$Happens(a, t) \;\equiv\; [[Perform(a, t) \;\wedge\; \neg Impossible(a, t)] \qquad\qquad\text{(EC7c)}$$
$$\vee\; Occurred(a, t)]$$

Finally, note that rules such as (R16) and (R17) partially defining $Impossible$ can have $Happens$ (and $Perform$ and $Occurred$) preconditions as well as $HoldsAt$ preconditions. This can be useful, for example, for expressing that it is impossible to perform certain combinations of actions simultaneously. For instance, the sentence

$$Impossible(a_1, t) \;\leftarrow\; [Perform(a_2, t) \;\wedge\; a_1 \neq a_2]$$

states that it is in general impossible to perform more than one action at a time. Like (R16) and (R17), such sentences must be placed within the scope of the circumscription of $Impossible$.

3.4   Categorisation of Fluents in the Event Calculus

For some domains, it is appropriate to categorise fluents into *frame* fluents and *non-frame* fluents (or *primitive* and *derived* fluents), and then to restrict the application of the principles of persistence encapsulated in axioms (EC3)-(EC6) to frame fluents only. To do this it is necessary to introduce a new predicate

$Frame \subseteq \mathcal{F}$, and alter (EC3)-(EC6) as follows:

$$HoldsAt(f, t_2) \leftarrow [Happens(a, t_1) \wedge Initiates(a, f, t_1) \quad\quad (EC3d)$$
$$\wedge \ Frame(f) \ \wedge \ t_1 < t_2$$
$$\wedge \ \neg Clipped(t_1, f, t_2)]$$

$$\neg HoldsAt(f, t_2) \leftarrow [Happens(a, t_1) \ \wedge \ Terminates(a, f, t_1) \quad\quad (EC4d)$$
$$\wedge \ Frame(f) \ \wedge \ t_1 < t_2$$
$$\wedge \ \neg Declipped(t_1, f, t_2)]$$

$$HoldsAt(f, t_2) \leftarrow [HoldsAt(f, t_1) \ \wedge \ t_1 < t_2 \quad\quad (EC5d)$$
$$\wedge \ Frame(f) \ \wedge \ \neg Clipped(t_1, f, t_2)]$$

$$\neg HoldsAt(f, t_2) \leftarrow [\neg HoldsAt(f, t_1) \ \wedge \ t_1 < t_2 \quad\quad (EC6d)$$
$$\wedge \ Frame(f) \ \wedge \ \neg Declipped(t_1, f, t_2)]$$

This axiom set can be useful when we want to include simple types of indirect effects in domain descriptions, since we are now free to write definitions or partial definitions of non-frame fluents (i.e. *state constraints*) in terms of *HoldsAt* and frame fluents. For example, as regards the robot we may wish to introduce a non-frame fluent *Happy* and state that, although the robot is not happy at time 0, it is in general happy if it is holding the key:

$$Frame(f) \equiv [f = Inside \ \vee \ f = HasKey \ \vee \ f = Locked] \quad\quad (R19)$$

$$\neg HoldsAt(Happy, 0) \quad\quad (R20)$$

$$HoldsAt(Happy, t) \leftarrow HoldsAt(HasKey, t) \quad\quad (R21)$$

Using (EC1), (EC2), (EC3d)-(EC6d), (R1)-(R4) and (R19)-(R21) we can now, for example, infer $HoldsAt(Happy, 5)$. Indeed, we can also infer $\neg HoldsAt(HasKey, 0)$ and therefore $\neg HoldsAt(HasKey, 1)$. But we can neither infer $HoldsAt(Happy, 1)$ nor $\neg HoldsAt(Happy, 1)$, since the non-frame fluent *Happy* has no intrinsic persistence of its own.

3.5   Trajectories, Delayed Actions and Gradual Change

Several techniques are available within the context of the Event Calculus for describing delayed effects. The simplest approach is to write rules in terms of *Happens*. For example, if setting an alarm clock causes it to ring 8 hours later, we can write

$$Happens(StartRing, t+8) \leftarrow Happens(Set, t) \quad\quad (A1)$$

$$Initiates(StartRing, Ringing, t) \quad\quad (A2)$$

A disadvantage of rules such as (A1) is that it is difficult to express that the occurrence of the later action might be prevented by some intervening action (e.g. somebody might switch off the alarm during the night).

A more flexible approach involves the use of *trajectories* [58]. It is convenient to illustrate this technique here by introducing a new sort $\mathcal{P}$ of *parameters* into the language. Like fluents, parameters are time-varying properties, but unlike (frame) fluents they have no associated default persistence. More precisely, parameters are names for arbitrarily-valued functions of time, and accordingly we introduce a new function $ValueAt : \mathcal{P} \times \mathcal{T} \mapsto \mathcal{X}$. For example, we might write $ValueAt(Countdown, 5) = 2$ to indicate that at time 5 the parameter *Countdown*, representing the time remaining before the alarm clock rings, has a value of 2. To represent delayed and triggered effects, as well as simple forms of gradual or continuous change, specific parameters are associated with specific fluents via the predicate $Trajectory \subseteq \mathcal{F} \times \mathcal{T} \times \mathcal{P} \times \mathcal{T} \times \mathcal{X}$. The intended meaning of $Trajectory(F, T_1, P, T_2, X)$ is that if fluent F is initiated at time T_1 and continues to hold until time $T_1 + T_2$, this results in parameter P having a value of X at time $T_1 + T_2$. For example, in the case of the alarm clock we might write

$$Trajectory(SwitchedOn, t_1, Countdown, t_2, 8 - t_2) \tag{A3}$$

We can translate this intended meaning into Event Calculus terms with the addition of a single extra domain independent axiom

$$
\begin{aligned}
ValueAt(p, t_1 + t_2) = x \leftarrow \\
[Happens(a, t_1) \land Initiates(a, f, t_1) \\
\land \; 0 < t_2 \land Trajectory(f, t_1, p, t_2, x) \\
\land \neg Clipped(t_1, f, t_1 + t_2)]
\end{aligned}
\tag{EC11}
$$

Continuing with our example, it is straightforward to express that when *Countdown* reaches 0 the alarm goes off:

$$Happens(StartRing, t) \leftarrow ValueAt(Countdown, t) = 0 \tag{A4}$$

We can complete our description of the domain by stating that switching on the alarm activates the timing mechanism (provided it is not already activated), that the ringing event switches off the timing mechanism, that when the timing mechanism is switched off the countdown is permanently fixed at 8, that the alarm is initially not switched on and that someone switches it on at time 2:

$$Initiates(Set, SwitchedOn, t) \leftarrow \neg HoldsAt(SwitchedOn, t) \tag{A5}$$

$$Terminates(StartRing, SwitchedOn, t) \tag{A6}$$

$$ValueAt(Countdown, t) = 8 \leftarrow \neg HoldsAt(SwitchedOn, t) \tag{A7}$$

$$\neg HoldsAt(SwitchedOn, 0) \tag{A8}$$

$$Happens(Set, 2) \tag{A9}$$

The theory consisting of (EC1)-(EC6), (EC11), (A3), (A7), (A8), $CIRC[(A4) \wedge$ (A9) ; $Happens]$ and $CIRC[(A2) \wedge (A5) \wedge (A6)$; $Initiates, Terminates]$ entails, for example, $Happens(StartRing, 10)$ and $HoldsAt(Ringing, 11)$.

The Event Calculus is symmetric as regards positive and negative $HoldsAt$ literals and as regards $Initiates$ and $Terminates$. Hence (EC11) has its counterpart in terms of $Terminates$:

$$
\begin{aligned}
ValueAt(p, t_1+t_2) = x \ \leftarrow \qquad\qquad\qquad\qquad & \\
[Happens(a, t_1) \ \wedge \ Terminates(a, f, t_1) & \\
\wedge \ 0 < t_2 \ \wedge \ AntiTrajectory(f, t_1, p, t_2, x) & \\
\wedge \ \neg Declipped(t_1, f, t_1+t_2)] &
\end{aligned}
\tag{EC12}
$$

This axiom uses the predicate $AntiTrajectory \subseteq \mathcal{F} \times \mathcal{T} \times \mathcal{P} \times \mathcal{T} \times \mathcal{X}$. The intended meaning of $AntiTrajectory(F, T_1, P, T_2, X)$ is that if fluent F is terminated at time T_1 and continues not to hold until time T_1+T_2, this results in parameter P having a value of X at time T_1+T_2. We can illustrate the use of anti-trajectories by representing the fact that a hot-air balloon rises when the air-heater is on, but falls when it is not:

$$
\begin{aligned}
Trajectory(HeaterOn, t_1, Height, t_2, x_1+t_2) & \\
\leftarrow ValueAt(Height, t_1) = x_1 &
\end{aligned}
\tag{H1}
$$

$$
\begin{aligned}
AntiTrajectory(HeaterOn, t_1, Height, t_2, x_1-t_2) & \\
\leftarrow ValueAt(Height, t_1) = x_1 &
\end{aligned}
\tag{H2}
$$

(Note that in the alarm clock example (A7) can also be expressed as $AntiTrajectory(SwitchedOn, t_1, Countdown, t_2, 8)$.)

Note that the functions captured in individual trajectories need not be continuous or even numerically valued. For example, we can use a trajectory to model the fact that the left indicator light of a car flashes once per second while the indicator switch is depressed:

$$Trajectory(IndicatorDepressed, t_1, Light, t_2, BlinkFunction(t_2)) \tag{L1}$$

$$AntiTrajectory(IndicatorDepressed, t_1, Light, t_2, Off) \tag{L2}$$

$$BlinkFunction(t) = On \ \leftarrow \ [t \bmod 2 < 1] \tag{L3}$$

$$BlinkFunction(t) = Off \ \leftarrow \ [t \bmod 2 \geq 1] \tag{L4}$$

In domains which include non-deterministic actions (in the sense that actions or combinations of actions can simultaneously initiate and terminate fluents) axioms (EC11) and (EC12) are too weak. For example, if the switching on mechanism is faulty in our alarm clock example, so that we have both (A5) and

$$Terminates(Set, SwitchedOn, t) \qquad\qquad (A10e)$$

axiom (EC11) will not inform us that the countdown is activated even in the circumstance where fluent *SwitchedOn* holds immediately after time 2. One solution is to replace (EC11) and (EC12) with equivalent axioms which have an extra *HoldsAt* condition in their right-hand sides, but use *StoppedIn* and *StartedIn* (see axioms (EC9b) and (EC10b)) instead of *Clipped* and *Declipped*:

$$ValueAt(p, t_1+t_2) = x \leftarrow \qquad\qquad (EC11e)$$
$$[Happens(a, t_1) \land Initiates(a, f, t_1)$$
$$\land\ 0 < t_2 \land Trajectory(f, t_1, p, t_2, x)$$
$$\land HoldsAt(f, t_1+t_2) \land \neg StoppedIn(t_1, f, t_1+t_2)]$$

$$ValueAt(p, t_1+t_2) = x \leftarrow \qquad\qquad (EC12e)$$
$$[Happens(a, t_1) \land Terminates(a, f, t_1)$$
$$\land\ 0 < t_2 \land AntiTrajectory(f, t_1, p, t_2, x)$$
$$\land \neg HoldsAt(f, t_1+t_2) \land \neg StartedIn(t_1, f, t_1+t_2)]$$

In Event Calculus axiomatisations where a distinction is made between fluents which are (temporarily or permanently) inside or outside the frame (such as in Section 3.4), we may dispense with the extra sort $\mathcal{P}$ in favour of non-frame fluents, and replace (EC11) and (EC12) with axioms such as

$$HoldsAt(f_2, t_1+t_2) \leftarrow [\neg Frame(f_2) \land Happens(a, t_1) \qquad (EC11f)$$
$$\land Initiates(a, f_1, t_1) \land 0 < t_2$$
$$\land Trajectory(f_1, t_1, f_2, t_2)$$
$$\land \neg Clipped(t_1, f_1, t_1+t_2)]$$

$$HoldsAt(f_2, t_1+t_2) \leftarrow [\neg Frame(f_2) \land Happens(a, t_1) \qquad (EC12f)$$
$$\land Terminates(a, f_1, t_1) \land 0 < t_2$$
$$\land AntiTrajectory(f_1, t_1, f_2, t_2)$$
$$\land \neg Declipped(t_1, f_1, t_1+t_2)]$$

Here *Trajectory* $\subseteq \mathcal{F} \times \mathcal{T} \times \mathcal{F} \times \mathcal{T}$, and the intended meaning of *Trajectory*(F_1, T_1, F_2, T_2) is that if fluent F_1 is initiated at time T_1 and continues to hold until time T_1+T_2, this results in F_2 holding at time T_1+T_2 (similarly for *AntiTrajectory*). In the alarm clock example *Countdown* would then be parameterised, (A3), (A4) and (A6) would be written

$$Trajectory(SwitchedOn, t_1, Countdown(8-t_2), t_2) \qquad \text{(A3f)}$$

$$Happens(StartRing, t) \leftarrow HoldsAt(Countdown(0), t) \qquad \text{(A4f)}$$

$$HoldsAt(Countdown(8), t) \leftarrow \neg HoldsAt(SwitchedOn, t) \qquad \text{(A7f)}$$

and the domain description would include the additional constraint

$$[HoldsAt(Countdown(x_1), t) \wedge HoldsAt(Countdown(x_2), t)] \qquad \text{(A10f)}$$
$$\rightarrow x_1 = x_2$$

3.6   The Event Calculus and Actions with Duration

The Event Calculus can be modified in various ways so that actions can be represented as occurring over intervals of time. To illustrate, we present here a simple modification in which actions are assigned a numerical duration using the function $Dur : \mathcal{A} \mapsto \mathcal{T}$. This avoids the need to introduce extra arguments of sort $\mathcal{T}$ in the predicates $Happens$, $Initiates$ and $Terminates$. For example, we will interpret the assertion $Happens(A, T)$ to mean "the action A starts to occur at T" (so that it finishes at $T + Dur(A)$).

We will be cautious in the assumptions we make about the effects of actions. We will assume that actions may affect relevant fluents from the moment they start, but the effects only become certain after the actions have finished. Hence the values of affected fluents should be undetermined by the axiomatisation during action occurrences. To incorporate these assumptions in the domain independent axioms (EC1)-(EC6) it is necessary only to modify the various inequality relations between the timepoint variables in (EC1)-(EC4):

$$Clipped(t_1, f, t_2) \stackrel{\text{def}}{=} \exists a, t[Happens(a, t) \wedge t_1 \leq (t + Dur(a)) \qquad \text{(EC1g)}$$
$$\wedge t < t_2 \wedge Terminates(a, f, t)]$$

$$Declipped(t_1, f, t_2) \stackrel{\text{def}}{=} \exists a, t[Happens(a, t) \wedge t_1 \leq (t + Dur(a)) \qquad \text{(EC2g)}$$
$$\wedge t < t_2 \wedge Initiates(a, f, t)]$$

$$HoldsAt(f, t_2) \leftarrow [Happens(a, t_1) \wedge Initiates(a, f, t_1) \qquad \text{(EC3g)}$$
$$\wedge (t_1 + Dur(a)) < t_2$$
$$\wedge \neg Clipped(t_1, f, t_2)]$$

$$\neg HoldsAt(f, t_2) \leftarrow [Happens(a, t_1) \wedge Terminates(a, f, t_1) \qquad \text{(EC4g)}$$
$$\wedge (t_1 + Dur(a)) < t_2$$
$$\wedge \neg Declipped(t_1, f, t_2)]$$

The issue of preconditions becomes more complex when actions have duration. We may for example wish to make a distinction between preconditions which must hold at the start of the action and those which must hold throughout the action. It is therefore often convenient to define auxiliary predicates such as $HoldsIn \subseteq \mathcal{F} \times \mathcal{T} \times \mathcal{T}$:

$$HoldsIn(f, t_1, t_3) \overset{\text{def}}{\equiv} \forall t_2 [t_1 \leq t_2 \leq t_3 \rightarrow HoldsAt(f, t_2)] \qquad \text{(EC13)}$$

To illustrate the use of $HoldsIn$, consider a simple description of an automated train which can move at a fixed speed S along a track running from West to East, provided its motor is engaged. Using the action term $MoveEast(T)$ to represent the action of moving east for T time units, we can for example write axioms such as

$$Dur(MoveEast(t)) = t \qquad \text{(T1)}$$

$$\begin{aligned}
Initiates&(MoveEast(t), Location(x_2), t_1) \leftarrow \qquad \text{(T2)}\\
&[HoldsAt(Location(x_1), t_1)\\
&\land x_2 = (x_1 + S \times Dur(MoveEast(t)))\\
&\land HoldsIn(MotorEngaged, t_1, (t_1 + Dur(MoveEast(t))))]
\end{aligned}$$

An alternative way of dealing with actions with duration is to split them into an (instantaneous) "start of action" (e.g. $StartMoveEast$), an "end of action" (e.g. $StopMoveEast$) and introduce an extra fluent representing the fact that the action is taking place (e.g. $MovingEast$). This approach is more easily integrated with the mechanisms described in Section 3.5 for dealing with gradual change, and allows straightforward description of interruptions of partly executed actions.

3.7   Dynamic Management of the Frame

We have already seen in Sections 3.4 and 3.5 how it can sometimes be advantageous to regard some fluents ("frame" fluents) as having an intrinsic (default) persistence, but regard other fluents as liable to change truth values between action occurrences. It can also be useful to be able to express that particular fluents have a default persistence during some intervals of time but not during others. This can, for example, help succinctly describe domains involving non-determinism, continuous change and indirect effects of actions (see [62] for details). In this section we illustrate how this facility for "dynamic management of the frame" can be incorporated into the Event Calculus by use of a new predicate $Releases \subseteq \mathcal{A} \times \mathcal{F} \times \mathcal{T}$. A form of this predicate was first introduced in [28] and it is related to Sandewall's idea of *occlusion* [56].

$Releases(A, F, T)$ expresses that if A occurs at T it will disable the fluent F's innate persistence. The truth value of F will then be free to fluctuate until the next action occurrence which initiates or terminates it. $Releases$ is defined in the domain-dependent part of the theory and circumscribed in parallel with $Initiates$ and $Terminates$. For example, in the alarm clock example of Section 3.5, we may write

$$Releases(Set, Countdown, t)$$

and if this is the only such statement in our theory, the circumscription will then give

$$Releases(a, f, t) \equiv [a = Set \land f = Countdown]$$

$Initiates(A, F, T)$ (respectively $Terminates(A, F, T)$) now expresses that if A occurs at T it will both initiate (respectively terminate) the fluent F and enable F's innate persistence. At any given time-point, therefore, a fluent can be in one of four states – true and persisting, false and persisting, true and released or false and released. To describe these states explicitly, we introduce a predicate $ReleasedAt \subseteq \mathcal{F} \times \mathcal{T}$ analogous to $HoldsAt$. Finally we need two new auxiliary predicates $ReleasedBetween \subseteq \mathcal{T} \times \mathcal{F} \times \mathcal{T}$ and $PersistsBetween \subseteq \mathcal{T} \times \mathcal{F} \times \mathcal{T}$. $ReleasedBetween(T_1, F, T_2)$ means "an action releases the fluent F between times T_1 and T_2" and $PersistsBetween(T_1, F, T_2)$ means "the fluent is not in a state of release at any time between T_1 and T_2."

The Event Calculus described in (EC1)–(EC6) needs fairly radical modifications to incorporate these extra concepts and predicates. The modified axiomatisation is as follows (for ease of reading (EC1) and (EC2) are listed again, although they are unmodified). The first three axioms are all similar and give definitions for $Clipped$, $Declipped$ and $ReleasedBetween$:

$$Clipped(t_1, f, t_2) \stackrel{\text{def}}{\equiv} \exists a, t[Happens(a, t) \land t_1 \leq t < t_2 \qquad \text{(EC1)}$$
$$\land \; Terminates(a, f, t)]$$

$$Declipped(t_1, f, t_2) \stackrel{\text{def}}{\equiv} \exists a, t[Happens(a, t) \land t_1 \leq t < t_2 \qquad \text{(EC2)}$$
$$\land \; Initiates(a, f, t)]$$

$$ReleasedBetween(t_1, f, t_2) \stackrel{\text{def}}{\equiv} \qquad\qquad\qquad \text{(EC14h)}$$
$$\exists a, t[Happens(a, t) \land t_1 \leq t < t_2$$
$$\land \; Releases(a, f, t)]$$

The next four axioms indicate how particular actions can put a fluent in one of the four states described above:

$$HoldsAt(f, t_2) \leftarrow [Happens(a, t_1) \land Initiates(a, f, t_1) \qquad \text{(EC3h)}$$
$$\land \; t_1 < t_2 \land \neg Clipped(t_1, f, t_2)$$
$$\land \neg ReleasedBetween(t_1, f, t_2)]$$

$$\neg HoldsAt(f, t_2) \leftarrow [Happens(a, t_1) \land Terminates(a, f, t_1) \qquad \text{(EC4h)}$$
$$\land \; t_1 < t_2 \land \neg Declipped(t_1, f, t_2)$$
$$\land \neg ReleasedBetween(t_1, f, t_2)]$$

$$ReleasedAt(f, t_2) \leftarrow [Happens(a, t_1) \land Releases(a, f, t_1) \qquad \text{(EC15h)}$$
$$\land \ t_1 < t_2 \ \land \ \neg Clipped(t_1, f, t_2)$$
$$\land \neg Declipped(t_1, f, t_2)]$$

$$\neg ReleasedAt(f, t_2) \leftarrow \qquad\qquad\qquad\qquad\qquad\qquad \text{(EC16h)}$$
$$[Happens(a, t_1) \land \ t_1 < t_2$$
$$\land \ [Initiates(a, f, t_1) \lor Terminates(a, f, t_1)]$$
$$\land \neg ReleasedBetween(t_1, f, t_2)]$$

A weakened version of the "commonsense law of inertia" is captured in the following three axioms:

$$PersistsBetween(t_1, f, t_2) \overset{\text{def}}{\equiv} \qquad\qquad\qquad\qquad\qquad \text{(EC17h)}$$
$$\neg \exists t[ReleasedAt(f, t) \ \land \ t_1 \leq t \leq t_2]$$

$$HoldsAt(f, t_2) \leftarrow [HoldsAt(f, t_1) \ \land \ t_1 < t_2 \qquad\qquad \text{(EC5h)}$$
$$\land \ PersistsBetween(t_1, f, t_2)$$
$$\land \neg Clipped(t_1, f, t_2)]$$

$$\neg HoldsAt(f, t_2) \leftarrow [\neg HoldsAt(f, t_1) \ \land \ t_1 < t_2 \qquad\qquad \text{(EC6h)}$$
$$\land \ PersistsBetween(t_1, f, t_2)$$
$$\land \neg Declipped(t_1, f, t_2)]$$

Finally, we need to state that the meta-property of being "released" is itself subject to a form of meta-persistence between action occurrences:

$$ReleasedAt(f, t_2) \leftarrow [ReleasedAt(f, t_1) \ \land \ t_1 < t_2 \qquad\qquad \text{(EC18h)}$$
$$\land \neg Clipped(t_1, f, t_2)$$
$$\land \neg Declipped(t_1, f, t_2)]$$

$$\neg ReleasedAt(f, t_2) \leftarrow [\neg ReleasedAt(f, t_1) \ \land \ t_1 < t_2 \qquad\qquad \text{(EC19h)}$$
$$\land \neg ReleasedBetween(t_1, f, t_2)]$$

Individual *ReleasedAt* literals can be included in the domain dependent part of the theory in the same way as *HoldsAt* literals (see for example axiom (R3) in Section 2.1).

The above axiomatisation is fairly complex – it replaces our original six axioms with twelve (longer) ones and introduces four new predicates. However, for practical and computational purposes (e.g. ease of translation into logic programs) and where we are using non-negative time, we can dispense with the predicates *ReleasedAt*, *ReleasedBetween* and *PersistsBetween* and simply incorporate *Releases* in the definitions of *Clipped* and *Declipped*. This gives rise to the following alternative (and complete) set of domain independent axioms:

$$Clipped(t_1, f, t_2) \stackrel{\text{def}}{\equiv} \tag{EC1i}$$
$$\exists a, t[Happens(a, t) \wedge t_1 \leq t < t_2$$
$$\wedge [Terminates(a, f, t) \vee Releases(a, f, t)]]$$

$$Declipped(t_1, f, t_2) \stackrel{\text{def}}{\equiv} \tag{EC2i}$$
$$\exists a, t[Happens(a, t) \wedge t_1 \leq t < t_2$$
$$\wedge [Initiates(a, f, t) \vee Releases(a, f, t)]]$$

$$HoldsAt(f, t_2) \leftarrow [Happens(a, t_1) \wedge Initiates(a, f, t_1) \tag{EC3}$$
$$\wedge t_1 < t_2 \wedge \neg Clipped(t_1, f, t_2)]$$

$$\neg HoldsAt(f, t_2) \leftarrow [Happens(a, t_1) \wedge Terminates(a, f, t_1) \tag{EC4}$$
$$\wedge t_1 < t_2 \wedge \neg Declipped(t_1, f, t_2)]$$

$$HoldsAt(f, t) \leftarrow [InitiallyP(f) \wedge \neg Clipped(0, f, t)] \tag{EC5a}$$

$$\neg HoldsAt(f, t) \leftarrow [InitiallyN(f) \wedge \neg Declipped(0, f, t)] \tag{EC6a}$$

Note that fluents which are *InitiallyP* or *InitiallyN* are initially in the frame, whereas those which are neither *InitiallyP* nor *InitiallyN* are effectively initially "released". Hence axiom (EC8a) (see Section 3.1) implies that all fluents are initially in the frame, and may or may not be appropriate for a given domain.

3.8   The Event Calculus, Continuous Change and Mathematical Modelling

The techniques using the *Trajectory* and *AntiTrajectory* predicates discussed in Section 3.5 are sufficient for modelling domains with very simple forms of continuous change, in particular where an explicit function of time is known for a particular parameter after a particular fluent has been initiated or terminated. However, this method is in general insufficient for integrating standard mathematical modelling techniques with the Event Calculus, for several reasons. First, the majority of mathematical models are expressed as sets of differential equations, and these cannot in general be solved so as to produce explicit functions of time for each parameter involved. Second, there might only be incomplete knowledge, expressed perhaps using inequalities, about the mathematical relationship between various parameters and/or their derivatives. Third, the circumstances under which various mathematical relationships hold between parameters might not be (easily) expressible in terms of a single fluent. Fourth, trajectories and antitrajectories do not provide mechanisms for describing continuous change in time intervals before any relevant initiating and/or terminating actions have occurred.

A more general approach is to include domain independent axioms which explicitly utilise the mathematical definitions of continuity and differentiability of real-valued functions of time. Under this approach, which is partly inspired by

Sandewall's work [54,55] and described in more detail in [42], continuity of real-valued parameters is regarded as a default analogous to default persistence of fluents, so that discontinuities arise only in particular parameters when specific actions occur. For this section, we will assume that time is represented either as the real numbers or as the non-negative real numbers. We will assume that some or all terms of sort $\mathcal{P}$ (introduced in Section 3.5) represent real-valued functions of time, and accordingly introduce two new function symbols $Value :$ $\mathcal{P} \times \mathcal{T} \mapsto \mathbb{R}$ and $\delta : \mathcal{P} \mapsto \mathcal{P}$. The term $Value(P, T)$ represents the numerical value of parameter P at time T, and the axiomatisation below ensures that the term $Value(\delta(P), T)$ represents the numerical value at time T of its first derivative (at all time-points where this exists).

To integrate the standard mathematical concepts of continuity and differentiability into the Event Calculus, we need to express them in terms of $Value$ and δ. It is also convenient to introduce the predicates $LeftContinuous \subseteq \mathcal{P} \times \mathcal{T}$ and $RightLimit \subseteq \mathcal{P} \times \mathcal{T}$ to capture the corresponding (standard) mathematical concepts[6]:

$$Continuous(p, t) \equiv \forall r \exists t_1 \forall t_2 [[|t - t_2| < t_1 \wedge 0 < r] \quad \text{(EC20j)}$$
$$\rightarrow |Value(p, t) - Value(p, t_2)| < r]$$

$$Differentiable(p, t) \equiv \quad \text{(EC21j)}$$
$$\forall r \exists t_1 \forall t_2 [[0 < |t - t_2| < t_1 \wedge 0 < r] \rightarrow$$
$$|(\frac{Value(p,t) - Value(p,t_2)}{t - t_2}) - Value(\delta(p), t)| < r]$$

$$LeftContinuous(p, t) \equiv \quad \text{(EC22j)}$$
$$\forall r \exists t_1 \forall t_2 [[t_2 < t \wedge (t - t_2) < t_1 \wedge 0 < r] \rightarrow$$
$$|Value(p, t) - Value(p, t_2)| < r]$$

$$RightLimit(p, t, r) \equiv \quad \text{(EC23j)}$$
$$\forall r_1 \exists t_1 \forall t_2 [[t < t_2 \wedge (t_2 - t) < t_1 \wedge 0 < r_1]$$
$$\rightarrow |Value(p, t_2) - r| < r_1]$$

To respect the convention that actions take effect immediately *after* they occur, it is necessary to axiomatise the mathematical constraint that, at every time-point (including those at which actions occur), the function associated with each parameter is left-hand continuous:

---

[6] A function is *left-continuous* if discontinuities occur only between successive intervals where the first is closed on the right and the second is open on the left. For example the function $f(t) = 0$ for all $t \leq 1$, $f(t) = 2$ otherwise, is left-continuous at all time-points, whereas the function $f'(t) = 0$ for all $t < 1$, $f'(t) = 2$ otherwise, is not. The *right-limit* of a function at a particular point is the limit value as the point is approached from the right. So, for example, the right-limit of both f and f' at 1 is 2.

$$LeftContinuous(p,t) \tag{EC24j}$$

To describe instantaneous changes in the values of parameters at times when actions occur, and discontinuities in their corresponding functions of time, the predicates $BreaksTo \subseteq \mathcal{A} \times \mathcal{P} \times \mathcal{T} \times \mathbb{R}$ and $Breaks \subseteq \mathcal{A} \times \mathcal{P} \times \mathcal{T}$ are introduced. Both are minimised (by circumscribing them in parallel). $BreaksTo(A, P, T, R)$ should be read as 'at time T, an occurrence of action A will cause parameter P to instantaneously take on value R'. More precisely, Axiom (EC27j) below states that if A does indeed occur at time T, then R is the value of the right-hand limit of P at T. $Breaks(A, P, T)$ can be read as 'at time T, action A potentially causes a discontinuity in parameter P'. The following domain-independent axioms make direct use of $BreaksTo$ and $Breaks$. Axioms (EC25j) and (EC26j) can be likened to 'frame axioms' for parameters. Axiom (EC28j) states the relationship between $BreaksTo$ and $Breaks$, and Axiom (EC29j) states that if an action potentially causes a discontinuity in a given parameter, it also potentially causes discontinuities in its higher derivatives.

$$\neg[Happens(a,t) \wedge Breaks(a,p,t)] \rightarrow Continuous(p,t) \tag{EC25j}$$

$$\neg[Happens(a,t) \wedge Breaks(a,\delta(p),t)] \rightarrow Differentiable(p,t) \tag{EC26j}$$

$$[BreaksTo(a,p,t,r) \wedge Happens(a,t)] \rightarrow RightLimit(p,t,r) \tag{EC27j}$$

$$BreaksTo(a,p,t,r) \rightarrow Breaks(a,p,t) \tag{EC28j}$$

$$Breaks(a,p,t) \rightarrow Breaks(a,\delta(p),t) \tag{EC29j}$$

To make useful derivations using this axiomatisation, for any given time point T it is useful to be able to refer to the next point after T at which an action occurs, if there is such a point. Axioms (EC30j), (EC31j) and (EC32j) state that if any action occurs at any time point after T, then the term $Next(T)$ refers to the least such time point. (Such points are somewhat analogous to the "least natural time points" discussed in [51].)

$$t < Next(t) \tag{EC30j}$$

$$[t < t_1 \wedge t_1 < Next(t)] \rightarrow \neg Happens(a,t_1) \tag{EC31j}$$

$$[Happens(a_1,t_1) \wedge t < t_1] \rightarrow \exists a.Happens(a, Next(t)) \tag{EC32j}$$

The above axiomatisation leaves us free to include (unsolved) sets of simultaneous differential equations in domain descriptions. As a simple illustration, suppose we wish to represent that the rate of change of the level of liquid in a tank is negatively proportional to the flow through a valve in its bottom, and that when the valve is open the flow is in turn proportional to the level (i.e. pressure). We need a single fluent $ValveOpen$, two parameters $Level$ and

Flow, and actions *Open Valve* and *Close Valve*. As well as *Happens*, *Initiates* and *Terminates* facts such as

$$Initiates(OpenValve, ValveOpen, t) \tag{V1}$$

$$Terminates(CloseValve, ValveOpen, t) \tag{V2}$$

we can represent information about the instantaneous effects of actions on parameters using *Breaks*,

$$Breaks(OpenValve, Flow, t) \tag{V3}$$

$$Breaks(OpenValve, \delta(Level), t) \tag{V4}$$

and include mathematical constraints (differential equations) which hold in different circumstances, e.g.

$$Value(\delta(Level), t) = -Value(Flow, t) \tag{V5}$$

$$HoldsAt(ValveOpen, t) \rightarrow \exists r [Value(Flow, t) = r.Value(Level, t)] \tag{V6}$$

In this case the full theory will include the circumscription $CIRC[(\text{EC28j}) \wedge (\text{EC29j}) \wedge (\text{V3}) \wedge (\text{V4}) ; Breaks, BreaksTo]$. The Event Calculus now allows us to infer new boundary conditions for sets of differential equations which become applicable when actions such as *Open Valve* and *Close Valve* occur. A variation of this example is discussed in more detail in [42].

The above axiomatisation lays a foundation for integrating the Event Calculus with representational and computational techniques from the field of Qualitative Reasoning [9] [34]. An Event Calculus based axiomatisation of some of the basic concepts in [34] is given in [42].

3.9   Other Issues and Extensions

Space limitations forbid a detailed summary of all work done on extending the classical logic Event Calculus in this article. In particular, three important topics we have not covered are *hierarchical actions*, *ramifications*, and *knowledge producing actions*.

Hierarchical or *compound actions* are non-instantaneous actions whose occurrence consists of the occurrence of a set of shorter actions. (For example, the "go to work" action might comprise a "walk to the station" action, a "get the train" action and a "walk to the office" action.) These can be formalised in the Event Calculus using "happens if happens" formulae. For more details, see [63] or [68]. Davila [10] has done related work on formulating programming constructs within an Event Calculus framework.

The ramification problem is the problem of representing permanent constraints between collections of fluents, and indirect effects of actions propagated via such constraints, whilst preserving a succinct and elaboration tolerant solution to the frame problem. Shanahan [66] has shown that a straightforward extension of the Event Calculus can handle many canonical examples of the ramification problem, including those in which concurrent events simultaneously affect the same fluent. In Section 4 we show an equivalence between the Event Calculus and the Language $\mathcal{E}$ [21], and $\mathcal{E}$ has been extended to deal with ramifications in [22] by using fixed point definitions to express how actions indirectly initiate and terminate fluents. It seems likely that this same technique can be described in the classical Event Calculus using inductive definitions similar to those in [69] and [70].

To our knowledge, little work has been done in the Event Calculus on representing the effects of knowledge producing actions. These are important, for example, in the context of planning. To catch a flight, an agent may plan to go to the airport and then look at the departures board to find out which gate the flight is boarding from. The action of looking at the board doesn't change the state of the external world but rather the agent's knowledge of it. To reason about such actions, the agent has to have a model about its own future knowledge state and how this will relate to the external world. Work on addressing these issues in the context of other action formalisms can be found for example in [36], [38], [46], [47] and [57].

4   A Correspondence Result

The focus of the previous sections has been on the development of Event Calculus axiomatisations written in standard predicate calculus to represent knowledge about the effects of actions. In this sense it follows the tradition established by McCarthy and others in developing the Situation Calculus [40]. Implicit in such work is the idea that such classical logic theories can act as specifications for computer programs that simulate various forms of reasoning about the domains represented. However, more recently there has been a trend towards the use of more specialised logics for representing and reasoning about the effects of actions, and in particular a growing body of work on the development and implementation of "action description languages" [16,17]. It is not our intention here to argue the merits and demerits of specialised as opposed to general purpose logics. (We do not for example subscribe to the view that formulations in classical or other general purpose logics require formulations in specialised logics to act as their "specification" or "semantics", or that specialised logics are at a "higher level" because they lack a proof theory.) However, it is clearly advantageous to explore correspondences between various types of representation, so that results and implementations for one approach can be more readily adapted to others.

While the majority of action description languages bear a resemblance to the Situation Calculus, the Language $\mathcal{E}$ [21,22] is inspired by, and inherits its ontology from, the Event Calculus. In this section we describe the circumstances under which Event Calculus theories correspond to Language $\mathcal{E}$ domain descriptions and may thus take advantage of the provably correct automated proof procedures that have been developed for $\mathcal{E}$ (see e.g. [23], [24], [26]).

4.1   The Language $\mathcal{E}$

The definition of the Language $\mathcal{E}$ given here corresponds to that in [21]. (This definition has subsequently been extended in various ways, in particular to deal with ramifications and the ramification problem [22,23].)

The Language $\mathcal{E}$ is really a collection of languages. The particular vocabulary of each language depends on the domain being represented, but always includes a set of *fluent constants*, a set of *action constants*, and a partially ordered set of *time-points*. A *fluent literal* may either be a fluent constant or its negation, as shown in the following definitions.

Definition 1 (Domain Language). *A domain language is a tuple $\langle \Pi, \preceq , \Delta, \Phi \rangle$, where $\preceq$ is a partial (possibly total) ordering defined over the non-empty set Π of time points, Δ is a non-empty set of action constants, and Φ is a non-empty set of fluent constants.*

Definition 2 (Fluent literal). *A fluent literal of $\mathcal{E}$ is an expression either of the form F or of the form $\neg F$, where $F \in \Phi$.*

Three types of statements are used to describe domains; *h-propositions* ("h" for "happens"), *t-propositions* ("t" for "time point") and *c-propositions* ("c" for "causes"). Their intended meanings are clear from their definitions:

Definition 3 (h-proposition). *An h-proposition in $\mathcal{E}$ is an expression of the form*

$$A \textbf{ happens-at } T$$

where $A \in \Delta$ and $T \in \Pi$.

Definition 4 (t-proposition). *A t-proposition in $\mathcal{E}$ is an expression of the form*

$$L \textbf{ holds-at } T$$

where L is a fluent literal of $\mathcal{E}$ and $T \in \Pi$.

Definition 5 (c-proposition). *A c-proposition in $\mathcal{E}$ is an expression either of the form*

$$A \textbf{ initiates } F \textbf{ when } C$$

or of the form

$$A \textbf{ terminates } F \textbf{ when } C$$

where $F \in \Phi$, $A \in \Delta$, and C is a set of fluent literals of $\mathcal{E}$.

C-propositions of the form "*A* **initiates** *F* **when** $\emptyset$" and "*A* **terminates** *F* **when** $\emptyset$" can be written more simply as "*A* **initiates** *F*" and "*A* **terminates** *F*" respectively. A *domain description* in $\mathcal{E}$ is a triple $\langle \gamma, \eta, \tau \rangle$, where γ is a set of c-propositions, η is a set of h-propositions and τ is a set of t-propositions.

The Event Calculus domain described in Section 2.1 might be described as an $\mathcal{E}$ domain description D_R as follows. For action and fluent constants we would have $\Delta = \{Insert, GoThrough, Pickup\}$ and $\Phi = \{Inside, HasKey, Locked\}$ respectively. For Π and $\preceq$ we would use the real numbers with the usual ordering relation. Axioms (R1)–(R4) would be expressed in D_R as:

$$Pickup \textbf{ initiates } HasKey$$
$$Insert \textbf{ initiates } Locked \textbf{ when } \{\neg Locked, HasKey\}$$
$$GoThrough \textbf{ initiates } Inside \textbf{ when } \{\neg Locked, \neg Inside\}$$
$$GoThrough \textbf{ terminates } Inside \textbf{ when } \{\neg Locked, Inside\}$$
$$Insert \textbf{ terminates } Locked \textbf{ when } \{Locked, HasKey\}$$
$$Locked \textbf{ holds-at } 0$$
$$Inside \textbf{ holds-at } 0$$
$$Pickup \textbf{ happens-at } 2$$
$$Insert \textbf{ happens-at } 4$$
$$GoThrough \textbf{ happens-at } 6$$

(The reader may also find it useful to compare this collection of propositions with axioms (R5)–(R12) in Section 2.2.)

The semantics of $\mathcal{E}$ is based on simple definitions of interpretations and models. Since the primary interest is in inferences about what holds at particular time-points in Π, it is sufficient to define an interpretation as a mapping of fluent/time-point pairs to *true* or *false* (i.e. a "*holds*" relation). An interpretation *satisfies* a fluent literal or set of fluent literals at a particular time-point if it assigns the relevant truth values to each of the corresponding fluent constants:

Definition 6 (Interpretation). *An interpretation of $\mathcal{E}$ is a mapping*

$$H : \Phi \times \Pi \mapsto \{true, false\}$$

Definition 7 (Point satisfaction). *Given a set of fluent literals C of $\mathcal{E}$ and a time point $T \in \Pi$, an interpretation H satisfies C at T iff for each fluent constant $F \in C$, $H(F, T) = true$, and for each fluent constant F' such that $\neg F' \in C$, $H(F', T) = false$.*

The definition of a model in $\mathcal{E}$ is parametric on the definitions of an *initiation point* and a *termination point*. Initiation and termination points are simply time-points where a c-proposition and an h-proposition combine to describe a direct effect on a particular fluent:

Definition 8 (Initiation/termination point). *Let H be an interpretation of $\mathcal{E}$, let $D = \langle \gamma, \eta, \tau \rangle$ be a domain description, let $F \in \Phi$ and let $T \in \Pi$. T is* *an initiation-point (respectively termination-point) for F in H relative to D iff there is an $A \in \Delta$ such that (i) there is both an h-proposition in η of the form "A **happens-at** T" and a c-proposition in γ of the form "A **initiates** F **when** C" (respectively "A **terminates** F **when** C") and (ii) H satisfies C at T.*

For an interpretation to qualify as a model, three basic properties need to be satisfied; (1) fluents change their truth values only via occurrences of initiating or terminating actions, (2) initiating a fluent establishes its truth value as *true*, and (3) terminating a fluent establishes its truth value as *false*. In addition, (4) every model must match with each of the t-propositions in the domain description:

Definition 9 (Model). *Given a domain description $D = \langle \gamma, \eta, \tau \rangle$ in $\mathcal{E}$, an interpretation H of $\mathcal{E}$ is a model of D iff, for every $F \in \Phi$ and $T, T', T_1, T_3 \in \Pi$ such that $T_1 \prec T_3$, the following properties hold:*

1. *If there is no initiation-point or termination-point T_2 for F in H relative to D such that $T_1 \preceq T_2 \prec T_3$, then $H(F, T_1) = H(F, T_3)$.*
2. *If T_1 is an initiation-point for F in H relative to D, and there is no termination-point T_2 for F in H relative to D such that $T_1 \prec T_2 \prec T_3$, then $H(F, T_3) = true$.*
3. *If T_1 is a termination-point for F in H relative to D, and there is no initiation-point T_2 for F in H relative to D such that $T_1 \prec T_2 \prec T_3$, then $H(F, T_3) = false$.*
4. *For all t-propositions in τ of the form "F **holds-at** T", $H(F, T) = true$, and for all t-propositions of the form "$\neg F$ **holds-at** T'", $H(F, T') = false$.*

Definition 10 (Consistency). *A domain description is consistent iff it has a model.*

Definition 11 (Entailment). *A domain description D entails the t-proposition "F **holds-at** T", written[7] "$D \models_{\mathcal{E}} F$ **holds-at** T", iff for every model H of D, $H(F, T) = true$. D entails the t-proposition "$\neg F$ **holds-at** T" iff for every model H of D, $H(F, T) = false$.*

As regards the robot example, using the above definitions it is easy to see that

$$D_R \models_{\mathcal{E}} \neg Inside \text{ \textbf{holds-at} } 8$$

More generally, if time is taken as the integers or reals, Definitions 8 and 9 indicate that the Language $\mathcal{E}$ corresponds to the "deterministic" Event Calculus described in Section 3.2, i.e. with domain independent axioms (EC1), (EC2), (EC3b), (EC4b), (EC5), (EC6), (EC9b) and (EC10b). Specifically condition 1 of Definition 9 mirrors axioms (EC1), (EC2), (EC5) and (EC6), condition 2 mirrors (EC3b) and (EC9b), and condition 3 mirrors (EC4b) and (EC10b). This correspondence is established more formally in the next section.

---

[7] The symbol $\models_{\mathcal{E}}$ is used here to distinguish Language $\mathcal{E}$ entailment from entailment in classical logic. It is identical in meaning to the symbol $\models$ used in other publications concerning the Language $\mathcal{E}$.

4.2   Translating Between the Event Calculus and $\mathcal{E}$

Clearly, for some domains (such as the robot example) translation from the Event Calculus to $\mathcal{E}$ (and vice versa) is straightforward. Equally clearly, for some other Event Calculus theories, perhaps with disjunctive or existentially quantified sentences partially defining *Initiates*, *Terminates*, *Happens* or *HoldsAt* (e.g. the robot example extended with (R13)), a translation into the restricted syntax of $\mathcal{E}$ is not possible. But it is difficult and cumbersome in general to describe necessary and sufficient syntactic conditions whereby an Event Calculus theory can be translated into an equivalent Language $\mathcal{E}$ domain description.

To illustrate, consider the following Event Calculus description of a "millennium counter" – a display of the minutes passed since 12 midnight on 31 December 2000. Time is taken as the integers, where each integer represents one second and 0 represents 12 midnight, 31 December 2000. An action *Tick* happens once every 60 seconds and increments the display by 1:

$$Initiates(a, f, t) \equiv [a = Tick \ \wedge \ \exists n.[f = Display(n)$$
$$\wedge \ HoldsAt(Display(n-1), t)]]$$

$$Terminates(a, f, t) \equiv [a = Tick \ \wedge \ \exists n.[f = Display(n)$$
$$\wedge \ HoldsAt(Display(n), t)]]$$

$$Happens(a, t) \equiv [a = Tick \ \wedge \ \exists t'.[t = (t' * 60)]]$$

$$HoldsAt(Display(0), 0) \ \wedge \ \forall n.[n \neq 0 \ \rightarrow \ \neg HoldsAt(Display(n), 0)]$$

This axiomatisation might at first seem problematic as regards translation into $\mathcal{E}$; it entails an infinite number of positive ground *Initiates*, *Terminates* and *Happens* literals and (even without augmentation with domain independent Event Calculus axioms) an infinite number of negative ground *HoldsAt* literals (at $t = 0$). All of these need explicit representation in $\mathcal{E}$. But the following (infinite) Language $\mathcal{E}$ domain description $\langle \gamma, \eta, \tau \rangle$ is well defined and clearly entails the same collection of "holds at" facts along the time line:

$$\gamma = \{ Tick \text{ terminates } Display(n) \text{ when } \{ Display(n) \} \mid n \in \mathbb{Z} \}$$
$$\cup$$
$$\{ Tick \text{ initiates } Display(n) \text{ when } \{ Display(m) \} \mid$$
$$n, m \in \mathbb{Z} \text{ and } n = m + 1 \}$$

$$\eta = \{ Tick \text{ happens-at } (t * 60) \mid t \in \Pi \}$$

$$\tau = \{ Display(0) \text{ holds-at } 0 \}$$
$$\cup$$
$$\{ \neg Display(n) \text{ holds-at } 0 \mid n \in \mathbb{Z} \text{ and } n \neq 0 \}$$

This example illustrates that any general syntactic constraints that we place on Event Calculus theories in order to ensure that they are translatable into $\mathcal{E}$ are likely to be over-restrictive. In what follows, we therefore instead concentrate on establishing a collection of sufficient (and intuitive) "semantic" constraints for a correct translation to be possible. Each of these will in most cases be straightforward to check from the form of the axiomatisation in question. Precisely what we mean by a "correct translation" is established in Proposition 1.

In Definitions 12 to 20 and Proposition 1 that follow, we will assume that $D = \langle \gamma, \eta, \tau \rangle$ is a Language $\mathcal{E}$ domain description written in the language $\langle \Pi, \leq , \Delta, \Phi \rangle$ (where Π is either $\mathbb{Z}$ or $\mathbb{R}$). We will also assume that T_{EC} is a collection of (domain dependent) axioms written in a sorted predicate calculus language of the type described in Section 2 that constrains the interpretation of the sort $\mathcal{T}$ to be Π, and that T_{EC} does not mention the predicates *Clipped*, *Declipped*, *StoppedIn* and *StartedIn*. Furthermore we will assume that the language of T_{EC} includes all symbols in Δ as ground terms of sort $\mathcal{A}$ and all symbols in Φ as ground terms of sort $\mathcal{F}$. **Notation:** We will denote as $\Phi^{\pm}$ the set of all (positive and negative) fluent literals that can be formed from the fluent constants in Φ. Given a model M of T_{EC}, $\|G\|_M$ will denote the interpretation (i.e. the denotation) of the ground term or symbol G in M. We will refer to the set of domain independent Event Calculus axioms $\{(\text{EC1}), (\text{EC2}), (\text{EC3b}), (\text{EC4b}), (\text{EC5}), (\text{EC6}), (\text{EC9b}), (\text{EC10b})\}$ (see Sections 2 and 3.2) as Det_{EC}.

The first condition to express is that (in all its models) T_{EC} establishes uniqueness of names for the fluents and actions referred to in D:

Definition 12 (Name-matches). *D name-matches T_{EC} iff for every model M of T_{EC}, for every $F, F' \in \Phi$ and for every $A, A' \in \Delta$,*

- *if $F \neq F'$ then $\|F\|_M \neq \|F'\|_M$, and*
- *if $A \neq A'$ then $\|A\|_M \neq \|A'\|_M$.*

Typically this name-matches property might be established by a collection of inequality statements in T_{EC} between ground fluent and action literals (e.g. *Inside* $\neq$ *HasKey*, etc. in the Robot example) or by universally quantified implications such as $\forall m, n.[Display(m) = Display(n) \rightarrow m = n]$.

The next condition to establish (Definitions 13 to 16 below) is that all interpretations of *Initiates*, *Terminates* and *Happens* licensed by T_{EC} are isomorphic to the unique interpretation (relative to the interpretation of *HoldsAt*) explicitly indicated by the c- and h-propositions in D:

Definition 13 (h-satisfies). *Given a model M of T_{EC}, a time-point $T \in \Pi$ and a set $C \subseteq \Phi^{\pm}$ of Language $\mathcal{E}$ fluent literals, M h-satisfies C at T iff for all $F \in \Phi$, if $F \in C$ then $\langle \|F\|_M, T \rangle \in \|HoldsAt\|_M$, and if $\neg F \in C$ then $\langle \|F\|_M, T \rangle \notin \|HoldsAt\|_M$.*

Definition 14 (Initiates-matches). *D initiates-matches T_{EC} iff for every model M of T_{EC}, every time-point T and every action α and fluent ϕ in the domain of discourse of M the following holds. $\langle \alpha, \phi, T \rangle \in \|Initiates\|_M$ if and* *only if there exist $F \in \Phi$, $A \in \Delta$ and $C \subseteq \Phi^{\pm}$ such that $\alpha = \|A\|_M$, $\phi = \|F\|_M$, M h-satisfies C at T, and "A initiates F when C" $\in \gamma$.*

Definition 15 (Terminates-matches). *D terminates-matches T_{EC} iff for every model M of T_{EC}, every time-point T and every action α and fluent ϕ in the domain of discourse of M the following holds. $\langle \alpha, \phi, T \rangle \in \|Terminates\|_M$ if and only if there exist $F \in \Phi$, $A \in \Delta$ and $C \subseteq \Phi^{\pm}$ such that $\alpha = \|A\|_M$, $\phi = \|F\|_M$, M h-satisfies C at T, and "A terminates F when C" $\in \gamma$.*

Definition 16 (Happens-matches). *D happens-matches T_{EC} iff for every model M of T_{EC}, every time-point T and every action α in the domain of discourse of M the following holds. $\langle \alpha, T \rangle \in \|Happens\|_M$ if and only if there exists $A \in \Delta$ such that $\alpha = \|A\|_M$ and "A happens-at T" $\in \eta$.*

Finally, it is necessary to establish that (without the domain independent Event Calculus axioms in Det_{EC}), T_{EC} imposes exactly the same collection of pointwise constraints on the interpretation of *HoldsAt* that are indicated by the t-propositions in D. To do this it is necessary to impose a domain closure property on fluent names (the first condition in Definition 19). It is also necessary to ensure that T_{EC} does not entail any extra "global dependencies" not captured by the t-propositions of D, either between two or more fluents (e.g. $\forall t.[HoldsAt(HasKey, t) \rightarrow HoldsAt(Inside, t)]$), or between fluents and other facts represented in T_{EC} (e.g. $\forall t.[HoldsAt(HasKey, t) \rightarrow SmallEnoughToHold(Key)]$). This is guaranteed by the third condition in Definition 19.

Definition 17 (t-model). *An interpretation H of $\mathcal{E}$ is a t-model of D iff, for every $F \in \Phi$ and $T, T' \in \Pi$, for all t-propositions in τ of the form "F holds-at T", $H(F, T) = true$, and for all t-propositions of the form "$\neg F$ holds-at T'", $H(F, T') = false$.*

Definition 18 ($\mathcal{E}$-projection). *The $\mathcal{E}$-projection of a model M of T_{EC} is defined as the following (Language $\mathcal{E}$) interpretation H_M:*

$$H_M(F, T) = \begin{cases} true & if \ \langle \|F\|_M, T \rangle \in \|HoldsAt\|_M \\ false & otherwise \end{cases}$$

Definition 19 (Holds-matches). *D holds-matches T_{EC} iff for every model M of T_{EC} the following conditions are satisfied:*

- *for every fluent ϕ in the domain of discourse of M there exists $F \in \Phi$ such that $\phi = \|F\|_M$,*
- *the $\mathcal{E}$-projection of M is a t-model of D,*
- *For every t-model H^t of D there is a model M^{H^t} of T_{EC} which differs from M only in the interpretation of HoldsAt and is such that H^t is the $\mathcal{E}$-projection of M^{H^t}.*

Definition 20 (matches). *D matches T_{EC} iff D name-matches, initiates-matches, terminates- matches, happens-matches and holds-matches T_{EC}.*

Proposition 1. *Let $F \in \Phi$ and let $T \in \Pi$. If T_{EC} is consistent and D matches T_{EC} then:*

- *$D \models_{\mathcal{E}} F$ holds-at T iff $T_{EC} \cup Det_{EC} \models HoldsAt(F, T)$*
- *$D \models_{\mathcal{E}} \neg F$ holds-at T iff $T_{EC} \cup Det_{EC} \models \neg HoldsAt(F, T)$*

Proof. It is sufficient to prove the following:

1. If there exists a model H of D such that $H(F, T) = true$ then there exists a model M^H of $T_{EC} \cup Det_{EC}$ such that $M^H \Vdash HoldsAt(F, T)$.
2. If there exists a model M of $T_{EC} \cup Det_{EC}$ such that $M \Vdash HoldsAt(F, T)$ then there exists a model H_M of D such that $H_M(F, T) = true$.
3. If there exists a model H of D such that $H(F, T) = false$ then there exists a model M^H of $T_{EC} \cup Det_{EC}$ such that $M^H \Vdash \neg HoldsAt(F, T)$.
4. If there exists a model M of $T_{EC} \cup Det_{EC}$ such that $M \Vdash \neg HoldsAt(F, T)$ then there exists a model H_M of D such that $H_M(F, T) = false$.

Proof of (1):

If there exists a model H of D such that $H(F, T) = true$ then by Definitions 9 and 17 H is a t-model of D. Hence, since T_{EC} is consistent, by Definition 19 there exists a model M^H of T_{EC} such that H is the $\mathcal{E}$-projection of M^H. Therefore $M^H \Vdash HoldsAt(F, T)$. Since T_{EC} does not mention the predicates *Clipped*, *Declipped*, *StoppedIn* and *StartedIn* then clearly we can assume that M^H is such that it satisfies (EC1), (EC2), (EC9b) and (EC10b). Since D name-matches, initiates-matches, terminates-matches and happens-matches T_{EC} then by condition 1 of Definition 9 M^H satisfies (EC5) and (EC6), by condition 2 of Definition 9 M^H satisfies (EC3b), and by condition 3 of Definition 9 M^H satisfies (EC4b). Therefore M^H is a model of $T_{EC} \cup Det_{EC}$.

Proof of (2):

If there exists a model M of $T_{EC} \cup Det_{EC}$ such that $M \Vdash HoldsAt(F, T)$, then by Definition 19 the $\mathcal{E}$-projection H_M of M is a t-model of D and $H_M(F, T) = true$. It remains to show that H_M satisfies conditions 1, 2 and 3 of Definition 9. Since D name-matches, initiates-matches, terminates-matches and happens-matches T_{EC}, it follows directly from the fact that $M \Vdash [(EC5) \wedge (EC6)]$ that H_M satisfies condition 1 of Definition 9, it follows directly from the fact that $M \Vdash (EC3b)$ that H_M satisfies condition 2 of Definition 9, and it follows directly from the fact that $M \Vdash (EC4b)$ that H_M satisfies condition 1 of Definition 9.

Proof of (3):

This is identical to the proof of (1), but substituting "$H(F, T) = false$" for "$H(F, T) = true$" and substituting "$M^H \Vdash \neg HoldsAt(F, T)$" for "$M^H \Vdash HoldsAt(F, T)$".

Proof of (4):

This is identical to the proof of (2), but substituting "$M \Vdash \neg HoldsAt(F, T)$" for "$M \Vdash HoldsAt(F, T)$" and substituting "$H_M(F, T) = false$" for "$H_M(F, T) = true$".

(end of proof of Proposition 1)

Proposition 1 is analogous in some respects to the results in [27], which show the equivalence of various classical logic formulations of the Situation Calculus to the Language $\mathcal{A}$. But whereas the conditions for the results in [27] are syntactic, those for Proposition 1 are semantic and so less restrictive. Although checking through all the conditions for Proposition 1 to hold might at first sight seem tedious, in many cases the fact that a collection of domain dependent axioms "matches" a Language $\mathcal{E}$ domain description will be obvious. In particular, it is clear that any Language $\mathcal{E}$ domain description written using only a finite number of action and fluent constants can be straightforwardly translated into an Event Calculus axiomatisation by formulating sentences analogous to (R1) – (R4) (see Section 2.1).

As stated earlier, Proposition 1 is useful because it allows the (deterministic) classical logic Event Calculus to take advantage of the provably correct automated reasoning procedures developed for $\mathcal{E}$ (see [21],[22],[23],[24]). Of these implementations, the most flexible is that described in [23,24], which is based on a sound and complete translation of $\mathcal{E}$ into an argumentation framework. The resulting implementation E-RES [24] [26] allows reasoning backwards and forwards along the time line even in cases where information about what holds in the "initial state" (i.e. before any action occurrences) is incomplete. E-RES has been further extended into an abductive planning system [25] able to produce plans and conditional plans even with incomplete information about the status of fluents along the time line.

5   Summary

In this article, we have described a basic, classical logic variation of the Event Calculus, and then summarised previous work on how this axiomatisation may be adapted and/or extended in various ways to represent various features of particular domains. In particular, we have described versions of the Event Calculus able to incorporate non-deterministic actions, concurrent actions, action preconditions and qualifications, delayed actions and effects, actions with duration, gradual and continuous change, and mathematical models using sets of simultaneous differential equations. We have also shown how one particular version of the basic Event Calculus may be given a sound and complete translation into the Language $\mathcal{E}$ and thus inherit $\mathcal{E}$'s provably correct automated reasoning procedures.

References

1. A. Baker, *Nonmonotonic Reasoning in the Framework of the Situation Calculus*, Artificial Intelligence, Vol 49(5-23), 1991.
2. I. Cervesato, L. Chittaro and A. Montanari, *A Modal Calculus of Partially Ordered Events in a Logic Programming Framework*, in Proceedings ICLP'95, MIT Press, pages 299-313, 1995.
3. I. Cervesato, L. Chittaro and A. Montanari, *A General Modal Framework for the Event Calculus and its Skeptical and Credulous Variants*, in in W. Wahlster, editor, Proceedings of the Twelfth European Conference on Artificial Intelligence (ECAI'96), pp. 33-37, John Wiley and Sons, 1996.
4. I. Cervesato, M. Franceschet and A. Montanari, *A Hierarchy of Modal Event Calculi: Expressiveness and Complexity*, in H.Barringer *et al*, Proceedings of the 2nd International Conference on Temporal Logic (ICTL'97, pp. 1-17, Kluwer Applied Logic Series, 1997.
5. I. Cervesato, M. Franceschet and A. Montanari, *Modal Event Calculi with Preconditions*, in R. Morris and L. Khatib, Proceedings of the Fourth International Workshop on Temporal Reasoning (TIME'97), pp. 38-45, IEEE Computer Society Press, 1997.
6. I. Cervesato, M. Franceschet and A. Montanari, *The Complexity of Model Checking in Modal Event Calculi with Quantifiers*, Journal of Electronic Transactions on Artificial Intelligence, Linköping University Electronic Press, http://www.ida.liu.se/ext/etai/, 1998.
7. L. Chittaro, A. Montanari and A. Provetti, *Skeptical and Credulous Event Calculi for Supporting Modal Queries*, in A. Cohn, Proceedings of the Eleventh European Conference on Artificial Intelligence (ECAI'94), pp. 361-365, John Wiley and Sons, 1994.
8. N. Chleq, *Constrained Resolution and Abductive Temporal Reasoning*, Computational Intelligence, vol. 12, no. 3, pp. 383?406, 1996.
9. J. M. Crawford and D. W. Etherington, *Formalizing Reasoning about Change: A Qualitative Reasoning Approach*, Proceedings AAAI'92, pp. 577-583, 1992.
10. J. Davila, *Reactive Pascal and the Event Calculus*, Proceedings FAPR'96 Workshop on Reasoning about Actions and Planning in Complex Environments, eds. U. Siegmund and M. Thielscher, vol. 11 of Technical Report AIDA, 1996.
11. M. Denecker, L. Missiaen and M. Bruynooghe, *Temporal Reasoning with Abductive Event Calculus*, in Proceedings ECAI 92, Vienna, 1992.
12. M. Denecker, K. Van Belleghem, G. Duchatelet, F. Piessens and D. De Schreye *A Realistic Experiment in Knowledge Representation in Open Event Calculus : Protocol Specification*, in Proceedings of the Joint International Conference and Symposium on Logic Programming, 1996.
13. M. Denecker, D. Theseider Dupré, and K. Van Belleghem, *An Inductive Definition Approach to Ramifications*, in Electronic Transactions on Artificial Intelligence, vol 2, 1998.
14. P. Doherty, *Reasoning about Action and Change Using Occlusion*, Proceedings ECAI'94, pp. 401-405, 1994.
15. K. Eshghi, *Abductive Planning with Event Calculus*, Proceedings of the 5th International Conference and Symposium on Logic Programming, ed.s Robert Kowalski and Kenneth Bowen, MIT Press, pp. 562-579, 1988.
16. M. Gelfond and V. Lifschitz, *Representing Actions in Extended Logic Programming*, JICSLP'92, ed. Krzysztof Apt, 560, MIT Press, 1992.

17. M. Gelfond and V. Lifschitz, *Representing Action and Change by Logic Programs*, JLP, 17 (2,3,4) 301–322, 1993.
18. R. C. Jeffrey, *Formal Logic: Its Scope and Limits*, McGraw-Hill, 1967.
19. C. G. Jung, K. Fischer and A. Burt, *Multi-Agent Planning Using an Abductive Event Calculus*, DFKI Report RR-96-04 (1996), DFKI, Germany, 1996.
20. C. G. Jung, *Situated Abstraction Planning by Abductive Temporal Reasoning*, Proceedings ECAI'98, pp. 383?387, 1998.
21. A. Kakas and R. Miller, *A Simple Declarative Language for Describing Narratives with Actions*, JLP 31(1–3) (Special Issue on Reasoning about Action and Change) 157–200, 1997.
22. A. Kakas and R. Miller, *Reasoning about Actions, Narratives and Ramifications*, Journal of Electronic Transactions on Artificial Intelligence 1(4), Linköping University Electronic Press, http://www.ida.liu.se/ext/etai/, 1998.
23. A. Kakas, R. Miller and F. Toni, *An Argumentation Framework for Reasoning about Actions and Change*, Proceedings of LPNMR'99, 1999.
24. A. Kakas, R. Miller and F. Toni, *E-RES – A System for Reasoning about Actions, Events and Observations*, Proceedings of NMR 2000, Special Session on System Demonstrations and Descriptions, http://xxx.lanl.gov/abs/cs.AI/0003034, 2000.
25. A. Kakas, R. Miller and F. Toni, *Planning with Incomplete Information*, Proceedings of NMR 2000, Special Session on Representing Actions and Planning, http://xxx.lanl.gov/abs/cs.AI/0003049, 2000.
26. A. Kakas, R. Miller and F. Toni, *E-RES - Reasoning about Actions, Events and Observations*, Proceedings of the 6th International Conference on Logic Programming and Nonmonotonic Reasoning (LPNMR'2001), September 17-19, 2001, Vienna, Austria, ed. T. Eiter, M. Truszczynski and W. Faber, pub. Springer-Verlag (LNCS/LNAI series), 2001.
27. G. N. Kartha, *Soundness and Completeness Theorems for Three Formalizations of Action*, Proceedings IJCAI'93, page 724, 1993.
28. G. N. Kartha and V. Lifschitz, *A Simple Formalization of Actions Using Circumscription*, Proceedings IJCAI'95, pp. 1970-1975, 1995.
29. R. A. Kowalski, *Database Updates in the Event Calculus*, Journal of Logic Programming, vol. 12, pp. 121-146, 1992.
30. R. A. Kowalski, *Legislation as Logic Programs*, Informatics and the Foundations of Legal Reasoning, Kluwer Academic Publishers, ed.s Z. Bankowski et al., pp. 325-356, 1995.
31. R. A. Kowalski and F. Sadri, *The Situation Calculus and Event Calculus Compared*, in Proceedings of the International Logic Programming Symposium (ILPS'94), 1994.
32. R. A. Kowalski and F. Sadri, *Reconciling the Event Calculus with the Situation Calculus*, Journal of Logic Programming, Special Issue on Reasoning about Action and Change, vol. 31, pp. 39-58, 1997.
33. R. A. Kowalski and M. J. Sergot, *A Logic-Based Calculus of Events*, New Generation Computing, vol. 4, pp. 67-95, 1986.
34. B. Kuipers, *Qualitative Reasoning: Modeling and Simulation with Incomplete Knowledge*, MIT Press, 1994.
35. F. Lévy and Joachim Quantz, *Representing Beliefs in a Situated Event Calculus*, Proceedings ECAI'98, pp. 547?551, 1998.
36. H. Levesque, *What is Planning in the Presence of Sensing?*, in Proceedings of AAAI'96, 1996.

37. V. Lifschitz, *Circumscription*, in The Handbook of Logic in Artificial Intelligence and Logic Programming, Volume 3: Nonmonotonic Reasoning and Uncertain Reasoning, ed. D. M. Gabbay, C .J. Hogger and J. A. Robinson, Oxford University Press, pp. 297-352, 1994.

38. J. Lobo, G. Mendez and S. Taylor, *Adding Knowledge to the Action Description Language A*, in Proceedings of AAAI'97, 1997.

39. J. McCarthy, *Circumscription N A Form of Non-Monotonic Reasoning*, Artificial Intelligence, vol. 13, pp. 27-39, 1980.

40. J. McCarthy and P. J. Hayes, *Some Philosophical Problems from the Standpoint of Artificial Intelligence*, in Machine Intelligence 4, ed. D. Michie and B. Meltzer, Edinburgh University Press, pp. 463-502, 1969.

41. R. Miller, *Situation Calculus Specifications for Event Calculus Logic Programs*, in Proceedings of the Third International Conference on Logic Programming and Non-monotonic Reasoning, Lexington, KY, USA, Springer Verlag, 1995.

42. R. S. Miller and M. P. Shanahan, *Reasoning about Discontinuities in the Event Calculus*, Proceedings 1996 Knowledge Representation Conference (KR'96), pp. 63?74, 1996.

43. R. S. Miller and M. P. Shanahan, *The Event Calculus in Classical Logic - Alternative Axiomatisations*, Journal of Electronic Transactions on Artificial Intelligence, Vol. 3 (1999), Section A, pages 77-105, http://www.ep.liu.se/ej/etai/1999/016/, 1999.

44. L. R. Missiaen, *Localized Abductive Planning for Robot Assembly*, Proceedings 1991 IEEE Conference on Robotics and Automation, pub. IEEE Robotics and Automation Society, pages 605-610, 1991.

45. L. R. Missiaen, M. Denecker and M. Bruynooghe, *An Abductive Planning System Based on Event Calculus*, Journal of Logic and Computation, volume 5, number 5, pages 579–602, 1995.

46. R. C. Moore, *A Formal Theory of Knowledge and Action*, In Hobbs and Moore, ed.s, Formal Theories of the Commonsense World, Ablex, Norwood, USA, 1985.

47. L. Morgenstern, *Knowledge Preconditions for Actions and Plans*, in Proceedings of the International Joint Conference in Artificial Intelligence 1987 (IJCAI'97), Morgan Kaufmann, 1987.

48. J. Pinto and R. Reiter, *Temporal Reasoning in Logic Programming: A Case for the Situation Calculus*, Proceedings ICLP 93, page 203, 1993.

49. A. Provetti, *Hypothetical Reasoning about Actions: From Situation Calculus to Event Calculus*, Computational Intelligence, volume 12, number 2, 1995.

50. R. Reiter, *The Frame Problem in the Situation Calculus: A Simple Solution (Sometimes) and a Completeness Result for Goal Regression*, in Artificial Intelligence and Mathematical Theory of Computation: Papers in Honor of John McCarthy, ed. V. Lifschitz, Academic Press, pp. 359-380, 1991.

51. R. Reiter, *Natural actions, concurrency and continuous time in the situation calculus*, in Principles of Knowledge Representation and Reasoning: Proceedings of the Fifth International Conference (KR'96), Cambridge, Massachusetts, U.S.A, November 5-8, 1996.

52. A. Russo, R. Miller, B. Nuseibeh and J. Kramer, *An Abductive Approach for Handling Inconsistencies in SCR Specifications*, in proceedings of the 3rd International Workshop on Intelligent Software Engineering (WISE3), Limerick, Ireland, June, 2000.

53. F. Sadri and R. Kowalski, *Variants of the Event Calculus*, Proceedings of the International Conference on Logic Programming, Kanagawa, Japan, Stirling L. (Ed), The MIT Press, pp. 67-81, 1995.

54. E. Sandewall, *Combining Logic and Differential Equations for Describing Real World Systems*, Proceedings KR'89, Morgan Kaufman, 1989.
55. E. Sandewall, *Filter Preferential Entailment for the Logic of Action in Almost Continuous Worlds*, Proceedings IJCAI'89, pages 894-899, 1989.
56. E. Sandewall, *The Representation of Knowledge about Dynamical Systems, Volume 1*, Oxford University Press, 1994.
57. R. Scherl and H. Levesque, *The Frame Problem and Knowledge-Producing Actions*, in Proceedings of AAAI'93, 1993.
58. M. P. Shanahan, *Representing Continuous Change in the Event Calculus*, Proceedings ECAI'90, pp. 598-603, 1990.
59. M. P. Shanahan, *A Circumscriptive Calculus of Events*, Artificial Intelligence, vol 77 (1995), pages 249-284, 1995.
60. M. P. Shanahan, *Robotics and the Common Sense Informatic Situation*, Proceedings ECAI'96, pp. 684-688, 1996.
61. M. P. Shanahan, *Noise and the Common Sense Informatic Situation for a Mobile Robot*, Proceedings AAAI'96, pp. 1098-1103, 1996.
62. M. P. Shanahan, *Solving the Frame Problem: A Mathematical Investigation of the Common Sense Law of Inertia*, MIT Press, 1997.
63. M. P. Shanahan, *Event Calculus Planning Revisited*, Proceedings 4th European Conference on Planning (ECP'97), Springer Lecture Notes in Artificial Intelligence no. 1348, pp. 390-402,1997.
64. M. P. Shanahan, *Noise, Non-Determinism and Spatial Uncertainty*, Proceedings AAAI'97, pp. 153-158, 1997.
65. M. P. Shanahan, *Reinventing Shakey*, Working Notes of the 1998 AAAI Fall Symposium on Cognitive Robotics, pp. 125-135, 1998.
66. M. P. Shanahan, *The Ramification Problem in the Event Calculus*, Proceedings IJCAI'99, 1999.
67. M. P. Shanahan, *A Logical Account of the Common Sense Informatic Situation for a Mobile Robot*, Electronic Transactions on Artificial Intelligence, 1999.
68. M. P. Shanahan, *The Event Calculus Explained*, in Artificial Intelligence Today, eds. M. J. Wooldridge and M. Veloso, Springer-Verlag Lecture Notes in Artificial Intelligence no. 1600, Springer-Verlag, pages 409-430, 1999.
69. E. Ternovskaia, *Inductive Definability and the Situation Calculus*, in "Transactions and Change in Logic Databases", Lecture Notes in Computer Science, volume 1472, Ed. Freitag B., Decker H., Kifer M. (Eds.), pub. Springer Verlag, 1997.
70. E. Ternovskaia, *Causality via Inductive Definitions*, in Working Notes of "Prospects for a Commonsense Theory of Causation", pages 94-100, AAAI Spring Symposium Series, March 23-28, 1998.
71. K. Van Belleghem, M. Denecker and D. De Schreye, *Representing Continuous Change in the Abductive Event Calculus*, in Proceedings 1994 International Conference on Logic Programming, ed. P. Van Hentenrijck, pages 225-240, 1994.
72. K. Van Belleghem, M. Denecker and D. De Schreye, *The Abductive Event Calculus as a General Framework for Temporal Databases*, Proceedings of the International Conference on Temporal Logic, 1994.
73. K. Van Belleghem, M. Denecker and D. De Schreye, *Combining Situation Calculus and Event Calculus*, in Proceedings of the International Conference on Logic Programming, 1995.
74. K. Van Belleghem, M. Denecker and D. De Schreye, *On the Relation Between Situation Calculus and Event Calculus*, Journal of Logic Programming, 31(1–3) (Special Issue on Reasoning about Action and Change), 1996.

Issues in Learning Language in Logic

James Cussens

Department of Computer Science, University of York
Heslington, York, Y010 5DD, UK
jc@cs.york.ac.uk

Abstract. Selected issues concerning the use of logical representations in machine learning of natural language are discussed. It is argued that the flexibility and expressivity of logical representations are particularly useful in more complex natural language learning tasks. A number of inductive logic programming (ILP) techniques for natural language are analysed including the CHILL system, abduction and the incorporation of linguistic knowledge, including active learning. Hybrid approaches integrating ILP with manual development environments and probabilistic techniques are advocated.

1 Introduction

The statistical natural language processing revolution has brought empirical methods of producing natural language resources to the fore. However, despite the long-established use of logic in NLP, most work in natural language learning (NLL) does not take place within a logical framework. Partly this is because many of the successful techniques (e.g. n-gram language models) derive from speech processing where logical approaches are not generally used. Also a logical representation is often seen as unnecessarily complex for NLL. Simpler representations, engineered for specific NLL tasks, are more common.

This chapter will argue that learning language in logic—using a logical representation for NLL—is both practical and desirable for a range of NLL problems. Our argument will fall into two parts. In Section 2 there is a high-level discussion of the role of logical representations in language learning where there are few specific examples. Section 3, in contrast, is much more detailed, examining particularly important issues which arise in actually existing LLL applications. We finish with some tentative predictions in Section 4.

2 Logical Representation in Natural Language Learning

Most logical learning can be defined as inductive logic programming (ILP), so for the sake of completeness, we begin by stating a highly simplified version of the ILP task in Table 1. The flexibility and expressivity of logic are amongst the prime reasons for using ILP. This flexibility and expressivity has also led to the development of a variety of logic-based resources for natural language

A.C. Kakas, F. Sadri (Eds.): Computat. Logic (Kowalski Festschrift), LNAI 2408, pp. 491–505, 2002.

Table 1. Highly simplified version of the ILP problem

Given background knowledge B and data E
Find a hypothesis H such that
$$B \wedge H \models E$$
where B, H and E are all logic programs.

---

processing (NLP). If we wish to *learn* such resources then this section argues that it is appropriate that the learning process also stays within this 'native' logical representation. I argue here that moving to a less expressive learning framework such as attribute-value learning will mean that some flexibility or expressivity will be lost.

In cases where the output of learning is required to be logical it is easy to motivate logical learning. Learning problems of this type include learning definite clause grammars [11, 21], semantic grammars [27], and learning rules to translate between logically represented semantic forms from different languages [5].

In other cases the form of the data motivates a logical representation. For example, text can be annotated with information such as part-of-speech (PoS) tags, syntactic parse trees or semantic interpretations. As we move from 'lower-level' information such as PoS tags to 'higher-level' information such as semantic annotation, the case for a logical representation becomes stronger. This is simply because logic has been designed to elegantly represent complex and structured information. A good example is the use of *quasi-logical forms (QLFs)* to provide semantic interpretation. Table 2 shows a pair of QLFs representing (at a semantic level) the English sentence *List the prices* and the French sentence *Indiquez les tarifs*. For details of QLF syntax and semantics see [3]. This sort of data is used by Boström [5] to learn transfer rules between French and English QLFs. It is difficult to see how the complex terms required to represent this sort of information could be adequately translated to a non-logical representation.

Sometimes the learning problem is such that a logical representation is not *essential* but is more convenient than simpler approaches. For example, when learning from unannotated text it is possible to represent sequences of words (n-grams) by feature vectors of length n. The same holds true of text annotated with information such as PoS tags. This is because, in practice, all such sequences will be of bounded length. If the longest sequence in the data were of length $n = 20$, then the sentence *James loves Gill* would be represented as the feature vector

$$(\text{W1} = \text{James}, \text{W2} = \text{loves}, \text{W3} = \text{Gill}, \text{W4} = \emptyset, \ldots, \text{W20} = \emptyset)$$

where $\emptyset$ is a null value. However, this 'flat' representation is unwieldy since all sequences except those of maximal length will involve superfluous null values. A much more compact representation would be to use lists, which are first-order terms, so that *James loves Gill* becomes

```
['James',loves,'Gill']
```

Table 2. A pair of QLFs for *List the prices* and *Indiquez les tarifs* reproduced from [5]

```
qlf_pair([imp, form(_,verb(no,no,no,imp,y),A,          %English
         B^[B,[list_Enumerate,A,
             term(_,ref(pro,you,_,1([])),_,
                         C^[personal,C],_,_),
             term(_,q(_,bare,plur),_,
                   D^[fare_Price,D],_,_)]],_)],
         [imp, form(_,verb(impera,no,no,impera,y),E, %French
          F^[F,[indiquer_Show,E,
              term(_,ref(pro,vous,_,1([])),G,
                          H^[personal,H],_,_),
              term(_,ref(def,le,plur,1([G-_])),_,
                      I^[tarif_Fares,I],_,_)]],_)]).
```

The logical representation becomes more attractive as the size (n) of the equivalent feature vector increases. Indeed when n is small then a feature vector representation is preferable and more efficient. For example, part-of-speech taggers can be efficiently learnt from bigrams ($n = 2$) and trigrams ($n = 3$). Note though that there is a natural logical representation of feature vectors of length n as n-ary terms, so if, say, $n = 5$, then a reasonable logical representation of *James loves Gill* would be

```
fivegram('James',loves,'Gill',empty,empty)
```

The vector representation is efficient because we can access elements in constant time (using `arg/3`) which is not the case with the list representation.

In other cases, n is so large as to make a feature vector approach impractical. This is the case in the learning problems addressed by the ASIUM system. In [15, 20], a principal goal is to cluster words into concepts and so to form an ontology of a domain. ASIUM begins by clustering words into "basic classes" to give a first level of clusters. ASIUM continues by building second level clusters from these, then third level clusters from the second level and so on. The final output is an acyclic directed graph (not necessarily a tree) representing a generality hierarchy between concepts.

> The nature of ASIUM input and output is intrinsically relational: it consists of relations between verbs and their complements and generality relations between concepts. [20]

Non-ILP clustering algorithms such as COBWEB or AUTOCLASS were rejected:

> Attributes of the input vector would be head words and values, their frequencies. As attributes would need to be the same in all vectors, very large vectors representing a whole dictionary would be required (about

2000 words in our experimentation) and most of their values would be equal to zero. [15]

Here we see the "exploding attribute space phenomenon" which occurs when unsuitable data is crow-barred into attribute value form. Another system which uses a first-order representation to handle unbounded context is the RAPIER system [7] which uses ILP to learn information extraction rules. RAPIER learns rules which use a word's context to decide whether it should fill one of a number of predefined slots. The context is defined to be the entire document within which a word appears, making the length (n) of any feature vector representation much too large to be practical.

3   Inductive Logic Programming Techniques for Natural Language Learning

3.1   Classification versus Analysis for Natural Language

In the standard presentation of ILP [17] the task is to learn a logic program H that correctly defines one, or more rarely several, predicates. Since a predicate symbol denotes a set this amounts to specifying which tuples are members of this set. So in this standard presentation, phrase structure grammar learning, for example, amounts to finding definitions for sentence, noun phrase, verb phrase, etc which correctly *classify* sequences of tokens into these categories.

This approach is in contrast with NLP where dividing up sequences of words between, say, sentences and non-sentences is much less important than finding *the correct linguistic analysis* for a piece of natural language. For example, the task of a semantic parser induced by the CHILL system [27] might be to translate a natural language query into a database query. It does not matter too much if the query is not grammatical. Indeed a semantic parser that can make sense of ungrammatical but meaningful natural language queries would be superior to one that rejected such queries.

3.2   Direct versus Indirect ILP for Linguistic Analysis

There is both a direct and indirect way to apply the standard ILP learning approach to linguistic analysis. In the *indirect* approach, we perform analysis via *proof*. For example, in the case of phrase structure grammars a proof that a word sequence is a sentence is a parse of that sentence and produces an analysis in the form of a parse tree. In the indirect approach the proofs are what we care about and this means we must differentiate between logic programs which are semantically equivalent (the same things get proved) but syntactically distinct (things get proved in different ways). For example, the two grammars in Table 3 define the same set of sentences, but only Grammar 2 can be used to analyse sentences.

In the *direct* approach to analysis we say that if the task is really to produce an analysis from text then this should be explicitly represented. Rather than

Table 3. Grammar 1 and Grammar 2 define the same S, NP and VP categories but provide different parses

---

Grammar 1

S	→ I went	S	→ He went	S	→ He goes	S	→ I go
NP → I		NP → He		VP → went		VP → goes	

Grammar 2

S	→ NP VP						
NP → I		NP → He		VP → went		VP → goes	

---

learn definitions of S, NP, VP, etc that indirectly encode analyses via proofs we should plainly state the problem as learning a binary predicate `parse(Sentence, Analysis)`.

To compare the two approaches, suppose that we have a training set of annotated sentences such as (*He goes*/s(np(He),vp(goes))). In the indirect case we would induce a grammar with rules such as

```
s --> np, vp.
```

whereas in the direct case we have to induce rules which explicitly represent the parse, for example, with an extra argument:

```
s(s(NP,VP)) --> np(NP), vp(VP).
```

An empirical comparison between the direct and indirect approaches can be found in [28]. Zelle and Mooney list a number of problems with the direct approach. Firstly, they state that

> We do not have a convenient set of negative examples [for the `parse(Sentence, Analysis)` predicate - JC] or a good theory of what the background relations should be. [28]

Considering first the problem of negative examples, it is clear that explicitly constructing all possible incorrect analyses for sentences in the data is intractable. Also any random sample of this set is likely not to contain "near-miss" examples of almost correct analyses which are crucial for learning. Although Zelle and Mooney do not favour the direct approach, they do apply it to the problem of parser construction for the purposes of comparison and hence need some way of getting round these problems and producing negative examples. Their solution is to use a preliminary over-general parser to parse training sentences and then label all incorrect analyses produced by this parser as negative examples. This is an example of the *output-completeness* assumption [29] which holds when we are learning an input-output relation and where those outputs not known to be correct are assumed incorrect.

Zelle and Mooney go on to argue that even with suitable examples (and background knowledge) the correct definition of `parse(Sentence,Analysis)` for problems of realistic size will be too complex to learn—the hypothesis space is too big. Their empirical results support this hypothesis. The direct approach was compared with their indirect approach (which we describe shortly) on a small artificial learning problem where case-role analyses are constructed from text. No significant difference in test-set accuracy was found. The two approaches were then applied to the problem of generating parse trees from text using the ATIS corpus as the training set. The parse trees with which sentences are annotated in the ATIS corpus are significantly more complex than the case-role analyses found in the first experiment. In this more complex situation the indirect approach was overwhelmingly superior, achieving 84% accuracy on test data where the direct approach only managed 20% accuracy. In both representations the same ILP algorithm (CHILLIN) was used to do the underlying induction, so the ATIS results provide strong evidence in favour of the indirect approach.

3.3   The CHILL System

We now describe Zelle and Mooney's indirect approach: the CHILL system [27]. CHILL views analysis as a sequential decision problem. At each stage in the process of parsing we must decide between several options. CHILL's task is to learn rules which choose only correct options. CHILL uses shift-reduce parsing to produce analyses. Shift-reduce parsing uses an input buffer (initially containing the text to be parsed) and a stack (initially empty). At each stage of parsing the parser can either shift or reduce. Shifting moves the word from the front of the buffer onto the top of the stack. There are a finite number of reduce operations each of which combine items at the top of the stack and replace them with a single combined item. For example, the REDUCE-agt operation would transform a stack

[ate,[man,det:the]]

to

[[ate,agt:[man,det:the]]]

leaving the buffer unaltered. The final analysis is represented by the contents of the stack once parsing is complete.

By parsing the annotated data with an initial overly general parser, CHILL determines which parser actions are permissible at each stage in parsing. The contents of the buffer and the stack are also recorded for each stage. This gives CHILL a set of positive examples for each of the possible parsing actions. For example,

op([ate,[man,det:the]],[the,pasta],A,B).

would be a positive example for the agt operator. The first two arguments were the state of the stack and buffer, respectively, when a REDUCE-agt operation was correctly applied. The final two arguments represent the stack and buffer after the parser operation, and are not important for induction. Negative examples are produced by mapping Stack,Buffer pairs to parsing operations that were not performed. In fact, since there are a finite number of parsing actions we can simplify by translating

`op([ate,[man,det:the]],[the,pasta],A,B).`

to

`reduce_agt([ate,[man,det:the]],[the,pasta]).`

These are the sorts of examples given to CHILLIN, CHILL's ILP algorithm, which outputs rules (implicitly) mapping parser states to parser operations. In essence, CHILL breaks down the complex problem of learning a mapping `parse(+Sentence,-Analysis)` into the much simpler problem of learning a finite number of `sr_op(+State)` predicates, one for each possible shift-reduce operation `sr_op`.

3.4 CHILL as a Multiple Predicate Learner

In [30] the initial over general parser used by CHILL is expressed as a logic program, which, viewed declaratively, is an over general grammar. (The induced control rules are 'folded' into this over general parser to produce the final appropriately specialised parser.) Table 4 reproduces this representation of the CHILL parser except that I have added guard literals (to be explained later) to each `op/4` clause.

Table 4 provides a partial definition of the `parse/2` predicate which was impossible to learn accurately using the direct representation in [28]. In that paper this partial definition was not used when learning with the direct representation, but was, in essence, supplied to the CHILL system. Table 4 is just the shift-reduce parsing architecture supplied to CHILL expressed as logical background knowledge. The definition is partial since the control rule predicates `reduce_agt_guard`, `reduce_det_guard`, `shift_guard`, etc need to be induced. This is, of course, what CHILL does. We can re-express what CHILL does by assuming that the initial definitions of `reduce_agt_guard`, `reduce_det_guard`, `shift_guard`, etc are maximally general and are then specialised appropriately.

In other words, the goal is to induce definitions for `reduce_agt_guard`, `reduce_det_guard`, `shift_guard`, etc which classify `Stack,Buffer` pairs correctly. This could also be done by a general-purpose ILP algorithm, *as long as it is capable of learning a definition of one predicate using examples of another one.* This is an example of *multiple predicate learning*. In this case, the various guard predicates must be learnt from examples of `parse/2`. In brief, we can map the control rule learning problem into a guard predicate learning problem. This is a mapping from an indirect approach (learning a parser) to a direct one (learning a definition of `parse/2` by modifying the predicates that define `parse/2`). Given the impressive results of the CHILL system there is no pressing reason to actually carry out the reformulation presented in Table 4; the goal here is to clarify the relationship between the CHILL parsing architecture and logically encoded background knowledge.

Table 4. Shift-reduce parsing architecture as background knowledge

```
parse(S,Parse) :- parse([],S,[Parse],[]).

parse(Stack,Input,Stack,Input).
parse(InStack,InInput,OutStack,OutInput) :-
    op(InStack,InInput,MidStack,MidInput),
    parse(MidStack,MidInput,OutStack,OutInput).

op([Top,Second|Rest],Input,[NewTop|Rest],Input) :-
    reduce_agt_guard([Top,Second|Rest],Input),
    reduce(Top,agt,Second,NewTop).


op([Top,Second|Rest],Input,[NewTop|Rest],Input) :-
    reduce_det_guard([Top,Second|Rest],Input),
    reduce(Top,det,Second,NewTop).


op([Top,Second|Rest],Input,[NewTop|Rest],Input) :-
    reduce_obj_guard([Top,Second|Rest],Input),
    reduce(Top,obj,Second,NewTop).

% ....

op(Stack,[Word|Words],[Word|Stack],Words) :-
    shift_guard(Stack,[Word|Words]).
```

3.5   Abduction

ILP approaches to learning a definition of one predicate using examples of another usually rest upon some form of *abduction*. This issue is discussed in some
detail in [12], so here we will be brief. Given (i) a positive example

```
s(['I',went],[]).
```

(ii) a known grammar rule in the background knowledge

```
s(A,B) :- np(A,C), vp(C,B).
```

and (iii) background knowledge such that

```
vp([went|T],T).
```

is entailed, we can abduce the fact (positive example)

```
np(['I'|T],T).
```

This amounts to guessing that I *went* is a sentence because I is a noun phrase. This is a perhaps misleadingly simple example of using abduction when we have unannotated data. In general, unannotated data is hard to abduce with. Consider doing abduction with unannotated data with the sort of background knowledge given in Table 4, i.e. suppose that only values for the Sentence variable and not the Analysis variable are given in examples of parse/2. In this case the annotation is not there to guide us towards the correct abduced facts, and the problem becomes highly unconstrained. In [11] abduction is possible with unannotated data, but only because it is assumed that only a single fact needs to be abduced. With annotated data we have enough information to abduce a whole series of facts the conjunction of which, when added to the rest of the grammar, allow the correct parse to be produced. In the case of shift-reduce parsing these facts correspond to the parser actions required for the correct parse.

3.6 Disambiguation and Probabilistic Approaches

One can view parser control rules such as those produced by CHILL as performing disambiguation. Given a grammar which produces too many parses for a sentence, each step in each parse can be checked against control rules to see if the parse is allowed. Here we compare this rule-based approach with those available via statistical language models.

In its most general application, statistical language modelling computes the probability of a particular analysis based on the 'features' the analysis has. Each feature has a weight which measures its importance. In a stochastic context-free grammar (SCFG) the features of a parse are simply the grammar rules used in that parse, and the weights are the probabilities attached to these rules. This same basic approach has also been applied in work on stochastic attribute-value grammars [1] and stochastic logic programs [18, 19, 8, 9, 10].

Although this choice of features is appealingly simple, there is no compelling linguistic argument that the grammar rules used to produce an analysis should be the linguistically important features of that analysis. This is the view developed by Riezler [22, 23] who defines log-linear distributions over analyses structured on arbitrary features including those "indicating the number of argument-nodes or adjunct-nodes in the tree, and features indicating complexity, parallelism or branching-behaviour" (Stefan Riezler, personal communication).

However, rule-based features are appealingly simple and probabilistic approaches using them connect more directly with existing ILP techniques. The basic idea is simple: rather than learn rules to select between parser operations or grammar rules we estimate conditional probabilities of the form $P(\text{Parser\_op}|\text{Parser\_state})$. Translating this to general logic programs we get $P(\text{Clause}|\text{Goal})$. Since each proof can be identified by the sequence of clauses it uses, we can derive a distribution over proofs, and hence linguistic analyses. Table 5 is a simple probabilistic version of Table 4 where the control rule predicates now compute probabilities rather than just succeeding or failing. A more sophisticated approach to probabilistic semantic parsing which integrates ILP with estimation of the correct probabilities can be found in [24].

Table 5. Probabilistic shift-reduce parsing

```
parse(S,Parse,Prob) :- parse([],S,1,[Parse],[],Prob).

parse(Stack,Input,Prob,Stack,Input,Prob).
parse(InStack,InInput,InProb,OutStack,OutInput,OutProb) :-
   op(InStack,InInput,InProb,MidStack,MidInput,MidProb),
   parse(MidStack,MidInput,MidProb,OutStack,OutInput,OutProb).

op([Top,Second|Rest],Input,InProb,[NewTop|Rest],Input,OutProb) :-
   reduce_agt_prob([Top,Second|Rest],Input,Prob),
   reduce(Top,agt,Second,NewTop),
   OutProb is Prob*InProb.

op([Top,Second|Rest],Input,InProb,[NewTop|Rest],Input,OutProb) :-
   reduce_det_prob([Top,Second|Rest],Input,Prob),
   reduce(Top,det,Second,NewTop),
   OutProb is Prob*InProb.

op([Top,Second|Rest],Input,InProb,[NewTop|Rest],Input,OutProb) :-
   reduce_obj_prob([Top,Second|Rest],Input,Prob),
   reduce(Top,obj,Second,NewTop),
   OutProb is Prob*InProb.

% ....

op(Stack,[Word|Words],InProb,[Word|Stack],Words,InProb) :-
   shift_guard(Stack,[Word|Words]).
```

As Abney [1] has pointed out estimating these probabilities is considerably harder in stochastic attribute-value grammars than in the context-free case. This is essentially because some clause choices lead to unification failure. Algorithms for parameter estimation in the non-context-free case can be found in [23, 10].

3.7 Building on Existing Linguistic Knowledge

Data driven methods of producing natural language resources are motivated by the difficulty of producing such resources manually. However, it would be wasteful not to draw on human linguistic knowledge when it is economical to do so. Brill argues that we should ...

> focus on ways of capitalizing on the relative strengths of people and machines, rather than simply viewing machine learning as another way to do the same thing. [6]

In this section we look at how this can be done in a logical learning framework, dividing the approaches into static (before learning) and active (during learning) approaches.

Static Incorporation of Linguistic Information A static approach to incorporating linguistic knowledge demands that the user presents linguistic information to the learning system before the learning process begins. In ILP this information is given by (i) defining the hypothesis space; often using extra-logical constraints on acceptable hypotheses and (ii) providing the background knowledge B as in Table 1. It is well known amongst ILP practitioners that "getting the background knowledge right" is crucial to the success of an ILP application. Logical learning techniques have most to offer where this background information (i.e. information other than data) has a logical representation. For example, in the case of learning grammars we can take whatever initial grammar we might have and add it to the background knowledge B—as in our reformulation of the CHILL system. This initial grammar will always be unsatisfactory, hence the need for learning which will revise it in some way, but starting from such knowledge that we do have is more efficient than *ab initio* techniques.

In [11] an initial grammar is provided as background knowledge, but the main focus is on constraining the hypothesis space. Although an inductive approach to grammar construction assumes that it is undesirable to do manual grammar writing, it is not unreasonable to expect a user to constrain the hypothesis space with general linguistic principles. In [11] the goal was to add sufficiently tight linguistic constraints such that no linguistically implausible grammar rule or lexical item gets past the constraints to be evaluated against the data. Constraints on headedness and gap-threading proved particularly useful, not only in filtering out implausible rules, but also in constructing rules. The great practical advantage of a logical approach here is that these constraints can be expressed declaratively (in Prolog) using a logical representation specifically devised to facilitate the expression of linguistic knowledge.

A more fundamental approach is offered by Adriaans and de Haas [2]. They note that

> If one wants to use logic to describe certain phenomena in reality there are in principle two options. 1) One takes some variant of predicate calculus, e.g. Horn clauses, and one tries to model the phenomena in this medium, or, 2) one tries to find a certain variant of logic in the substructural landscape that has characteristics that intrinsically model the target concepts. The latter route is to the knowledge of the authors hardly taken by researchers in ILP. [2]

Adriaans and de Haas argue for the latter option:

> we show that in some areas, especially grammar induction, the substructural approach has specific advantages. These advantages are: 1) a knowledge representation that models the target concepts intrinsically,

2) of which the complexity issues are well known, 3) with an expressive power that is in general weaker than the Horn-clause or related representations that are used in more traditional ILP research, 4) for which explicit learnability results are available. [2]

This approach underlies the EMILE algorithm which learns categorial grammars from unannotated data and queries to the user. In general, there are different benefits from using a logic just expressive enough for a particular learning problem (e.g. grammar learning) and using problem specific constraints within a more expressive logic. This is, at base, a practical problem and further work is required to compare the hard-coded restrictions of Adriaans and de Haas with the problem-specific restrictions commonly used in ILP.

Active Learning An active learning system seeks out information, usually data, *during the course of learning.* For example, in the ASIUM system the user is called upon in two ways. One is to give comprehensible names to predicates invented by ASIUM, the other more important way is to check each stage of generalisation to prevent over-generalisation. Thompson and Califf [25] use a *selective sampling* approach to active learning where the learning systems (in their case the ILP systems CHILL and RAPIER) ask the user to annotate particularly informative examples.

We finish this section by considering early work by Wirth [26] where abduction is used to guess missing facts and the user is asked whether these guesses are correct. For example, in Wirth's grammar learning example the user is asked whether the abduced facts

```
intransitive_verb([loves,a,man],[])
```

and

```
verb_phrase([loves,a,man],[])
```

are true. The user also has to evaluate the conjectured rule, a practice which Wirth defends as follows:

> A system that learn[s] concepts or rules from looking at the world is useless as long as the results are not verified because a user who feels responsible for his knowledge base rarely use these concepts or rules. [26]

The counter-argument to this is that an output of, say, a large lexicon is too big for a user to check, and so the best verification in such a case is against out-of-sample data. The desirability of user interaction is a quantitative matter; one needs to weigh up the effort required of the user against the gains in quality of output. In the systems discussed in this section the user is required to give a yes/no answer to a hypothesis produced by the system or provide an annotation for a particular example. Both these interactions, particularly the former, do not put a heavy burden on the user as long as they are not required too frequently. In any case, they are vastly less burdensome than a purely manual non-inductive approach, and so, given the valuable information which users can supply seem likely to be used extensively in future work on language learning.

4 Conclusions

This paper has not examined all aspects of learning language in logic (LLL). For example, there is almost no discussion of LLL work in morphology or PoS tagging, overviews of which are given by [16] and [13], respectively. However, hopefully some key issues have been discussed in sufficient detail to back up the argument that LLL is both practical and desirable for a number of NLL tasks.

Looking ahead, it seems likely that hybrid approaches will be important for LLL. One important hybridisation is between manual development environments and inductive techniques. I have previously argued that LLL is attractive because logic is often the native representation for NLP—this should also make integrated systems easier to build. Also such an environment is the right one for active learning. An existing LLL system that takes user interaction seriously is ASIUM (see http://www.lri.fr/~faure/Demonstration.UK/Presentation_Demo.html).

However, the most important hybridisation is between logic and probability, an enterprise which has been continuing since the very beginning of symbolic logic [4]. NLL is considerably more difficult than many other machine learning tasks, so it is inconceivable that NLL outputs will not have residual uncertainty. In much ILP work uncertainty is left unquantified or dealt with in a statistically unsophisticated manner. The statistical NLP revolution has demonstrated the advantages of (i) recognising the inevitability of uncertainty and (ii) modelling it properly using probabilistic models. There is no reason, in principle, why these elementary observations can not be applied to LLL. However, the very flexibility of logic that makes it so attractive for NLP gives rise to complex probabilistic models. Nonetheless there has been progress in this area [14, 18, 1, 22, 23, 8, 9, 10].

References

[1] Steven Abney. Stochastic attribute-value grammars. *Computational Linguistics*, 23(4):597–618, 1997.

[2] Pieter Adriaans and Erik de Haas. Grammar induction as substructural inductive logic programming. In James Cussens and Sašo Džeroski, editors, *Learning Language in Logic*, volume 1925 of *LNAI*. Springer, 2000.

[3] H. Alshawi, editor. *The Core Language Engine*. MIT Press, Cambridge, Mass, 1992.

[4] George Boole. *An Investigation of the Laws of Thought, on which are founded the Mathematical Theories of Logic and Probabilities*. Dover, 1854.

[5] Henrik Boström. Induction of recursive transfer rules. In James Cussens and Sašo Džeroski, editors, *Learning Language in Logic*, volume 1925 of *LNAI*. Springer, 2000.

[6] Eric Brill. A closer look at the automatic induction of linguistic knowledge. In James Cussens and Sašo Džeroski, editors, *Learning Language in Logic*, volume 1925 of *LNAI*. Springer, 2000.

[7] M.E. Califf and R.J. Mooney. Relational learning of pattern-match rules for information extraction. In *Proceedings of the Sixteenth National Conference on Artificial Intelligence*, pages 328–334, Orlando, FL, July 1999.

[8] James Cussens. Loglinear models for first-order probabilistic reasoning. In *Proceedings of the Fifteenth Annual Conference on Uncertainty in Artificial Intelligence (UAI-99)*, pages 126–133, San Francisco, CA, 1999. Morgan Kaufmann Publishers.

[9] James Cussens. Stochastic logic programs: Sampling, inference and applications. In *Proceedings of the Sixteenth Annual Conference on Uncertainty in Artificial Intelligence (UAI-2000)*, pages 115–122, San Francisco, CA, 2000. Morgan Kaufmann.

[10] James Cussens. Parameter estimation in stochastic logic programs. *Machine Learning*, 2001. To appear.

[11] James Cussens and Stephen Pulman. Incorporating linguistics constraints into inductive logic programming. In *Proceedings of CoNLL2000 and LLL2000*, pages 184–193, Lisbon, September 2000. ACL.

[12] Sašo Džeroski, James Cussens, and Suresh Manandhar. An introduction to inductive logic programming and learning language in logic. In James Cussens and Sašo Džeroski, editors, *Learning Language in Logic*, volume 1925 of *LNAI*. Springer, 2000.

[13] Martin Eineborg and Nikolaj Lindberg. ILP in part-of-speech tagging — an overview. In James Cussens and Sašo Džeroski, editors, *Learning Language in Logic*, volume 1925 of *LNAI*. Springer, 2000.

[14] Andreas Eisele. Towards probabilistic extensions of constraint-based grammars. Contribution to DYANA-2 Deliverable R1.2B, DYANA-2 project, 1994. Available at ftp://moon.philo.uva.nl/pub/dekker/dyana/R1.2.B.

[15] D. Faure and C. Nédellec. A Corpus-based Conceptual Clustering Method for Verb Frames and Ontology Acquisition. In Paola Velardi, editor, *LREC workshop on Adapting lexical and corpus ressources to sublanguages and applications*, pages 5–12, Granada, Spain, May 1998.

[16] Dimitar Kazakov. Achievements and prospects of learning word morphology with inductive logic programming. In James Cussens and Sašo Džeroski, editors, *Learning Language in Logic*, volume 1925 of *LNAI*. Springer, 2000.

[17] S. Muggleton and L. De Raedt. Inductive logic programming: Theory and methods. *Journal of Logic Programming*, 20:629–679, 1994.

[18] Stephen Muggleton. Stochastic logic programs. In Luc De Raedt, editor, *Advances in Inductive Logic Programming*, volume 32 of *Frontiers in Artificial Intelligence and Applications*, pages 254–264. IOS Press, Amsterdam, 1996.

[19] Stephen Muggleton. Semantics and derivation for stochastic logic programs. In Richard Dybowski, editor, *Proceedings of the UAI-2000 Workshop on Fusion of Domain Knowledge with Data for Decision Support*, 2000.

[20] Claire Nedellec. Corpus-based learning of semantic relations by the ILP system, Asium. In James Cussens and Sašo Džeroski, editors, *Learning Language in Logic*, volume 1925 of *LNAI*. Springer, 2000.

[21] Miles Osborne. DCG induction using MDL and parsed corpora. In James Cussens and Sašo Džeroski, editors, *Learning Language in Logic*, volume 1925 of *LNAI*. Springer, 2000.

[22] Stefan Riezler. *Probabilistic Constraint Logic Programming*. PhD thesis, Universität Tübingen, 1998. AIMS Report 5(1), 1999, IMS, Universität Stuttgart.

[23] Stefan Riezler. Learning log-linear models on constraint-based grammars for disambiguation. In James Cussens and Sašo Džeroski, editors, *Learning Language in Logic*, volume 1925 of *LNAI*. Springer, 2000.

[24] Lappoon R. Tang and Raymond J. Mooney. Automated construction of database interfaces: Integrating statistical and relational learning of semantic parsing. In *Proceedings of the Joint SIGDAT Conference on Empirical Methods in Natural Language Processing and Very Large Corpora(EMNLP/VLC-2000)*, pages 133–141, Hong-Kong, October 2000.

[25] Cynthia A. Thompson and Mary Elaine Califf. Improving learning by choosing examples intelligently in two natural language tasks. In James Cussens and Sašo Džeroski, editors, *Learning Language in Logic*, volume 1925 of *LNAI*. Springer, 2000.

[26] Ruediger Wirth. Learning by failure to prove. In Derek Sleeman, editor, *Proceedings of the 3rd European Working Session on Learning*, pages 237–251, Glasgow, October 1988. Pitman.

[27] J. M. Zelle and R. J. Mooney. Learning semantic grammars with constructive inductive logic programming. In *Proceedings of the Eleventh National Conference on Artificial Intelligence*, pages 817–822, Washington, D.C., July 1993.

[28] J. M. Zelle and R. J. Mooney. Comparative results on using inductive logic programming for corpus-based parser construction. In S. Wermter, E. Riloff, and G. Scheler, editors, *Connectionist, Statistical, and Symbolic Approaches to Learning for Natural Language Processing*, pages 355–369. Springer, Berlin, 1996.

[29] J.M. Zelle, C.A. Thompson, M.E. Califf, and R.J. Mooney. Inducing logic programs without explicit negative examples. In L. De Raedt, editor, *Proceedings of the 5th International Workshop on Inductive Logic Programming*, pages 403–416. Department of Computer Science, Katholieke Universiteit Leuven, 1995.

[30] John M. Zelle and Raymond J. Mooney. An inductive logic programming method for corpus-based parser construction. Unpublished Technical Report, 1997.

On Implicit Meanings

Veronica Dahl

Logic and Functional Programming Group
School of Computing Science
Simon Fraser University
Burnaby, B.C. Canada V5A 1S6
veronica@cs.sfu.ca

Abstract. We present a logic programming parsing methodology which we believe especially interesting for understanding implicit human-language structures. It records parsing state constituents through linear assumptions to be consumed as the corresponding constituents materialize throughout the computation. Parsing state symbols corresponding to implicit structures remain as undischarged assumptions, rather than blocking the computation as they would if they were subgoals in a query. They can then be used to glean the meaning of elided structures, with the aid of parallel structures. Word ordering inferences are made not from symbol contiguity as in DCGs, but from invisibly handling numbered edges as parameters of each symbol. We illustrate our ideas through a metagrammatical treatment of coordination, which shows that the proposed methodology can be used to detect and resolve parallel structures through syntactic and semantic criteria.

Keywords: elision, parallel structures, logic grammars, datalog grammars, hypothetical reasoning, bottom-up parsing, left-corner parsing, chart parsing, linear affine implication, prediction, coordination.

1   Introduction

Work on implicit meaning reconstruction has typically centered around the notion of parallelism as a key element in the determination of implicit meanings. [1] defines parallelism as

> a pairing of constituents ... and their parts, such that each pair contains two semantically and structurally similar objects

For instance, in *Bob likes tea and Alain coffee.* we can recognize two parallel verb phrases, one complete (likes tea) and one incomplete (coffee), in which the verb's meaning is implicit and can be inferred from that of the verb in the parallel, complete verb phrase.

Because the parallel structures can be any phrase at all, it would be highly inefficient to try to code all possible cases explicitly, even if we did not have the added complication of possible elision. Instead, we can use the metarule

```
X --> X conj X
```

A.C. Kakas, F. Sadri (Eds.): Computat. Logic (Kowalski Festschrift), LNAI 2408, pp. 506–525, 2002.
© Springer-Verlag Berlin Heidelberg 2002

to express the large number of its specific instances (where X = noun phrase, X = adjective, X = sentence, etc.).

A parser dealing with conjunction metagrammatically must keep this meta-rule "in mind" to try to parse a structure of same category on either side of the conjunction (the conjoints), and then identify the string covering both conjoints as being of the same category, its meaning an appropriate combination of the meanings of both conjoints.

In this article we present a bottom-up, left-corner rendition of datalog grammars [8] in which expected constituents are expressed as continuation- based affine linear assumptions [6, 11, 17]. These can be consumed at most once, are backtrackable, and remain available during the entire continuation. Parsing state symbols that correspond to implicit structures can then remain as undischarged assumptions, rather than blocking the computation as they would if they were subgoals in a query. By examining the undischarged assumptions with the aid of the parallel structures concerned, we can recover the meaning of elided strings.

As proof of concept, we present a parser which charts, or memoes, the theorems obtained at each level of iteration, and examines them when searching for parallel structures.

Given that memoization has been born from Computational Linguistics [26, 4, 15, 29], it is poetic justice that related ideas boomerang back into CL to help solve a long-time, very interesting problem, and that they should do so in a context that honours Robert Kowalski,a source of so many beautiful and lasting ideas.

Datalog grammars themselves, partly inspired from database theory, follow an assertional representation of parsing which was first cast in logic programming terms by Kowalski [18]–in which sentences are coded with numbered word edges rather than as lists of words, e.g.

```
the(1,2).
resolution(2,3).
principle(3,4).
```

rather than

```
[the,resolution,principle]
```

The two representations are equivalent, though it is faster to look up words in an assertionally represented sentence than to repeatedly pick a list apart.

After an Introduction and a Background section, we present our parsing methodology: predictive left-corner datalog. Section 5 examines our treatment of elipsis in the context of coordinated sentences, and section 6 discusses our results and related work.

2   Background

As we saw in the Introduction, the notion of parallelism is central to work on ellipsis. [12], following [23], also postulates the necessity, within a feature-structure setting, of combining elements which exhibit a degree of syntactico-semantic parallelism in order to determine the way in which some kinds of anaphora are resolved, and argue that the use of default unification (or priority union) improves on Prust's operation for combining the parallel structures. Intuitively, default unification [3] takes two feature structures, one of which (called the TARGET) is identified as "strict", while the other one (called the SOURCE) is "defeasible", and combines the information in both such that the information in the strict structure takes priority over that in the defeasible structure.

For instance, the combination of the feature structures shown below for sentences 1a and 1b:

1a. Hannah likes beetles.

```
        [ AGENT Hannah
          PATIENT beetle ]
                        likes
```

1b. So does Thomas.

```
        [ AGENT Thomas ]
                    agentive
```

results in the priority union:

```
        [ AGENT Thomas
          PATIENT beetle ]
                        likes
```

Thus, the implicit constituent in the second sentence is reconstituted from the first by using a generally applicable procedure on the representations of the parallel structures.

[10] postulated a similar analysis, but it was based on λ-calculus semantic representations, and used higher order unification. For instance, in their example:

Dan likes golf, and George does too.

they identify the antecedent or source as the complete structure ("Dan likes golf"), whereas the target clause ("George does too") is either missing, or contains only vestiges of, material found overtly in the source.

Their analysis of such structures consist of:

a) determining the parallel structure of source and target;

b) determining which are parallel elements in source and target (e.g., "Dan" and "George" are parallel elements in the example);

c) using Huet's higher-order unification algorithm [14] for finding a property P such that $P(s_1,...,s_n) = S$, where s_1 through s_n are the interpretations of the parallel elements of the source, and S is the interpretation of the source itself.

Only solutions which do not contain a primary occurrence of the parallel elements are considered (occurrences are primary if they arise directly from the parallel elements, as opposed to those arising, for instance, from a pronoun). In the example,

```
P(dan) = likes(dan,golf)
```

is solved by equating P with λx. likes(x,golf) given that the other possible solution, λx. likes(dan,golf) contains a primary occurrence of the parallel element, "dan", and must therefore be discarded;

d) applying the property on the representation of the target, e.g. P(george)= [λx.likes(x,golf)] george = likes(george,golf);

e) conjoining the meanings of the source and of the target thus completed, e.g.:

```
likes(dan,golf) & likes(george,golf)
```

Unlike previous analyses, both [10] and [12] provide ambiguous readings of discourses such as

```
Jessy likes her brother. So does Hannah.
```

without having to postulate ambiguity in the source (this is achieved in [12] by allowing for priority union to either preserve or not preserve structure-sharing information in the source, and in [10] by the distinction between primary and secondary occurrences of parallel elements).

Another notable point in both these approaches is that they address the issue of semantic parallelism, which in most previous approaches was understressed in favour of syntactic parallelism.

However, both methods share the following limitations:

a) neither method formulates exactly how parallelism is to be determined- it is just postulated as a prerequisite to the resolution of ellipsis (although [12] speculates on possible ways of formulating this)

b) both approaches stress semantic parallelism, while pointing out that this is not sufficient in all cases.

We shall provide a method for detecting parallel structures through syntactic and semantic criteria in the context of coordinated structures.

3   Our Parsing Methodology: Predictive Left-Corner Datalog

A left corner is the first symbol of a rewriting rule's right hand side. Our bottom-up parser records a left corner's expectations through affine linear assumptions. A rule of the form

```
q --> p1,..., pn.
```

which, with numbered edges, would read

`q(E0,En+1) :- p1(E0,E1),..., pn(En,En+1).`

is transformed (through compilation) into:

`rule(p1(E0,E1),q(E0,En+1)):- +p2(E0,E1),..., +pn(En,En+1).`

(We use BinProlog's [1] notation for linear assumptions ("+" for assuming, "-" for consuming):)
For instance, the rule

`np(X,Z):- det(X,Y), adj(Y,W), noun(W,Z).`

compiles into a rule which, upon recognizing a determiner between X and Y, hypothesizes an adjective between Y and some point W, and a noun between W and Z, and tentatively recognizes a noun phrase between X and Z, subject to the assumptions being (later) satisfied . If they turn out not to be, backtracking will occur.

`rule(det(X,Y),np(X,Z)):- +adj(Y,W), +noun(W,Z).`

Assumptions, combined with sentence boundaries, allow us to distinguish and manage incomplete parses on the fly. The numbered edges give us information as to which constituents might be missing where. The parser charts the results of each level of iteration, and pertinent rules are examined top-down to complete the elided parts.

Appendix I shows our core parser, not including implicit meaning determination.

3.1   Terminology

Symbols that have been assumed are called *predictions*

Symbols whose start and end edges are both known are called *completed*. They are charted at each iteration of our parser.

Symbols placed in the parse state (i.e., the head symbols of applied rules) are called *pending symbols*, because they are subject to the remainder of the right hand side materializing.

A grammar symbol is said to be *implicit* if either:

a) it has been predicted and both its edges coincide, denoting an empty string, or

b) it corresponds to an elided left corner element

---

[1] BinProlog's online manual, http://www.binnetcorp.com/BinProlog/index.html

4 An Interesting Case Study-Coordination

Coordination is one of the most difficult phenomena in language processing. Typically, any two constituents (even of different kind) can be coordinated, and often some substring that is explicit in one of the conjuncts is missing in the other, as in Wood's well-known example

John drove the car through and completely demolished a window.

where the first conjunct is missing an object ("the window") and the second one, a subject ("John").

4.1 Coordinating Complete Constituents

Conjoining complete constituents involves finding two constituents of same category to the left and to the right of the conjunction- e.g. "John and Mary laugh" and "Mary sings and dances" could be respectively represented as

```
laughs(and(john,mary))
```

```
and(sings(mary),dances(mary))
```

Our parser keeps track of the start and end points of the conjunction through a linear assumption which will be consulted every time we generate a new level of parsing, to check whether the two conjoints can be recognized at this level.

For instance, for "John and Mary laugh" the conjoints can be recognized at level two, when we know that "john" and "mary" are names, and we can simply conjoin the names. For "the walrus and the carpenter ate all the oysters", instead, we have to wait until level 3, where we know that both "the walrus" and "the carpenter" are noun phrases, and we can then create a new noun phrase whose meaning will be the conjunction of both noun phrases' meanings. The parser consults just the latest level of iteration to produce the next, with the exception of its conjoining procedure, which consults the entire history of known facts.

4.2 Coordinating Incomplete Constituents

As for the case of complete constituents, we also assume that there are two coordinating constituents, C1 and C2, surrounding the conjunction, which must in general be of the same category. If there are any missing elements, we must moreover require that they have parallel parses. Thus any missing elements in either C1 or C2 can be reconstructed from the other. We also adopt the heuristic that closer scoped coordinations will be attempted before larger scoped ones. Thus in Woods' example, "John drove his car through and completely demolished a window","vp conj vp" is tried before "sent conj sent". We have

```
C1= drove his car through
C2= completely demolished a window
```

As the conjunction is reached before the first verb phrase's parsing is finished ("through" analyses as a preposition introducing a prepositional phrase- i.e., expecting a noun phrase to follow, when there is none), the unfulfilled expectation of a noun phrase is postponed until it can be equated with a parallel noun phrase in C2.

What we mean by C1 and C2 having parallel parses is not that they must necessarily follow the same structure to the last details, but that their structures must complement each other so that missing constituents in one may be reconstructed from the other.

We further assume, for the purposes of this article, that they both must have the same root (in this case, a verb phrase root), although this assumption is not necessary in general.

The Parallel-Structures Constraint In a sentence where a conjunction stretches between points P1 and P2, we predict that our parser will derive (excusing the abuse of notation) at least one of the two structures Cat(A,P1) and Cat(P2,Z); complete the incomplete one (if any) in parallel with the completed one; and derive the new theorem Cat(A,Z). We call this constraint the parallel-structures constraint, and note it as follows:

```
{Cat(A,P1) parallel to Cat(P2,Z) ==> chart Cat(A,Z)}
```

As soon as one of these predictions is fulfilled, we can further specify the other prediction to follow the same structure as that of the found noun phrase, which will allow us to reconstruct any missing elements.

For instance, for "Jean mange une pomme rouge et une verte", stretching between edges 0 and 8, once we have identified "une pomme rouge" as a np, we can expect a np of similar structure after "et".

Of course, backtracking can occur. For instance, the parser's first guess is that the conjoined categories must be "adjective", and that A=4 (this would be a good guess for: "Jean mange une pomme *rouge et verte*"). But this first try will fail to find an adjective after the conjunction, so backtracking will undo the bindings and the parallel-structures constraint will suspend until other suitable candidates for "Cat" and "A" have been derived.

Determining the Parallel Structure of Source and Target By examining ellipsis in the context of coordinated structures, which are parallel by definition, and by using constraints on sentence boundaries, we provide a method in which parallel structures are detected and resolved through syntactic and semantic criteria, and which can be applied to grammars using different semantic representations- feature structure, λ-calculus, etc.

Note that in the parallel structures constraint, P1 and P2 are known, being the edges of the conjunction, while A and Z may or may not be known. We identify the source as the symbol with known edges, and the target as that with an unknown edge.

To complete the target, rules whose head matches the target are tried on it top down, until one is found whose different elements can be found among the completed constituents (i.e., in the chart), the predicted ones (i.e., those assumed) or, in the case of the target itself, the pending symbols(i.e., those in the parse state).

An undischarged prediction becomes a new target, which also needs to be completed with respect to a matching new source within the present source's range, as we shall see later.

Implicit targets (i.e. those corresponding to non-overt constituents) are filled in through their corresponding source, as we shall also see.

An Example Going back to our example *Bob likes tea and Alain coffee*, we have the iteration levels:

```
L1= {bob(0,1),likes(1,2),tea(2,3),and(3,4),alain(4,5),coffee(5,6)}
L2= {name(0,1),verb(1,2),noun(2,3),conj(3,4),name(4,5),noun(5,6)}
L3= {np(0,1),vp(1,3),np(2,3),np(4,5),np(5,6)} L4={s(0,3)}
```

After backtracking on failed choices, the parallel structures constraint instantiates to

```
s(0,3) parallel to s(4,Z)
```

where s(0,3) is the source and s(4,Z) the target.

To complete the target, we now try rules for s top-down, e.g.:

```
s(E1,E2) --> np(E1,E2), vp(E2,E3)
```

which for our target instantiates to

```
s(4,Z) --> np(4,E2), vp(E2,Z).
```

Notice that the charted left-corner np(4,5) has generated (using the same rule in predictive datalog notation) the tentative constituent s(4,Z'), to be found in the parse state, as well as the assumption +vp(5,Z'). The presence of all these elements confirms the adequacy of this s rule, and further instantiates it to

```
s(4,Z) --> np(4,5), vp(5,Z).
```

The non-discharged assumption vp(5,Z) becomes the next target, with source vp(1,3). Trying rules for the new target top-down yields:

```
vp(5,Z) --> verb(5,P), np(P,Z).
```

The charted np in the target's range (namely, np(5,6)) fixes P=5, Z=6, which postulates an empty verb (an implicit symbol) between edge 5 and itself, to be determined through the source vp's overt verb (namely, verb(1,2)).

Determining the Meaning of Ellided Structures As seen in the Background section, after having determined the parallel structure of source and target, we need to determine which are parallel elements in them, and using for instance higher-order unification, we must void the source's meaning from those elements which are not shared with the target and then apply the resulting property on the representation of the target. For our example, we must from the meaning representation of "Bob likes tea" reconstruct the more abstract property $[\lambda y.\lambda x.\text{likes}(x,y)]$, which can then be applied on "coffee" and "alain" to yield likes(alain,coffee).

We bypass this need through a simple heuristic for choosing our source constituent.

The Most Ancient Source Heuristic We adopt the heuristic of preferring as source symbol, among those in the appropriate range, the most ancient (i.e., least instantiated) symbol of same category as the target.

This ensures that any variables extraneous to this constituent's strict meaning representation are not bound by the remaining context in the source.

To exemplify, we add meaning representations to our sample grammar above (where $\wedge$ is a binary operator in infix notation, used to build semantic structure):

```
s(Sem) --> np(X), vp(X^Sem).
```

```
vp(X^Sem) --> verb(Y^X^Sem), np(Y).
```

```
np(X) --> name(X); noun(X).
```

```
verb(Y^X^likes(X,Y)) --> likes.
```

Top-down application of the (datalog) vp rule to complete the target now yields:

```
vp(X^Sem,5,Z) :- v(Y^X^Sem,5,P), np(Y,P,Z).
```

which generates the new target

```
verb(coffee^X^Sem),5,5)
```

to be parsed in parallel with the source verb:

```
verb(Y'^X'^likes(X',Y'),1,2)
```

rather than the more recently charted:

```
verb(tea^bob^likes(bob,tea),1,2)
```

This heuristic is easily implemented by looking for candidate sources in the chart's lowest possible level. Thus, we neither need to explicitly determine which are parallel elements (e.g., "Bob" and "Alain", "tea" and "coffee"), nor abstract them away from the property to be applied, nor apply the property on the target's meaning.

Conjoining the Meaning of the Source and the Completed Target We now need to conjoin the meanings of the completed target and the source, e.g. likes(bob,tea) and likes(alain,coffee). However, this is not enough in the general case. In Wood's example, representing the coordinated sentences "John drove the car through a window" and "John demolished a window", say in some logical form, we must take care of not re-quantifying "a window" when we reconstitute its meaning at the missing point: the window driven through must be equated with the one demolished.

The input string is labeled as follows:

```
john(0,1), drove(1,2),the(2,3),car(3,4),through(4,5),and(5,6),
demolished(6,7),a(7,8),window(8,9).
```

Appendix III shows the grammar we use, in ordinary rather than left-corner notation. Appendix IV shows the subsequent target/source pairs and top-down rule applications.

The parallel structures to be conjoined, in this three-branched quantification example, are:

```
TARGET=
vp(X'^a(W,window(W),the(Y,car(Y),drove_through(X,Y,W))),1,5)

SOURCE=
vp(X^a(V,window(V),demolished(X,V)),6,9)}
```

We use what we call *c-unification*: unify those parts in the parallel structures which are unifiable, and conjoin those that are not (i.e., the parallel elements) (with the exception of the last two arguments, of course, which become the start point of the first conjoint and the end point of the second).

We obtain:

```
vp(X^a(W,window(W),and(the(Y,car(Y),drove_through(X,Y,W)),
                        demolished(X,W))),1,9)
```

After this theorem's addition in the chart, the sentence rule can apply to derive

```
sent(a(W,window(W),and(the(Y,car(Y),drove_through(john,Y,W)),
                        demolished(john,W))),0,9)
```

5   Related Work

Some of our ideas regarding implicit meaning understanding were sketched in embryonic form in [7].

Early metagrammatical approaches to coordination ([28, 9]) treat the appearance of a coordinating word, or conjunction, as a demon (a demon is a procedure automatically triggered by the apearance of some run-time condition- in this case, the recognition of a conjunction). When a conjunction appears in a sentence of the form

```
A X conj Y B
```

a process is triggered which backs up in the parse history in order to parse Y parallel to X, while B is parsed by merger with the state interrupted by the conjunction.

Thus, in Wood's example we have

```
A= John
X= drove his car through
conj= and
Y= completely demolished
B= a window
```

The reconstructed phrase is then A X B and A Y B, with the provision already made regarding requantification.

It is interesting that, whereas in Woods' analysis we end up with two conjoined *sentences*, in ours we first produce one sentence having a verb phrase composed of two conjoined verb phrases. Linguistically speaking, it is arguable whether one analysis is preferable over the other one. But computationally speaking, our analysis is more general in that it accommodates sentences for which Woods' analysis would fail. Our known example "Jean mange une pomme rouge et une verte", for instance, cannot be split into A X conj B Y to reconstitute an unreduced structure following previous analyses. On the other hand, using our approach we can postulate

```
C1= une pomme rouge
C2= une verte
```

and require that C2 follow a structure parallel to that of C1, which then allows us to reconstitute the missing noun in C2.

Recent work, e.g. [16] discusses an alternative approach to anaphoric dependencies in ellipsis, in which the dependence between missing elements in a target clause and explicit elements in a source clause does not follow from some uniform relation between the two clauses, but follows indirectly from independently motivated discourse principles governing pronominal reference.

While containing linguistically deep discussions, the literature on discourse-determined analysis also focusses mostly on ellipsis resolution, and still leaves unresolved the problem of automatically determining which are the parallel structures.

On another line of research, Steedman's CCGs [25] provide an elegant treatment of a wide range of syntactic phenomena, including coordination, which does not resort to the notions of movement and empty categories, instead using limited combinatory rules such as type raising and functional composition . However, these are also well known to increase the complexity of parsing, originating spurious ambiguity- that is, the production of many irrelevant syntactic analyses as well as the relevant ones. Extra work for getting rid of such ambiguity seems to be needed, e.g. as proposed in [22].

6 Discussion

We have introduced a parsing methodology, Predictive Left-Corner Datalog, which makes it syntactically obvious which constituents have been completely parsed (those charted), which ones contain incomplete constituents (those corresponding to symbols with an unknown edge), and which constituents are missing (those corresponding to implicit symbols). The first two of these properties are standard in chart-based approaches with active edges. The latter represents a significant advantage with respect to these approaches.

We have also shown that introducing syntactic as well as semantic parallelism within our Predictive Datalog methodology can help automatically determine which are the parallel structures, and we have exemplified this for the case of coordination. Our analysis of parallelism, inspired in that of [10], complements it in various ways.

Several observations are in order. In the first place, we must note that a simple conjoining of the representations obtained for the parallel structures as proposed in [10] may not, as we have seen, suffice. Since these structures may be quite dissimilar, we must conjoin only the parallel elements. We postulate that, in compositionally defined semantics, the parallel elements will be represented by those subterms which are not unifiable.

Secondly, we do not need to commit to higher-order unification, property reconstructions, etc. Again for compositionally defined semantics, the parallel structures constraint together with top-down target determination and most ancient source heuristics ensures that the correct meanings are associated to the elided constituents as a side effect of parsing.

In the third place, we should note that our analysis allows for the source clause to not necessarily be the first one- as we have also seen, we can have structures in which the incomplete substructure does not antecede the complete one. Thus our analysis can handle more cases than those in the previous related work.

Note that some special cases allow us to use unification between isomorphic objects to obtain the proper quantification. By slightly modifiying the grammar as

```
np(X^Scope^Sem) --> np0(X^Scope^Sem).

np(X^Scope^and(Sem1,Sem2)) --> np0(X^Scope^Sem1),
                               conj(and),
                               np(X^Scope^Sem2).
```

we can handle directly phrases like:

Each man ate an apple and a pear.

Clearly this works only for a class of particular constituents exhibiting strong isomorphism in the constructed meaning. For instance, noun groups of the form *np1*, *np2* and *np3* do have this property.

We must note, however, that in some cases we will need to complement our analysis with a further phase which we shall call "reshaping". Take, for instance, the sentence "Each man and each woman ate an apple". Here we need to reshape the result of the analysis through distribution, thus converting

```
each(X,man(X)&woman(X),exists(Z,apple(Z),ate(X,Z)))
```

into

```
and(each(X,man(X),exists(Z,apple(Z),ate(X,Z))),
    each(X,woman(X),exists(Z,apple(Z),ate(X,Z))))
```

Reshaping operations have been used in [9], and are useful in particular to decide on appropriate quantifier scopings where coordination is involved. It would be interesting to study how to adapt these operations to the present work. Another interesting observation is that the results in [10] concerning the use of the distinction between primary and secondary occurrences of parallel elements in order to provide ambiguous readings of discourses such as "Jessie likes her brother. So does Hannah." could, in principle, be transferred into our approach as well.

Let us also note that, as observed in [1], the notion of compositional semantics of the two clauses (on which the related previous work, and ours to some extent, is based) is not enough in some cases.

For instance, consider:

If Fred drinks, half the bottle is gone. But if Sam drinks too, the bottle is empty.

In the first sentence, the conclusion which holds if Fred drinks *but Sam does not*, does not hold if both Fred and Sam drink. The implicit information that the first conclusion holds only if the premise of the second sentence does not hold must be inferred. Using our approach, we could use the re-shaping phase to deal with cases such as this one, in which the presence of words such as "too" would trigger the generation of the full reading. A sentence of the form

If Fred drinks, C1, but if Sam drinks too, C2.

would roughly generate a representation such as

```
but(if(drink(fred),C1),if(too(drink(sam)),C2))
```

which, after reshaping would become:

```
and(if(and(drink(fred),not(drink(sam))),C1),
    if(and(drink(fred),drink(sam),C2)))
```

Finally, let us note that the methodology presented here is also useful for other applications. Preliminary work re. error detection and correction through datalog grammars has shown good promise [2, 27] in this respect.

Acknowledgements

Thanks are due to Kimberly Voll and Tom Yeh for their help with testing and debugging the parser, and to Kimberly Voll, Paul Tarau and Lidia Moreno for useful comments on a first draft. This research was made possible by NSERC research grant 611024, NSERC Equipment grant 31-613183, and NSERC's Summer Study Program.

Appendix I: The Core Parser

```
p(L,List):- parse(L,List,List1), L1 is L+1, +level(L1,List1),
endcheck(L1,List1).

endcheck(L,[s(0,P)]):- !. endcheck(Level,List):- p(Level,List).

parse(L,[C|Cs],[D|Ds]):- rule(C,D), parse(L,Cs,Ds).
parse(L,[C|Cs],Ds):- rule(C), !,
parse(L,Cs,Ds). % for assumed constituents
parse(L,[C|Cs],[C|Ds]):- parse(L,Cs,Ds).
    % no rule applies to C at this level parse(_,[],[]).
```

The topmost predicate, p/2, will be called for L=1 and List containing the (automatically constructed) datalog representation of the input string. The procedure parse then builds one more level of iteration at a time, and records it through assumption rather than assertion. This way, tentative levels proposed can be backtracked upon, and we only assert the most recent level information for each level, once we have reached the end of the parsing (i.e., once the start symbol is all that remains in the current level).

The equivalent of Matsumoto et al's chart parser's termination rules (see previous section) is also needed here, and can likewise be transparently created by compilation. From our sample np rule above, the following termination clauses are generated:

```
% Rules for assumed constituents

rule(noun(X,Y)):- -noun(X,Y). rule(adj(X,Y)):- -adj(X,Y).
```

Appendix II: A Complete Constituent Coordination Demon and Sample Grammar

The following grammar constructs three-branched quantification representations of the input sentence. The parser metagrammatically treats complete constituent coordination, e.g. *John smiled and Mary laughed, John drove the car through a window and Mary laughed, John drove a car through the window and smiled.*

```
% PARSER

:- dynamic level/2.

go:- abolish(known/1), input_sentence(List), +known([]),
     p(List), nl, listing(known/1).

input_sentence(Level1):- sent(S), datalog(S,0,Level1).

datalog([C|CS],Pos,[MC|MCS]) :- NextPos is Pos + 1,
                                MC =.. [C,Pos,NextPos],
                                datalog(CS,NextPos,MCS).
datalog([],_,[]).

p(L):- write('Parsing: '), write(L), nl, nl,
       parse(L,L1),
       write('L1: '), write(L1), nl,
       endcheck(L1),!.

endcheck([s(_,0,_)]):- !.
endcheck(List):- -known(Old),update(List,Old,K),+known(K),p(List).

update(_,_,K):- -newK(K), !. % if we've conjoined
update(List,OldK,K):- append(List,OldK,K).

parse(Cs,L):- -conj_demon(P1,P2),
              -known(K), +known(K), % get known info
              find(K,Pred,P1,P2,K1,C1,C2),
                  % update current level with conjoining new info:
              remove(conj(_,_),Cs,L1),
              remove(C1,L1,L2),
              remove(C2,L2,L3),
              append(L3,[Pred],L),
              +newK(K1). %marker for endcheck to update K properly

parse([C|Cs],[D|Ds]):- rule(C,D), parse(Cs,Ds).
parse([C|Cs],Ds):- rule(C), parse(Cs,Ds).
          % for assumed constituents parse([],[]).

find(K,Pred,X,Y,[Pred|K],C1,C2):-member(C1,K),C1=..[Cat,Sem1,A,X],
                                  nonvar(A),C2=..[Cat,Sem2,Y,Z],
                                  member(C2,K),nonvar(Z),
                                  c_unify(Sem1,Sem2,Sem3),
                                  Pred=..[Cat,Sem3,A,Z].
```

```
c_unify(X^M1,X^M2,X^and(M1,M2)).
c_unify(X,Y,and(X,Y)).
```

% Utilities

```
remove(X,[X|Ls],List) :- remove(X,Ls,List).
remove(X,[L|Ls],[L|Ds]) :- member(X,Ls), remove(X,Ls,Ds).
remove(_X,List,List).
```

% THE COMPILED GRAMMAR (see Appendix III for source rules)

% Lexicon:

```
rule(john(P1,P2),name(john,P1,P2)).
rule(mary(P1,P2),name(mary,P1,P2)).
rule(the(P1,P2), det(X^S^R^def(X,S,R),P1,P2)).
rule(a(P1,P2), det(X^S^R^exists(X,S,R),P1,P2)).
rule(car(P1,P2), noun(X^car(X),P1,P2)).
rule(window(P1,P2), noun(X^window(X),P1,P2)).
rule(smiled(P1,P2), verb0(X^smile(X),P1,P2)).
rule(laughed(P1,P2), verb0(X^laughed(X),P1,P2)).
rule(drove(P1,P2), verb2(X^Y^Z^drove_through(X,Y,Z),P1,P2).
rule(demolished(P1,P2), verb1(X^Y^demolished(X,Y),P1,P2)).
rule(name(X^P1^P2),np(X^S^S),P1,P2)).
rule(through(P1,P2), prep(P1,P2)).
rule(and(P1,P2), conj(P1,P2)):- +conj_demon(P1,P2).
        % sets up the conjunction demon
```

% Syntactic Rules:

```
rule(verb0(X^S,P1,P2),vp(X^S,P1,P2)).
rule(verb1(X^Y^S0,P1,P2),vp(X^S,P1,P3)) :-
   +np(Y^S0^S,P2,P3).
rule(verb2(X^Y^Z,^S0,P1,P2),vp(X^S,P1,P4)) :-
   +np(Y^S0^S1,P2,P3),
   +pp(Z^S1^S,P3,P4).
rule(prep(P1,P2),pp(X^S0^S,P1,P3)) :- +np(X^S0^S,P2,P3).
rule(det(X^Res^Sco^Sem,P1,P2),np(X^Sco^Sem,P1,P3)):-
   +noun(X^Res,P2,P3).
rule(np(X^Sco^Sem,P1,P2),s(Sem,P1,P3)):- +vp(X^Sco,P2,P3).
rule(conj(P1,P2),conj(P1,P2)).
   % Demon was unable to conjoin
   % at previous level- will try again in the next level
```

```
% Rules for assumed constituents:

rule(noun(Sem,P1,P2)):- -noun(Sem,P1,P2).
rule(vp(Sem,P1,P2)):- -vp(Sem,P1,P2).
rule(np(Sem,P1,P2)):- -np(Sem,P1,P2).
rule(pp(Sem,P1,P2)):- -pp(Sem,P1,P2).
```

Appendix III: A Toy Grammar

```
sent(S) --> np(X^Scope^S), vp(X^Scope).

np(X^Scope^S) --> det(X^Restriction^Scope^S),
                  noun(X^Restriction).
np(X^S^S) --> name(X).

vp(X^S) --> verb0(X^S).
vp(X^S) --> verb1(X^Y^S0), np(Y^S0^S).
vp(X^S) --> verb2(X^Y^Z^S0), np(Y^S0^S1), pp(Z^S1^S)).

pp(X^S0^S) --> prep, np(X^S0^S).

verb0(X^laugh(X)) --> [laughed].
verb1(X,^Y^demolished(X,Y)) --> [demolished].
verb2(X^Y^Z^drove_through(X,Y,Z)) --> [drove].

det(X^Restriction^Scope^exists(X,Restriction,Scope)) --> [a].
det(X^Restriction^Scope^def(X,Restriction,Scope)) --> [the].

noun(X^window(X)) --> [window].
noun(X^car(X)) --> [car].

name(john) --> [john].
name(mary) --> [mary].

prep --> [through].

conj --> [and].
```

Appendix IV: Target Completion Follow-Up for Example in 4.2.6

The parallel structures constraint: Cat(...,A,5) parallel to Cat(...,6,Z) suspends until the following new theorems have been derived at level four:

```
L4= {np(john^S^S),0,1),np(Y^Sc^the(Y,car(Y),Sc)),2,4),
     np(W^Sc1^a(W,window(W),Sc1)),7,9),vp(X^a(W,window(W),
        demolished(X,W)),6,9)}
```

Since there is a source vp starting at point 6, we can now postulate Cat=vp and try to derive top-down a (possibly incomplete) vp parallel to this one and ending at point 5 (the target vp). When trying the third rule for vp , the rule instance

```
vp(X^Sem,P,5) --> verb2(X^Y^Z^SO,P,P1),
                  np(Y^SO^S1,P1,P2),
                  pp(Z^S1^Sem,P2,5).
```

is generated, whose diverse parts can be found among the known theorems or assumptions, thus confirming that this is the right rule, and further instantiating the vp rule into:

```
vp(X^Sem,1,5) --> verb2(X^Y^Z^drove_through(X,Y,Z),1,2),
                  np(Y^drove_through(X,Y,Z),
                     the(Y,car(Y),drove_through(X,Y,Z)))),2,4),
                  pp(Z^the(Y,car(Y),
                     drove_through(X,Y,Z))),Sem),4,5).
```

New target:

```
pp(Z^the(Y,car(Y),drove_through(X,Y,Z))^Sem),4,5)
```

Since this pp prediction remains unfulfilled, the parser tries and fails to find a parallel pp within the source vp, so pp rules are now applied top-down, generating the rule instance

```
pp(Z^the(Y,car(Y),drove_through(X,Y,Z))^Sem),4,5) -->
        prep(4,5),
        np(Z^the(Y,car(Y),drove_through(X,Y,Z))^Sem),5,5).
```

This pp's np becomes the target np. Its parallel np (the window, between points 2 and 4) within the source vp is then identified as its source (to be found at the oldest possible level of iteration).

We have:

```
SOURCE NP: np(W^Scope^a(W,window(W),Scope)),2,4)
TARGET NP: np(Z^the(Y,car(Y),drove_through(X,Y,Z))^Sem),5,5)
```

Unifying the source and target np yields the

```
COMPLETED TARGET NP:
np(W^the(Y,car(Y),drove_through(X,Y,W)^
        a(W,window(W),the(Y,car(Y), drove_through(X,Y,W))),5,5)
```

This is added to the current level of iteration, which will cause the undischarged np assumption to be consumed, which in turn completes the target vp:

```
COMPLETED TARGET VP:
vp(X'^a(W,window(W),the(Y,car(Y),drove_through(X,Y,W))),1,5)
```

References

[1] N. Asher. *Reference to Abstract Objects in Discourse.* Studies in Linguistics and Philosophy, 50, 1992.

[2] J. Balsa, V. Dahl, and J.G. Pereira Lopes. Datalog Grammars for Abductive Syntactic Error Diagnosis and Repair. *Proceedings of the Natural Language Understanding and Logic Programming Workshop*, Lisbon, 1995.

[3] J. H. R. Calder. *An Interpretation of Paradigmatic Morphology.* PhD thesis, University of Edinburgh, 1990.

[4] J. Cocke, and J. I. Schwartz. *Programming Languages and Their Compilers.* Courant Institute of Mathematical Sciences, New York University, 1970.

[5] A. Colmerauer. *Metamorphosis Grammars*, pages 133–189. Lecture Notes in Computer Science, Springer-Verlag, 63, 1978.

[6] V. Dahl, P. Tarau and R. Li. Assumption Grammars for Processing Natural Language. *Proceedings of the International Conference on Logic Programming'97*, pages 256–270, 1997.

[7] V. Dahl, P. Tarau, L. Moreno and M. Palomar. Treating Coordination with Datalog Grammars. *COMPULOGNET/ELSNET/EAGLES Workshop on Computational Logic for Natural Language Processing*, Edinburgh, April 3-5, 1995, pages 1–17.

[8] V. Dahl, P. Tarau and Y. N. Huang. Datalog Grammars. *Proceedings of the 1994 Joint Conference on Declarative Programming*, Peniscola, Spain, 1994.

[9] V. Dahl and M. McCord. *Treating Coordination in Logic Grammars.* American Journal of Computational Linguistics, 9:69–91, 1983.

[10] M. Darlymple, S. Shieber, and F. Pereira. *Ellipsis and Higher-Order Unification* Linguistics and Philosophy, 14(4):399–452, 1991.

[11] J.-Y. Girard. *Linear Logic.* Theoretical Computer Science, (50):1-102, 1987.

[12] C. Grover, C. Brew, S. Manandhar, and M. Moens. Priority Union and Generalization in Discourse Grammars. *Proceedings of the 32nd ACL Conference*, New Mexico, 1994.

[13] J. Hodas. Specifying Filler-Gap Dependency Parsers in a Linear-Logic Programming Language. In Krzysztof Apt, editor, *Logic Programming Proceedings of the Joint International Conference and Symposium on Logic programming*, pages 622–636, Cambridge, Massachusetts London,England, 1992. MIT Press.

[14] G. Huet. *A Unification Algorithm for Typed Lambda-Calculus.* Theoretical Computer Science, 1:27–57, 1975.

[15] T. Kasami. *An efficient recognition and syntax algorithm for context-free languages.* Technical Report AF-CRL-65-758, Air Force Cambridge Research Laboratory, Bedford, MA., 1965.

[16] A. Kehler, and S. Shieber. *Anaphoric Dependencies in Ellipsis.* Computational Linguistics, 23(3), 1997.

[17] A.P. Kopylov. Decidability of Linear Affine Logic. In *Proceedings of the 10th Annual IEEE Symposium on Logic in Computer Science*, pages 496–504, 1995.

[18] R.A.K. Kowalski. *Logic for Problem Solving*. North-Holland, 1979.

[19] Y. Matsumoto, H. Tanaka, H. Hirakawa, H. Miyoshi,and H. Yasukawa. BUP: a bottom-up parser embedded in Prolog. *New Generation Computing*, 1:pages 145–158, 1983.

[20] D.A. Miller and G Nadathur. Some uses of higher -order logic in computational linguistics. In *Proceedings of the 24th Annual Meeting of the Association for Computational Linguistics*, pages 247–255, 1986.

[21] R. Pareschi and D. Miller. *Extending definite clause grammars with scoping constructs* Warren, David H. D. and Szeredi, P. (eds.) International Conference in Logic Programming, MIT Press, pages 373–389, 1990.

[22] J.C. Park and H.J. Cho. Informed Parsing for Coordination with Combinatory Categorial Grammar. *Proceedings of the International Conference on Computational Linguistics (COLING)*, pages 593–599, 2000.

[23] H. Prust. *On Discourse Structuring, Verb Phrase Anaphora and Gapping*. PhD thesis, Universiteit van Amsterdam, 1992.

[24] D. Srivastava and R. Ramakrishnan. *Pushing Constraint Selections*. The Journal of Logic Programming, 16:361–414, 1993.

[25] M. Steedman. *Gapping as Constituent Coordination*. Linguistics and Philosophy, 1990.

[26] D. S. Warren. *Memoing for logic programs*. Communications of the ACM, 35(3):94–111, 1992.

[27] K. Voll, T. Yeh, and V. Dahl. *An Assumptive Logic Programming Methodology for Parsing*. International Journal of Artificial Intelligence Tools, Vol. 10, No. 4, pages 573–588, 2001.

[28] W. Woods. *An Experimental Parsing System for Transition Network Grammars*. In R. Rustin, editor, Natural Language Processing, pages 145–149, Algorithmic Press, New York, 1973.

[29] D. H. Younger. *Recognition and Parsing of Context-free Languages in Time*. Information and Control, 10(2):189–208, 1967.

Data Mining as Constraint Logic Programming[*]

Luc De Raedt

Institut für Informatik, Albert-Ludwig-University
Georghes Koehler Allee 79, D-79110 Freiburg, Germany
`deraedt@informatik.uni-freiburg.de`

Abstract. An inductive database allows one to query not only the data but also the patterns of interest. A novel framework, called RDM, for inductive databases is presented. It is grounded in constraint logic programming. RDM provides a small but powerful set of built-in constraints to query patterns. It is also embedded in the programming language Prolog. In this paper, the semantics of RDM is defined and a solver is presented. The resulting query language allows us to *declaratively* specify the patterns of interest, the solver then takes care of the *procedural* aspects.

1 Introduction

Imielinski and Mannila [17] present a database perspective on knowledge discovery. From this perspective, data mining is regarded as a querying process to a database mining system. Querying for knowledge discovery requires an extended query language (w.r.t. database languages), which supports primitives for the manipulation, mining and discovery of rules, as well as data. The integration of such rule querying facilities provides new challenges for database technology.

The view of Imielinski and Mannila is very much in the spirit of Kowalski's celebrated equation "Algorithm = Logic + Control" [19]. Indeed, inductive database queries declaratively specify the *logic* of the problem and the inductive database management system should provide the procedures (i.e. the *control*) for solving the query. Thus the hope is that inductive databases will allow us to nicely separate the declarative from the procedural aspects.

Mannila and Toivonen [23] formulate the general pattern discovery task as follows. Given a database r, a language $\mathcal{L}$ for expressing patterns, and a constraint q, find the theory of r with respect to $\mathcal{L}$ and q, i.e. $Th(\mathcal{L}, r, q) = \{\phi \in \mathcal{L} \mid q(r, \phi) \ is \ true\}$. This formulation of pattern discovery is generic in that it makes abstraction of several specific tasks including the discovery of association rules, frequent patterns, inclusion dependencies, functional dependencies, frequent episodes, ... Also, efficient algorithms for solving these tasks are known [23].

So far, the type of constraint that has been considered is rather simple and typically relies on the frequency of patterns. However, knowledge discovery is

---

[*] This paper significantly extends [8,9].

A.C. Kakas, F. Sadri (Eds.): Computat. Logic (Kowalski Festschrift), LNAI 2408, pp. 526–547, 2002.

often regarded as a cyclic and iterative process [12], where one considers a number of different selection predicates, patterns, example sets and versions of the database. Therefore, there is a need for considering and combining different theories $Th(\mathcal{L}_i, q_i, r_i)$. The practical necessity to combine various theories $Th(\mathcal{L}_i, q_i, r_i)$ forms one of the main motivations underlying the work on inductive databases [24] and database mining query languages [17,18,26,14,13]. Indeed, a number of extensions to database languages such as SQL have been introduced with the aim of supporting the generation and manipulation of theories $Th(\mathcal{L}_i, q_i, r_i)$.

The need to combine various constraints and answers also arises in constraint programming [15]. We pursue this analogy further by developing a constraint logic programming language RDM to support data mining. RDM stands for Relational Database Mining. There are different versions of RDM, which depend on the pattern domain under consideration. The pattern domains considered here include item-sets [1], sequences [20,21], graphs [30] and Datalog queries [7]. Nevertheless, the key primitives and the execution mechanism of RDM are presented at a domain independent level. RDM supports a larger variety of constraints than previous inductive databases. In addition to providing an operator to manipulate patterns, it also allows the user to manipulate sets of examples, and provides selection predicates that impose a minimum or maximum frequency threshold. This allows RDM to address descriptive as well as predictive induction. In descriptive induction one is interested in frequent patterns, whereas in predictive induction, one aims at discovering patterns that are frequent on the positive examples but infrequent on the negative ones. Further primitives in RDM are based on the notion of generality (among patterns) and on coverage (w.r.t. specific examples). Queries in RDM consist of a number of primitives which specify constraints on the patterns of interest. Each constraint results in a theory $Th(\mathcal{L}_i, q_i, r_i)$. RDM then efficiently computes the intersection of these theories. RDM's execution mechanism is based on the level-wise version space algorithm of De Raedt and Kramer [10], which integrates the version space approach by Tom Mitchell [28] with the level-wise algorithm [23]. Directly computing the intersection of the theories contrasts with the approach taken in the MINE RULE operator by Meo et al. [26] and the approach of Boulicaut et al. [3] in that these latter approaches typically generate one theory $Th(\mathcal{L}_i, q_i, r_i)$ and then repeatedly modify it.

Embedding RDM within a programming language such as Prolog [4] puts database mining on the same methodological grounds as constraint programming. Indeed, each of the different selection predicates or primitives in a query or program imposes certain constraints on the patterns. As in constraint programming, we have to specify the semantics of these primitives as well as develop efficient solvers for queries and programs. In this paper, an operational semantics for RDM is specified and an efficient solver presented.

The paper is organised as follows: in Section 2, we define the basic RDM primitives, in Section 3, we show RDM at work through a number of database mining queries, in Section 4, we present a solver for simple queries, in Section 5,

528   Luc De Raedt

we discuss how to extend the language and the solver, and finally, in Section 6, we conclude and touch upon related work.

2   Data Mining Domains

2.1   Data Mining Domains

In data mining various pattern domains are employed. A pattern domain consists of three components: a set of patterns, a set of observations or examples and a set of constraints. The most fundamental constraint is the coverage constraint, which specifies when a pattern matches (covers) an example.

For illustration, consider the following pattern domains:

Definition 1. (Item-Sets IS [1]) Consider the a set of all possible items $\mathcal{I}$. An item-set i is then a subset of $\mathcal{I}$. Examples (sometimes called transactions) and patterns are item-sets. A pattern p covers an example e if and only if $p \subseteq e$.

Items are elements of interest. Typical examples of items are products. The set $\mathcal{I}$ then corresponds to the set of all products in a given supermarket and an item-set to a possible transaction.

Definition 2. (Sequences SEQ) Examples and patterns are sequences. A pattern covers an example if it is a subsequence of the example.

Definition 3. (Graphs G [30]) Examples as well as patterns are labeled graphs. A pattern p covers an example e if and only if p is isomorphic to a subgraph of e.

Definition 4. (Datalog Queries DQ [7]) A database $\mathcal{D}$ is given. The database contains the predicate $key/1$. An example is then a substitution $\{K \leftarrow k\}$ such that k is a constant and $\mathcal{D} \models key(k)$. A pattern is a Datalog query of the form $? - key(K), l_1, ..., l_n$, where the l_i are literals. A pattern p covers an example θ in $\mathcal{D}$ if $\mathcal{D} \models p\theta$.

Throughout the paper we will employ the domain of item-sets to illustrate our inductive database formalism RDM. However, we wish to stress that all of the concepts introduced equally apply to the other domains (SEQ, G, and DQ).

2.2   Data Types

The RDM framework distinguishes three basic data types:

- $\mathcal{E}$: denotes the example data type; depending on the application domain this can be item-sets, graphs, sequences, keys, ...
- $\mathcal{P}$: denotes the pattern data type; depending on the application domain this can be item-sets, graphs, sequences, Datalog queries, ...

- $\mathcal{D}(\mathcal{T})$: denotes the data set type over the type $\mathcal{T}$; datasets of type $\mathcal{D}(\mathcal{E})$ are sets of examples of type $\mathcal{E}$; datasets of type $\mathcal{D}(\mathcal{P})$ are sets of patterns of type $\mathcal{P}$.

In many situations the data types $\mathcal{E}$ and $\mathcal{P}$ will be identical. Throughout the rest of the paper we will make abstraction of how specific data types are internally represented. This will leave a maximum degree of freedom to the implementer. We do however assume that the membership predicate is available:

- ?Element in +Set : succeeds when Element $\in$ Set.

We employ the usual conventions where we list the modes of the predicates using '$-$","$+$' and '?': '$-$' arguments must not be instantiated, '$+$' arguments must be instantiated to a non-variable term, and '?' arguments may but need not be instantiated at the time of calling.

Let us illustrate the types on the domain of item-sets IS. $\mathcal{E}$ as well as $\mathcal{P}$ could be represented as ordered lists of atoms (in which each item appears at most once). Furthermore, the sets of examples and patterns could then easily be represented using lists of elements of $\mathcal{P}$ or $\mathcal{E}$.[1]

2.3   Basic Querying Primitives in RDM

The following primitives are supported by the inductive database language RDM. We provide generic definitions of the primitives that are meaningful across different pattern domains. However, we illustrate them mainly on item-sets, which results in the language RDM(IS). Throughout the paper we employ a Prolog like style and syntax. Consider the following predicates:

- +Pattern covers +Example: succeeds whenever the Pattern covers the Example.
- ?Pattern1 <<= +Pattern2: succeeds whenever Pattern1 is 'more general than' Pattern2, i.e. whenever Pattern1 covers an example e, Pattern2 covers e as well[2]. Also, the usual variant 'strictly more general' is <<.

It will be convenient to refer to the most specific pattern within the domain as bottom and to the most general one as top.

In the domain of item-sets IS (with the above sketched data types), both covers and <<= correspond to the subset relation. Indeed, for item-sets $P, P1, P2$ and E, P *covers* E if and only if $P \subseteq E$, and $P1$ <<= $P2$ if and only if $P1 \subseteq P2$.

---

[1] In practical implementations, it is likely that sets would be represented differently, e.g. using files.

[2] The reason for employing the notation <<= to denote the 'is more general than' relation is that this relation often coincides with the subset relation $\subseteq$ (or a variant thereof). The reader has to keep this interpretation in mind when reasoning about <<=.

Although for item-sets, covers and $<<=$ coincide this is not the case for some of the more complex domains such as DQ. Indeed, for Datalog queries, the typical 'more general than' notion corresponds to a form of θ-subsumption, whereas coverage would be tested by instantiating the query with the example and answering the resulting query on the database.

The following properties of primitives will turn out to be crucial for efficiency reasons.

Definition 5. Let $f : \mathcal{D}(\mathcal{P}) \to R$ be a function from patterns to real numbers. We say that f is monotonic (resp. anti-monotonic) whenever $P <<= Q$ implies $f(P) \leq f(Q)$ (resp. $(f(P) \geq f(Q))$) for two patterns P and Q.

Let us now extend these notions of monotonicity and anti-monotonicity to the case where f is a unary predicate taking patterns as argument. The value $f(P)$ of the predicate f is then 1 for those patterns P for which $f(P)$ is true, and 0 for the other patterns. Under this definition the predicate f defined by the clause

```
f(P) :- P covers ex.
```

where ex is a specific example, is anti-monotonic.

Abusing terminology, we will sometimes talk about monotonic or anti-monotonic queries. These queries then implicitly define a unary predicate over patterns.

Sometimes it will be useful to relax the condition on coverage. For instance, one might be interested in patterns that almost cover the example. This can be realized using the following primitive.

– match(+Pattern,+Example) denotes the degree to which the Pattern matches the Example. It is required that matches(P,ex) for any specific example ex is monotonic w.r.t. $<<=$.

For instance, the degree to which an item-set P considered as a pattern matches an item-set E considered as an example could be defined as follows.

$$match(P, E) = \mid P \mid - \mid P \cap E \mid$$

This notion of matching might appear unnatural at first sight because it yields the value 0 when there is a perfect match and a positive integer otherwise. This notion of matching is however motivated by the monotonicity requirement, which is as we shall see, crucial for efficiency reasons.

For some applications it might also be more natural to work with a dual notion of matching, called anti-matching. The function anti-match(P,E) for item-sets could be defined as $\mid P \cap E \mid$. Anti-matching should (and in this case does) satisfy the anti-monotonicity requirement.

The typical use of the primitive match (as well of the primitives frequency, anti-match and similarity introduced below) will be in a literal of the form match (P,E) op Num where op is a comparison operator such as $<, >, \leq, \geq$, and P, E and Num are a pattern, example and a number, respectively. Notice that for fixed E, Num and op the corresponding query behaves either monotonically or non-monotonically.

Another desirable primitive concerns similarity.

- similarity(+Element1,+Element2): denotes the similarity between the two elements Element1 and Element2.

Similarity among two item-sets I and J can be defined as

$$similarity(I, J) = \frac{2 \times |I \cap J|}{|I| + |J|}$$

This definition has the property that the similarity between I and J is 1 if and only if I and J are identical. Similarity could be used to perform similarity based reasoning such as required by the k-nearest neighbor algorithm or clustering algorithm, where the basic operation is the computation of the similarity of one example to another. Unfortunately similarity is neither monotonic nor anti-monotonic. This will make its efficient implementation hard.

The true data miner's favourite primitive is:

- frequency(-E, +Set,+Query): denotes the number of all elements E in Set for which Query succeeds. It is required that the variable E occurs in Query. The frequency corresponds to the cardinality of the set NewSet when the predicate defineset(E,Set,Query,NewSet) (cf. below) succeeds.

Now that we have defined all the basic operations on examples and patterns, we still need to define primitives that allow us to manipulate sets of examples and of patterns.

- defineset(-E,+Set,+Query,-NewSet): succeeds when NewSet is the set of elements E for which Query succeeds. It is mandatory that E occurs in Query.

For instance, the query defineset(E, DataSet, anti-match([beer,mustard,cheese], E) ≥ 2), Set), succeeds if Set is the list of all examples in DataSet that have at least two items in common with [beer,mustard,cheese].

The predicate defineset could - for the domain of item-sets - be implemented using Prolog's setof0 predicate.

```
defineset(El,Set,Query,NewSet) :-
    sefof0(El,(member(El,Set), call(Query)), NewSet).
```

The predicate defineset is crucial to the framework as it allows us to manipulate sets of patterns and data. This predicate is RDM's way to realize the so called closure property (cf. [3]).[3]

---

[3] An inductive database consists of data and patterns. Furthermore there are inductive queries that can be posed to an inductive database. The closure property states that the result of an inductive query is again an inductive database.

2.4   Queries, Modes, and Safety

It should be noted that not all queries are safe. For instance, the query ?- Pat1
<<= Pat2 is unsafe because it has - in general - an infinite number of answers.
To avoid this problem, queries involving patterns will typically be bounded from
below (using e.g. ?- Pat <<= bottom).

We address the problem in the usual Prolog manner. The above definitions
of the primitives include the usual mode-declarations. Mode conform queries are
safe.

3   Example Queries

In this section, we provide a number of queries to illustrate the power of the
querying approach. For simplicity, the examples are illustrated over the domain
IS. However, the example queries generalize to the other domains as well.

Throughout the paper we employ databases d1 and d2 which are defined as
the following lists of item-sets:

```
d1 = [[beer,mustard,sausage,bread,cheese],
      [beer,mustard,bread,cheese,wine],
      [coke,beer,bread,cheese,wine],
      [fries,mayo,cheese,beer],
      [bread,cheese,wine]]
d2 = [[coke,bread,cheese],
      [fries,mayo,sausage],
      [bread,cheese]]
```

Given only these databases,

```
bottom = [beer,bread,cheese,coke,fries,mayo,mustard,sausage,wine]
top = []
```

3.1   Simple Queries

Let us first specify a data mining query that returns all frequent item-sets together with their associated frequency.

(1) ?- P <<= bottom, F is frequency(E,d1,covers(P,E)), 3 < F.

The answers to this query are

```
F = 5, P = [] ? ;
F = 4, P = [beer] ? ;
F = 4, P = [beer,cheese] ? ;
F = 4, P = [bread] ? ;
F = 4, P = [bread,cheese] ? ;
F = 5, P = [cheese] ? ;
no
```

Notice that in displaying the answers, we use the traditional Prolog convention in which an answer substitution is listed explicitly and followed ended by a '?' when more answers exist. When the user inputs a ';' the next answer subsitution is listed.

A variant of query (1) succeeds only for those frequent item-sets in which beer occurs. It only returns the second and third answer.

```
(2) ?- P <<= bottom, [beer]<<= P, 3 < frequency(E,d1,covers(P,E)).
```

Another type of query generates patterns with high frequency on positives and low frequency on negatives. This is especially useful in predictive data mining.

```
(3) ?- P <<= bottom, 3 < frequency(E,d1,covers(P,E)),
frequency(E,d2,covers(P,E))< 1.
```

```
P = [beer] ? ;
P = [beer,cheese] ? ;
no
```

The induced patterns can directly or indirectly be used for classification, i.e. to discriminate among examples from d1 and d2.

Alternatively one might impose a constraint on the accuracy directly. The following query generates patterns that are at least 80 per cent accurate. Such queries will however be harder to answer efficiently (cf. Section 5.2).

```
(4) ?- P <<= bottom, F1 is frequency(E,d1,covers(P,E)),
F2 is frequency(E,d2,covers(P,E)), Acc is F1/(F1 + F2), Acc > 0.7.
```

```
P = [beer], F1 = 4, F2 = 0, Acc = 1  ? ;
P = [mustard], F1 = 2, F2 = 0, Acc = 1  ? ;
P = [cheese], F1 = 5, F2 = 2, Acc = 0.71  ? ;
...
```

Queries about similarity can be used to find those elements that are similar to a specified example.

```
(5) ?- E in d1, similarity(E,[beer,cheese,wine,port])> 0.5 .
```

```
E = [beer,mustard,bread,cheese,wine] ? ;
E = [coke,beer,bread,cheese,wine] ? ;
E = [bread,cheese,wine] ? ;
no
```

At this point it would also be interesting if the query language sorted its answers, e.g. according to the degree of matching or similarity. E.g. the substitutions for E could be sorted in such a way that the most similar examples were returned first. This is implemented in William Cohen's query language Whirl [5] which extends Datalog with a unique primitive called a 'soft join', which is related to our notion of similarity. Alternatively one might use the optimization primitives which are introduced in Section 5.1.

Let us consider also the match construct. Matching can be useful in sequence analysis in biological data (such as proteins). One can then discover all patterns that frequently match with the example to a certain extent, e.g.

```
(6) ?- P <<= [bread,cheese,coke], match(P,[beer,coke]) < 2 .

P = [] ? ;
P = [bread] ? ;
P = [bread,coke] ? ;
P = [cheese] ? ;
P = [cheese,coke] ? ;
P = [coke] ? ;
no

(7) ?- P <<= bottom, [wine] <<= P,
4 < frequency(E,d1,(match(P,E) < 2)).

P = [beer,cheese,wine] ? ;
P = [beer,wine] ? ;
P = [cheese,wine] ? ;
P = [wine] ? ;
no
```

3.2   Embedding within Prolog

Part of the attractiveness of constraint logic programming is that constraint processing abilities are embedded within a general programming language (such as Prolog). The same is true for database languages. These can often be embedded within other programming languages as well.

In this section, we show how the above introduced primitives can be embedded within Prolog and how this allows us to easily formulate some queries of interest.

As a first illustration we show how to query for association rules over a given database (cf. [1]). It is assumed that MinAcc and MinSupport are instantiated at the time of querying to the minimum required accuracy and support, respectively.

```
asso_rule(DataSet, Conclusion, Condition, MinAcc, MinSupport) :-
    F1 is frequency(E,DataSet,covers(Conclusion,E)),
    F1 > MinSupport,
    propersubset(Condition, Conclusion),
```

```
F2 is frequency(E,DataSet,covers(Condition,E)),
Acc is F2 / F1,
Acc > MinAcc.
```

(8) ?- P <<= bottom, asso_rule(d1, P, C, 0.3, 3).

The above program can easily be extended to also include a statistical criterion for deciding whether the association rule is interesting.

Suppose that we are interested in classifying the example [beer,cheese,wine,
port] as belonging to d1 or to d2. We could use the following query

(9) ?- E1 in d1, S1 is similarity(E1,[beer,cheese,wine,port]),
    not(E2 in d2, S2 is similarity(E2,[beer,cheese,wine,port]),
    S1<S2).

This query implements the simple nearest neighbor algorithm. If the query succeeds the system predicts d1, otherwise d2.

Finally consider also the following complex query, which queries for two interrelated patterns.

(10) ?- P1 <<= bottom, P2 <<= bottom,
frequency(E,d1,P1) > 3, defineset(E,d1,(not covers(P1,E)),D2),
frequency(E,D2,P2) > 1.

Query (10) first generates pattern P1 which has minimum frequency of 4 on
the dataset d1. It then looks for pattern P2 which has minimum frequency of 2 on
the examples not covered by pattern P1. Thus pattern P2 depends on pattern P1.
Though such queries are in principle interesting (e.g. when learning concepts),
it is - at this point - an open question as how these queries can efficiently be
answered (cf. Section 4.5).

4   Solving Queries

So far, we have defined the semantics of the language RDM and embedded it
within Prolog. It is straightforward to implement the *specification* of the primitives within Prolog. However, a direct implementation would not work efficiently
because it would simply generate all patterns and test whether they satisfy the
constraints. Below we sketch more efficient algorithms to answer queries.

The key insights that lead to efficient algorithms for answering database
mining queries are 1) that the space of patterns is partially ordered by the 'is
more general than' relation, and 2) that all primitives (with the exception of
similarity) are either monotonic or anti-monotonic (cf. below). From now on, we
focus on such monotonic and anti-monotonic queries and will therefore ignore
the similarity primitive in the rest of this section.

4.1 Various Types of Queries

Let us now investigate the introduced primitives more closely. First, the *basic* atoms are atoms involving the predicates covers, $<<=$ and match(P,E) op ct. Atoms of the form frequency(E,D,Query) op ct are called frequency literals.

The following basic atoms are anti-monotonic :

- Arg1 covers Arg2 : where Arg1 is a variable and Arg2 is ground
- Arg1 $<<=$ Arg2 : where Arg1 is a variable and Arg2 is ground
- match(Arg1,Arg2) op Arg3 : where Arg1 is a variable and Arg2 and Arg3 are ground and op is either $<$ or $\leq$

The basic monotonic atoms can be obtained dually:

- Arg1 covers Arg2 : where Arg2 is a variable and Arg1 is ground
- Arg1 $<<=$ Arg2 : where Arg2 is a variable and Arg1 is ground
- match(Arg1,Arg2) *op* Arg3 : where Arg1 is a variable and Arg2 and Arg3 are ground and *op* is either $>$ or $\geq$

Furthermore, the following properties hold:

- The negation *not A* of monotonic atom A is anti-monotonic and vice versa.
- The conjunction of a set of monotonic (resp. anti-monotonic) literals querying for the same pattern variable is monotonic (resp. anti-monotonic).
- The frequency literals of the form frequency(Arg1,Arg2,Arg3) op Arg4 are anti-monotonic for a pattern variable P when Arg3 is an anti-monotonic query in P, Arg1 is a variable occurring in Arg3, Arg2 and Arg4 are ground, and op is $>$ or $\geq$. For instance, the literal frequency(E,d1,covers(P,E)) $>$ 3 is anti-monotonic for P. Frequency literals can change their status from monotonic to anti-monotonic by negating the query Arg3 or by inverting the operator op.

4.2 The Search Space

One of the most popular algorithms in data mining is the so-called level-wise search algorithm (cf. [23]). The level-wise algorithm generates all patterns p that are a solution to an anti-monotonic query $query(p)$.

In order to introduce the level-wise algorithm, we need some terminology. First, let us define the notions of minimally general and maximally general elements of a set S w.r.t. $<<=$.

Definition 6. Let S be a set of patterns :

- $min(S) = \{p \in S \mid \neg \exists q \in S : q << p\}$
- $max(S) = \{p \in S \mid \neg \exists q \in S : p << q\}$

Second, we need operators on patterns.

Definition 7. Let p be a pattern, then

- A refinement operator $\rho_s(p) = max\{p'\ is\ a\ pattern\ |\ p << p'\}$.
- A generalization operator $\rho_g(p) = min\{p'\ is\ a\ pattern\ |\ p' << p\}$.

For the domain IS, $\rho_s(p)$ contains all item-sets obtained by adding a single item to p; $\rho_g(p)$ contains all item-sets obtained by deleting a single item from p.

The level-wise algorithm is shown below. It works iteratively and alternates between candidate evaluation (the F step) and candidate generation (the L step). The candidate evaluation step evaluates which of the candidates satisfy the query; the candidate generation step generates those candidates that may still satisfy the query. To this aim it employs the anti-monotonicity property together with the solutions at the previous level F_i. Notice that if $\rho_g(p) \not\subseteq F_i$ then p cannot satisfy the query because of the anti-monotonicity property.

$L_0 := \{top\};$
$i := 0;$
while $L_i \neq \emptyset$ **do**
$\qquad F_i := \{p\ |\ p \in L_i\ and\ query(p) = true\}$
$\qquad\qquad$ *i.e. determine the elements in L_i that satisfy the query*
$\qquad L_{i+1} := \{p\ |\ \exists q \in F_i : p \in \rho_s(q)\ and\ \rho_g(p) \subseteq F_i\}$
$\qquad\qquad$ *generation: find candidate patterns at next level*
$\qquad i := i + 1$
endwhile
output $\cup_j F_j$

A well-known fact (cf. [23]) is that the space of solutions for anti-monotonic queries is bounded by a border BD. [4]

Definition 8. $BD^+(query) = S(query) = min\{p\ |\ p\ is\ a\ pattern\ and\ query(p)\}$

This set corresponds to the S-set in Mitchell's version space framework [28]. For convenience, we use Mitchell's terminology. For an anti-monotonic query, the S-set completely characterizes the set of all solutions. Indeed, all patterns that are more general than a pattern in S is a solution to the query as well.

Because Mitchell considers not only anti-monotonic but also monotonic constraints, Mitchell introduces also the dual of the S-set, which is the so called G-set.

Definition 9. $G(query) = max\{p\ |\ p\ is\ a\ pattern\ and\ query(p)\}$

Because of the duality, the G-set completely characterizes the set of solutions to a monotonic query. Furthermore, if one works with a conjunctive query that

---

[4] Manilla and Toivonen [23] also introduce a negative border. The negative border BD^- contains all patterns whose strict generalizations are a solution. Though it might be interesting to use the negative border in our framework, we choose not to do this for convenience.

involves both monotonic and non-monotonic literals, the space of solutions is completely characterized by the two boundaries S and G.

The space of solutions is called the version space $VS(query)$ by Mitchell.

Definition 10. $VS(query) = \{p \mid p \text{ is a pattern and } query(p) \}$

The following property can be proven.

Property 1. $VS(query) = \{p \mid p \text{ is a pattern and } \exists s \in S(query), \exists g \in G(query) : g <<= p <<= s\}$

Mitchell's original formulation of the version space considered only constraints of the form $P <<= ex$ and $not(P <<= ex)$. It has been extended by various researchers, see [27,29,16]. For instance, Mellish extended it to also take into account constraints of the form $p <<= P$ and $not(p <<= P)$. Mellish provides an algorithm that constructively computes the S and G-sets for these 4 types of constraints. In this paper, we elaborate on the version space framework by also taking into account the frequency constraints.

4.3   Solving Simple RDM Queries

The version space framework is important in our context because it can be adapted to solve simple RDM queries.

Definition 11. An RDM query $? - l_1, ..., l_n$ is simple if all literals l_i 1) concern the same pattern P, and 2) are either monotonic or non-monotonic.

For simple queries, Property 1 holds and the space of solutions can be represented by the S and G-sets. To illustrate this, we reformulate the answers to the above simple queries in terms of G and S :

```
(1) G = {[]} ;      S = {[beer,cheese],[bread,cheese]}
(2) G = {[beer]} ;  S = {[beer,cheese]}
(3) G = {[beer]} ;  S = {[beer,cheese]}
(6) G = {[]} ;      S = {[bread,coke],[cheese,coke]}
(7) G = {[wine]} ;  S = {[beer,cheese,wine]}
```

The naive way of solving a simple query would be to first split the query q in two parts q_a and q_m corresponding to the anti-monotonic and monotonic parts respectively, and then to use the two dual versions of the level-wise algorithm. Though this approach would work it is clear that one can do better by adopting the version space algorithm.

When analyzing simple queries, the most expensive literals are those concerning frequency, because computing the frequency requires access to the data(bases). For the other literals, concerning covers, match, $<<=$, this is not necessary. Therefore, a good strategy is to first compute the G and S boundaries using the constraints mentioning covers, match, $<<=$ and then further shrink the version space using the frequency constraints. By doing this the hope is that the first step results in a small version space to be explored in the second step, and hence in a small number of passes through the data.

Let us first outline the algorithm for the first step. The literals for $<<=$ can be processed using Mellish's description identification algorithm. This algorithm employs the following operations patterns:

Definition 12. Let a, b and d be patterns :

- the greatest lower bound
  $glb(a, b) = max\{d \mid a <<= d \text{ and } b <<= d\}$
- the least upper bound
  $lub(a, b) = min\{d \mid d <<= a \text{ and } d <<= b\}$
- the most general specialisations of a w.r.t. b
  $mgs(a, b) = max\{d \mid a <<= d \text{ and } not(d <<= b)\}$
- the most specific generalisations of a w.r.t. b
  $msg(a, b) = min\{d \mid d <<= a \text{ and } not(b <<= d)\}$

function versionspace($i_1 \wedge ... \wedge i_n$: conjunctive query)
    returns S and G defining the versionspace of $i_1 \wedge ... \wedge i_n$

$S := \{top\}$; $G := \{bottom\}$;
for all basic literals i **do**
    case i of $q <<= Pattern$:
        $S := \{s \in S \mid q <<= s\}$
        $G := max \{glb(q, g) \mid g \in G \text{ and } \exists s \in S : glb(q, g) <<= s\}$
    case i of $Pattern <<= q$:
        $G := \{g \in G \mid g <<= q\}$
        $S := min \{lub(q, s) \mid s \in S \text{ and } \exists g \in G : g <<= lub(q, s)\}$
    case i of $not\ Pattern <<= q$:
        $S := \{s \in S \mid not(s <<= q)\}$
        $G := max \{m \mid \exists g \in G : m \in mgs(g, q) \text{ and } \exists s \in S : m <<= s\}$
    case i of $not\ q <<= Pattern$:
        $G := \{g \in G \mid not(q <<= g)\}$
        $S := min \{m \mid \exists s \in S : m \in msg(s, q) \text{ and } \exists g \in G : g <<= m\}$
    case i of $Pattern\ covers\ ex$
        $G := \{g \in G \mid g\ covers\ ex\}$
        $S := min \{s' \mid s'\ covers\ ex \text{ and } \exists s \in S : s' <<= s \text{ and }$
            $\exists g \in G : g <<= s\}$
    case i of $not\ Pattern\ covers\ ex$:
        $S := \{s \in S \mid not\ s\ covers\ ex\}$
        $G := max \{g' \mid not\ g'\ covers\ ex \text{ and }$
            $\exists g \in G : g <<= g' \text{ and } \exists s \in S : g' <<= s\}$
    case i of $match(Pattern, ex) \leq n$
        $G := \{g \in G \mid match(g, ex) \leq n\}$
        $S := min \{s' \mid match(s', ex) \leq n \text{ and } \exists s \in S : s' <<= s \text{ and }$
            $\exists g \in G : g <<= s\}$
    case i of $match(Pattern, ex) \geq n$

$$S := \{s \in S \mid match(s, ex) \geq n\}$$
$$G := \max \{g' \mid match(g', ex) \geq n \text{ and}$$
$$\exists g \in G : g <<= g' \text{ and } \exists s \in S : g' <<= s\}$$

The above algorithm can be specialized according to the pattern domain under consideration. For the domain IS the specialization is rather straightforward and results in an efficient algorithm. For other domains such as DQ, the implementation of the steps for matching is more complicated. The key point about this algorithm is however that it does not require to access the data and that - depending on the constraints - it results in a reduced version space.

The second step of the algorithm then deals with the frequency literals. The general outline of the algorithm is shown below. The efficient implementation of this algorithm is less straightforward. However, it turns out that we can integrate the level-wise algorithm with that of version spaces.

for all frequency literals $freq$ **do**
    case i is anti-monotonic :
        $G := \{g \in G \mid freq(g)\}$
        $S := \min \{s' \mid freq(s') \text{ and}$
            $\exists s \in S : s' <<= s \text{ and } \exists g \in G : g <<= s'\}$
    case i is monotonic :
        $S := \{s \in S \mid freq(s)\}$
        $G := \max \{g' \mid freq(g') \text{ and}$
            $\exists g \in G : g <<= g' \text{ and } \exists s \in S : g' <<= s\}$

The first case of the second step can be implemented as follows (we assume an anti-monotonic frequency constraint $freq$):

$L_0 := G$
$i := 0$
while $L_i \neq \emptyset$ **do**
    $F_i := \{p \mid p \in L_i \text{ and } freq(p)\}$
    $I_i := L_i - F_i$ the set of infrequent patterns considered
    $L_{i+1} := \{p \mid \exists q \in F_i : p \in \rho_s(q) \text{ and } \exists s \in S : p <<= s \text{ and}$
        $\rho_g(p) \cap (\cup_{j \leq i} I_j) \neq \emptyset \}$
    $i := i + 1$
endwhile
$G := F_0$
$S := min(\cup_j F_j)$

To explain the algorithm, let us first consider the case where $S = \{bottom\}$ and $G = \{top\}$ and where we work with itemsets. In this case the refinement operator will merely add a single item to a query and the generalization operator will delete a single item from the itemset (in all possible manners). In this case, the above algorithm will behave roughly as the level-wise algorithm presented earlier. The only difference is that we keep track also of the infrequent item-sets I_i. L_i will contain only itemsets of size i. The algorithm will then repeatedly compute a set of candidate refinements L_{i+1}, delete those item-sets that cannot be frequent by looking at the frequency of its generalizations, and evaluate the resulting possibly frequent itemsets on the database. This process continues until L_i becomes empty.

The basic modifications to run it in our context are concerned with the fact that we need not consider any element that is not in the already computed version space (i.e. any element not between an element of the G and the S set). Secondly, we have to compute the updated S set, which should contain those frequent elements whose refinements are all infrequent.

Finding the updated G and S sets can also be realized in the dual manner. In this case, one will initialize L_0 with the elements of S and proceed otherwise completely dual. The resulting algorithm is shown below.

Whether the top down or bottom up version will work more efficiently is likely to depend on the application and query under consideration. At this point it remains an open question as to when which strategy will work more efficiently.

$L_0 := S$
$i := 0$
$G := \{g \in G \mid freq(g)\}$
while $L_i \neq \emptyset$ **do**
    $F_i := \{p \mid p \in L_i \text{ and } freq(p)\}$
    $I_i := L_i - F_i$ the set of infrequent patterns considered
    $L_{i+1} := \{p \mid \exists q \in I_i : p \in \rho_g(q) \text{ and } \exists g \in G : g <<= p \text{ and }$
        $\rho_s(p) \cap (\cup_{j \leq i} F_j) \neq \emptyset \}$
    $i := i + 1$
endwhile
$S := min(\cup_j F_j)$

Finally, it is also easy to modify the above algorithms (exploiting the dualities) in order to handle *monotonic* frequency atoms (i.e. the second case in the algorithm for the second step).

Whereas in this section we have adopted the standard level-wise algorithm to search for the borders, it would also be possible to adopt more efficient algorithms such as e.g. the randomized ones proposed in [22].

4.4   The Proof of the Concept

In [10,20,21] we present an implementation, called MolFea, of the level-wise version space algorithm for use in molecular applications. The examples in MolFea correspond to the 2D structure of chemical compounds, i.e. they are essentially graphs. Patterns in MolFea are molecular fragments, i.e. linearly connected sequences of atoms and bonds.

Furthermore, we have performed various experiments with MolFea on challenging data sets. In one of the experiments [21], we have discovered fragments of interest in a database containing information on the HIV activity of over 40000 chemicals. MolFea did find long patterns (of about 25 atom/bonds) in relatively short time (a couple of hours on Linux workstation - Pentium III). As compared to state-of-the-art inductive logic programming systems (such as e.g. Dehaspe's Warmr [7]), which have also been applied to this type of problem, MolFea compares favourable in that it handles much more data in much less time and also discovers longer patterns. Indeed, typical experiments reported in the literature indicate that inductive logic programming systems handle a few hundred compounds, find patterns of length 7 or 8, and require several days of cpu-time.

This provides evidence that the level-wise version space algorithm is an effective solver for database mining.

4.5 Optimising the Algorithms

Various optimisations to the algorithm seem possible and worthy of further investigation.

Instead of processing the frequency literals independent of each other, one might rather combine these so that only one pass through the search space is necessary. Indeed, consider the following query :

(11) ?- P <<= bottom, frequency(E,d1,covers(P,E)) > 3,
frequency(E,d1,(match(P,E) < 2)) > 4 .

This query could be answered by taking as constraint the conjuction of the two frequency literals and then performing one pass through the search space. Executing the query in this way may be more efficient.

This example illustrates that reasoning about queries and their execution will be beneficial. To this aim, one could employ notions such as query simplification, equivalence, implication, redundancy, etc. All of these standard notions could essentially be applied to the framework of RDM. Some of these notions have already been worked out for related languages such as MINESQL (cf. e.g. [26,13]). For RDM, this remains however a topic for further research.

4.6 Solving More Complex RDM Queries

So far we have only dealt with simple RDM queries. We have not considered the similarity literals, because they are neither monotonic nor anti-monotonic. Therefore, it seems natural (though inefficient) to compute these using a generate-and-test strategy after the G and S borders have been computed for the rest of the query.

What also remains an open question is how to process queries such as query (10), which involves two patterns. The problem is how to capture the dependencies among the two patterns within the version space model. Extensions to the basic version space model are necessary to realize this. These are also a topic for further research.

5   Extensions to the Basic Engine and Language

5.1   Optimization Primitives

Two primitives that seem especially useful are min and max. Indeed, one could imagine being interested in those patterns that satisfy a number of constraints and in addition have maximum frequency on a certain dataset or are minimally general. Let us therefore define:

– max(p(P), term(P)) (resp. min) takes as argument a query p(P) where P is a pattern variable and a term term(P) to be optimized. The predicate p imposes constraints on the pattern P and term(P) specifies the criterion that should be maximized (resp. minimized). term(P) should be a monotonic or anti-monotonic term or literal in P. The optimization literal then succeeds for those patterns satisfying the constraints imposed by p and being optimal w.r.t. term(P).

As an example consider the following query :

```
patt(P) :-
     P <<= [beer,mustard,wine,yoghurt],
     frequency(E,d5,covers(P,E)) > 100.

?-min(pat(P), frequency(E,d8,covers(P,E))).
```

The predicate patt succeeds for all patterns involving beer, mustard, wine and yoghurt with a frequency on d5 that is larger than 100. The optimisation literal then selects among the patterns that satisfy patt, those with minimal frequency in d8.

Within the sketched solver for simple queries, it is relatively easy to accommodate this type of optimization constructs. Indeed, because the optimization terms should be monotonic or anti-monotonic, one only needs to consider the maximum (resp. minimum) elements in the solutions to p. So, to answer a query containing an optimization literal, one first computes the S and G sets for p. If the criterion to be optimized is generality, then the solutions are given directly by either S or G (depending on whether one wishes to minimize or to maximize). If on the other hand one wishes to optimize w.r.t. frequency then one needs to compute the frequency of all elements in either G or S. The answers to the queries are then those patterns within either G or S that are optimal.

5.2   Heuristic Solvers

So far, we have described *complete* solvers, which compute all solutions within the specified constraints. However, completeness often comes at a (computational) cost. Therefore, complete solvers may not always be desirable. There are at least two situations already encountered where this might be the case.

First, the provided primitives for optimisation were so far quite simple. Also, the criterion one may want to optimize is not necessarily frequency. Indeed, one is often more interested in accuracy, or entropy, etc. To optimize w.r.t. accuracy (as in query (4)) one cannot employ the above sketched method because accuracy involves combining maximum frequency on positives and minimum frequency on negatives. Hence, optimal patterns might lie in the middle of the version space.

Second, there is a discrepancy between answering simple queries and answering queries over multiple inter-related patterns (as in query (11)). There is good hope and evidence that the former solver is reasonably efficient, but it is also clear that the latter one is much less efficient, because it merely enumerates all the possibilities.

Therefore, we need to extend the current solver with *heuristic* methods. This situation is again akin to what happens in constraint logic programming (cf. [25]). The effect of heuristic methods would be that queries get a heuristic answer, that some solutions might not be found, and that suboptimal solutions might be generated. Of course, in such cases the user should be aware of this. When allowing for heuristic methods, it becomes possible to extend the database mining engine with various well-known database mining algorithms, such as a beam-search procedure to greedily find the most interesting clauses in predictive modelling. In this context, any-time algorithms would also be quite effective.

5.3   The Knowledge Discovery Cycle

Knowledge discovery in databases typically proceeds in an iterative manner. Data are selected, possibly cleaned, formatted, input in a data mining engine, and results are analysed and interpreted. As first results often can be improved, one would typically re-iterate this process until some kind of a local optimum has been reached (cf. [12]).

Because knowledge discovery is an iterative process data mining tools should support this process. One consequence of the iterative nature of knowledge discovery in our context is that many of the queries formulated to the database mining engine will be related. Indeed, one can imagine that various queries are similar except possibly for some parameters such as thresholds, data sets, pattern syntax, etc. The relationships among consecutive queries posed to the data mining engine should provide ample opportunities for optimization. The situation is - to some extent - akin to the way that views are dealt with in databases (cf. [11]). Views in databases are similar to patterns in data mining in that both constructs are virtual data structures, i.e. they do not physically exist in the database. Both forms of data can be queries and it is the task of the engines to efficiently answer questions concerning these virtual constructs.

Answering queries involving views can be realized essentially in two different ways. First, one can *materialize* views, which means that one generates the tuples in the view relation explicitly, and then processes queries as if a normal relation were queried. Second, one can perform *query modification*, which means that any query to a view is 'unfolded' into a query over the base relations. The advantage of materialization is that new queries are answered much faster whereas the disadvantage is that one needs to recompute or update the view whenever something changes in the underlying base relations. At a more general level, this corresponds to the typical computation versus storage trade-off in computer science.

These two techniques also apply to querying patterns in data mining. Indeed, if consecutive queries are inter-related, it would be useful to store the results (and possibly also the intermediate results) of one query in order to speed up the computation of the latter ones. This corresponds to materializing the patterns (together with accompanying information). Doing this would result in effective but fairly complicated solvers.

6 Related Work

RDM is related to other proposals for database mining query languages' such as e.g. [26,17,14,13,?]. However, it differs from these proposals in a number of aspects. First, due to the use of deductive databases as the underlying database model, RDM allows - in principle - to perform pattern discovery over various domains, such as e.g. item-sets, sequences, graphs, datalog queries, ... Secondly, a number of new and useful primitives are foreseen. Using RDM one is not restricted to finding frequent patterns, but one may also look for infrequent ones with regards to certain sets of (negative) examples. One can also require that certain examples are (resp. are not) covered by the patterns to be induced. Thirdly and most importantly, RDM allows to combine different primitives when searching for patterns. Finally, its embedding within Prolog puts database mining on the same methodological grounds as constraint programming.

As another contribution, we have outlined an efficient algorithm for answering complex database mining queries. This algorithm integrates the principles of the level-wise algorithm with those of version spaces and thus provides evidence that RDM can be executed efficiently. It also provides a generalized theoretical framework for data mining. The resulting framework extends the borders in the level-wise techniques sketched by [23], who link the level-wise algorithm to the S set of Mitchell's version space approach but do not further exploit the version space model as we do here. An implementation of the level-wise versionspace algorithm for use in molecular applications has been implemented [20,21] and the results obtained are promising.

Finally, the author hopes that this work provides a new perspective for data mining, which is grounded in the methodology of computational logic. The hope is that this will result in a clear separation of the declarative from the procedural aspects in data mining.

Acknowledgements

This work was partially supported by the EU IST project cInQ. The author is grateful to Stefan Kramer, Jean-Francois Boulicaut and the anonymous reviewers for comments, suggestions and discussions on this work. Finally, he would like to thank the editors for their patience.

References

1. R. Agrawal, T. Imielinski, A. Swami. Mining association rules between sets of items in large databases. In *Proceedings of ACM SIGMOD Conference on Management of Data*, pp. 207-216, 1993.
2. E. Baralis, G. Psaila. Incremental Refinement of Mining Queries. In Mukesh K. Mohania, A. Min Tjoa (Eds.) *Data Warehousing and Knowledge Discovery, First International Conference DaWaK '99* Proceedings. Lecture Notes in Computer Science, Vol. 1676, Springer Verlag, pp. 173-182, 1999.
3. Jean-Francois Boulicaut, Mika Klemettinen, Heikki Mannila: Querying Inductive Databases: A Case Study on the MINE RULE Operator. In *Proceedings of PKDD-98*, Lecture Notes in Computer Science, Vol. 1510, Springer Verlag, pp. 194-202, 1998.
4. I. Bratko. *Prolog Programming for Artificial Intelligence*. Addison-Wesley, 1990. 2nd Edition.
5. W. Cohen, Whirl : a word-based information representation language. *Artificial Intelligence*, Vol. 118 (1-2), pp. 163-196, 2000.
6. L. Dehaspe, H. Toivonen and R.D. King. Finding frequent substructures in chemical compounds, in *Proceedings of KDD-98*, AAAI Press, pp. 30-36, 1998.
7. L. Dehaspe, H. Toivonen. Discovery of Frequent Datalog Patterns, in *Data Mining and Knowledge Discovery Journal*, Vol. 3 (1), pp. 7-36, 1999.
8. L. De Raedt, An inductive logic programming query language for database mining (Extended Abstract), in *Proceedings of Artificial Intelligence and Symbolic Computation*, Lecture Notes in Artificial Intelligence, Vol. 1476, Springer Verlag, pp. 1-13, 1998.
9. L. De Raedt, A Logical Database Mining Query Language. in *Proceedings of the 10th Inductive Logic Programming Conference*, Lecture Notes in Artificial Intelligence, Vol. 1866, Springer Verlag, pp. 78-92, 2000.
10. L. De Raedt, S. Kramer, The level-wise version space algorithm and its application to molecular fragment finding, in *Proceedings of the Seventeenth International Joint Conference on Artificial Intelligence*, Morgan Kaufmann, pp. 853-862, 2001.
11. R. Elmasri, S. Navathe. Fundamentals of database systems. Benjamin Cummings. 1994.
12. Usama M. Fayyad, Gregory Piatetsky-Shapiro, Padhraic Smyth, Ramasamy Uthurusamy (Eds.). Advances in Knowledge Discovery, The MIT Press, 1996.
13. F. Giannotti, G. Manco: Querying Inductive Databases via Logic-Based User-Defined Aggregates. In *Proceedings of PKDD 99*, Lecture Notes in Artificial Intelligence, Vol. 1704, Springer Verlag, pp. 125-135, 1999.
14. J. Han, Y. Fu, K. Koperski, W. Wang, and O. Zaiane, DMQL: A Data Mining Query Language for Relational Databases, in SIGMOD'96 Workshop on Research Issues on Data Mining and Knowledge Discovery, Montreal, Canada, June 1996.
15. J. Han, L. V. S. Lakshmanan, and R. T. Ng, Constraint-Based, Multidimensional Data Mining, *Computer*, Vol. 32(8), pp. 46-50, 1999.
16. H. Hirsh. Generalizing Version Spaces. *Machine Learning*, Vol. 17(1), pp. 5-46 (1994).

17. T. Imielinski and H. Mannila. A database perspectivce on knowledge discovery. *Communications of the ACM*, Vol. 39(11), pp. 58–64, 1996.
18. T. Imielinski, A. Virmani, and A. Abdulghani. Application programming interface and query language for database mining. In *Proceedings of KDD 96*. AAAI Press, pp. 256-262, 1996.
19. Robert A. Kowalski. Algorithm = Logic + Control. *Communications of the ACM*, 22(7), pp. 424-436, 1979.
20. S. Kramer, L. De Raedt. Feature Construction with Version Spaces for Biochemical Applications, in *Proceedings of the Eighteenth International Conference on Machine Learning*, Morgan Kaufmann, 2001.
21. S. Kramer, L. De Raedt, C. Helma. Molecular Feature Mining in HIV Data, in *Proceedings of the Seventh ACM SIGKDD International Conference on Knowledge Discovery and Data Mining*, ACM Press, pp. 136-143, 2001.
22. D. Gunopulos, H. Mannila, S. Saluja: Discovering All Most Specific Sentences by Randomized Algorithms. In Foto N. Afrati, Phokion Kolaitis (Eds.): *Database Theory - ICDT '97, 6th International Conference*, Lecture Notes in Computer Science, Vol. 1186, Springer Verlag, pp. 41-55, 1997.
23. H. Mannila and H. Toivonen, Levelwise search and borders of theories in knowledge discovery, *Data Mining and Knowledge Discovery*, Vol. 1(3), pp. 241-258, 1997.
24. H. Mannila. Inductive databases. in *Proceedings of the International Logic Programming Symposium*, The MIT Press, pp. 21-30, 1997.
25. Marriott, K. and Stuckey, P. J. Programming with constraints : an introduction. The MIT Press. 1998.
26. R. Meo, G. Psaila and S. Ceri, An extension to SQL for mining association rules. *Data Mining and Knowledge Discovery*, Vol. 2 (2), pp. 195-224, 1998.
27. C. Mellish. The description identification algorithm. *Artificial Intelligence*, Vol. 52 (2), pp,. 151-168, 1990.
28. T. Mitchell. Generalization as Search, *Artificial Intelligence*, Vol. 18 (2), pp. 203-226, 1980.
29. G. Sablon, L. De Raedt, and Maurice Bruynooghe. Iterative Versionspaces. *Artificial Intelligence*, Vol. 69(1-2), pp. 393-409, 1994.
30. A. Inokuchi, T. Washio, H. Motoda. An Apriori-based algorithm for mining frequent substructures from graph data. in D. Zighed, J. Komorowski, and J. Zyktow (Eds.) *Proceedings of PKDD 2000*, Lecture Notes in Artificial Intelligence, Vol. 1910, Springer-Verlag, pp. 13-23, 2000.

DCGs: Parsing as Deduction?

Chris Mellish

Division of Informatics
University of Edinburgh
80 South Bridge, Edinburgh EH1 1HN, Scotland
C.Mellish@ed.ac.uk,
http://www.dai.ed.ac.uk/homes/chrism/

Abstract. The idea of viewing parsing as *deduction* has been a powerful way of explaining formally the foundations of natural language processing systems. According to this view, the role of grammatical description is to write logical axioms from which the well-formedness of sentences in a natural language can be deduced.

However, this view is at odds with work on unification grammars, where categories are given complex descriptions and the process of building satisfying *models* is at least as relevant as that of building deductive proofs. In some work feature logics are even used to replace the context-free component of grammars. From this work emerges the view that grammatical description is more like writing down a set of constraints, with well-formed sentences being the possible solutions to these constraints.

In this paper, we concentrate on Definite Clause Grammars (DCGs), the paradigm example of "parsing as deduction". The fact that DCGs are based on using deduction (validity) and unification grammar approaches are based on constructing models (satisfiability) seems to indicate a significant divergence of views. However, we show that, under some plausible assumptions, the computation involved in using deduction to derive consequences of DCG clauses produces exactly the same results as would be produced by a process of model building using a set of axioms derived syntactically from the original clauses.

This then suggests that there is a single view of parsing (and generation) that reconciles the two approaches. This is a view of parsing as model-building, not a view of parsing as deduction. Even in the original paradigm case there is some doubt as to whether "parsing as deduction" is the best, or only, explanation of what is happening.

1   Parsing as Deduction?

The idea of viewing parsing as *deduction*, which goes back to the work of Colmerauer [Colmerauer 1978] and Kowalski [Kowalski 1979], has been a powerful way of explaining formally the foundations of natural language processing systems. According to this view, the role of grammatical description is to write logical axioms from which the well-formedness of sentences in a natural language can be deduced. Pereira and Warren [Pereira and Warren 1983] cite a number of benefits that arise from investigating the connection between the two, including

A.C. Kakas, F. Sadri (Eds.): Computat. Logic (Kowalski Festschrift), LNAI 2408, pp. 548–566, 2002.
© Springer-Verlag Berlin Heidelberg 2002

the transfer of useful techniques between theorem-proving and computational linguistics. Shieber [Shieber 1988] and others have used similar arguments for also considering *generation* as deduction.

The paradigm examples of parsing as deduction have used Definite Clause Grammars (DCGs [Pereira and Warren 1980]). Demonstrating that a sentence is well-formed according to a DCG grammar is achieved using the theorem-proving approach known as SLD resolution. Part of the resolution model involves having a *unification* operation to establish when a category required to be present could be decomposed by one of the grammar rules. In general, unification involves applying rewrite rules acting on sets of constraints in such a way as to build representations of possible models of those constraints. In the DCG case, the constraints are so simple (equality in the Herbrand universe of the terms which can be constructed from the constants and function symbols in the grammar) that the unification operation almost goes unnoticed as part of the definition of valid inference. However later work has introduced the possibility of describing categories by complex feature descriptions expressed in a feature logic (e.g. [Kasper and Rounds 1986], [Smolka 1992]). In such cases, unification can be doing a significant part of the real work in a parser. When such complex feature descriptions are used to annotate context-free phrase structure rules, as in PATR [Shieber 1986], a hybrid model such as Höhfeld and Smolka's model of constraint logic programming [Höhfeld and Smolka 1988] is needed to provide a way of reconciling the use of a model-building component within an inference system. The simple view that parsing is deduction has now become more complex.

The situation unfortunately becomes different again when the feature logic is also used also to replace the context-free skeleton present in DCGs (as in [Manaster-Ramer and Rounds 1987],[Emele and Zajac 1990],[Manandhar 1993], and many approaches based on HPSG or Categorial Grammar). In this case, unification becomes more or less all there is in a parser, which leads to the view that parsing is really model-building. For instance, in typed unification grammars parsing is implemented as a process of type checking (in the presence of a type theory expressing the constraints of the grammar) which rewrites an input term to possible normal forms corresponding to models of it ([Emele 1994], [Aït-Kaci and Podelski 1993]). According to this view, grammatical description is more like writing down a set of constraints, with well-formed sentences being the possible solutions to these constraints.

The fact that DCGs are based on using deduction (validity) and feature logic approaches are based on constructing models (satisfiability) seems to indicate a significant divergence of views about what parsing "is" [Johnson 1992].

2   Definite Clause Grammars - The Usual Account

In this section, we briefly give the standard account of DCGs and show how they illustrate the idea of parsing as deduction. This section contains nothing original, but we wish to go through the steps fairly carefully in order that we can later show that a different account explains the same phenomena.

Figure 1 shows a simple example of a definite clause grammar (DCG) in various forms (to be discussed below).

1. Original DCG:

```
vp(Num) --> vtr(Num), np(Num1).
vp(Num) --> vintr(Num).
vtr(sing) --> [hates].
vtr(plur) --> [hate].
np(Num) --> [sheep].
```

2. Context-free skeleton:

vp → vtr np
vp → vintr
vtr → hates
vtr → hate
np → sheep

3. Prolog translation Π:

```
vp(Num,P0,P1) :- vtr(Num,P0,P1), np(Num1,P1,P2).
vp(Num,P0,P1) :- vintr(Num,P0,P1).
vtr(sing,[hates|P],P).
vtr(plur,[hate|P],P).
np(Num,[sheep|P],P).
```

4. Horn Clause interpretation Π^{if}:

$\forall Num, Num1, P0, P1, P2. \quad vp(Num, P0, P2) \quad \subset \quad vtr(Num, P0, P1) \ \wedge$
$np(Num1, P1, P2)$
$\forall Num, P0, P1. \ vp(Num, P0, P1) \subset vintr(Num, P0, P1)$
$\forall P. \ vtr(sing, [hates|P], P)$
$\forall P. \ vtr(plur, [hate|P], P)$
$\forall Num, P. \ np(Num, [sheep|P], P)$

Fig. 1. Definite Clause Grammar in various forms

Basically, the notation allows for the expression of context-free rules where the nonterminal symbols can be associated with values for particular features (using a fixed positional notation for each nonterminal). Feature values in the grammar rules can be given as constants (e.g. **sing**) or by variables (whose names begin with upper case letters, e.g. Num). Where variables are used, the intent is that every time a rule is used the same value must be used consistently for each occurrence of a given variable in the rule. If the feature annotations are stripped away from a DCG, the result is (modulo trivial syntactic differences) a context-free grammar, the *context-free skeleton*. The context-free skeleton in general generates a larger language than the DCG because it ignores all the feature constraints. The context-free skeleton for the above example is shown in the figure. hates, hate and sheep are terminals, and all other symbols are nonterminals. [1]

A DCG can be viewed as an abbreviation for a Prolog program Π, which makes explicit the relation between the phrases and portions of the string by using a threading technique on two extra arguments added to each nonterminal. In the figure, $[X|Y]$ is the usual Prolog syntactic sugar for $cons(X,Y)$, for some function symbol *cons* used to construct lists. [2] The two extra arguments represent a *difference list* of a string and a (not necessarily proper) tail of that string (encoded as lists), the given category then being taken to describe the portion of the string which is the *difference* between these. This translation Π is standardly interpreted as a set of Horn clauses of logic Π^{if} which states a set of "if" definitions (also shown in the figure).

In this example, "hates sheep" is a valid VP because

$$vp(sing, [hates, sheep], [])$$

is a logical consequence of the above axioms. [3] In general, the set of strings α making up the language generated by a DCG is the set of strings for which $s(f_1, \ldots f_n, \alpha', [])$ is a valid logical consequence of the appropriate Π^{if} axioms, where s is the initial symbol, $f_1, \ldots f_n$ can be any values for the features associated with that category by the grammar and α' is the encoding of α as a list structure as illustrated by the example. That is, we are interested in the situation where

$$\Pi^{if} \vdash s(f_1, \ldots f_n, \alpha', [])$$

SLD resolution, which is the basis of the execution mechanism of Prolog, is one way in which logical consequences can be derived from Π^{if}. The operation of

---

[1] Note that, although the word "sheep" is ambiguous as to number, in a good grammar one would only want to allow plural nouns to stand alone as NPs. This formulation has been chosen here to make a particular formal point later. The examples in this paper are not intended to make any real claims about any natural languages. The reader is asked to imagine that the example grammars really do make plausible claims.

[2] We also assume the Prolog syntactic sugar $[a, b, c]$ for $cons(a, cons(b, cons(c, nil)))$ and $[]$ for *nil*, for some constant symbol *nil*.

[3] In this section, and in the rest of the paper, we will actually concentrate on the *recognition* problem, rather than the parsing problem, for DCGs. Given, however, that we *will* be concerned with the possible feature values for the categories that are recognised and that there are standard ways to express parse trees in the features of the categories [Pereira and Warren 1980], this represents no limitation.

a successful proof of an atom ϕ from axioms Π^{if} is characterised by an SLD refutation of $\Pi^{if} \cup \{\leftarrow \phi\}$. An example successful SLD refutation showing the grammaticality of "hates sheep" is shown in Figure 2. SLD resolution operates on

Goals	Renamed Clause/ Subst
← vp(Num,P0,P1)	vp(Num1,P01,P11):-vtr(Num1,P01,P11),np(Num11,P11,P21) [Num/Num1,P0/P01,P1/P11]
← vtr(Num1,P01,P11),np(Num11,P11,P21)	vtr(sing,[hates|P22],P22) [Num1/sing,P01/[hates|P22],P11/P22]
np(Num11,P22,P21)	np(Num3,[sheep|P33],P33) [Num11/Num3,P22/[sheep|P33],P21/P33]

Fig. 2. SLD refutation

the "Prolog" representation of the grammar rules, not the DCG representation or the Predicate Calculus version. Each line of the refutation starts with a sequence of goals to be proved. Initially, the only goal is to find some true instance of the predicate for the initial symbol. As well as the goals, each line must mention a clause whose left hand side matches the first goal in the sequence (this clause has its variables renamed so as not accidentally to clash with those in the goal) and the minimal substitution (computed by unification) required to make the left hand side of the clause the same as that first goal. The next line of the refutation starts with the remaining goals preceded by the right hand side of the chosen clause, to all of which the just-computed substitution has been applied. The last line must be empty (indicating that all goals have been proved).

The completeness and soundness of SLD resolution [Lloyd 1987] ensure that a ground atom ϕ' is a logical consequence of Π^{if} if and only if there is a finite SLD refutation of $\Pi \cup \{\leftarrow \phi'\}$. This justifies regarding a Prolog implementation of DCGs as doing "parsing by deduction". Note that in this paper, we will assume that our primary interest is in possible ground conclusions that can be drawn from a grammar and some input data. For instance, we may wish to know which particular logical forms can be associated with a particular sentence, or vice versa. Thus if a particular atom ϕ is of interest to us (e.g. vp(plur,X,[])) then that interest can be expressed as an interest in those ground instances ϕ' of ϕ that are true. Restricting attention to the ground case simplifies the presentation and does not sacrifice generality.

3   Limitations

DCGs allow one to describe a language in terms of the Horn clause subset of Predicate Logic, but the expressive limitations of this are well-known. Horn clauses do not allow for arbitrary occurrences of negation and disjunction, and yet these (and other extensions) are well-motivated from a linguistic point of view [Wedekind 1990].

If "parsing as deduction" is an appropriate model for DCGs then the account should be able to be extended to the case where axioms of other forms are allowed. Unfortunately, with even modest extensions to the Horn clause subset the parsing as deduction model does not always seem to fit naturally. The following examples are intended to indicate such unnaturalness. Obviously what is "natural" is to some extent subjective.

1. We might want to add information about typing. A natural idea would be to add to our DCG grammar additional statements describing what are legal "number" values:

   $\forall Num, P0, P1. (Num = sing \vee Num = plur \subset np(Num, P0, P1))$

   Unfortunately, this does not prevent $np(christmas, [sheep], [])$ from being inferred from the grammar. Indeed, our original DCG formulation has *already* stated that *anything* is a legal number value for "sheep" (and so the addition renders the axioms inconsistent). It is not possible to add this new information without changing many individual references to number in the grammar. This example shows a place where perhaps the original interpretation of the DCG is not what a linguist would have intended, or at least where adding new information is not as easy as one might expect.

2. We might want to express partial knowledge about a word (replacing the existing information about "hates" with the following, which expresses ignorance as to whether it is a transitive verb or a plural noun):

   $(\forall P. \; vtr(sing, [hates|P], P)) \vee (\forall P. \; np(plur, [hates|P], P))$

   If used to replace the existing information about *hates*, the proposition

   $vp(Num, [hates, sheep], [])$

   no longer follows from the grammar (which is exactly what would happen if there was *no* provided information about the word). There does seem to be an argument to suggest that possible analyses requiring either of the two categories (but not both simultaneously) should be produced.

In each of these two examples, there are other ways in which the desired effect might be achieved, but the fact that such logically seemingly innocuous ways introduce problems gives some cause to question whether the basic account corresponds to what one really wants.

4 Model Construction

It seems plausible to us that viewing parsing (or generation) as deduction only works if one accepts an impoverished notion of what a grammar can be. A more general approach would be to view parsing and generation as instances of model construction. According to this view, a grammar specifies a set of *constraints* that must be satisfied by well-formed sentences (in the spirit of modern grammatical formalisms such as GB and HPSG). Parsing then involves establishing that a string is compatible with the constraints and what else must be true if it is to be well-formed. This view is supported by the work discussed above that already implicitly takes a model-building view (at least in part) by using complex unification operations. It is also suggested by approaches to interpretation that use abduction (for instance [Hobbs et al 1993]) – a closely related technique. Finally, Reiter and Mackworth [Reiter and Mackworth 1990] have presented an elegant model-construction characterisation of the process of visual interpretation, whose adaptation to language tasks would look promising.

A *model* of a set of logical axioms is an interpretation of the vocabulary of those axioms which interprets terms as denoting individuals in some domain and predicates as denoting relations on that domain, such that the axioms all correspond to true statements about that domain. Such an interpretation is completely specific about what each possible ground term denotes and whether or not each relation applies to any given combination of entities from the domain. As a special case, a *Herbrand model* is a model where the domain is the *Herbrand universe* of the program, that is, the set of ground terms that can be constructed using the constants and function symbols in the program itself, and where, amongst other things, each ground term is taken to denote itself. For instance, Figure 3 shows a small set of axioms and one possible Herbrand model for them. A Herbrand model is *normal* if it includes an interpretation

$$\exists X. p(X)$$
$$q(a, f(a)) \vee r(a)$$
$$\forall X. p(X) \supset p(f(X))$$
$$a = f(a) \vee p(a)$$

Predicate	Denotation
p	$\{< a >, < f(a) >, < f(f(a)) >, < f(f(f(a))) >, \ldots\}$
q	$\{< a, f(a) >\}$
r	$\{\}$
$=$	$\{< a, a >, < f(a), f(a) >, < f(f(a)), f(f(a)) >, \ldots\}$

Fig. 3. Axioms and one Herbrand model

for the equality predicate $=$ as the identity relation on the Herbrand universe. The above model $\mathcal{M}$ is normal. Linguistic objects are finite and with DCGs we are using predicates to characterise categories of phrases that exist in a given situation. It thus seems appropriate to concentrate on models with a certain finiteness property:

Definition: A *finite category model* is a normal Herbrand model where the denotation of each predicate apart from $=$ is finite.

The above model is not finite category, because p has an infinite denotation.

If $\mathcal{M}$ is a model of some axioms, a formula ϕ is said to be *true* in $\mathcal{M}$,

$$\mathcal{M} \models \phi$$

if, using the denotation of the symbols of ϕ given by $\mathcal{M}$, it represents a proposition that holds in that particular world. Thus, for instance, if $\mathcal{M}$ is the model indicated in the figure,

$$\mathcal{M} \models (\exists X . q(X, f(X)))$$

We claim that it is productive to regard parsing as a model construction process in which the aim is to determine certain formulae that are true in models of the grammatical axioms. But if parsing (and generation) are best conceived of as model construction tasks, how is it that the parsing as deduction explanation fits DCGs so well? In this paper, we will show that there is an equally good model construction explanation for the DCG phenomenon. It is because DCGs have a particular constrained form that the deduction and model construction explanations are equivalent. But, whereas the deduction model considers Π^{if} to be the underlying intention behind the DCG whose Prolog translation is Π, the model construction explanation assumes a different underlying logical intention, $\Pi^{only-if}$, which we now explain.

5   The Only-If Interpretation

Frequently with Prolog programs there is an intended "if and only if" semantics, which is obtained if a closed world assumption is made. This can be thought of in terms of there being implicit extra "only if" axioms [Kowalski 1979]. Let the corresponding "only if" axioms from Π be called $\Pi^{only-if}$. (The Clark completion of Π^{if} [Clark 1978] [Lloyd 1987] is then $\Pi^{if} \cup \Pi^{only-if}$). For the above Π, $\Pi^{only-if}$ has two parts, shown in Figure 4.

The first part (not shown completely here) is a set of axioms for equality which in the Herbrand universe of the DCG force = to be the identity relation (and hence any Herbrand model to be normal). The equality axioms ([Lloyd 1987], page 79) are straightforwardly constructed from the function symbols, predicates and constants that occur in the DCG.

The second part of $\Pi^{only-if}$ is obtained from Π by repeating the following for each predicate p:

1. Collect all clauses for p
2. Make the new axiom:
   $$\forall X_1 X_2 ... X_n . p(X_1, ... X_n) \supset \beta_1 \vee \beta_2 ... \vee \beta_m$$
   where n is the arity of p, X_1 to X_n are new variables and each β_i is derived from the ith clause of p and takes the form:
   $$\exists Y_1 ... Y_k : X_1 = a_1 \wedge ... X_n = a_n \wedge B$$
   where Y_1 to Y_k are the variables in the ith clause, a_j is the term in the jth argument position of the head of that clause and B is the body of the clause (translated into a logical conjunction).

Importantly, $\Pi^{only-if}$ is a straightforward syntactic translation of Π, in the same way that Π^{if} is.

Consider a DCG, its Prolog translation Π and an atom ϕ whose true instances we are interested in (e.g. vp(Num,[hates,poetry],[])). The standard way is

$hates \neq hate$
$hates \neq sheep$
$\forall X, Y.[X|Y] \neq hates$
$\forall X, Y.[X|Y] \neq X$
$\forall X1, Y1, X2, Y2.[X1|Y1] = [X2|Y2] \iff X1 = X2 \wedge Y1 = Y2$
$\dots$

$\forall X, Y, Z.$
$\quad vp(X, Y, Z) \supset$
$\qquad (\exists Num, Num1, P0, P1, P2 :$
$\qquad\quad X = Num \wedge Y = P0 \wedge Z = P2 \wedge vtr(Num, P0, P1) \wedge np(Num, P1, P2)) \vee$
$\qquad (\exists Num, P0, P1 :$
$\qquad\quad X = Num \wedge Y = P0 \wedge Z = P1 \wedge vintr(Num, P0, P1))$
$\forall X, Y, Z.$
$\quad vtr(X, Y, Z) \supset$
$\qquad (\exists P : X = sing \wedge Y = [hates|P] \wedge Z = P) \vee$
$\qquad (\exists P : X = plur \wedge Y = [hate|P] \wedge Z = P)$
$\forall X, Y, Z.$
$\quad np(X, Y, Z) \supset$
$\qquad \exists Num, P :$
$\qquad\quad X = Num \wedge Y = [sheep|P] \wedge Z = P$

Fig. 4. $\Pi^{only-if}$ for the example (equality axioms incomplete)

to investigate what ground instances of ϕ are logical consequences of Π^{if} (for instance, by SLD resolution). A different approach is to formulate the additional axiom

$$\exists Num : vp(Num, [hates, poetry], [])$$

and to see what is true in models of this combined with the "only if" axioms $\Pi^{only-if}$. That is, we could investigate what the vp relation might look like in possible models of the extended $\Pi^{only-if}$. Here we explore the relation between these two approaches. The reader who is prepared to take the relevant proofs on trust can skip to section 9, apart from taking in the statement of Theorem 1.

6   Analysis Trees

Before we show the equivalence of the deduction and model construction approaches for DCGs, it is useful to introduce a notational variant for a ground instance of a finite SLD refutation which makes the arguments simpler. An analysis tree is a ground instance of a "trace" of a DCG "execution". Alternatively, it can be regarded as a possible phrase structure analysis of some phrase of the language described by the grammar (possibly further instantiated with ground feature values).

Definition: Let Π be the Prolog translation of a DCG and ϕ' a ground atom. Then an *analysis tree* for ϕ' based on Π is a finite tree with every node labelled by a pair $< \rho, \Gamma >$, ρ a clause of Π (the *clause label*), Γ a ground atom (the *atom label*), with the following properties:

1. The root node is labelled with $< \rho, \phi' >$, for some clause ρ.
2. For every node, with label $< \rho, \Gamma >$, with n children labelled (in left-right order) with $< \rho_1, \Gamma_1 >$, $< \rho_2, \Gamma_2 >$, ... $< \rho_n, \Gamma_n >$, the form 'Γ :- Γ_1, Γ_2, ... Γ_n.' is a ground instance of clause ρ.

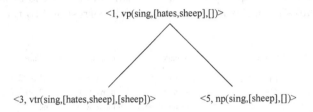

<1, vp(sing,[hates,sheep],[])>

<3, vtr(sing,[hates,sheep],[sheep])>   <5, np(sing,[sheep],[])>

Fig. 5. Analysis tree

Figure 5 shows an example analysis tree for the example DCG grammar. Here clauses have been numbered for compactness. Note that the first argument of the np node is instantiated (to sing) even though nowhere in the grammar is this value constrained. In fact, any ground term could appear in this position without affecting the correctness of the analysis tree.

From the definition of analysis tree it is clear that if we restrict attention to only the predicate names at the nodes then an analysis tree corresponds to a phrase structure analysis (minus the terminal symbols) of a sentence according to the context-free skeleton of the DCG[4]. We now indicate the fairly obvious result that analysis trees correspond to SLD refutations.

Lemma 1a:

Let Π be a Prolog program and ϕ' a ground atom. Then:

If there is an analysis tree for ϕ' based on Π then there is a finite SLD refutation of $\Pi^{if} \cup \{\leftarrow \phi'\}$.

Proof: The following method will construct the appropriate SLD refutation. First of all, the first line of the derivation will start off with ϕ' as the only goal. The rest of the derivation is built from a depth-first, left-right traversal of the nodes of the analysis tree, starting from the top node (each node being passed through exactly once, just before its leftmost child). As a node n is encountered

---

[4] This assumes that the goals in the Prolog clauses are ordered in the same way as those in the DCG, which is not strictly necessary, as the necessary ordering is already conveyed in the Prolog program by the string arguments

in the traversal, a renamed version of the clause in n's label is entered into the current line of the derivation. The substitution entered into this line is the result of unifying the first goal with the left hand side of this renamed clause. Then we move to the next line of the derivation and set up the goals in the way that an SLD refutation requires (fully determined by the previous line), before moving on to the next node.

To see that the above is possible and generates an SLD refutation, it suffices to see that whenever a given node n is visited the first goal in the current line of the refutation is an equally or more general version of the atom label of n. Figure 6 indicates for an example that the goals correspond to the nodes about

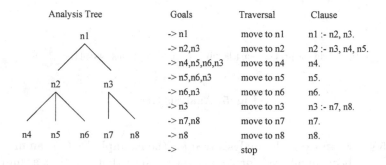

Fig. 6. Analysis tree and goals sequence

to be visited (since when new goals are added the right hand side of the clause used corresponds to the left-right order of the children of the current node). This correspondence reflects the fact that a depth-first, left-right tree traversal can be implemented via a certain kind of stack. In general, by induction on the line number of the refutation, the goals are equally or more general than the atom labels of the nodes that they will correspond to. This is because the goals arise from the unification of one of the previous goals with the left hand side of a clause and the analysis tree records a ground instance of such a unification. Therefore at each stage the first goal will unify with the left hand side of the appropriate clause and so it will always be possible to continue constructing the refutation. The refutation finishes with an empty line when the last node in the analysis tree has been left.

Lemma 1b:

Let Π be a Prolog program and ϕ' a ground atom. Then:

If there is a finite SLD refutation of $\Pi^{if} \cup \{\leftarrow \phi'\}$ then there is an analysis tree for ϕ' based on Π.

Proof: The analysis tree can be constructed in the following way. First of all, if all of the substitutions recorded in the refutation are applied to all of the goals, the result is that a given goal always has exactly the same form wherever it appears throughout the "goals" column of the refutation and that every goal taken together with the set of new goals that it introduces is an instance of the clause chosen to reduce that goal. The same applies if an arbitrary grounding substitution is subsequently applied to the goals. All that is then necessary is to map the goals into a tree, in the converse of what was done for Lemma 1a.

The completeness and soundness of SLD resolution give:

Lemma 1:

Let Π be the Prolog program arising from a DCG and ϕ' a ground atom. Then:

There is an analysis tree for ϕ' based on Π if and only if $\Pi^{if} \vdash \phi'$.

7   DCGs and Model Construction

In this section, we show, for Prolog programs Π derived from DCGs, the equivalence between logical consequences of Π^{if} and propositions that are true in models of $\Pi^{only-if}$ augmented with an assumption that a "solution" exists. First of all, however, it is necessary to make some assumptions.

7.1   Assumptions

The following results will only apply to DCG grammars which have the following property of *irredundancy*:

- For every category (predicate) **p** mentioned in the grammar. there is at least one clause of the form **p(....)** --> .... That is, there are no "undefined" categories.
- The grammar is has the property that in the context-free skeleton there is no possible derivation of the form $c \rightarrow^+ c$ for a category c. This corresponds closely to the notion of "offline parsibility" ([Pereira and Warren 1983] and [Kaplan and Bresnan 1982]) and hence is independently motivated. The existence of such a possible derivation would mean either that the context-free skeleton was infinitely ambiguous or that the context-free skeleton actually allowed no phrases of category c.

It seems plausible that any "reasonable" DCG can be transformed into an "equivalent" irredundant program.

Just as Lemma 1 shows that analysis trees correspond to proofs using Π^{if}, we now build up to Lemma 2, which shows that they also correspond to models of an augmented $\Pi^{only-if}$. Combining these results will give the main equivalence result, Theorem 1.

Lemma 2a:

Consider the Prolog translation Π of an irredundant DCG and a ground atom ϕ'. Then:

If $\mathcal{M}$ is a Herbrand model of $\Pi^{only-if}$ and $\mathcal{M} \models \phi'$, then there is a clause ρ of Π with head H and body B, and a grounding substitution σ such that:
- $\sigma(H)$ is identical to ϕ'.
- the literals in $\sigma(B)$ are all true in $\mathcal{M}$.

Proof: Let the predicate of ϕ' be p and its arguments be $\bar{t}$. Since Π is irredundant, there is a clause in it "defining" p. In $\Pi^{only-if}$ this corresponds to an axiom of the form

$$\forall \bar{X}. \; p(\bar{X}) \supset \beta_1 \vee \beta_2 \vee \ldots \beta_n$$

(where each disjunct corresponds to one clause of Π). Since ϕ' is in the model and instantiates p, we have an instance of this axiom for which the antecedent is true in the model (this is obtained by substituting the arguments of ϕ' for $\bar{X}$). Therefore the consequence is also true in the model. That is:

$$(\beta_1 \vee \beta_2 \vee \ldots \beta_n)[\bar{X}/\bar{t}]$$

is true in the model. At least one of the disjuncts must be true in the model, so pick one which is true (say the ith). Thus:

$$\beta_i[\bar{X}/\bar{t}]$$

is true in the model. From the form of β_i, it follows that:

$$(\exists \bar{Y} : X_1 = a_1 \wedge \ldots X_n = a_n \wedge B) \; [\bar{X}/\bar{t}]$$

is true, where each a_k is the kth argument in the head of the clause from which β_i was derived (a term whose free variables are taken from $\bar{Y}$) and B is the body of that clause.

Since this disjunct is true in the model, choose ground values $\bar{y}$ for $\bar{Y}$ which make it true. Then:

$$(X_1 = a_1 \wedge \ldots X_n = a_n \wedge B) \; [\bar{X}/\bar{t}][\bar{Y}/\bar{y}]$$

is true in the model. From the fact that equality is identity in the model, it follows that t_i must be identical to $a_i[\bar{Y}/\bar{y}]$. Also, since the variables $\bar{X}$ do not appear in the B literals,

$$B[\bar{Y}/\bar{y}]$$

is true in the model.

We claim that the clause corresponding to the chosen β_i satisfies the requirements for ρ, with the substitution σ being the one which assigns the values $\bar{y}$ to $\bar{Y}$. The fact that the instances of the literals in the body of ρ are true has just been shown (since these literals are B). It remains to show that σ applied to the head of the clause yields ϕ'. But the jth argument of the head of ρ is a_j, and we have shown that $a_j[\bar{Y}/\bar{y}]$ is identical to t_j, the jth argument of ϕ'.

Lemma 2b:

Consider the Prolog translation Π of an irredundant DCG and a ground atom ϕ'. Then:

If $\mathcal{M}$ is a finite category model of $\Pi^{only-if}$ and $\mathcal{M} \models \phi'$, then there is an analysis tree τ for ϕ' based on Π.

Proof: Consider the following method of constructing the tree τ. We start off with a trivial tree with a root node labelled with ϕ', but without a clause label, and with no daughters. We then extend the tree "downwards" as follows. By Lemma 2a, there is a ground instance of a clause ρ whose head is identical to ϕ' and whose body literals are all true in $\mathcal{M}$. The tree can thus be extended as follows. First of all, the existing leaf node is given the clause label ρ. Then m daughters are added, corresponding to the literals of the body of the clause, each labelled with the ground instance which has just been selected as being in the model, and each not yet assigned a clause label.

What we have just done is extend the initial tree by picking a leaf with an unassigned clause label, assigning it a clause label and adding a (possibly empty) set of daughters each of which was labelled with a ground atom true in $\mathcal{M}$. The leaf and daughters come from a ground instance of clause ρ.

The tree construction procedure is to repeat this operation until there are no more such leaves. In terms of the original program Π, what we are doing at each stage in extending the tree is taking a ground instance of a clause ρ whose head is the existing leaf node label and whose body goals are used to label the daughters. Thus what we are building is a well-formed analysis tree. In addition, we can show, by induction on the size of the tree, that every atom label in the tree is an atom that is in $\mathcal{M}$. At each point in the growth of the tree, if there is a leaf with an unassigned clause label, we are able to continue growing because the selected leaf will always be labelled with an atom that is in $\mathcal{M}$ and so Lemma 2a will apply.

If the above procedure terminates (i.e. reaches a tree that has no leaf nodes with unassigned clause labels) then the result is an analysis tree for ϕ' based on Π. For the proof of the Lemma, we need only to worry about what happens if the procedure does not terminate. This could only be because an infinite tree is built. Since the tree is finitely-branching (the number of daughters of a node is the number of goals in a clause of Π), this would be because some infinite branch can be built. Each node in this branch would be labelled with an atom true in $\mathcal{M}$. But since $\mathcal{M}$ is finite category, there are only finitely many instances of each predicate (apart from $=$) true in it. There are also only finitely many predicates in the DCG. Therefore some ground atom would have to appear more than once on the same branch. Since we are dealing with a DCG, this means that an atom in the tree dominates another atom with the same predicate (context-free category) and string arguments. If this predicate is c, from the way that the tree (so far) has been constructed, it follows that there is a derivation $c \rightarrow^{+} c$ in the context-free skeleton (anything else in the derivation from c must be empty because of the identity of the string arguments). This contradicts the assumption of irredundancy of the DCG.

Lemma 2c:

Consider a Prolog program Π and atom ϕ with ground instance ϕ'. Then:

> If there is an analysis tree τ for ϕ' based on Π then there is a finite category model $\mathcal{M}$ of $\Pi^{only-if} \cup \{\bar{\exists}\phi\}$ such that $\mathcal{M} \models \phi'$. [5]

> **Proof:** Consider the Herbrand interpretation that makes true:

- every atom appearing in τ
- = when its two arguments are identical

and which makes every other atom false. We claim that this is a finite category model of $\Pi^{only-if} \cup \{\bar{\exists}\phi\}$. First of all, it is a model. It clearly satisfies $\bar{\exists}\phi$. Now consider an axiom in $\Pi^{only-if}$, "defining" the predicate p. This axiom could only possibly be false in the interpretation if there was an instance of p with the antecedent true and the consequence false. For the consequence to be false, each disjunct would have to be false. But every instance of p true in the interpretation is in the analysis tree τ, and its presence there (with its daughters) guarantees that an instance of one of the disjuncts is true (which one is indicated by the clause label). Therefore there is no way that the interpretation could make the axiom false. Also trivially the equality axioms are satisfied. Thus the interpretation is indeed a Herbrand model. Because of the way equality is interpreted, it is also a normal model. It is finite category because the finite analysis tree provides all the true instances of the predicates. Finally it supports ϕ' because ϕ' is in the tree.

Lemma 2:

Let Π be the Prolog translation of an irredundant DCG, ϕ an atom and ϕ' a ground instance of ϕ. Then:

> There is an analysis tree τ for ϕ' based on Π if and only if there is a finite category model $\mathcal{M}$ of $\Pi^{only-if} \cup \{\bar{\exists}\phi\}$ and $\mathcal{M} \models \phi'$.

This follows immediately from Lemma 2b and Lemma 2c.

Theorem 1:

Let Π be the Prolog translation of an irredundant DCG, ϕ an atom and ϕ' a ground instance of ϕ. Then:

> There is a finite category model $\mathcal{M}$ of $\Pi^{only-if} \cup \{\bar{\exists}\phi\}$ and $\mathcal{M} \models \phi'$ if and only if $\Pi^{if} \vdash \phi'$.

This follows immediately from Lemma 1 and Lemma 2. This theorem justifies the idea of constructing models of an augmented $\Pi^{only-if}$ as a way of generating logical consequences of Π^{if}, and vice versa.

---

[5] $\bar{\exists}$ is existential closure.

8 Minimality

It would be useful if it was sufficient to construct *minimal* models of the augmented $\Pi^{only-if}$. However, there are versions of Π with logical consequences of Π^{if} that do not correspond to such models. Consider, for instance, in an augmented version of the example grammar that allowed S complements of some verbs, what would happen if we were interested in all true instances of vp(X,Y,[]). An analysis tree for "insists he hates sheep" would contain a subtree that was an analysis tree for "hates sheep". The model of

$$\Pi^{only-if} \cup \{\bar{\exists}vp(X, Y, [])\}$$

that supported "insists he hates sheep" would have as a proper part a model of the same axioms that supported "hates sheep", and hence would not be a minimal model of these axioms.

If, for some irredundant Π there is a minimal finite Herbrand model of an augmented $\Pi^{only-if}$ which supports ϕ' then since that is a model it follows from Theorem 1 that ϕ' is a logical consequence of Π^{if}.

For the converse, we need to make extra assumptions about Π or about the queries ϕ that we wish to present to it. If ϕ' follows from Π^{if} then there is a minimal SLD-refutation, which corresponds to a minimal analysis tree for ϕ' based on Π (an analysis tree such that no strict subset of its atom labels could be rearranged into an analysis tree for the same root atom label). It would be nice to say the set of atoms in such a minimal analysis tree (together with the instances of equality) corresponded to a minimal model of $\Pi^{only-if} \cup \{\bar{\exists}\phi\}$. Clearly (by the proof of Lemma 2c) it is a model, but is it minimal? Certainly none of the instances of = could be missed out. If some strict subset of the atoms in the tree, including the root atom label, was a model then by Lemma 2b it would be possible to form an analysis tree with root atom label ϕ', and this would correspond to a "smaller" SLD-refutation, which is a contradiction. It is, however, possible that a strict subset, obtained by omitting the root atom label and possibly other atoms (possibly with a reduced set of = instances), could be a smaller model of $\Pi^{only-if} \cup \{\bar{\exists}\phi\}$. This would correspond to there being a possible analysis tree for ϕ' with an instance of ϕ labelling a non-root node. If we can disallow this, then it can be seen that the set of atoms in the analysis tree constitutes a minimal finite Herbrand model. This motivates the following definition and theorem.

A DCG with Prolog translation Π is *nonrecursive* with respect to an atom ϕ if there is no analysis tree for a ground instance ϕ' of ϕ with respect to Π such that an instance of ϕ labels a non-root node. The nonrecursiveness condition amounts to a strengthening of the requirements imposed for irredundancy, but the assumption only has to be made for the "goal atom" ϕ. This is actually a reasonable assumption to be made in many cases. For instance, in a traditional parsing situation ϕ already includes complete information about the goal category (at least its context-free part) and the string. In this case an analysis tree involving the same category and string as a proper subpart would amount to a violation of irredundancy (this is the case that was considered in Lemma 2b). As another example, for generation if ϕ were a proposition expressing the semantic content of some sentence but with only the context-free part of the category instantiated, it would be reasonable to assume that a derivation yielding the desired syntax would not include as a part a derivation of the syntactic structure of another phrase with the same category and semantics. Thus we have:

Theorem 2:

Let Π be the Prolog translation of an irredundant DCG, ϕ an atom, with respect to which Π is nonrecursive, and ϕ' a ground instance of ϕ. Then:

There is a minimal finite category model $\mathcal{M}$ of $\Pi^{only-if} \cup \{\exists \phi\}$ such that $\mathcal{M} \models \phi'$ if and only if $\Pi^{if} \vdash \phi'$.

9   Discussion

The results about the two equivalent ways of interpreting DCGs have a number of wider implications.

9.1   Extensions

Although the results of this paper are oriented towards the understanding of DCGs, the notions of Π^{if} and $\Pi^{only-if}$ are relevant for any Prolog program Π. Underneath the results are therefore some more general results about interpretations of Prolog programs and constraint logic programs. Nevertheless the assumptions on which the proofs rested, i.e. irredundancy (and nonrecursiveness), whilst reasonable for DCGs, are not necessarily natural for arbitrary Prolog programs.

Within Computational Linguistics, a natural idea would be to extend the results to give an account of PATR and other unification grammar formalisms with richer constraint structures which no longer relies on logical deduction. Recasting the semantics of the context-free backbone in terms of the "only-if" interpretation may also make it easier to consider extensions.

9.2   Algorithms for Computing Models

Although we treated a particular algorithm for generating logical consequences from Π^{if} (SLD resolution), because of the completeness results the results actually extend (in terms of *what* results are produced by deduction and model-building, not *how*) to other complete proof procedures. This means, for instance, that (in the case of well-behaved DCGs) Earley parsing[Pereira and Warren 1983] can also be regarded as a way of constructing a certain kind of model. One of the contributions that the result makes is to show that, given sufficient restrictions on the form of axioms, certain algorithms for *deduction* can be used for constructing models. Note, however, that we have never discussed constructing *all* models, merely those (minimal) finite category models that support a particular kind of "conclusion".

9.3   Grammar Interpretations

If the same results can be computed (albeit by different mechanisms) regardless of whether a DCG is interpreted in the Π^{if} or the $\Pi^{only-if}$ sense, the question naturally arises as to what a linguist thinks they are saying when they write a DCG. For instance, in writing:

```
s --> np, vp.
```

which of the following do they intend?

$$\Pi^{if}: \qquad \forall P1, P2, P3.\; s(P1, P3) \subset np(P1, P2) \wedge vp(P2, P3)$$
$$\Pi^{only-if}: \forall P1, P2, P3.\; s(P1, P3) \supset np(P1, P2) \wedge vp(P2, P3)$$

In the former case, a *sufficient* condition for sentencehood is expressed. This could be used, for instance, in a situation where it was accepted that there may be other kinds of sentences currently unaccounted for by the grammar. In the latter case, the condition is *necessary*, which suggests that one is partially *defining* a sentence as something with the stated decomposition (i.e. that "sentence" is a technical term not influenced by any real-world constraints).

Possibly the linguist intends *both* interpretations (i.e. "if and only if"). But if a deductive point of view is taken, axioms in $\Pi^{only-if}$ allow no further inferences to be made if the other axioms are just Π^{if}. Similarly, if a model-building view is taken then the Π^{if} axioms seem to make no interesting difference to the models obtained from the augmented $\Pi^{only-if}$ axioms. It seems that one is forced to ignore one half.

To determine which approach is better motivated (e.g. in a given application) and hence what the logical reading of a DCG "should be", we believe it will be necessary to consider the implications of the deduction/ model-building distinction for more complex types of grammars where the syntactic restrictions of DCGs are relaxed. Our expectation is that it is the model-building view that will be supported by this.

10   Acknowledgements

This paper was largely written when the author was visiting IMS at the University of Stuttgart in the summer of 1994 and he would like to thank IMS for giving him this opportunity. The ideas here have benefited from many useful discussions, for instance with Mike Reape, Mark Johnson and Jochen Dörre.

References

[Aït-Kaci and Podelski 1993] Aït-Kaci, H. and Podelski, A., "Towards a Meaning of LIFE", *Journal of Logic Programming* Vol 16, Nos 3,4, 1993.
[Clark 1978] Clark, K. L., "Negation as Failure" in Gallaire, H. and Minker, J. (Eds), *Logic and Databases*, Plenum Press, 1978.

[Emele and Zajac 1990] Emele, M. and Zajac, R., "Typed Unification Grammars", in *Procs of the 13th International COLING*, 1990.

[Emele 1994] Emele, M., "Die TFS Sprache und ihre Implementierung", IMS, Universität Stuttgart, Germany, 1994.

[Höhfeld and Smolka 1988] Höhfeld, M. and Smolka, G., "Definite Relations over Constraint Languages", LILOG report 53, IBM Deutschland, 1988.

[Johnson 1992] Johnson, M., "Two Ways of Formalizing Grammars", Cognitive and Linguistic Sciences, Brown University, 1992.

[Kaplan and Bresnan 1982] Bresnan, J. and Kaplan, R., "Lexical-Functional Grammar: A Formal System for Grammatical Representation", in Bresnan, J., Ed., *The Mental Representation of Grammatical Relations*, MIT Press, 1982.

[Colmerauer 1978] Colmerauer, A., "Metamorphosis Grammars", in Bolc, L., Ed., *Natural Language Communication with Computers*, Springer Verlag, 1978.

[Hobbs et al 1993] Hobbs, J., Stickel, M., Appelt, D. and Martin, P., "Interpretation as Abduction", , *Artificial Intelligence* Vol 63, Nos 1-2, pp69-142, 1993.

[Kasper and Rounds 1986] Kasper, R. and Rounds, W., "A Logical Semantics for Feature Structures", in *Proceedings of the 24th Annual Meeting of the ACL*, 1986.

[Kowalski 1979] Kowalski, R., *Logic for Problem Solving*, North Holland, 1979, Chapter 11.

[Lloyd 1987] Lloyd, J. W. *Foundations of Logic Programming*, Springer Verlag, Second Edition, 1987.

[Manandhar 1993] Manandhar, S., *Relational Extensions to Feature Logic: Applications to Constraint-Based Grammars*, PhD thesis, University of Edinburgh, 1993.

[Manaster-Ramer and Rounds 1987] Manaster-Ramer, A. and Rounds, W., "A Logical Version of Functional Grammar", in *Procs of the 25th Annual Meeting of the ACL*, 1987.

[Pereira and Warren 1980] Pereira, F. and Warren, D., "Definite Clause Grammars for Language Analysis - a Survey of the Formalism and a Comparison with Augmented Transition Networks", *Artificial Intelligence* Vol 13, pp231-278, 1980.

[Pereira and Warren 1983] Pereira, F. and Warren, D., "Parsing as Deduction", in *Proceedings of the 21st Annual Meeting of the ACL*, 1983.

[Reiter and Mackworth 1990] Reiter, R. and Mackworth, A. K., "A Logical Framework for Depiction and Image Interpretation", *Artificial Intelligence* Vol 41, pp125-155, 1989/90.

[Shieber 1986] Shieber, S., *An Introduction to Unification-Based Theories of Grammar*, CSLI Lecture Notes Series, University of Chicago Press, 1986.

[Shieber 1988] Shieber, S., "A uniform architecture for parsing and generation", *Procs of 12th COLING*, 1988.

[Smolka 1992] Smolka, G., "Feature Constraint Logics for Unification Grammars", *Journal of Logic Programming* Vol 12, Nos 1&2, pp51-88, 1992.

[Wedekind 1990] Wedekind, J., (Ed.), "A Survey of Linguistically Motivated Extensions to Unification-Based Formalisms", Deliverable R3.1.A, ESPRIT Basic Research Action BR3175 (DYANA), 1990.

Statistical Abduction with Tabulation[*]

Taisuke Sato and Yoshitaka Kameya

Dept. of Computer Science, Graduate School of Information
Science and Engineering, Tokyo Institute of Technology
2-12-1 Ookayama Meguro-ku Tokyo Japan 152-8552
sato@mi.cs.titech.ac.jp, kame@mi.cs.titech.ac.jp

Abstract. We propose *statistical abduction* as a first-order logical framework for representing, inferring and learning probabilistic knowledge. It semantically integrates logical abduction with a parameterized distribution over abducibles. We show that statistical abduction combined with tabulated search provides an efficient algorithm for probability computation, a Viterbi-like algorithm for finding the most likely explanation, and an EM learning algorithm (the graphical EM algorithm) for learning parameters associated with the distribution which achieve the same computational complexity as those specialized algorithms for HMMs (hidden Markov models), PCFGs (probabilistic context-free grammars) and sc-BNs (singly connected Bayesian networks).

1 Introduction

Abduction is a form of inference that generates the best explanation for observed facts. For example, if one notices that the grass is wet in the yard, he/she might abduce that it rained last night, or the sprinkler was on, by using general rules such as "if it rains, things get wet." Abduction has been used for diagnosis systems [30], planning [14, 41], natural language processing [5, 15], user modeling [9] etc in AI.

It is possible to formalize (part of) abduction in logic programming as follows [16, 17]. We have a background theory T consisting of clauses and an observed fact G (usually a ground atom) to be explained, and the task of abduction is to search for an explanation $E = \{a_1, \ldots, a_n\}$ by choosing ground atoms a_is from a particular class of primitive hypotheses called *abducibles*[1] such that $T \cup E \models G$ and $T \cup E$ is consistent.[2] The quality of E, the abduced explanation, is evaluated by various criteria such as precision, simplicity, abduction cost, and so on [15, 16, 40].

---

[*] This paper is based on a workshop paper presented at the UAI-2000 workshop on Fusion of Domain Knowledge with Data for Decision Support, Stanford, 2000.

[1] The term "explanation" is henceforth used as a synonym of a conjunction (or set) of abducibles.

[2] Sometimes $T \cup E$ is required to satisfy integrity constraints, but in this paper, we do not consider them.

A.C. Kakas, F. Sadri (Eds.): Computat. Logic (Kowalski Festschrift), LNAI 2408, pp. 567–587, 2002.
© Springer-Verlag Berlin Heidelberg 2002

While the above framework is simple and logically sound, it is obviously incomplete. Especially it entirely ignores the problem of uncertainty in the real world. Our observations are often partial, inconsistent or contaminated by noise. So the abduced explanation should be treated as being true only to some degree. Also it must be noticed that our observations are always finite but potentially infinite (we may have another observation indefinitely), and it is often critical to evaluate how far our explanation holds on average. Since these problems are certainly not in the realm of logic, but belong to statistics, it is natural and desirable to build an interdisciplinary framework that unifies the logical aspects and the statistical aspects of abduction.

There are many ways of doing this, but one of the simplest ways is to introduce a *parameterized probability distribution over abducibles*. We term the resulting logical-statistical framework *statistical abduction*, in which we calculate the probabilities of explanations from the parameters associated with the distribution and determine the most likely explanation among possible ones as the one with the highest probability. Parameters are statistically learnable from observations if they are unknown.

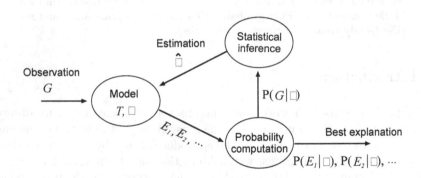

Fig. 1. Statistical abduction

The idea of statistical abduction is illustrated above. We have abducibles $a_1, a_2, \ldots$ with a probability distribution parameterized by θ, and a clausal theory T. For a given G, observation, we search for possible explanations $E_1, E_2, \ldots$ each of which is a conjunction of finitely many abducibles. $P(E_1 \mid \theta), P(E_2 \mid \theta), \ldots$, probabilities of explanations, are computed from marginal distributions for these constituent abducibles. Their probabilities are used to select the best explanation and also to compute $P(G \mid \theta)$. The parameter θ is estimated by applying ML (maximum likelihood) estimation to $P(G \mid \theta)$.

We would like to first emphasize that statistical abduction is not merely abduction of logical explanations but aims at the inference of their distribution. Second, it has wide coverage, as we will see, ranging from logic programming to popular symbolic-statistical frameworks. By popular symbolic-statistical frameworks, we mean for instance HMMs (hidden Markov Models), PCFGs (probabilistic context free grammars) and BNs (Bayesian networks) explained below.

An HMM is a special type of Markov chain in which a symbol is emitted at a state and we can only observe a sequence of emitted symbols whereas state transitions are not observable, i.e. they are hidden. HMMs are used as a modeling tool for speech recognition, genome informatics etc [22, 32]. Also a PCFG is a context free grammar with probabilities assigned to each rule in such a way that if there are n production rules $A \rightarrow B_1, \ldots, A \rightarrow B_n$ for a non-terminal symbol A, probabilities p_i is assigned to $A \rightarrow B_i$ $(1 \leq i \leq n)$ where $\sum_i p_i = 1$. The probability of a parse tree is the product of probabilities assigned to rules appearing in the tree, and that of a sentence is the sum of probabilities of possible parse trees for the sentence [6, 22, 43]. PCFGs form a basis of stochastic natural language processing. Finally, a Bayesian network means an acyclic directed graph consisting of nodes of random variables where a child node conditionally depends on the parent nodes, and the dependency is specified by a CPT (conditional probability table) when nodes take discrete values. BNs are used to model probabilistic-causal relationships. A *singly connected* BN is one that does not include loops when directions are ignored in the graph [4, 28].

Turning back to statistical abduction, we note that all statistical techniques from fitting test to random sampling and to parameter learning are applicable. They provide us with powerful means for the statistical analysis of abduction logically formalized.

There are however two fundamental problems; one is theoretical and the other is practical. First of all, statistical abduction must deal with infinitely many objects sanctioned by the language of first-order logic and their joint distributions, which raises the mathematical question of defining a probability space consistently giving a joint distribution over a set of arbitrarily chosen objects. It goes beyond probabilistic semantics often seen in AI that deals only with finite domains and finitely many random variables.

Secondly, to apply it in practice, we need to know *all* values of statistical parameters, but determining a large number of statistical parameters is a hard task, known as the where-do-the-numbers-come-from problem. Although one might hope that the problem is mitigated by learning, there has been little work on parameter learning in the literature of logical framework of abduction.

The objective of this paper is to make it clear that there exists a firm theoretical basis for statistical abduction and we have an efficient algorithm for computing probabilities, thereby being able to efficiently determine the most likely hypothesis and an efficient EM algorithm for parameter learning.[3] Since the subject is broad and the space is limited, we concentrate on putting the major ideas across and details are left to the related literature [18, 33, 34, 35, 36, 37].[4] In what follows, after reviewing some historical background in Section 2, we

---

[3] The EM algorithm is an iterative algorithm which is a standard method for ML estimation of statistical parameters from incomplete data [23].

[4] We submitted a comprehensive paper on the subject [38].

sketch our probabilistic semantics in Section 3, and explain in Section 4 PRISM, a symbolic-statistical modeling language implementing distribution semantics as an embodiment of statistical abduction. Section 5 is a main section. We first describe three basic computational tasks required for statistical abduction. We then propose the use of tabulated search and combine it with general algorithms for PRISM programs to perform the three tasks, and finally state the time complexity of PRISM programs for the case of HMMs, PCFGs and sc-BNs (singly connected Bayesian networks), which indicates that the proposed algorithms run as efficiently as specialized algorithms for HMMs, PCFGs and sc-BNs. Section 6 is a conclusion.

2 Background

Looking back on the role of probabilities in logic programming, two approaches, "constraint approach" and "distribution approach," are distinguishable. The former focuses on the inference of probability intervals assigned to atoms and clauses as constraints, whereas the prime interest of the latter is to represent a single probability distribution over atoms and formulas, from which various statistics are calculated.

The constraint approach is seen in the early work of Nilsson [27] where he tried to compute, by using linear programming technique, the upper and lower bound of probability of a target sentence consistent with a first-order knowledge base in which each sentence is assigned a probability. In logic programming, Ng and Subrahmanian took the constraint approach to formulate their probabilistic logic programming [25] (see [12] for recent development). Their program is a set of annotated clauses of the form $A : \mu_0 \leftarrow F_1 : \mu_1, \ldots, F_n : \mu_n$ where A is an atom, F_i $(1 \leq i \leq n)$ a basic formula, i.e. a conjunction or a disjunction of atoms, and the annotation μ_j $(0 \leq j \leq n)$ a sub-interval in $[0, 1]$ indicating a probability range. A query $\leftarrow \exists (F_1 : \mu_1, \ldots, F_n : \mu_n)$ is answered by an extension of SLD refutation. Their language contains only a finite number of constant and predicate symbols, and no function symbols are allowed.

A similar probabilistic framework was proposed by Lakshmanan and Sadri under the same syntactic restrictions (finitely many constant, predicate symbols, no function symbols) in a different uncertainty setting [21]. They used annotated clauses of the form $A \xleftarrow{c} B_1, \ldots, B_n$ where A and B_i are atoms and $c = \langle [\alpha, \beta], [\gamma, \delta] \rangle$, the confidence level, represents the belief interval $[\alpha, \beta]$ $(0 \leq \alpha \leq \beta \leq 1)$ and doubt interval $[\gamma, \delta]$ $(0 \leq \gamma \leq \delta \leq 1)$ which an expert has in the clause [21].

By comparison, the distribution approach has been actively pursued outside logic programming. In particular, researchers in the Bayesian network community have been using definite clauses with probabilities attached to express probabilistic events such as gene inheritance. In the framework of KBMC (knowledge-based model construction) [1, 3, 19, 26] for instance, clauses are used as a macro language to compactly represent similar Bayesian networks. Basically a knowledge base KB contains clauses representing general rules and CPTs (conditional probability tables). Every time a set of evidence and context is given as ground atoms, a specialized Bayesian network is constructed by tracing logical/probabilistic dependencies in KB to compute the probability of a query atom. Uncertain parameters associated with CPTs can be learned by applying the EM learning algorithm for Bayesian networks [4] to the constructed network [19]. It can be said that KBMC implicitly defines a collection of local distributions in the form of Bayesian networks, each corresponding to a pair of evidence and context. The question of whether there exists a single distribution compatible with these implicitly defined (and infinitely many) local distributions or not remains open.

In contrast to KBMC, statistical abduction explicitly defines a single distribution (probability measure) over ground atoms. It was begun by Poole as "probabilistic Horn abduction" [31]. In his approach, a program is comprised of non-probabilistic definite clauses and probabilistic disjoint declarations. A disjoint declaration is of the form disjoint([h_1:p_1,...,h_n:p_n]). It says h_i, an abducible atom, becomes exclusively true with probability p_i $(1 \leq i \leq n)$. Abducibles in different declarations are independent. The probability of a non-abducible ground atom is then calculated by reduction in a top-down manner through program clauses to a DNF formula made out of abducibles in the disjoint declarations. The probabilistic Horn abduction is able to represent Bayesian networks [31].

While the probabilistic Horn abduction opened a new vista on extending Bayesian networks to first-order languages, it makes various assumptions on programs such as the acyclicity condition[5] and the covering property.[6] These assumptions are not easy to verify and could be severe restrictions in programming. For example, under the acyclicity condition, when a clause includes local variables like Y in $p(X) \leftarrow q(X, Y),\ldots$ one cannot write recursive clauses about q such as $member(X, cons(H, Y)) \leftarrow member(X, Y)$. Also the defined probability measure is not proved to be completely additive either. In other words, the continuity $\lim_{n \to \infty} P(p(t_1) \vee \ldots \vee p(t_n)) = P(\exists X p(X))$ where t_is are ground terms, is not necessarily guaranteed. More serious is the problem of determining parameters in disjoint declarations. How can we get them? It remained unanswered.

SLP (stochastic logic programming) proposed by Muggleton [24] is another attempt to define probabilities over ground atoms. He associated, analogously to PCFGs, probabilities p_i's with range-restricted clauses[7] C_i's like p_i : C_i $(1 \leq i \leq n)$. The probability of a ground atom G is defined as the product

---

[5] It says that every ground atom A must be assigned a unique integer $n(A)$ such that $n(A) > n(B_1), \ldots, n(B_n)$ for every ground instance of a clause of the form $A \leftarrow B_1, \ldots, B_n$.

[6] It requires that when there are finite ground instances $A \leftarrow \alpha_i$ $(1 \leq i \leq m)$ about a ground atom A in the program, $A \leftrightarrow \alpha_1 \vee \ldots \vee \alpha_m$ holds. Intuitively the property ensures every observation has an explanation. Logically it is equivalent to assuming the iff completion [13].

[7] A clause is range-restricted if variables appearing in the head also appear in the body. So, a unit clause must be ground.

of such p_is appearing in G's SLD refutation, but with a modification such that if a subgoal g can invoke n clauses, $p_i : C_i$ ($1 \leq i \leq n$) at some derivation step, the probability of choosing k th clause is normalized to $p_k / \sum_{i=1}^{n} p_i$. More recently, Cussens extended SLP by introducing the notion of loglinear models for SLD refutations and defined probabilities of ground atoms in terms of their SLD-trees and "features" [10]. To define the probability of a ground atom $s(a)$, he first defines the probability $P(R)$ of an SLD refutation R for the most general goal $\leftarrow s(X)$ as $P(R) \stackrel{\text{def}}{=} Z^{-1} \exp\left(\sum_i \log(\lambda_i) f(R, i)\right)$. Here λ_i is a number (parameter) associated with a clause C_i and $f(R, i)$ is a feature such as the number of occurrences of C_i in R. Z is a normalizing constant. The probability assigned to a ground atom $s(a)$ is the sum of probabilities of all possible SLD refutations for $\leftarrow s(a)$ [10]. An EM algorithm for inferring parameters taking failures into account is proposed in [11]. Presently, assigning probabilities to arbitrary quantified formulas is out of the scope of both of SLPs.

Looking at the distribution approach to probabilistic functional languages, we notice that Koller et al. proposed a probabilistic functional language which can represent HMMs, PCFGs and BNs [20], but neither the problem of defining declarative semantics nor that of learning parameters in a program was not discussed. Later Pfeffer developed it into another functional language with declarative semantics which is based on the products of countably infinite uniform distributions over the unit interval $[0, 1)$. EM learning is sketched [29].

3   Distribution Semantics: An Overview

Aiming at providing a broader theoretical basis and a learning algorithm for statistical parameters of statistical abduction, Sato proposed *distribution semantics* [33] and developed a first-order statistical modeling language PRISM (http://mi.cs.titech.ac.jp/prism/) [33, 34]. The proposed semantics rigorously defines a *probability measure* over the set of Herbrand interpretations as the denotation of a PRISM program. It is exactly a probabilistic extension of the least Herbrand model semantics to the possible world semantics with a probability measure, but eliminates extraneous assumptions made in the previous approaches. For example, there is no need for the covering assumption or the acyclicity condition [31] (because every definite program has a least Herbrand model and the iff completion [13] holds in it.). Similarly, neither the range-restrictedness condition nor normalization in SLPs [10, 24] is necessary. What is more, there is no vocabulary restriction. We may use as many constant symbols, function symbols and predicate symbols as we need, and can write whatever program we want, though in actual programming, we have to care about efficiency, termination, etc.

Syntactically, our program DB is a set $F \cup R$ where F is a set of atoms (abducibles) and R is a set of definite clauses such that no clause head in R is unifiable with an atom in F. In the theoretical context however, we always consider DB as a set of ground clauses made up of all possible ground instances of the original clauses in DB. F then is a set of infinitely many ground atoms.

We associate with F a *basic probability measure* P_F. It is defined over the set of Herbrand interpretations of F and makes every atom A in F a random variable taking on 1 when A is true and 0 otherwise. Hence atoms in F are probabilistically true and random sampling determines a set of true atoms F'. Then think of a new definite clause program $DB' = F' \cup R$ and its least Herbrand model $M(DB')$ [13]. $M(DB')$ determines the truth values of all ground atoms in DB, which implies that every ground atom in DB is a random variable. Therefore a probability measure P_{DB} over the set of Herbrand interpretations of DB is definable [33]. P_F mentioned above is constructed from a collection of finite joint distributions $P_F^{(n)}(A_1 = x_1, \ldots, A_n = x_n)$ $(n = 1, 2, \ldots)$ where A_is $(\subset F)$ are random variables (abducibles) such that

$$P_F^{(n)}(A_1 = x_1, \ldots, A_n = x_n) = \sum_{x_{n+1} \in \{0,1\}} P_F^{(n+1)}(A_1 = x_1, \ldots, A_{n+1} = x_{n+1}).$$

In the following, for the sake of intuitiveness, we use a joint distribution and a probability measure interchangeably. This is because the probability measure P_F behaves as if it were an infinite probability distribution whose marginal distribution is $P_F^{(n)}(\cdot)$ $(n = 1, 2, \ldots)$.[8]

PRISM, an implementation of the distribution semantics with P_F chosen to be a specific form (the direct products of infinitely many random switches) has been developed as a symbolic-statistical modeling language for complex phenomena governed by rules and probabilities [33, 34, 35]. It is a general logic programming language equipped with a built-in EM algorithm by which we can learn parameters associated with P_F from observations represented by ground atoms. As PRISM allows us to use programs to specify distributions (*programs as distributions*), we have an enormous degree of freedom and flexibilities in modeling complex symbolic-statistical phenomena. Actually, we have found it rather easy to write a PRISM program modeling complicated interactions between gene inheritance and social rules (bi-lateral cross cousin marriage) observed in the Kariera tribe, an anthropological tribe which lived 80 years ago in the west Australia [35, 44].

4 PRISM Programs

In this section, we explain PRISM programs by examples. In our framework, observations are represented by ground atoms and the role of PRISM programs is to specify their joint distributions in terms of built-in probabilistic atoms (abducibles).

A PRISM program is a definite clause program $DB = F \cup R$ such that R, a set of definite clauses, represents non-probabilistic rules such as Mendel's law whereas F, a set of ground atoms, represents basic probabilistic events and has an infinite joint distribution P_F. F and P_F must satisfy the following.

---

[8] This also applies to P_{DB}.

1. F is a set of probabilistic atoms of the form $\mathtt{msw}(i,n,v)$. They are random variables taking on 1 (resp. 0) when true (resp. false). The arguments i and n are ground terms called *switch name* and *trial-id*, respectively. We assume that V_i, a finite set of ground terms, is associated with each i, and $v \in V_i$ holds. V_i is called the *value set* of i.

2. Let V_i be $\{v_1, v_2, \ldots, v_{|V_i|}\}$. Then, one of the ground atoms $\mathtt{msw}(i,n,v_1)$, $\mathtt{msw}(i,n,v_2), \ldots, \mathtt{msw}(i,n,v_{|V_i|})$ becomes exclusively true on each trial. For each i, $\theta_{i,v} \in [0,1]$ is a *parameter* of the probability of $\mathtt{msw}(i,\cdot,v)$ being true ($v \in V_i$), and $\sum_{v \in V_i} \theta_{i,v} = 1$ holds.

3. For arbitrary i, i', n, n', $v \in V_i$ and $v' \in V_{i'}$, random variable $\mathtt{msw}(i,n,v)$ is independent of $\mathtt{msw}(i',n',v')$ if $n \neq n'$ or $i \neq i'$.

A ground atom $\mathtt{msw}(i,n,v)$ represents an event "a probabilistic switch named i takes on v as a sample value on the trial n" ($\mathtt{msw}$ stands for *multi-valued switch*). The second and the third condition say that a logical variable $\mathtt{V}$ in $\mathtt{msw}(i,n,\mathtt{V})$ behaves like a random variable which is realized to v_k with probability θ_{i,v_k} ($k = 1 \ldots |V_i|$). Moreover, from the third condition, the logical variables $\mathtt{V1}$ and $\mathtt{V2}$ in $\mathtt{msw}(i,n_1,\mathtt{V1})$ and $\mathtt{msw}(i,n_2,\mathtt{V2})$ can be seen as *independent and identically distributed* (i.i.d.) random variables if n_1 and n_2 are different ground terms. From an abductive point of view, $\mathtt{msw}$ atoms are abducibles.[9]

To get a feel for PRISM programs, we first take a look at a non-recursive PRISM program. Imagine a lawn beside a road and their observations such as "the road is dry but the lawn is wet." Assume that the lawn is watered by a sprinkler that (probabilistically) works only when it does not rain. The process that generates an observation $\mathtt{observed(road(X),lawn(Y))}$ ("the road is X and the lawn is Y") where $\mathtt{X}, \mathtt{Y} \in \{\mathtt{wet}, \mathtt{dry}\}$ is described by the program DB_{rs} in Figure 2.

```
(1) target(observed/2).
(2) values(rain,[yes,no]).
(3) values(sprinkler,[on,off]).
(4) observed(road(X),lawn(Y)):-
        msw(rain,once,A),
        ( A = yes, X = wet, Y = wet
        ; A = no,  msw(sprinkler,once,B),
            ( B = on,  X = dry, Y = wet
            ; B = off, X = dry, Y = dry ) ).
```

Fig. 2. DB_{rs}

This program first declares $\mathtt{observed/2}$ as a target predicate corresponding to our observations by clause (1). (2) and (3) declare the use and value sets

---

[9] The second and the third condition correspond to the disjoint declaration in Poole's framework [31]: $\mathtt{disjoint}([\mathtt{msw}(i,\mathtt{N},v_1):\theta_{i,v_1}, \ldots, \mathtt{msw}(i,\mathtt{N},v_{|V_i|}):\theta_{i,v_{|V_i|}}])$.

of msw atoms. For example (2) declares a probabilistic multi-ary switch named rain whose values are {yes, no}. (4), the main clause defining observed/2 is read like an ordinary Prolog clause. The difference between (4) and usual clauses is two usages of built-in msw atoms in the body. msw(rain, once, A) for example returns in A one of {yes, no} sampled according to a parameterized distribution $P_{F_r}(\cdot \mid \theta_r)$ described below. msw(sprinkler, once, B) behaves similarly.[10] If disjunctions look messy, by the way, it is possible to split the clause into three clauses each of which has a conjunctive body. By doing so however, we will have a multiple occurrences of the same msw atom.

Write the program as $DB_{rs} = F_{rs} \cup R_{rs}$ where $F_{rs} = \{$msw(rain, once, yes), msw(rain, once, no), msw(sprinkler, once, on), msw(sprinkler, once, off)$\}$ and R_{rs} is the set of ground instantiations of (4). To define a basic distribution $P_{F_{rs}}$ over F_{rs}, put $F_r = \{$msw(rain, once, yes), msw(rain, once, no)$\}$ and introduce a distribution $P_{F_r}(\cdot, \cdot)$ over F_r parameterized by θ_r ($0 \leq \theta_r \leq 1$) such that[11]

$$P_{F_r}(\text{msw(rain, once, yes)} = 1, \text{msw(rain, once, no)} = 1) = 0$$
$$P_{F_r}(\text{msw(rain, once, yes)} = 1, \text{msw(rain, once, no)} = 0) = \theta_r$$
$$P_{F_r}(\text{msw(rain, once, yes)} = 0, \text{msw(rain, once, no)} = 1) = 1 - \theta_r$$
$$P_{F_r}(\text{msw(rain, once, yes)} = 0, \text{msw(rain, once, no)} = 0) = 0.$$

Introduce analogously another distribution $P_{F_s}(\cdot, \cdot)$ parameterized by θ_s over the set $F_s = \{$msw(sprinkler, once, on), msw(sprinkler, once, off)$\}$. The basic distribution $P_{F_{rs}}$ is then defined as the products of P_{F_r} and P_{F_s}. Hereafter for simplicity, we use $P(A)$ as a synonym for $P(A = 1)$, and $P(\neg A)$ for $P(A = 0)$. Accordingly we write $P_{DB_{rs}}(\text{observed(road(dry), lawn(wet))}) = (1 - \theta_r)\theta_s$ etc.

PRISM provides the user with not only various built-ins to set statistical parameter values and compute probabilities of atoms using them, but a built-in EM learning routine for ML (maximum likelihood) estimation to infer parameter values from observed atoms. That is if we have a random sample such as

observed(road(wet), lawn(wet)), observed(road(dry), lawn(wet)), ...

we can statistically infer θ_r and θ_s from them as the maximizers of the likelihood of the sample (we further discuss EM learning later).

Now we turn to another feature of PRISM, recursion. The existence of recursion in a program potentially introduces a countably infinite number of random variables and the construction of an underlying probability space is an absolute necessity for their joint distributions to be consistently defined, but presents some technical difficulties. Distribution semantics however achieves it through the least model semantics [33].

---

[10] If a ground msw atom such as msw(rain, once, yes) is called, we first execute msw(rain, once, A) and then execute A = yes. So the goal fails if the sampled value returned in A is no.

[11] "once" in msw(rain, once, yes) is a constant to identify a trial that is attempted only once in the program.

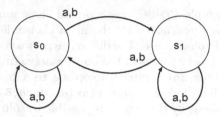

Fig. 3. Two state HMM

```
(1) target(hmm/1).
(2) table([hmm/1,hmm/3]).
(3) values(init,[s0,s1]).
(4) values(out(_),[a,b]).
(5) values(tr(_),[s0,s1]).
(6) hmm(Cs):- msw(init,once,Si),hmm(1,Si,Cs).
(7) hmm(T,S,[C|Cs]):- T=<3,
        msw(out(S),T,C),msw(tr(S),T,NextS),
        T1 is T+1,hmm(T1,NextS,Cs).
(8) hmm(T,_,[]):- T>3.
```

Fig. 4. PRISM program DB_{hmm} for the two state HMM

As an example of recursive PRISM program, we look at an HMM program DB_{hmm} in Figure 4 describing a two state HMM in Figure 3 that generates strings $\{a, b\}^*$ (of finite length, 3 in this case). In the program, clause (1) declares that only ground atoms containing hmm/1 are observable. (2) is concerned with tabulated search which will be explained later. Since msw atoms that can appear as goals during execution have similar patterns, (4) and (5) declare them by terms containing "_" that matches anything. Clauses (6)~(8) specify the probabilistic behavior of the HMM. T is a time step and S and NextS are states. Cs represents a list of output symbols. Clause (7) probabilistically chooses an output symbol C and the next state NextS. To represent switches sampled at each state S, it uses non-ground terms out(S) and tr(S). T is used to guarantee independence among choices at different time steps. DB_{hmm} as a whole describes a process of stochastic generation of strings such as hmm([a,b,a]). The program is procedurally understandable by Prolog reading except msw atoms. That is, given ground S and T, C in msw(out(S),T,C) behaves like a random variable taking discrete values $\{a, b\}$ declared by clause (4).

5   Three Computational Tasks

To apply statistical abduction to the real world, we need computational tractability in addition to expressive power. We here consider three basic computational tasks based on the analogy of HMMs [32]:

(1) computing $P_{DB}(G \mid \theta)$,[12] the probability of an atom G representing an observation,
(2) finding E^*, the most likely explanation for G, and
(3) adjusting the parameters so that the probability of a given sequence $\mathcal{G} = \langle G_1, G_2, \ldots, G_T \rangle$ of observations is maximized.
All solutions should be computationally tractable.

As for HMMs, these methods correspond to the forward procedure, the Viterbi algorithm and the Baum-Welch algorithm respectively [32, 22].

Poole [31] described a method for the first task. Let us consider a program $DB = F \cup R$ and the following if-and-only-if (iff) relation under $comp(R)$, the Clark's completion of the rules R [7]:

$$comp(R) \models G \leftrightarrow E^{(1)} \vee \cdots \vee E^{(m)}. \tag{1}$$

He assumes that there exist finitely many explanations $\psi_{DB}(G) = \{E^{(1)}, \cdots, E^{(m)}\}$ for G as above each of which is a finite conjunction of independent abducibles and they are mutually exclusive, i.e. $P_{DB}(E^{(i)} \wedge E^{(j)}) = 0$ $(1 \leq i \neq j \leq m)$ (we say DB satisfies the *exclusiveness condition*). Let I be the set of switch names and $\sigma_{i,v}(E)$ the number of occurrences of $\mathtt{msw}(i, \cdot, v)$ in an explanation E. His method for solving the first task is formulated in our notation as follows:

$$P_{DB}(G \mid \theta) = \sum_{E \in \psi_{DB}(G)} P_F(E) = \sum_{E \in \psi_{DB}(G)} \prod_{i \in I, v \in V_i} \theta_{i,v}^{\sigma_{i,v}(E)}. \tag{2}$$

A little modification of the above formula would give one for the second task:

$$E^* = \arg\max_{E \in \psi_{DB}(G)} \prod_{i \in I, v \in V_i} \theta_{i,v}^{\sigma_{i,v}(E)}. \tag{3}$$

Unfortunately, $|\psi_{DB}(G)|$, the number of explanations for G, often grows exponentially in the complexity of the model (e.g. the number of states in an HMM), or in the complexity of each observation (e.g. the string length).

5.1   OLDT Search and Support Graphs

For the three basic computational tasks to be practically achievable, it is a must to suppress computational explosions. Define anew $\psi_{DB}(G)$, the set of all explanations for a goal G, by $\psi_{DB}(G) \overset{\text{def}}{=} \{E \mid \text{minimal } E \subset F, R \cup E \vdash G\}$.

Statistical abduction has two potential sources of computational explosions. One is a search phase searching for $\psi_{DB}(G)$. It will be explosive if the search is done by backtracking as can be easily confirmed by the HMM program DB_{hmm}. The other is a probability computation phase corresponding to (2) and/or (3). They would be explosive without factoring computations. Suppose we have a program $\{\mathtt{g:-m1,m2,m3.\ g:-m4,m2,m3.}\}$ in which g is a goal. (2) leads us to $P(\mathtt{g}) = P(\mathtt{m1})P(\mathtt{m2})P(\mathtt{m3}) + P(\mathtt{m4})P(\mathtt{m2})P(\mathtt{m3})$. However this computation repeats the same computation $P(\mathtt{m2})P(\mathtt{m3})$ twice.

---

[12] θ is the vector consisting of parameters associated with all abducibles which forms the explanations for the observed fact G or an observation in $\mathcal{G}$.

It is possible to avoid these computational redundancies all at once by adopting tabulated search that results in a compact graphical representation of all solutions, which in turn makes it possible to factor probability computations. The point is to introduce intermediate atoms between a goal and abducibles called *table atoms* to factor out common probability computations and let them store the computed results. In the above case, we should write {g:-m1,h. g:-m4,h. h:-m2,m3.} using a new table atom h. We compute $P(h) = P(m2)P(m3)$ once and use the result twice later in the computation of $P(g) = P(m1)P(h) + P(m4)P(h)$. The remaining of this subsection and the next subsection detail the idea sketched above.

In OLDT search [42] which is a complete tabulated search method for logic programs that adopts the tabling of goals, we store explanations for a goal G in a global table called a *solution table* while letting them hierarchically share common sub-explanations [18, 42]. Such hierarchical sharing reflects on factoring probability computations carried out after search. From the solution table, a graph called *support graph* representing $\psi_{DB}(G)$ is extracted as an ordered set of disconnected subgraphs. Once the support graph is extracted, it is relatively easy to derive algorithms for the three computational tasks that run as efficiently as specialized ones such as the forward procedure, the Viterbi algorithm and the Baum-Welch algorithm in HMMs, as we see later.

Mathematically we need some assumptions to validate the derivation of these efficient algorithms. Namely we assume that the number of explanations for a goal is finite (we say DB satisfies the *finite support condition*) and there exists a linearly ordered set[13] $\tau_{DB}(G) \overset{\text{def}}{=} \langle \tau_0, \tau_1, \ldots, \tau_K \rangle$ $(\tau_0 = G)$ of *table atoms*[14] satisfying the following conditions:

- Under *comp*(R), a table atom τ_k $(0 \leq k \leq K)$ is equivalent to a disjunction $E_{k,1} \vee \cdots \vee E_{k,m_k}$. Each disjunct $E_{k,h}$ $(1 \leq h \leq m_k)$ is called a *tabled-explanation* for τ_k and made up of msw atoms and other table atoms. The set $\tilde{\psi}_{DB}(\tau_k) = \{E_{k,1}, \ldots, E_{k,m_k}\}$ is called the *tabled-explanations* for τ_k. Logically, it must hold that

$$comp(R) \models (G \leftrightarrow E_{0,1} \vee \cdots \vee E_{0,m_0}) \qquad (4)$$
$$\wedge (\tau_1 \leftrightarrow E_{1,1} \vee \cdots \vee E_{1,m_1})$$
$$\wedge \cdots \wedge (\tau_K \leftrightarrow E_{K,1} \vee \cdots \vee E_{K,m_K}).$$

Also we require that table atoms be layered in the sense that atoms appearing in the right hand side of $\tau_k \leftrightarrow E_{k,1} \vee \cdots \vee E_{k,m_k}$ belong in $F \cup \{\tau_{k+1}, \ldots, \tau_K\}$

---

[13] Here we use a vector notation to emphasize the set is ordered.

[14] *Table atoms* mean atoms containing a *table predicate*. They make an entry for a table to store search results. Table predicates are assumed to be declared by the programmer in advance like table([hmm/1,hmm/3]) in DB_{hmm}. We treat the top goal G as a special table atom τ_0.

(*acyclic support condition*). In other words, τ_k can only refer to $\tau_{k'}$ such that $k < k'$ in the program.

- $P_{DB}(E_{k,i}, E_{k,j}) = 0$ if $i \neq j$ for $E_{k,i}, E_{k,j} \in \tilde{\psi}_{DB}(\tau_k)$ $(0 \leq k \leq K)$ (*t-exclusiveness condition*) and each tabled-explanation in $\tilde{\psi}_{DB}(\tau_k)$ is comprised of statistically independent atoms (*independent condition*).

$\tau_{DB}(G)$ satisfying these conditions (the finite support condition, the acyclic support condition, the t-exclusiveness condition and the independent condition) is obtained, assuming due care is taken by the programmer, by (a specialization of) OLDT search [42] as follows. We look at the HMM program DB_{hmm} in Section 4 as a running example. We first translate DB_{hmm} to a Prolog program similarly to DCGs (*definite clause grammars*). The Prolog program DB_{hmm}^t in Figure 5 is a translation of DB_{hmm}. Clauses (Tj) and (Tj') are generated from the clause (j) in DB_{hmm}. In translation, we add two arguments as difference-list

```
(T1)   top_hmm(Cs,Ans):- tab_hmm(Cs,Ans,[]).
(T2)   tab_hmm(Cs,[hmm(Cs)|X],X):- hmm(Cs,Ans,[]).
(T2')  tab_hmm(T,S,Cs,[hmm(T,S,Cs)|X],X):- hmm(T,S,Cs,Ans,[]).
(T3)   e_msw(init,T,s0,[msw(init,T,s0)|X],X).
(T3')  e_msw(init,T,s1,[msw(init,T,s1)|X],X).
   :
(T6)   hmm(Cs,X0,X1):-
              e_msw(init,once,Si,X0,X2),
              tab_hmm(1,Si,Cs,X2,X1).
(T7)   hmm(T,S,[C|Cs],X0,X1):- T=<3,
              e_msw(out(S),T,C,X0,X2),
              e_msw(tr(S),T,NextS,X2,X3),
              T1 is T+1, tab_hmm(T1,NextS,Cs,X3,X1).
(T8)   hmm(T,S,[],X,X):- T>3.
```

Fig. 5. Translated program DB_{hmm}^t

to atoms to hold a tabled-explanation. Table predicates do not change, so table predicates in DB_{hmm}^t are hmm/3 and hmm/5. We rename the abducibles declared by (3)∼(5) in DB_{hmm} so that they are placed in the callee's difference-list. We treat table atoms just like msw atoms except that they invoke subsequent calls to search for their tabled-explanations returned in Ans. For this purpose, we add clauses with the head of the form tab_...() like (T2).

After translation, we fix the search strategy of OLDT to *multi-stage depth-first strategy* [42] (it is like Prolog execution) and run DB_{hmm}^t for a top-goal, for instance :- top_hmm([a,b,a],Ans,[]) to search for all tabled-explanations of the tabled atom $\tau_0 = $ hmm([a,b,a]) corresponding to our observation. They are returned in Ans as answer substitutions.

The top goal invokes a table atom hmm([a,b,a],Ans,[]) through clause (T2). Generally in OLDT search, when a table atom hmm(t,Ans,[]) is called for the first time, we create a new entry hmm(t) in the solution table. Every time hmm(t,Ans,[]) is solved with answer substitutions $t = t'$ and Ans $= e$, we store e (= a tabled-explanation for hmm(t')) in the solution table under the sub-entry hmm(t') (in the current case however, t and t' are ground, so they coincide). If the entry hmm(t) already exists, hmm(t,Ans,[]) returns with one of the unused solutions. The OLDT search terminates exhausting all solutions for hmm([a,b,a],Ans,[]) and yields a solution table in Figure 6.

```
hmm([a,b,a]):
    [hmm([a,b,a]): [[msw(init,once,s0),hmm(1,s0,[a,b,a])],
                    [msw(init,once,s1),hmm(1,s1,[a,b,a])]]]
hmm(1,s0,[a,b,a]):
    [hmm(1,s0,[a,b,a]):[[msw(out(s0),1,a),msw(tr(s0),1,s0),hmm(2,s0,[b,a])],
                        [msw(out(s0),1,a),msw(tr(s0),1,s1),hmm(2,s1,[b,a])]]]
hmm(1,s1,[a,b,a]):
    [hmm(1,s1,[a,b,a]):[[msw(out(s1),1,a),msw(tr(s1),1,s0),hmm(2,s0,[b,a])],
                        [msw(out(s1),1,a),msw(tr(s1),1,s1),hmm(2,s1,[b,a])]]]
    :
```

Fig. 6. Part of solution table for hmm([a,b,a]).

A list

$$[[\text{msw(init,once,s0)},\text{hmm(1,s0,[a,b,a])}],$$
$$[\text{msw(init,once,s1)},\text{hmm(1,s1,[a,b,a])}]]$$

under the sub-entry hmm([a,b,a]) in Figure 6 means

$$comp(R_{\text{hmm}}) \vdash \text{hmm}([a,b,a]) \leftrightarrow$$
$$(\text{msw}(\text{init}, \text{once}, \text{s0}) \wedge \text{hmm}(1, \text{s0}, [a, b, a]) \vee$$
$$(\text{msw}(\text{init}, \text{once}, \text{s1}) \wedge \text{hmm}(1, \text{s1}, [a, b, a])$$

where $R_{\text{hmm}} = \{(6), (7), (8)\}$ in DB_{hmm} in Section 4.

After OLDT search, we collect all tabled-explanations from the solution table and topologically sort them to get linearly ordered table atoms $\tau_{DB}(G) = \langle \tau_0, \tau_1, \ldots, \tau_K \rangle$ ($\tau_0 = G$) together with their tabled-explanations $\tilde{\psi}_{DB}(\tau_k)$ ($0 \leq k \leq K$) satisfying Equation (4).

Since all the data we need in the subsequent computation of $P_{DB}(G \mid \boldsymbol{\theta})$ is $\tau_{DB}(G)$ (and $\tilde{\psi}_{DB}(\cdot)$), and since it is much more natural from a computational point of view to look upon $\tau_{DB}(G)$ as a graph than a set of atoms logically connected, we introduce a graphical representation of $\tau_{DB}(G)$, and call it a *support graph*. Namely, a *support graph* $\tau_{DB}(G)$ for G is a *linearly ordered* set $\langle \tau_0, \tau_1, \ldots, \tau_K \rangle$ ($\tau_0 = G$) of disconnected subgraphs. Each subgraph τ_k ($0 \leq k \leq K$) (we identify a subgraph with the table atom labeling it) is comprised of linear graphs of the form start-e_1-$\cdots$-e_M-end representing some tabled-explanation $e_1 \wedge \cdots \wedge e_M$ for τ_k. Here start is a fork node and end is a join node and e_h ($1 \leq h \leq M$) is either a msw atom or a table atom labeling the corresponding subgraph in a lower layer. Part of the support graph for $\tau_{DB}(\text{hmm}([a, b, a]))$ is described in Figure 7.

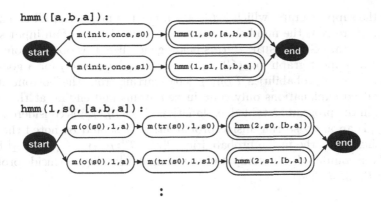

Fig. 7. Part of the support graph for hmm([a,b,a])

5.2 Computing the Observation Probability and the Most Likely Explanation

Given a support graph for G, an efficient algorithm for computing $P_{DB}(G \mid \boldsymbol{\theta})$ (the first task) is derived based on the analogy of the inside probabilities in Baker's Inside-Outside algorithm [2]. In our formulation, the inside probability of a table atom τ (sometimes called the *generalized inside probabilitiy* of τ) is $P_{DB}(\tau \mid \boldsymbol{\theta})$. Recall that the computation of $P_{DB}(G \mid \boldsymbol{\theta})$ by Equation (2) completely ignores the fact that $P_F(E)$ and $P_F(E')$ $(E \neq E')$ may have common computations, and hence always takes time proportional to the number of explanations in $\psi_{DB}(G)$.

The use of the support graph $\tau_{DB}(G) = \langle \tau_0, \tau_1, \ldots, \tau_K \rangle$ $(\tau_0 = G)$ enables us to factor out common computations. First note that distribution semantics ensures that $P_{DB}(\tau_k) = \sum_{E_{k,h} \in \tilde{\psi}_{DB}(\tau_k)} P_{DB}(E_{k,h})$ holds for every k $(0 \leq k \leq K)$ [33, 38]. Consequently from the support graph

$$\tau_{DB_{\text{hmm}}}(\text{hmm}([a,b,a])) = \langle \text{hmm}([a,b,a]), \text{hmm}(1,s0,[a,b,a]), \ldots, \text{hmm}(4,s1,[]) \rangle$$

in Figure 7, we have

$$\begin{cases} P(\text{hmm}([a,b,a])) \\ \quad = \theta_{(\text{init},s0)} P(\text{hmm}(1,s0,[a,b,a])) + \theta_{(\text{init},s1)} P(\text{hmm}(1,s1,[a,b,a])) \\ P(\text{hmm}(1,s0,[a,b,a])) \\ \quad = \theta_{(\text{out}(s0),a)} \theta_{(\text{tr}(s0),s0)} P(\text{hmm}(2,s0,[b,a])) \\ \quad\quad + \theta_{(\text{out}(s0),a)} \theta_{(\text{tr}(s0),s1)} P(\text{hmm}(2,s1,[b,a])) \\ \cdots \\ P(\text{hmm}(4,s1,[])) = 1 \end{cases}$$

Here $P(\cdot) = P_{DB_{\text{hmm}}}(\cdot)$. Second note that by computing inside probabilities sequentially from the bottom table atom hmm(4,s1,[]) to the top table atom hmm([a,b,a]), we can obtain $P(\text{hmm}(1,s1,[a,b,a]))$ in time proportional to the size of the support graph which is $O(N^2 L)$, not the number of all explanations $O(N^L)$, where N is the number of states and L the length of an input string.

The program GET-INSIDE-PROBS below generalizes this observation. It takes as input a support graph $\tau_{DB}(G) = \langle \tau_0, \tau_1, \ldots, \tau_K \rangle$ ($\tau_0 = G$) for a goal G and computes inside probabilities $P_{DB}(\tau_k \mid \boldsymbol{\theta})$ starting from the bottom atom τ_K whose tabled explanations only contain msw atoms and ending at the top goal $\tau_0 = G$. In the program, a tabled explanation $E \in \tilde{\psi}_{DB}(\tau_k)$ is considered as a set and $\mathcal{P}[\cdot]$ is an array storing inside probabilities. It should be noted that, when computing $\mathcal{P}[\tau_k]$, the inside probabilities $\mathcal{P}[\tau_K]$, $\mathcal{P}[\tau_{K-1}]$, $\ldots$, $\mathcal{P}[\tau_{k+1}]$ have already been computed. The computation terminates leaving the inside probability of G in $\mathcal{P}[\tau_0](= \mathcal{P}[G])$.

1: **procedure** GET-INSIDE-PROBS$(\tau_{DB}(G))$ **begin**
2:    **for** $k := K$ **downto** 0 **do**
3:      $\mathcal{P}[\tau_k] := \sum_{E \in \tilde{\psi}_{DB}(\tau_k)} \prod_{\mathrm{msw}(i, \cdot, v) \in E} \theta_{i,v} \prod_{\tau \in E \cap \{\tau_{k+1}, \ldots, \tau_K\}} \mathcal{P}[\tau]$
4: **end**.

Similarly, an efficient algorithm for computing the most likely explanation E^* for G (i.e. the second task) is derived. The algorithm GET-ML-EXPL below first computes $\delta[\tau_k]$, the maximum probability of tabled-explanations for each table atom $\tau_k \in \tau_{DB}(G)$. $\mathcal{E}[\tau_k]$, the most likely tabled-explanation for τ_k, is simultaneously constructed. Finally, we construct E^* from $\mathcal{E}[\cdot]$.

1: **procedure** GET-ML-EXPL$(\tau_{DB}(G))$ **begin**
2:    **for** $k := K$ **downto** 0 **do begin**
3:      **foreach** $E \in \tilde{\psi}_{DB}(\tau_k)$ **do**
4:        $\delta'[\tau_k, E] :=$
5:          $\prod_{\mathrm{msw}(i, \cdot, v) \in E} \theta_{i,v} \prod_{\tau \in E \cap \{\tau_{k+1}, \ldots, \tau_K\}} \delta[\tau];$
6:      $\delta[\tau_k] := \max_{E \in \tilde{\psi}_{DB}(\tau_k)} \delta'[\tau_k, E];$
7:      $\mathcal{E}[\tau_k] := \arg\max_{E \in \tilde{\psi}_{DB}(\tau_k)} \delta'[\tau_k, E]$
8:    **end**;
9:    $s := \{G\}$; $E^* := \emptyset$;
10:   **while** $s \neq \emptyset$ **do begin**
11:     Select and remove A from s;
12:     **if** $A = \mathrm{msw}(\cdot, \cdot, \cdot)$ **then** add A to E^*
13:     **else** $s := s \cup \mathcal{E}[A]$
14:   **end**
15: **end**

We remark that GET-ML-EXPL is equivalent to the Viterbi algorithm, a standard algorithm for finding the most likely state-transition path of HMMs [22, 32]. Furthermore, in case of PCFGs (probabilistic context-free grammars), it is easily shown that a probabilistic parser for a PCFG in PRISM combined with GET-ML-EXPL can find the most likely parse of a given sentence.

5.3   Graphical EM Algorithm

The third task, i.e. the parameter learning of PRISM programs means ML estimation of statistical parameters $\boldsymbol{\theta}$ associated with msws in a program DB, for which the EM algorithm is appropriate [23]. A new EM algorithm named *graphical EM algorithm* that runs on support graphs has been derived based on the analogy of computation of the *outside probabilities* in the Inside-Outside algorithm [18, 37]. We added an assumption that all observable atoms are exclusive to each other and their probabilities sum up to one to guarantee the mathematical correctness of the algorithm (we say DB satisfies the *uniqueness condition*).

Although details of the graphical EM algorithm are left to [18, 37], we give a brief account. The algorithm takes as input a set of support graphs $\{\tau_{DB}(G_1), \ldots, \tau_{DB}(G_T)\}$ generated from a random sample of goals $\langle G_1, \ldots, G_T \rangle$. It first initializes the parameters $\boldsymbol{\theta}$ randomly and then iterates the simultaneous update of $\boldsymbol{\theta}$ by executing the E(expectation) step followed by the M(aximization) step until the likelihood of the observed goals saturates. Final values $\boldsymbol{\theta}^*$ become the learned ones that locally maximize $\prod_{t=1}^{T} P_{DB}(G_t \mid \boldsymbol{\theta})$. The crux in the graphical EM algorithm is the E step, i.e. the computation of the expected counts of occurrences of $\mathtt{msw}(i, \cdot, v)$ in a proof of the top goal G:

$$\sum_{E \in \psi_{DB}(G)} \sigma_{i,v}(E) P_F(E \mid G, \boldsymbol{\theta})$$

to update a parameter $\theta_{i,v}$ associated with $\mathtt{msw}(i, \cdot, v)$. Naive computation of the expected counts as above causes computation time to become proportional to the number of explanations for G, which must be avoided. We compute it indirectly using the *generalized outside probabilities* which are recursively (and efficiently) computed from the support graph $\tau_{DB}(G) = \langle \tau_0, \tau_1, \ldots, \tau_K \rangle$ for G like the inside probabilities but from τ_0 to τ_K. Update of parameters per iteration completes by scanning the support graph twice, once for inside probabilities and once for outside probabilities, and hence update time is linear in the size of $\tau_{DB}(G)$ [18, 37].

5.4   Complexity

In this subsection, we first analyze the time complexity of our methods for the first and the second task. The method for the first (resp. the second) task comprises two phases – OLDT search to generate a support graph for a goal and a subsequent computation by GET-INSIDE-PROBS (resp. by GET-ML-EXPL). Hence, we should estimate each phase separately. First assuming that table access can be performed in $O(1)$ time,[15] the computation time of OLDT search is measured by the size of the search tree which depends on a class of models.

---

[15]   In reality $O(1)$ can be subtle. The worst case should be $O(\log n)$ for n data by using balanced trees [8]. In the case of PCFGs, n is $N^3 L^3$ with N non-terminals for a sentence of length L, so $O(\log n) = O(\log \max\{N, L\})$. On the other hand, if we consider the abundance of memory available nowadays, it seems technically and

As for the computation time of GET-INSIDE-PROBS and GET-ML-EXPL, since both algorithms scan a support graph $\tau_{DB}(G)$ only once, it is linear in the size of $\tau_{DB}(G)$, or $O(\xi_{num}\xi_{maxsize})$ in notation where $\xi_{num} \overset{def}{=} |\Delta|$, $\xi_{maxsize} \overset{def}{=} \max_{E \in \Delta} |E|$, and $\Delta \overset{def}{=} \bigcup_{\tau \in \tau_{DB}(G)} \tilde{\psi}_{DB}(\tau)$.

Now we examine concrete models. For an HMM program like DB_{hmm} in Section 4, OLDT time and the size of a support graph are both $O(N^2L)$ where N is the number of states and L the length of an input string. This is because for a ground top-goal $hmm([w_1, \ldots, w_L])$ (we are thinking of DB_{hmm}), there are at most NL goal patterns of table atom $hmm(t,s,l)$ during the execution. Each goal causes N recursive calls in the body of clause (7) in DB_{hmm}. Thanks to OLDT search, each table atom is computed once (we assume in the programming, the arguments t and s in $hmm(t,s,l)$ are numbers and $O(1)$ table access is available). Therefore the size of the search tree is $O(N^2L)$ and so is the search time for all solutions. Also as each tabled-explanation is a conjunction of at most three atoms (see Figure 7), we conclude that $\xi_{num} = O(N^2L)$ and $\xi_{maxsize} = O(1)$. Hence, the time complexity of GET-INSIDE-PROBS and GET-ML-EXPL for HMMs becomes $O(N^2L)$. This is the same order as that of the forward procedure and the Viterbi algorithm. So GET-INSIDE-PROBS (resp. GET-ML-EXPL) is a generalization of the forward procedure (resp. the Viterbi algorithm).

For PCFGs, we assume grammars are in Chomsky normal form. Then it is shown that the time complexity of OLDT search is $O(M^3L^3)$ and so is the size of a support graph (see [18]), and hence GET-INSIDE-PROBS and GET-ML-EXPL run in time $O(M^3L^3)$. Here M is the number of non-terminals in the grammar, L the length of an input sentence.

Compared to HMMs and PCFGs, Bayesian networks present harder problems as computing marginal probabilities in a Bayesian network is NP-hard. So we focus on the sub-class, singly connected Bayesian networks [4, 28], though expressing general Bayesian networks by PRISM programs is straightforward [34]. By writing an appropriate PRISM program for a singly connected Bayesian network which has a clause corresponding to each node in the singly connected network, it is relatively easy to show that OLDT time for GET-INSIDE-PROBS and GET-ML-EXPL is linear in the number of nodes in the network [38]. We here assumed that the maximum number of parent nodes is fixed.

Our method for the third task (EM learning) comprises OLDT search and the graphical EM algorithm. For the latter, time complexity is measured by the re-estimation time per iteration (since we do not know how many times it iterates until convergence in advance). It is shown, analogously to GET-INSIDE-PROBS and GET-ML-EXPL however, to be $O(\xi_{num}\xi_{maxsize})$ for one goal. The reader is referred to [18, 37, 38] for details.

---

economically reasonable to employ an array in order to ensure $O(1)$ data access time, as has been traditionally assumed in parsing algorithms. Also we note that hashing achieves average $O(1)$ data access time under a certain assumption [8].

Model	OLDT time	GIP/GMLE	GEM
HMMs	$O(N^2 L)$	$O(N^2 L)$	$O(N^2 L T)$
PCFGs	$O(M^3 L^3)$	$O(M^3 L^3)$	$O(M^3 L^3 T)$
sc-BNs	$O(|V|)$	$O(|V|)$	$O(|V|T)$

Table 1. Time complexity for the three computational tasks

We summarize computation time w.r.t. popular symbolic-statistical models in Table 1. In the table, the second column "OLDT time" indicates that computation time of OLDT search (assuming $O(1)$ table access) for the models in the first column. The third column "GIP/GMLE" means the time complexity of GET-INSIDE-PROBS (the first task) and GET-ML-EXPL (the second task) respectively corresponding to the model in the first column. The fourth column "GEM" is the time complexity of (one iteration of) the graphical EM algorithm (the third task, parameter estimation by EM learning). N, L, M, $|V|$ and T are respectively the number of states of the target HMM, the maximum length of input strings, the number of non-terminal symbols in the target PCFG, the number of nodes in the target singly connected Bayesian network and the size of training data. In statistical abduction with OLDT search, time complexity for each of the three computational tasks is the sum of OLDT time and the subsequent probability computations which is linear in the total size of support graphs.

Table 1 exemplifies that our general framework can subsume specific algorithms in terms of time complexity. For HMMs, $O(N^2 L)$ is the time complexity of the forward algorithm, the Viterbi algorithm and one iteration of the Baum-Welch algorithm [32, 22]. $O(M^3 L^3)$ is the time complexity of one iteration of the Inside-outside algorithm for PCFGs [2, 22]. $O(|V|)$ is the time complexity of a standard algorithm for computing marginal probabilities in singly connected Bayesian networks [28, 4] and that of one iteration of the EM algorithm for singly connected Bayesian networks [4].

6   Conclusion

We have proposed statistical abduction as a combination of abductive logic programming and a distribution over abducibles. It has first-order expressive power and integrates current most powerful probabilistic knowledge representation frameworks such as HMMs, PCFGs and (singly connected) Bayesian networks. Besides, thanks to a new data structure, support graphs which are generated from OLDT search, our general algorithms developed for the three computational tasks (probability computation, the search for the most likely explanation, and EM learning) accomplish the same efficiency as specialized algorithms for above three frameworks. On top of that, recent learning experiments with PCFGs by the graphical EM algorithm using two Japanese corpora of moderate size[16] suggest that the graphical EM algorithm can run much (orders of magnitude) faster than the Inside-Outside algorithm [37, 39].

There remains however a lot to be done including finishing the implementation of OLDT search and the proposed algorithms in PRISM and the development of various applications of statistical abduction. Also a theoretical extension to programs containing negation is an important research topic.

References

[1] Bacchus, F., Using First-Order Probability Logic for the Construction of Bayesian Networks, Proc. of UAI'93, pp219-226, 1993.
[2] Baker,J.K., Trainable Grammars for Speech Recognition, Proc. of Spring Conference of the Acoustical Society of America, pp547-550, 1979.
[3] Breese,J.S., Construction of Belief and Decision Networks, J. of Computational Intelligence, Vol.8 No.4 pp624-647, 1992.
[4] Castillo,E., Gutierrez,J.M., and Hadi,A.S., Expert Systems and Probabilistic Network Models, Springer-Verlag, 1997.
[5] Charniak,E., A neat theory of marker passing, Proc. of AAAI'86, pp584-588, 1986.
[6] Charniak,E., Statistical Language Learning, The MIT Press, 1993.
[7] Clark, K., Negation as failure, In Gallaire, H., and Minker, J. (eds), Logic and Databases, pp293-322, Plenum Press, 1978.
[8] Cormen,T.H., Leiserson,C. E. and Rivest,R.L., Introduction to Algorithms, MIT Press, 1990.
[9] Csinger,A., Booth,K.S. and Poole,D., AI Meets Authoring: User Models for Intelligent Multimedia, Artificial Intelligence Review 8, pp447-468, 1995.
[10] Cussens,J., Loglinear models for first-order probabilistic reasoning, Proc. of UAI'99, pp126-133, 1999.
[11] Cussens,J., Parameter estimation in stochasitc logic programs, Machine Learning 44, pp245-271, 2001.
[12] Dekhtyar,A.and Subrahmanian,V.S., Hybrid Probabilistic Programs, Proc. of ICLP'97, pp391-405, 1997.
[13] Doets,K., From Logic to Logic Programming, MIT Press, Cambridge, 1994.
[14] Eshghi,K. Abductive Planning with Event Calculus, Proc. of ILCP'88, pp562-579, 1988.
[15] Hobbs,J.R., Stickel,M.E., Appelt,D.E. and Martin,P., Interpretation as abduction, Artificial Intelligence 63, pp69-142, 1993.
[16] Kakas,A.C., Kowalski,R.A. and Toni,F., Abductive Logic Programming, J. Logic Computation, Vol.2 No.6, pp719-770, 1992.
[17] Kakas,A.C., Kowalski,R.A. and Toni,F., The role of abduction in logic programming, Handbook of Logic in Artificial Intelligence and Logic Programming, Oxford University Press, pp235-324, 1998.
[18] Kameya,Y. and Sato,T., Efficient EM learning with tabulation for parameterized logic programs, Proc. of CL2000, LNAI 1861, Springer-Verlag, pp269-284, 2000.

---

[16] One corpus contains 9,900 sentences. It has a very ambiguous grammar (2,687 rules), generating 3.0×10^8 parses/sentence at the average sentence length 20. The other corpus consists of 10,995 sentences, and has a much less ambiguous grammar (860 rules) that generates 958 parses/sentence.

[19] Koller,D. and Pfeffer,A., Learning probabilities for noisy first-order rules, Proc. of IJCAI'97, Nagoya, pp1316-1321, 1997.

[20] Koller,D., McAllester,D. and Pfeffer,A., Effective Bayesian Inference for Stochastic Programs, Proc. of AAAI'97, Rhode Island, pp740-747, 1997.

[21] Lakshmanan,L.V.S. and Sadri,F., Probabilistic Deductive Databases, Proc. of ILPS'94 pp254-268, 1994.

[22] Manning, C. D. and Schütze, H., Foundations of Statistical Natural Language Processing, The MIT Press, 1999.

[23] McLachlan, G. J. and Krishnan, T., The EM Algorithm and Extensions, Wiley Interscience, 1997.

[24] Muggleton,S., Stochastic Logic Programs, in Advances in Inductive Logic Programming (Raedt,L.De ed.) OSP Press, pp254-264, 1996.

[25] Ng,R. and Subrahmanian,V.S., Probabilistic Logic Programming, Information and Computation 101, pp150-201, 1992.

[26] Ngo,L. and Haddawy,P., Answering Queries from Context-Sensitive Probabilistic Knowledge Bases, Theoretical Computer Science 171, pp147-177, 1997.

[27] Nilsson,N.J., Probabilistic Logic, Artificial Intelligence 28, pp71-87, 1986.

[28] Pearl,J., Probabilistic Reasoning in Intelligent Systems, Morgan Kaufmann, 1988.

[29] Pfeffer,A., IBAL:A Probabilistic Programming Language, Proc. of IJCAI'01, pp733-740, 2001.

[30] Poole,D., Goebel,R. and Aleliunas,R., Theorist: a logical reasoning system for default and diagnosis, In Cercone,N., and McCalla., eds., The Knowledge Frontier, Springer, pp331-352, 1987.

[31] Poole,D., Probabilistic Horn abduction and Bayesian networks, Artificial Intelligence 64, pp81-129, 1993.

[32] Rabiner, L. and Juang, B. *Foundations of Speech Recognition*, Prentice-Hall, 1993.

[33] Sato,T., A Statistical Learning Method for Logic Programs with Distribution Semantics, Proc. of ICLP'95, pp715-729, 1995.

[34] Sato,T. and Kameya,Y., PRISM:A Language for Symbolic-Statistical Modeling, Proc. of IJCAI'97, pp1330-1335, 1997.

[35] Sato,T., Modeling Scientific Theories as PRISM Programs, ECAI Workshop on Machine Discovery, pp37-45, 1998.

[36] Sato,T., Parameterized Logic Programs where Computing Meets Learning, Proc. of FLOPS2001, LNCS 2024, 2001, pp40-60.

[37] Sato,T., Kameya,Y., Abe,S. and Shirai,K., Fast EM learning of a Family of PCFGs, Titech Technical Report (Dept. of CS) TR01-0006, Tokyo Institute of Technology, 2001.

[38] Sato,T. and Kameya, Y., Parameter Learning of Logic Programs for Symbolic-statistical Modeling, submitted for publication.

[39] Sato,T., Abe,S., Kameya,Y. and Shirai,K., A Separate-and-Learn Approach to EM Learning of PCFGs, Proc. of NLPRS2001, Tokyo, 2001.

[40] Sakama,T. and Inoue,K., Representing Priorities in Logic Programs, Proc. of JICSLP'96, MIT Press, pp82-96, 1996.

[41] Shanahan,M., Prediction is Deduction but Explanation is Abduction, Proc. of IJCAI'89, pp1055-1060,1989.

[42] Tamaki, H. and Sato, T., OLD resolution with tabulation, Proc. of ICLP'86, LNCS 225, pp84-98, 1986.

[43] Wetherell,C.S., Probabilistic Languages: A Review and Some Open Questions, Computing Surveys, Vol.12,No.4, pp361-379, 1980.

[44] White,H.C., An Anatomy of Kinship, Prentice-Hall INC., 1963.

Logicism and the Development of Computer Science

Donald Gillies

Department of Philosophy,
King's College,
London, UK

Abstract. This paper argues for the thesis that ideas originating in the philosophy of mathematics have proved very helpful for the development of computer science. In particular, logicism, the view that mathematics can be reduced to logic, was developed by Frege and Russell, long before computers were invented, and yet many of the ideas of logicism have been central to computer science. The paper attempts to explain how this serendipity came about. It also applies Wittgenstein's later theory of meaning to human-computer interaction, and draws the conclusion that computers do understand the meaning of the symbols they process. The formal language of logic is suitable for humans trying to communicate with computers.

1 Introduction

Philosophy is often thought of as an activity, which may have considerable theoretical interest, but which is of little practical importance. Such a view of philosophy is, in my opinion, profoundly mistaken. On the contrary, I would claim that philosophical ideas and some kind of philosophical orientation are necessary for many quite practical activities. Bob Kowalski's researches are an excellent example of this thesis, since they have been characterised by an explicit and productive use of philosophical ideas. His work, therefore, naturally suggests looking at the general question of how far philosophy has influenced the development of computer science. My own view is that the influence of philosophy on computer science has been very great. In the first three or four decades of the computer, this influence came mainly from earlier work in the philosophy of mathematics. In the last two decades, however, there has been an increasing influence of ideas from the philosophy of science, particularly ideas connected with probability, induction, and causality. In this paper, however, I will focus on the philosophy of mathematics. In section 2, I will give a brief sketch of the development of philosophy of mathematics during the so-called 'foundational' period (c. 1879 - 1939). This period saw the emergence of three main schools: logicism, formalism, and intuitionism. As a matter of fact, all three subsequently influenced the development of computer science, but in this paper I will concentrate on logicism, partly for reasons of space, and partly because it is the philosophical position most relevant to Kowalski's work. Section 3 therefore is devoted to logicism and

A.C. Kakas, F. Sadri (Eds.): Computat. Logic (Kowalski Festschrift), LNAI 2408, pp. 588–604, 2002.

computer science, and I try to show two things. First of all that the ideas of logicism were developed (particularly by Frege and Russell) for purely philosophical reasons, and second that these ideas proved very fruitful in computer science. This naturally raises a problem. Why did concepts and theories developed for philosophical motives before computers were even invented, prove so useful in the practice of computing? I will attempt to sketch the beginnings of a possible answer to this question. In section 4, however, I will turn to an influence in the opposite direction. The logic invented by the logicists proved to be useful in computer science, but the application of logic in computer science changed logic in many ways. In section 4, therefore, I will examine some of the ways in which applications in computing have changed the nature of logic. Section 5 closes the paper by considering some ideas of Wittgenstein. During his later ('ordinary language') period, which began around 1930, Wittgenstein developed a criticism of logicism. I am very far from accepting this criticism in its entirety, but it does raise some interesting points. In particular, in conjunction with some of Wittgenstein's later ideas on meaning, it suggests some further reasons why formal logic has proved so fruitful in computer science.

2   Philosophy of Mathematics in the Foundational Period

The foundational period in the philosophy of mathematics (c. 1879 - 1939) is characterised by the emergence and development of three different schools, each of which aimed to give a satisfactory foundation for mathematics. These schools were:

(i)   logicism (the view that mathematics is reducible to logic),
(ii)   formalism (mathematics as the study of formal systems),
(iii) intuitionism (mathematics based on the intuitions of the creative
       mathematician).

Logicism was started by Frege. Strictly speaking his aim was not to show that the whole of mathematics was reducible to logic, but only that arithmetic was reducible to logic. Frege adopted a non-logicist, Kantian view of geometry. To accomplish his goal, Frege devised a way of defining number in terms of purely logical notions. The existing Aristotelian logic was not adequate for his purpose. So he devised a new kind of formal logic which he published in his *Begriffsschrift* [7] (literally concept writing) of 1879. This is essentially the same as the formal logic taught today, except that Frege used a curious two dimensional notation, which has been abandoned in favour of the more usual one dimensional manner of writing. Frege then went on to set up a complicated formal system with what were intended to be purely logical axioms, and tried to show that the whole of arithmetic could be logically deduced within this system using his definition of number. The first volume of this formal system [10] took Frege 9 years to complete and it appeared in 1893. By the summer of 1902 , Frege had worked for another 9 years on the second volume, which was nearing completion, and it must have seemed to him that he had successfully completed the project to which he had devoted almost his whole adult life. At this moment, however, disaster struck. Frege received a letter dated 16 June 1902 from a young logician named Bertrand Russell who showed that it was possible to derive a contradiction from what Frege had taken as the basic axioms of logic. This is what is now known as Russell's paradox. Here is an extract from Frege's reply to Russell dated 22 June 1902 ([11], 1902, pp. 127-8):

> 'Your discovery of the contradiction caused me the greatest surprise and, I would almost say, consternation, since it has shaken the basis on which I intended to build arithmetic. It seems, then, ... that my Rule V ... is false ... . I must reflect further on the matter. It is all the more serious since, with the loss of my Rule V, not only the foundations of my arithmetic, but also the sole possible foundations of arithmetic seem to vanish. ... In any case your discovery is very remarkable and will perhaps result in a great advance in logic, unwelcome as it may seem at first glance.'[1]

Russell's discovery of the paradox did not cause Russell to give up logicism. On the contrary, Russell tried to provide logicism with new foundations. He invented what is known as *the theory of types* to resolve his paradox, and, using this theory, he constructed with A.N.Whitehead a new massive system of formal logic in which it was hoped that the whole of mathematics could be derived. When the three huge volumes of this system, known as *Principia Mathematica* [22], were published in 1913, it looked as if the logicist programme had been brought to a successful conclusion. However, once again, this apparent success proved short-lived. In 1931 Kurt Gödel, a logician and member of the Vienna Circle, published his two incompleteness theorems. The first of these, in its modern form, shows that if *Principia Mathematica* is consistent, then there is an arithmetical statement which cannot be proved within the system, but which can be shown to be true by an informal argument outside the system. In effect not all the truths of arithmetic can be derived in *Principia Mathematica* which thus fails in its logicist goal of reducing arithmetic to logic. If *Principia Mathematica* were inconsistent, the situation would be no better - indeed it would be worse. In that event any statement whatever could be proved in the system which would therefore be useless. Gödel showed that the results of his paper applied not just to *Principia Mathematica* but to any similar logicist system. He had thus demonstrated that it was impossible to carry out the logicist programme of Frege and Russell.

Let us now turn to *formalism*. The formalist philosophy of mathematics was developed by the German mathematician David Hilbert. Hilbert took over the concept of formal system from the logicists. The logicists tried to construct a single formal system based on the axioms of logic within which the whole of mathematics (or in Frege's case the whole of arithmetic) could be derived. Hilbert, however, suggested that a different axiomatic formal system could be constructed for each branch of mathematics, e.g. arithmetic, geometry, algebra,

---

[1] For further details of Frege's logicism, and the impact on it of Russell's paradox, see Gillies (1982) [12].

set theory, probability theory, etc. Frege's work had shown that there was a danger of a contradiction appearing in a formal system. To avoid this difficulty, Hilbert suggested that the formal systems of mathematics should be proved to be consistent using only the simple informal methods of finite arithmetic. Unfortunately Gödel's second incompleteness theorem showed that such consistency proofs could not be given for nearly all the significant branches of mathematics. Thus Gödel had shown in a single paper published in 1931 that two of the three major positions in the philosophy of mathematics were untenable.

This leaves us with the last of the three major schools - *intuitionism*. This was not in fact refuted by Gödel's incompleteness theorems, but it had other difficulties which made it unacceptable to most mathematicians. A systematic working out of the idea that mathematics was the intuitive construction of creative mathematicians seemed to indicate that some of the logical laws assumed in standard mathematics, notably the law of the excluded middle, had no proper justification. The intuitionists therefore created a new kind of mathematics not involving the law of the excluded middle and other suspect laws. Unfortunately this new mathematics turned out to be more involved and intricate than standard mathematics, and, as a result, it was rejected by most mathematicians as just too complicated to be acceptable.

The Wall Street crash of 1929 ushered in the depression of the 1930's. One could say that the Gödel crash of 1931 initiated a period of depression in the philosophy of mathematics. The three main schools all appeared to have failed. Not one had carried out its promise of providing a satisfactory foundation for mathematics. Yet fate was preparing an odd turn of events. In the post-War period the ideas of these philosophical programmes turned out, surprisingly, to be of the greatest possible use in the new and rapidly expanding field of computer science. In the next section I will examine how this came about in the case of the logicist programme. For reasons of space I cannot analyse the contributions of all three programmes, and I have chosen to concentrate on logicism, as it is the programme most closely connected to Bob Kowalski's work.

3   Logicism and Computer Science

Let us begin with the predicate calculus introduced by Frege in his *Begriffsschrift* of 1879 which opened the foundational period in the philosophy of mathematics. This has become one of the most commonly used theoretical tools of computer science. One particular area of application is in automated theorem proving. In his 1965 paper [19], Alan Robinson developed a form of the predicate calculus (the clausal form) which was specifically designed for use in computer theorem proving, and which has also proved useful in other applications of logic to computing. At the beginning of his paper, Robinson has an interesting section in which he discusses how a logic designed for use by a computer may differ from one suitable for human use. I will now expound his ideas on this point as they will be very helpful in dealing with the issues raised in the present paper.

Robinson begins by pointing out that in a logic designed for humans, the rules of inference have usually been made very simple. As he says ([19], pp. 23):

'Traditionally, a single step in a deduction has been required, for pragmatic and psychological reasons, to be simple enough, broadly speaking, to be apprehended as correct by a human being in a single intellectual act. No doubt this custom originates in the desire that each single step of a deduction should be indubitable, even though the deduction as a whole may consist of a long chain of such steps. The ultimate conclusion of a deduction, if the deduction is correct, follows logically from the premises used in the deduction; but the human mind may well find the unmediated transition from the premises to the conclusion surprising, hence (psychologically) dubitable. Part of the point, then, of the logical analysis of deductive reasoning has been to reduce complex inferences, which are beyond the capacity of the human mind to grasp as single steps, to chains of simpler inferences, each of which is within the capacity of the human mind to grasp as a single transaction.'

If the logic is to be used by a computer, then the requirement that the rules of inference be simple no longer applies. A rule of inference which requires a great deal of computation for its application poses no problem for a computer, as it would for a human. On the other hand, for computer applications, it might well be desirable to reduce the number of rules of inference as much as possible. If a system has a large number of simple rules of inference, a human endowed with some intuitive skill could see which of these rules would be the appropriate one to employ in a particular situation. A computer, lacking this intuitive skill, might have to try each of the rules of the list in turn before hitting on the appropriate one. So we could say, that a logic for humans could have a large number of simple rules of inference, while a logic for computers would be better with fewer but more complicated rules. In fact Robinson introduced a system with a single rule of inference - the *resolution principle*. As he says ([19], pp. 24):

'When the agent carrying out the application of an inference principle is a modern computing machine, the traditional limitation on the complexity of inference principles is no longer very appropriate. More powerful principles, involving perhaps a much greater amount of combinatorial information-processing for a single application, become a possibility.
In the system described in this paper, one such inference principle is used. It is called the resolution principle, and it is machine-oriented, rather than human-oriented, in the sense of the preceding remarks. The resolution principle is quite powerful, both in the psychological sense that it condones single inferences which are often beyond the ability of the human to grasp (other than discursively), and in the theoretical sense that it alone, as sole inference principle, forms a complete system of first-order logic. ...
The main advantage of the resolution principle lies in the ability to allow

us to avoid one of the major combinatorial obstacles to efficiency which have plagued earlier theorem-proving procedures.'

The important point to note here, and to which we shall return later in the paper, is that, as regards logico-linguistic systems, the requirements of a computer may be very different from those of a human.

Alan Robinson's version of the predicate calculus has indeed been used with great success in automated theorem proving. It also led through the work of Kowalski, and of Colmerauer and his team, to the logic programming language PROLOG (for historical details, see Gillies [13], **4.1**, pp. 72-5). Muggleton's concept of *inductive logic programming*, originated from the idea of inverting Robinson's deductive logic to produce an inductive logic. PROLOG has been an essential tool in the development of Muggleton's approach, which has resulted in some very successful machine learning programs (for some further details see Muggleton [17], and Gillies, [13], **2.4**, pp. 41-44).

The examples just given, and some further examples which will be mentioned below, show that Frege's invention of the predicate calculus provided a useful, perhaps indeed essential, tool for computer science. Yet Frege's motivation was to establish a particular position in the philosophy of mathematics, namely that arithmetic could be reduced to logic. Indeed in the entire body of his published and unpublished writings, Frege makes only one reference to questions of computation. His predecessor Boole had also introduced a system of formal logic, and Jevons, influenced by Babbage, had actually constructed a machine to carry out logical inferences in his own version of Boolean logic. Jevons had the machine constructed by a clockmaker in 1869, and describes it in his paper of 1870 [14]. Frege made a number of comments on these developments in a paper written in 1880-1, although only published after his death. He wrote ([8], 1880-1, pp. 34-5):

'I believe almost all errors made in inference to have their roots in the imperfection of concepts. Boole presupposes logically perfect concepts as ready to hand, and hence the most difficult part of the task as having been already discharged; he can then draw his inferences from the given assumptions by a mechanical process of computation. Stanley Jevons has in fact invented a machine to do this.'

Frege, however, made clear in a passage occurring a little later that he did not greatly approve of these developments. He wrote ([8], 1880-1, pp. 35):

'Boolean formula-language only represents a part of our thinking; our thinking as a whole can never be coped with by a machine or replaced by purely mechanical activity.'

On the whole it seems that Jevons' attempts to mechanise logical inference had only a slight influence on Frege's thinking. So we can say that considerations of computing had almost no influence on Frege's development of the predicate calculus, and yet the predicate calculus has proved a very useful tool for computer science.

Let us now move on from Frege to Russell. Bertrand Russell devised the theory of types in order to produce a new version of the logicist programme (the programme for reducing mathematics to logic) when Frege's earlier version of the programme had been shown to be inconsistent by Russell's discovery of his paradox. Thus Russell's motivation, like Frege's, was to establish a particular position in the philosophy of mathematics (logicism), and there is no evidence that he even considered the possibility of his new theory being applied in computing. Indeed Russell's autobiographical writings show that he was worried about devoting his time to logicism rather than to useful applied mathematics. Thus in his 1959 *My Philosophical Development* [20], he writes of the years immediately following the completion of his first degree (pp. 39):

'I was, however, persuaded that applied mathematics is a worthier study than pure mathematics, because applied mathematics - so, in my Victorian optimism, I supposed - was more likely to further human welfare. I read Clerk Maxwell's *Electricity and Magnetism* carefully, I studied Hertz's *Principles of Mechanics*, and I was delighted when Hertz succeeded in manufacturing electro-magnetic waves.'

Moreover in his autobiography, Russell gives a letter which he wrote to Gilbert Murray in 1902 which contains the following passage ([21], 1967, pp. 163):

'Although I denied it when Leonard Hobhouse said so, philosophy seems to me on the whole a rather hopeless business. I do not know how to state the value that at moments I am inclined to give it. If only one had lived in the days of Spinoza, when systems were still possible ...'

In view of Russell's doubts and guilt feelings, it is quite ironical that his work has turned out to be so useful in computer science.

Russell's theory of types failed in its original purpose of providing a foundation for mathematics. The mathematical community preferred to use the axiomatic set theory developed by Zermelo and others. Indeed type theory is not taught at all in most mathematics departments. The situation is quite different in computer science departments where courses on type theory are a standard part of the syllabus. This is because the theory of types is now a standard tool of computer science.

Let us now examine how Russell's ideas about types came in to computer science. A key link in the chain was Church who worked for some of his time on Russell's programme. Indeed Church's invention of the $\lambda - calculus$ arose out of his attempts to develop the logicist position of Russell and Whitehead (1910-13). Russell and Whitehead had written the class of all x's such that $f(x)$ as $\hat{x}f(x)$. Church wished to develop a calculus which focused on functions rather than classes, and he referred to the function by moving the symbol down to the left of x to produce $\wedge x f(x)$. For typographic reasons it was easier to write this as $\lambda x f(x)$, and so the standard notation of the $\lambda - calculus$ came into being. (cf. [18] Rosser, 1984, pp. 338)

Church had intended his first version of the $\lambda-calculus$ in 1932 [1] to provide a new foundation for logic in the style of Russell and Whitehead. However it turned out to be inconsistent. This was first proved by Kleene and Rosser in 1935 using a variation of the Richard paradox, while Curry in 1942 [4] provided a simpler proof based on Russell's paradox. Despite this set-back the $\lambda-calculus$ could be modified to make it consistent, and turned out to be very useful in computer science. It became the basis of programming languages such as LISP, Miranda, and ML, and indeed is used as a basic tool for the analysis of other programming languages. Functional programming languages such as Miranda and ML are usually typed, and indeed some form of typing is incorporated into most programming languages. It is desirable when specifying a function e.g. $f(x,y)$ to specify also the types of its variables x, y. If this is not done, errors can be produced by substituting something of the wrong type for one of the variables. Type mistakes of this sort will often lead to a nonsensical answer. Of course the type theories used in contemporary computer science are *not* the same as Russell's original type theory, but they are descendants nonetheless of Russell's original system. An important link in the chain was Church's 1940 version [2] of the theory of types which was developed from Russell's theory, and which influenced workers in computer science. Davis sums up the situation very well as follows ([6], 1988, pp. 322):

'Although the role of a hierarchy of types has remained important in the foundations of set theory, strong typing has not. It has turned out that one can function quite well with variables that range over sets of whatever type. So, Russell's ultimate contribution was to programming languages!'

Robinson's ideas about the different requirements of humans and computers regarding logico-linguistic systems help to explain what happened here. A system whose variables may be of a variety of different types is awkward and inconvenient for humans to handle, nor does it really confer any advantages. Humans can easily in most cases avoid making type errors in formulae, since their intuitive grasp of the meaning which the formula is supposed to convey will prevent them from writing down nonsense. The situation is almost exactly the opposite as regards computers. Computers have no problem at all about handling variables belonging to many different types. On the other hand, without the guidance provided by a strictly typed syntax, a computer can easily produce nonsensical formulae, since it lacks any intuitive grasp of the intended meaning of the formula. Once again different systems are suitable for human and computers, so that it is not so inappropriate after all that set theory but not type theory is taught in mathematics departments, and type theory in computer science departments.

I have mentioned so far quite a number of uses of logic in computer science, but in fact there are several more. Logic is a fundamental tool for both program and hardware verification. As regards programming, the influence of logic is not restricted to the specifically logical programming languages such as PROLOG and LISP mentioned above. In fact logic has provided the syntactic core for ordinary programming languages[2]. At an even more fundamental level, the *Begriffsschrift* is the first example of a fully formalised language, and so, in a sense, the precursor of all programming languages[3].

We must now try to tackle the problem which has arisen from the preceding discussion. The research of Frege and Russell was motivated by philosophical considerations, and they were influenced either not at all, or to a negligible extent, by considerations to do with computing. Why then did their work later on prove so useful in computer science?

Before the work of Frege and Russell, mathematics might be described as semi-formal. Of course symbolism was used, but the symbols were embedded in ordinary language. In a typical proof, one line would not in general follow from the previous ones using some simple logical rule of inference. On the contrary, it would often require a skilled mathematician to 'see' that a line followed from the previous ones. Moreover even skilled mathematicians would sometimes 'see' that a line in a proof followed from earlier lines when it did not in fact follow. As a result mistaken proofs were often published, even by eminent mathematicians. Moreover the use of informal language often resulted in ambiguities in the concepts employed, which could create confusions and errors.

Of course mathematics is still done today in this semi-formal style, but Frege, in his quest for certainty, thought that he could improve things by a process of formalisation. Concepts would have to be precisely defined to avoid ambiguities and confusions. The steps in a proof would have to be broken down, so that each individual step involved the application of a simple and obviously correct logical rule. By this process, which Frege thought of as the elimination of anything intuitive, he hoped to eliminate the possibility of error creeping in. As he put it ([9], 1884, p. 2): 'The aim of proof is ... to place the truth of a proposition beyond all doubt ...' It was this approach which led him to develop a formal system of logic, his *Begriffsschrift* (or concept writing), which is equivalent to present day predicate calculus.

It is now easier to see how the methods which Frege used in his search for certainty in mathematics created a system suitable for use in computer science. What Frege was doing was in effect mechanising the process of checking the validity of a proof. If a proof is written out in the characteristic human semi-formal style, then its validity cannot be checked mechanically. One needs a skilled human mathematician to apply his or her intuition to 'see' whether a particular line follows from the previous ones. Once a proof has been formalised, however, it is a purely mechanical matter to check whether the proof is valid using the prescribed set of rules of inference. Thus Frege's work can be seen as replacing the craft skills of a human mathematician with a mechanical process[4].

---

[2] I owe this point to Mark Priestley who is researching into this topic at the moment.

[3] I owe this point to Martin Davis. See [6], 1988, p. 316.

[4] It should be stressed that this is my way of viewing Frege's work, and that Frege himself would not have seen things in this light. (I owe this point to Carlo Cellucci.)

The process of mechanisation in general takes place in something like the following manner. The starting point is handicraft production by skilled artisans. The next step is the division of labour in the workshop in which the production process is broken down into smaller and simpler steps, and an individual worker carries out only one such step instead of the process as a whole. Since the individual steps are now quite simple and straightforward, it becomes possible to get them carried out by machine, and so production is mechanised.

Frege and his successors in the logicist tradition were carrying out an analogous process for mathematics. Mathematical proofs were broken down into simple steps which at a later stage could be carried out by a machine. From a general philosophical point of view, Frege and Russell were engaged in the project of mechanising thought. Since they lived in a society in which material production had been so successfully mechanised and in which there was an ever increasing amount of mental (white collar) labour, this project for mechanising thought was a natural one. Moreover it was equally natural that mathematics should be the area chosen to begin the mechanisation process, since mathematics was already partially formalised, unlike other areas of thought.

These considerations perhaps explain why the philosophy of mathematics has assumed such importance within the philosophy of our time. Naturally as well as the thinkers who have pressed forward with the mechanisation of mathematics, there have been those who have objected to this mechanisation, and stressed the human and intuitive aspects of mathematics. Poincaré, Brouwer, Gödel, the later Wittgenstein, and, more recently, Penrose all belong to this trend. Although this line of thought is in many ways reactionary and of course has not halted the advances of mechanisation, there is nonetheless some truth in it, for, as long as mathematics continues to be done by humans at all, it will evidently retain some intuitive characteristics. This is another reason why the logicists, although they thought they were building a secure foundation for mathematics and rendering its results certain, were in fact creating a form of mathematics suitable for computer science.

4   How Computer Science Has Affected Logic

So far we have examined how logical ideas, originating in the logicist programme for the philosophy of mathematics, proved useful in computer science. However the application of these logical concepts to computer science resulted in changes in the concepts themselves. We will next examine some of these changes. The earlier theoretical work of Robinson, Kowalski and others had been concerned with the problem of adapting ordinary classical first-order logic for the computer. In the course of actually implementing PROLOG it turned out that use had to made not of classical negation, but of a different type of negation called *negation as failure*. This issue was clarified by Clark in his 1978 paper [3], which contains a study of this new type of negation. A logic with negation as failure is just one example of a new type of logic known as *non-monotonic logic*. Non-monotonic logic has been developed by computer scientists since the early 1980's, and is an example of an entirely new kind of logic which was introduced as the result of applying logic to computer science.

PROLOG, because of its negation as failure, turned out to be a non-monotonic logic. We must next examine what is a much more profound change - namely PROLOG's introduction of control into deductive logic. As we shall see, negation as failure is really just one consequence of PROLOG's control elements. We can perhaps most easily introduce the topic of logic and control by comparing a passage from Frege with one from Kowalski. In the conclusion of his 1884 book [9] on *The Foundations of Arithmetic*, Frege claims to have made it probable that his logicist programme can be carried out. He goes on to describe what this means as follows ([9], 1884, §87, pp. 99):

> 'Arithmetic thus becomes simply a development of logic, and every proposition of arithmetic a law of logic, albeit a derivative one. ... calculation becomes deduction.'

Let us compare Frege's statement: 'calculation becomes deduction' with the following statement from Kowalski's (1979) *Logic for Problem Solving* [16], p. 129: 'computation = controlled deduction'. It is clear that Kowalski has added control to Frege's deduction. Let us now try to see what this means.

Suppose we have a PROLOG database (including programs). If the user inputs a query e.g. ? − $p(a)$. (i.e. is p(a) true?), PROLOG will automatically try to construct a proof of p(a) from the database. If it succeeds in proving p(a), the answer will be: 'yes', while, if it fails to prove p(a), the answer will be: 'no' (negation as failure). In order to construct these proofs, PROLOG contains a set of instructions (often called the PROLOG interpreter) for searching systematically through various possibilities. The instructions for carrying out such searches are clearly part of a control system which has been added to the inference procedures of the logic.

One symptom of the addition of control is that logic programs often contain symbols relating to control which would not occur in ordinary classical logic. An example of this is the cut facility, written !. The PROLOG interpreter when conducting its searches automatically backtracks in many situations. In some problems, however, we may not wish the program to carry out so much backtracking which could result in a waste of time, the provision of unnecessary solutions etc. The facility ! controls, in a precise though somewhat complicated way, the amount of backtracking which occurs.

Negation as failure can be defined in terms of !, and another of PROLOG's control elements: fail, a primitive which simply causes the interpreter to fail. A logic program which defines negation as failure is the following:

$$notX \leftarrow X, !, fail.$$
$$notX.$$

The interesting point here is that negation as failure is defined using the control elements !, and fail. Thus PROLOG's non-classical negation arises out of its control elements, and the difference between PROLOG and classical logic regarding negation can be seen as a symptom of the more profound difference that PROLOG introduces control into deductive logic.

I will now argue that these developments in PROLOG are a natural extension of the mechanisation process which gave rise to modern logic in the first place. In the previous section I claimed that the work of Frege and Russell can be seen as a mechanisation of the process of checking the validity of a proof. Still their classical logic leaves the construction of the proof entirely in the hands of the human mathematician who has to use his or her craft skills to carry out the task. PROLOG carries the mechanisation process one stage further by mechanising the construction of proofs. In this respect, then, it goes beyond classical logic, and this is also why PROLOG has to introduce control into logic.

A major theme of this paper so far has been the different conceptual requirements of a computer and of a human mathematician. Further light will be cast on this issue by considering an argument against logicism which Wittgenstein formulated in his later period. This will be the subject of the fifth and final section of the paper.

5   A Criticism of Logicism by Wittgenstein and Its Significance

Wittgenstein began his career in philosophy as a student of Russell's, and his first published book [23], the *Tractatus* of 1921, is full of enthusiasm for Russell's logic. Indeed Wittgenstein claims that the new logic reveals the underlying structure of language. After finishing the Tractatus, Wittgenstein gave up philosophy for about a decade, and engaged in a variety of other activities. He was a village schoolmaster for several years, and also helped with the construction of his sister's mansion in Vienna. Perhaps partly because of these experiences, when he returned to philosophy he developed new views about language which were very different from those of the Tractatus. These were eventually published in 1953, after his death, in the *Philosophical Investigations* [24]. Wittgenstein's later theory is that the meaning of a word is given by its use in a language-game. By a 'language-game' he means some kind of rule-guided social activity in which the use of language plays an essential part. He himself introduces the concept as follows: 'I shall also call the whole, consisting of language and the actions into which it is woven, the "language-game."' ([24], 1953, §7, p. 5). And again: 'Here the term "language-*game*" is meant to bring into prominence the fact that the speaking of language is part of an activity, or of a form of life.' ([24], 1953, §23, p. 11)

Wittgenstein illustrates his concept of language-game by his famous example involving a boss and a worker on a building site. The boss shouts 'slab', for example, and the worker has to fetch a slab. Wittgenstein's point is that the meaning of the word 'slab' is given by its use in the activity carried out by boss and worker.

Wittgenstein also devoted a great deal of thought to the philosophy of mathematics during his later period. His reflections on this subject were eventually published as *Remarks on the Foundations of Mathematics* [25] in 1956, though they were written much earlier. In these remarks, Wittgenstein displays great hostility both to logicism and the use of logic in mathematics. He speaks of 'The disastrous invasion of mathematics by logic.([25], 1956, V-24, p. 281), and of 'The curse of the invasion of mathematics by mathematical logic ...' ([25], 1956, V-46, p. 299)

These harsh words about logic are of course connected with his new views of language and meaning. Wittgenstein now thought that it was absurd to claim that the whole of mathematics could be reduced to a single system such as *Principia Mathematica*. On the contrary mathematics consists of a whole variety (or motley) of techniques carried out in different language-games; as he says: ' ... what we call mathematics is a family of activities with a family of purposes ...' ([25], 1956, V-15, pp. 273). These mathematical language-games are also connected with the language-games of everyday life, as, for example, arithmetic may be used on the building site.

From this point of view, Russell's *Principia Mathematica* does not provide a foundation for mathematics, but is simply a new piece of mathematics, a new mathematical language-game. As Wittgenstein says ([25], 1956, III-4, pp. 146):

> 'But still for small numbers Russell does teach us to add; for then we take the groups of signs in the brackets in at a glance and we can take them as numerals; for example 'xy', 'xyz', 'xyzuv'.
> Thus Russell teaches us a new calculus for reaching 5 from 2 and 3; and that is true even if we say that a logical calculus is only - frills tacked on to the arithmetical calculus.'

In my view this is partly right and partly wrong. I agree with Wittgenstein that mathematical logic is a new mathematical calculus but does not provide a foundation for the rest of mathematics as the logicists thought it would. On the other hand Wittgenstein clearly thought that this new mathematical calculus was useless, and that 'a logical calculus is only - frills tacked on to the arithmetical calculus.' The passages I have quoted from Wittgenstein were written in the period 1939-44, and it was not unreasonable at that time to think that the formal systems produced by the logicists would be useless. Contrary, however, to Wittgenstein's expectations, these same logicist systems turned out to be very useful for computer science. I next want to argue that Wittgenstein's later theory of meaning, with which I largely agree, helps to explain why formal logic has proved valuable in computer science.

Let us return to the example of the boss and the worker on the building site. If the boss shouts 'slab', and the worker fetches a slab, then we can surely say that the worker has understood the meaning of the word 'slab', because he has acted appropriately, or, in Wittgenstein's terminology, has made the right move in the language-game. It is interesting in this context to consider the historical example of the Norman conquest of England. The Normans spoke French and the serfs on the estates which they had conquered spoke English. This must have created difficulties for the Norman overlords in giving orders to their serfs.

Thus the lord might have said: 'Donnez-moi un de vos moutons', while the serf would only have understood: 'Give me one of your sheep'. Now the serfs would have lacked the educational facilities to learn French, and it might indeed have been in their interest to pretend to understand less French than they really did. Thus the Norman overlords must have been forced to learn English to be able to give orders to their serfs. This may perhaps explain why the speaking of French disappeared in England over the centuries, though not before it had modified the English language in many ways. Let us now see how all this might be applied to computers.

Several philosophers have denied that computers can understand language, but, if we adopt Wittgenstein's later theory of meaning, it looks as if they were wrong to do so. In Wittgenstein's example, we have only to replace the worker by a computer. I can certainly give orders to my computer, by, for example, typing in a program. If the computer carries out my instructions, surely it is sensible to say, just as in the human case, that it has understood those instructions. The computer and I are playing a language-game. Both of us are using the symbols involved correctly, and so, by Wittgenstein's criterion, we both understand the meaning of those symbols. In a similar fashion, we can say that dogs understand at least a few words of human language. Thus if my dog performs the appropriate actions when I say: 'sit', 'beg' and 'fetch', we can say that he understands the meaning of these three words. There is, however, a very significant difference between dogs and computers as regards language. Dogs can only understand commands consisting of essentially of one symbol (which may in practice be composed of a few words, e.g. sit down). Grammar is quite beyond them. Computers by contrast are much more finicky about grammar than humans. Humans often speak ungrammatically, and their utterances can usually be understood nonetheless. This applies even to the greatest of writers. Thus Shakespeare in describing the wound which Brutus gave Caesar wrote: 'This was the most unkindest cut of all' (Julius Caesar, act III, scene ii, line 188). Shakespeare's line is surely ungrammatical, and yet it is perfectly comprehensible to us. By contrast my computer has, all too frequently, failed to understand one of my instructions merely because that instruction has contained some trivial syntactical error!

This brings us back to the central theme of the different linguistic requirements of computers and humans. Computers find it easiest to understand very precise formal languages which are difficult for humans. The language which is easiest for computers is machine code which is quite opaque to all but a few highly trained humans. Conversely humans find loose informal natural (for humans) languages very easy to understand, and these cannot be understood at all by computers. This is the point of the analogy with the French-speaking Norman lords, and their English-speaking serfs. We humans are in the position of the Norman lords with regard to our computer serfs. These computers will do wonderful things for us, but we have to give them their orders in a language they can understand. This is a difficult task since computers cannot cope with languages which are easy and natural for us. This is where the language of formal logic has proved to be helpful. This language is intermediate between the machine code which is natural for computers, and an everyday language such as English which is natural for humans. Formal logic has the precise syntax which makes its sentences accessible to computers, while it has sufficient resemblance to ordinary language to be comprehensible to humans after a little training. Even within logic itself, there are, as Robinson pointed out in the passages quoted above in section 3, some formulations which are more suitable for computers and others that are more suitable for humans. Thus the clausal form of logic with its single, but complicated, rule of inference is more suitable for computers, whereas other systems of logic with several, but much simpler, rules of inference are more suitable for humans. In general terms, however, formal logic is a language system somewhat intermediate between those which are most suitable for computers, and those which are most suitable for humans. It is thus very helpful in facilitating human-computer interaction, and this I would see as the fundamental reason why it has proved so useful in computer science.

Frege in the *Begriffsschrift* where he introduces a formal system for logic for the first time explains the differences between his system and ordinary language by means of a striking analogy ([7], 1879, p. 6):

> 'I believe that I can best make the relation of my ideography to ordinary language clear if I compare it to that which the microscope has to the eye. Because of the range of its possible uses and the versatility with which it can adapt to the most diverse circumstances, the eye is far superior to the microscope. Considered as an optical instrument, to be sure, it exhibits many imperfections, which ordinarily remain unnoticed only on account of its intimate connection with our mental life. But, as soon as scientific goals demand greater sharpness of resolution, the eye proves to be insufficient. The microscope, on the other hand, is perfectly suited to precisely such goals, but that is just why it is useless for all others.'

Similarly the language of formal logic is suited to the scientific goal of communicating with computers, since this task demands great precision of expression. It is less suited, however, to the task of communicating with other human beings.

The idea that different languages are suited to different purposes is already to be found in a reputed saying of the multi-lingual emperor Charles V. He is supposed to have said that he found French the most suitable language for talking to men, Italian for women, Spanish for God, and German for horses. If he had lived today, he could have added that the language of formal logic was the most suitable for talking to computers.

Acknowledgements I have been researching into the connections between philosophy and computer science for several years now. The specific focus on philosophy of mathematics arose out of discussions with Yuxin Zheng during his visit to King's College London from April to September 1997. This visit was made possible by Yuxin Zheng's receipt of a British Academy K.C.Wong Fellowship, and a travel grant from the Open Society Institute. I would like to thank the British Academy and the Open Society Institute for the support, which made this collaborative research possible, as well as Yuxin Zheng for many helpful suggestions.

Earlier versions of some of the ideas in this paper were presented at the Annual Conference of the British Society for the Philosophy of Science in September 1998, at the Logic Club, Department of Philosophy, University of California, Berkeley in November 1998, at a conference on Philosophy and Computing at King's College London in February 1999, and at the Applied Logic Colloquium at Queen Mary College London in November 1999. I am very grateful for the comments received on these occasions, and particularly for some points made by Martin Davis at Berkeley, one of which is mentioned in footnote 3.

I would also like to thank a number of computer scientists with whom I discussed this problem and who made many helpful suggestions, which have been incorporated in the paper. These include James Cussens, Mark Gillies, Stephen Muggleton, David Page, and Ashwin Srinivasan.

References

1. Church, A., A Set of Postulates for the Foundation of Logic, Annals of Mathematics, 33, pp. 346-66, 1932.
2. Church, A. A Formulation of the Simple Theory of Types, Journal of Symbolic Logic, 5, pp. 56-68, 1940.
3. Clark, K., Negation as Failure. In H.Gallaire and J. Minker (eds.), Logic and Data Bases, Plenum Press, pp. 293-322, 1978.
4. Curry, H.B., The Inconsistency of Certain Formal Logics, Journal of Symbolic Logic, 7, pp. 115-17, 1942.
5. Davis, M., Mathematical Logic and the Origin of Modern Computing. In Rolf Herken (ed.), The Universal Turing Machine. A Half-Century Survey, Oxford University Press, pp. 149-74, 1988.
6. Davis, M., Influences of Mathematical Logic on Computer Science. In Rolf Herken (ed.), The Universal Turing Machine. A Half-Century Survey, Oxford University Press, pp. 315-26, 1988.
7. Frege, G. Begriffsschrift, Eine der arithmetischen nachgebildete Formelsprache des reinen Denkens, English translation in Jean van Heijenoort (ed.), From Frege to Gödel: A Source Book in Mathematical Logic, 1879-1931, Harvard University Press, pp. 1-82, 1967.
8. Frege, G., Boole's Logical Calculus and the Concept-Script. English translation in Gottlob Frege: Posthumous Writings, Blackwell, pp. 9-52.
9. Frege, G., The Foundations of Arithmetic: A Logico-Mathematical Enquiry into the Concept of Number. English translation by J.L.Austin, Blackwell, 1968.
10. Frege, G., Grundgesetze der Arithmetik, Begriffsschriftlich abgeleitet. Vol. I. (1893) and Vol. II. (1903). Reprinted by G.Olms, 1962.
11. Frege, G., Letter to Russell. English translation in J. van Heijenoort (ed.) From Frege to Gödel, Harvard University Press, pp. 127-8, 1967.
12. Gillies, D.A., Frege, Dedekind, and Peano on the Foundations of Arithemtic, Van Gorcum, 1982.

13. Gillies, D.A., Artificial Intelligence and Scientific Method, Oxford University Press, 1996.
14. Jevons, W.S., On the Mechanical Performance of Logical Inference, Philosophical Transactions of the Royal Society, 160, pp. 497-518.
15. Kleene, S.C. and Rosser, J.B., The Inconsistency of Certain Formal Logics, Annals of Mathematics, 36, pp. 630-36, 1935.
16. Kowalski, R.A., Logic for Problem Solving, North-Holland, 1979.
17. Muggleton, S. (ed.), Inductive Logic Programming, Academic Press, 1992.
18. Rosser, J.B., Highlights of the History of the Lambda-Calculus, Annals of the History of Computing, 6(4), pp. 337-9, 1984.
19. Robinson, J.A., A Machine-Oriented Logic Based on the Resolution Principle, Journal for the Association for Computing Machinery, 12, pp. 23-41, 1965.
20. Russell, B., My Philosophical Development, George Allen and Unwin, 1959.
21. Russell, B., Autobiography. Volume 1, George Allen and Unwin, 1967.
22. Russell, B., and Whitehead, A.N., Principia Mathematica, Cambridge University Press, 1910-13.
23. Wittgenstein, L., Tractatus Logico-Philosophicus. English translation by D.F.Pears and B.F.McGuinness, Routledge and Kegan Paul, 1963.
24. Wittgenstein, L., Philosophical Investigations. English translation by G.E.M.Anscombe, Blackwell, 1967.
25. Wittgenstein, L., Remarks on the Foundations of Mathematics. English translation by G.E.M.Anscombe, Blackwell, 3rd edition, revised and reset, 1978.

Simply the Best: A Case for Abduction*

Stathis Psillos

Department of Philosophy and History of Science, University of Athens,
37 John Kennedy Str. 16121 Athens, Greece.

Abstract. This paper formulates what I think is *the basic problem* of any
attempt to characterise the abstract structure of scientific method, viz., that it
has to satisfy two conflicting desiderata: it should be ampliative (content-
increasing) and it should confer epistemic warrant on its outcomes. Then, after
two extreme solutions to the problem of the method, viz., Enumerative
Induction and the Method of Hypothesis, are examined, the paper argues that
abduction, suitably understood as Inference to the Best Explanation, offers the
best description of scientific method and solves the foregoing problem in the
best way: it strikes the best balance between ampliation and epistemic warrant.

1 Introduction

In the last decade there has been a lot of work on abduction, both among philosophers
and researchers in Artificial Intelligence (AI). Philosophers have mostly tried to
unravel the conceptual problems that this mode of reasoning faces[1], whereas workers
in AI have looked into its computational modelling.[2] Pioneering among the
researchers in AI has been Bob Kowalski. Together with his collaborators, Kowalski
has attempted to offer a systematic treatment of both the syntax and the semantic of
abduction, with an eye to how Logic Programming can offer the appropriate
framework to deal with these issues. It is this primarily theoretical work that will be, I
think, the lasting influence of Kowalski's work on our thinking about abduction. In

---

* This essay is dedicated to Bob Kowalski for his very generous help and the long time we
spent in London discussing about philosophy of science and Artificial Intelligence. His
inquisitive mind and sharp criticism made me think harder about the philosophical problems of
abduction. Many thanks are due to two anonymous readers for this volume and John Norton for
useful comments. An earlier version of this paper was presented at the NORDPLUS Intensive
Programme on Inference to the Best Explanation in Iceland. Comments made by Jan Faye,
Olav Gjelsvik, Mikael Karlsson and Bengt Hansson were particularly useful.

[1] Some recent philosophical work includes [2], [6], and [36]. A fresh approach to abduction
has been presented in [11] where Fodor uses the very fact that reasoners employ abduction to
raise some important worries against computational theories of mind.

[2] For appraisals of the recent work on abduction in AI, see [1], [25], [28], [37], [49] and the
papers in [10].

A.C. Kakas, F. Sadri (Eds.): Computat. Logic (Kowalski Festschrift), LNAI 2408, pp. 605-625, 2002.
□ Springer-Verlag Berlin Heidelberg 2002

particular, Kowalski saw very clearly that a number of tangles in the foundations of AI could be dealt with successfully by taking abduction seriously and by incorporating it within AI.[3] In this article, however, my aim is not to deal with the philosophical implications and the possible problems of the analysis of abduction within Logic Programming. I have tried to do this in [42], which can usefully be seen as a companion to the present article. Instead, in this paper I will do two things. First, I shall formulate what I think is *the basic problem* of any attempt to characterise the abstract structure of scientific method, viz., that it has to satisfy two conflicting desiderata: it should be ampliative (content-increasing) and it should confer epistemic warrant on its outcomes (cf. [13], [41]). Second, and after I have examined two extreme solutions to the problem of the method, viz., Enumerative Induction and the Method of Hypothesis, I will try to show that abduction, suitably understood as Inference to the Best Explanation (henceforth, IBE), offers the best description of scientific method and solves the foregoing problem in the best way: it strikes the best balance between ampliation and epistemic warrant. So, the paper to follow will aim to offer a philosophical vindication of the recent interest in abduction among researchers in AI.

The general framework I will follow is John Pollock's [38] analysis of defeasible reasoning in terms of the presence or absence of *defeaters*. This framework makes possible to investigate the conditions under which defeasible reasoning can issue in warranted beliefs. I shall also raise and try to answer some general philosophical questions concerning the epistemic status of abduction.

In what follows, I shall deliberately leave aside all the substantive issues about the nature of explanation.[4] This is partly because they are just too many to be dealt with in this article and partly because I think that--barring some general platitudes about the nature of explanation--my claims about IBE should be neutral vis-à-vis the main theories of explanation.[5] At any rate, I think that the very possibility of Inference to the Best Explanation as a warranted ampliative method must be examined independently of specific models of the explanatory relationship between hypotheses and evidence. Ideally, IBE should be able to accommodate different conceptions of what explanation is. This last thought implies that abduction (that is, IBE) is not usefully seen as a species of ampliative reasoning, but rather as a *genus* whose several species are distinguished by plugging assorted conceptions of explanation in the reasoning schema that constitutes the genus. So, for instance, if the relevant notion of explanation is revealing of causes, then IBE becomes an inference to the best causal explanation. Or, if the relevant notion of explanation is subsumption under laws, then IBE becomes as a kind of inference to the best Deductive-Nomological explanation, and so forth. Given that there is too much disagreement on the notion of explanation, and given that no account offered in the literature so far seems to cover fully all aspects of explanation, it seems to me methodologically useful to treat the reference to explanation in IBE as a 'placeholder' which can be spelled out in different ways in

---

[3] The paper [24] in this volume contains a very useful analysis of Abductive Logic Programming.

[4] These are dealt in detail in my [43].

[5] The relevant literature is really massive. Some important recent items include [27], [30] and [45].

different contexts. Some philosophers may think that this approach to IBE renders it an unnatural agglomeration of many different types of reasoning where explanatory considerations are involved. But I think it is at least premature to call this agglomeration 'unnatural'. After all, as I hope to show in this piece, the general ways in which explanatory considerations can enter into defeasible reasoning can be specified without a prior commitment to the nature of the explanatory relation.

2 Ampliation and Epistemic Warrant

Any attempt to characterise the abstract structure of scientific method should make the method satisfy two general and intuitively compelling desiderata: it should be ampliative and epistemically probative. Ampliation is necessary if the method is to deliver informative hypotheses and theories, viz., hypotheses and theories which exceed in content the observations, data, experimental results and, in general, the experiences which prompt them. This 'content-increasing' aspect of scientific method is indispensable, if science is seen, at least prima facie, as an activity which purports to extend our knowledge (and our understanding) beyond what is observed by means of the senses. But this ampliation would be merely illusory, qua increase of *content*, if the method was not epistemically probative: if, that is, it did not convey epistemic warrant to the excess content produced thus (viz., hypotheses and theories). To say that the method produces--as its output--more information than what there is in its input is one thing. To say that this extra information can reasonably be held to be warranted is quite another. Now, the real problem of the scientific method is that these two plausible desiderata are not jointly satisfiable. Or, to weaken the claim a bit, the problem is that there seems to be good reason to think that they are not jointly satisfiable. The tension between them arises from the fact that ampliation does not carry its epistemically probative character on its sleeves. When ampliation takes place, the output of the method can be false while its input is true. The following question then arises: what makes it the case that the method conveys epistemic warrant to the intended output rather than to any other output which is consistent with the input? Notice that ampliation has precisely the features that deduction lacks. Suppose one thought that a purely deductive method is epistemically probative in the following (conditional) sense: if the input (premises) is warranted, then the method guarantees that the output cannot be less warranted than the input. No ampliative method can be epistemically probative in the above sense. But can there be any other way in which a method can be epistemically probative? If the method is not such that the input excludes all but one output, in what sense does it confer any warrant on a certain output?

'In no sense', is the strong sceptical (Humean) answer. The sceptic points out that any attempt to strike a balance between ampliation and epistemic warrant is futile for the following reason. Given that ampliative methods will fail to satisfy the aforementioned conditional, they will have to base any differential epistemic treatment of outputs which are consistent with the input on some *substantive and contingent assumptions*, (e.g., that the world has a natural-kind structure, or that the world is governed by universal regularities, or that observable phenomena have unobservable causes, etc.). It is these substantive assumptions that will do all the work in conferring epistemic warrant on some output rather than another. But, the sceptic goes on, what else, other than ampliative reasoning itself, can possibly establish that these substantive and contingent assumptions are true of the world? Arguing in a circle, the sceptic notes, is inevitable and this simply means, he concludes, that the alleged balance between ampliation and epistemic warrant carries no rational compulsion with it. In other words, the sceptic capitalises on the fact that in a purely deductive (non-ampliative) method, the transference of the epistemic warrant from the premises to the conclusion is parasitic on their formal (deductive) relationship, whereas in an ampliative method the alleged transference of the epistemic warrant from the premises to the conclusion depends on substantive (and hence challengeable) background beliefs and considerations.[6]

A standard answer to the problem of method is to grant that the sceptic has won. But I think this is too quick. Note that the sceptical challenge is far from intuitively compelling. It itself relies on a substantive *epistemic* assumption: that any defence of an ampliative but epistemically probative method should simply mirror some formal relations between the input and the output of the method and should depend on no substantive and contingent assumptions whose truth cannot be established by independent means. This very assumption is itself subject to criticism.[7] First, if it is accepted, it becomes *a priori* true that there can be no epistemically probative ampliative method. Yet, it may be reasonably argued that the issue of whether or not there can be an ampliative yet epistemically probative method should hinge on information about the actual world and its structure (or, also on information about those possible worlds which have the same nomological structure as the actual). A proof that a method could be both ampliative and epistemically probative in *all* possible worlds (that is, a proof which we have *no* reasons to believe is forthcoming) would certainly show that it can have these features in the actual world. But the very request of such a proof (one that could persuade the sceptic) relies on the substantive assumption that an epistemically probative method should be totally insensitive to the actual features (or structure) of the world. This request is far from compelling. After all, we primarily need our methods to be the right ones for the world we live in. If the range of their effectiveness is larger, then that's a pleasant bonus. But we can live without it. Second, if the sceptical assumption is accepted, even the possibility of epistemically probative demonstrative reasoning becomes dubious. For truth-transmission, even though it is guaranteed by deductive reasoning, requires some truths to start with. Yet, the truth of any substantive claims that feature in the premises of a deductive argument can only be established by ampliative reasoning, and hence it is equally open to the sceptical challenge.[8] Naturally, the point here is not that relations of deductive entailment between some premise P and a conclusion Q fail to offer an epistemic warrant for accepting Q, *if one already warrantedly accepts P*. Rather, the point is that coming to accept as true a premise P with any serious content will typically involve some ampliative reasoning. The sceptical challenge is not

---

[6] Philosophical attempts to offer circular justifications of ampliative modes of reasoning have been analysed in [40, chapter 4] and in [32].

[7] For a rather compelling criticism of the sceptical challenge to induction and of its philosophical presuppositions, see [35].

[8] It might be claimed that some self-evident beliefs are ampliative and yet certain enough to be the deductive foundations of all knowledge. But a) it is contentious whether there are such beliefs; and b) even if there were, they would have to be implausibly rich in content, since deduction cannot create any new content.

incoherent. But if its central assumption is taken seriously, then what is endangered is not just the very possibility of any kind of learning from experience, but also any kind of substantive reasoning.

There is, however, something important in a mild reading of the sceptical answer to the problem of method: if we see it as a challenge to offer a satisfactory account of method which is both ampliative and epistemically probative, then we can at least make some progress in our attempt to understand under what conditions (and under what substantive assumptions) the two desiderata can co-exist.

3 Between Two Extremes

In order to start making this progress, we need to see how the two standard accounts of scientific method fare vis-à-vis the two desiderata. So, we'll look at Enumerative Induction (EI) and crude hypothetico-deductivism (HD) (or, the 'method of hypothesis') and compare them in terms of the strength of ampliation and the strength of the epistemic warrant. But let me first make an important note.

3.1 Defeasibility and Defeaters

The very idea of ampliation implies that the outcome of the application of an ampliative method (or of a mode of ampliative reasoning) can be defeated by new information or evidence. So, unlike deductive methods, ampliative methods are *defeasible*. The issue here is not just that further information can make the output not to logically follow from the input. It is rather that further information can remove the *warrant* for holding the output of the method. So, further information can make the previous input not be strong enough to warrant the output. Following Pollock ([38] chapter 2, section 3; [39]), we can call "prima facie" or "defeasible" any type of reason which is not conclusive (in the sense that it is not deductively linked with the output it is a reason for). Given that ampliative reasoning is defeasible, we can say that such reasoning provides *prima facie warrant* for an output (belief). What Pollock has rightly stressed is that to call a warrant (or a reason) prima facie is not to degrade it, *qua* warrant or reason. Rather, it is to stress that a) it can be defeated by further reasons (or information); and b) its strength, *qua* reason, is a function of the presence or absence of "defeaters". "Defeaters" are the factors (generally, reasons or information) that, when they are taken into account, can remove the prima facie warrant for an outcome (belief). On Pollock's insightful analysis of reasoning and warrant, the presence or absence of defeaters is directly linked with the degree to which one is warranted to hold a certain belief. Suppose that a subject S has a prima facie (nonconclusive) reason R to believe Q. Then S is warranted to believe that Q on the basis of R, *unless* either there are further reasons R' such that, were they to be taken into account, they would lead S to doubt the integrity of R as a reason for Q, or there are strong (independent) reasons to hold not-Q. Generalising this idea to the problem of method, we may say that the presence or absence of defeaters is directly linked with the degree to which an ampliative method can confer epistemic warrant on an outcome, that is, the degree to which it can be epistemically probative. So, to say that S is prima facie warranted to accept the outcome Q of an ampliative method is to say that although it is possible that there are defeaters of the outcome Q, such defeaters are not actual. In particular, it is to say that S has considered several possible defeaters of the reasons offered for this outcome Q and has shown that they are not present. If this is done, we can say that there are no *specific* doubts about the outcome of the method and, that belief in this outcome is prima facie warranted.

This talk of defeaters is not abstract. There are general *types* of defeater that one can consider. Hence, when it comes to considering whether an outcome is warranted, there are certain things to look at such that, if present, they would remove the warrant for the outcome. Even if it is logically possible that there could be considerations that would undercut the warrant for the outcome (a possibility that follows from the very idea of defeasibility), the concrete issue is whether or not there actually are such considerations (actual defeaters).[9] Besides, if the reasoner has done whatever she can to ensure that such defeaters are not present in a particular case, there is a strong sense in which she has done what it can plausibly be demanded of her in order to be epistemically justified. Pollock ([38], 38-39) has identified two general types of defeater: "rebutting" and "undercutting". Suppose, for simplicity, that the ampliative method offers some prima facie reason P for the outcome Q. A factor R is called a rebutting defeater for P as a reason for Q if and only if R is a reason for believing not-Q. And a factor R is called an undercutting defeater for P as a reason for Q if and only if R is a reason for denying that P offers warrant for Q.[10] So, considering whether or not Q is warranted on the basis of P one has to consider whether or not there are rebutting and undercutting defeaters. Taking all this into account, let us look at the two extreme cases of ampliative method.

3.2 Enumerative Induction

Enumerative Induction (EI) is based on the following: if one has observed *n* As being B and *no* As being not-B, and if the evidence is enough and variable, then one should infer that (with high probability) 'All As are B'. The crux of EI is that ampliation is effected by *generalisation*. We observe a pattern among the data (or, among the instances of two attributes), and then generalise it so that it covers all the values of the relevant variables (or all instances of the two attributes). For obvious reasons, we can call EI, the "more-of-the-same" method (cf. [31], 16). The prime advantage of EI is that it is content-increasing in a, so to speak, 'horizontal way': it allows the acceptance of generalisations based on observed evidence in a way that stays close to what has been actually observed. In particular, no new entities (other then those referred to in (descriptions of) the data) are introduced by the ampliation. Let me call this *minimal ampliation*. The basic substantive assumptions involved in this ampliation are that a) there are projectable regularities among the data; and b) the pattern detected among the data (or the observations) in the sample is representative of the pattern (regularity) in the whole relevant population. The prima facie warrant that EI confers on its

---

[9] As Pollock ([38], 39) notes the mere presence of a defeater R' is not enough to remove the prima facie warrant for a belief Q. For, being itself a reason, R' might also be subject to defeaters. Hence, faced with a possible defeater R', we should examine whether R' can itself be (or actually is) defeated by other reasons (what Pollock calls "defeater defeaters").

[10] Pollock frames this in terms of the subjunctive conditional: R is a reason to deny that *P would be* true unless Q *were* true.

outcomes is based on these substantive assumptions. But this warrant--and the assumptions themselves--are subject to evaluation. EI admits of both undercutting and rebutting defeaters. If there are specific reasons to doubt that the pattern among the data can be projected to a lawful regularity in the population, then the projection is not warranted.[11] If there are specific reasons to doubt the fairness of the sample, then the projection is also no longer warranted. Note that although the sample may be unfair (e.g., the sample might involve only ravens in a certain region), the conclusion (viz., that all ravens are black) may well be true. Yet, knowing that the sample was unfair does remove the warrant for the conclusion. These are cases of undercutting defeaters. Besides, EI admits of rebutting defeaters. If we find a negative instance (e.g. a black swan) the warrant (e.g. for the conclusion that all swans are white) is completely removed.[12] So, in EI we know precisely what kind of defeaters can remove the prima facie warrant for making the ampliation (generalisation). And, on very many occasions, we a) can certify the presence or absence of defeaters; and b) we can withhold the conclusion until we have reasons to believe that the potential defeaters are not present (e.g. by making meticulous search for cases which would rebut the conclusion). Given the very specific character of defeaters in EI, and the general feasibility of the search for defeaters, we can say that EI can be *maximally epistemically probative* (among ampliative methods). Here again, the point is not that the sceptic loses. Nor is it that EI *is* maximally epistemically probative. Rather, the point is that if ampliative--and hence defeasible--methods can be warranted at all based on the presence or absence of defeaters, and given that in the case of EI we know exactly what defeaters we should look for and how to do it, EI fares best in terms of how warranted an outcome of a successful (undefeated) application of EI can be.

So, EI is minimally ampliative and maximally epistemically probative. But this is precisely the problem with EI: that what we gain in (epistemic) austerity we lose in strength (of ampliation). EI is too restrictive. It cannot possibly yield any hypothesis about the causes of the phenomena. Nor can it introduce new entities. The basic problem is that the input and the output of EI are couched in the same vocabulary: conclusions that state generalisations are necessarily couched in the vocabulary of the premises. Hence, EI cannot legitimately introduce new vocabulary. Hence, it cannot possibly be used to form ampliative hypotheses that refer to entities whose descriptions go beyond the expressive power of the premises.[13]

---

[11] This is essentially what Goodman [12] observed in his notorious "new riddle of induction".

[12] This may be a bit too strong, since we know that we can always fault the observation. We may, for instance, insist that the observed swan was not really black. Or we may make it part of the meaning of the term 'swan' that all swans are white. On this last move, a black swan cannot really be a swan. But such manoeuvres, though logically impeccable, do not always have the required epistemic force to save the generalisation from refutation. In any case, in EI we know exactly what sort of manoeuvres we have to block in order to render a generalisation rebutted.

[13] Goodman-type stories of the form 'All observed emeralds are green. Therefore, all emeralds are grue' involve a different vocabulary between premises and conclusion only in a trivial way. For predicates such as 'grue' are fully definable in terms of the vocabulary of the premises (plus other antecedently understood vocabulary). So, for instance, 'grue' is defined as: 'green if observed before 2001 and blue thereafter'.

3.3 The Method of Hypothesis

Let us turn to the crude version of the 'method of hypothesis' (HD). This is based on the following: Form a hypothesis H and derive some observational consequences from it. If the consequences are borne out, then the hypothesis is confirmed (accepted). If they are not borne out, then the hypothesis is disconfirmed (rejected). So, the crux of the method is that a hypothesis is warrantedly accepted on the basis of the fact that it entails all available relevant evidence. In HD, ampliation is effected by *confirmation*. An ampliative hypothesis H is accepted because it gets confirmed by the relevant evidence. To be sure, the operation of HD is more complicated. The observational consequences follow from the conjunction of H with some statements of initial conditions, other auxiliary assumptions and some bridge-principles which connect the vocabulary in which H is couched and the vocabulary in which the observational consequences are couched. It is this bridge-principles that make HD quite powerful, since they allow for what I shall call 'vertical extrapolation'--to be contrasted with the 'horizontal extrapolation' characteristic of EI. The content of H may well be much richer than the content of the relevant observational consequences and the deductive link between the two contents is guaranteed by the presence of bridge-principles. The prime attraction of HD is precisely that is can be content-increasing in a, so to speak, 'vertical way': it allows the acceptance of hypotheses about the, typically unobservable, causes of the phenomena. In particular, new entities (other then those referred to in the data) are introduced by the ampliation. So, in contrast to EI, let me call this *maximal ampliation*. The basic substantive assumptions involved in this type of ampliation are that a) there are causally and explanatory relevant entities and regularities behind the observed data or phenomena; and b) the pattern detected among the data (or the observations) is the causal-nomological outcome of entities and processes behind the phenomena. What about the warrant that HD confers on its outcomes? As in the case of EI, we should look at the possible defeaters of the reasons offered by HD for the acceptance of a hypothesis H. The rebutting defeaters seem to be clear-cut: if the predicted observation is not borne out, then--by modus tollens--the hypothesis is refuted. This seems quite compelling, yet there are well-known problems. As we have just seen, it is typically the case that, in applications of HD, the predictions follow from the conjunction of the hypothesis with other auxiliary assumptions and initial and boundary conditions. Hence, when the prediction is *not* borne out, it is the whole cluster of premises that gets refuted. But HD alone cannot tell us how to apportion praise and blame among them. At least one of them is false but the culprit is not specified by HD. It might be that the hypothesis is wrong, or some of the auxiliaries were inappropriate. So, a possible rebutting defeater (a negative prediction) does not carry with it the epistemic force to defeat the hypothesis and hence to remove the warrant for it. (This is a version of the well-known Duhem-Quine problem.) In order for the rebutting defeater to do its job, we need further information, viz., whether the hypothesis is warranted enough to be held on, or whether the auxiliaries are vulnerable to substantive criticism etc. But all these considerations go a lot beyond the deductive link between hypotheses and data that forms the backbone of HD and are not incorporated by the logical structure of HD. What about the undercutting defeaters? Here, it's not clear what these are. It seems a good idea to say that an undercutting defeater for a hypothesis H which does conform to the observations is another hypothesis H* which also conforms to the observations.

For if we know that there is another H*, then it seems that our confidence about H is negatively affected. The prima facie warrant for H (based as it is on the fact that H entails the evidence) may not be totally removed, but our confidence that H is correct will surely be undermined. To put the same point in a different way, if our warrant for H is solely based on the fact that it entails the evidence, then insofar as there is another hypothesis H* which also entails the evidence, H and H* will be equally warranted. It may be that H* entails H, which means that, on probabilistic considerations, H will be at least as probable as H*. But this is a special case. The general case is that H and the alternative hypothesis H* will be mutually inconsistent. Hence, HD will offer no way to discriminate between them in terms of warrant. The existence of each alternative hypothesis will act as an undercutting defeater for the rest of them. Given that, typically, for any H there will be alternative hypotheses which also entail the evidence, HD suffers from the existence of just too many undercutting defeaters. All this can naturally lead us to the conclusion that HD is *minimally epistemically probative*, since it does not have the resources to show how the undercutting defeaters can be removed.[14]

So, HD is maximally ampliative and minimally epistemically probative. But this is precisely the problem with it: that what we gain in strength (of ampliation) we lose in (epistemic) austerity. Unlike EI, it can lead to hypotheses about the causes of the phenomena. And it can introduce new entities. That is, it can also be 'vertically ampliative'. But, also unlike EI, HD is epistemically too permissive. Since there are, typically, more than one (mutually incompatible) hypothesis which entail the very same evidence, if a crude 'method of hypothesis' were to license any of them as probably true, it would also have to license all of them as probably true. But this permissiveness leads to absurdities. The crude 'method of hypothesis' simply lacks the discriminatory power that scientific method ought to have.[15]

4 A Case for Abduction

Faced with these two extreme solutions to the problem of the scientific method, the question is whether there can be a characterisation of the method that somehow moves in-between them. So far, we have noted that ampliation is inversely proportional to epistemic warrant. This is clearly not accidental, since ampliation amounts to risk and the more the risk taken, the less the epistemic security it enjoys. But it is an open issue whether or not there can be a way to strike a balance between ampliation and epistemic warrant, or (equivalently) between strength and austerity. In particular, it is an open issue whether there can be a characterisation of the method which strikes a balance between EI's restrictive ampliation and HD's epistemic permissiveness. I want to explore the suggestion that abduction, if suitably understood as Inference to the Best Explanation (IBE), can offer the required trade-off. But first, what is abduction?

---

[14] For a telling critique of hypothetico-deductivism see [29]. However, Laudan wrongly assimilates Inference to the Best Explanation to hypothetico-deductivism.

[15] It may be objected that EI is equally epistemically permissive since, on any evidence, there will be more than one generalisation which entails it. Yet in order to substantiate this claim for the case of EI, one is bound to produce alternative generalisations which either are non-projectible or restate merely sceptical doubts (e.g., that all ravens are black when someone observes them).

4.1 What Is Abduction?

I am going to leave aside any attempt to connect what follows with Peirce's views on abduction.[16] Rather, I shall take Harman's [15] as the *locus classicus* of the characterisation of IBE. "In making this inference", Harman notes, "one infers, from the fact that a certain hypothesis would explain the evidence, to the truth of that hypothesis. In general, there will be several hypotheses that might explain the evidence, so one must be able to reject all such alternative hypotheses before one is warranted in making the inference. Thus one infers, from the premise that a given hypothesis would provide a 'better' explanation for the evidence than would any other hypothesis, to the conclusion that the given hypothesis is true" (1965, 89). Following Josephson ([22], 5), IBE can be put schematically thus (A):

> D is a collection of data (facts, observations, givens).
> H explains D (would, if true, explain D)
> No other hypothesis can explain D as well as H does.
>
> ---
>
> Therefore, H is probably true.[17]

It is important to keep in mind that, on IBE, it is not just the semantic relation between the hypothesis and the evidence which constitutes the prima facie warrant for the acceptance of the hypothesis. Rather, it is the *explanatory quality* of this hypothesis, on its own but also taken in comparison to others, which contributes essentially to the warrant for its acceptability. So, what we should be after here is a kind of measure of the explanatory power of a hypothesis. Explanatory power is connected with the basic function of an explanation, viz., providing understanding. Whatever the formal details of an explanation, it should be such that it enhances our understanding of why the explanandum-event happened. This can be effected by incorporating the explanandum into the rest of our background knowledge by providing some link between the explanandum and other hypotheses that are part of our background knowledge. Intuitively, there can be better and worse ways to achieve this incorporation--and hence the concomitant understanding of the explanandum. For instance, an explanation which does not introduce gratuitous hypotheses in the explanatory story it tells, or one that tallies better with the relevant background knowledge, or one that by incorporating the explanandum in the background knowledge it enhances its unity, offers a better understanding and, hence has more explanatory power.

I think the evaluation of explanatory power takes place in two directions. The *first* is to look at the specific background information (beliefs) which operate in a certain application of IBE. The *second* is to look at a number of structural features (standards) which competing explanations might possess. The prime characteristic of IBE is that it cannot operate in a "conceptual vacuum", as Ben-Menahem ([2], 330) put it. Whatever else one thinks of an explanation, it must be such that it establishes

---

[16] For Peirce's views the interested reader should look at [4], [8], [14], [47] and [9].

[17] Here I am using the word 'probably' with no specific interpretation of the probability calculus in mind. Its use implies only that the conclusion does not follow from the premises in the way that a deductive argument would have it.

some causal-nomological connection between the explanandum and the explanans. The details of this connection--and hence the explanatory story that they tell--will be specified relative to the available background knowledge. So, to say that a certain hypothesis H is the best explanation of the evidence is to say, at least in part, that the causal-nomological story that H tells tallies best with background knowledge. This knowledge must contain all relevant information about, say, the types of causes that, typically, bring about certain effects, or the laws that govern certain phenomena etc. At least in non-revolutionary applications of IBE, the relevant background knowledge can have the resources to discriminate between better and worse potential explanations of the evidence. So, the explanatory power of a potential explanation depends on what other substantive information there is available in the background knowledge.[18] Let me call 'consilience' this feature of IBE which connects the background knowledge with the potential explanation of the evidence.

Consilience: Suppose that there are two potentially explanatory hypotheses H_1 and H_2 but the relevant background knowledge favours H_1 over H_2. Unless there are specific reasons to challenge the background knowledge, H_1 should be accepted as the best explanation.

Yet, to a certain extent, there is room for a structural specification of the best explanation of a certain event (or piece of evidence). That is, there are structural standards of explanatory merit which mark the explanatory power of a hypothesis and which, when applied to a certain situation, rank competing explanations in terms of their explanatory power. These standards operate crucially when the substantive information contained in the relevant background knowledge cannot forcefully discriminate between competing potential explanations of the evidence. The following list, far from being complete, is an indication of the relevant standards.[19]

Completeness: Suppose that only one explanatory hypothesis H explains all data to be explained. That is, all other competing explanatory hypotheses fail to explain some of the data, although they are not refuted by them. H should be accepted as the best explanation.
Importance: Suppose that two hypotheses H_1 and H_2 do not explain all relevant phenomena, but that H_1, unlike H_2, explains the most salient phenomena. Then H_1 is to be preferred as a better explanation.
Parsimony: Suppose that two composite explanatory hypotheses H_1 and H_2 explain all data. Suppose also that H_1 uses fewer assumptions than H_2. In particular, suppose that the set of hypotheses that H_1 employs to explain the data is a proper subset of the hypotheses that H_2 employs. Then H_1 is to be preferred as a better explanation.

---

[18] A reader has pressed me to explain how the background knowledge can discriminate among competing hypotheses that, if true, would explain a certain explanandum. I don't think there is a deep mystery here. In a lot of typical cases where reasoners employ IBE, there is just one 'best explanation' that the relevant background knowledge makes possible. Finding it consists in simply searching within the relevant background knowledge. For more on this issue, and for an interesting scientific example, see [40], 217-219.
[19] For a fuller discussion see [48].

Unification: Suppose that we have two composite explanatory hypotheses H^k and H^j a body of data $e_1,...,e_n$. Suppose that for every piece of data e_i ($i=1,...,n$) to be explained H^j introduces an explanatory assumption $H^j{}_i$ such that $H^j{}_i$ explains e_i. H^k, on the other hand, subsumes the explanation of all data under a few hypotheses, and hence it unifies the explananda. Then H^k is a better explanation than H^j.

Precision: Suppose that H_1 offers a more precise explanation of the phenomena than H_2, in particular an explanation that articulates some causal-nomological mechanism by means of which the phenomena are explained. Then H_1 is to be preferred as a better explanation.

Such standards have a lot of intuitive pull. Besides, they can characterise sufficiently well several instances of application of IBE in scientific practice (cf. [46], [48]). But even if one granted that these standards have some genuine connection with explanatory quality or merit, one could question their epistemic status: why are they anything more than pragmatic virtues? (cf. [51]) If to call a certain virtue 'pragmatic' is to make it non-cognitive, to relegate it to a merely self-gratifying 'reason' for believing things, then it should be clear that the foregoing explanatory virtues (standards) are not pragmatic. For they possess a straight cognitive function. As Thagard [49] has persuasively argued, such standards safeguard the explanatory coherence of our total belief corpus as well as the coherence between our belief corpus and a new potential explanation of the evidence. To say that a hypothesis that meets these standards has the most explanatory power among its competitors is to say that it has performed best in an explanatory coherence test among its competitors. Explanatory coherence is a cognitive virtue because, on some theories of justification at least, it is a prime way to confer justification on a belief or a corpus of beliefs (cf. [3], [17]). Naturally, the warrant conferred on the chosen hypothesis, viz., that it fares better than others in an explanatory-quality test and that, as a result of this, it enhances the explanatory coherence of the belief corpus, is a defeasible warrant. But this is as it should be. The problem might be thought to be that there is no algorithmic way to connect all these criteria (with appropriate weights) so that they always engender a clear-cut ranking. And the obvious rivalries among some of the criteria suggest that a lot of judgement should be exercised in this ranking. Such problems would be fatal only for those who thought that a suitable description of the method would have to be algorithmic, and in particular that it would have to employ a simple and universal algorithm. This aspiration should not have been taken seriously in the first place. Note also that although a simple and universal algorithm for IBE is not possible, there have been implementations of IBE, e.g., by Thagard [49] which employ a variety of algorithms. Besides, although IBE may be characterised at a very general and abstract level in the way presented above, there is good reason to think that many specific applications (e.g., in medical diagnosis) may employ important domain-specific criteria which require more careful empirical study.

4.2 Some Philosophical Issues

Some philosophers have expressed doubts about IBE which are based on the following worry: why should the information that a hypothesis is the best explanation of the evidence be a prima facie *reason* to believe that this hypothesis is true (or likely to be true)? Cartwright ([5], 4) for instance, has argued that the foregoing question cannot be successfully answered.[20] Meeting this challenge will have to engage us in a proper understanding of the interplay between substantive background knowledge and considerations of explanatory coherence in rendering IBE a legitimate mode of inference. Those readers who feel that these doubts are ill-motivated or just philosophical can skip the rest of this section.

So, what sort of *inference* is IBE? There are two broad answers to this. (1) We infer to the probable truth of the likeliest explanation insofar as and because it is the *likeliest* explanation. On this answer, what matters is how likely the explanatory hypothesis is. If it is likely we infer it; if it isn't we don't. (2) The best explanation, *qua* explanation, is likely to be true (or, at least more likely to be true than worse explanations). That is, the fact that a hypothesis H is the *best* explanation of the evidence issues a warrant that H is likely. In his ([31], 61-65), Lipton has noted that the first answer views IBE as an inference to the Likeliest Potential Explanation, while the second views it as an inference to the Loveliest Potential Explanation. The loveliest potential explanation is "the one which would, if correct, be the most explanatory or provide the most understanding" (op.cit., p.61). If we go for the Likeliest Potential Explanation, then Cartwright's challenge evaporates. For, best explanation and epistemic warrant are linked *externally* via some considerations of likelihood.[21] If there are reasons to believe that a certain hypothesis is likely (or the likeliest available), then there is no further issue of epistemically warranted acceptance. But if we go for the Likeliest Potential Explanation (i.e., the first answer above) then, IBE loses all of its excitement. For what is particularly challenging with IBE is the suggestion--encapsulated in answer (2) above--that the fact that a hypothesis is the *best* explanation (i.e. the loveliest one) *ipso facto* warrants the judgement that it is likely. If the loveliness of a potential explanation is shown to be a symptom of its truth, then Cartwright's challenge is met in a significant and *internal* way.[22] Lipton's own strategy has been to impose two sorts of filters on the choice of hypotheses. One selects a relatively small number of potential explanations as plausible, while the other selects the best among them as the actual explanation. Both filters should operate with explanatory considerations. That is, both filters should act as explanatory-quality tests. Still, although plausibility might have to do with explanatory considerations, why should plausibility have anything to do with likelihood? Here, Lipton's answer is to highlight the *substantive assumptions* that need

---

[20] She does believe however in a special case of IBE, viz., inference to the most likely cause (cf. [5], 6).

[21] Note that here I am using the term "likelihood" informally and not in the statistical sense of it. An attentive reader has pressed me to elaborate on the possible relation between IBE and Bayesianism. I have attempted to offer a few thoughts on this matter in [42]. Suffice it to say here that I take IBE to be a way to assign a kind of objective prior probabilities to hypotheses whose posterior degree of confirmation--in light of further evidence for them--can be calculated by Bayesian techniques.

[22] Failure to discriminate between the Likeliest and the Loveliest Explanation seems to be the reason why Ben-Menahem ([2], 324) claims that "[t]here is nothing particularly deep about the inference to the best explanation. At least there is nothing particularly deep about it qua type of inference".

to be in place for IBE (as Inference to the Loveliest Potential Explanation) to be possible. Explanatory considerations enter into the first filter (that of selecting a small number of hypotheses) by means of our substantive background knowledge that favours hypotheses that cohere well with (or are licensed by) our background beliefs (cf. [31], 122). Insofar as these background beliefs are themselves likely, then IBE operates within an environment of likely hypotheses. Given that the background beliefs themselves have been the product of past applications of IBE, they have been themselves imputed by explanatory considerations. So, the latter enter implicitly in the first filter and explicitly in the second (that of choosing the best among the competing hypotheses that are licensed by the background beliefs). We can see the crux of all this by looking at Josephson's aforementioned schema (A) for IBE. The crucial judgement for the inference to take place is that no other hypothesis explains D as well as H. This judgement is the product of a) filtering the competing hypotheses according to substantive background knowledge and b) choosing among *them* by explanatory considerations. The upshot of all this is that the application of IBE relies on substantive background knowledge. Without it, IBE as an *inference* is simply impotent.[23] But notice that the structural features that make an explanation better than another are part and parcel of the background knowledge. They are just this more abstract part of it which tells us how to evaluate potential explanations. Notice also that these general structural features are complemented by particular ones when it comes to specific applications of IBE. As Josephson ([22], 14) has noted, in specific cases the likelihood of the chosen 'best explanation' H will depend on considerations such as "how decisively H surpasses the alternatives" and "how much confidence there is that all plausible explanations have been considered (how thorough was the search for alternative explanations)".

But suppose that all this is not convincing. Suppose, that is, that we haven't made a case for the claim that the best (loveliest) explanation and the likeliest explanation may reasonably be taken to coincide in light of the relevant background knowledge. There is still an indirect answer available to Cartwright's challenge. Note that we are concerned with the prima facie warrant for accepting a hypothesis H. The question then is: is the fact that H is rendered the best explanation of the evidence a prima facie reason for its acceptance? If, following Pollock ([38], 124), we view justification as "epistemic permissibility", it is obvious that the answer to the foregoing question can only be positive. For to say that the fact that H is the best explanation of the evidence is a *reason* for the acceptance of H is to say that a) it is all right (i.e., it is permissible) to believe in H on this basis; and b) that this permissibility is grounded on the explanatory connection between H and the evidence. It is this explanatory connection which makes the acceptance of H prima facie reasonable since it enhances the coherence of our total belief corpus. By incorporating H in our belief corpus BC as the best explanation of the evidence we enhance the capacity of BC to deal with new information and we improve our understanding not just of why the evidence is the way it is but also of how this evidence gets embedded in our belief corpus. To see how all this works out, note the following. It is explanatory (causal-nomological) connections which hold our belief corpus together. It is such connections which organise the individual beliefs that form it and make the corpus useful in understanding, planning, anticipating etc. (cf. [16]). Faced with a choice among competing explanatory

---

[23] I have defended the reliability of IBE in some detail in my ([40], 81-90 & 212-2).

hypotheses of some event, we should appeal to reasons to eliminate some of them.[24] Subjecting these hypotheses to an explanatory-quality test is the prime way to afford these reasons. Those hypotheses which fare badly in this test get eliminated. For, by having done badly in the test, they have failed at least some of the intuitively compelling criteria of explanatory power. So, they have either failed to cohere well with the relevant background information, or have left some of the data unaccounted for, or have introduced gratuitous assumptions into the explanatory story, or what have you. If this test has a clear winner (the best explanation), then this is the only live option for acceptance. In the end, what IBE does is to enhance the explanatory coherence of a background corpus of belief by choosing a hypothesis which brings certain pieces of evidence into line with this corpus. And it is obviously reasonable to do this enhancement by means of the best available hypotheses. This coherence-enhancing role of IBE, which has been repeatedly stressed by Harman ([16], [17], [18]), Lycan [33] and Thagard ([46], [49]), is ultimately the warrant-conferring element of IBE.

Some philosophers think that there may be a tension between the two prime aspects of IBE that I have described above, viz., its reliance on considerations of explanatory coherence and its dependence on substantive background beliefs. Day and Kincaid ([6], 275) for instance, argue that if IBE is primarily seen as relying on considerations of explanatory coherence, it becomes "redundant and uninformative". For it reduces to "nothing more than a general admonition to increase coherence ([6], 279). And if IBE is primarily seen as being dependent on substantive background knowledge, it "does not name a fundamental pattern of inference" ([6], 282). Rather, they argue, it is an instance of a strategy "that infers to warranted beliefs from background information and the data", without necessarily favouring an explanatory connection between hypotheses and the data (cf. ibid.). Day and Kincaid favour a *contextual* understanding of IBE, since, they say, it has "no automatic warrant" and its importance "might well differ from one epistemic situation to the next" ([6], 282). I think, however, that a) the two aspects of IBE are not in any tension; and b) they engender a rather general and exciting mode of ampliative reasoning. Certainly, more work needs to be done on the notion of coherence and its link with explanation. But if we adopt what Lycan [33] has called "explanationism", it should be clear that explanatory coherence is a vehicle through which an inference is performed and justified. IBE is the mode of inference which effects ampliation via explanation and which licenses conclusions on the basis of considerations which increase explanatory coherence. Yet, as I have noted above, it is wrong to think that the achievement (or enhancement) of explanatory coherence is just a formal-structural matter. Whatever else it is, the best explanation of the evidence (viz., the one that is the best candidate for an enhancement of the explanatory coherence of a belief corpus) has some substantive content which is constrained (if not directly licensed) by the relevant substantive background knowledge. So, substantive background information is not just the material on which some abstract considerations of explanatory coherence should be imposed. It is also the means by which this coherence is achieved. To infer to the best explanation H of the evidence is to search within the relevant background knowledge for explanatory hypotheses and to select the one (if there is one) which

---

[24] Normally, we need to eliminate all but one of them (insofar as they are mutually incompatible, of course), but we should surely allow for ties.

makes the incorporation of the evidence into this background corpus the most explanatorily coherent one. The selection, as we have seen, will be guided by both the substantive background knowledge and some relatively abstract structural standards. That this process is not an inference can be upheld only if one entertains the implausible views that to infer is to deduce and that to infer is to have "an automatic warrant" for the inference. Not all changes in the background knowledge will be based on explanatory considerations. But given that some (perhaps most) are, IBE will have a distinctive (and exciting) role to play.

To sum up, the prima facie reasonableness of IBE cannot be seriously contested. Even if one can question the link between best explanation and truth, one cannot seriously question that the fact that a hypothesis stands out as the best explanation of the evidence offers defeasible reasons to warrantedly accept this hypothesis.[25]

4.3 Abduction and the Two Desiderata

This preliminary defence of the reasonableness of IBE was necessary in order to dispel some natural doubts towards it.[26] Now, we need to see how IBE fares vis-à-vis EI and HD. I will suggest that both EI and HD are extreme cases of IBE, but while EI is an interesting limiting case, HD is a degenerate one whose very possibility shows why IBE is immensely more efficient. Besides, I will argue that IBE has all the strengths and none of the weaknesses of either EI or HD.

That proper inductive arguments are instances of IBE has been argued by Harman [16] and been defended by Josephson ([22], [23]) and Psillos [42]. The basic idea is that good inductive reasoning involves comparison of alternative potentially explanatory hypotheses. In a typical case, where the reasoning starts from the premise that 'All As in the sample are B', there are (at least) two possible ways in which the reasoning can go. The first is to withhold drawing the conclusion that 'All As are B', even if the relevant predicates are projectable, based on the claim that the observed correlation in the sample is due to the fact that the sample is biased. The second is to draw the conclusion that 'All As are B' based on the claim that that the observed correlation is due to the fact that there is a nomological connection between being A and being B such that All As are B. This second way to reason implies (and is supported by) the claim that the observed sample is not biased. What is important in any case is that which way the reasoning should go depends on explanatory considerations. Insofar as the conclusion 'All As are B' is accepted, it is accepted on the basis it offers a better explanation of the observed frequencies of As which are B in the sample, in contrast to the (alternative potential) explanation that someone (or something) has biased the sample. And insofar as the generalisation to the whole population is not accepted, this judgement will be based on providing reasons that the biased-sample hypothesis offers a better explanation of the observed correlations in the sample. Differently put, EI is an extreme case of IBE in that a) the best

---

[25] Here I am leaving aside van Fraassen's [52] claim that the reasons for acceptance are merely pragmatic rather than epistemic. For a critical discussion of his views see ([40] 171-76) and ([20] chapter 4).

[26] Van Fraassen ([50], 160-70) suggested that IBE--conceived as a rule--is incoherent. Harman [19] and Douven [7] have rebutted this claim.

explanation has the form of a nomological generalisation of the data in the sample to the whole relevant population and b) the nomological generalisation is accepted, if at all, on the basis that it offers the best explanation of the observed correlations on the sample. HD, on the other hand, is a limiting but degenerate case of IBE in the following sense: if the only constraint on an explanatory hypothesis is that it deductively entails the data, then any hypothesis which does that is a potential explanation of the data. If there is only one such hypothesis, then it is automatically the 'best' explanation. But it is trivially so. The very need for IBE is suggested by the fact that HD is impotent, as it stands, to discriminate between competing hypotheses which entail (and hence explain in this minimal sense) the evidence.

How, then, does IBE fare vis-à-vis the two desiderata for the method, viz. ampliation and epistemic warrant? Remember that EI is minimally ampliative and maximally epistemically probative, whereas HD is maximally ampliative and minimally epistemically probative. Like HD, IBE is maximally ampliative: it allows for the acceptance of hypotheses which go far beyond the data not just in a horizontal way but also in a vertical one. And given that EI is a special case of IBE, IBE can-- under certain circumstances--be as epistemically probative as EI. But unlike HD, IBE can be epistemically probative in circumstances that HD becomes epistemically too permissive. For IBE has the resources to deal with the so-called 'multiple explanations' problem (cf. [42], 65). That is, IBE can rank competing hypotheses which all, prima facie, explain the evidence in terms of their explanatory power and therefore evaluate them.[27] In order to see how this evaluative dimension of IBE can issue in epistemic warrant, let us examine the types of defeaters to the reasons offered by IBE.

Recall from section 3 that to say that one is prima facie warranted to accept the outcome of an ampliative method is to say that one has considered several possible defeaters of the reasons offered for this outcome and has shown that they are not present. If this is done, we noted there, there are no *specific* doubts about the warrant for the outcome of the method. Recall also that there are two general types of defeater, rebutting and undercutting ones. Naturally, if there is an observation which refutes the best explanation of the evidence so far, then this is a rebutting defeater of the best explanation. But IBE fares better than HD vis-à-vis the Duhem-Quine problem. For, although any hypothesis can be saved from refutation by suitable adjustments to some auxiliary assumptions (and hence although any rebutting defeater can be neutralised), IBE can offer means to evaluate the impact of a recalcitrant piece of evidence on the conclusion that the chosen hypotheses is the best explanation of the evidence. HD does not have the resources to perform this evaluation. If the sole constraint on the acceptance of the hypothesis is whether or not it entails the evidence, it is clear that a

---

[27] As one of the anonymous readers observed, abduction, as this is typically used in Logic Programming, does not require ranking of competing hypotheses in terms of their explanatory power. In particular, it does not require that no other hypothesis be a better explanation than the one actually chosen. This is indeed so. But, as I have argued [42], this is precisely the problem that suggests that the computational modelling of abduction in Logic Programming should be more complicated than it actually is. In many cases of abductive Logic Programming it is already a difficult (and valuable) task to generate an explanation of a certain event. But, as many advocates of abductive Logic Programming are aware, there will typically be competing explanations of the event to be explained (cf. [25]). So there is bound to be need to discriminate between them in terms of their explanatory power. This point of view is also entertained by [24] in this volume.

negative observation can only refute the hypothesis. If the hypothesis is to be saved, then the blame should be put on some auxiliaries, but--staying within HD--there is no independent reason to do so. In IBE, the required independent reasons are provided by the relevant explanatory considerations: if there are strong reasons to believe that a hypothesis is the best explanation of the evidence, there is also reason to stick to this hypothesis and make the negative observation issue in some changes to the auxiliary assumptions. After all, if a hypothesis has been chosen as the best explanation, then it has fared best in an explanatory-quality test with its competing rivals. So unless there is reason to think that it is superseded by an even better explanation, or unless there is reason to believe that the recalcitrant evidence points to one of the rivals as a better explanation, to stick with the best explanatory hypothesis is entirely reasonable. This last thought brings us to the role of undercutting defeaters in IBE. Recall that in the case of HD, any other hypothesis which entails the same evidence as H is an undercutting defeater for (the warrant for) H. And given that there are going to be a lot of such alternative hypotheses, the warrant for H gets minimised. But in IBE it is simply not the case that any other hypothesis which entails the evidence offers an explanation of it. For it is not required that the explanatory relation between the evidence and the hypothesis be deductive (cf. [31], 96).[28] Even if we focus on the special case in which this relation is deductive, IBE dictates that we should look beyond the content of each potential explanatory hypothesis and beyond the relations of deductive entailment between it and the evidence in order to appraise its explanatory power. Two or more hypotheses may entail the same evidence, but one of them may be a better explanation of it. So, the presence of a worse explanation cannot act as a possible undercutting defeater for the acceptance of the best explanatory hypothesis. The choice of the best explanation has already involved the consideration of possible undercutting defeaters (viz., other potential explanations of the evidence) and has found them wanting. The judgement that a certain hypothesis is the best explanation of the evidence is warranted precisely because it has rested on the examination and neutralisation of possible undercutting defeaters. To be sure, IBE is defeasible. And the discovery of an even better explanation of the evidence will act as an undercutting (sometimes even as a rebutting defeater) of the chosen hypothesis. But this is harmless for two reasons. First, given the information available at a time t, it is reasonable to infer to the best available explanation H of the present evidence even if there may be even better possible explanations of it. The existence of hitherto unthought of explanations is a contingent matter. H has fared in the explanatory-quality test better than its extant competitors. Hence it has neutralised a number of possible undercutting defeaters. That there may be more possible undercutting defeaters neither can be predicted, nor can it retract from the fact that it is prima facie reasonable to accept H. In any case, if the search for other potential explanations has been thorough, and if the present information does not justify a further exploration of the logical space of potentially explanatory hypotheses, there is no *specific* reason to

---

[28] A hypothesis might explain an event without entailing it. It might make it occurrence probable; or it might be such that it makes the occurrence of the event more probable than it was before the explanatory hypothesis was taken into account. More generally, IBE should be able to take the form of statistical explanation either in the form of the Hempelian Inductive-Statistical model (cf. [21]) or in the form of Salmon's Statistical-Relevance model (cf. [44]).

doubt that the current best explanation is simply the best explanation. If such doubts arise later on they are welcome, but do not invalidate our present judgement.[29]

The natural conclusion of all this is that IBE admits of clear-cut undercutting defeaters, but unlike HD it has the resources to show when a potential undercutting defeater can be neutralised. And it also admits of clear-cut rebutting defeaters, but unlike HD it can explain how and why such a possible defeater can be neutralised. So, when its comes to its epistemically probative character, IBE can reach the maximal epistemic warrant of EI (since EI is an extreme case of IBE), but it goes far beyond the minimal epistemic warrant of HD (since it offers reasons to evaluate competing hypotheses in an explanatory-quality test). And when it comes to ampliation, like HD and unlike EI, it reaches up to maximal ampliation (cf. the following chart).

	EI	HD	IBE
Ampliation	Minimal	Maximal	Maximal
Epistemic Warrant	Maximal	Minimal	Far more than minimal and up to maximal

5 Conclusion

I have argued that abduction, understood as Inference to the Best Explanation, satisfies in the best way the two desiderata of ampliation and epistemic warrant and also strikes the best balance between the role that background knowledge plays in ampliative reasoning and the role that explanatory considerations (as linked with the demand of explanatory coherence) plays in justifying an inference. I will then conclude with a couple of issues that need more attention in future work.

One such issue is the connection between Kowalski's work on argumentation and the approach to IBE suggested in this paper. Kowalski and Toni [26] have suggested that practical reasoning can be understood as a "dialectic process" in which two reasoners present defeasible arguments in favour of their respective positions. Part of the reasoning process is, then, for each side to present defeaters for the other side's arguments. The possibility is then open that we can think of cases where the best explanation of an event is sought as cases in which reasoners argue for their favoured hypotheses being the 'best explanation' and defend it against the defeaters offered by the other side. It may indeed be useful to see how the abstract framework for argumentation that Kowalski and Toni have put forward, and which makes heavy use of defeaters, can be enlarged (or customised) to incorporate cases of conclusions reached by IBE. Obviously, more work needs to be done on the notion of explanatory coherence and also on the role of coherence in justification. But the good news so far seems to be that IBE can emerge as the general specification of scientific method which promises to solve in the best way its central philosophical problem.

---

[29] In his [37], Pereira makes some interesting observations as to how defeasibility considerations can be captured within Logic Programming, especially in connection with the role that negation plays within this framework.

References

1. Aliseda, A.: Seeking Explanations: Abduction in Logic. Philosophy of Science and Artificial Intelligence. ILLC Dissertation Series (1997) Amsterdam: University of Amsterdam
2. Ben-Menahem, Y.: The Inference to the Best Explanation. Erkenntnis 33 (1990) 319-344
3. BonJour, L.: The Structure of Empirical Knowledge. (1985) Cambridge MA: Harvard University Press
4. Burks A.: 'Peirce's Theory of Abduction. Philosophy of Science 13 301-306
5. Cartwright, N.: How the Laws of Physics Lie. (1983) Oxford: Clarendon Press
6. Day, T. & Kincaid, H.: Putting Inference to the Best Explanation in its Place. Synthese 98 (1994) 271-295
7. Douven, I.: Inference to the Best Explanation Made Coherent. Philosophy of Science 66 (Proceedings) (1999) S424-435
8. Fann, K.T.: Peirce's Theory of Abduction. (1970) Martinus Nijhoff
9. Flach, P. & Kakas, A.: Abductive and Inductive Reasoning: Background and Issues. In Flach, P. & Kakas, A. (eds.): Abduction and Induction: Essays on their Relation and Integration. (2000) Dordrecht: Kluwer Academic Publishers
10. Flach, P. & Kakas, A. (eds.): Abduction and Induction: Essays on their Relation and Integration. Dordrecht: Kluwer Academic Publishers
11. Fodor, G.: The Mind Doesn't Work That Way. (2000) MIT Press
12. Goodman, N.: Fact, Fiction and Forecast. (1954) Cambridge MA: Harvard University Press
13. Gower, B.: Scientific Method: An Historical and Philosophical Introduction. (1998) London: Routledge.
14. Hanson, N.R.: Notes Towards a Logic of Discovery. In Bernstein, R. J. (ed.): Critical Essays on C. S. Peirce. (1965) Yale University Press.
15. Harman, G.: Inference to the Best Explanation. The Philosophical Review 74 (1965) 88-95
16. Harman, G.: Reasoning and Explanatory Coherence. American Philosophical Quarterly 17 (1979) 151-157
17. Harman, G.: Change in View: Principles of Reasoning. (1986) Cambridge MA: MIT Press
18. Harman, G.: Rationality. In Smith, E. E. & Osherson, D. N. (eds.) An Invitation to Cognitive Science Vol. 3 (Thinking) (1995) Cambridge MA: MIT Press
19. Harman, G.: Pragmatism and the Reasons for Belief. In Kulp, C. B. (ed.) Realism/Anti-realism and Epistemology. (1996) New Jersey: Rowan & Littlefield
20. Harman, G.: Reasoning, Meaning and Mind. (1999) Oxford: Oxford University Press
21. Hempel, C.: Aspects of Scientific Explanation. (1965) New York: Basic Books
22. Josephson, J. et al.: Abductive Inference. (1994) Cambridge: Cambridge University Press
23. Josephson, J.: Smart Inductive Generalisations are Abductions. In Flach, P. & Kakas, A. (eds.) Abduction and Induction: Essays on their Relation and Integration. (2000) Dordrecht: Kluwer Academic Publishers
24. Denecker, M & A.C. Kakas.: Abduction in Logic Programming. This volume
25. Kakas, A.C., Kowalski, R.A., & Toni, F.: Abductive Logic Programming. Journal of Logic and Computation 2 (1992) 719-770
26. Kowalski, R. A. & Toni, F.: Abstract Argumentation. Artificial Intelligence and Law 4 (1996) 275-296
27. Kitcher, P.: Explanatory Unification. Philosophy of Science 48 (1981) 251-81
28. Konolige, K.: Abductive Theories in Artificial Intelligence. In Brewka, G. (ed.) Principles of Knowledge Representation. (1996) CSLI Publications

29. Laudan, L.: Damn the Consequences. The Proceedings and Addresses of the American Philosophical Association 6 (1995) 27-34
30. Lewis, D.: Causal Explanation. In his Philosophical Papers, Vol.2, (1986) Oxford University Press
31. Lipton, P.: Inference to the Best Explanation. (1991) London: Routledge
32. Lipton, P.: Tracking Track Records. Proceedings of the Aristotelian Society Suppl. Volume 74 (2000) 179-205
33. Lycan, W.: Judgement and Justification. (1988) Cambridge: Cambridge University Press
34. Lycan, W.: Explanationism, ECHO, and the Connectionist Paradigm. Behavioural and Brain Sciences 12 (1989) 480
35. Mellor, D. H.: The Warrant of Induction. (1988) Cambridge: Cambridge University Press
36. Niiniluoto, I.: Defending Abduction. Philosophy of Science 66 (Proceedings) (1999) S436-S451
37. Pereira, L. M.: Philosophical Impingement of Logic Programming. In Gabbay, D. & Woods, J. (eds) Handbook of History and Philosophy of Logic. (2001) Kluwer Academic Press
38. Pollock, J.: Contemporary Theories of Knowledge. (1986) New Jersey: Rowan & Littlefield
39. Pollock, J.: Defeasible Reasoning. Cognitive Science 11 (1987) 481-518
40. Psillos, S.: Scientific Realism: How Science Tracks Truth. (1999) London: Routledge
41. Psillos, S.: Review of Gower, B: Theories of Scientific Method. Ratio XII (1999) 310-316
42. Psillos, S.: Abduction: Between Conceptual Richness and Computational Complexity. In Flach, P. & Kakas, A. (eds.) Abduction and Induction: Essays on their Relation and Integration. (2000) Dordrecht: Kluwer Academic Publishers
43. Psillos, S.: Causation and Explanation. (forthcoming) Acumen
44. Salmon, W.: Scientific Explanation and the Causal Structure of the World. (1984) Princeton: Princeton University Press
45. Salmon, W.: Four Decades of Scientific Explanation. (1989) Minnesota University Press
46. Thagard, P.: Best Explanation: Criteria for Theory Choice. Journal of Philosophy 75 (1978) 76-92
47. Thagard, P.: Peirce on Hypothesis and Abduction. In C. S. Peirce Bicentennial International Congress. (1981) Texas University Press
48. Thagard, P.: Computational Philosophy of Science. (1988) Cambridge MA: MIT Press
49. Thagard, P.: Explanatory Coherence. Behavioural and Brain Sciences 12 (1989) 435-502
50. Thagard, P. & Shelley, C.: Abductive Reasoning: Logic, Visual Thinking and Coherence. In Dalla Chiara, M. L. (ed.) Logic and Scientific Methods. (1997) Kluwer Academic Publishers
51. van Fraassen, B.C.: The Scientific Image. (1980) Oxford: Clarendon Press
52. van Fraassen, B.C.: Laws and Symmetry. (1989) Oxford: Clarendon Press

Author Index